Safety Evaluation Report
Related to the License Renewal of
Palisades Nuclear Plant

I0488143

Docket No. 50-255

Nuclear Management Company, LLC

Manuscript Completed: September 2006
Date Published: January 2007

Division of License Renewal
Office of Nuclear Reactor Regulation
U.S. Nuclear Regulatory Commission
Washington, DC 20555-0001

ABSTRACT

This safety evaluation report (SER), documents the technical review of the Palisades Nuclear Plant (PNP) license renewal application (LRA) by the U.S. Nuclear Regulatory Commission (NRC) staff (the staff). By letter dated March 22, 2005, Nuclear Management Company, LLC (NMC or the applicant) submitted the LRA in accordance with Title 10, Part 54, of the *Code of Federal Regulations* (10 CFR Part 54). NMC requests renewal of the operating license for Palisades Nuclear Plant (Facility Operating License Number DPR-20) for a period of 20 years beyond the current expiration date at midnight March 24, 2011.

PNP is located approximately five miles south of South Haven, Michigan. The NRC issued the construction permit for PNP on March 14, 1967, and the operating license on February 21, 1991. The plant's nuclear steam supply system consists of a pressurized water reactor (PWR-DRYAMB) with two closed loops. The nuclear steam supply system was supplied by Combustion Engineering and the balance of the plant was originally designed and constructed by Bechtel. PNP operates at a licensed power output of 2565 megawatt thermal, with a gross electrical output of approximately 767 megawatt electric.

This SER presents the status of the staff's review of information submitted to the NRC through July 5, 2006, the cutoff date for consideration in the SER. The staff identified no open items and one confirmatory item that had to be resolved before the staff could make a final determination on the application. Sections 1.5 and 1.6 of this report summarize these items. Section 6 provides the staff's final conclusion on the review of the PNP LRA.

TABLE OF CONTENTS

ABBREVIATIONS

AC	alternating current
ACI	American Concrete Institute
ACRS	Advisory Committee on Reactor Safeguards
ACSR	aluminum conductor steel reinforced
ADAMS	Agencywide Document Access and Management System
AERM	aging effects requiring management
AFW	auxiliary feedwater
AISC	American Institute of Steel Construction
ALEX	advanced license extension
AMP	aging management program
AMR	aging management review
ANSI	American National Standards Institute
ASME	American Society of Mechanical Engineers
ASME III	American Society of Mechanical Engineers Boiler and Pressure Vessel Code, Section III
ASTM	American Society for Testing of Materials
ATWS	anticipated transient without scram
B&W	Babcock and Wilcox
B&PV	boiler and pressure vessel
BTP	branch technical position
BWR	boiling water reactor
CAS	compressed air system
CASS	cast austentitic stainless steel
CCCW	closed cycle cooling water
CCS	component cooling system
CCW	component cooling water
CDS	condensate system
CE	combustion engineering
CEOG	Combustion Engineering Owners Group
CF	chemistry factor
CFR	*Code of Federal Regulations*
CHECWORKS	The suite of products developed to evaluate power plants for the most common forms of corrosion that degrade their performance and shorten the operating life of critical components.
CI	confirmatory item
CLB	current licensing basis
CMAA	Crane Manufactures Association of America
CP Co	Consumers Power Company
CRD	control rod drive
CRDM	control rod drive mechanism
CRS	containment air and recirculation and cooling system
CSS	containment spray system
CST	condensate storage tank
CUF	cumulative usage factor

CVCS	chemical and volume control system
CWS	circulating water system
DBA	design-basis accident
DBD	design-basis document
DBE	design-basis event
DC	direct current
DMW	demineralized makeup water
ECT	eddy current testing
EDB	equipment database
EDG	emergency diesel generator
EER	electrical equipment room
EEQ	electrical equipment qualification
EFPY	effective full-power years
EHC	electrohydraulic control
EMA	equivalent margins analysis
EPR	ethylene propylene rubber
EPRI	Electric Power Research Institute
EQ	environmental qualification
ESF	engineered safety feature
ESS	engineered safeguards system
F	Fahrenheit
FAC	flow-accelerated corrosion
FatiguePro	automated cycle counting and fatigue monitoring program
FE	flow element
F_{en}	environmental fatigue factor
FERC	Federal Energy Regulatory Commission
FOS	fuel oil system
FP	fire protection
FPS	fire protection system
FSAR	final safety analysis report
FW	feedwater
GALL	generic aging lessons learned
GEIS	Generic Environmental Impact Statement
GL	generic letter
GSI	generic safety issues
HAZ	heat-affected zone
HELB	high-energy line break
HPSI	high pressure safety injection
HSS	high safety significance
HVAC	heating, ventilation, and air conditioning
HX	heat exchanger
I&C	instrumentation and controls
IASCC	irradiation assisted stress corrosion cracking

ID	identification or inside diameter
IEB	inspection and enforcement bulletin
IGA	intergranular attack
IGSCC	intergranular stress corrosion cracking
IN	information notice or inch
INPO	Institute of Nuclear Power Operations
IPA	integrated plant assessment
IR	insulation resistance or inspection report
ISG	interim staff guidance
ISI	inservice inspection
IWB	ASME Boiler and Pressure Vessel Code, Section XI, Requirements for Class 1 Components of Light-Water Cooled Power Plants
IWC	ASME Boiler and Pressure Vessel Code, Section XI, Requirements for Class 2 Components of Light-Water Cooled Power Plants
IWD	ASME Boiler and Pressure Vessel Code, Section XI, Requirements for Class 3 Components of Light-Water Cooled Power Plants
IWE	ASME Boiler and Pressure Vessel Code, Section XI, Requirements for Class MC and Metallic Liners of Class CC Components of Light-Water Cooler Power Plants
IWF	ASME Boiler and Pressure Vessel Code, Section XI, Requirements for Class 1, 2, 3, and MC Component Supports of Light-Water Cooled Power Plants
IWL	ASME Boiler and Pressure Vessel Code, Section XI, Requirements for Class CC Concrete Components of Light-Water Cooled Power Plants
K_{IC}	reference stress intensity factor as a function of the metal temperature (T) and the metal references nil-ductility temperature (RT_{NDT})
KIP	1000 lb; or 1 kilo-pound
ksi	one KIP per square inch, 1000 psi
Lb	pound
LER	licensee event report
LLC	limited liability company
LOCA	loss-of-coolant accident
LPSI	low pressure safety injection
LPCI	low pressure coolant injection
LR	license renewal
LRA	license renewal application
LTOP	low-temperature overpressure protection
LWR	light water reactor
MEL	master equipment list
MIC	microbiologically induced corrosion
MRP	EPRI materials reliability program
MS	main steam
MSIV	main steam isolation valve
MT	magnetic particle test
NA or N/A	not applicable

NDE	non-destructive examination
NEI	Nuclear Energy Institute
NEPA	National Environmental Policy Act of 1969
NESC	National Electrical Safety Code
NFPA	National Fire Protection Association
NMC	Nuclear Management Company
NPS	nominal pipe size
NRC	U.S. Nuclear Regulatory Commission
NSAC	nuclear safety analysis center
NSR	nonsafety-related
NSSS	nuclear steam supply system
NUMARC	Nuclear Utility Management and Resource Council
NUREG	nuclear regulation document
ODSCC	outside-diameter stress-corrosion cracking
OBE	operating-basis earthquake
OI	open item
P&ID	piping and instrumentation diagram
P-T	pressure -temperature
PCP	primary coolant pump
PCS	primary coolant system
PLL	predicted lower limits
PNP	Palisades Nuclear Plant
PORV	power-operated relief valve
PT	penetrant testing
PTS	pressurized thermal shock
PVC	polyvinyl chloride
PWR	pressurized water reactor
PWSCC	primary water stress-corrosion cracking
QA	quality assurance
RAI	request for additional information
RCP	reactor coolant pump
RCPB	reactor coolant pressure boundary
RCS	reactor coolant system
RG	regulatory guide
RI-ISI	risk-informed inservice inspection
RPS	reactor protection system
RPV	reactor pressure vessel
RTD	resistance temperature detectors
RT_{NDT}	reference temperature nil ductility transition
RT_{PTS}	reference temperature pressurized thermal shock
RVI	reactor vessel internals
RVID	Reactor Vessel Internals Database
RWS	radwaste system
SBO	station blackout

SC	structure and component
SCC	stress corrosion cracking
SCS	shield cooling system
SDC	shutdown cooling
SER	safety evaluation report
SFP	spent fuel pool
SG	steam generator
SIR	safeguards implementation report
SIRW	safety injection and refueling water
SOC	statement of considerations
SR	safety-related
SRP	Standard Review Plan
SRP-LR	Standard Review Plan for Review of License Renewal Applications for Nuclear Power Plants
SS	stainless steel or sampling system
SSC	system, structure, or component
SWS	service water system
t	thickness
TGS	turbine generator system
TLAA	time-limited aging analysis
TR	technical report
TRM	training requirements manual
TS	technical specification
TSC	technical support center
U.S.	United States
UFSAR	updated final safety analysis report
UHS	ultimate heat sink
USE	upper-shelf energy
USI	unresolved safety issue
UT	ultrasonic testing
UTS	ultimate tensile strength
UV	ultraviolet
VAC	Volts - alternating current
VCT	volume control tank
VDC	Volts - direct current
VT	visual examination
WCAP	Westinghouse Commercial Atomic Power
XLPE	cross-linked polyethylene
Zn	Zinc

SECTION 1

INTRODUCTION AND GENERAL DISCUSSION

1.1 Introduction

This document is a safety evaluation report (SER) on the license renewal application (LRA) for Palisades Nuclear Plant (PNP), as filed by Nuclear Management Company, LLC (NMC or the applicant). By letter dated March 22, 2005, NMC submitted its application to the to U.S. Nuclear Regulatory Commission (NRC) for renewal of the PNP operating license for an additional 20 years. The NRC staff (the staff) prepared this report, which summarizes the results of its safety review of the renewal application, for compliance with the requirements of Title 10, Part 54, of the *Code of Federal Regulations* (10 CFR Part 54), "Requirements for Renewal of Operating Licenses for Nuclear Power Plants." The NRC license renewal project manager for the PNP license renewal review is Juan Ayala. Mr. Ayala can be contacted by telephone at 301-415-4063 or by electronic mail at jxa3@nrc.gov. Alternatively, written correspondence may be sent to:

License Renewal and Environmental Impacts Program
U.S. Nuclear Regulatory Commission
Washington, D.C. 20555-0001
Attention: Juan Ayala, Mail Stop 0-11F1

In its March 22, 2005, submission letter, the applicant requested renewal of the operating license issued under Section 104b (Operating License No. DPR-20) of the Atomic Energy Act of 1954, as amended, for PNP, for a period of 20 years beyond the current license expiration at midnight March 24, 2011. PNP is located approximately five miles south of South Haven, Michigan. The NRC issued the construction permit for PNP on March 14, 1967 and the operating license on February 21, 1991. The plant's nuclear steam supply system consists of a pressurized water reactor (PWR) with two closed loops. The nuclear steam supply system was supplied by Combustion Engineering and the balance of the plant was originally designed and constructed by Bechtel. PNP operates at a licensed power output of 2565 megawatt thermal, with a gross electrical output of approximately 767 megawatt electric. The final safety analysis report (FSAR) contains details of the plant and the site.

The license renewal process consists of two concurrent reviews–a technical review of safety issues and an environmental review. The NRC regulations in 10 CFR Parts 54 and 51, respectively, set forth requirements for these reviews. The safety review for the PNP license renewal is based on the applicant's LRA and on the responses to the staff's requests for additional information (RAIs). The applicant supplemented and clarified its responses to the LRA and RAIs in audits, meetings, and docketed correspondence. Unless otherwise noted, the staff reviewed and considered information submitted through April 26, 2006. The staff reviewed the information received after that date on a case-by-case basis, depending on the stage of the safety review and the volume and complexity of the information. The public may view the LRA and all pertinent information and materials, including the FSAR, at the NRC Public Document Room, One White Flint North, 11555 Rockville Pike (First Floor), Rockville, MD 20852-2738 (301-415-4737/800-397-4209), and at the South Haven Memorial Library, 314 Broadway, South

Haven, Michigan. In addition, the public may find the LRA, as well as materials related to the license renewal review, on the NRC Web Site at www.nrc.gov.

This SER summarizes the results of the staff's safety review of the LRA and describes the technical details considered in the evaluation of safety aspects of the unit's proposed operation for an additional 20 years beyond the term of the current operating license. The staff reviewed the LRA in accordance with NRC regulations and the guidance of NUREG-1800, "Standard Review Plan for Review of License Renewal Applications for Nuclear Power Plants" (SRP-LR), dated July 2001.

SER Sections 2 through 4 address the staff's evaluation of license renewal issues considered during its review of the application. SER Section 5 is reserved for the report of the Advisory Committee on Reactor Safeguards (ACRS). The conclusions of this report are in SER Section 6.

SER Appendix A is a table that identifies the applicant's commitments for the renewal of the operating license. SER Appendix B is a chronology of the principal correspondence between the NRC and the applicant related to the review of the application. SER Appendix C is a list of principal contributors to the SER. SER Appendix D is a bibliography of the references in support of the review.

In accordance with 10 CFR Part 51, the staff prepared a draft plant-specific supplement to the Generic Environmental Impact Statement (GEIS). This supplement discusses the environmental considerations related to license renewal for PNP. The staff issued draft Supplement 27 to NUREG-1437 "Generic Environmental Impact Statement for License Renewal of Nuclear Plants, Regarding Palisades Nuclear Plant, Draft Report for Comment," on February 14, 2006.

1.2 License Renewal Background

Pursuant to the Atomic Energy Act of 1954, as amended, and NRC regulations, operating licenses for commercial power reactors are issued for 40 years. These licenses can be renewed for up to 20 additional years. The original 40-year license term was selected on the basis of economic and antitrust considerations, rather than on technical limitations; however, some individual plant and equipment designs may have been engineered based on an expected 40-year service life.

In 1982, the staff anticipated interest in license renewal and held a workshop on nuclear power plant aging. This workshop led the NRC to establish a comprehensive program plan for nuclear plant aging research. On the basis of the results of that research, a technical review group concluded that many aging phenomena are readily manageable and pose no technical issues that would preclude life extension for nuclear power plants. In 1986, the staff published a request for comment on a policy statement that would address major policy, technical, and procedural issues related to license renewal for nuclear power plants.

In 1991, the staff published the license renewal rule in 10 CFR Part 54 (the Rule). The staff participated in an industry-sponsored demonstration program to apply the Rule to a pilot plant and to gain experience necessary to develop implementation guidance. To establish a scope of review for license renewal, the Rule defined age-related degradation unique to license renewal; however, during the demonstration program, the staff found that many aging mechanisms occur to plant systems and components with effects managed during the initial license period. In addition, the staff found that the scope of the review did not allow sufficient credit for existing programs, particularly the implementation of the Maintenance Rule, which also manages plant-aging phenomena. As a result, the staff amended the Rule in 1995. As amended, 10 CFR Part 54 established a regulatory process that is simpler, more stable, and more predictable than the previous Rule. In particular, as amended, 10 CFR Part 54 focused on management of adverse aging effects rather than on identification of age-related degradation unique to license renewal. The staff initiated these rule changes to ensure that important systems, structures, and components (SSCs) will continue to perform their intended functions during the period of extended operation. In addition, the revised Rule clarified and simplified the integrated plant assessment (IPA) process for consistency with the revised focus on passive, long-lived structures and components (SCs).

In parallel with these efforts, in a separate rulemaking effort, the staff amended 10 CFR Part 51 to focus the scope of the review of environmental impacts of license renewal and fulfill the NRC's responsibilities under the National Environmental Policy Act of 1969 (NEPA).

1.2.1 Safety Review

License renewal requirements for power reactors are based on two key principles:

(1) The regulatory process is adequate to ensure that the licensing bases of all currently operating plants maintain an acceptable level of safety, with the possible exception of the detrimental aging effects on the function of certain SSCs, as well as a few other safety-related (SR) issues, during the period of extended operation.

(2) The plant-specific licensing basis must be maintained during the renewal term in the same manner and to the same extent as during the original licensing term.

In implementing these two principles, 10 CFR 54.4 defines the scope of license renewal as including SSCs (1) that are SR, (2) whose failure could affect SR functions, and (3) that are relied on to demonstrate compliance with NRC regulations for fire protection (FP), environmental qualification (EQ), pressurized thermal shock (PTS), anticipated transient without scram (ATWS), and station blackout (SBO).

Pursuant to 10 CFR 54.21(a), an applicant for a renewed license must review all SSCs within the scope of the Rule to identify SCs subject to an aging management review (AMR). Those SCs subject to an AMR perform an intended function without moving parts or without a change in configuration or properties, and are not subject to replacement based on a qualified life or specified time period. As required by 10 CFR 54.21(a), an applicant for a renewed license must demonstrate that aging effects will be managed in such a way that the intended function(s) of those SCs will be maintained, consistent with the current licensing basis (CLB), for the period of extended operation; however, active equipment is considered adequately monitored and maintained by existing programs. In other words, detrimental aging effects that may affect active equipment are readily detectable and can be identified and corrected through routine surveillance, performance monitoring, and maintenance. Surveillance and maintenance programs for active equipment, as well as other maintenance aspects of plant design and licensing basis, are required throughout the period of extended operation.

Pursuant to 10 CFR 54.21(d), each LRA is required to include an FSAR supplement that must have a summary description of the applicant's programs and activities for managing aging effects and the evaluation of time-limited aging analyses (TLAAs) for the period of extended operation.

License renewal also requires TLAA identification and updating. During the plant design phase certain assumptions are made about the length of time the plant can operate. These assumptions are incorporated into design calculations for several plant SSCs. In accordance with 10 CFR 54.21(c)(1), the applicant must show that these calculations will remain valid for the period of extended operation, project the analyses to the end of the period of extended operation, or demonstrate that effects of aging on these SSCs can be adequately managed for the period of extended operation.

In 2001, the staff developed and issued Regulatory Guide (RG) 1.188, "Standard Format and Content for Applications to Renew Nuclear Power Plant Operating Licenses." This RG endorses Nuclear Energy Institute (NEI) 95-10, Revision 3, "Industry Guideline for Implementing the Requirements of 10 CFR Part 54 - The License Renewal Rule," issued in March 2001 by the NEI. NEI 95-10 details an acceptable method of implementing the Rule. The staff also used the SRP-LR to review this application.

In its LRA, the applicant fully utilized the process defined in NUREG-1801, "Generic Aging Lessons Learned (GALL) Report," issued in July 2001. The GALL Report provides a summary of staff-approved aging management programs (AMPs) for the aging of many SCs subject to an AMR. If an applicant commits to implementing these staff-approved AMPs, the time, effort, and resources to review an applicant's LRA can be greatly reduced, thereby improving the efficiency and effectiveness of the license renewal review process. The GALL Report summarizes the aging management evaluations, programs, and activities credited for managing aging for most SCs used throughout the industry. The report is also a reference for both applicants and staff reviewers to quickly identify AMPs and activities that can provide adequate aging management during the period of extended operation.

1.2.2 Environmental Review

In December 1996, the staff revised the environmental protection regulations to facilitate the environmental review for license renewal. The staff prepared a "Generic Environmental Impact Statement (GEIS) for License Renewal of Nuclear Plants" (NUREG-1437, Revision 1) to document its evaluation of the possible environmental impacts associated with renewing licenses of nuclear power plants. For certain types of environmental impacts, the GEIS establishes generic findings applicable to all nuclear power plants. These generic findings are codified in Appendix B to Subpart A of 10 CFR Part 51. Pursuant to 10 CFR 51.53(c)(3)(i), an applicant for license renewal may incorporate these generic findings in its environmental report. In accordance with 10 CFR 51.53(c)(3)(ii), an environmental report must also include analyses of environmental impacts that must be evaluated on a plant-specific basis (i.e., Category 2 issues).

In accordance with NEPA and the requirements of 10 CFR Part 51, the staff performed a plant-specific review of the environmental impacts of license renewal, including whether the GEIS had not considered new and significant information. As part of its scoping process, the staff held a public meeting on July 28, 2005, in South Haven, Michigan to identify plant-specific environmental issues. The NRC's draft plant-specific GEIS Supplement 27, issued on February 14, 2006, documents the results of the environmental review and includes a preliminary recommendation as to license renewal action. The staff held another public meeting on April 5, 2006, in South Haven, Michigan, to discuss the draft plant-specific GEIS Supplement 27. After considering comments on the draft, the staff will prepare and publish, a final plant-specific supplement to the GEIS separately from this report.

1.3 Principal Review Matters

Part 54 of 10 CFR describes the requirements for renewing operating licenses for nuclear power plants. The staff performed its technical review of the LRA in accordance with NRC guidance and 10 CFR Part 54 requirements. Section 54.29 of 10 CFR sets forth the standards for renewing a license. This SER describes the results of the staff's safety review.

Under 10 CFR 54.19(a), the NRC requires a license renewal applicant to submit general information. The applicant provided this general information in LRA Section 1, which it submitted, by letter dated March 22, 2005. The staff reviewed LRA Section 1 and found that the applicant had submitted the information required by 10 CFR 54.19(a).

Under 10 CFR 54.19(b), the NRC requires that each LRA include "conforming changes to the standard indemnity agreement, 10 CFR 140.92, Appendix B, to account for the expiration term of the proposed renewed license." The applicant stated the following in the LRA on this issue:

> The current indemnity agreement No. B-40 for Palisades states, in Article VII, that the agreement shall terminate at the time of expiration of that license specified in Item 3 of the attachment to the agreement, which is the last to expire. Item 3 of the attachment to the indemnity agreement, as revised by Amendment No. 1, lists DPR-20 as the applicable operating license number. Should the operating license number be changed upon issuance of the renewed license, NMC requests that conforming changes be made to Item 3 of the attachment, and any other sections of the indemnity agreement as appropriate.

The staff intends to maintain the original license number upon issuance of the renewed license, if approved. Therefore, conforming changes to the indemnity agreement need not be made and the requirements of 10 CFR 54.19(b) have been met.

Under 10 CFR 54.21, the NRC requires that each LRA contain (a) an IPA, (b) a description of any CLB changes during the staff's review of the LRA, (c) an evaluation of TLAAs, and (d) an FSAR supplement. LRA Sections 3 and 4 and Appendix B address the license renewal requirements of 10 CFR 54.21(a), (b), and (c). LRA Appendix A satisfies the license renewal requirements of 10 CFR 54.21(d).

Under 10 CFR 54.21(b), the NRC requires that each year following submission of the LRA, and at least three months before the scheduled completion of the staff's review, the applicant submit an LRA amendment identifying any CLB changes of the facility that materially affect the contents of the LRA, including the FSAR supplement. The applicant submitted an update to the LRA by letter dated March 21, 2006, summarizing the CLB changes that have occurred at PNP during the staff's review of the LRA. This submission satisfies the requirements of 10 CFR 54.21(b) and is still under staff review.

Under 10 CFR 54.22, the NRC requires that an applicant's LRA include changes or additions to the technical specifications (TS) necessary to manage aging effects during the period of extended operation. In LRA Appendix D, the applicant stated that it had not identified any TS changes necessary to support issuance of the renewed operating license. This statement adequately addresses the 10 CFR 54.22 requirement.

The staff evaluated the technical information required by 10 CFR 54.21 and 10 CFR 54.22 in accordance with NRC regulations and the guidance of the SRP-LR. SER Sections 2, 3, and 4 document the staff's evaluation of the technical information in the LRA.

As required by 10 CFR 54.25, the ACRS will issue a report to document its evaluation of the staff's LRA review and associated SER. SER Section 5 will incorporate the ACRS report once it is issued. SER Section 6 will document the findings required by 10 CFR 54.29.

The final plant-specific GEIS supplement will document the staff's evaluation of the environmental information required by 10 CFR 54.23 and will specify the considerations for renewing the PNP license. The staff will prepare the supplement separately from the SER.

1.4 Interim Staff Guidance

License renewal is a living program. The staff, industry, and other interested stakeholders gain experience and develop lessons learned with each renewed license. The lessons learned address the staff's performance goals of maintaining safety, improving effectiveness and efficiency, reducing regulatory burden, and increasing public confidence. Interim staff guidance (ISG) is documented for use by the NRC staff, industry, and other interested stakeholders until incorporated into such license renewal guidance documents as the SRP-LR and the GALL Report.

The following table provides the current set of ISGs as well as the SER sections in which the staff addresses ISG issues.

ISG Issue (Approved ISG No.)	Purpose	SER Section
GALL Report presents one acceptable way to manage aging effects (ISG-1)	This ISG clarifies that the GALL Report contains one acceptable way, but not the only way, to manage aging for license renewal.	N/A

ISG Issue (Approved ISG No.)	Purpose	SER Section
Station Blackout (SBO) Scoping (ISG-2)	The license renewal rule 10 CFR 54.4(a)(3) includes 10 CFR 50.63(a)(1)—SBO. The SBO rule requires that a plant must withstand and recover from an SBO event. The recovery time for offsite power is much faster than that of emergency diesel generators (EDGs). The offsite power system should be included within the scope of license renewal.	2.1.3.3.2 2.5.1.5 3.0.3.1.7
Concrete AMP (ISG-3)	Lessons learned from the GALL demonstration project indicated that GALL is not clear on whether concrete requires an AMP.	3.0.3.2.10 3.5.2.1.4 3.5.2.1.5 3.5.2.1.6 3.5.2.1.10 3.5.2.2.1 3.5.2.2.2
Fire Protection (FP) System Piping (ISG-4)	This ISG clarifies the staff position for wall-thinning of the FP piping system in GALL AMPs XI.M26 and XI.M27. The staff's new position is that there is no need to disassemble FP piping, as disassembly can introduce oxygen to FP piping, which can accelerate corrosion. Instead, use a non-intrusive method, such as volumetric inspection. Testing of sprinkler heads should be performed at year 50 of sprinkler system service life, and every 10 years thereafter. This ISG eliminates the Halon/carbon dioxide system inspections for charging pressure, valve line-ups, and the automatic mode of operation test from GALL; the staff considers these test verifications to be operational activities.	3.0.3.2.5

ISG Issue (Approved ISG No.)	Purpose	SER Section
Identification and Treatment of Electrical Fuse Holders (ISG-5)	This ISG includes electrical fuse holders AMR and AMP (i.e., same as terminal blocks and other electrical connections). The position includes only fuse holders that are not inside the enclosure of active components (e.g., inside of switchgears and inverters). Operating experience finds that metallic clamps (spring-loaded clips) have a history of age-related failures from aging stressors such as vibration, thermal cycling, mechanical stress, corrosion, and chemical contamination. The staff finds that visual inspection of fuse clips is not sufficient to detect the aging effects from fatigue, mechanical stress, and vibration.	2.1.4.4.2 3.6.2.3.1
The ISG Process (ISG-8)	This ISG provides clarification and update to the ISG process on Improved License Renewal Guidance Documents.	N/A
Standardized Format for License Renewal Applications (ISG-10)	The purpose of this ISG is to provide a standardized license renewal application format for applicants.	N/A

1.5 Summary of Open Items

After its review of the LRA, including additional information submitted through April 26, 2006, the staff has identified no open items (OIs). An item is considered open if the applicant has not presented a sufficient basis for issue resolution. Each open item OI has been assigned a unique identifying number.

1.6 Summary of Confirmatory Items

Following the staff's review of the LRA, including additional information and clarifications submitted to the NRC through April 26, 2006, the staff identified the following confirmatory items (CIs). An item was considered confirmatory if the staff and the applicant had reached a satisfactory resolution, but the resolution had not been formally submitted to the staff. Each CI has been assigned a unique identifying number. The items identified in this section have been properly closed by the technical staff.

CI 4.7.6-1 (Section 4.7.6 Reactor Vessel Underclad Cracking)

The staff reviewed the LRA changes related to potential underclad cracking. By letter, dated April 26, 2006, the applicant added LRA Section 4.7.6 that described the mechanisms responsible for underclad cracking and indicated the effects of aging could be managed by performing a fracture mechanics analysis. The staff agrees that the effects of aging on underclad cracks can be evaluated and managed by a fracture mechanics analysis. The applicant indicated that the staff had reviewed and approved a bounding fracture mechanics analysis that assessed underclad cracking for Westinghouse RPVs. The applicant stated that although PNP is not a Westinghouse plant, it had been determined that if underclad cracking were postulated to exist in the PNP RPV, the bounding fracture mechanics analysis in "A Review of Cracking Associated with Weld Deposited Cladding in Operating PWR Plants," Westinghouse WCAP-15338-A, October 2002, would be applicable to the issue and would be used as a reference to support a 60 year life. The applicant made similar changes to LRA Sections 3.1.2.2.5 and A4.5.6 and to the LRA Section 4.7 references to add the references to WCAP-15338-A and the staff approval letter. The staff determined the changes were acceptable pending verification of the applicability of the WCAP-15388-A. This is Confirmatory Item CI 4.7.6-1.

The staff reviewed the July 5, 2006, information for closing this confirmatory item and determined that WCAP-16605-NP is a plant-specific version of WCAP-15338-A but with PNP plant-specific data (including design transients) for the fatigue crack growth analysis. The staff found that for underclad flaws with aspect ratio (flaw length to flaw depth) of 2 to 6, the amounts of flaw growth cited in WCAP-15338-A are about the same as those of WCAP-16605-NP for all assumed flaw locations. As fatigue crack growth analysis uses design transients and applied stress intensity factors as input, the bounding nature of this analysis in WCAP-15338-A indicates that the ASME Code Section XI flaw evaluation provided by Westinghouse in 2001 to supplement WCAP-15338-A also applies to PNP. This flaw evaluation demonstrates that after 60 years of fatigue crack growth significant margins (2 for normal, upset, and test conditions and 3 for emergency and faulted conditions) remain for underclad flaws of realistic shapes at different locations in addition to the ASME Code-specified structural factors. Hence, this TLAA is acceptable and Confirmatory Item CI 4.7.6-1 is closed.

1.7 Summary of Proposed License Conditions

Following the staff's review of the LRA, including subsequent information and clarifications provided by the applicant, the staff identified three proposed license conditions.

The first license condition requires the applicant to include the FSAR supplement required by 10 CFR 54.21(d) in the next FSAR update, as required by 10 CFR 50.71(e), following the issuance of the renewed license.

The second license condition requires that the list of commitments in Appendix A to this SER be completed in accordance with the schedule in Appendix A and tracked in the FSAR supplement.

The third license condition requires that all capsules in the reactor vessel, that are removed and tested, must meet the requirements of American Society for Testing and Materials (ASTM) E 185-82 to the extent practicable for the configuration of the specimens in the capsule. Any changes to the capsule withdrawal schedule, including spare capsules, must be approved by the staff prior to implementation. All capsules placed in storage must be maintained for future insertion. Any changes to storage requirements must be approved by the staff, as required by 10 CFR Part 50, Appendix H.

SECTION 2

STRUCTURES AND COMPONENTS SUBJECT TO AGING MANAGEMENT REVIEW

2.1 Scoping and Screening Methodology

Title 10, Section 54.21 of the *Code of Federal Regulations* (10 CFR 54.21), "Contents of Application Technical Information," requires that each license renewal application (LRA) contain an integrated plant assessment (IPA). Furthermore, the IPA must list and identify those structures and components (SCs) subject to an aging management review (AMR) for those systems, structures, and components (SSCs) within the scope of license renewal in accordance with 10 CFR 54.4.

In LRA Section 2.1, "Scoping and Screening Methodology," Nuclear Management Company, LLC (NMC or the applicant) described the scoping and screening methodology used to identify the SSCs at Palisades Nuclear Plant (PNP) within the scope of license renewal and the SCs subject to an AMR. The staff reviewed the applicant's scoping and screening methodology to determine if it meets the requirements of 10 CFR 54.4(a) and 10 CFR 54.21 for scoping and screening, respectively.

In developing its scoping and screening methodology for the LRA, the applicant considered the requirements of 10 CFR Part 54, "Requirements for Renewal of Operating Licenses for Nuclear Power Plants" (the Rule), statements of consideration (SOCs) related to the Rule, and the guidance provided in Nuclear Energy Institute (NEI) 95-10, Revision 4, "Industry Guideline for Implementing the Requirements of 10 CFR Part 54 - The License Renewal Rule." Additionally, in developing this methodology, the applicant considered the correspondence between the staff of the U.S. Nuclear Regulatory Commission (NRC) (the staff) and other applicants, and/or NEI.

2.1.1 Introduction

In LRA Sections 2 and 3, the applicant provided the technical information required by 10 CFR 54.21(a). In LRA Section 2.1, the applicant described the process used to identify the SSCs that meet the license renewal scoping criteria of 10 CFR 54.4(a) and the process used to identify the SCs subject to an AMR, as required by 10 CFR 54.21(a)(1). In addition, the applicant provided the results of the process used to identify the SCs subject to an AMR in the following LRA sections:

- Section 2.2, "Plant Level Scoping Results"

- Section 2.3, "Scoping and Screening Results: Mechanical Systems"

- Section 2.4, "Scoping and Screening Results: Containments, Structures, and Component Supports"

- Section 2.5, "Scoping and Screening Results: Electrical and Instrumentation and Controls"

LRA Section 3, "Aging Management Review Results," contains the applicant's aging management results in the following LRA sections:

- Section 3.1, "Aging Management of the Reactor Coolant System"

- Section 3.2, "Aging Management of Engineered Safety Features Systems"

- Section 3.3, "Aging Management of Auxiliary Systems"

- Section 3.4, "Aging Management of Steam and Power Conversion System"

- Section 3.5, "Aging Management of Containments, Structures and Component Supports"

- Section 3.6, "Aging Management of Electrical and Instrumentation and Controls"

LRA Section 4, "Time-Limited Aging Analyses," contains the applicant's identification and evaluation of time-limited aging analyses (TLAAs).

2.1.2 Scoping and Screening Program Review

The staff evaluated the scoping and screening methodology in the LRA in accordance with the guidance contained in NUREG-1800, "Standard Review Plan for Review of License Renewal Applications for Nuclear Power Plants" (SRP-LR), dated July 2001, Section 2.1, "Scoping and Screening Methodology." The following regulations form the basis for the acceptance criteria for the scoping and screening methodology review:

- 10 CFR 54.4(a), as it relates to the identification of plant SSCs within the scope of the Rule

- 10 CFR 54.4(b), as it relates to the identification of the intended functions of plant SSCs determined to be within the scope of the Rule

- 10 CFR 54.21(a)(1) and (2), as they relate to the methods used by the applicant to identify plant SCs subject to an AMR

As part of the review of the applicant's scoping and screening methodology, the staff reviewed the activities described in LRA Sections 2.1 using the guidance contained in SRP-LR Section 2.1 to ensure that the applicant described a process for identifying SSCs within the scope of license renewal, in accordance with the requirements of 10 CFR 54.4(a).

In addition, the staff reviewed the activities described in LRA Sections 2.2 through 2.5 using the guidance described in SRP-LR Section 2.2, "Plant Level Scoping Results;" Section 2.3, "Scoping and Screening Results: Mechanical Systems;" Section 2.4, "Scoping and Screening Results: Containments, Structures, and Component Supports;" and Section 2.5, "Scoping and Screening Results: Electrical and Instrumentation and Controls," to ensure that the applicant described a process for determining structural, mechanical, and electrical components at PNP subject to an AMR, in accordance with the requirements of 10 CFR 54.21(a)(1) and (2).

In addition, the staff conducted a scoping and screening methodology audit at PNP in South Haven, Michigan during the week of June 27 through July 1, 2005. The audit focused on ensuring that the applicant had developed and implemented adequate guidance to conduct the scoping and screening of SSCs in accordance with the methodologies described in the LRA and the requirements of the Rule. The staff reviewed implementation procedures and technical reports describing the applicant's scoping and screening methodology. In addition, the staff conducted detailed discussions with the applicant on implementation and control of the license renewal program and reviewed administrative control documentation and selected design documentation used by the applicant during the scoping and screening process. The staff reviewed the applicant's processes for quality assurance (QA) with respect to development of the LRA and training and qualification of the LRA development team. The staff also reviewed a sample of system scoping and screening results reports for the main steam system (MSS) and shutdown cooling/low pressure safety injection (SDC/LPSI) system to ensure that the applicant had appropriately implemented the methodology outlined in the administrative controls and that the results were consistent with the current licensing basis (CLB) documentation. The staff documented its review in an audit trip report issued July 19, 2005. The report identifies several issues which require additional information from the applicant prior to completion of the review effort.

2.1.2.1 Implementation Procedures and Documentation Sources Used for Scoping and Screening

2.1.2.1.1 Summary of Technical Information in the Application

In LRA Section 2.1.1.1, "Plant Information Sources," the applicant cited the following information sources during the license renewal scoping and screening process:

- CLB for license renewal purposes, as defined in 10 CFR 54.3

- PNP equipment database (EDB)

- design-basis documents (DBDs) - The DBDs include information from the final safety analysis report (FSAR), technical specifications (TSs), industry codes, standards and regulations, regulatory correspondence, technical correspondence, analyses and reports, calculations, drawings, specifications, modifications, and vendor reports, specifications, and drawings.

- controlled plant reference drawings

- license renewal scoping boundary drawings

- Maintenance Rule information

- license renewal tools: license renewal database (LRDB)

The applicant stated that the in-scope boundary is depicted on the license renewal boundary drawings showing the in-scope components highlighted in color. The screening process evaluated the in-scope SCs to determine which were passive and long-lived. The result was a list of passive, long-lived components subject to an AMR.

2.1.2.1.2 Staff Evaluation

The staff reviewed the applicant's scoping and screening implementation procedures to verify that the process used to identify SCs subject to an AMR was consistent with the LRA and the SRP-LR. Additionally, the staff reviewed the scope of CLB documentation sources and the process used by the applicant to ensure that CLB commitments were appropriately considered and that the applicant adequately implemented the procedural guidance during the scoping and screening process.

Scoping and Screening Implementation Procedures. The staff reviewed the following scoping and screening methodology implementation procedures and engineering reports:

- License Renewal Project Guideline (LRPG) 1, Revision 2, "License Renewal Project Guidance"

- LRPG 3, Revision 3, "IPA Scoping and Screening"

- LR-TR-012, Revision 2, "Mechanical and Electrical Scoping and Screening Methodology and Summary Report"

- LR-TR-022-CS, Revision 2, "Civil Structural (C/S) Integrated Plant Assessment-Scoping/Screening and Aging Management Review Methodology and Results"

During the review of these procedures, the staff focused on the consistency of the detailed procedural guidance with information in the LRA, including the implementation of staff positions documented in the SRP-LR and interim staff guidance (ISG) documents.

The staff found overall direction for implementing 10 CFR 54 requirements in LRPG 1. The staff found guidance for determining plant systems and structures within the scope of the Rule, including guidelines for determining which components of the in-scope systems and structures were subject to an AMR. The applicant documented individual scoping and screening methodology results for civil/structural components in accordance with LRPG 3. The mechanical and electrical scoping and screening methodology was described in LR-TR-012.

After reviewing the LRA and supporting documentation the staff found that the scoping and screening methodology instructions were generally consistent with LRA Section 2.1. The applicant's methodology contained sufficient detail to provide concise guidance on the scoping and screening implementation process to be followed during the LRA activities.

Sources of Current Licensing Basis Information. The staff reviewed the scope and depth of the applicant's CLB review to verify that the methodology was sufficiently comprehensive to identify SSCs within the scope of license renewal, as well as SCs that require an AMR. As defined in 10 CFR 54.3(a), the CLB is (a) a set of NRC requirements applicable to a specific plant, (b) a licensee's written commitments for ensuring compliance with, and operation within, applicable NRC requirements, and (c) the plant-specific design bases that are docketed and in effect. The CLB included certain NRC regulations, orders, license conditions, exemptions, TSs, design-basis information documented in the most recent updated final safety analysis report (UFSAR), and licensee commitments remaining in effect that were made in docketed licensing correspondence, such as licensee responses to NRC bulletins, generic letters (GLs), and enforcement actions, as well as licensee commitments documented in NRC safety evaluations or licensee event reports (LERs).

During the audit the staff reviewed pertinent information sources utilized by the applicant. The staff reviewed samples of information utilized by the applicant including the PNP EDB, DBDs, controlled plant reference drawings, license renewal scoping boundary drawings, Maintenance

Rule information, and the LRDB reviewed by the applicant during the scoping and screening process. The DBDs include information from the FSAR, TSs, industry codes, standards and regulations, regulatory correspondence, technical correspondence, analyses and reports, calculations, drawings, specifications, modifications, and vendor reports, specifications, and drawings. The applicant stated that, although some of these documents were not considered to be a part of the CLB (such as Maintenance Rule scoping documents and the DBD), they were used to identify potential CLB functions and additional CLB references.

The staff determined that the LRA Section 2.1 description of the CLB and related documents used during the scoping and screening process is consistent with the guidance contained in the SRP-LR. In addition, the staff reviewed technical reports utilized to support identification of systems and structures relied upon to demonstrate compliance with the requirements of 10 CFR 54.4(a)(1) through (3) for safety-related (SR) criteria and nonsafety-related (NSR) criteria as well as compliance with the five regulations referenced. Project guidance documentation LRPG 1 and LRPG 3 comprehensively list documents used to support scoping and screening evaluations. The staff found these design documentation sources useful for ensuring that the initial scope of SSCs identified by the applicant was consistent with the plant's CLB.

2.1.2.1.3 Conclusion

On the basis of a review of information provided in LRA Section 2.1, a review of the applicant's detailed scoping and screening implementation procedures, and the results from the scoping and screening audit; the staff concludes that the applicant's scoping and screening methodology considered CLB information consistent with the guidance contained in the SRP-LR and is, therefore, acceptable.

2.1.2.2 Quality Controls (QA) Applied to LRA Development

The staff reviewed the QA controls used by the applicant to assure that the scoping and screening methodologies in the LRA were adequately implemented. Although the applicant did not develop the LRA under a 10 CFR 50 Appendix B QA program the staff determined that, during the LRA development, the applicant utilized the following QA processes:

- Implementation of the scoping and screening methodology was governed by written procedures, guidelines, and scoping position papers.

- The LRA was reviewed and approved by the applicant's Off-Site Review Committee and the Plant Review Committee prior to submittal to the staff.

- The applicant planned to retain certain license renewal documents as quality records or controlled documents.

- The applicant performed an industry peer review of license renewal activities.

- Nuclear Oversight performed two self-assessments in the implementation of license renewal procedures and position papers.

On the basis of its review of pertinent LRA development guidance, discussion with the applicant's license renewal personnel, and review of the two Nuclear Oversight quality audit reports, the staff concludes that these QA activities provided additional assurance that LRA development activities were performed consistently with the LRA descriptions.

2-5

2.1.2.3 Training

The staff reviewed the applicant's training process to ensure the guidelines and methodology for the scoping and screening activities were performed in a consistent and appropriate manner. The applicant developed several license renewal lesson plans used to train all technical leads, site personnel, and contractors performing license renewal activities. The license renewal project manager or discipline lead was responsible for determining training requirements for a given assignment. LRPG 2, Revision 2, "Staff Training Requirements and Qualifications," provides guidelines for exemptions from training. License renewal personnel were also required to review applicable license renewal regulations, NEI 95-10, and associated procedures. The staff reviewed the training records for the applicant's license renewal personnel and noted no discrepancies.

Based on discussions with the applicant's license renewal personnel responsible for the scoping and screening process and a review of selected documentation in support of the process, the staff concludes that the applicant's personnel understood the requirements and adequately implemented the scoping and screening methodology documented in the LRA. The staff concludes that license renewal personnel were adequately trained and qualified for license renewal activities.

2.1.2.4 Conclusion for Scoping and Screening Program Review

On the basis of its review of information in LRA Section 2.1, review of the applicant's detailed scoping and screening implementation procedures, discussions with the applicant's LRA personnel, and review of results from the scoping and screening audit, the staff finds that the applicant's scoping and screening program is consistent with the guidance contained in the SRP-LR and is, therefore, acceptable.

2.1.3 Plant Systems, Structures, and Components Scoping Methodology

In LRA Section 2.1 the applicant described the methodology used for scoping SSCs, in accordance with the requirements of the 10 CFR 54.4(a) scoping criteria. The applicant described the scoping process for the plant in terms of systems and commodity groups and then identified their system-level functions. Next, these functions were evaluated against the scoping criteria in 10 CFR 54.4(a)(1) - (3) to determine whether they performed or supported a license renewal system intended function. Finally, the applicant evaluated components that performed or supported system intended functions. The in-scope boundary was depicted on the license renewal boundary drawings. The applicant's scoping methodology described in the LRA is addressed in the following sections.

2.1.3.1 Application of the Scoping Criteria in 10 CFR 54.4(a)(1)

2.1.3.1.1 Summary of Technical Information in the Application

In LRA Section 2.1.2.1.1, "Scoping Criteria 1 - Safety Related Systems and Structures," the applicant described the scooping methodology required by 10 CFR 54, as it relates to SR criteria in accordance with 10 CFR 54.4(a)(1). With respect to the SR criteria, the applicant stated that the first scoping category in 10 CFR 54.4 involves SR SSCs. The license renewal criteria for SR SSCs are consistent with the applicant's CLB, Quality List, and Maintenance Rule classifications criteria.

The CLB definition for SR is not identical to the definition in the Rule. The CLB definition of SR is the safe shutdown earthquake definition described in 10 CFR Part 100, Appendix A. Part 54 of 10 CFR cites 10 CFR 50.67(b)(2) and 10 CFR 50.34(a)(1), in addition to 10 CFR Part 100.11. The CLB cites only 10 CFR 100.11. The applicant analyzed the differences in the SR definitions and the results are included in the following section.

2.1.3.1.2 Staff Evaluation

In accordance with 10 CFR 54.4(a)(1), an applicant must consider all SR SSCs relied upon to remain functional during and following a design-basis event (DBE) to ensure the following functions: (i) the integrity of the reactor coolant pressure boundary (RCPB), (ii) the capability to shut down the reactor and maintain it in a safe shutdown condition, or (iii) the capability to prevent or mitigate the consequences of accidents that could result in potential offsite exposures comparable to those in 10 CFR 50.34(a)(1), 10 CFR 50.67(b)(2), or 10 CFR 100.11.

With regard to identification of DBEs SRP-LR Section 2.1.3, "Review Procedures," states:

> The set of design basis events as defined in the rule is not limited to Chapter 15 (or equivalent) of the UFSAR. Examples of design basis events that may not be described in this chapter include external events, such as floods, storms, earthquakes, tornadoes, or hurricanes, and internal events, such as a high energy line break. Information regarding design basis events as defined in 10 CFR 50.49(b)(1) may be found in any chapter of the facility UFSAR, the Commission's regulations, NRC orders, exemptions, or license conditions within the CLB. These sources should also be reviewed to identify systems, structures, and components that are relied upon to remain functional during and following design basis events (as defined in 10 CFR 50.49(b)(1)) to ensure the functions described in 10 CFR 54.4(a)(1).

The applicant's program for satisfying the scoping requirements of 10 CFR 54.4(a)(1) required the identification of all major SR and NSR plant SSCs and the function(s) that each major SSC is required to accomplish.

The applicant performed scoping of SSCs for 10 CFR 54.4(a)(1) in accordance with procedures LRPG 3 and LR-TR-001-SR, Revision 2, "Components Identification and Data Processing for Safety Related SSC within Scope of 10 CFR 54.4(a)(1)." The staff reviewed these documents to identify the major SSCs and their respective functions, which were compared with the scoping criteria described in 10 CFR 54.4(a) to identify those functions that are within the scope of license renewal. SSC functions that meet the scoping criteria requirements of 10 CFR 54.4(a) are identified as intended functions in the Advanced License Extension (ALEX) license renewal database. During the document review, the staff noted that the CLB definition of "SR" was not identical to the definition in the Rule. As discussed above, 10 CFR 54.4(a)(1)(iii) cites 10 CFR 50.67(b)(2) and 10 CFR 50.34(a)(1), in addition to 10 CFR 100.11, whereas the CLB definition of SR cites only 10 CFR 100.11.

The staff reviewed the analysis of the differences in the SR definition performed by the applicant and documented in LR-TR-001-SR. In the analysis, the applicant stated that 10 CFR 50.67(b)(2) does not apply because PNP is not revising its accident source term. For the case of the 10 CFR 50.34(a)(1) reference, the 10 CFR 50.34(a)(1)(ii) dose limits only pertain to applicants for construction permits who apply on or after January 10, 1997. This information, if applicable, could have impacted the designation of components as SR or NSR. In addition, the applicant stated that 10 CFR 50.34(a)(1)(i) references only 10 CFR Part 100, which is consistent with the CLB. The applicant's analysis concluded that the portions of 10 CFR 50.34(a)(1) that could impact the PNP definition of SR, in a way that could potentially affect SSCs within the scope of license renewal, are not a part of the CLB; therefore, the staff finds that the differences in the applicant's definition did not result in any additional components being considered SR beyond those identified using the CLB.

To provide additional assurance that the applicant adequately implemented its SR scoping methodology, the staff reviewed a sample of the license renewal scoping report results and discussed the methodology and results with the applicant's personnel who were responsible for these evaluations. The staff verified that the applicant has identified and used pertinent engineering and licensing information to identify the SSCs required to be in scope in accordance with the 10 CFR 54.4(a)(1) criteria.

2.1.3.1.3 Conclusion

Based on this sample review, discussions with the applicant, and review of the applicant's scoping process, the staff determines that the applicant's methodology for identifying systems and structures meets the scoping criteria of 10 CFR 54.4(a)(1) and is, therefore, acceptable.

2.1.3.2 Application of the Scoping Criteria in 10 CFR 54.4(a)(2)

2.1.3.2.1 Summary of Technical Information in the Application

In LRA Section 2.1.2.1.2, "Scoping Criterion 2 - Non Safety Related Systems and Structures Affecting Safety Related Functions," the applicant described the scoping methodology as related to the NSR criteria, in accordance with 10 CFR 54.4(a)(2). The applicant evaluated the components that met 10 CFR 54.4(a)(2) using the following three categories: (1) NSR SSCs identified in the CLB, (2) NSR SSCs directly connected to SR SSCs, and (3) NSR SSCs not directly connected to SR SSCs. Categories 2 and 3 apply to NSR SCs that may not be specifically identified in the CLB. These two categories were part of an expanded scoping effort by the applicant in the application development process. A summary description of the three categories follows.

(1)   Category 1 NSR SSCs identified in the CLB include SSCs used to mitigate and protect SR SSCs from high-energy line break (HELB), internal/external flooding, and internal/external missiles. Category 1 also includes heavy load lifting equipment that could drop on and damage SR equipment. Category 1 SSCs that support operator habitability and access to SR equipment are also included within the scope of license renewal.

(2)   For Category 2 NSR SSCs directly connected SR SSCs (typically piping systems), the NSR piping and supports, up to and including the first equivalent anchor beyond the

safety/non-safety interface, were within the scope of license renewal in accordance with 10 CFR 54.4(a)(2). An actual seismic anchor is defined as a physical six-way barrier support (restraint against forces and movements in each of three orthogonal directions) that can be a fabricated, or a base-mounted component (e.g., pump, heat exchanger, tank, etc.). An equivalent seismic anchor is typically defined as at least two rigid supports in each of the three orthogonal directions. Components of the NSR piping segment up to and including the first seismic or equivalent anchor are included within the scope of license renewal.

(3) For Category 3 NSR SSCs not directly connected to SR SSCs, or connected downstream of the first seismic or equivalent anchor, the applicant concluded that the NSR SSC may be within scope of license renewal if its failure could prevent the performance of an SR SSC safety function, including (a) NSR piping failures on adjacent SSCs (e.g., pipe whip, jet impingement, spray, flooding, etc.) and (b) loss of NSR piping supports resulting in piping falling on SR SSCs (Seismic II/I). To determine which in-scope NSR SSCs may be included, the applicant initially evaluated two options - mitigative or preventive.

The mitigative option considers the effects of NSR SSCs failures on SR SSCs controlled by some features to protect the SR SSCs (e.g., whip restraints, spray shields, supports, barriers, etc). This mitigation is such that the failure of the NSR SSC will not prevent the performance of an SR SSC intended function identified in 10 CFR 54.4(a)(1). If the mitigative option is used, the mitigative feature (whip restraints, spray shields, supports, barriers, etc.) needs to be included within the scope of license renewal, in accordance with 10 CFR 54.4(a)(2). The NSR SSCs can be excluded from within the scope of license renewal, provided that adequate mitigative features address all potential failure locations that could result from aging.

If mitigative features are not installed or cannot adequately protect SR SSCs, then the preventive option is used. For the preventive option, vulnerable SR components in proximity to the NSR systems are identified by review of plant documentation or plant walk-downs to identify NSR systems, or portions of systems, that have the potential for spatial interaction (pipe whip, spray, flooding, etc.) with vulnerable SR equipment, assuming a failure occurs anywhere along the length of the SR system.

The applicant initially conducted scoping under the preventive option using engineering judgment to exclude certain components from within the scope of license renewal, but subsequently expanded the license renewal scope under this category by removing the exclusion guidelines (i.e., no limitation to exposure duration from spray/leakage and no limitation to distance) in assessing whether SR components may be excluded from within the scope of license renewal if it exists in the same general area of NSR piping systems. Thus, an augmented methodology was implemented requiring all pressurized liquid/steam systems in the general area of SR components, passive or active, within the scope of license renewal.

The LRA states that potential spatial interactions of nonsafety piping systems that may fall on or otherwise physically impact SR SSCs is considered seismic II/I. Palisades was not originally designed to seismic II/I. To address seismic II/I considerations at PNP, guidance from NEI 95-10 was utilized. The applicant determined that because NSR piping system supports are managed, falling of piping sections is not credible and the piping section itself would not be within the scope of license renewal in accordance with 10 CFR 54.4(a)(2) due to any physical impact hazard (although leakage/spray/flooding hazards may still apply). All supports for nonseismic piping systems with a potential for spatial interaction with SR SSCs are included within the scope of license renewal. These supports are addressed in a commodity fashion within the civil/structural area utilizing a "spaces" approach, in which all pipe supports that exist in areas that contain SR components are included within the scope of license renewal.

Air and gas systems are included within the scope of license renewal utilizing a "spaces' approach and are not considered hazardous to other plant equipment. A site-specific review of air/gas systems operating experience verified that they have not negatively affected other plant equipment. The applicant's review of industry operating experience also revealed no events of this nature; therefore, the applicant concluded that the air/gas systems are not within the scope of license renewal.

Most heating, ventilation, and air conditioning (HVAC) ducts and supports are NSR and located overhead many SR SSCs. Similar to air/gas pipe systems, HVAC ducts are not hazardous to other equipment; however, there is a potential falling spatial interaction. The in-scope portion was determined by the same "Scoping Criterion 2" methodology used for piping systems. The boundaries of the in-scope portions of the HVAC system include HVAC components and ducting.

2.1.3.2.2 Staff Evaluation

In accordance with 10 CFR 54.4(a)(2), the applicant must consider all NSR SSCs whose failure could prevent satisfactory accomplishment of SR SSCs relied upon to remain functional during and following a DBE to ensure the following functions: (i) the integrity of the RCPB, (ii) the ability to shut down the reactor and maintain it in a safe shutdown condition, or (iii) the capability to prevent or mitigate the consequences of accidents that could result in potential offsite exposures comparable to those referenced in 10 CFR 50.34(a)(1), 10 CFR 50.67(b)(2), or 10 CFR 100.11.

By letters dated December 3, 2001 (ADAMS Accession No. ML013380013), and March 15, 2002 (ADAMS Accession No. ML020770026), the staff issued a staff position to the NEI that provided staff expectations for determining what SSCs meet the 10 CFR 54.4(a)(2) criterion (i.e., all NSR SSCs whose failure could prevent satisfactory accomplishment of any SR functions identified in 10 CFR 54.4(a)(1)(i) - (iii)).

The letter dated December 3, 2001, provides specific examples of operating experience that identified pipe failure events (summarized in NRC Information Notice (IN) 2001-09, "Main Feedwater System Degradation in Safety Related ASME Code Class 2 Piping Inside the Containment of a Pressurized Water Reactor") and the approaches the staff considers acceptable to determine which piping systems should be included within the scope of license renewal based on the 10 CFR 54.4(a)(2) criterion. The letter dated March 15, 2002, further describes the staff's expectations for the evaluation of non-piping SSCs to determine which additional NSR SSCs are within the scope of license renewal. The position states that applicants should not consider hypothetical failures, but rather should base their evaluation on the plant's CLB, engineering judgement and analyses, and relevant operating experience. The paper further describes operating experience as all documented plant-specific and industry-wide experience that can be used to determine the plausibility of a failure.

Documentation would include NRC generic communications and event reports, plant-specific condition reports, industry reports like safety operational event reports, and engineering evaluations.

The staff reviewed LRA Section 2.1.2.1.2, LR-TR-002-NSAS, Revision 3, "Components Identification and Data Processing for NonSafety Related Affecting Safety Related SSCs within the Scope of License Renewal," and LRPG 3.

As part of the applicant's licensing basis, the term "NSR affecting SR" was not applied to the plant when it was licensed under 10 CFR Part 50; therefore, the applicant has no comprehensive listing of NSR components affecting SR components. The applicant utilized the scoping process to identify NSR components affecting SR components, as defined under 10 CFR Part 54.

The applicant evaluated the components that met 10 CFR54.4(a)(2) using the three categories set forth in the NRC guidance to the industry on identification and treatment of SSCs meeting 10 CFR 54.4(a)(2). The evaluations of each of the categories are detailed below.

Nonsafety-Related SSCs Identified in the CLB (Category 1). In LRA Section 2.1.2.1.2, the applicant stated that NSR SSCs that directly support the function of an SR SSC or whose failure could prevent the performance of an SR function were included within the scope of license renewal. LR-TR-002-NSAS further states that, for an SSC to be within the scope of license renewal, it must either be in the plant CLB or identified for inclusion as a special case by NRC/NEI guidance relevant to license renewal.

In the CLB, the applicant identified a number of event-related issues that met the 10 CFR 54.4(a)(2) criteria, including HELBs, internal/external flooding, internal/external missiles, heavy load lifting equipment, and operator habitability and access. The applicant described each issue in LR-TR-002-NSAS.

The applicant used FSAR Section 5.2, "Classification of Structures, Systems and Components," as the primary source of CLB information for determining which system portions or functions were SR. In order to identify NSR SSCs, the applicant defines NSR equipment as that not meeting the definition of SR equipment. NSR SSCs were then determined using all existing SSCs not identified as SR. Another approach used by the applicant was a review of the CLB for events initiated by NSR SSCs. The use of NRC guidance and a review of the CLB gave the applicant the list of event-related issues described above for consideration and evaluation against the 10 CFR 54.4(a)(2) criteria.

The staff finds that the applicant has considered an adequate scope of CLB information and implemented a reasonable approach for scoping of NSR SSCs that provide support functions to SR intended functions.

Nonsafety-Related SSCs Directly Connected to Safety-Related SSCs (Category 2). The applicant identified NSR SSCs that directly affect the ability of attached SR SSCs to perform functions. The applicant's review addressed two areas in considering NSR SSCs directly connected to SR SSCs:

(1) The possibility that the pressure boundary related spatial interactions of the NSR piping could damage adjacent SR components.

(2) Ensure that the attached NSR piping and supports maintain their structural integrity to ensure the structural integrity of the attached SR piping.

For the pressure boundary related spatial interactions, pipe whip, jet impingement, spray, and flood were included as part of the analysis of the NSR SSCs not directly connected to SR SSCs (refer to the Category 3 discussion below).

To ensure that the attached NSR piping and supports maintain their structural integrity, the applicant included pipe supports up to and including the first seismic or equivalent seismic anchor beyond the SR/NSR interface as within the scope of license renewal. Specifically, the applicant located the first equivalent seismic anchor on the NSR piping, and included all of the intermediate piping components and supports between the anchor and the nonsafety to safety-related piping interface as within the scope of license renewal. The applicant used piping and instrumentation diagrams (P&IDs), piping stress reports, stress isometrics documents, and plant walk-downs to determine the systems or system portions located in each building or structure.

During its review of the LRA and the license renewal procedures, the staff identified areas in which additional information was necessary to answer questions associated with the applicant's methodology for performing license renewal scoping of NSR SSCs, pursuant to the requirements of 10 CFR 54.4(a)(2). The applicant responded to the staff's requests for additional information (RAIs) as discussed below.

In RAI 2.1-1, dated July 27, 2005, regarding use of alternatives to seismic anchors or equivalent anchors, that staff stated that LRPG 3 Section 6.1.3 provides guidance for establishing system boundaries for NSR piping systems connected directly to SR piping systems. The guideline states, in part, that for NSR SSCs directly connected to SR SSCs, the NSR piping and supports up to and including the first anchor, or equivalent anchor, beyond the SR/NSR interface, are within the scope of license renewal. An alternative to specifically identifying a seismic anchor or series of supports comprising equivalent anchors that support the SR/NSR piping interface is to include enough of the NSR piping run to conservatively encompass the anchors and ensure that piping and anchor intended functions are maintained. The guideline states that the following examples are typically used to establish the end of pipe stress analysis models and can be used to define conservative end points in the license renewal boundary:

- a flexible connection is generally considered a pipe stress analysis model end point because it does not support loads or transfer loads across it on to connecting piping

- a point where buried piping enters the ground because the ground acts like an anchor

The staff requested that the applicant provide the technical basis for establishing a flexible connection and a point where buried piping enters the ground as adequate end points for determining piping within the scope of license renewal.

In its response, by letter dated August 27, 2005, the applicant provided information concerning flexible connections and buried piping used as adequate endpoints. The applicant included NEI guidance and PNP's current analysis criteria, which state that flexible connections are adequate endpoints due to the de-coupling function of the flexible connection. The use of a flexible connection prevents piping system loads to be transferred across the SR/NSR connection. According to NEI guidance, establishing an endpoint at the flexible connection is a conservative endpoint in the license renewal boundary.

In its response, the applicant further stated that the use of the point where buried piping enters the ground as the anchor point is a common practice in stress analysis modeling in the industry. Contrary to the established NEI guidance, PNP did not include the buried portion of the piping as within the scope of license renewal. A justification for this assumption is the stiffness contribution of the buried piping and the restraint provided by the grouted penetration where the piping typically runs.

After subsequent discussions with the staff, the applicant determined to include the buried portion of the piping within the scope of license renewal and to subject them to aging management.

The staff finds the applicant's response to RAI 2.1-1 acceptable after review of the additional information. Therefore, the staff's concern described in RAI 2.1-1 is resolved.

In RAI 2.1-2, dated July 27, 2005, regarding 10 CFR 54.4(a)(2) scoping criteria for NSR SSCs, the staff stated that, by letter dated December 3, 2001, the staff issued a staff position to the NEI describing areas to be considered and options to be used in determining which SSCs meet the 10 CFR 54.4(a)(2) criterion (i.e., all NSR SSCs whose failure could prevent satisfactory accomplishment of any of the SR functions identified in 10 CFR 54.4(a)(1)(i) - (iii)). Specifically, the staff's concern is that Seismic II/I piping, though seismically supported, would be subject to the same plausible aging effects as SR piping. For example, depending on piping material, geometrical configuration, and operating conditions like water chemistry, temperature, flow velocity, and external environment, erosion and corrosion may be plausible aging effects for some Seismic II/I piping. Those effects, if not properly managed, could result in age-related failures and adversely impact the safety functions of SR SSCs.

Based on a review of the LRA, the applicant's scoping and screening implementation procedures, and discussions with the applicant, the staff determined that additional information was required with respect to certain aspects of the applicant's evaluation of the 10 CFR 54.4(a)(2) criteria.

In RAI 2.1-2, dated July 27, 2005, the staff requested that the applicant provide information in two areas:

(1) LRA Section 2.1.2.1.2, item (3)(b) states that as long as the supports for these piping systems are managed, falling of piping sections is not credible, and the piping section itself would not be within the scope of license renewal, in accordance with 10 CFR 54.4(a)(2), due to physical impact hazard (although the leakage/spray/flooding hazard may still apply). Therefore, the staff requested that the applicant clarify whether it considered flow-accelerated corrosion (FAC) piping failures, as demonstrated in Information Bulletin 2001-09, regardless of whether the piping supports remain intact.

(2) LRA Section 2.1.2.1.2, item (3)(a) states that all pressurized liquid/steam systems in the general area of SR components, passive or active, be considered within the scope of

license renewal. The staff requested that the applicant clarify if it used system pressure as a means to exclude any liquid or steam piping systems or system portions from the scope of license renewal. Specifically, the staff requested the applicant to clarify if it considered nonpressurized liquid or steam systems within the scope of license renewal.

In its response, by letter dated August 27, 2005, the applicant addressed each area of RAI 2.1-2, as described below.

(1) Concerning FAC, the applicant stated that dynamic and spatial effects due to potential FAC failures were considered as part of the analysis for NSR piping whose failure could affect SR SSC. LRA Section B2.1.11 describes the FAC Program and includes procedures and administrative controls to maintain structural integrity of carbon steel and low-alloy steel lines containing high-energy fluids. To clarify the seismic II/I considerations and FAC piping failures, the applicant revised LRA Section 2.1.2.1.2(3)(b) to include the exception of FAC failures in the assumption used for falling of piping segments.

(2) The applicant stated that system pressure did not exclude piping systems, with one exception. The exception was nonpressurized chemical addition tanks and piping with spill retention dikes. These were excluded from within the scope of license renewal if there were no SR SSCs within the dike-enclosed area.

The staff finds the applicant's response to RAI 2.1-2 acceptable after review of the additional information. Therefore, the staff's concern described in RAI 2.1-2 is resolved.

Nonsafety-Related SSCs Not Directly Connected to Safety-Related SSCs (Category 3). As stated in LRA Section 2.1.2.1.2, item 3, the applicant evaluated NSR SSCs not directly connected to SR SSCs to determine if their failure could prevent the performance of the safety function of an SR SSC. The applicant evaluated two failure scenarios:

- Consequences due to the loss of pressure boundary of the NSR piping on adjacent SSCs (e.g., pipe whip, jet impingement, spray, flooding, etc.)

- Loss of NSR piping support resulting in a potential fall onto adjacent SSCs (Seismic II/I).

The applicant followed guidance from the NRC and NEI 95-10 to evaluate pressure boundary spatial interactions to determine NSR SSCs within the scope of license renewal, including the mitigative and preventive options described in the March 15, 2002, letter. The applicant identified and included in-scope mitigative features during the review of the CLB. Per NRC guidance, the applicant needed to demonstrate that the identified mitigative features are adequate to protect SR SSCs from failures of NSR piping segments. With that demonstration and the inclusion of these mitigative features within the scope of license renewal, the piping segments themselves need not be included within the scope of license renewal. Where the applicant found no piping mitigative features installed, or could not demonstrate that the installed mitigative features safely protected SSCs, the preventive option was used. The staff noted that the preventive option brought the entire NSR piping system into the scope of license renewal and that an AMR was performed on the components within the piping system to ensure that age-related degradation of nonsafety SSCs does not lead to interactions with SR SSCs not previously considered and might create additional failures of the SR SSCs.

2.1.3.2.3 Conclusion

Based on a review of sample systems, discussions with the applicant, and review of the applicant's scoping process, the staff determines that the applicant's methodology for identifying systems and structures meets the scoping criteria of 10 CFR 54.4(a)(2) and is, therefore, acceptable.

2.1.3.3 Application of the Scoping Criteria in 10 CFR 54.4(a)(3)

2.1.3.3.1 Summary of Technical Information in the Application

In LRA Section 2.1.2.1.3, "Scoping Criterion 3 - Systems and Structures Required by Other Regulations Identified in 10 CFR 54," the applicant described the scoping methodology for SSCs relied upon in safety analyses or plant evaluation reports performing an intended function that demonstrates compliance with NRC regulations. SSCs for fire protection (FP), environmental qualification (EQ), pressurized thermal shock (PTS), anticipated transient without scram (ATWS), and station blackout (SBO) have been included within the scope of license renewal in accordance with the criteria of 10 CFR 54.4(a)(3). For each of these five applicable regulations, the applicant dedicated an LRA subsection that describes the methodology and reports used to determine the scope of SSCs required by 10 CFR 54.4(a)(3). The applicant utilized technical reports and equipment lists for input to the scoping process.

Fire Protection (FP). In LRA Section 2.1.2.1.3, item 1, the applicant described the scoping of SSCs required to demonstrate compliance with fire protection requirements of 10 CFR 50.48. The technical reports and equipment lists for fire protection include: (a) Fire Protection Program Report, (b) Post Fire Safe Shutdown Analysis Report, (c) Safe Shutdown Equipment (Appendix R), (d) Color Coded Appendix R Safe Shutdown Equipment Drawings, and (e) Fire Hazard Analysis. Using these information sources, the applicant reviewed the components in its EDB to determine those required to support the safe shutdown functions for fire protection. Components supporting safe shutdown were added to the LRDB, as necessary. Those SSCs which are relied upon in safety analyses or plant evaluations to perform functions demonstrating compliance with the Commission's regulations for fire protection were included within the scope for license renewal.

Environmental Qualification. The applicant selected the electrical equipment required for EQ from the EQ master equipment list (MEL), which defines all equipment within the scope of 10 CFR 50.49. Identified systems were placed within the scope of license renewal and the components from the MEL define the boundary.

Pressurized Thermal Shock. The applicant addressed the PTS requirement in a time-limited aging analysis (TLAA) in LRA Section 4.2.2. For this requirement the only component within the scope of license renewal is the reactor vessel.

Anticipated Transients Without Scram. The applicant described the scoping of SSCs required to demonstrate compliance with the ATWS requirements of 10 CFR 50.62 in LRA Section 2.1.2.1.3, item 4, and FSAR Section 7.4. PNP has circuitry independent of the reactor protection system (RPS) to shutdown the reactor, trip the turbine, and start the turbine driven auxiliary feedwater (AFW) pump following a transient in which the RPS does not trip in the normal fashion. The LRA states that those SSCs relied upon in safety analyses or plant evaluations to perform a function that demonstrates compliance with the Commission's regulations for ATWS are within the scope of license renewal.

Station Blackout. The applicant described the scoping criteria in LRA Section 2.1.2.1.3, item 5 and in technical paper LR-TR-007-SBO, "Component Identification and Data Processing for SSC Within Scope of 10 CFR 54.4(a)(3) for Station Blackout." The applicant stated that SCs identified are required to restore offsite power consistent with GALL. System boundaries and components required to support system functions were included within the scope of license renewal due to SBO requirements.

2.1.3.3.2 Staff Evaluation

Pursuant to 10 CFR 54.4(a)(3), the applicant must consider all SSCs relied on in safety analyses or plant evaluations to perform functions demonstrating compliance with the Commission's regulations for fire protection, EQ, PTS, ATWS, and SBO.

SRP-LR Section 2.1.3.1.3, "Regulated Events," states that all SSCs relied upon in the plant's CLB (as defined in 10 CFR 54.3), plant-specific operating experience, industry-wide operating experience (as appropriate), and safety analyses or plant evaluations to perform functions demonstrating compliance with 10 CFR 54.4(a)(3), are required to be included within the scope of the Rule. However, consideration of hypothetical failures that could result from system interdependencies not part of the CLB and not previously experienced is not required.

The staff reviewed the applicant's approach for identifying SSCs relied upon to perform functions related to the five regulations described in 10 CFR 54.4(a)(3). As part of this review, the staff discussed the methodology with the applicant's LRA team, reviewed the documentation developed to support the review, and evaluated a sample of the resultant SSCs identified as within the scope of license renewal in accordance with the 10 CFR 54.4(a)(3) criteria.

For the fire protection regulated event, the staff reviewed the LRA, LR-TR-003-FP, Revision 3, "Component Identification and Data Processing for SSCs within Scope of 10 CFR 54.4(a)(3) for Fire Protection," and LR-TR-015-FP-APPR Motors, Revision 0, "Fire Protection - APPR Spare Motors." Components selected by the applicant to satisfy the post-fire safe shutdown requirements of Appendix R are listed in the Post-fire shutdown analysis and the safe shutdown equipment list. The LRA and LR-TR-003-FP identify the equipment within the scope of license renewal for fire protection.

The scoping methodology also applied to the Appendix R post-fire repair equipment maintained in storage. This information is also identified in LR-TR-015-FP-APPR. The Appendix R stored equipment restores power to pre-selected plant components and cools certain areas after a fire to attain cold shutdown. This scoping methodology was in accordance with the guidelines in draft ISG-07.

For the EQ regulated event, the staff reviewed the LRA and LR-TR-004-EQ, Revision 2, "Component Identification and Data Processing for SSCs within the Scope of 10 CFR 54.4(a)(3) for Environmental Qualification." The applicant's EQ MEL identifies all equipment within the scope of 10 CFR 50.49. The types of components subject to the EQ Program are listed in the EQ MEL.

The list of EQ equipment is also provided in LR-TR-004-EQ, Attachment 1. The list does not include a list of individual components, such as breakers, fuses, and cables associated with this equipment; however, the cables, connectors, and terminations were included in an electrical AMR that treats these components as commodities. The staff determines that the applicant's approach to scoping SSCs needed to meet EQ requirements was acceptable.

For the PTS regulated event, the staff reviewed the LRA and LR-TR-005-PTS, Revision 2, "Component Identification and Data Processing for SSCs within the Scope of 10 CFR 54.4(a)(3) for Pressurized Thermal Shock." LR-TR-005-PTS states that the applicant had performed a review of the applicability of 10 CFR 50.61, "Fracture Toughness Requirements for Protection Against Pressurized Thermal Shock Events." The applicant determined that only the reactor vessel is within the scope of license renewal, in accordance with the current NRC regulatory guidance. The staff found the applicant's scoping methodology acceptable.

For the ATWS regulated event, the staff reviewed the LRA and LR-TR-006-ATWS, Revision 2, "Component Identification and Data Processing for SSCs within the Scope of 10 CFR 54.4(a)(3) for Anticipated Transients Without Scram." LR-TR-006-ATWS provided ATWS electrical and P&IDs from DBD 2.05 and a table that identified components within the scope of 10 CFR 54.4(a)(3) for ATWS.

The staff reviewed FSAR Section 7.4.3, which contains the design description of the AFW controls used for ATWS. The applicant stated that it has a diverse circuitry independent of the RPS to shut down the reactor, trip the turbine, and start the turbine driven AFW pump following a transient in which the RPS does not trip in the normal fashion. This circuitry was added during the 1990-91 refueling outage to satisfy the 10 CFR 50.62 ATWS rule. Upon a loss of direct current electrical power, the turbine driven pump will start automatically via the diverse start system added for ATWS. The applicant stated that these components were determined to be within the scope of license renewal.

The staff noted that the cables and connectors associated with ATWS signal detection were included in the Cables and Terminations commodity group. Additional SSCs that support ATWS (e.g. control drive system, reactor protective system) were also included within the scope of license renewal. The staff determines that the applicant's approach to scoping SSCs relied on to demonstrate compliance with the ATWS Rule (10 CFR 50.62) was adequate.

For the SBO regulated event, the staff reviewed the LRA and LR-TR-007-SBO, Revision 3, "Component Identification and Data Processing for SSCs within the Scope of 10 CFR 54.4(a)(3) for Station Blackout." The staff reviewed the applicant's methodology for identifying systems and components meeting the scoping criteria of 10 CFR 54.4(a)(3). As part of this review, the staff sampled the LRDB scoping results and the analyses and supporting documentation.

The staff discussed the methodology and results with the applicant's personnel responsible for these evaluations. The staff verified that the applicant had identified and used pertinent engineering and licensing information to identify SSCs required to be within the scope of license renewal in accordance with the 10 CFR 54.4(a)(3) criteria. However, the staff's review of the LRA and license renewal procedures found an area in which additional information was necessary, associated with the applicant's methodology for performing license renewal scoping and screening of SBO related equipment in accordance with the requirements of 10 CFR 54.4(a)(3).

In RAI 2.5-1, dated July 26, 2005, the staff stated that the NRC had issued a staff position (ISG-02) to the NEI, by letter dated April 1, 2002, describing the plant system portion of the offsite power system connecting to the offsite power source that should be included within the scope of license renewal. Consistent with the staff position described in the aforementioned letter, in RAI 2.5-1, the staff requested that the applicant provide additional information regarding the scoping methodology associated with the 10 CFR 54.4(a)(3) evaluation. Specifically, the staff requested that the applicant describe the scoping methodology for determining the license renewal boundary within the switchyard. This request was to ensure that both coping and recovery portions of the SBO mitigation strategy were evaluated as part of the scoping evaluation.

In its response, by letter dated August 27, 2005, the applicant provided information concerning the SBO methodology. The applicant stated that the scoping of SBO equipment was performed in accordance with ISG-2. The applicant provided a drawing of the offsite power source connections to the plant distribution system and components within the scope of license renewal, as described in LRA Section 2.1.2.1.3. The applicant stated that the scoping boundary starts at the motor operated disconnects 24F1 and 24R2, which are the preferred offsite power source and alternate offsite power source connection points, respectively and includes the disconnects, the offsite transformers (startup transformers, safeguards transformer), overhead and buried cables, buses, and isolation breakers. Upstream of the motor operating disconnects are the 345kV buses, switchyard components, and the incoming lines, all parts of the transmission (grid) system, not the plant system.

As part of the staff's scoping methodology review, the Electrical Engineering Branch (EEEB) evaluated the switchyard components and the license renewal boundaries established by the applicant. EEEB concluded that the switchyard configuration provides a reliable source of power to the preferred and alternate offsite circuits. In addition, EEEB concluded that the established boundaries are adequate for switchyard equipment within the scope of license renewal.

The staff finds the applicant's response to RAI 2.5-1 acceptable based on the additional information provided by the applicant. Therefore, the staff's concern described in RAI 2.5-1 is resolved.

2.1.3.3.3 Conclusion

Based on the sample review, discussions with the applicant, and review of the applicant's scoping process, the staff determines that the applicant's methodology for identifying systems and structures meets the scoping criteria of 10 CFR 54.4(a)(3), and is, therefore, acceptable.

2.1.3.4 Plant-Level Scoping of Systems and Structures

2.1.3.4.1 Summary of Technical Information in the Application

Plant and System Level Scoping. In LRA Section 2.1.2.1, the applicant described the scoping methodology for SR and NSR systems, structures, and commodity groups relied upon to perform a function for any of the five regulations described in 10 CFR 54.4(a)(3). The scoping methodology is consistent with NRC guidance in the SR-LR and industry guidance in NEI 95-10. In LRA Section 2.2, the applicant evaluated systems, structures, and commodities to determine whether they were within the scope of license renewal by using the methodology described in

LRA Section 2.1. In LRA Table 2.2-1, "Plant Level Scoping Results," the results of plant scoping for all major SSCs and commodity groups are provided.

The applicant's system-level function information was used to create a system function list in the LRDB. CLB references were added to each system function. In LRA Sections 2.1.2.1.1 through 2.1.2.1.3, the applicant described how all three 10 CFR 54.4 criteria were used to generate the list of SSCs considered within the scope of license renewal. Not all components of each system and structure within the scope of license renewal were themselves within the scope because not all components support system intended functions. Therefore, where possible, boundaries were depicted on drawings with color overlays to indicate those SSCs within the scope of license renewal.

Component Level Scoping. After the applicant identified the intended functions of systems or structures within the scope of license renewal, a review determined which components of each in-scope system and structure support license renewal intended functions. The components that support intended functions were considered within the scope of license renewal and screened to determine if an AMR was required. The applicant considered three component classifications during this stage of the scoping methodology: (1) mechanical, (2) civil/structural, and (3) electrical.

Commodity Groups Scoping. The applicant described the use of commodity groups for mechanical, civil/structural, and electrical components in LRPG 3. The applicant grouped components with similar design, functions, materials of construction, aging management practices, locations, and/or environments. These groups were then evaluated against the 10 CFR 54.4(a)(1) - (3) criteria to determine whether they support a license renewal intended function. The applicant stated that even if only a portion of a system or commodity group met the scoping criteria of 10 CFR 54.4, the system or commodity group was identified as within the scope of license renewal.

Descriptions of the commodity groups were developed based on the commodity group functions; a detailed listing of the types of structural components (e.g., columns, beams, walls, ceilings, etc.) required to support those functions were included in the description. Next, the commodity groups' intended functions were evaluated to determine which were within the scope of license renewal; boundary descriptions were developed to detail the structural components required to support the in-scope functions.

In summary, for civil/structural components, the applicant based the grouping criteria primarily on design function. Choice of design-based commodity groups were made considering the component group's design functions as well as materials, environments, locations, and expected aging management programs (AMPs). The applicant established boundaries at the building column lines and/or physical barriers.

For mechanical systems, some components within a system were removed from the mechanical system scoping and were further generically addressed in commodity groups. The applicant left electrical components within mechanical systems in their respective electrical systems; however, some electrical components were left in their respective mechanical systems because they served a passive pressure boundary function.

As most electrical and instrumentation and controls (I&C) components are active, they were screened out and therefore not subject to AMR. For several passive and long-lived electrical and I&C components, commodity groups were established. In LRA Section 2.5.1, the applicant described the commodity groups used to evaluate all in-scope electrical and I&C components subject to AMR. During the scoping process, boundaries for each system and commodity group based on system function were identified and electrical and I&C components were assigned to commodity groups.

Insulation. LRA Section 2.1.3.4 addresses screening of thermal insulation. The applicant determined from the CLB that only a small portion of the thermal insulation is within the scope of license renewal. During review of insulation for AMR, the applicant further determined that insulation is not subject to any aging effects requiring management (AERMs).

In two cases insulation was determined to be within the scope of license renewal: (1) main feedwater and main steam penetrations with the intended function to limit adjacent concrete temperatures to less than 179 °F and (2) stainless steel mirror insulation on the reactor vessel supplied as part of the seismically qualified nuclear steam supply system for spatial impact of SR components.

There were no other plant locations within PNP where insulation on piping and components was credited to reduce heat transfer for individual room heat load calculations in support of accident analysis or safe shutdown for regulated events.

Consumables. In SRP-LR Table 2.1-3, guidance is given for screening consumables. According to this table, consumables may be divided into the following four categories for the purpose of license renewal: (a) packing, gaskets, component seals, and O-rings, (b) structural sealants, (c) oil, grease, and component filters, and (d) system filters, fire extinguishers, fire hoses, and air packs. For category (b) these sub-components may perform a function without moving parts or configuration changes and they are not typically replaced.

2.1.3.4.2 Staff Evaluation

The staff reviewed the applicant's methodology for scoping plant systems and components for consistency with 10 CFR 54.4(a). The methodology used to determine the scope of systems and components was documented in LRPG 3, LR-TR-012, and LR-TR-022-CS IPA; plant-level scoping results are identified in LRA Table 2.2-1. The scoping process defined the entire plant in terms of major systems, structures, and commodity groups, as documented in LRPG 3. Specifically, LRPG 3 determines which systems and structures are included within the scope of license renewal. The process was completed for all systems and structures to ensure that the entire plant was addressed. The applicant's personnel initially evaluated systems and structures identified in the FSAR, the Maintenance Rule Scoping Document, and the EDB. The ALEX database was then populated with all components from the EDB and systems, structures and functions from the Maintenance Rule database. The following information was documented in the ALEX database: (1) the system or structure name and designated abbreviation, (2) a concise but complete description of the system or structure, (3) a complete list of system or structure functions (which may or may not be intended functions), and (4) applicable design or CLB references for the information obtained. The staff found the applicant's approach for system and structure scoping consistent with the methodology in LRA Sections 2.1.1, 2.1.2, and 2.2.

The staff noted that a system or component was presumed to be within the scope of license renewal if it performed one or more SR functions or met the other scoping criteria per the Rule, as determined by CLB review. Components supporting intended functions also were considered within the scope of license renewal. In discussions with the applicant's license renewal personnel the staff noted that the identified SSC functions were compared to a list of scoping criteria to determine functions that met the scoping criteria of 10 CFR 54.4(a). Components supporting intended functions were further screened to determine if they required AMR. The staff did not identify any discrepancies in the applicant's methodology.

The staff reviewed the applicant's methodology to generate commodity groups. LRPG 3 states that commodity groups were developed based on SCs' similar characteristics, environment, design, functions, or aging management practices. The staff reviewed the commodity group-level functions identified and evaluated by the applicant against criteria provided in 10 CFR 54.4(a)(1) - (3). This process determined whether the commodity group was considered within the scope of license renewal. The staff found the methodology acceptable.

The staff reviewed the results of the scoping process documented in LR-TR-012 and LR-TR-022-CS IPA. This documentation includes descriptions of the systems or structures, the 10 CFR 54.4(a) scoping criteria met, and references to the license renewal scoping and screening report for each system or structure. The staff also reviewed a sample of the applicant's scoping documentation and concluded that it contains an appropriate level of detail to document the scoping process.

The applicant reviewed the CLB and determined that only a small portion of the thermal insulation was within the scope of license renewal. The staff noted that through AMR, the applicant determined that insulation is not subject to any AERMs. The applicant's methods and conclusions are consistent with the precedents established by past license renewal applicants.

During the staff's discussions with the applicant, it stated that there are no locations where insulation on piping and components is credited to reduce heat transfer for individual room heat load calculations in support of accident analyses or safe shutdown for regulated events.

The staff questioned why the AFW system pump room piping insulation was not within the scope of license renewal. To support this determination, the applicant provided an engineering analysis that calculated AFW room temperature with a loss of ventilation, but with the piping insulation installed. The applicant however, stated that other rooms containing SR equipment had analyses performed that did not credit insulation to reduce heat transfer in support of accident analyses. There was no documentation presented to the staff to prove that the AFW system thermal insulation is not credited to reduce room temperatures and prevent a loss of system intended function.

In RAI 3.4-1, dated July 26, 2005, the staff requested that the applicant provide additional justification that the pipe insulation in the AFW pump room is not required to ensure temperatures remain below the values that could cause SR equipment in the room to fail.

In its response, by letter dated August 27, 2005, the applicant provided information concerning thermal insulation. The applicant determined that the insulation on the steam supply and exhaust piping for the steam-driven AFW pump within the AFW pump room is within the scope of license renewal. The applicant stated that it would update the appropriate sections of the LRA to include the AFW pump steam piping insulation within the scope of license renewal with AMR results.

In its response, by letter dated October 31, 2005, the applicant provided additional information about thermal insulation. The applicant added pipe insulation as a component subject to an AMR.

The staff found the applicant's response to RAI 3.4-1 acceptable after review of the additional information provided by the applicant. Therefore, the staff's concern described in RAI 3.4-1 is resolved.

The staff reviewed LRA Section 2.1.3.2 for consumables. This section states that, "Consumables are a special class of short-lived items that can include packing, gaskets, component seals, O-rings, oil grease, component filters, system filters, fire extinguishers, fire hoses, and air packs."

In RAI 2.1-3, dated July 27, 2005, the staff stated that, based on the above statement and evaluation of implementing guidance, engineering evaluations, and discussions with the applicant's license renewal personnel additional information was necessary to complete the review. Therefore, the staff requested that the applicant clarify how it considered, in the screening process, structural sealants in the identification of short-lived components and consumables.

In its response, by letter dated August 27, 2005, the applicant provided the information concerning structural sealants. The applicant stated that structural sealants were included within the scope of license renewal and an AMR was performed, as part of the miscellaneous structural and bulk commodities in LRA Section 2.4.8. This commodity includes miscellaneous component types of various materials that were identified as requiring an AMR, but did not fit into other structural commodity groups. A summary of aging management evaluations for the miscellaneous structural and bulk commodities is presented in LRA Table 3.5.2-8.

The staff finds the applicant's response to RAI 2.1-3 acceptable after review of the additional information provided by the applicant. Therefore, the staff's concern described in RAI 2.1-3 is resolved.

2.1.3.4.3 Conclusion

Based on a review of the LRA, ALEX database, scoping and screening implementation procedures, and a sampling review of system scoping results during the audit, the staff concludes that the applicant's scoping methodology for plant systems, components, commodity groups, insulation, and consumables is acceptable. In particular, the staff determines that the applicant's methodology reasonably identified SCs within the scope of license renewal and their associated intended functions in accordance with 10 CFR 54.4(a).

2.1.3.5 Mechanical Component Scoping

2.1.3.5.1 Summary of Technical Information in the Application

In LRA Section 2.1.2.2, "Mechanical Discipline-Specific Scoping Methodology," the applicant discussed the methods used to identify mechanical components in piping systems as within the scope of license renewal. The applicant based mechanical component identification on the EDB and Maintenance Rule system designations and boundaries. The Maintenance Rule system functions were augmented to include functions to address license renewal criteria as necessary. Some components within a system were moved to commodity groups. For example, system pipe supports were moved to the component supports commodities in the civil/structural discipline. Electrical components in the mechanical system were usually left in the associated mechanical system since the majority of the electrical components were active and eventually screened out. Within most systems, new assets or sub-components were created in the LRDB to ensure that all necessary assets/components were accurately described and addressed in the license renewal process. The applicant identified the following asset/component examples used in the mechanical area: instrument tubing, valves and manifolds, fasteners in each system utilizing pressure boundary bolting, heat exchanger sub-components, and piping assets created to identify combinations of material.

The applicant depicted mechanical system boundaries on drawings with color overlays. In some instances, components were reviewed as part of another interfacing system in order to more accurately portray system functional boundaries or to streamline the license renewal process. In a few cases, all in-scope components for a single system could be reviewed as part of another interfacing system. In this manner, some systems were de-populated of in-scope functions and, therefore, shown as not within the scope of license renewal. LRA Table 2.2-1 provides a list of license renewal systems and indicates whether they are within the scope of license renewal.

2.1.3.5.2 Staff Evaluation

In LRA Section 2.1.2.2 and LR-TR-012, the applicant discussed the methods used to identify components in mechanical piping systems as within scope of license renewal. The scoping methodology focused on evaluation of plant mechanical piping systems. License renewal system designators were based on the EDB and Maintenance Rule system designations and boundaries. The Maintenance Rule system functions were augmented to include functions to address license renewal criteria as necessary.

To determine whether a system was within the scope of license renewal, all system-level functions were evaluated against the criteria specified in 10 CFR 54.4(a)(1) - (3). Systems that met any of the criteria are within the scope of license renewal. There were unique clarifications and exceptions for mechanical components within the scope of license renewal:

- When all in-scope components are active, a system was considered within the scope of license renewal, but with no components that require an AMR.

- If the only in-scope portion of a system was comprised of components that were part of a commodity group evaluation (e.g., fire barrier, equipment support, cables, etc.), that system or structure could be identified as not within the scope of license renewal. These instances and the commodity group evaluation were documented.

As an example, pipe supports were removed from the respective mechanical systems because they were handled as component support commodities in the civil/structural discipline. Electrical components in mechanical systems were usually left in their respective systems as a majority of the components were active and eventually screened out. Some electrical components were also kept in their mechanical systems when they served a passive pressure boundary function. Within most systems, new assets or subcomponents were created in the LRDB to ensure that all necessary assets/components were accurately described and addressed in the license renewal process. The applicant identified the following asset/component examples used in the mechanical area:

- instrument tubing, valves and manifolds
- asset for fasteners used in each system that utilizes pressure boundary bolting
- heat exchanger subcomponents
- piping assets created to identify combinations of material

The applicant depicted mechanical system boundaries on drawings with color overlays. In some instances, components were reviewed as part of another interfacing system in order to more accurately portray system functional boundaries or to streamline the license renewal process. In a few cases, all in-scope components for a single system could be reviewed as part of another interfacing system. In this manner, some systems were de-populated of in-scope functions and therefore shown as not within the scope of license renewal. LRA Table 2.2-1 provides a list of systems either within, or not within, the scope of license renewal. LR-TR-012, Attachment 2, "Mechanical and Electrical System Level Scoping Summary and Definition," identified the same mechanical systems within the scope of license renewal. LR-TR-012, Attachment 1, "Component Intended Function Summary and Definition," comprehensively listed passive mechanical component intended functions subject to an AMR.

The staff sampled the scoping methodology for the MSS and SDC/LPSI systems to verify that systems and components were accurately identified as within scope for license renewal. The results of the review are discussed below.

The staff extensively reviewed two systems to verify the adequacy of the applicant's scoping methodology. In LRA Section 2.3.4.6, "Main Steam System," the applicant provided the scoping methodology results for the SSCs within the MSS. The scoping results indicated that the MSS contained sixteen system functions along with eleven license renewal system functions. In LRA Section 2.3.2.1, "Engineered Safeguards System," for the SDC/LPSI subsystem, the applicant provided the scoping methodology results for the SSCs within the SDC subject to an AMR. The scoping results indicated that the SDC/LPSI system contains three system functions within the scope of license renewal and two system functions not within the scope of license renewal. The staff did not identify any issues associated with the scoping results for the MSS and SDC/LPSI systems. The staff reviewed the applicant's methodology for identifying systems and structures meeting mechanical scoping criteria, as defined in the Rule. The staff also reviewed a sample of the scoping methodology implementation procedures and discussed the methodology and results with the applicant. The staff verified that the applicant had identified and used pertinent engineering and licensing information in order to determine the SSCs required to be within the scope of license renewal, in accordance with the 10 CFR 54.4(a) criteria.

2.1.3.5.3 Conclusion

Based on the staff's review of the information in the LRA, the systems sample review, and discussions with the applicant, the staff concludes that the applicant's methodology for identifying mechanical systems and structures meets the scoping criteria of 10 CFR 54.4(a), and is, therefore, acceptable.

2.1.3.6 Civil/Structural Component Scoping

2.1.3.6.1 Summary of Technical Information in the Application

The applicant described the methodology used for civil/structural scoping in LRA Section 2.1.2.3, "Civil/Structural Discipline Specific Scoping Methodology," LRPG 3, Revision 3, "IPA Scoping and Screening," and license renewal project implementing procedure LR-TR-022-CS IPA, "Civil Structural Integrated Plant Assessment (IPA) - Scoping/Screening and Aging Management Review Methodology and Results."

Civil/structural scoping was performed in a manner to ensure that all plant buildings, yard structures, and their constituent parts were considered for license renewal. The scoping was performed using a structures and commodity approach to ensure that all civil/structural components and commodities that serve a license renewal intended function were identified and evaluated for aging management. Descriptions of the commodity groups were based on functions and a detailed listing of the types of structural components (e.g., columns, beams, walls, ceilings, etc.) required to support those functions was included in the description. Next, the commodity group intended functions were evaluated to determine which were within the scope of license renewal and boundary descriptions were developed to detail the structural components required to support the in-scope functions.

2.1.3.6.2 Staff Evaluation

The staff found that the civil/structural scoping was performed in a manner to ensure all plant buildings, yard structures, and their constituent parts were considered for license renewal. The scoping used a structures and commodities approach to identify all civil/structural components and commodities that serve a license renewal intended function and evaluate them for aging management.

The staff noted that several steps were involved in civil/structural scoping. The first step identified all site structures based on the review of CLB sources of information. This list of plant structures, substructures, or groups of structures was loaded into the LRDB for scoping evaluation. From the LRDB, commodity groups, including identification of components and component types, were developed based primarily on design functions of the structures (i.e., structural support, fuel storage and handling, cranes, mechanical component supports, flooding protection, etc.).

For structural NSR SSCs identified in the CLB, the staff found that the applicant had used the same methodology applied to piping systems for identifying structures within the scope of license renewal because a protective feature was typically associated with the structure. These protective features for HELB include: whip restraints, jet impingement shields, blowout panels, etc. The staff found this methodology acceptable.

Once the structural commodity groups were defined, the commodity group intended functions were developed and evaluated to determine which were within the scope of license renewal in accordance with the 10 CFR 54.4 criterion. The staff sampled boundary descriptions that detailed the structural components required to support the in-scope functions.

2.1.3.6.3 Conclusion

Based on the staff's review of information in the LRA, the applicant's detailed scoping implementation procedures, and a sampling review of civil/structural scoping results, the staff concludes that the applicant's methodology for identifying civil/structural SSCs within the scope of license renewal meets the requirements of 10 CFR 54.4(a) and is, therefore, acceptable.

2.1.3.7 Electrical and I&C Component Scoping

2.1.3.7.1 Summary of Technical Information in the Application

LRA Section 2.1.2.4, "Electrical and I&C Discipline-Specific Scoping Methodology," describes the scoping process associated with electrical and I&C systems and components. Electrical and I&C systems or commodity groups were identified by specific system intended functions. The applicant then evaluated the system and commodity groups against the scoping criteria in 10 CFR 54.4(a)(1) - (3) to determine if they perform license renewal intended functions. Relevant plant documentation was used during the scoping process.

2.1.3.7.2 Staff Evaluation

The applicant described the methodology for electrical and I&C scoping in LRA Section 2.1.2.4 and LR-TR-012. The audit staff reviewed the LRA, procedures, drawings, ALEX database, and a sample of the results of the application of the scoping methodology for selected systems. During the scoping process, the applicant established evaluation boundaries for each system or commodity group based on system functions. Following the system functions review, the applicant evaluated each system or commodity group against the scoping criteria in 10 CFR 54.4(a)(1) through (3) to determine whether it performs or supports a license renewal system intended function.

The staff noted that the applicant added to the ALEX database system descriptions, system functions, applicable design-basis references, and reference documentation which identified the system function that placed the system within the scope of license renewal. The applicant used multiple information sources that include CLB documents, controlled drawings, the EDB, and Maintenance Rule Program documents. Various electrical and I&C system components identified during the scoping process and requiring an AMR were assigned to commodity groups due to similarities across systems in materials, environments, and aging management strategies.

The staff discussed the electrical scoping methodology with the applicant's cognizant engineers and reviewed several plant electrical packages to verify proper implementation of the scoping process.

2.1.3.7.3 Conclusion

Based on its review of information contained in the LRA, the applicant's detailed scoping implementation procedures, and a sampling review of electrical and I&C scoping results, the staff concludes that the applicant's methodology for identification of electrical and I&C SSCs within the scope of license renewal meets the requirements of 10 CFR 54.4(a) and is, therefore, acceptable.

2.1.3.8 Scoping Methodology Conclusion

Based on a review of the LRA and the scoping implementation procedures, the staff determines that the applicant's scoping methodology is consistent with the guidance contained in the SRP-LR and identified SSCs (1) that are SR, (2) whose failure could affect SR functions, and (3) that are necessary to demonstrate compliance with NRC regulations for fire protection, EQ, PTS, ATWS, and SBO. Therefore, the staff concludes that the applicant's methodology meet the requirements of 10 CFR 54.4(a).

2.1.4 Screening Methodology

After identifying the SSCs within the scope of license renewal, the applicant implemented a process for determining which SSCs were subject to an AMR, in accordance with 10 CFR 54.21.

2.1.4.1 General Screening Methodology

2.1.4.1.1 Summary of Technical Information in the Application

In LRA Section 2.1.3, "Screening Methodology," the applicant discussed the method of identifying in-scope components of systems and commodity groups subject to an AMR. The screening process consisted of the following steps:

- Identification of components subject to an AMR (passive and long-lived) for each in-scope system, structure, or commodity.

- Identification of the component-level intended functions for all components subject to an AMR. A component level intended function is one that supports the system/structure/commodity-level intended function.

- Identification of applicable references used during the screening process to make the determinations.

Most of the electrical and I&C system components were determined to be active. Active components were screened and, therefore, did not require AMR. For passive, long-lived electrical components, a commodity group was established for non-EQ electrical cables and connections, inaccessible medium voltage cables and connections, non-EQ connections in instrument circuits, non-EQ electrical and I&C penetration assemblies, fuse holders, non-segregated phase bus and connectors, high voltage transmission conductors, high voltage switchyard bus connections, and high voltage insulators.

All temperature element components were reviewed to determine whether pressure boundary functions were performed (either thermowell or other fitting if it was a direct immersion temperature element). All solenoid valves (SVs) were reviewed to determine whether they performed a passive pressure boundary function. All instruments were reviewed to determine whether pressure retaining boundary functions were performed. Fan and damper housings were considered passive components subject to AMR. With the exception of some crane components, all civil/structural components are considered passive, long-lived and subject to an AMR.

The screening process also identified short-lived components and consumables. The short-lived components are not subject to an AMR. Consumables are a special class of short-lived items that include packing, gaskets, component seals, O-rings, oil, grease, component filters, system filters, fire extinguishers, fire hoses, and air packs. Consumables are considered subcomponents of identified components not subject to performance monitoring.

The screening process identified some stored equipment that is reserved for installation in the plant in response to a DBE and requires an AMR, including Appendix R stored equipment used to restore power to preselected plant components and to provide cooling in certain areas after a fire in order to attain cold shutdown. The stored Appendix R cables are included in the non-EQ electrical cables and connections commodity.

2.1.4.1.2 Staff Evaluation

In accordance with 10 CFR 54.21, each license renewal application must contain an IPA that identifies SCs within the scope of license renewal that are subject to an AMR. The IPA must identify components that perform an intended function without moving parts or a change in configuration or properties (passive), as well as components not subject to periodic replacement based on a qualified life or specified time period (long-lived). The IPA describes and justifies the methodology used to determine the passive and long-lived SCs, and a demonstration that the effects of aging on those SCs will be adequately managed so that the intended functions will be maintained under all design conditions imposed by the plant-specific CLB for the period of extended operation.

The staff reviewed the applicant's methodology to determine whether mechanical, civil/structural, and electrical and I&C components within the scope of license renewal should be subject to an AMR. The applicant implemented a process for determining which SCs were subject to an AMR in accordance with the requirements of 10 CFR 54.21(a)(1). In LRA Section 2.1.3, the applicant described these screening activities as they related to SCs within the scope of license renewal.

The screening process evaluated these in-scope SCs to determine which were passive and long-lived and, therefore, subject to an AMR. The staff reviewed LRA Sections 2.3, 2.4, and 2.5, which provide the results of the process used to identify SSCs subject to an AMR. The staff also reviewed the screening results reports for the MSS and SDC/low pressure coolant injection (LPCI) systems.

The applicant provided the staff with details of the processes used for each discipline and administrative documentation that described the screening methodology. Specific methodology for mechanical, electrical, and civil/structural classifications is discussed below.

2-28

2.1.4.1.3 Conclusion

On the basis of its review of the LRA, screening implementation procedures, and a sampling of screening results, the staff determines that the applicant's screening methodology is consistent with the guidance contained in the SRP-LR and is capable of identifying passive, long-lived components within the scope of license renewal that are subject to an AMR. The staff determines that the applicant's process for determining which SSCs were subject to an AMR meets the requirements of 10 CFR 54.21 and is, therefore, acceptable.

2.1.4.2 Mechanical Component Screening

2.1.4.2.1 Summary of Technical Information in the Application

Following system-level scoping for mechanical systems, the applicant screened SCs to identify those within the scope of license renewal. The screening process evaluated these in-scope SCs to determine which were passive, long-lived and, therefore, subject to an AMR. The intended functions from the documentation sources cited in SER Section 2.1.2.1.2, were used as input to the screening process to identify the passive and long-lived components within the scope of license renewal. Passive and long-lived component determinations were made in accordance with 10 CFR 54.21(a)(1)(i) and guidance in NEI 95-10.

The applicant depicted evaluation boundaries on the license renewal boundary drawings for mechanical systems. The screening process evaluates the in-scope components for active/passive and long-lived properties. For those components identified as passive and long-lived (i.e., subject to an AMR), the component function was documented.

The applicant's mechanical screening process was in accordance with LRPG 3 and LR-TR-012, Revision 2, "Mechanical and Electrical Scoping and Screening Methodology and Summary Report." The applicant performed and documented the screening process evaluations in the LRDB. The results for major component screening of mechanical systems are presented in LRA Sections 2.3.1 through 2.3.4.

2.1.4.2.2 Staff Evaluation

In LRA Section 2.1.3 and 2.3, the applicant discussed the screening methodology for passive and long-lived mechanical system components in accordance with the requirements of 10 CFR 54.21. The applicant implemented the screening methodology for mechanical system components using LRPG 3 and LR-TR-012. The staff reviewed the screening methodology for passive mechanical component intended functions, as documented in LR-TR-012, Attachment 1, "Component Intended Function Summary and Definition." The passive and long-lived mechanical components subject to an AMR were identified in the system/structure scoping and screening results report for each individual system.

The staff sampled the screening methodology of the MSS and SDC/LPSI systems to verify that systems and components were accurately identified as within the scope of license renewal and subject to an AMR. The results of this review are discussed below.

The staff reviewed the system/structure results report LR-SS-MSS, Revision 2, "Palisades Nuclear Plant System/Structure Scoping and Screening Results for the Main Steam System," and license renewal boundary P&IDs. In LR-SS-MSS Section 4, the applicant evaluated the entire list of active and passive components within the MSS. The section identified passive individual components subject to an AMR. Component types determined assignments of passive commodity groups screened into license renewal and subject to an AMR.

In LR-SS-MSS Section 5, the applicant listed equipment moved from the MSS to other systems due to component functions that provide an I&C function for that system or other cross-over functions completed by other systems. The MSS components were also moved to pseudo system commodity groups such as hangers and supports, compressed air components, radiation monitors, and fuses for station power. Components were moved to these cross-over systems and commodity groups due to the existing AMPs established for these components.

The staff also reviewed LR-SS-ESS, Revision 2, "Palisades Nuclear Plant System/Structure Scoping and Screening Results for the Engineering Safeguards," and the associated color-coded boundary P&IDs. In LR-SS-ESS Section 4, the applicant evaluated the entire list of passive and active components within the system. The section identified individual passive components subject to an AMR or AMP. As with the MSS system, the component type determined the assignments for passive commodity groups screened into license renewal and subject to an AMR or AMP.

2.1.4.2.3 Conclusion

Based on its review of the LRA, screening implementation procedures, and a sample review of two systems' screening results, the staff determines that the applicant's mechanical component screening methodology is consistent with the guidance contained in the SRP-LR and identified those passive, long-lived components that are within the scope of license renewal and subject to an AMR. The staff also concludes that the applicant's methodology for identifying mechanical components subject to an AMR meets the requirements of 10 CFR 54.21(a)(1).

2.1.4.3 Civil/Structural Component Screening

2.1.4.3.1 Summary of Technical Information in the Application

The applicant described the methodology used for civil/structural screening in LRA Section 2.1.3.1, "Active/Passive Determination," LRPG 3, and LR-TR-022-CS IPA. Civil/structural components and commodities (assets) were created to cover all in-scope structures and commodity groups. With the exception of certain crane components, all civil/structural components included within the scope of license renewal received an AMR. The applicant performed and documented the screening process evaluations in the ALEX database. The screening results for civil/structural components are presented in LRA Sections 2.4.1 through 2.4.10.

2.1.4.3.2 Staff Evaluation

The staff reviewed the applicant's methodology for civil/structural screening, as described in LRA Section 2.1.3.1 and in implementing procedure LR-TR-022-CS IPA.

The staff found that, in general, the screening process for civil/structural components and commodities was identical to that applied for mechanical and electrical components. The staff found that generic commodity components (assets) were created based on design function, structure (or yard), material, and environment to aid in the AMR. The staff noted that the structural commodity assets were regrouped into five AMR groups for more effective AMR evaluations and alignment to NUREG-1801, Volume 2, "Generic Aging Lessons Learned (GALL) Report," dated July 2001. The five groups developed included (1) containment pressure boundary, (2) component supports, (3) structural concrete, (4) structural steel, and (5) miscellaneous/bulk commodities. The staff reviewed components samples in each group. No deficiencies were noted.

Finally, the staff noted that AMR results were regrouped and presented by individual structure, component support commodity group, or the miscellaneous/bulk commodity group for alignment with the standard LRA format.

2.1.4.3.3 Conclusion

Based on its review of information in the LRA, the applicant's detailed screening implementation procedures, and a sampling review of civil/structural screening results, the staff concludes that the applicant's methodology for identification of civil/structural components and commodities subject to an AMR meets the requirements of 10 CFR 54.21(a)(1).

2.1.4.4 Electrical/I&C Component Screening

2.1.4.4.1 Summary of Technical Information in the Application

LRA Section 2.1.3 describes the methodology used for screening electrical and I&C components. The applicant applied the screening methodology for electrical and I&C components consistently with the guidance in NEI 95-10. The screening process identified the components that performed intended functions from the scoping evaluation previously addressed. The applicant identified components performing intended functions without moving parts or change in configuration properties (passive) and not subject to replacement based on a qualified life or specified time period (long-lived).

2.1.4.4.2 Staff Evaluation

The staff reviewed the applicant's screening methodology for electrical and I&C in LRA Section 2.1.2.4, "Electrical and I&C Discipline-Specific Scoping Methodology," and in LR-TR-012. The applicant used the screening process described in LR-TR-012 to identify the electrical and I&C components subject to AMR for each in-scope system, component, or commodity. Component-level intended functions for all components subject to AMR and applicable references used during the screening process to make the determinations were also identified.

The commodity groups established for passive and long-lived components were evaluated to identify whether they were subject to replacement based on a qualified life or specified time period (short-lived) or not subject to replacement based on a qualified life or specified time period (long-lived). The staff noted that components within electrical systems determined to require AMR were addressed as part of the defined commodity groups.

In the LRA, the applicant stated that most electrical and I&C system components were active. Using NEI 95-10 Appendix B as guidance, the applicant screened out active components since they do not require an AMR in accordance with 10 CFR 54.21(a)(1).

The staff reviewed the applicant's approach for scoping and screening of electrical fuse holders in accordance with ISG-05, "Identification and Treatment of Electrical Fuse Holders for License Renewal." ISG-05 states that, consistent with the requirements specified in 10 CFR 54.4(a), fuse holders (including fuse clips and fuse blocks) are considered passive electrical components. Fuse holders should be scoped, screened, and included in the AMR in the same manner as terminal blocks and other types of electrical connections treated in the process. ISG-05 also states that fuse holders that are part of an active component assembly (i.e., switchgear, power supplies, power inverters, battery chargers and circuit boards) are not considered within the scope of license renewal.

The staff reviewed and discussed the evaluations of fuse holders made by the applicant. The applicant evaluated fuse holders not inside active equipment and determined that they are located in environmentally controlled areas and, therefore, no further AMR evaluation was required.

2.1.4.4.3 Conclusion

The staff reviewed the LRA, procedures, drawings, LRDB, and a sample of the results of the screening methodology. The staff determines that the applicant's methodology is consistent with the description in LRA Section 2.1.2.4 and the applicant's implementation procedures. Based on its review of information in the LRA, the applicant's detailed screening implementation procedures, and a sampling review of electrical and I&C screening results, the staff concludes that the applicant's methodology for identification of electrical and I&C SCs subject to an AMR meets the requirements of 10 CFR 54.21(a)(1).

2.1.4.5 Screening Methodology Conclusion

Based on its review of the LRA, screening implementation procedures, discussions with the applicant's staff, and a sample review of screening results, the staff determines that the applicant's screening methodology is consistent with the guidance contained in the SRP-LR and identified those passive and long-lived components within the scope of license renewal that are subject to an AMR. The staff concludes that the applicant's methodology meets the requirements of 10 CFR 54.21(a)(1) and is, therefore, acceptable.

2.1.5 Summary of Evaluation Findings

The staff formed the basis of its safety determination on its review of the information presented in LRA Section 2.1, supporting information in the scoping and screening implementation procedures and reports, information presented during the scoping and screening methodology audit, and the applicant's responses to the staff's RAIs. The staff verified that the applicant's scoping and screening methodology is consistent with the requirements of the Rule. From this review, the staff concludes that there is reasonable assurance that the applicant's methodology for identifying SSCs within the scope of license renewal and SCs requiring an AMR is consistent with the requirements of 10 CFR 54.4 and 10 CFR 54.21(a)(1).

2.2 Plant-Level Scoping Results

2.2.1 Introduction

The SOC for the license renewal rule (60 *Federal Register* 22478) indicates that an applicant is flexible in determining the set of SSCs for which an AMR is performed. In LRA Section 2.1, the applicant described the methodology for identifying the SSCs within the scope of license renewal. In LRA Section 2.2, the applicant used the scoping methodology to determine which of the SSCs are required to be included within the scope of license renewal. The staff reviewed the plant-level scoping results to determine whether the applicant had properly identified all plant-level systems and structures relied upon to mitigate DBEs, as required by 10 CFR 54.4(a)(1), or whose failure could prevent satisfactory accomplishment of any of the SR functions, as required by 10 CFR 54.4(a)(2), as well as the systems and structures relied on in safety analysis or plant evaluations to perform a function required by any of the regulations referenced in 10 CFR 54.4(a)(3).

The staff reviewed the SSCs that the applicant did not identify as within the scope of license renewal to verify whether the systems and structures have any intended functions that would require their inclusion within the scope of license renewal. The staff also reviewed selected SSCs that the applicant identified as being within the scope of license renewal to verify that the applicant had properly identified their components, within the evaluation boundaries, that are subject to an AMR in accordance with 10 CFR 54.21(a)(1). To determine whether the applicant identified the SSCs that are subject to an AMR, the staff reviewed the components that the applicant did not identify as being subject to an AMR.

2.2.2 Summary of Technical Information in the Application

In LRA Table 2.2-1, the applicant provided a list of the plant-level scoping results, identifying those systems and structures that are within the scope of license renewal. Based on the DBEs considered in the plant's CLB, other CLB information relating to NSR systems and structures, and certain regulated events, the applicant identified those plant-level systems and structures that are within the scope of license renewal, as defined by 10 CFR 54.4.

2.2.3 Staff Evaluation

In LRA Section 2.1, the applicant described its methodology for identifying the systems and structures that are within the scope of license renewal and subject to an AMR. The staff reviewed the scoping and screening methodology and provided its evaluation in SER Section 2.1. To verify that the applicant had properly implemented its methodology, the staff focused its review on the implementation results, as shown in LRA Table 2.2-1 to confirm that there were no omissions of plant-level systems and structures within the scope of license renewal.

The staff determined whether the applicant had properly identified the systems and structures within the scope of license renewal in accordance with 10 CFR 54.4. The staff reviewed selected systems and structures that the applicant did not identify as falling within the scope of license renewal to verify whether the systems and structures have any intended functions that would require their inclusion within the scope of license renewal. The staff's review of the applicant's implementation was conducted in accordance with the guidance described in SRP-LR Section 2.2, "Plant-Level Scoping Results."

In reviewing LRA Section 2.2, the staff identified areas in which additional information was necessary to complete the evaluation of the applicant's plant-level scoping results to determine whether the applicant properly applied the scoping criteria of 10 CFR 54.4. The applicant responded to the staff's RAIs as discussed below.

In RAI 2.2-3, dated September 21, 2005, the staff requested that the applicant explain how the following component types, subject to an AMR, serve the pressure boundary intended function:

- electric heaters and motors in LRA Table 2.3.3-6
- transmitters in LRA Tables 2.3.4-2 and 2.3.4-4

In addition, the staff requested that the applicant identify any other electrical components included within the mechanical systems. The staff also requested that the applicant identify the mechanical systems with which these components are evaluated and explain how these electrical components serve their intended functions.

In its response, by letter dated October 21, 2005, the applicant stated:

> The electrical heaters on LRA Table 2.3.3-6 are the emergency diesel generators (EPS) lube oil (EH-25A/B) and jacket water (EH-27A-D) heaters that penetrate the lube oil and jacket water pressure boundaries, respectively, and require an AMR. The heaters connect to the fluid pressure boundaries with screwed and/or flanged connections (i.e., are not in thermowells).
>
> The motors on LRA Table 2.3.3-6 are the Emergency Diesel Generator (EPS) crankcase exhausters (EMG-0501, 0502, 0503 & 0504). These motors have a pressure boundary intended function and require an AMR.
>
> The transmitter/element on LRA Table 2.3.4-2 are the demineralized makeup water (DMW) restricting orifices. These have a pressure boundary intended function and require an AMR.
>
> The transmitter/element on LRA Table 2.3.4-4 are pressure transmitters. These should not have been listed since they are "Active Components" per NEI 95-10 and not subject to AMR. Accordingly, LRA Tables 2.3.4-4 and 3.4.2-4 are revised to delete the Transmitter/Element line items.
>
> Other electrical component types that were evaluated with the mechanical components include the following: Differential Pressure Indicator, Differential Pressure Switch, Level Indicating Alarm, Level Switch, Level Transmitter, Pressure Indicator, Pressure Transmitter, Pressure Switch, Temperature Element, Temperature Indicator, Temperature Switch. Consistent with NEI 95-10, these component types are active and do not require an AMR.
>
> The Primary Coolant System pressurizer heaters and backup heaters penetrate the pressurizer, provide fluid pressure boundary intended functions and are subject to AMR.

Based on its review, the staff finds the applicant's response to RAI 2.2-3 acceptable. The applicant specifically identified the electrical component types that were evaluated with associated mechanical systems and their passive, mechanical, intended function of fluid pressure boundary. Therefore, the staff's concerns described in RAI 2.2-3 are resolved.

In RAI 2.2-6, dated September 21, 2005, the staff stated that LRA Section 2.1.3.1 states that solenoid valves (SVs) are typically active components; however, in some cases, the solenoid valve body actually performs a pressure boundary intended function. All SVs were reviewed against this criteria and those needed to maintain a pressure boundary were identified as passive. LRA Section 2.3 does not identify SVs within the scope of license renewal, but are excluded as subject to an AMR because they do not serve a pressure boundary intended function. To determine whether the SVs had been properly evaluated in the LRA, the staff requested that the applicant list the SVs within the scope of license renewal, but not subject to an AMR in accordance with 10 CFR 54.4(a) and 10 CFR 54.21(a)(1), respectively.

In its response, by letter dated October 21, 2005, the applicant stated:

> The following SV's are in scope of license renewal only for their seismic mounting and are not subject to AMR as mechanical components: SV-0101, 0148, 0150, 1501, 1502, 1503, 1553, 1553A, 1553B, 1555, 1555A, 1555B, 1683A, 1684A, 1834, 1843, 1893, 1894, 2001, 2002A, 2002B, 2003, 2004, and 2005. The hangers/supports for these components are in scope for license renewal and evaluated separately within a civil/structural commodity.
>
> The following SV's are in scope of License Renewal and not subject to AMR due to being managed by the Palisades Electrical Qualification Program: SV-0338, 0342, 0346, 0347, 0522A, 0738, 0739, 0767, 0768, 0821, 0824, 0825, 0844, 0845, 0846, 0847, 0857, 0861, 0862, 0864, 0865, 0867, 0869, 0870, 0873, 0876, 0877, 0878, 0879, 0880, 0910, 0911, 0913, 0937, 0938, 0939, 0940, 0944A, 0945, 0946, 0947, 0948, 0949, 0950, 0951, 1002, 1007, 1036, 1038, 1044, 1045, 1103, 1104, 1805, 1806, 1807, 1808, 1901, 1902, 1903, 1904, 1905, 2113, 2115, 2117, 2413A, 2413B, 2415A, 2415B, 2418L, 2419L, 2420L, 3001, 3002, 3018, 3027A, 3027B, 3029A, 3029B, 3030A, 3030B, 3031A, 3031B, 3036, 3037, 3055A, 3055B, 3056A, 3056B, 3057A, 3057B, 3059, 3069, 3070, 3071, 3084, 3085, 3212A, 3212B, 3213A, 3213B, 3223A, 3223B, 3224A and 3224B.
>
> The following SV's are in scope of License Renewal and not subject to AMR due to being Active: SV-1819, 1820, 1821, 1822. These four SVs are in containment, associated with the containment air cooling system, and provide flow to a radiation monitor (RE-1817) that is not in scope for LR. These SVs are in scope because they are installed on branch lines that are attached to the main air ductwork for the containment air cooler fans. The lines associated with these SVs are 3/4" and non-safety related. Leakage through these lines, in comparison with the flow through the main ductwork, will not keep the containment air coolers from performing their system functions.

Based on its review, the staff finds the applicant's response to RAI 2.2-6 acceptable because the applicant specifically identified SVs that are within the scope of license renewal, but are not subject to an AMR because they are either managed by the Electrical Qualification Program or active per NEI 95-10 Appendix B guidance. Therefore, the staff's concerns described in RAI 2.2-6 are resolved.

2.2.4 Conclusion

The staff reviewed LRA Section 2.2, accompanying scoping boundary drawings, RAI responses, and supporting information in the FSAR to determine whether any systems and structures within the scope of license renewal had not been identified by the applicant. The staff's review did not identify any omissions. On the basis of its review, the staff concludes that the applicant had properly identified the systems and structures that are within the scope of license renewal in accordance with 10 CFR 54.4.

2.3 Scoping and Screening Results: Mechanical Systems

This section documents the staff's review of the applicant's scoping and screening results for mechanical systems. Specifically, this section discusses the following mechanical systems:

- reactor vessel, internals, and reactor coolant system
- engineered safety features
- auxiliary systems
- steam and power conversion systems

In accordance with the requirements of 10 CFR 54.21(a)(1), the applicant identified and listed passive, long-lived SCs that are within the scope of license renewal and subject to an AMR. To verify that the applicant properly implemented its methodology, the staff focused its review on the implementation results. This approach allowed the staff to confirm that there were no omissions of mechanical system components that meet the scoping criteria and are subject to an AMR.

Staff Evaluation Methodology. The staff's evaluation of the information provided in the LRA was performed in the same manner for all mechanical systems. The objective of the review was to determine if the components and supporting structures for a specific mechanical system, that appeared to meet the scoping criteria specified in the Rule, were identified by the applicant as within the scope of license renewal, in accordance with 10 CFR 54.4. Similarly, the staff evaluated the applicant's screening results to verify that all long-lived, passive components were subject to an AMR in accordance with 10 CFR 54.21(a)(1).

Scoping. To perform its evaluation, the staff reviewed the applicable LRA section and associated component drawings, focusing its review on components that had not been identified as within the scope of license renewal. The staff reviewed relevant licensing basis documents, including the final safety analysis report (FSAR), for each mechanical system to determine if the applicant had omitted components with intended functions delineated under 10 CFR 54.4(a) from the scope of license renewal. The staff also reviewed the licensing basis documents to determine if all intended functions delineated under 10 CFR 54.4(a) were specified in the LRA. If omissions were identified, the staff requested additional information to resolve the discrepancies.

Screening. Once the staff completed its review of the scoping results, the staff evaluated the applicant's screening results. For those systems and components with intended functions, the staff sought to determine: (1) if the functions are performed with moving parts or a change in configuration or properties, or (2) if they are subject to replacement based on a qualified life or specified time period, as described in 10 CFR 54.21(a)(1). For those that did not meet either of these criteria, the staff sought to confirm that these mechanical systems and components were subject to an AMR as required by 10 CFR 54.21(a)(1). If discrepancies were identified, the staff requested additional information to resolve them.

2.3.1 Reactor Vessel, Internals, and Reactor Coolant System

In LRA Section 2.3.1, the applicant identified the structures and components of the reactor vessel, internals, and reactor coolant system that are subject to an AMR for license renewal.

The applicant described the supporting structures and components of the reactor vessel, internals, and reactor coolant system in the following sections of the LRA:

- 2.3.1.1   primary coolant system
- 2.3.1.2   reactor vessel
- 2.3.1.3   reactor vessel internals
- 2.3.1.4   replacement steam generators

The staff's review findings regarding LRA Sections 2.3.1.1 – 2.3.1.4 are presented in SER Sections 2.3.1.1 – 2.3.1.4, respectively.

2.3.1.1 Primary Coolant System

2.3.1.1.1 Summary of Technical Information in the Application

In LRA Section 2.3.1.1, the applicant described the primary coolant system (PCS). The PCS is designed to remove heat from the reactor core and internals and transfer it to the secondary (steam generating) system by the controlled circulation of pressurized borated water which serves both as a coolant and a neutron moderator. The PCS serves as a barrier to the release of radioactive material to the containment building, and is equipped with controls and safety features that assure safe conditions within the system. The system contains two steam generators to transfer the heat generated in the PCS to the secondary system. The steam generators operate with the primary coolant in the tube side and the secondary fluid on the shell side. System pressure is maintained by regulation of the water temperature in the pressurizer where steam and water are held in thermal equilibrium. Overpressure protection is provided by spring-loaded safety valves connected to the top of the pressurizer. Steam discharged from the valves is cooled and condensed by water in a quench tank. A subsystem of the PCS is the pressurizer pressure and level control system. The pressurizer pressure and level control is a redundant control system designed to maintain pressurizer level and pressure during heatup, cooldown and load changes. Level is accomplished by controlling the rate of water added to or removed from the PCS. Pressure is controlled by energizing electrical heaters or activating steam space spray water to the pressurizer.

The PCS contains SR components that are relied upon to remain functional during and following DBEs. The failure of NSR SSCs in the PCS could potentially prevent the satisfactory accomplishment of an SR function. In addition, the PCS performs functions that support fire protection, EQ, ATWS, and SBO.

The intended functions within the scope of license renewal include the following:

- provides fluid pressure boundary
- provides shelter/protection to SR components
- provides spray pattern at discharge nozzle

In LRA Table 2.3.1-1, the applicant identified the following PCS component types that are within the scope of license renewal and subject to an AMR:

- alloy 600 cladding
- alloy 600 safe ends
- alloy 600 thermal sleeves
- bolting and fasteners
- carbon steel nozzles
- carbon steel pipe (30" and 42")
- flow element (PCP controlled bleed)
- non-CASS valves in PCS and connected systems
- PCS spray and drain nozzles
- PORV isolation, quench tank spray manual valves
- PORV isolation valves
- pressurizer alloy 600 instrument penetrations
- pressurizer heater sleeves
- pressurizer heaters
- pressurizer integral support weld
- pressurizer manway and flange bolting
- pressurizer manway and flanges
- pressurizer quench tank
- pressurizer quench tank shell and heads
- pressurizer spray head
- primary coolant pump casing
- primary coolant sample heat exchanger shell
- reactor head vent
- reactor head vent orifice
- sample point (quench tank liquid, loop 2 hot leg, pressurizer surge line)
- small bore stainless steel pipe (PCS and connected systems)
- SS cladding
- stainless steel pipe (PCS and connected systems)
- stainless steel safe ends (pressurizer and connected systems)
- stainless steel thermal sleeves
- stainless steel tubing
- vessels, pressure (pressurizer)

2.3.1.1.2 Staff Evaluation

The staff reviewed LRA Section 2.3.1.1 and the FSAR using the evaluation methodology described in SER Section 2.3. The staff conducted its review in accordance with the guidance described in SRP-LR Section 2.3, "Scoping and Screening Results: Mechanical Systems."

In conducting its review, the staff evaluated the system functions described in the LRA and FSAR in accordance with the requirements of 10 CFR 54.4(a) to verify that the applicant had not omitted from the scope of license renewal any components with intended functions delineated under 10 CFR 54.4(a). The staff then reviewed those components that the applicant had identified as being within the scope of license renewal to verify that the applicant had not omitted any passive and long-lived components that should be subject to an AMR in accordance with the requirements of 10 CFR 54.21(a)(1).

The staff's review of LRA Section 2.3.1.1 identified areas in which additional information was necessary to complete the review of the applicant's scoping and screening results. The applicant responded to the staff's RAIs as discussed below.

In RAI 2.3.1-1, dated June 3, 2005, the staff requested the applicant to verify whether the vessel flange leak detection lines, which provide a pressure boundary, have been included within scope of license renewal and subject to an AMR or to explain the exclusion.

In its response, by letter dated July 1, 2005, the applicant stated:

> NMC has treated the between-the-seals portion of the reactor flange leak detection tap and the downstream piping as having intended functions and being in-scope for license renewal. The in-scope head leak detection piping is depicted on drawing LR-M-201-1. LRA Table 2.3.1-2 for the reactor vessel provides a specific line item for reactor vessel head O-ring leakage monitoring, with an intended function of pressure boundary/fission product retention. The outer flange seal leak detection portion is not in scope.
>
> The reactor vessel head leak detection tube (or tap) was evaluated with the reactor vessel and included in the summary of reactor vessel aging management evaluations provided in LRA Table 3.1.2-2. As indicated in the table, this component of head leak detection is managed using the Alloy 600 Program.
>
> The stainless steel leak detection piping downstream of the integral alloy 600 tube (or tap) was evaluated with the Primary Coolant System (not the reactor pressure vessel). Therefore, the stainless steel flange leak detection lines are included in LRA Table 3.1.2-1 under the small-bore stainless steel pipe line item. NMC grouped the leak detection piping in this manner to prevent having an incongruous "piping" component in the reactor vessel group.

With inclusion of this component the staff finds the applicant's response to RAI 2.3.1-1 acceptable. Therefore, the staff's concern described in RAI 2.3.1-1 is resolved.

In RAI 2.3.1-2, dated August 23, 2005, the staff stated that scoping boundary drawing LR 201, sheet 2 indicates that pressurizer safety valves are within the scope of license renewal; however, LRA Table 2.3.1-1 does not identify pressurizer safety valves separately as within the scope of license renewal. Therefore, the staff requested that the applicant indicate which line item in LRA Table 2.3.1-1 includes the subject component.

In its response, by letter dated September 16, 2005, the applicant stated that "In LRA Table 2.3.1-1, Pressurizer relief [safety] valves (RV-1039, 1040 & 1041) are included in component type 'Non-CASS Valves in PCS and Connected Systems'."

With the inclusion of this component, the staff finds the applicant's response to RAI 2.3.1-2 acceptable. Therefore, the staff's concern described in RAI 2.3.1-2 is resolved.

2.3.1.1.3 Conclusion

The staff reviewed the LRA, accompanying scoping boundary drawings, and RAI responses to determine whether any SSCs that should be within the scope of license renewal had not been identified by the applicant. No omissions were identified. In addition, the staff performed a review to determine whether any components that should be subject to an AMR had not been identified by the applicant. No omissions were identified. On the basis of its review, the staff concludes that the applicant adequately identified the PCS components that are within the scope of license renewal, as required by 10 CFR 54.4(a), and the PCS components that are subject to an AMR, as required by 10 CFR 54.21(a)(1).

2.3.1.2 Reactor Vessel

2.3.1.2.1 Summary of Technical Information in the Application

In LRA Section 2.3.1.2, the applicant described the reactor vessel (RVG). The RVG is the pressure vessel used to contain the core, core supports, and nuclear fuel in a pressurized water reactor. The reactor vessel is an integral part of the PCS, which provides the means of removing heat from the fuel and generating steam in the steam generator. The reactor vessel is operated at a pressure high enough to ensure that the bulk primary coolant remains in a liquid phase. The reactor vessel and top head are designed in accordance with American Society of Mechanical Engineers (ASME) B&PV Code, Section III, Class A, 1965, W65a.

The RVG contains SR components that are relied upon to remain functional during and following DBEs. In addition, the RVG performs functions that support PTS.

The intended functions within the scope of license renewal include the following:

- provides structural support to safety-related components
- provides pressure boundary or fission product retention barrier
- prevents excessive core displacement
- reduces core inlet flow inequalities
- provides structural and/or functional support to SR equipment

In LRA Table 2.3.1-2, the applicant identified the following RVG component types that are within the scope of license renewal and subject to an AMR:

- control rod drive mechanism (CRDM) seal pressure housing
- CRDM upper pressure housing & flange
- CRDM/Incore instrument bolting
- incore instrument closure flanges
- internal SS cladding
- reactor vessel column support
- reactor vessel bottom head
- reactor vessel closure head
- reactor vessel closure head lifting lugs

- reactor vessel core stabilizer lugs
- reactor vessel CRDM nozzles
- reactor vessel flow skirt
- reactor vessel head o-ring leakage monitoring
- reactor vessel head vent nozzle
- reactor vessel incore instrument nozzles
- reactor vessel intermediate shell
- reactor vessel lower shell
- reactor vessel nozzle safe ends
- reactor vessel primary coolant nozzles
- reactor vessel seal ledge ring
- reactor vessel stop lugs
- reactor vessel studs, nuts, washers
- reactor vessel surv. capsule holder
- reactor vessel upper shell
- reactor vessel upper shell flange
- under head CRDM support

2.3.1.2.2 Staff Evaluation

The staff reviewed LRA Section 2.3.1.2 and the FSAR using the evaluation methodology described in SER Section 2.3. The staff conducted its review in accordance with the guidance described in SRP-LR Section 2.3.

In conducting its review, the staff evaluated the system functions described in the LRA and FSAR in accordance with the requirements of 10 CFR 54.4(a) to verify that the applicant had not omitted from the scope of license renewal any components with intended functions delineated under 10 CFR 54.4(a). The staff then reviewed those components that the applicant had identified as being within the scope of license renewal to verify that the applicant had not omitted any passive and long-lived components that should be subject to an AMR in accordance with the requirements of 10 CFR 54.21(a)(1).

2.3.1.2.3 Conclusion

The staff reviewed the LRA to determine whether any SSCs that should be within the scope of license renewal had not been identified by the applicant. No omissions were identified. In addition, the staff performed a review to determine whether any components that should be subject to an AMR had not been identified by the applicant. No omissions were identified. On the basis of its review, the staff concludes that the applicant adequately identified the RVG components that are within the scope of license renewal, as required by 10 CFR 54.4(a), and the RVG components that are subject to an AMR, as required by 10 CFR 54.21(a)(1).

2.3.1.3 Reactor Vessel Internals

2.3.1.3.1 Summary of Technical Information in the Application

In LRA Section 2.3.1.3, the applicant described the reactor vessel internals (RVI). The RVI support and orient the reactor core fuel bundles and control rods, absorb the control rod dynamic loads and transmit these and other loads to the reactor vessel flange, provide a passageway for the reactor coolant, and support incore instrumentation. The reactor vessel internals safely perform their functions during all steady-state conditions and during normal operating transients. The internals safely withstand the forces due to deadweight, handling, system pressure, flow impingement, temperature differential, shock and vibration. All reactor components are considered Class 1 for seismic design.

The RVI contain SR components that are relied upon to remain functional during and following DBEs.

The intended function, within the scope of license renewal, is to provide structural and/or functional support to SR equipment.

In LRA Table 2.3.1-3, the applicant identified the following RVI component types that are within the scope of license renewal and subject to an AMR:

- IV.B3.2 - control element assembly shroud assemblies: control rod shroud, shroud support lug, fuel guide pin, fuel guide pin nuts, shroud top support, control rod support lug, fuel plate cap screw

- IV.B3.4 - core shroud assembly: anchor block, centering plate, core shroud plate, anchor screw & pin, centering screw & pin, positioning screw, shroud bolt & pin

- IV.B3.3 - core support barrel assembly core support barrel, core support barrel integral upper flange

- incore instrument guide tube (not addressed in GALL IV.B3) instrument guide tube, guide tube bracket, guide tube plugs, guide tube plug screw, guide tube support

- IV.B3.5 - lower internal assembly core support plate, core support column, core support barrel snubber lug, core support barrel cap screws, fuel alignment pins, core support column support beams and tie rods

- upper guide structure - not in GALL spacer shim, instrument sleeve

- IV.B3.1 - upper internal assembly fuel alignment plate, fuel plate align lug, fuel plate cap screw, fuel plate guide pin, holddown ring plunger, holddown ring strap, holddown ring, brace grid beam, cross brace screw, shroud grid ring

2.3.1.3.2 Staff Evaluation

The staff reviewed LRA Section 2.3.1.3 and the FSAR using the evaluation methodology described in SER Section 2.3. The staff conducted its review in accordance with the guidance described in SRP-LR Section 2.3.

In conducting its review, the staff evaluated the system functions described in the LRA and FSAR in accordance with the requirements of 10 CFR 54.4(a) to verify that the applicant had not omitted from the scope of license renewal any components with intended functions delineated under 10 CFR 54.4(a). The staff then reviewed those components that the applicant had identified as being within the scope of license renewal to verify that the applicant had not omitted any passive and long-lived components that should be subject to an AMR in accordance with the requirements of 10 CFR 54.21(a)(1).

2.3.1.3.3 Conclusion

The staff reviewed the LRA to determine whether any SSCs that should be within the scope of license renewal had not been identified by the applicant. No omissions were identified. In addition, the staff performed a review to determine whether any components that should be subject to an AMR had not been identified by the applicant. No omissions were identified. On the basis of its review, the staff concludes that the applicant adequately identified the RVI components that are within the scope of license renewal, as required by 10 CFR 54.4(a), and the RVI components that are subject to an AMR, as required by 10 CFR 54.21(a)(1).

2.3.1.4 Replacement Steam Generators

2.3.1.4.1 Summary of Technical Information in the Application

In LRA Section 2.3.1.4, the applicant described the replacement steam generators. The two replacement steam generators are of the vertical pressurized water type wherein the heating surface and steam separation equipment are in the same vessel. The steam generators have been designed and fabricated as Class 1 vessels, as defined in ASME Boiler and Pressure Vessel Code, Section III, 1977 Edition. Materials of construction are as specified in ASME Code, Section III. Each steam generator is a vertical U-tube heat exchanger, which operate with the primary coolant on the tube side and secondary coolant on the shell side. A vertical divider plate separates the inlet and outlet plenums of the primary head. Secondary system feedwater enters the steam generator through the feed ring, mixes with the recirculating water from the moisture separators, and flows down the annulus between the tube bundle wrapper plate and the steam generator shell. Here it turns, flows inward around the base of the tube bundle and then upward around the tubes. As it flows up past the tubes, it is heated to saturation temperature and boils. The wet steam passes through sets of moisture separators and steam driers in the upper portion of the steam generator and leaves as dry steam through the outlet flow restricting nozzle at the top. Manways and handholes are provided for access to the steam generator internals for inspection or repairs.

The replacement steam generators contain SR components that are relied upon to remain functional during and following DBEs.

The intended functions within the scope of license renewal include the following:

* provides spray shield or curbs for directing flow
* provides fluid pressure boundary
* provides heat transfer
* provides structural and/or functional support to SR components

In LRA Table 2.3.1-4, the applicant identified the following replacement steam generators component types that are within the scope of license renewal and subject to an AMR:

* alloy 690 tube plugs
* handhole cover
* tube bundle support assembly
* tube bundle wrapper
* lower head

- primary manway cover
- manway cover diaphragm
- nozzle safe ends
- primary divider plate
- primary inlet and outlet nozzles
- feedwater inlet nozzles and thermal sleeves
- steam outlet nozzle and flow limiter, blowdown nozzle
- secondary side inspection port cover
- wide and narrow range water level nozzles, sampling and instrument nozzles
- shells (lower, upper, transition)
- tubesheet
- upper head
- U-tubes
- fasteners

2.3.1.4.2 Staff Evaluation

The staff reviewed LRA Section 2.3.1.4 and the FSAR using the evaluation methodology described in SER Section 2.3. The staff conducted its review in accordance with the guidance described in SRP-LR Section 2.3.

In conducting its review, the staff evaluated the system functions described in the LRA and FSAR in accordance with the requirements of 10 CFR 54.4(a) to verify that the applicant had not omitted from the scope of license renewal any components with intended functions delineated under 10 CFR 54.4(a). The staff then reviewed those components that the applicant had identified as being within the scope of license renewal to verify that the applicant had not omitted any passive and long-lived components that should be subject to an AMR in accordance with the requirements of 10 CFR 54.21(a)(1).

The staff's review of LRA Section 2.3.1.4 identified an area in which additional information was necessary to complete the review of the applicant's scoping and screening results. The applicant responded to the staff's RAI as discussed below.

In RAI 2.3.4-1, dated June 3, 2005, the staff stated that the steam generator feedwater inlet ring is not included within the scope of license renewal. The staff believed that this component has intended functions, as delineated in 10 CFR 54.4(a); therefore, the staff requested that the applicant justify the exclusion of the steam generator feedwater inlet ring from within the scope of license renewal.

In its response, by letter dated July 1, 2005, the applicant stated:

> The methodology for identifying the SSCs within the scope of license renewal and their intended functions is provided in Section 2 of the LRA, and is consistent with that contained in Section 2.1.2.1 of NUREG-1800 and in NEI 95-10. As indicated in NEI 95-10 and NUREG-1800, the fact that a system or component serves an intended license renewal function in accordance with 10 CFR 54.4(b) does not mean that all of its subcomponents also support that function. In this case, as shown on drawing LR-M-201, the replacement steam generators (RSG)

are components within the scope of license renewal; however, the feedwater inlet ring subcomponents do not support the RSG intended functions.

The intended functions of the RSG itself are listed in Section 2.3.1.4 of the LRA. The feed rings are not discussed in that section because the feed rings do not support any of the listed RSG functions, and, therefore, do not require an AMR.

The forged carbon steel feedwater nozzle is welded into the upper shell of the RSG. Feedwater is introduced into the secondary side of the RSG through the feedwater nozzle, the welded thermal sleeve assembly and fittings, and the feedwater distribution ring. Since the distribution function is performed within the steam generator, the feed ring does not support the RSG secondary side fluid pressure boundary function, nor does it perform any other license renewal intended function.

The feedwater ring support is included in GALL, with the ring itself, as Volume 2 Line Item IVD.1.3.1. The aging mechanism of concern is flow-accelerated corrosion (FAC). The Palisades response to that issue is provided in LRA Section 3.1.2.2.14. At Palisades, the carbon steel feed ring support is considered an integral part of the RSG carbon steel upper shell. Accordingly, the feed ring support function is subsumed into and managed as indicated under GALL Volume 2 line item IVD.1.1-c in LRA Table 3.1.2-4 for the RSG shell.

Based on its review of the applicant's response, the staff determines that the feed ring along with the feed ring supports should be included within the scope of license renewal because the feed ring provides structural and/or functional support for in-scope equipment. Loss of one or more J-tubes could make the feed ring assembly more susceptible to water hammer. In addition, there is the possibility that failure of the feed ring or one or more of the J-tubes might damage the steam generator tube sheet and prevent any of the 10 CFR 54.4(a)(1)(i), (ii), or (iii) functions.

In response to the staff's concerns, the applicant provided a supplemental response, by letter dated November 18, 2005. In the supplemental response the applicant stated:

NMC has added the Palisades steam generator feedwater rings into the scope of license renewal under criterion (a)(2). Conforming changes are made to the LRA as follows:

In LRA Table 2.3.1-4 on page 2-68, a new Subcomponent "Feedwater Ring" is hereby added, with the Intended Function of "Structural Support for Safety Related."

In LRA Table 3.1.2-4 on page 3-61, a new line item for Component Type "Feedwater Ring" is hereby added with Intended Function, Material, Environment, and Aging Effect Requiring Management, of "Structural Support for Safety Related," "Carbon Steel," Treated Water (Int and Ext)," and "Loss of Material," respectively. For this aging effect two separate sets of entries for Aging Management Program, NUREG-1801 Volume 2 Line Item, Table 1 Item and Notes are, "Steam Generator Tube Integrity Program," "IV.D1.3-a," 3.1.1-21," and "A, 126;" and "Water Chemistry Program," IV.D1.3-a," 3.1.1-21," and "A, 126," respectively.

On page 3-63, new plant-specific note 126 is added to read, "The Steam Generator Tube Integrity Program is credited for managing wall thinning/flow-accelerated corrosion. Palisades also credits the Water Chemistry Program for other loss of material mechanisms."

The feedwater rings are managed for wall thinning/flow-accelerated corrosion by the Water Chemistry Program and the Steam Generator Tube Integrity Program (SGTIP). The Steam Generator Tube Integrity Program includes secondary side inspections in the area of the feedwater rings for degradation such as the following:

- Feedwater Distribution Box - For signs of broken or damaged supports, thermal liner, gusset welded attachments, and outlet nozzles.

- Entire Length of Feedwater Ring - For signs of broken, damaged, or eroded J-Nozzles; broken, loose, or missing feedwater ring U-bolts and supports.

- Auxiliary Feedwater Inlet Nozzle and Piping - For signs of damaged, eroded, or broken piping.

- J-Nozzle Internal Video Probe - For signs of broken or eroded J-Nozzle to feedwater weld joint.

Based on its review, the staff finds the applicant's response to RAI 2.3.4-1 acceptable because the applicant agreed to include the feedwater inlet ring within the scope of license renewal. Therefore, the staff's concerns described in RAI 2.3.4-1 are resolved.

2.3.1.4.3 Conclusion

The staff reviewed the LRA, accompanying scoping boundary drawing, and RAI response to determine whether any SSCs that should be within the scope of license renewal had not been identified by the applicant. No omissions were identified. In addition, the staff performed a review to determine whether any components that should be subject to an AMR had not been identified by the applicant. No omissions were identified. On the basis of its review, the staff concludes that the applicant adequately identified the replacement steam generators components that are within the scope of license renewal, as required by 10 CFR 54.4(a), and the replacement steam generators components that are subject to an AMR, as required by 10 CFR 54.21(a)(1).

2.3.2 Engineered Safety Features

2.3.2.1 Engineered Safeguards System

2.3.2.1.1 Summary of Technical Information in the Application

In LRA Section 2.3.2.1, the applicant described the engineered safeguards system (ESS). The ESS is a two train independent and diverse system designed to identify inadequate core cooling, and then start the necessary pumps and open the needed valves to establish adequate core cooling conditions. The system is subdivided functionally into seven mechanical subsystems: (1) high pressure safety injection (HPSI) (2) low pressure safety injection (LPSI), (3) containment spray (CSS), (4) safety injection tanks (SIT), (5) safety injection and refueling water (SIRW) tank and containment sump suction, (6) shutdown cooling, (7) reactor cavity flood (RCF); and two electrical subsystems: (1) engineered safety features (ESF) actuation, and (2) normal shutdown (NSD) and design basis accident (DBA) sequencers.

Except for reactor cavity flood, the mechanical subsystems use most of the same system components for the various subsystem functions, and those components are hydraulically interconnected. Therefore, ESS is most accurately presented as a single system, and the groups of components that provide each major function are characterized as subsystems for license renewal purposes. The HPSI subsystem is designed to supply high pressure cooling water to the PCS under accident conditions. The LPSI subsystem is designed to supply low pressure cooling water to the PCS under accident conditions. The CSS subsystem is designed to limit post-accident containment pressure and remove heat from the containment atmosphere under accident conditions. The four SIT are designed to flood the core with borated water following a depressurization of the PCS. The SIRW and containment sump suction subsystem delivers water to the engineered safeguards system pumps through one of four paths. It provides the suction source for high pressure injection, low pressure injection, containment spray subsystems, and chemical and volume control system. The shutdown cooling subsystem is designed to transfer heat from the PCS to the component cooling system after reactor shutdown and after an accident, and to maintain a suitable temperature for refueling and maintenance during plant shutdowns. The reactor cavity flood subsystem consists of a network of floor drain piping designed to: (a) collect normal expected floor drains and transport these drains to the containment sump for subsequent disposal outside the containment, and (b) collect a portion of the containment spray water during accident conditions and transport this water into the cavity (annulus) between the biological shield and reactor vessel for flooding and cooling the outside of the reactor vessel bottom head. Neither function is SR or credited in the FSAR accident analyses.

The ESF actuation subsystem (electrical) consists of two independent and isolated circuits that initiate operation of redundant engineered safeguards equipment. These control circuits monitor whether offsite and/or emergency power is available and select load groups in accordance with the available power supply. The normal shutdown and design basis accident sequencer (electrical) is designed to sequentially load the safe shutdown equipment on to the emergency buses and diesel generators. Sequencing of loads ensures that the equipment is energized when needed while preventing excessive step loads from being imposed on the diesel generator which could result in loss of the diesel generator.

The ESS contains SR components that are relied upon to remain functional during and following DBEs. The failure of NSR SSCs in the ESS could potentially prevent the satisfactory accomplishment of an SR function. In addition, the ESS performs functions that support fire protection and EQ.

The intended functions within the scope of license renewal include the following:

- provides fluid pressure boundary
- provides heat transfer

In LRA Table 2.3.2-1, the applicant identified the following ESS component types that are within the scope of license renewal and subject to an AMR:

- SIRW tank
- safety injection tank
- SDC heat exchanger (HX) shell
- SIRWT HX shell
- SDC, SIRWT HX shell
- SIRWT HX shell
- SDC HX shell
- SDC HX channel head
- SDC HX tube sheet shell side
- SDC HX channel head shell side
- cont. spray pump HX shell, LPSI pump HX shell
- SIRWT HX tubes
- cont. spray, LPSI pump coils
- PCP seal cooler coils, cont. Spray pump coils, LPSI pump coils, SDC HX tubes
- SDC HX tubes
- PCP seal cooler coils, SIRWT HX tubes
- PCP seal cooler coils
- SIRWT HX tubes
- cont. spray pump coils, SDC HX tubes
- SIRWT HX tubes, cont. spray pump coils, LPSI pump coils, SDC HX tubes
- fasteners
- containment spray system fasteners
- containment spray pump bolting
- containment spray system valves bolting
- containment spray system valves header bolting
- HPSI, LPSI pumps bolting
- SIRWT HX bolting
- SIRWT bolting
- HPSI check valves, SCDC from PCS MOVs
- HPSI check valves
- hot leg injection check valves
- HPSI check valves loops 1A, 2A

2.3.2.1.2 Staff Evaluation

The staff reviewed LRA Section 2.3.2.1 and the FSAR Sections 6.1 and 6.2 using the evaluation methodology described in SER Section 2.3. The staff conducted its review in accordance with the guidance described in SRP-LR Section 2.3.

In conducting its review, the staff evaluated the system functions described in the LRA and FSAR in accordance with the requirements of 10 CFR 54.4(a) to verify that the applicant had not omitted from the scope of license renewal any components with intended functions delineated under 10 CFR 54.4(a). The staff then reviewed those components that the applicant had identified as being within the scope of license renewal to verify that the applicant had not omitted any passive and long-lived components that should be subject to an AMR in accordance with the requirements of 10 CFR 54.21(a)(1).

The staff's review of LRA Section 2.3.2.1 identified areas in which additional information was necessary to complete the review of the applicant's scoping and screening results. The applicant responded to the staff's RAIs as discussed below.

In RAI 2.3.2-1, dated June 3, 2005, the staff requested that the applicant provide a separate system description, system function listing, FSAR reference, scoping boundary drawings, and components subject to an AMR for each of the nine systems that have been integrated in LRA Section 2.3.2, "Engineered Safety Features."

In its response, by letter dated July 1, 2005, the applicant stated:

> Section 2.3.2 of the LRA describes Palisades' Engineered Safeguards System (ESS) in a manner consistent with the current licensing basis (CLB) and Final Safety Analysis Report (FSAR). The LRA statement, "The system is divided functionally into seven mechanical subsystems: ..." was not intended to imply that ESS consists of, or can be subdivided into discrete subsystems. The statement was intended to show that the ESS System, with one exception, is a single mechanical system (not a compilation of separate systems), which provides multiple functions. The exception is the Reactor Cavity Flood subsystem. Section 2.3.2 goes on to state, "Except for Reactor Cavity Flood, the mechanical subsystems use most of the same system components for the various subsystem functions, and those components are hydraulically interconnected. Therefore, ESS is most accurately presented as a single system, and the groups of components that provide each major function are characterized as subsystems for license renewal purposes."
>
> NMC's use of the term "subsystem" in the descriptions of the ESS functions, and the inclusion of electrical items in the mechanical ESS description, may have caused some confusion. The following discussion will address the listed "subsystems" in three parts: ESS Mechanical (High Pressure Safety Injection (HPSI), Low Pressure Safety Injection (LPSI), Containment Spray (CSS), Safety Injection Tanks (SIT), Safety Injection and Refueling Water Tank and Containment Sump Suction (SIRW), and Shutdown Cooling (SDC)); Reactor Cavity Flood (RCF); and ESS Electrical.

ESS Mechanical (HPSI, LPSI, CSS, SIT, SIRW, SDC)

These mechanical functions are provided by a single system using shared components, as described in FSAR Chapter 6, Engineered Safety Features, Sections 6.1 and 6.2. In practice, these in-scope mechanical functions of ESS are not independent of each other, but are provided by changing the operating equipment lineups. Since the mechanical components supporting these ESS functions are nominally made of the same materials and subject to the same environments, the aging management results do not vary by function. It was judged most logical for license renewal, therefore, to review the components providing these functions as a single system.

To facilitate NRC reviewer understanding of ESS design and operation for these functions, and to facilitate assignment of a separate reviewer for CSS, additional descriptive information and training materials were provided on April 12, April 15, and April 27, 2005. The information provided did not revise the application, but supported improved understanding and a suggested boundary definition for reviewer assignments.

For these functions, therefore, the requested information is provided in LRA Section 2.3.2.

Based on the discussion above and additional descriptive information provided on April 12, 15, and 27, 2005, the staff finds the applicant's response to RAI 2.3.2-1 acceptable. Therefore, the staff's concern described in RAI 2.3.2-1 is resolved.

In RAI 2.3.2-3, dated August 23, 2005, the staff stated that LRA Page 2-71 shows the major components of the seven mechanical subsystems of the engineered safeguards system; however, in LRA Table 2.3.2-1 and supplemental letter dated May 5, 2005, the following components are not identified as within the scope of license renewal: (a) LPCI control valves, (b) LPCI injection header, and (c) containment spray headers. Therefore, the staff requested that the applicant either justify why these components were not within the scope license renewal or submit an AMR for them.

In its response, by letter dated September 16, 2005, the applicant stated:

Palisades does not have 'LPCI control valves' or an 'LPCI injection header' but it does have Containment Spray headers.

The most comparable component to LPCI control valves would be the Low and High Pressure Safety Injection valves. These valves have motor operators. The motor operators were screened out as active components. The valve bodies are in scope and were evaluated as part of Engineered Safeguards. In the LRA these valves were included in the line items of Table 2.3.2-1 on page 2-80 and in Table 3.2.2-1 on page 3-76. However, to provide improved clarity for NRC review, Tables 2.3.2-1 and 3.2.2-1 were revised and submitted to the NRC in an NMC letter dated August 27, 2005. These clarified tables have consolidated the various valve components into the new consolidated component type "Valves" which includes the subject valve bodies.

2-50

The most comparable component to an LPCI Injection header would be the safety injection piping that supplies low and high pressure safety injection flow to the Primary Coolant System. This piping is included in the Piping & Fittings line item of Table 2.3.2-1.

The containment Spray header piping is included in the Piping & Fittings line item of Table 2.3.2-1.

Based on the applicant's inclusion of the above components, the staff finds the applicant's response to RAI 2.3.2-3 acceptable. Therefore, the staff's concern described in RAI 2.3.2-3 is resolved.

In RAI 2.3.2.1, dated August 26, 2005, the staff stated that, for the containment spray subsystem, valves and flow elements included within the scope of license renewal on drawings LR-M-203, sheet 2 and LR-M-204, sheets 1 and 1A, are not listed in LRA Table 2.3.2.1. Therefore, the staff requested that the applicant clarify whether these components are within the scope of license renewal and subject to an AMR. If excluded from within the scope of license renewal and not subject to an AMR, the staff requested that the applicant justify the exclusion.

In its response, by letter dated September 16, 2005, the applicant stated:

Valves and flow elements that are described in LRA Section 2.3.2.1, are included in the scope of license renewal, in accordance with 10CFR54.4(a), and are subject to an AMR in accordance with 10CFR54.21(a)(1).

As discussed in the response to NRC RAI 2.3.2-3 above, revised Tables 2.3.2-1 and 3.2.2-1 were provided to the NRC in an NMC letter dated August 27, 2005. These clarified tables have the various valve components consolidated into the component type 'Valves' which includes the subject valve bodies.

Flow elements are included in the 'Piping and Fittings' line items within these tables.

Based on its review, the staff finds the applicant's response to RAI 2.3.2.1 acceptable because the applicant had included the valves and flow element components within the scope of license renewal and subject to an AMR. Therefore, the staff's concern described in RAI 2.3.2.1 is resolved.

2.3.2.1.3 Conclusion

The staff reviewed the LRA, license renewal drawings, licensing-basis information, and RAI responses to determine whether any SSCs that should be within the scope of license renewal had not been identified by the applicant. No omissions were identified. In addition, the staff performed a review to determine whether any components that should be subject to an AMR had not been identified by the applicant. No omissions were identified. On the basis of its review, the staff concludes that the applicant adequately identified the ESS components that are within the scope of license renewal, as required by 10 CFR 54.4(a), and the ESS components that are subject to an AMR, as required by 10 CFR 54.21(a)(1).

2.3.3 Auxiliary Systems

In LRA Section 2.3.3, the applicant identified the structures and components of the auxiliary systems that are subject to an AMR for license renewal.

The applicant described the supporting structures and components of the auxiliary systems in the following sections of the LRA:

- 2.3.3.1   chemical and volume control system
- 2.3.3.2   circulating water system
- 2.3.3.3   component cooling water system
- 2.3.3.4   compressed air system
- 2.3.3.5   containment air recirculation and cooling system
- 2.3.3.6   emergency power system
- 2.3.3.7   fire protection system
- 2.3.3.8   fuel oil system
- 2.3.3.9   heating, ventilation, and air conditioning system
- 2.3.3.10   miscellaneous gas system
- 2.3.3.11   radwaste system
- 2.3.3.12   service water system
- 2.3.3.13   shield cooling system
- 2.3.3.14   spent fuel pool cooling system
- 2.3.3.15   waste gas system
- 2.3.3.16   domestic water system
- 2.3.3.17   chemical addition system

The staff's review findings regarding LRA Sections 2.3.3.1 – 2.3.3.17 are presented in SER Sections 2.3.3.1 – 2.3.3.17, respectively.

2.3.3.1 Chemical and Volume Control System

2.3.3.1.1 Summary of Technical Information in the Application

In LRA Section 2.3.3.1, the applicant described the chemical and volume control (CVC) system. The CVC system design basis is to maintain the required volume of water in the PCS over the range of full to zero reactor power, maintain the chemistry and purity of the primary coolant, maintain the desired boric acid concentration in the PCS, and pressure test the PCS. The boric acid concentration and chemistry of the primary coolant are maintained by the CVC system. Chemicals are introduced to the PCS by means of a metering pump which pumps the chemical solution from a chemical addition tank and introduces it to the charging pumps suction header.

The PCS may be pressure tested for leaks by means of the variable speed charging pump. The CVC system is not credited in the Chapter 14 accident analyses with any mitigating actions; however, the system responds to safety injection actuation signal and the charging pumps inject concentrated boric acid into the primary coolant system.

The CVC system contains SR components that are relied upon to remain functional during and following DBEs. The failure of NSR SSCs in the CVC system could potentially prevent the satisfactory accomplishment of an SR function. In addition, the CVC system performs functions that support fire protection, EQ, and SBO.

The intended function, within the scope of license renewal, is to provide fluid pressure boundary.

In LRA Table 2.3.3-1, the applicant identified the following CVC system component types that are within the scope of license renewal and subject to an AMR:

- boric acid storage tanks
- oil cooler shell
- oil cooler tubes
- letdown heat exchanger shell
- letdown heat exchanger channel head, tubes, tube sheet
- letdown heat exchanger tubes, tube sheet
- letdown heat exchanger tubes
- letdown heat exchanger channel head, tubes, tube sheet
- fasteners
- pipe - CVC cooler
- nozzle - CVC spray
- tubing - CVC oil
- pipe and pressure test fittings
- flow elements
- regenerative heat exchanger tubes, tube sheet
- regenerative heat exchanger channel head, shell, tubes, tube sheet
- control valves
- letdown stop valve CV-2001
- check, control, manual & relief valves; instrument assemblies

2.3.3.1.2 Staff Evaluation

The staff reviewed LRA Section 2.3.3.1 and the FSAR using the evaluation methodology described in SER Section 2.3. The staff conducted its review in accordance with the guidance described in SRP-LR Section 2.3.

In conducting its review, the staff evaluated the system functions described in the LRA and FSAR in accordance with the requirements of 10 CFR 54.4(a) to verify that the applicant had not omitted from the scope of license renewal any components with intended functions delineated under 10 CFR 54.4(a). The staff then reviewed those components that the applicant had identified as being within the scope of license renewal to verify that the applicant had not omitted any passive and long-lived components that should be subject to an AMR in accordance with the requirements of 10 CFR 54.21(a)(1).

The staff's review of LRA Section 2.3.3.1 identified an area in which additional information was necessary to complete the review of the applicant's scoping and screening results. The applicant responded to the staff's RAI as discussed below.

In RAI 2.3.3-1, dated August 23, 2005, the staff noted that scoping boundary drawing LR 202, sheets 1 and 1A indicate boric acid pumps and filters as within the scope of license renewal;

however, LRA Table 2.3.3-1 does not identify boric acid pumps and filters as within the scope of license renewal. Therefore, the staff requested that the applicant indicate which line item in LRA Table 2.3.3-1 includes the subject components.

In its response, by letter dated September 16, 2005, the applicant stated:

> NMC's LRA Supplement Letter, dated May 5, 2005, added the line items 'Pumps' to Tables 2.3.3-1 and 3.3.2-1 with the intended function of Fluid Pressure Boundary. This line item includes the Concentrated Boric Acid Pumps. Table 2.3.3-1 does not itemize filters as a separate line item. Therefore, a new line item for filters is hereby added to Table 2.3.3-1 with intended functions of Fluid Pressure Boundary and Filtration.

In its response, the applicant also added corresponding entries to LRA Table 3.3.2-1.

Based on the applicant's inclusion of the above components, the staff finds the applicant's response to RAI 2.3.3-1 acceptable. Therefore, the staff's concern described in RAI 2.3.3-1 is resolved.

2.3.3.1.3 Conclusion

The staff reviewed the LRA, accompanying scoping boundary drawing, and RAI response to determine whether any SSCs that should be within the scope of license renewal had not been identified by the applicant. No omissions were identified. In addition, the staff performed a review to determine whether any components that should be subject to an AMR had not been identified by the applicant. No omissions were identified. On the basis of its review, the staff concludes that the applicant adequately identified the CVC system components that are within the scope of license renewal, as required by 10 CFR 54.4(a), and the CVC system components that are subject to an AMR, as required by 10 CFR 54.21(a)(1).

2.3.3.2 Circulating Water System

2.3.3.2.1 Summary of Technical Information in the Application

In LRA Section 2.3.3.2, the applicant described the circulating water system (CWS). The CWS is a closed cycle system using two mechanical draft cooling towers. Each loop supplies one-half of the main condenser with cooling water by gravity flow from the two 18-cell, induced draft cross-flow cooling towers. The cooling towers are erected to the south of the plant over concrete basins. Two half-capacity vertical wet pit cooling tower pumps receive heated circulating water from the condenser pump suction makeup basin. The cooling tower pumps return the circulating water to the cooling tower distribution headers through two 96-inch pipes. Improved cooling efficiency and reduced system scaling are obtained by injection of dilution water pump discharge into the condenser inlet.

The failure of NSR SSCs in the CWS could potentially prevent the satisfactory accomplishment of an SR function.

The intended function, within the scope of license renewal, is to provide fluid pressure boundary.

In LRA Table 2.3.3-2, the applicant identified the following CWS component types that are within the scope of license renewal and subject to an AMR:

- fasteners
- pipe & fittings
- pumps
- valves & dampers

2.3.3.2.2 Staff Evaluation

The staff reviewed LRA Section 2.3.3.2 and the FSAR using the evaluation methodology described in SER Section 2.3. The staff conducted its review in accordance with the guidance described in SRP-LR Section 2.3.

In conducting its review, the staff evaluated the system functions described in the LRA and FSAR in accordance with the requirements of 10 CFR 54.4(a) to verify that the applicant had not omitted from the scope of license renewal any components with intended functions delineated under 10 CFR 54.4(a). The staff then reviewed those components that the applicant had identified as being within the scope of license renewal to verify that the applicant had not omitted any passive and long-lived components that should be subject to an AMR in accordance with the requirements of 10 CFR 54.21(a)(1).

The staff's review of LRA Section 2.3.3.2 identified areas in which additional information was necessary to complete the review of the applicant's scoping and screening results. The applicant responded to the staff's RAIs as discussed below.

In RAI 2.3.3.2-2, dated September 21, 2005, the staff stated that LRA Section 2.3.3.2 identifies the following valves as boundaries of portions of the circulating water system within the scope of license renewal: (1) valves MOV-5315 and MO-5316, which lead from dilution pumps P-40 A/B to cooling towers E-30 A/B makeup/fill located outside of the intake structure pump house and (2) valves MOV-5326A (basin 'A' cooling tower blowdown line isolation) and MOV-5326B (basin 'B' cooling tower blowdown line isolation). However, valves MOV-5315, MOV-5316, MOV-5326A, and MOV-5326B are not shown within the scope of license renewal on LR drawing LR-653, sheet 1 at locations D-2, F-2, D-1, and G-1, respectively. Therefore, the staff requested that the applicant clarify whether these valves are within the scope of license renewal and, if not, justify their exclusion in accordance with 10 CFR 54.4(a).

In its response, by letter dated October 21, 2005, the applicant stated:

> The words 'portion of the piping' in the referenced descriptions, and the highlighting on the drawing, were intended to indicate that only the indoor portions of the listed lines, and not the outdoor piping and valves, are in scope. The Circulating Water System piping in the screen house and turbine building is in scope for an (a)(2) spray concern only. The subject valves and some of the piping are outdoors (Screen House Roof or nearby ground level) and not required for isolation for license renewal, so they are not in scope.

For additional clarification, LRA Section 2.3.3.2 on page 2-88, third paragraph, items 2) and 3) are hereby revised to read as follows, '2) the indoor portions of the piping from ... house), and 3) the indoor portions of the piping from ... house).

Based on its review, the staff finds the applicant's response to RAI 2.3.3.2-2 acceptable because it adequately clarified that the LRA description identifying valves as boundaries of the portions of the circulating water system is for the pipe up to, but not including the valves, as indicated on the license renewal drawings. The applicant sufficiently explained that the piping in the screen house and turbine building is within the scope of license renewal in accordance with 10 CFR 54.4(a)(2). The valves in question are outside and not required for isolation; therefore, they perform no license renewal intended function. The staff concluded that the valves in question were correctly excluded from the scope of license renewal. Therefore, the staff's concern described in RAI 2.3.3.2-2 is resolved.

In RAI 2.3.3.2-3, dated September 21, 2005, the staff stated that license renewal drawing, LR-653, sheet 1, shows metering orifices (FE-5327A/B) at locations C-1 and G-1 not within the scope of license renewal. These metering orifices appear to be located in an area where their failure could cause the failure of SR components; therefore, these components should be within the scope of license renewal in accordance with 10 CFR 54.4(a)(2). Also, metering orifices are listed in LRA Table 2.3.3-2 as component groups subject to an AMR. These components serve a fluid pressure boundary intended function and are passive and long-lived. Therefore, the staff requested that the applicant justify the exclusion of metering orifices from the scope of license renewal and subject to an AMR in accordance with 10 CFR 54.4(a) and 10 CFR 54.21(a)(1), respectively.

In its response, by letter dated October 21, 2005, the applicant stated:

> FE-5327A/B are not located in the Screen House or Turbine Building, or near safety-related components, and are not in scope of license renewal. They are located outside after the circulating water piping exits the Screen House. For the FE's that are in scope for this system, they are included in component type 'Pipe & Fittings' which is included in Table 3.3.2-2.

Based on its review, the staff finds the applicant's response to RAI 2.3.3.2-3 acceptable because it adequately clarified that the metering orifices in question (FE-5327A/B) are not located in the screen house, turbine building, or near SR components; therefore, the components are not within the scope of license renewal in accordance with 10 CFR 54.4(a)(2). The applicant further explained that other flow elements within the scope of license renewal for the circulating water system are included in the LRA table under the "Pipe & Fittings" component group as subject to an AMR in accordance with 10 CFR 54.21(a)(1). Therefore, the staff's concerns described in RAI 2.3.3.2-3 are resolved.

2.3.3.2.3 Conclusion

The staff reviewed the LRA, accompanying scoping boundary drawings, and RAI responses to determine whether any SSCs that should be within the scope of license renewal had not been identified by the applicant. No omissions were identified. In addition, the staff performed a review to determine whether any components that should be subject to an AMR had not been identified by the applicant. No omissions were identified. On the basis of its review, the staff concludes that the applicant adequately identified the CWS components that are within the scope of license renewal, as required by 10 CFR 54.4(a), and the CWS components that are subject to an AMR, as required by 10 CFR 54.21(a)(1).

2.3.3.3 Component Cooling Water System

2.3.3.3.1 Summary of Technical Information in the Application

In LRA Section 2.3.3.3, the applicant described the component cooling water system (CCS). The CCS is designed to cool components carrying radioactive and potentially radioactive fluids. It provides a monitored intermediate barrier between these fluids and the service water system which transfers the heat to the outside environment. Thus, the probability of leakage of contaminated fluid into the lake is greatly reduced. System components are rated for the maximum heat removal requirements that occur during normal, shutdown or accident operation as applicable. The parts of the system located inside containment are isolated in the event of a containment high-pressure signal. The component cooling water to the radwaste evaporators and spent fuel cooling system are isolated on safety injection actuation signal. The system is a closed loop consisting of three motor-driven circulating pumps, two heat exchangers, a surge tank, associated valves, piping, instrumentation and controls. The system is continuously monitored by a process monitor which detects radioactivity which may have leaked into the system from the fluids being cooled. The component cooling water system uses demineralized water to which an inhibitor is added for corrosion control. Makeup to the system is automatically supplied from the primary system makeup storage tank. Heat is transferred from the system to plant service water by means of two component cooling heat exchangers. Service water from the critical service water header is provided to the tube side of the heat exchangers and the rejected heat from the system is discharged by service water into the cooling tower pump makeup basin.

The CCS contains SR components that are relied upon to remain functional during and following DBEs. The failure of NSR SSCs in the CCS could potentially prevent the satisfactory accomplishment of an SR function. In addition, the CCS performs functions that support fire protection and EQ.

The intended functions within the scope of license renewal include the following:

- provides fluid pressure boundary
- provides heat transfer

In LRA Table 2.3.3-3, the applicant identified the following CCS component types that are within the scope of license renewal and subject to an AMR:

- accumulators
- bistable/switch (in-line flow indicator)
- component cooling heat exchanger
- cooler
- heat exchanger
- fasteners
- primary coolant pump motor oil cooler
- pipe & fittings

2-57

- pumps
- valves & dampers
- waste gas compressor cooler

2.3.3.3.2 Staff Evaluation

The staff reviewed LRA Section 2.3.3.3 and the FSAR using the evaluation methodology described in SER Section 2.3. The staff conducted its review in accordance with the guidance described in SRP-LR Section 2.3.

In conducting its review, the staff evaluated the system functions described in the LRA and FSAR in accordance with the requirements of 10 CFR 54.4(a) to verify that the applicant had not omitted from the scope of license renewal any components with intended functions delineated under 10 CFR 54.4(a). The staff then reviewed those components that the applicant had identified as being within the scope of license renewal to verify that the applicant had not omitted any passive and long-lived components that should be subject to an AMR in accordance with the requirements of 10 CFR 54.21(a)(1).

The staff's review of LRA Section 2.3.3.3 identified an area in which additional information was necessary to complete the review of the applicant's scoping and screening results. The applicant responded to the staff's RAI as discussed below.

In RAI 2.3.3.3-1, dated September 21, 2005, the staff stated that LR drawing LR-223, sheet 1B shows component cooling water going to the component cooling water radiation monitor RE-0915 to be monitored for activity at location D-4. The radiation monitor and the component cooling lines entering and exiting it are shown as within the scope of license renewal; however, LRA Tables 2.3.3-3 and 3.3.2-3 do not list radiation monitor as a component group subject to an AMR. Therefore, the staff requested that the applicant clarify whether radiation monitors are included in a component group already listed in LRA Table 2.3.3-3, or, if not, justify their exclusion from an AMR in accordance with 10 CFR 54.21(a)(1).

In addition, the staff noted an apparent inconsistency between different systems related to the scoping of the radiation monitoring instrument. The radiation monitor associated with the component cooling water system is included in this system; however, radiation monitors associated with the radwaste system and steam generator blowdown systems are included in the radiation monitoring system. Therefore, the staff requested that the applicant explain this inconsistency.

In its response, by letter dated October 21, 2005, the applicant stated:

> RE-0915 is 'Active' per NEI 95-10, Appendix B analysis, and not subject to AMR.
>
> It is correct that the RE's could have been included consistently in either their parent system or in a consolidated radiation monitor system. However, at Palisades, radiation elements are active components (not subject to AMR) regardless of which system they are assigned to, so system assignments were based on convenience. No change is needed to the Palisades LRA.

Based on its review, the staff finds the applicant's response to RAI 2.3.3.3-1 acceptable. The applicant sufficiently clarified that radiation element RE-0915 had been screened out as an active component according to NEI 95-10 Appendix B guidance; therefore, it is not subject to an AMR in accordance with 10 CFR 54.21(a)(1). The applicant also agreed that the radiation elements could have been included consistently in either their parent system or in a consolidated radiation monitor system. Therefore, the staff's concerns described in RAI 2.3.3.3-1 are resolved.

2.3.3.3.3 Conclusion

The staff reviewed the LRA, accompanying scoping boundary drawings, and RAI response to determine whether any SSCs that should be within the scope of license renewal had not been identified by the applicant. No omissions were identified. In addition, the staff performed a review to determine whether any components that should be subject to an AMR had not been identified by the applicant. No omissions were identified. On the basis of its review, the staff concludes that the applicant adequately identified the CCS components that are within the scope of license renewal, as required by 10 CFR 54.4(a), and the CCS components that are subject to an AMR, as required by 10 CFR 54.21(a)(1).

2.3.3.4 Compressed Air System

2.3.3.4.1 Summary of Technical Information in the Application

In LRA Section 2.3.3.4, the applicant described the compressed air systems (CAS). The CAS consist of the instrument air system, the high pressure air system, various backup systems, and the feedwater purity air system. The instrument air system is an NSR system that is required for normal plant operation. The system is designed to provide a reliable supply of dry, oil-free air for instruments and controls, and for service air requirements. The design of the system is based on an estimated instrument air consumption rate of 80 scfm (standard cubic feet per minute) for the nuclear steam supply system and 115 scfm for the remainder of the plant. The high pressure air system consists of three oil-lubricated air compressors, each with its own dryer and air receiver. One of these high pressure air compressors resides in the turbine building and is NSR, while the other two are located in the east and west safeguards rooms. Moisture is removed from the high pressure air by dryers that are in series with the compressors' air-cooled aftercoolers. Backup systems consist of bottled nitrogen stations, bottled air station, bulk nitrogen, local accumulators, and manual valve actuators. The feedwater purity system is not an SR system. When manually aligned the system is capable of supplying air to the instrument and service air systems.

The CAS contains SR components that are relied upon to remain functional during and following DBEs. The failure of NSR SSCs in the CAS could potentially prevent the satisfactory accomplishment of an SR function. In addition, the CAS performs functions that support fire protection and EQ.

The intended functions within the scope of license renewal include the following:

- provides filtration
- provides fluid pressure boundary

In LRA Table 2.3.3-4, the applicant identified the following CAS component types that are within the scope of license renewal and subject to an AMR:

- accumulators
- air dryers
- blowers fans compressors vacuum
- filters/strainers
- heat exchangers
- fasteners
- pipe & fittings
- pumps
- traps (steam)
- valves & dampers

2.3.3.4.2 Staff Evaluation

The staff reviewed LRA Section 2.3.3.4 and the FSAR using the evaluation methodology described in SER Section 2.3. The staff conducted its review in accordance with the guidance described in SRP-LR Section 2.3.

In conducting its review, the staff evaluated the system functions described in the LRA and FSAR in accordance with the requirements of 10 CFR 54.4(a) to verify that the applicant had not omitted from the scope of license renewal any components with intended functions delineated under 10 CFR 54.4(a). The staff then reviewed those components that the applicant had identified as being within the scope of license renewal to verify that the applicant had not omitted any passive and long-lived components that should be subject to an AMR in accordance with the requirements of 10 CFR 54.21(a)(1).

The staff's review of LRA Section 2.3.3.4 identified areas in which additional information was necessary to complete the review of the applicant's scoping and screening results. The applicant responded to the staff's RAIs as discussed below.

In RAI 2.3.3.4-1, dated September 21, 2005, the staff stated that FSAR Section 9.5.1.5.c for the compressed air system states, "The safety positions and position on a loss of air supply for significant safety-related or important to safety air-operated valves are listed in FSAR Table 9-9. No failure of valves due to degraded instrument air precludes maintaining the plant in a safe condition provided the backup systems are available." FSAR Table 9-9 for valve CV-2191 indicates that the safety position of this valve is open and the position after loss of air is closed with a note also indicating that an accumulator is installed to open the valve upon loss of normal air supply. However, the air supply line and components between valve CV-2191 and the air reservoir are shown as not within the scope of license renewal on drawing LR-202, sheet 1, at location H-5. Failure of the air supply line, its associated in-line components, and the air reservoir will cause the valve to close when the safety position of the valve is open. As such, the air supply line and components between valve CV-2191 and the air reservoir should be within the scope of license renewal. Therefore, the staff requested that the applicant provide the basis for excluding the above-mentioned components from within the scope of license renewal and subject to an AMR in accordance with the requirements of 10 CFR 54.4(a) and 10 CFR 54.21(a)(1), respectively.

In its response, by letter dated October 21, 2005, the applicant stated:

> SV-2191, MV-CA2191, PCV-2191, CK-CA498, and the associated tubing and air reservoir are hereby included in scope of license renewal. Aging management of these components will be consistent with aging management of the other similar air system components as reported in the LRA.

Based on its review, the staff finds the applicant's response to RAI 2.3.3.4-1 acceptable. The applicant adequately explained that the air supply line and components between valve CV-2191 and the air reservoir are within the scope of license renewal because failure of the air supply line, its associated in-line components, and the air reservoir will cause the valve to close when the safety position of the valve is open. The applicant further clarified that aging management of these components will be consistent with the aging management of similar air system components, as reported in the LRA. Therefore, the staff's concern described in RAI 2.3.3.4-1 is resolved.

In RAI 2.3.3.4-2, dated September 21, 2005, the staff stated that compressed air license renewal drawings show portions of the air line connecting to the control valve/damper operators as within the scope of license renewal; however, the following portions of the control valve/damper operators are not shown as within the scope of license renewal. Therefore, the staff requested that the applicant explain how the valves/dampers perform their functions with a failure (loss of air) in the portions excluded from the scope of license renewal:

(a) the air supply line and solenoid valve to valve CV-2165 on LR-202, sheet 1A (location F-5)

(b) the air supply line beyond the solenoid valve to CV-0522A, which is shown with red tick marks on LR-205, sheet 2 (location G-3)

(c) the air supply line connecting to valve CV-0736 up to POC-0736 on LR-207, sheet 2 (location C4)

(d) the air supply line connecting to valve CV-1061 up to POC-1061 on LR-210, sheet 2 (location G-3)

(e) F-354 and the air supply line beyond SV-1768, F-353 and F-355 and the connecting air supply lines on LR-218, sheet 6; also, F-352 and F-351 and the connecting air supply lines on LR-218, sheet 6A

(f) the air supply lines and associated components on both sides of valves CV-0735, CV-1221 and CV-0734 on LR-212, sheet 2 (locations F-8, G-8, and H-6) and valves MV-PC-161, MV-PC-162, and MV-PC-163 on LR-219, sheet 1B (locations D-5, D-6)

In its response, by letter dated October 21, 2005, the applicant addressed each area of the staff's RAI:

(a) SV-2165 and associated tubing are hereby included in scope of license renewal under criterion (a)(3) (Appendix R).

(b) These components are retired in place and are isolated by locked valves and a blind flange.

(c) This air supply tubing is hereby in scope of license renewal.

(d) CV-1061 is in scope for non-safety related SSC attached to safety related SCC. POC-1061 is in scope because it is seismically mounted. The air supply tubing does not have operational safety significance and does not meet license renewal scoping criteria.

(e) The solenoid valves associated with F-351, F-352, F-353, F-354, and F-355 fail to their safe positions. The supply filters have no affect on the solenoid valves. These items do not meet license renewal scoping criteria. During this review it was noticed that F-356, on LR-M-218 Sheet 6 (E-2), was erroneously highlighted. Filter F-356 does not meet license renewal scoping criteria and the highlighting on the drawing should be disregarded.

(f) CV-0734 and CV-0735 are normally closed, their safety position is closed, and they fail as-is (i.e., closed) on loss of supply air. CV-1221 is normally open and fails open on loss of air supply. These items do not meet license renewal scoping criteria. MV-PC161, 162, and 163 were shown as in scope solely for administrative reasons because they have no Q-list interpretation and are, therefore, treated as safety-related. The associated air lines and control valves do not have operational safety significance and do not meet license renewal scoping criteria.

Based on its review, the staff finds the applicant's response to RAI 2.3.3.4-2 acceptable for each area because:

(a) The applicant adequately clarified that the air supply line and solenoid valve to valve CV-2165 on drawing LR-202, sheet 1A are within the scope of license renewal in accordance with 10 CFR 54.4(a)(3), as credited in Appendix R.

(b) The applicant adequately clarified that the air supply line beyond the solenoid valve to valve CV-0522A is retired in-place and isolated by locked valves and a blind flange; therefore, it is not within the scope of license renewal in accordance with 10 CFR 54.4(a).

(c) The applicant adequately clarified that the air supply line connecting to valve CV-0736 up to POC-0736 on drawing LR-207, sheet 2 performs an intended function of pressure boundary in accordance with 10 CFR 54.4(a); therefore, it is within the scope of license renewal.

(d) The applicant adequately clarified that valve CV-1061 is within the scope of license renewal in accordance with 10 CFR 54.4(a)(2) (i.e., NSR SSC attached to an SR SSC) and that POC-1061 is within the scope of license renewal because it is seismically mounted. The applicant further clarified that the air supply line connecting to valve CV-1061, up to POC-1061 on LR-210, sheet 2 does not meet any license renewal scoping criteria; therefore, it is not within the scope of license renewal in accordance with 10 CFR 54.4(a).

(e) The applicant adequately explained that the solenoid valves associated with F-351, F-352, F-353, F-354, and F-355 fail to their safe positions and that the supply filters have no affect on the solenoid valves. The applicant further clarified that these components meet no license renewal scoping criteria; therefore, they are not within the scope of license renewal in accordance with 10 CFR 54.4(a).

(f)   The applicant adequately clarified that the air supply lines and associated components on both sides of valves CV-0735, CV-1221, and CV-0734 on drawing LR-212, sheet 2, and valves MV-PC-161, MV-PC-162, and MV-PC-163 on drawing LR-219, sheet 1B do not meet any license renewal scoping criteria; therefore, they are not within the scope of license renewal in accordance with 10 CFR 54.4(a).

Therefore, the staff's concerns described in RAI 2.3.3.4-2 are resolved.

In RAI 2.3.3.4-4, dated September 21, 2005, the staff stated that drawing LR-212, sheet 3 (locations D-7, F-5), shows air lines within the scope of license renewal; however, the continuing portions of these lines are shown as not within the scope of license renewal. Therefore, the staff requested that the applicant explain how the portions within the scope of license renewal will be isolated from the portions not within the scope of license renewal without a valve or other component to isolate them.

In its response, by letter dated October 21, 2005, the applicant stated:

Note 1 on LR-M-212, Sheet 3 states: 'Each individual component is served by a local isolation valve.' These valves are not shown on LR-M-212, Sheet 3 but are available to isolate components if necessary.

Based on its review, the staff finds the applicant's response to RAI 2.3.3.4-4 acceptable. The applicant adequately explained that each individual component on drawing LR-M-212, sheet 3, is served by a local isolation valve and that these valves are not shown on the drawing, but are available to isolate components, if necessary. Therefore, the staff's concern described in RAI 2.3.3.4-4 is resolved.

In RAI 2.3.3.4-6, dated September 21, 2005, the staff stated that drawing LR-225, sheet 1, shows a silencer (S-966) at location A-6 as within the scope of license renewal; however, silencers are not listed in LRA Table 2.3.3-4 as component groups subject to an AMR. Silencers serve a fluid pressure boundary intended function and are passive and long-lived. The silencer performs no mechanical function; this component appears to be within the scope of license renewal for structural considerations. Therefore, the staff requested that the applicant clarify whether silencers are already included in LRA Table 2.3.3-4 as part of any other component group. If not, the staff requested that the applicant justify the exclusion of silencers as subject to an AMR in accordance with 10 CFR 54.21(a)(1).

In its response, by letter dated October 21, 2005, the applicant stated, "S-966 is located on air dryer M-9B's filter drain line. Failure of the silencer will not prevent the air dryer from performing its intended function. However, the silencer is seismically supported. Therefore, the hangers/supports for the silencer are in-scope for license renewal and are evaluated with the appropriate structural commodity."

Based on its review, the staff finds the applicant's response to RAI 2.3.3.4-6 acceptable. The applicant clarified that the silencer (S-966) is seismically supported, but the failure of the silencer will not prevent the air dryer from performing its intended function. The applicant further clarified that, as a consequence of their seismic support intended function, the hangers/supports for the silencer are within the scope of license renewal, in accordance with the requirements of 10 CFR 54.4(a), and are evaluated with the appropriate structural commodity. Therefore, the staff's concern described in RAI 2.3.3.4-6 is resolved.

In RAI 2.3.3.4-7, dated September 21, 2005, the staff stated that drawing LR-212, sheet 1A, shows P&IDs for air compressors C-2A and C-2C. In the RAI, the staff requested that the applicant provide additional information for the following areas:

(a) The first and second stages of air compressor C-2A and C-2C are shown as not within the scope of license renewal. Also, a note on the drawing states, "Per NEI 95-10, air compressors are excluded from the scope of license renewal. Therefore all components located inside C-2A and C-2C are excluded from license renewal scope." However, compressors are listed in LRA Table 2.3.3-4 as subject to an AMR. Therefore, the staff requested that the applicant explain the contradiction between the LRA table and the license renewal drawing.

(b) Fans for the air compressors C-2A and C-2C are shown as within the scope of license renewal and listed in LRA Table 2.3.3-4 with the fluid pressure boundary intended function. The staff requested that the applicant explain how these air compressor fans serve a fluid pressure boundary.

(c) The positive displacement pumps for air compressors C-2A and C-2C are shown as within the scope of license renewal. The pumps component type is listed in LRA Table 2.3.3-4 as subject to an AMR with a fluid pressure boundary intended function. Therefore, the staff requested that the applicant explain how the pumps, with an internal fluid of oil, are within the scope of license renewal. The staff also requested that the applicant explain how the oil sump and oil manifold, which also maintain fluid pressure boundary and are passive, are excluded from within the scope of license renewal in accordance with 10 CFR 54.4(a).

(d) The oil cooler and aftercooler tube and shell sides for air compressors C-2A and C-2C are shown as within the scope of license renewal. The tube side of each intercooler, for air compressors C-2A and C-2A, is shown as within the scope of license renewal; however, the shell side of the intercoolers is shown as not within the scope of license renewal. Therefore, the staff requested that the applicant explain why the shell side of the compressor intercoolers is excluded from within the scope of license renewal in accordance with 10 CFR 54.4(a).

In its response, by letter dated October 21, 2005, the applicant responded to each of area of the staff's RAI:

(a) The first and second stages of air compressors, including the load/unload valves, are hereby in scope and highlighted to indicate that they are in the scope of license renewal. The note on LR-M-212, Sheet 1A is hereby deleted. However, in accordance with NEI 95-10, compressors are active and do not require aging management review; therefore, the line item 'Blowers, Fans, Compressors, Vacuum' is hereby removed from Table 2.3.3-4 on page 2-100 and Table 3.3.2-4 on page 3-130.

(b)   The fans should not be highlighted on LR-M-212, Sheet 1A; the highlighting of the fans should be disregarded. The Component Group 'Blowers Fans Compressors Vacuum' in Table 2.3.3-4 only includes compressors. In accordance with NEI 95-10, compressors are active and do not require aging management review; this component group is hereby removed from LRA Tables 2.3.3-4 and 3.3.2-4.

(c)   The oil sump and oil manifold are in scope and should be highlighted on the drawing. However, they are parts of the compressor which is active per NEI 95-10, and do not require aging management review.

(d)   The skid mounted coolers shown on the drawing are air cooled and have fins rather than shells. The tubing is in scope for its pressure boundary function but the external fins represented by the surrounding boxes have no license renewal intended function. Therefore, the highlighting on the rectangular boxes around the oil coolers and aftercoolers should be disregarded.

Based its review, the staff finds the applicant's response to RAI 2.3.3.4-7 acceptable for each area because:

(a)   The applicant adequately clarified that the first and second stages of air compressor C-2A and C-2C, including the load/unload valves, are within the scope of license renewal in accordance with 10 CFR 54.4(a). The applicant further clarified that compressors are "Active" per NEI 95-10, Appendix B guidance; therefore, they are not subject to an AMR in accordance with 10 CFR 54.21(a)(1) and that the line item "Blowers, Fans, Compressors, Vacuum" is hereby removed from LRA Tables 2.3.3-4 and 3.3.2-4.

(b)   The applicant adequately clarified that the fans for air compressors C-2A and C-2C were highlighted in error. The applicant further clarified that the component group "Blowers Fans Compressors Vacuum" in LRA Table 2.3.3-4 only includes compressors and not the fans for air compressors C-2A and C-2C because these fans do not perform the intended function of pressure boundary; therefore, they are not within the scope of license renewal in accordance with 10 CFR 54.4(a).

(c)   The applicant adequately clarified that the oil sump and oil manifold serve the intended function of pressure boundary and are within the scope of license renewal in accordance with 10 CFR 54.4(a); however, the oil sump and oil manifold are "Active" per NEI 95-10 Appendix B guidance. Therefore, they are not subject to an AMR in accordance with 10 CFR 54.21(a)(1).

(d)   The applicant adequately explained that the skid mounted coolers shown on the drawing are air-cooled and have fins rather than shells. The applicant further clarified that the external fins represented by the surrounding boxes do not meet any license renewal scoping criteria; therefore, they are not within the scope of license renewal in accordance with 10 CFR 54.4(a) and the highlighting on the rectangular boxes around the oil coolers and aftercoolers should be disregarded.

Therefore, staff's concerns described in RAI 2.3.3.4-7 are resolved.

In RAI 2.3.3.4-8, dated September 21, 2005, the staff stated that LRA Table 2.3.3-4 of the compressed air system lists blowers as a component group subject to an AMR, with the intended function of fluid pressure boundary in accordance with 10 CFR 54.21(a)(1). However, the staff could not identify any blowers within the scope of license renewal on the license renewal drawings for the compressed air system. Therefore, the staff requested that the applicant provide drawings or other documents that indicate the location of the compressed air system blowers within the scope of license renewal in accordance with 10 CFR 54.4(a).

In its response, by letter dated October 21, 2005, the applicant stated that

> The components in the group 'Blowers, Fans, Compressors, Vacuum' that are in scope for LR for CAS are the compressors C-2A, B and C and C-6A and B. These compressors are active per NEI 95-10 and an AMR is not required. As noted in response to RAI-2.3.3.4-7, the group 'Blowers, Fans, Compressors, Vacuum' is being removed from Tables 2.3.3-4 and 3.3.2-4.

Based on its review, the staff finds the applicant's response to RAI 2.3.3.4-8 acceptable. The applicant adequately clarified that the component group, "Blowers Fans Compressors Vacuum," in LRA Table 2.3.3-4 only includes compressors. The applicant further clarified that compressors are "Active" per NEI 95-10, Appendix B guidance and as such not subject to an AMR in accordance with 10 CFR 54.21(a)(1). Therefore, the staff's concern described in RAI 2.3.3.4-8 is resolved.

2.3.3.4.3 Conclusion

The staff reviewed the LRA, accompanying scoping boundary drawings, and RAI responses to determine whether any SSCs that should be within the scope of license renewal had not been identified by the applicant. No omissions were identified. In addition, the staff performed a review to determine whether any components that should be subject to an AMR had not been identified by the applicant. No omissions were identified. On the basis of its review, the staff concludes that the applicant adequately identified the CAS components that are within the scope of license renewal, as required by 10 CFR 54.4(a), and the CAS components that are subject to an AMR, as required by 10 CFR 54.21(a)(1).

2.3.3.5 Containment Air Recirculation and Cooling System

2.3.3.5.1 Summary of Technical Information in the Application

In LRA Section 2.3.3.5, the applicant described the containment air recirculation and cooling system (CRS). The CRS includes four air handling and cooling units located entirely within the containment building. Plant service water from the critical service water is circulated through the air cooling units. Air is drawn through the coils by two matched vaneaxial fans with direct connected motors. One fan motor is rated for normal operating conditions and the second fan motor is rated for post-DBE conditions. The fan motors rated for the post-DBE condition are fed from the emergency power buses. Four units are normally in operation with two fans in each unit operating. The coolers are automatically changed to the emergency mode by a safety injection actuation signal. This signal will trip the normal rated fan motor in each unit.

The CRS contains SR components that are relied upon to remain functional during and following DBEs. In addition, the CRS performs functions that support fire protection and EQ.

The intended functions within the scope of license renewal include the following:

- provides filtration
- provides orifice for purpose of fluid flow measurement
- provides fluid pressure boundary
- provides heat transfer
- provides pressure-retaining boundary
- provides structural and/or functional support to safety-related equipment

In LRA Table 2.3.3-5, the applicant identified the following CRS component types that are within the scope of license renewal and subject to an AMR:

- containment air cooler coils
- containment air cooler filter
- containment air cooler flow element
- containment air cooler housing
- containment air cooler recirculation fans
- dampers
- drip pans
- duct
- fasteners
- manual and instrumentation valves

2.3.3.5.2 Staff Evaluation

The staff reviewed LRA Section 2.3.3.5 and the FSAR Sections 6.3 and 14.18 using the evaluation methodology described in SER Section 2.3. The staff conducted its review in accordance with the guidance described in SRP-LR Section 2.3.

In conducting its review, the staff evaluated the system functions described in the LRA and FSAR in accordance with the requirements of 10 CFR 54.4(a) to verify that the applicant had not omitted from the scope of license renewal any components with intended functions delineated under 10 CFR 54.4(a). The staff then reviewed those components that the applicant had identified as being within the scope of license renewal to verify that the applicant had not omitted any passive and long-lived components that should be subject to an AMR in accordance with the requirements of 10 CFR 54.21(a)(1).

2.3.3.5.3 Conclusion

The staff reviewed the LRA, license renewal drawings, and licensing-basis information to determine whether any SSCs that should be within the scope of license renewal had not been identified by the applicant. No omissions were identified. In addition, the staff performed a review to determine whether any components that should be subject to an AMR had not been identified by the applicant. No omissions were identified. On the basis of its review, the staff concludes that the applicant adequately identified the CRS components that are within the scope of license renewal, as required by 10 CFR 54.4(a), and the CRS components that are subject to an AMR, as required by 10 CFR 54.21(a)(1).

2.3.3.6 Emergency Power System

2.3.3.6.1 Summary of Technical Information in the Application

In LRA Section 2.3.3.6, the applicant described the emergency power system (EPS). The EPS has four major subsystems: (1) emergency diesel generator, (2) 125 Volt vital DC, (3) 120 Volt preferred AC system and 4) emergency lighting. The emergency power sources are designed to furnish onsite power to reliably shut down the plant and maintain it in a safe shutdown condition under all conditions, including DBE, upon loss of normal and standby power. The emergency power sources are part of the engineered safeguards electrical system and are identified as Class 1E systems. The diesel fire pump engines have also been included in the EPS system for convenience to permit fire pump diesels to be evaluated in conjunction with the diesel generator diesel engines. The emergency diesel generator subsystem consists of two independent, physically separate, diesel-engine driven generators of equal size. Support systems associated with each diesel generator include a fuel oil system, air starting system, lube oil system, jacket water system, crankcase exhauster, two starting circuits and a load sequencer. Supply of electric power for these support subsystems is obtained from the generator they are supporting. The 125-Volt vital DC power subsystem consists of two independent and redundant safety-related Class 1E DC power sources. On loss of normal and standby AC power, the batteries will supply power to required preferred AC and DC loads until one of the diesel generators has started and can supply power for the chargers. Assuming that neither diesel emergency generator is available, the batteries have ample capacity to supply required DC loads and preferred AC loads during a complete loss of AC power for at least four hours. The 120-Volt preferred AC subsystem is part of the engineered safeguards electrical system, and provides a Class 1E service. The inverters are the normal source of power for the four preferred AC buses. The function of each inverter is to convert the 125-Volt vital DC power from the batteries and provide continuous uninterruptible preferred AC power for the instrumentation required to shut down the reactor and maintain it in a safe condition after an anticipated operational occurrence or a postulated DBE. The preferred AC system buses supply power to the four reactor protection system channels and other ESF controls and instrumentation.

The emergency lighting subsystem is composed of three subsystems: (1) emergency AC lighting (the portion of the normal lighting system which is supplied by the alternating current portion of the Class 1E engineered safeguards electrical system), (2) emergency DC lighting (supplied by the 125-VOLT vital DC battery), and (3) emergency lighting units (supplied by internal battery packs which are continuously maintained in the charged condition when normal lighting power is available).

The EPS contains SR components that are relied upon to remain functional during and following DBEs. The failure of NSR SSCs in the EPS could potentially prevent the satisfactory accomplishment of an SR function. In addition, the EPS performs functions that support fire protection and SBO.

The intended functions within the scope of license renewal include the following:

- provides filtration
- provides fluid pressure boundary
- provides heat transfer

In LRA Table 2.3.3-6, the applicant identified the following EPS component types that are within the scope of license renewal and subject to an AMR:

- accumulators
- blowers fans compressor vacuum
- cooler
- fasteners
- filters/strainers
- heat exchangers
- heaters, electric
- misc mechanical (mufflers, oil pans)
- motors
- pipe & fittings
- pumps
- traps (steam)
- valves & dampers

2.3.3.6.2 Staff Evaluation

The staff reviewed LRA Section 2.3.3.6 and the FSAR using the evaluation methodology described in SER Section 2.3. The staff conducted its review in accordance with the guidance described in SRP-LR Section 2.3.

In conducting its review, the staff evaluated the system functions described in the LRA and FSAR in accordance with the requirements of 10 CFR 54.4(a) to verify that the applicant had not omitted from the scope of license renewal any components with intended functions delineated under 10 CFR 54.4(a). The staff then reviewed those components that the applicant had identified as being within the scope of license renewal to verify that the applicant had not omitted any passive and long-lived components that should be subject to an AMR in accordance with the requirements of 10 CFR 54.21(a)(1).

2.3.3.6.3 Conclusion

The staff reviewed the LRA and accompanying scoping boundary drawings to determine whether any SSCs that should be within the scope of license renewal had not been identified by the applicant. No omissions were identified. In addition, the staff performed a review to determine whether any components that should be subject to an AMR had not been identified by the applicant. No omissions were identified. On the basis of its review, the staff concludes that the applicant adequately identified the EPS components that are within the scope of license renewal, as required by 10 CFR 54.4(a), and the EPS components that are subject to an AMR, as required by 10 CFR 54.21(a)(1).

2.3.3.7 Fire Protection System

2.3.3.7.1 Summary of Technical Information in the Application

In LRA Section 2.3.3.7, the applicant described the fire protection system (FPS). The FPS includes the diverse design and operational features intended to prevent and mitigate the effects of fires. Building structures have been designed and arranged to prevent the spread of fire and to ensure integrity of redundant safe shutdown systems and areas. Fire suppression is provided by fixed water spray systems, such as sprinkler systems and deluge systems, fire hose reels and cabinets, portable fire extinguishers, fire barriers and fire detection systems. These fire suppression provisions are found throughout the plant site. Fixed fog deluge systems protect the main, start-up and station auxiliary transformers. Fire detection is provided in the form of smoke and ultraviolet detectors. Portable fire extinguishers are provided at convenient and accessible locations. The extinguishing media are pressurized water, CO_2 or dry chemical as appropriate for the service requirements of the area. Water for the fire suppression system is supplied by one of three full-capacity fire pumps. Each pump is capable of providing water to the largest system demand plus fire hose streams in the area of demand. A jockey pump with local controls is provided to maintain the fire suppression system full and pressurized. The building structure has been designed and arranged to prevent the spread of fire and to ensure integrity of redundant safe shutdown systems and areas.

The FPS contains SR components that are relied upon to remain functional during and following DBEs. The failure of NSR SSCs in the FPS could potentially prevent the satisfactory accomplishment of an SR function. In addition, the FPS performs functions that support fire protection.

The intended functions within the scope of license renewal include the following:

- provides fluid pressure boundary
- provides spray pattern at discharge nozzle

In LRA Table 2.3.3-7, the applicant identified the following FPS component types that are within the scope of license renewal and subject to an AMR:

- accumulators
- fasteners
- filters/strainers
- pipe & fittings
- pumps
- sprinkler heads
- valves & dampers

2.3.3.7.2 Staff Evaluation

The staff reviewed LRA Section 2.3.3.7 and the FSAR using the evaluation methodology described in SER Section 2.3. The staff conducted its review in accordance with the guidance described in SRP-LR Section 2.3.

In conducting its review, the staff evaluated the system functions described in the LRA and FSAR in accordance with the requirements of 10 CFR 54.4(a) to verify that the applicant had not omitted from the scope of license renewal any components with intended functions delineated under 10 CFR 54.4(a). The staff then reviewed those components that the applicant had identified as being within the scope of license renewal to verify that the applicant had not omitted any passive and long-lived components that should be subject to an AMR in accordance with the requirements of 10 CFR 54.21(a)(1).

The staff also reviewed the Fire Protection Program Report Volumes 1 & 2, Revision 6, April 2005, approved fire protection SER, dated September 1, 1978, and its fire protection SER supplements. These reports are referenced in the fire protection CLB, which summarizes the Fire Protection Program and its commitments to 10 CFR 50.48 using the guidance of Branch Technical Position (BTP) Auxiliary and Power Conversion Systems Branch (APCSB) 9.5-1, Appendix A, dated August 23, 1976.

The staff's review of LRA Section 2.3.3.7 identified areas in which additional information was necessary to complete the review of the applicant's scoping and screening results. The applicant responded to the staff's RAIs as discussed below.

In RAI 2.3.3.7-1, dated August 31, 2005, the staff stated that drawing LR-216, sheet 2 shows fire hose stations FHS 1, FHS 6, FHS 2, and FHS 4 (locations F-3 and F-4) as not within the scope of license renewal. From the information available in the CLB the staff could not determine whether these fire hose stations should be excluded from the scope of license renewal. Therefore, the staff requested that the applicant provide the fire area and the room number in which these fire hose stations are located. The staff also requested that the applicant justify the exclusion of these fire hose stations from within the scope of license renewal in accordance with 10 CFR 54.4(a)(3).

In its response, by letter dated September 16, 2005, the applicant stated:

> FHS 1 is located in the Auxiliary Building, elevation 649', in room 405 and is not located in a Palisades Fire Zone Area. It is located in an administrative area over the Instrument & Control Shop, which is not located near equipment in scope of License Renewal.

> FHS 2 is located in the Auxiliary Building, elevation 611', in room 210 and is not located in a Palisades Fire Zone Area. It is located in the Auxiliary Building Access Control Area, which is not located near equipment in scope of License Renewal.

> FHS 4 is located in the Auxiliary Building, elevation 611', in room 216 and is not located in a Palisades Fire Zone Area. It is located in an Auxiliary Building hallway, which is not located near equipment in scope of License Renewal.

> FHS 6 is located in the Auxiliary Building, elevation 625', in room 731 and is located in Palisades Fire Zone Area #27. It is located in the Volume Reduction Area (equipment retired in place), and is not located near equipment in scope of License Renewal.

Based on its review, the staff finds the applicant's response to RAI 2.3.3.7-1 acceptable. Hose stations not within the scope of license renewal have been located and discussed. Therefore, the staff's concern described in RAI 2.3.3.7-1 is resolved.

In RAI 2.3.3.7-3, dated August 31, 2005, the staff stated that LRA Section 2.4.7 lists building framing-concrete, below grade (wall, foundation, slab, grout, reinforcement, trenches, cable pits, tunnels, etc.) as a component group subject to an AMR. Although fire barrier was identified as one of the intended functions for this component group, it was not clear whether all of the fire barriers and the radiant energy shield wall in the screen house (intake structure) described in the fire hazards analysis report (fire area 9) are included within the scope of license renewal. Therefore, the staff requested that the applicant confirm that all fire barriers and components such as doors, dampers, and penetration seals described in the plant Fire Hazards Analysis Report are included within the scope of license renewal or to state why any of the fire barriers or the radiant energy shield are excluded from the scope of license renewal.

In its response, by letter dated September 16, 2005, the applicant stated:

> Fire barrier commodities are in scope of license renewal and are included in the Miscellaneous and Bulk Commodities section of the LRA. See LRA Table 2.4.8-1, Miscellaneous Structural and Bulk Commodities on page 2-244 and 2-245, and Table 3.5.2-8 on page 3-362 to 3-370. The radiant energy shield in the intake structure is specifically included in component type 'Fire Barrier - Intake Structure Bldg, Carbon Steel, Protected' and is shown on pages 2-245 and 3-365.

Based on its review, the staff finds the applicant's response to RAI 2.3.3.7-3 acceptable. Fire barriers and the associated opening protective components were identified as within the scope of license renewal and subject to an AMR under the specified section of the LRA. Therefore, the staff's concern described in RAI 2.3.3.7-3 is resolved.

In RAI 2.3.3.7-4, dated August 31, 2005, the staff stated that drawing LR-216, sheet 2, depicted sprinkler systems in the following areas as not within the scope of license renewal: boiler rooms, office (elevation 607'), cold chemical and IC labs (elevation 590'), and office (elevation 590') at location F-7 and maintenance storage (elevation 590'), and tech and maintenance office (elevation 590'). Therefore, the staff requested that the applicant provide the basis for excluding the sprinkler systems for those rooms from the scope of license renewal or to revise the license renewal drawings accordingly.

In its response, by letter dated September 16, 2005, the applicant stated:

> The subject rooms do not contain safety related equipment, and they are not required to achieve Safe Shutdown for a regulated event. These rooms have sprinklers for loss prevention/property protection purposes only. Therefore, the sprinkler systems in these rooms are not in the scope of license renewal.

Based on its review, the staff finds the applicant's response to RAI 2.3.3.7-4 acceptable. The sprinkler systems in the identified rooms are not within the scope of license renewal because the protected rooms do not contain any equipment required for safe shutdown in the event of a fire. Therefore, the staff's concern described in RAI 2.3.3.7-4 is resolved.

In RAI 2.3.3.7-5, dated August 31, 2005, the staff stated that LRA Section 2.4.3, system function listing S0100-FP, has a comment that containment exterior wall is an Appendix R fire barrier; however, LRA Table 2.4.3-1 does not include fire barrier/protection as an intended function.

Therefore, the staff requested that the applicant provide information to clarify that LRA Table 2.4.3-1 does not need to describe fire protection as an intended function, or to revise the table to define fire protection as an intended function.

In its response, by letter dated September 16, 2005, the applicant stated:

> Table 2.4.3-1 addresses GALL Section IIA Containment concrete and steel commodities. GALL Section VIIG Fire Protection commodities, including concrete fire barriers/walls, are included in LRA Section 2.4.8, Miscellaneous and Bulk Commodities. The 'Fire Barrier' intended function for the Containment exterior wall is shown as part of component type 'Fire Barrier - Containment Bldg - Concrete, Exposed' in Table 2.4.8-1 (page 2-245 of the LRA).
>
> Therefore, it is not necessary for Table 2.4.3-1 to include Fire Barrier as an Intended Function, since this function of the Containment exterior wall is included in Table 2.4.8-1.

Based on its review, the staff finds the applicant's response to RAI 2.3.3.7-5 acceptable. The intended function of the containment exterior wall as a fire barrier was clarified through cross reference. Therefore, the staff's concern described in RAI 2.3.3.7-5 is resolved.

In RAI 2.3.3.7-6, dated August 31, 2005, the staff stated that LRA Section 2.4.6 states that the water purity building is within the scope of license renewal based on fire protection requirements to achieve safe shutdown. The section also states that the feedwater purity boiler room in the south end of the building houses a diesel fuel oil transfer pump and piping and that this is the only area in this structure that is within the scope of license renewal. Further, system function S0800-FP states that the building provides structural and/or functional support to fire protection-related components; however, LRA Table 2.4.6-1 does not identify fire protection/barrier as an intended function. Therefore, the staff requested that the applicant to clarify that fire barrier is not an intended function for the feedwater purity building, or to revise LRA Table 2.4.6-1 to define fire barrier as an intended function.

In its response, by letter dated September 16, 2005, the applicant stated:

> The safety related fuel oil transfer pump that is credited with providing makeup fuel oil to the emergency diesel generators (EDG) is located in the Intake Structure. A non-safety related backup pump is also located adjacent to the safety-related pump in the same room. In the event of a fire in this room, coincident with a loss of offsite power, there is non-safety related fuel oil transfer piping located in the boiler room of the Feedwater Purity Building that serves as an alternate source of fuel oil for the EDGs. The feedwater purity building only provides structural and functional support for these fuel oil components since simultaneous fires in both areas in conjunction with a loss of offsite power event do not need to be postulated. If there were a fire that would affect the fuel oil components in the feedwater purity building, the primary and backup pumps in the Intake Structure, which is a separate structure remote from the Feedwater Purity Building, would still be available. Based on this description, the water purity building does not have a 'Fire Barrier' intended function. The appropriate intended function is shown in LRA Table 2.4.6-1 as 'Structural Support for

Regulated Events' (i.e., provides structural and/or functional support to fire protection-related components).

Based on its review, the staff finds the applicant's response to RAI 2.3.3.7-6 acceptable. The water purity building was clarified to not have the intended function of a fire barrier. Therefore, the staff's concern described in RAI 2.3.3.7-6 is resolved.

In RAI 2.3.3.7-7, dated August 31, 2005, the staff stated that LRA Section 2.4.10 states that the turbine building also contains fire barrier concrete commodities credited in fire protection requirements for achieving safe shutdown. Further, the system function listing states that this section includes fire protection as one of the functions; however, LRA Table 2.4.10-1 does not include fire protection as an intended function. Therefore, the staff requested that the applicant provide information to clarify that fire barrier is not an intended function for the turbine building, or to revise LRA Table 2.4.10-1 to define fire barrier as an intended function.

In its response, by letter dated September 16, 2005, the applicant stated:

> LRA Table 2.4.10-1 addresses GALL Section IIIA Turbine Building concrete and steel commodities GALL Section VIIG Fire Protection commodities, including concrete fire barriers/walls, are included in LRA Section 2.4.8, Miscellaneous and Bulk Commodities. The 'Fire Barrier' intended function for the turbine building walls is shown as part of component type 'Fire Barrier - Turbine Bldg - Concrete, Exposed' in Table 2.4.8-1 (page 2-245 of the LRA) with the aging management evaluation summary provided in Table 3.5.2-8 (page 3-367 of the LRA).

Based on its review, the staff finds the applicant's response to RAI 2.3.3.7-7 acceptable. The intended function of the turbine building walls as a fire barrier was clarified through cross reference. Therefore, the staff's concern described in RAI 2.3.3.7-7 is resolved.

In RAI 2.3.3.7-9, dated August 31, 2005, the staff stated that flexible connections on drawing LR-216, sheet 1, at locations B-4, C-4, and E-5 and lube oil coolers on drawing LR-216, sheet 1, at locations B-4 and F-5 are shown on the fire protection license renewal drawings as within the scope of license renewal; however, LRA Table 2.3.3-7 does not list these components as subject to an AMR. These components are passive, long-lived, and serve a pressure boundary intended function. Therefore, the staff requested that the applicant clarify whether these components are included in LRA Table 2.3.3-7 as part of any other component group and, if not, to justify the exclusion of these components from being subject to an AMR in accordance with 10 CFR 54.21(a)(1).

In its response, dated September 16, 2005, the applicant stated:

> (a) The flexible connections for fire pump diesel drivers K-5 and K-10 are not subject to AMR because they are short-lived. They are periodically replaced under the preventive maintenance program every 3 years.

> (b) As discussed in LRA Section 2.3.3.6 on page 2-105, the fire pump diesel drivers are addressed as part of the Emergency Power System. The diesel fire pump lube oil coolers are shown as subject to AMR in Table 2.3.3-6 on page 2-112, and AMR results are included in Table 3.3.2-6 on page 3-142.

Based on its review, the staff finds the applicant's response to RAI 2.3.3.7-9 acceptable. The flexible connectors for the fire pump drivers were identified as short-lived components and not subject to an AMR. Also, the AMR for the diesel fire pump lube coolers was identified. Therefore, the staff's concern described in RAI 2.3.3.7-9 is resolved.

2.3.3.7.3 Conclusion

The staff reviewed the LRA, RAI responses, and the accompanying scoping boundary drawings to determine whether any SSCs that should be within the scope of license renewal had not been identified by the applicant. No omissions were identified. In addition, the staff performed a review to determine whether any components that should be subject to an AMR had not been identified by the applicant. No omissions were identified. On the basis of its review, the staff concludes that the applicant adequately identified the FPS components that are within the scope of license renewal, as required by 10 CFR 54.4(a), and the FPS components that are subject to an AMR, as required by 10 CFR 54.21(a)(1).

2.3.3.8 Fuel Oil System

2.3.3.8.1 Summary of Technical Information in the Application

In LRA Section 2.3.3.8, the applicant described the fuel oil system (FOS). The FOS is designed to provide storage of an adequate volume of fuel oil for accident conditions, to transfer fuel to the diesel engines or boilers at an adequate rate, and to stop delivery to the unit storage tank when filled. The primary loads on the fuel oil system are the two emergency diesel engines and the two fire pump engines.

The FOS contains SR components that are relied upon to remain functional during and following DBEs. The failure of NSR SSCs in the FOS could potentially prevent the satisfactory accomplishment of an SR function. In addition, the FOS performs functions that support fire protection.

The intended functions within the scope of license renewal include the following:

- provides rated fire barrier
- provides flow restriction (throttle)
- provides fluid pressure boundary

In LRA Table 2.3.3-8, the applicant identified the following FOS component types that are within the scope of license renewal and subject to an AMR:

- accumulators
- filters/strainers
- indicators/recorders (level glasses)
- fasteners
- pipe & fittings
- pumps
- valves & dampers

2.3.3.8.2 Staff Evaluation

The staff reviewed LRA Section 2.3.3.8 and the FSAR using the evaluation methodology described in SER Section 2.3. The staff conducted its review in accordance with the guidance described in SRP-LR Section 2.3.

In conducting its review, the staff evaluated the system functions described in the LRA and FSAR in accordance with the requirements of 10 CFR 54.4(a) to verify that the applicant had not omitted from the scope of license renewal any components with intended functions delineated under 10 CFR 54.4(a). The staff then reviewed those components that the applicant had identified as being within the scope of license renewal to verify that the applicant had not omitted any passive and long-lived components that should be subject to an AMR in accordance with the requirements of 10 CFR 54.21(a)(1).

2.3.3.8.3 Conclusion

The staff reviewed the LRA and accompanying scoping boundary drawings to determine whether any SSCs that should be within the scope of license renewal had not been identified by the applicant. No omissions were identified. In addition, the staff performed a review to determine whether any components that should be subject to an AMR had not been identified by the applicant. No omissions were identified. On the basis of its review, the staff concludes that the applicant adequately identified the FOS components that are within the scope of license renewal, as required by 10 CFR 54.4(a), and the FOS components that are subject to an AMR, as required by 10 CFR 54.21(a)(1).

2.3.3.9 Heating, Ventilation, and Air Conditioning System

2.3.3.9.1 Summary of Technical Information in the Application

In LRA Section 2.3.3.9, the applicant described the heating, ventilation, and air conditioning system (VAS). The VAS provides air flow to spaces in the plant. Various supply and exhaust fan combinations provide ventilation air for breathing, heated air to prevent equipment freezing and for personnel comfort in cold weather, and cooled air to selected locations to remove heat from lights, equipment, etc. This system is a collection of independent ventilation subsystems, with the major ones being: (1) control room HVAC; (2) containment purge; (3) engineered safeguards room HVAC; (4) emergency diesel generator room fans; (5) electrical equipment room HVAC, including battery room; and (6) fuel handling area ventilation.

The control room HVAC subsystem provides conditioned air to the control room, the technical support center (TSC), the viewing gallery and the mechanical equipment room. The subsystem has separate and redundant air handling units, air filtering units, condensing units, steam humidifiers, and continuous air monitors. The containment purge subsystem supplies air to the air room area of the containment building and provides an exhaust line that connects the containment building to the main exhaust fans. The engineered safeguard rooms HVAC subsystem normally supplies ventilation air to the east and west engineered safeguard rooms via the radwaste area HVAC subsystem. The ductwork to these rooms is automatically isolated if airborne radiation in the exhaust ductwork exceeds preset levels. The emergency diesel generator room fans provide ventilation air to remove heat generated by the diesel generators. The fans are powered by the particular emergency diesel generator for which they provide cooling. The electrical equipment, switchgear, & cable spreading room HVAC subsystem (includes battery rooms) draws outside air through a filter and heating coil and distributes the air to these spaces, which contain mainly electrical equipment. The fuel handling area ventilation subsystem provides the capability of filtering potential airborne radioactive particulates from the area of the spent fuel pool following a fuel handling accident or a fuel cask drop accident. The penetration and fan room HVAC subsystem was installed in conjunction with modifications to protect essential structures, systems and components from the effects of high energy line breaks.

The VAS contains SR components that are relied upon to remain functional during and following DBEs. The failure of NSR SSCs in the VAS could potentially prevent the satisfactory accomplishment of an SR function. In addition, the VAS performs functions that support fire protection, EQ, and SBO.

The intended functions within the scope of license renewal include the following:

- provides containment isolation
- provides filtration
- provides fluid pressure boundary
- provides heat transfer

In LRA Table 2.3.3-9, the applicant identified the following VAS component types that are within the scope of license renewal and subject to an AMR:

- blowers fans compressor vacuum
- ductwork
- filters/strainers
- heat exchangers
- heaters, electric
- muffler (CRHVAC refrig. condensing units)
- elastomers in flexible connections and seals inside/outside of containment
- CRHVAC duct silencer
- fasteners
- dampers
- pipe & fittings
- traps (steam)
- valves & dampers

2.3.3.9.2 Staff Evaluation

The staff reviewed LRA Section 2.3.3.9 and the FSAR Sections 5.8, 6.5, 7.2, 9.6, 9.8, 14.11, 14.19, 14.20, 14.21, 14.23, and 14.24 using the evaluation methodology described in SER Section 2.3. The staff conducted its review in accordance with the guidance described in SRP-LR Section 2.3.

In conducting its review, the staff evaluated the system functions described in the LRA and FSAR in accordance with the requirements of 10 CFR 54.4(a) to verify that the applicant had not omitted from the scope of license renewal any components with intended functions delineated under 10 CFR 54.4(a). The staff then reviewed those components that the applicant

had identified as being within the scope of license renewal to verify that the applicant had not omitted any passive and long-lived components that should be subject to an AMR in accordance with the requirements of 10 CFR 54.21(a)(1).

The staff's review of LRA Section 2.3.3.9 identified an area in which additional information was necessary to complete the review of the applicant's scoping and screening results. The applicant responded to the staff's RAI as discussed below.

In RAI 2.3.3.9, dated August 26, 2005, the staff stated that the VAS, as described in LRA Section 2.3.3.9, and Table 2.3.3.9, identifies the component groups that require an AMR and their intended functions; however, LRA Table 2.3.3.9 does not list all of the components of the systems as highlighted in drawings LR-M-208, sheets 1, 1A, and 1B; drawing LR-M-218, sheets 1, 2, 4, 5, 6A, and 7; and drawing LR-M-658, sheet 1. For example, LRA Table 2.3.3.9 does not list the associated components, such as filter housings, damper housings, fan housings, valve bodies etc. Therefore, the staff requested that the applicant clarify whether these and all other applicable components of the system are within the scope of license renewal and subject to an AMR in accordance with 10 CFR 54.4(a) and 10 CFR 54.21(a)(1), respectively, and, if excluded from within the scope of license renewal and not subject to an AMR, the staff requested that the applicant provide justification for the exclusion.

In its response, by letter dated September 16, 2005, the applicant stated:

> LRA Table 2.3.3.9 does include the components of the VAS System highlighted in drawings LR-M-208, sheets 1, 1A, and 1B,;LR-M-218, Sheets 1,2,4,5, 6A, 7; and LR-M- 658, sheet 1. As examples, filter housings are included in the LRA Table 2.3.3.9 line item 'Filters/Strainers' on LRA page 2-131, damper housings are included in line item 'Dampers' on LRA page 2-131, fan housings are included in line item 'Blowers Fans Compressor Vacuum' on LRA page 2-130, and valve bodies are included in line item 'Valves & Dampers' on LRA page 2-131.
>
> These components and other applicable components of the VAS System, highlighted on the drawings listed on LRA page 2-130, are within the scope of license renewal, in accordance with 10 CFR 54.4(a), and subject to an AMR in accordance with 10 CFR 54.21(a)(1).

Based on its review, the staff finds the applicant's response to RAI 2.3.3.9 acceptable because the applicant had included the applicable VAS components within the scope of license renewal and subject to an AMR. Therefore, the staff's concern described in RAI 2.3.3.9 is resolved.

2.3.3.9.3 Conclusion

The staff reviewed the LRA, license renewal drawings, licensing-basis information, and RAI response to determine whether any SSCs that should be within the scope of license renewal had not been identified by the applicant. No omissions were identified. In addition, the staff performed a review to determine whether any components that should be subject to an AMR had not been identified by the applicant. No omissions were identified. On the basis of its review, the staff concludes that the applicant adequately identified the VAS components that are within the scope of license renewal, as required by 10 CFR 54.4(a), and the VAS components that are subject to an AMR, as required by 10 CFR 54.21(a)(1).

2.3.3.10 Miscellaneous Gas System

2.3.3.10.1 Summary of Technical Information in the Application

In LRA Section 2.3.3.10, the applicant described the miscellaneous gas system (MGS). The MGS is a collection of all the compressed bottles and liquid storage of gases used in various plant process and equipment. The stored gases are nitrogen (liquid and gaseous), hydrogen, carbon dioxide (liquid), propane, helium, argon, acetylene, and air bottles (air bottles are included in the compressed air system scoping/screening). Each system has headers to attach the gaseous source, pressure regulators, monitoring gauges, valving and piping. The Appendix R evaluations identified the need for a backup to the air supply to air operated valves. This evaluation resulted in numerous nitrogen supply stations being strategically located throughout the plant.

The MGS contains SR components that are relied upon to remain functional during and following DBEs. The failure of NSR SSCs in the MGS could potentially prevent the satisfactory accomplishment of an SR function. In addition, the MGS performs functions that support fire protection, EQ, and SBO.

The intended function, within the scope of license renewal, is to provide fluid pressure boundary.

In LRA Table 2.3.3-10, the applicant identified the following MGS component types that are within the scope of license renewal and subject to an AMR:

- accumulators
- filters/strainers
- misc mechanical (fasteners, manifold, monitor)
- pipe & fittings
- valves & dampers
- tanks

2.3.3.10.2 Staff Evaluation

The staff reviewed LRA Section 2.3.3.10 and the FSAR using the evaluation methodology described in SER Section 2.3. The staff conducted its review in accordance with the guidance described in SRP-LR Section 2.3.

In conducting its review, the staff evaluated the system functions described in the LRA and FSAR in accordance with the requirements of 10 CFR 54.4(a) to verify that the applicant had not omitted from the scope of license renewal any components with intended functions delineated under 10 CFR 54.4(a). The staff then reviewed those components that the applicant had identified as being within the scope of license renewal to verify that the applicant had not omitted any passive and long-lived components that should be subject to an AMR in accordance with the requirements of 10 CFR 54.21(a)(1).

The staff's review of LRA Section 2.3.3.10 identified areas in which additional information was necessary to complete the review of the applicant's scoping and screening results. The applicant responded to the staff's RAIs as discussed below.

In RAI 2.3.3.10-1, dated September 21, 2005, the staff stated that the following components are shown on the license renewal drawings for the miscellaneous gas system as within the scope of license renewal:

(a) containment sampling pumps on drawing LR-224, sheet 2, at locations C-6 and G-5
(b) moisture separators on drawing LR-224, sheet 2, at locations C-5, F-5

However, LRA Table 2.3.3-10 does not list these component groups as subject to an AMR. These components perform a pressure boundary intended function, are passive and long-lived, and are in the hydrogen monitoring system. LRA Section 2.3.3.10 states that the hydrogen monitoring system is a portion of the miscellaneous gas system within the scope of license renewal. LRA Section 2.3.3.10 also states that "HYM-01" is an intended function of the hydrogen monitoring system, which provides continuous hydrogen monitoring of the containment atmosphere during post-accident conditions. Therefore, the staff requested that the applicant clarify if these components are already included in LRA Table 2.3.3-10 as part of any other component group. If not, the staff requested that the applicant justify the exclusion of these components as subject to an AMR in accordance with 10 CFR 54.21(a)(1).

In its response, by letter dated October 21, 2005, the applicant addressed each item of the staff's RAI:

(a) LRA Table 2.3.3-10 is hereby revised to include the new line item 'Pumps' with an Intended Function of Fluid Pressure Boundary. Table 3.3.2-10 is also revised to include the companion line item of Component Type 'Pumps,' with Intended Function, Material, Environment, Aging Effect Requiring Management, and Aging Management Program entries of Fluid Pressure Boundary, Stainless Steel, Air (Int) and Plant Indoor air (Ext), None, and None Required, respectively.

(b) The moisture separators are associated with the hydrogen monitors. LRA Table 2.3.3-10 is hereby revised to include a new line item 'Monitor,' with an -Intended Function of Fluid Pressure Boundary. Table 3.3.2-10 is also revised to include the new companion line item of Component Type 'Monitor,' with Intended Function, Material, Environment, Aging Effect Requiring Management, and Aging Management Program entries of Fluid Pressure Boundary, Stainless Steel, Air (Int) and Plant Indoor air (Ext), None, and None Required, respectively.

Based on its review, the staff finds the applicant's response to each item in RAI 2.3.3.10 acceptable because:

(a) The applicant agreed that the containment sampling pumps are passive and long-lived and perform the intended function of pressure boundary. In LRA Table 2.3.3-10, the applicant also added the component group "pumps" as subject to an AMR in accordance with 10 CFR 54.21(a)(1).

(b)   The applicant clarified that the moisture separators are associated with the hydrogen monitors. The applicant also agreed that these components are passive and long-lived with the intended function of pressure boundary and, in LRA Table 2.3.3-10, added the component group "monitor" to represent these moisture separators as subject to an AMR in accordance with 10 CFR 54.21(a)(1).

Therefore, the staff's concern described in RAI 2.3.3.10 is resolved.

In RAI 2.3.3.10-2, dated September 21, 2005, the staff stated that drawing LR-222, sheet 1 (location B-1), shows the nitrogen supply lines to the spent fuel pool gate as within the scope of license renewal; however, the symbol for the spent fuel pool gate and the inner and outer seals are shown as not within the scope of license renewal. Therefore, the staff requested that the applicant explain why the spent fuel pool gate seals are not within the scope of license renewal while the nitrogen gas, which is required to inflate the seals to perform their intended function, is within the scope of license renewal.

In its response, by letter dated October 21, 2005, the applicant stated:

LR-M-222, Sheet 1 depicts mechanical components in the scope of license renewal. The spent fuel pool gate and seals are addressed as civil/structural components.

The spent fuel pool gate seals perform no license renewal intended function. Therefore, the nitrogen gas supply (stainless steel tubing and related components from the HBD/SS Tubing reducer (located at C-3) to the gate seals) is not in the scope of license renewal. The highlighting on these components should be disregarded.

Based on its review, the staff finds the applicant's response to RAI 2.3.3.10-2 acceptable. The applicant adequately clarified that the spent fuel pool gate seals perform no license renewal intended function and are not within the scope of license renewal in accordance with 10 CFR 54.4(a). Therefore, the staff's concern described in RAI 2.3.3.10-2 is resolved.

In RAI 2.3.3.10-3, dated September 21, 2005, the staff stated that on drawing LR-222, sheets 2 and 3, almost all of the drawing is shown as within the scope of license renewal, except for the nitrogen bottles and air containers required to supply nitrogen and air backup to the SR systems. Therefore, the staff requested that the applicant explain why the nitrogen and air supply bottles are not within the scope of license renewal in accordance with 10 CFR 54.4(a). Similarly, hydrogen bottles on drawing LR-222, sheet 1A (location E-8), and their piping and components to line HB-21-1," are shown as not within the scope of license renewal. Line HB-21-1" from upstream of valve 1"-130-WE-88 to the continuation flag to the volume control tank, is shown as within the scope of license renewal for the chemical and volume control system. These hydrogen bottles appeared to supply backup hydrogen to the volume control tank T-54, as shown on drawing LR-202, sheet 1A, at location F-7. Therefore, the staff requested that the applicant justify the exclusion of the nitrogen bottles from within the scope of license renewal and subject to an AMR in accordance with 10 CFR 54.4(a) and 10 CFR 54.21(a)(1), respectively.

In its response, by letter dated October 21, 2005, the applicant stated:

> The nitrogen bottles and air containers on LR-M-222, Sheet 2, are within the scope of license renewal, and are consumables in accordance with NEI 95-10. The nitrogen bottles for the electrical penetrations on LR-M-222, Sheet 3, are not required to support any intended function of the penetrations or any license renewal scoping criteria, and are not in scope.

> The Room 139 hydrogen bottles on LR-M-222, Sheet 1A, provide a non-safety related backup supply, and do not meet any of the license renewal scoping criteria. Line HB- 21-1' from upstream of valve 1'-130-WE-88 to the continuation flag to the volume control tank is in scope of license renewal in accordance with 10 CFR 54.4(a)(2), nonsafety related SSC connected to safety related SSC. Equivalent anchors are located where the highlighting ends on the drawing.

Based on its review, the staff finds the applicant's response to RAI 2.3.3.10-3 acceptable. The applicant adequately explained that the nitrogen bottles and air containers on drawing LR-M-222, sheet 2, are treated as consumables, per NEI 95-10 guidance, and are within the scope of license renewal. The applicant further clarified that the nitrogen bottles for the electrical penetrations on drawing LR-M-222, sheet 3, and the hydrogen bottles on drawing LR-M-222, sheet 1A, do not support any license renewal intended function and are not within the scope of license renewal in accordance with 10 CFR 54.4(a). Therefore, the staff's concerns described in RAI 2.3.3.10-3 are resolved.

In RAI 2.3.3.10-5, dated September 21, 2005, the staff stated that drawing LR-224, sheet 1 (location C-7), shows pump P-2402 as within the scope of license renewal for the miscellaneous gas system; however, the license renewal drawing indicates that the pump is part of a portion of the system no longer in service. Therefore, the staff requested that the applicant identify the license renewal criterion that this pump supports while no longer in service. The staff also requested that the applicant explain why the pipes leading to this pump are not within the scope of license renewal. In addition, pumps are not listed in LRA Table 2.3.3-10 as a component group subject to an AMR. Therefore, the staff requested that the applicant justify the exclusion of pump P-2402 as subject to an AMR in accordance with 10 CFR 54.21(a)(1).

In its response, by letter dated October 21, 2005, the applicant stated:

> P-2402 shown on M-224, Sheet 1, is retired in-place and is in the process of being removed from LR-M-224-1 because the number duplicates a pump number on another drawing. The pump P-2402 shown on M-224, Sheet 2, (D-6) is in scope for license renewal.

> The NMC response to NRC RAI-2.3.3.10-1 above added line items for 'Pumps' to LRA Tables 2.3.3-10 and 3.3.2-10.

Based on its review, the staff finds the applicant's response to RAI 2.3.3.10-5 acceptable because it adequately clarified that pump P-2402, depicted on drawing LR-M-224, sheet 1, is retired in place and is not within the scope of license renewal. The applicant further explained that this pump duplicates a pump number on drawing LR-M-224, sheet 2, which is within the scope of license renewal. The applicant also added the component group "Pumps" to the LRA tables as subject to an AMR in accordance with 10 CFR 54.21(a)(1). Therefore, the staff's concern described in RAI 2.3.3.10-5 is resolved.

In RAI 2.3.3.10-6, dated September 21, 2005, the staff stated that drawing LR-224, sheet 2 (locations B-6 and D-6), shows the lines beyond two closed valves continuing on to drawing LR-219, sheet 2, as within the scope of license renewal. On drawing LR-219, sheet 2, the continuation of the lines from drawing LR-224, sheet 2, are also shown as within the scope of license renewal. However, these lines are shown as not within the scope of license renewal before entering the post-accident sampling monitoring panel C1 03-1. Therefore, the staff requested that the applicant explain why these lines were not within the scope of license renewal prior to entering the post accident sampling monitoring panel C103-1.

In its response, by letter dated October 21, 2005, the applicant stated, "The lines are in the scope of license renewal in accordance with 10 CFR 54.4 (a)(2) as an NSR SSC attached to an SR SSC. Where the highlighting stops represents the location of an equivalent anchor."

Based on its review, the staff finds the applicant's response to RAI 2.3.3.10-6 acceptable because it adequately clarified that the lines shown within the scope of license renewal meet the requirements of 10 CFR 54.4(a)(2) (i.e., NSR SSC attached to an SR SSC) and that the portion of the line before entering the post-accident sampling monitoring panel C103-1 represents the location of an equivalent anchor and, therefore, is not shown within the scope of license renewal. Therefore, the staff's concern described in RAI 2.3.3.10-6 is resolved.

In RAI 2.3.3.10-7, dated September 21, 2005, the staff stated:

> LRA Table 2.3.3-10 lists accumulators and tanks as component groups subject to an AMR with a fluid pressure boundary intended function. Clarify whether nitrogen bottles on license renewal Drawing LR-222, Sheet 1 (Locations B-2, E-7, and E-8), or reagent gas and calibration gas bottles on LR-224, Sheet 2, are included in the accumulators/tanks component groups. If these bottles are excluded from being subject to an AMR because they are considered as consumable components, then provide the frequency or condition of their replacement. Also, identify waste gas system accumulators/tanks that are within the scope of license renewal and are subject to an AMR in accordance with the requirements of 10 CFR 54.4(a) and 10 CFR 54.21(a)(1), respectively.

In a subsequent discussion, the staff requested clarification of the applicant's response contained in the letter dated October 21, 2005. The applicant stated that: "The previous response to RAI 2.3.3.10-7 is hereby revised in its entirety to incorporate the requested clarifications."

LRA Table 2.3.3-10 component group "accumulators" consists of selected nitrogen bottles shown on LR-M-222-1, LR-M-222-2, and LR-M-224-2, that have been determined to be in scope for license renewal.

These accumulators (replaceable gas bottles) are in scope of license renewal. However, because they are replaced after use, they are considered consumables and are screened out as not requiring aging management review in accordance with NEI 95-10. There are no components "accumulators" in Table 2.3.3-10 that require an AMR. Therefore, component type "accumulators" need not have been listed in either LRA Tables 2.3.3-10 or 3.3.2-10. LRA Table 3.3.2-10 provides the results of aging management reviews that have been completed, and does not include in-scope component types which did not require an AMR. Therefore, the line items for component type "accumulators" in LRA Table 2.3.3-10 on page 2-136 and 3.3.2-10 on 3-171 are hereby deleted.

The component group "tank" in LRA Table 2.3.3-10 is the bulk nitrogen tank (located outdoors and not shown on drawing). The bulk nitrogen tank is within the scope of license renewal and subject to AMR.

Based on its review, the staff finds the applicant's response to RAI 2.3.3.10-7 acceptable because it adequately clarified that the component group "accumulators" consists of selected nitrogen calibration and reagent gas bottles which are within the scope of license renewal. However, because these accumulators (replaceable gas bottles) are replaced after use, they are considered consumables and, therefore, are not subject to an AMR in accordance with 10 CFR 54.21(a)(1). The applicant further clarified that the line items for component type "accumulators" in LRA Tables 2.3.3-10 and 3.3.2-10 need not have been listed and are hereby deleted. Therefore, the staff's concern described in RAI 2.3.3.10-7 is resolved.

2.3.3.10.3 Conclusion

The staff reviewed the LRA, accompanying scoping boundary drawings, and RAI responses to determine whether any SSCs that should be within the scope of license renewal had not been identified by the applicant. No omissions were identified. In addition, the staff performed a review to determine whether any components that should be subject to an AMR had not been identified by the applicant. No omissions were identified. On the basis of its review, the staff concludes that the applicant adequately identified the MGS components that are within the scope of license renewal, as required by 10 CFR 54.4(a), and the MGS components that are subject to an AMR, as required by 10 CFR 54.21(a)(1).

2.3.3.11 Radwaste System

2.3.3.11.1 Summary of Technical Information in the Application

In LRA Section 2.3.3.11, the applicant described the radwaste system (RWS). The RWS contains two major subsystems: (1) liquid radwaste and (2) solid radwaste. The systems are designed and operated to achieve a near-zero discharge to the environment. The liquid radioactive waste system is divided into three sections: (a) the clean waste section which processes high-activity, high-purity (low solids) liquid waste; (b) the dirty waste section which processes low-activity, low-purity (high solids) liquid waste; and (c) the laundry waste. The solid waste management system is designed to collect, process, package and store for future offsite disposal low-level liquid and solid wastes, evaporator concentrates, spent ion-exchange resins and assorted solid wastes. Solid wastes are, as applicable, stored on site, shipped to contractors for incineration, immobilized by the addition of additives, and (resins) dewatered and compacted. There are two onsite storage/processing facilities.

The RWS contains SR components that are relied upon to remain functional during and following DBEs. The failure of NSR SSCs in the RWS could potentially prevent the satisfactory accomplishment of an SR function. In addition, the RWS performs functions that support EQ.

The intended functions within the scope of license renewal include the following:

- provides flood protection barrier
- provides fluid pressure boundary

In LRA Table 2.3.3-11, the applicant identified the following RWS component types that are within the scope of license renewal and subject to an AMR:

- accumulators
- demineralizer
- fasteners
- filters/strainers
- heat exchangers
- pipe & fittings
- pumps
- valves & dampers

2.3.3.11.2 Staff Evaluation

The staff reviewed LRA Section 2.3.3.11 and the FSAR using the evaluation methodology described in SER Section 2.3. The staff conducted its review in accordance with the guidance described in SRP-LR Section 2.3.

In conducting its review, the staff evaluated the system functions described in the LRA and FSAR in accordance with the requirements of 10 CFR 54.4(a) to verify that the applicant had not omitted from the scope of license renewal any components with intended functions delineated under 10 CFR 54.4(a). The staff then reviewed those components that the applicant had identified as being within the scope of license renewal to verify that the applicant had not omitted any passive and long-lived components that should be subject to an AMR in accordance with the requirements of 10 CFR 54.21(a)(1).

The staff's review of LRA Section 2.3.3.11 identified areas in which additional information was necessary to complete the review of the applicant's scoping and screening results. The applicant responded to the staff's RAIs as discussed below.

In RAI 2.3.3.11-1, dated September 21, 2005, the staff stated that drawing LR-650, sheet 1A, shows an instrument diaphragm at location F-7 as within the scope of license renewal; however, LRA Table 2.3.3-11 does not list instrument diaphragm as a component group subject to an AMR. This instrument diaphragm serves a pressure boundary intended function and is passive and long-lived. Therefore, the staff requested that the applicant clarify whether this component was already included in LRA Table 2.3.3-11 as part of any other component group and, if so, identify its intended function. If not, the staff requested that the applicant justify the exclusion of this instrument diaphragm from being subject to an AMR in accordance with 10 CFR 54.21(a)(1).

In its response, by letter dated October 21, 2005, the applicant stated:

> The instrument diaphragm on M-650-IA (F7) is not listed in the Radwaste System
> LRA Table 2.3.3-11. The referenced instrument diaphragm is addressed in

Chemical and Volume Control System LRA Tables 2.3.3-1 on page 2-87 and 3.3.2-1 on page 3-121 under Component Group 'Check, Control, Manual & Relief Valves; Instrument Assemblies.'

Based on its review, the staff finds the applicant's response to RAI 2.3.3.11-1 acceptable. The applicant sufficiently clarified that the instrument diaphragm is addressed in the chemical and volume control system in LRA Table 2.3.3-1, under the component group "Check, Control, Manual & Relief Valves; Instrument Assemblies," as subject to an AMR in accordance with 10 CFR 54.21(a)(1). Therefore, the staff's concern described in RAI 2.3.3.11-1 is resolved.

In RAI 2.3.3.11-4, dated September 21, 2005, the staff stated that drawing LR-654 shows a section of piping HCD-1-1/2" going to the controlled chemical lab drain tank T-76, at location B-6, as within the scope of license renewal; however, the continuation of this piping to drawing LR-210, sheet 1, at location H-8, is not shown as within the scope of license renewal. Therefore, the staff requested that the applicant resolve this inconsistency between the drawings and provide the basis for the resolution.

In its response, by letter dated October 21, 2005, the applicant stated:

> This line is in scope of License Renewal per 10CFR54.4(a)(2) due to being in proximity of SR equipment (spray). The Controlled Chemistry Laboratory Drain Tank is located in the Laundry Drain Tank Room. Once the line penetrates the Laundry Drain Tank Room wall, there is no safety related equipment in the area of the piping. Therefore, this section of piping on LR-M-210 located in the Laundry Drain Tank Room is not in scope of License Renewal.

Based on its review, the staff finds the applicant's response to RAI 2.3.3.11-4 acceptable. The applicant adequately explained that the controlled chemistry laboratory drain tank is located in the laundry drain tank room with no proximity to SR equipment; therefore, the section of piping on LR-M-210, located in the laundry drain tank room wall, does not meet the requirements of 10 CFR 54.4(a)(2) and is not within the scope of license renewal. Therefore, the staff's concern described in RAI 2.3.3.11-4 is resolved.

In RAI 2.3.3.11-5, dated September 21, 2005, the staff stated that drawing LR-210, sheet 1, shows a section of piping HC-1-1" coming from the controlled chemical lab drain tank T-76, at location F-8, continuing to the dirty waste drain header, as not within the scope of license renewal; however, the continuation of this piping on drawing LR-211, sheet 1, at location E-6, is shown as within the scope of license renewal. Therefore, the staff requested that the applicant resolve this inconsistency between the above-mentioned drawings and provide the basis for the resolution.

In its response, by letter dated October 21, 2005, the applicant stated:

> While the Controlled Chemistry Laboratory Drain Tank drain piping is located in the Laundry Drain tank Room, it is not in scope of License Renewal because no safety related equipment is located in the room. Once the pipe exits the room, as shown on the continuation drawing, it is in scope of License Renewal per 10CFR54.4(a)(2) due to being in proximity to safety related equipment (spray).

Based on its review, the staff finds the applicant's response to RAI 2.3.3.11-5 acceptable because it adequately clarified that the controlled chemistry laboratory drain tank piping is located in the laundry drain tank room with no proximity to SR equipment. This piece of piping cannot prevent an SR component from performing its intended function due to spatial interactions (i.e., spray). As such, it does not meet 10 CFR 54.4(a)(2) requirements, and is not within the scope of license renewal. Therefore, the staff's concern described in RAI 2.3.3.11-5 is resolved.

In RAI 2.3.3.11-7, dated September 21, 2005, the staff stated that drawing LR-210, sheet 1B, shows the clean resin transfer tank (T-61), at location D-1, as within the scope of license renewal in accordance with 10 CFR 54.4(a); however, the lines from the clean resin transfer tank to the purification and deborating ion exchangers and to the spent fuel demineralizer at location B-1 are excluded from within the scope of license renewal. Therefore, the staff requested that the applicant justify the exclusion of these lines (including flow gauge FG-1054) from within the scope of license renewal in accordance with 10 CFR 54.4(a).

In its response, by letter dated October 21, 2005, the applicant stated:

> The Clean Resin Transfer Tank (T-61) is in scope of License Renewal due to 10CFR54.4(a)(2) (seismic II attached to I), and is an anchor point for the line coming from Primary Make-up water (Line HC-9-2), including lines HC-32-1. T-61 is not in scope of License Renewal due to I0CFR54.4(a)(2) for spatial orientation (spray) due to being located in its own room with no safety related components being in the area. Therefore, all lines connecting to T-61, such as those discussed above, are not necessarily in scope.

Based on its review, the staff finds the applicant's response to RAI 2.3.3.11-7 acceptable. The applicant adequately explained that the clean resin transfer tank is within the scope of license renewal, as seismic category II attached to seismic category I, and not for spatial orientation (i.e., spray), because it is located in its own room with no SR components in the area; therefore, the clean resin transfer tank is within the scope of license renewal in accordance with 10 CFR 54.4(a)(2). The lines from the clean resin transfer tank to the purification and deborating ion exchangers and to the spent fuel demineralizer do not meet the requirements of 10 CFR 54.4(a)(2) and are not within the scope of license renewal. Therefore, the staff's concern described in RAI 2.3.3.11-7 is resolved.

In RAI 2.3.3.11-8, dated September 21, 2005, the staff stated that drawing LR-650, sheet 1B, shows clean waste transfer pumps P-94 and P-97B, at location C/D-6, as within the scope of license renewal; however, the lines from these clean waste transfer pumps to the continuation drawing LR-650, sheet 1, are shown as not within the scope of license renewal. Therefore, the staff requested that the applicant justify the exclusion of these lines, including the restricting orifices RO-5026 and RO-5027, from within the scope of license renewal in accordance with 10 CFR 54.4(a).

In its response, by letter dated October 21, 2005, the applicant stated:

> P-94 and P-97B are in scope of License Renewal due to 10CFR54.4(a)(2) (seismic II attached to I), and is an anchor point for the line coming from the Treated Waste Monitor Tanks (line HCC-49-3) on drawing LR-M-650-1B (G2).

The lines going to LR-M- 650-1 are truncated based on a moment of inertia truncation criteria.

Based on its review, the staff finds the applicant's response to RAI 2.3.3.11-8 acceptable. The applicant adequately explained that the clean waste transfer pumps are within the scope of license renewal as seismic category II attached to seismic category I, in accordance with 10 CFR 54.4(a)(2). These pumps are an anchor point for the line coming from the treated waste monitor tanks on drawing LR-M-650-1B. Therefore, the lines from the clean waste transfer pumps to the continuation drawing LR-650, sheet 1, are not within the scope of license renewal in accordance with 10 CFR 54.4(a). Therefore, the staff's concern described in RAI 2.3.3.11-8 is resolved.

In RAI 2.3.3.11-9, dated September 21, 2005, the staff stated that LRA Section 2.3.3.11 states that some SSCs are considered within the scope of license renewal due to EQ requirements, in accordance with 10 CFR 54.4(a)(3). LRA Section 2.3.3.11 states that the radwaste system contains components required by the current design-basis for EQ in accordance with 10 CFR 50.49; however, from the information provided in the FSAR and LRA the staff was unable to determine which mechanical components were within the scope of license renewal in accordance with 10 CFR 54.4(a)(3). Therefore, the staff requested that the applicant provide information (e.g., EQ database or reports) so that the staff could identify the radwaste system mechanical components within the scope of license renewal in accordance with the requirements of the 10 CFR 54.4(a)(3) EQ regulated event.

In its response, by letter dated October 21, 2005, the applicant stated:

> The Radwaste System at Palisades does not have mechanical components included in the EQ Program. However the Radwaste System does have electrical/I&C components in the EQ Program as follows: POS-1002, 1007, 1036, 1038, 1044, 1045, 1103, 1104 are EQ position switches evaluated in the radwaste system. These switches are active and not subject to AMR. SV-1002, 1007, 1036, 1038, 1044, 1045, 1103, 1104 are EQ air solenoid valves associated with radwaste control valves. These are evaluated in the Compressed Air System (CAS) and age managed by the EQ Program.

Based on its review, the staff finds the applicant's response to RAI 2.3.3.11-9 acceptable. The applicant sufficiently clarified that the radwaste system does not have any mechanical components within the scope of license renewal in accordance with the requirements of 10 CFR 54.4(a)(3). Therefore, the staff's concern described in RAI 2.3.3.11-9 is resolved.

2.3.3.11.3 Conclusion.

The staff reviewed the LRA, accompanying scoping boundary drawings, and RAI responses to determine whether any SSCs that should be within the scope of license renewal had not been identified by the applicant. No omissions were identified. In addition, the staff performed a review to determine whether any components that should be subject to an AMR had not been identified by the applicant. No omissions were identified. On the basis of its review, the staff concludes that the applicant adequately identified the RWS components that are within the scope of license renewal, as required by 10 CFR 54.4(a), and the RWS components that are subject to an AMR, as required by 10 CFR 54.21(a)(1).

2.3.3.12 Service Water System

2.3.3.12.1 Summary of Technical Information in the Application

In LRA Section 2.3.3.12, the applicant described the service water system (SWS). The SWS supplies Lake Michigan water as the cooling medium (ultimate heat sink) for removal of waste heat from the nuclear and steam plant auxiliary systems during normal, shutdown, or emergency conditions. Three half-capacity electric motor-driven pumps draw screened and intermittently chlorinated Lake Michigan water from the intake structure. Each service water pump discharges through a simplex strainer into a common header. The common header has a full-capacity takeoff at each end, which supplies critical plant systems. A third takeoff at one end of the common header supplies the noncritical auxiliary systems. The noncritical service water (NSW) header is isolated on a safety injection signal, thus ensuring that all available service water is routed to the critical systems. The SWS, also includes the ultimate heat sink (UHS) subsystem. Basically, UHS subsystem of SWS consists of the components that take water from Lake Michigan to the suction of the SWS system (intake crib, intake structure, and connecting pipe), take the water from the SWS system and discharge it to the lake (discharge structure), and transfer water from the discharge structure to the intake structure (P-5 and associated components) for the purpose of deicing or supplying alternate SWS supply water should the intake crib collapse.

The SWS contains SR components that are relied upon to remain functional during and following DBEs. The failure of NSR SSCs in the SWS could potentially prevent the satisfactory accomplishment of an SR function. In addition, the SWS performs functions that support fire protection and EQ.

The intended functions within the scope of license renewal include the following:

- provides filtration
- provides flow restriction (throttle)
- provides fluid pressure boundary
- provides spray pattern at discharge nozzle

In LRA Table 2.3.3-12, the applicant identified the following SWS component types that are within the scope of license renewal and subject to an AMR:

- accumulator
- fasteners in containment
- fasteners not in containment
- filters/stainers
- heat exchanger
- pipe & fittings
- pumps
- traveling screen spray nozzles
- valves & dampers

2.3.3.12.2 Staff Evaluation

The staff reviewed LRA Section 2.3.3.12 and the FSAR using the evaluation methodology described in SER Section 2.3. The staff conducted its review in accordance with the guidance described in SRP-LR Section 2.3.

In conducting its review, the staff evaluated the system functions described in the LRA and FSAR in accordance with the requirements of 10 CFR 54.4(a) to verify that the applicant had not omitted from the scope of license renewal any components with intended functions delineated under 10 CFR 54.4(a). The staff then reviewed those components that the applicant had identified as being within the scope of license renewal to verify that the applicant had not omitted any passive and long-lived components that should be subject to an AMR in accordance with the requirements of 10 CFR 54.21(a)(1).

2.3.3.12.3 Conclusion

The staff reviewed the LRA and accompanying scoping boundary drawings to determine whether any SSCs that should be within the scope of license renewal had not been identified by the applicant. No omissions were identified. In addition, the staff performed a review to determine whether any components that should be subject to an AMR had not been identified by the applicant. No omissions were identified. On the basis of its review, the staff concludes that the applicant adequately identified the SWS components that are within the scope of license renewal, as required by 10 CFR 54.4(a), and the SWS components that are subject to an AMR, as required by 10 CFR 54.21(a)(1).

2.3.3.13 Shield Cooling System

2.3.3.13.1 Summary of Technical Information in the Application

In LRA Section 2.3.3.13, the applicant described the shield cooling system (SCS). The SCS is designed to remove heat from the biological shield surrounding the reactor vessel thereby limiting thermal stresses in the structural concrete. It is not an SR system. The system is designed to maintain structural concrete temperature below 165 degrees F. The system assures the concrete in the reactor cavity does not overheat and develop excessive thermal stress. The shield cooling system is a closed loop system consisting of two full-capacity sets of cooling coils, two full-capacity pumps, a heat exchanger, a surge tank, associated piping, valves, instrumentation and controls. Each set of shield cooling coils is composed of individual cooling coils embedded in the concrete shield. The closed loop system transfers heat to the component cooling water system by means of the shield cooling heat exchanger. Demineralized water with a corrosion inhibitor is used in the shield cooling loop.

The SCS contains SR components that are relied upon to remain functional during and following DBEs. The failure of NSR SSCs in the SCS could potentially prevent the satisfactory accomplishment of an SR function. In addition, the SCS performs functions that support EQ.

The intended function, within the scope of license renewal, is to provide fluid pressure boundary.

In LRA Table 2.3.3-13, the applicant identified the following SCS component types that are within the scope of license renewal and subject to an AMR:

- accumulators
- fasteners
- filters/strainers
- pipe & fittings
- valves & dampers

2.3.3.13.2 Staff Evaluation

The staff reviewed LRA Section 2.3.3.13 and the FSAR using the evaluation methodology described in SER Section 2.3. The staff conducted its review in accordance with the guidance described in SRP-LR Section 2.3.

In conducting its review, the staff evaluated the system functions described in the LRA and FSAR in accordance with the requirements of 10 CFR 54.4(a) to verify that the applicant had not omitted from the scope of license renewal any components with intended functions delineated under 10 CFR 54.4(a). The staff then reviewed those components that the applicant had identified as being within the scope of license renewal to verify that the applicant had not omitted any passive and long-lived components that should be subject to an AMR in accordance with the requirements of 10 CFR 54.21(a)(1).

The staff's review of LRA Section 2.3.3.13 identified areas in which additional information was necessary to complete the review of the applicant's scoping and screening results. The applicant responded to the staff's RAIs as discussed below.

In RAI 2.3.3.13-2, dated September 21, 2005, the staff stated that LRA Section 2.3.3.13 states that some SSCs are considered within the scope of license renewal due to EQ in accordance with 10 CFR 54.4(a)(3). LRA Section 2.3.3.13 states that the shield cooling system contains equipment in compliance with the EQ requirements of 10 CFR 50.49; however, based on the information in the FSAR and LRA, the staff was not able to determine which mechanical components are within the scope of license renewal, in accordance with 10 CFR 54.4(a)(3), due to an EQ regulated event. Therefore, the staff requested that the applicant provide information (e.g., EQ database or reports) so that the staff could identify the mechanical components of the shield cooling system within the scope of license renewal in accordance with the EQ regulated event in 10 CFR 54.4(a)(3).

In its response, by letter dated October 21, 2005, the applicant stated:

> The Shield Cooling System at Palisades does not have mechanical components included in the EQ Program. However the Shield Cooling System does have electrical/I&C components in the EQ Program as follows: POS-0939, is an EQ position switch in this system. The switch is active and not subject to AMR. SV-0939 is a EQ air solenoid valve associated with a Shield Cooling System control valve. This SV is evaluated in the Compressed Air System (CAS) and is age managed by the EQ Program.

Based on its review, the staff finds the applicant's response to RAI 2.3.3.13-2 acceptable. The applicant sufficiently clarified that the shield cooling system has no mechanical components within the scope of license renewal in accordance with the requirements of the EQ regulated event in 10 CFR 54.4(a)(3). Therefore, the staff's concern described in RAI 2.3.3.13-2 is resolved.

In RAI 2.3.3.13-3, dated September 21, 2005, the staff stated that drawing LR-221, sheet 1, shows the shield cooling surge tank as within the scope of license renewal. The tank appears to be within the scope of license renewal in accordance with 10 CFR 54.4(a)(2) to protect SR components from spray, flooding, and seismic II/I considerations; however, it could not be determined from the information in the LRA and FSAR why the Y-strainers and the motor control switches had been included within the scope of license renewal whereas the piping between the heat exchanger (E-64), including the shield cooling pumps P-77A and P-77B and the shield cooling surge tank, was not within the scope of license renewal. Therefore, the staff requested that the applicant justify the exclusion of the above-mentioned piping from within the scope of license renewal in accordance with 10 CFR 54.4(a).

In its response, by letter dated October 21, 2005, the applicant stated:

> The surge tank, T-62, is in scope of License Renewal per 10CFR54.4(a)(2) because it serves as the anchor point for the attached Containment penetration piping (seismic II attached to I). The other piping and components associated with the surge tank do not meet license renewal scoping criteria.
>
> The control switches are in scope of LR because they are Q-listed for seismic reasons as they are mounted on a safety related panel. The control switches were not subject to AMR due to being active. The hangers/supports for these components are in-scope for license renewal and are evaluated in the appropriate structural commodity.
>
> The two Y-strainers are only in scope of LR and subject to AMR because the Q-list identified them as safety related for seismic support. The strainers did not have an operational function such as filtration. The hangers/supports for these components are in-scope for license renewal and are evaluated in the appropriate structural commodity.

Based on its review, the staff finds the applicant's response to RAI 2.3.3.13-3 acceptable. The applicant sufficiently clarified that the motor control switches and Y-strainers are SR components that provide the seismic support license renewal intended function; therefore, they are within the scope of license renewal. The applicant also clarified that the piping between the heat exchanger (E-64), including the shield cooling pumps P-77A and P-77B and the shield cooling surge tank, does not meet the scoping and screening criteria of 10 CFR 54.4(a) and 10 CFR 54.21(a)(1), respectively, and is not within the scope of license renewal. Therefore, the staff's concerns described in RAI 2.3.3.13-3 are resolved.

2.3.3.13.3 Conclusion

The staff reviewed the LRA, accompanying scoping boundary drawing, and RAI responses to determine whether any SSCs that should be within the scope of license renewal had not been identified by the applicant. No omissions were identified. In addition, the staff performed a review to determine whether any components that should be subject to an AMR had not been identified by the applicant. No omissions were identified. On the basis of its review, the staff concludes that the applicant adequately identified the SCS components that are within the scope of license renewal, as required by 10 CFR 54.4(a), and the SCS components that are subject to an AMR, as required by 10 CFR 54.21(a)(1).

2.3.3.14 Spent Fuel Pool Cooling System

2.3.3.14.1 Summary of Technical Information in the Application

In LRA Section 2.3.3.14, the applicant described the spent fuel pool cooling (SFP) system. The SFP system removes decay heat from spent fuel stored in the spent fuel pool. The system is a closed loop system consisting of two pumps, a heat exchange unit consisting of two heat exchangers in series, a bypass filter, a bypass demineralizer, a booster pump, piping, valves and instrumentation.

The SFP system contains SR components that are relied upon to remain functional during and following DBEs. The failure of NSR SSCs in the SFP system could potentially prevent the satisfactory accomplishment of an SR function.

The intended functions within the scope of license renewal include the following:

- provides fluid pressure boundary
- provides heat transfer

In LRA Table 2.3.3-14, the applicant identified the following SFP system component types that are within the scope of license renewal and subject to an AMR:

- accumulator
- demineralizers
- fasteners
- filters/strainers
- pipe & fittings
- pumps
- SFP heat exchanger shell and channel head
- SFP heat exchanger tube and tubesheet
- valves & dampers

2.3.3.14.2 Staff Evaluation

The staff reviewed LRA Section 2.3.3.14 and the FSAR using the evaluation methodology described in SER Section 2.3. The staff conducted its review in accordance with the guidance described in SRP-LR Section 2.3.

In conducting its review, the staff evaluated the system functions described in the LRA and FSAR in accordance with the requirements of 10 CFR 54.4(a) to verify that the applicant had not omitted from the scope of license renewal any components with intended functions delineated under 10 CFR 54.4(a). The staff then reviewed those components that the applicant

had identified as being within the scope of license renewal to verify that the applicant had not omitted any passive and long-lived components that should be subject to an AMR in accordance with the requirements of 10 CFR 54.21(a)(1).

The staff's review of LRA Section 2.3.3.14 identified areas in which additional information was necessary to complete the review of the applicant's scoping and screening results. The applicant responded to the staff's RAIs as discussed below.

In RAI 2.3.3.14-1, dated September 21, 2005, the staff stated that LRA Section 2.3.3.14 states that "SFP-01" is an intended function meeting Criterion 1 for inclusion of the spent fuel pool cooling system within the scope of license renewal. The SFP system removes decay heat from the fuel stored in the spent fuel pool and cools reactor cavity water during spent fuel transfer. Further, SFP-01 states that cooling the reactor cavity water is not an intended function of the spent fuel pool cooling system that meets the requirements of 10 CFR 54.4. Consequently, drawing LR-221-2 sheet 2, for the SFP system, shows piping that supports the cooling of the reactor cavity as not within the scope of license renewal and not subject to an AMR. The piping starts at location D-1 on drawing LR-221 sheet 2.

Based on the information provided in the LRA and the license renewal drawings for the spent fuel pool cooling system, it appeared that although cooling of the reactor cavity water had been correctly identified as not supporting an intended function, failure of the piping supporting this function might affect the intended function of removing decay heat from the fuel stored in the spent fuel pool. Therefore, the staff requested that the applicant justify exclusion of the spent fuel pool cooling system piping portions from the scope of license renewal in accordance with 10 CFR 54.4(a).

In its response, by letter dated October 21, 2005, the applicant stated:

> The two lines that go to the top of the reactor cavity were used to clean up the reactor cavity for water purity. The plant now uses filters that float on the top of the reactor cavity pool so these lines are not used. During refueling, with the reactor vessel head removed, heat removal for the reactor cavity is provided by the shutdown cooling system. Therefore these pipes are not required to be in scope for license renewal as they do not support any system intended function.

> Also, the normal position of the containment isolation valves (MV-SFP117, 118, 120 & 121) is locked closed providing isolation of the reactor cavity from the SFP system. Therefore, any failure of the portion of the lines not in scope would not affect the intended function of removing decay heat from the fuel stored in the spent fuel pool.

Based on its review, the staff finds the applicant's response to RAI 2.3.3.14-1 acceptable. The applicant adequately explained that the piping lines used to clean up the reactor cavity for water purity are no longer in use and are isolated from the spent fuel pool system during normal operation by locked closed containment isolation valves. Further, any failure of the portion of those piping lines would not affect the intended function of removing decay heat from the fuel stored in the spent fuel pool and these piping lines are not within the scope of license renewal in accordance with 10 CFR 54.4(a). Therefore, the staff's concern described in RAI 2.3.3.14-1 is resolved.

In RAI 2.3.3.14-2, dated September 21, 2005, the staff stated that LRA Table 2.3.3-14 identifies "Component Groups" and their intended functions. Within this table is the component group "Filters/Strainers" with the intended function of fluid pressure boundary. This LRA section also states that "SFP-01" is an intended function meeting criteria Criterion 1 for inclusion of the system within the scope of license renewal. The system intended function is to remove decay heat from fuel stored in the spent fuel pool.

Drawing LR-221-2 shows basket strainers BS-2100 and BS-2101 on the suction side of fuel pool cooling pumps P-51A and P-51B at location B-5. NEI-95-10 Table 4.1-1 states that "filtration" is an example of a component intended function. From the information in the LRA and license renewal drawings for the spent fuel pool cooling system in addition to the fluid pressure boundary function, the basket strainers appear to have a filtration function. Therefore, the staff requested that the applicant justify the exclusion of the filtration function of the spent fuel pool cooling systems basket strainers.

In its response, by letter dated October 21, 2005, the applicant stated:

> For Palisades the spent fuel pool water chemistry is maintained by the water chemistry program. It uses the Spent Fuel Pool Demineralizer and some filters that float on the top of the spent fuel pool to maintain the water chemistry. Strainers designed on pump suctions are for pump protection, which is not an intended function of license renewal. The basket strainers provide the intended function of fluid pressure boundary and are not required for filtration.

Based on its review, the staff finds the applicant's response to RAI 2.3.3.14-2 acceptable. The applicant adequately clarified that the strainers designed on pump suctions are solely for pump protection, which is not a license renewal intended function. The applicant further clarified that basket strainers perform the intended function of fluid pressure boundary and not the intended function of filtration. Therefore, the staff's concern described in RAI 2.3.3.14-2 is resolved.

In RAI 2.3.3.14-3, dated September 21, 2005, the staff stated that LRA Section 2.3.3.14 states in SFP-03, that maintaining spent fuel pool boron concentration at or greater than required concentrations meets the 10 CFR 54.4(a)(1) criteria and is therefore a system intended function. It further implies that there are components associated with performing this intended function but that they are not within the scope of license renewal. Therefore, the staff requested that the applicant provide more information about the method and components used to maintain boron concentration at or greater than its required concentrations. The staff requested the applicant to identify the method and the components that perform the intended function. In addition, the staff requested that the applicant justify the exclusion of those components from within the scope of license renewal and subject to an AMR in accordance with 10 CFR 54.4(a) and 10 CFR 54.21(a)(1), respectively.

In its response, by letter dated October 21, 2005, the applicant stated:

> The concentration of boron in the Spent Fuel Pool System is maintained by water exchange with the Safety Injection and Refueling Water Tank (SIRW). The SIRW Tank boron concentration is maintained by blending concentrated boric acid from the Recycled Boric Acid Storage Tank (T-96), thru the Concentrated Boric Acid Tanks (T-53A/B), Concentrated Boric Acid Pumps (P-56A/B), the Boric Acid

Blender (M-51) and demineralized water from the Primary System Makeup Water Tank (T-90) to the Boric Acid Blender (M-51). The referenced comment is in error; the associated components are in scope of license renewal, although they are not in the Spent Fuel Pool Cooling system.

In LRA Section 2.3.3.14, System Function SFP-03, on page 2-152, the comment is hereby revised to read as follows: 'This system function provides reactivity control. The components associated with the addition of boron concentration to the Spent Fuel Pool are evaluated in the Engineered Safeguards, Chemical and Volume Control and Demineralized Water Systems, as applicable.'

Based on its review, the staff finds the applicant's response to RAI 2.3.3.14-3 acceptable. The applicant adequately explained the method and components used to perform the intended function of maintaining spent fuel pool boron concentration at or greater than its required concentrations. The applicant further clarified that the components associated with performing this intended function are within the scope of license renewal and evaluated in the engineered safeguards, chemical and volume control, and demineralized water systems, as applicable. Therefore, the staff's concerns described in RAI 2.3.3.14-3 are resolved.

2.3.3.14.3 Conclusion

The staff reviewed the LRA, accompanying scoping boundary drawings, and RAI responses to determine whether any SSCs that should be within the scope of license renewal had not been identified by the applicant. No omissions were identified. In addition, the staff performed a review to determine whether any components that should be subject to an AMR had not been identified by the applicant. No omissions were identified. On the basis of its review, the staff concludes that the applicant adequately identified the SFP system components that are within the scope of license renewal, as required by 10 CFR 54.4(a), and the SFP system components that are subject to an AMR, as required by 10 CFR 54.21(a)(1).

2.3.3.15 Waste Gas System

2.3.3.15.1 Summary of Technical Information in the Application

In LRA Section 2.3.3.15, the applicant described the waste gas system (WGS). The WGS stores gaseous isotopes for a time period, which will permit sufficient radioactive decay, prior to their discharge to the environment. The waste gas system is divided into two sections: (1) the gas collection header which collects low-activity gases from liquids, which have been previously degassed and/or vented in other waste handling steps, and (2) the gas processing section which collects gases from potentially high-activity sources. The waste gas system also contains components to provide containment isolation. The hydrogen recombiners are a subsystem to the WGS. The recombiner units are natural convection, thermal reactor type, hydrogen/oxygen recombiners. The subsystem consists of two 100 percent recombiner units, each containing an electric heater bank, power supply panel, and control panel. The recombiner units are located inside the containment building and the power supplies and control panels are in the cable spreading room. These units rely on the containment air coolers to mix the air for optimal performance.

The WGS contains SR components that are relied upon to remain functional during and following DBEs. The failure of NSR SSCs in the WGS could potentially prevent the satisfactory accomplishment of an SR function. In addition, the WGS performs functions that support EQ.

The intended function, within the scope of license renewal, is to provide fluid pressure boundary.

In LRA Table 2.3.3-15, the applicant identified the following WGS component types that are within the scope of license renewal and subject to an AMR:

- accumulators
- fasteners
- filters/strainers
- pipe & fittings
- valves & dampers

2.3.3.15.2 Staff Evaluation

The staff reviewed LRA Section 2.3.3.15 and the FSAR using the evaluation methodology described in SER Section 2.3. The staff conducted its review in accordance with the guidance described in SRP-LR Section 2.3.

In conducting its review, the staff evaluated the system functions described in the LRA and FSAR in accordance with the requirements of 10 CFR 54.4(a) to verify that the applicant had not omitted from the scope of license renewal any components with intended functions delineated under 10 CFR 54.4(a). The staff then reviewed those components that the applicant had identified as being within the scope of license renewal to verify that the applicant had not omitted any passive and long-lived components that should be subject to an AMR in accordance with the requirements of 10 CFR 54.21(a)(1).

The staff's review of LRA Section 2.3.3.15 identified areas in which additional information was necessary to complete the review of the applicant's scoping and screening results. The applicant responded to the staff's RAIs as discussed below.

In RAI 2.3.3.15-1, dated September 21, 2005, the staff stated that the following components shown on the waste gas license renewal drawings as within the scope of license renewal:

(a)   drain traps at several locations on drawing LR-211, sheets 2 and 3
(b)   a flow indicator on drawing LR-211, sheet 3, at location G4.

LRA Table 2.3.3-15 does not list these components as component groups subject to an AMR. These components serve a pressure boundary intended function and are passive and long-lived. Therefore, the staff requested that the applicant clarify whether these components were already included in LRA Table 2.3.3-15 as part of another component group and, if not, justify exclusion of these components as subject to an AMR in accordance with 10 CFR 54.21(a)(1).

In its response, by letter dated October 21, 2005, the applicant stated:

(a) The drain traps are in scope of license renewal due to 10CFR54.4(a)(2) spatial orientation (spray) and have a intended function of 'provide fluid pressure boundary' only. AMR of the drain trap pressure boundaries is addressed as part of Component Type 'Pipe & Fittings' because of similar materials, environments, AERMs and AMPs.

(b) FI-1120 was conservatively evaluated as in scope even though it does not serve a pressure boundary function; however, it is not subject to AMR due to being 'Active' per NEI 95-10.

Based on its review, the staff finds the applicant's response to RAI 2.3.3.15-1 acceptable because it adequately clarified that drain traps are included in LRA Table 2.3.3-15 under the component group "pipe & fittings" with the intended function of fluid pressure boundary as subject to an AMR in accordance with 10 CFR 54.21(a)(1). In addition, the applicant adequately clarified that the flow indicator on drawing LR-211, sheet 3, is an active component, per NEI 95-10, Appendix B guidance and not subject to an AMR in accordance with 10 CFR 54.21(a)(1). Therefore, the staff's concern described in RAI 2.3.3.15-1 is resolved.

In RAI 2.3.3.15-2, dated September 21, 2005, the staff stated that drawing LR-211, sheet 3, shows pressure indicator/alarms associated with the waste gas decay tanks as within the scope of license renewal; however, pressure transmitters and piping to these transmitters are not shown as within the scope of license renewal. LRA Section 2.1.3.1 states that all instruments are considered active unless they form an integral part of the pressure retaining boundary. These instruments serve a pressure boundary intended function and should be subject to an AMR. Therefore, the staff requested that the applicant justify the exclusion of these instruments from within the scope of license renewal and subject to an AMR in accordance with 10 CFR 54.4(a) and 10 CFR 54.21(a)(1), respectively.

In its response, by letter dated October 21, 2005, the applicant stated:

The pressure indicator alarms are in scope of license renewal because they are identified as safety related for their seismic mounting in the Q-list as they are mounted on a safety related panel. The hangers/supports for these components are in-scope for license renewal and are evaluated in the appropriated structural commodity. The components do not support any other license renewal criterion and are, therefore, not included for further evaluation. The Waste Gas Decay Tanks are in scope of license renewal due to 10CFR54.4(a)(2) (seismic II attached to I), and are anchor points for the Waste Gas process piping to and from the tanks. The pressure transmitters and respective piping are not in scope of license renewal due to not meeting the requirements of 10CFR54.4(a)(1), (2) or (3).

Based on its review, the staff finds the applicant's response to RAI 2.3.3.15-2 acceptable because it adequately clarified that the pressure transmitters and piping to these transmitters on drawing LR-211, sheet 3, do not meet any license renewal scoping criteria and are not within the scope of license renewal in accordance with 10 CFR 54.4(a). Therefore, the staff's concern described in RAI 2.3.3.15-2 is resolved.

In RAI 2.3.3.15-3, dated September 21, 2005, the staff stated that LRA Section 2.3.3.15 states that the hydrogen recombiners are a subsystem of the waste gas system and are within the scope of license renewal in accordance with 10 CFR 54.4(a)(1). Drawing LR-218, sheet 2, shows hydrogen recombiners inside the containment at location H-8 as within the scope of license renewal; however, LRA Table 2.3.3-15 does not list hydrogen recombiners as a component group subject to an AMR. Therefore, the staff requested that the applicant clarify if the components of the hydrogen recombiners had been scoped and screened as complex assemblies. As complex assemblies, SRP-LR Table 2.1-2 states, "some structures and components, when combined, are considered a complex assembly... An applicant should establish the boundaries for each assembly by identifying each structure and component that makes up the complex assembly and determining whether or not each structure and component is subject to an AMR." The staff further requested that the applicant clarify if hydrogen recombiners are already included in LRA Table 2.3.3-15 as part of any other component group and, if not, the staff requested that the applicant justify the exclusion of hydrogen recombiners as subject to an AMR, in accordance with 10 CFR 54.21(a)(1).

In its response, by letter dated October 21, 2005, the applicant stated, "The hydrogen recombiners (M-69A and B) are in scope of license renewal and are age managed by the EQ program. The hydrogen recombiners were not scoped and screened as complex assemblies."

Based on its review, the staff finds the applicant's response to RAI 2.3.3.15-3 acceptable because it adequately clarified that the hydrogen recombiners (M-69A and B) were not evaluated for scoping and screening as complex assemblies. The hydrogen recombiners are within the scope of license renewal in accordance with 10 CFR 54.4(a) and aging management is performed by the EQ Program. Therefore, the staff's concern described in RAI 2.3.3.15-3 is resolved.

In RAI 2.3.3.15-4, dated September 21, 2005, the staff stated that FSAR Section 11.3.2.2 states, "if the surge tank is discharging directly to the ventilation stack, a high-radiation condition (as identified by a continuously operating monitoring system taking samples from the discharge line) will automatically close the discharge valve which is upstream of the stack. On occurrence of high surge tank pressure, a waste gas compressor starts automatically and, taking suction from the surge tank, discharges to the decay tanks." Drawing LR-211, sheet 2, shows waste gas compressors C-54, at location C-3, as within the scope of license renewal; however, LRA Table 2.3.3-15 does not list compressors as a component group subject to an AMR. In addition, a filter upstream and a cooler downstream of this compressor, with their associated piping, are shown as not within the scope of license renewal. Waste gas compressors C-50 A/B, at locations D-3 and E-3, are shown as not within the scope of license renewal. Therefore, the staff requested that the applicant:

(a) justify the exclusion of the waste gas compressor C-54 from being subject to an AMR in accordance with the requirements of 10 CFR 54.21(a)(1)

(b) explain why the filter and cooler associated with C-54 are not within the scope of license renewal in accordance with the requirements of 10 CFR 54.4(a).

(c) explain how waste gas compressors C-50 A/B function differently from C-45 and why C-50 A/B are not within the scope of license renewal in accordance with the requirements of 10 CFR 54.4(a)

In its response, by letter dated October 21, 2005, the applicant addressed each staff request as follows:

(a) C-54 is in scope of license renewal and subject to AMR due to 10 CFR54.4(a)(2) (seismic II attached to I), and is an anchor point. The hangers/supports for C-54 are in scope for license renewal, and are evaluated in the appropriate structural commodity.

(b) Filters (F-100,101 & 102) are in scope of license renewal and subject to AMR due to 10CFR54.4(a)(2) (seismic II attached to I), and is an anchor point. The compressor skid filter and cooler are not in license renewal scope due to not providing an intended function.

(c) As discussed in (a) above, C-54 is in scope of license renewal due only to 10CFR54.4(a)(2) (seismic II attached to I), because it is an anchor point. C-50A and B do not provide any license renewal intended function, and are, therefore, not in scope.

Based on its review, the staff finds the applicant's response to RAI 2.3.3.15-4 acceptable regarding each item:

(a) The applicant adequately clarified that the waste gas compressor C-54 is an anchor point and is within the scope of license renewal in accordance with 10 CFR 54.4(a)(2) due to being seismic category II attached to a seismic category I. The applicant further clarified that the hangers/supports for the waste gas compressor C-54 are also within the scope of license renewal and evaluated in the appropriate structural commodity.

(b) The applicant adequately clarified that the filters (F-100, 101, & 102) are an anchor point, within the scope of license renewal in accordance with 10 CFR 54.4(a)(2) due to being seismic category II attached to seismic category I, and subject to an AMR in accordance with 10 CFR 54.21(a)(1); however, the compressor skid filter and cooler do not meet any license renewal scoping criteria because they do not provide an intended function. Therefore, they are not within the scope of license renewal in accordance with 10 CFR 54.4(a).

(c) The applicant adequately clarified that the waste gas compressors C-50A/B do not meet any license renewal scoping criteria; therefore, they are not within the scope of license renewal in accordance with 10 CFR 54.4(a).

Therefore, the staff's concerns described in RAI 2.3.3.15-4 are resolved.

In RAI 2.3.3.15-5, dated September 21, 2005, the staff stated that LRA Section 2.3.3.15 states that the boundaries of the portions of the waste gas system within the scope of license renewal include the piping and valves for containment isolation located between containment penetration #52 to CV-1104; however, the staff was unable to find this section of piping on the license renewal drawings for the waste gas system. Therefore, the staff requested that the applicant Identify where this section of piping was located on the license renewal drawings for the waste gas system.

In its response, by letter dated October 21, 2005, the applicant stated:

> This statement is in error. The Containment isolation piping and valves located between Containment penetration #52 to CV-1104 were evaluated as part of the Radioactive Waste System (RWS) in Section 2.3.3.11 as shown on color-coded drawing LR-M-211, Sheet 1 (F-7). LRA Section 2.3.3.15, System Description, fifth paragraph, on page 2-155, is hereby revised to delete '2) the piping and valves for Containment isolation located between Containment penetration #52 to CV-1104.' LRA Section 2.3.3.11, System Description, fifth paragraph, on page 2-137, item 4), is hereby revised to read, '4) Containment isolation components for containment penetrations to and from drain tanks.'

Based on its review, the staff finds the applicant's response to RAI 2.3.3.15-5 acceptable because it adequately clarified that the description of the waste gas system boundaries was incorrect. The applicant explained that the containment isolation piping and valves located between containment penetration #52 to CV-1104 had been evaluated as part of the radioactive waste system in LRA Section 2.3.3.11, which was revised to reflect the change. Therefore, the staff's concern described in RAI 2.3.3.15-5 is resolved.

2.3.3.15.3 Conclusion

The staff reviewed the LRA, accompanying scoping boundary drawings, and RAI responses to determine whether any SSCs that should be within the scope of license renewal had not been identified by the applicant. No omissions were identified. In addition, the staff performed a review to determine whether any components that should be subject to an AMR had not been identified by the applicant. No omissions were identified. On the basis of its review, the staff concludes that the applicant adequately identified the WGS components that are within the scope of license renewal, as required by 10 CFR 54.4(a), and the WGS components that are subject to an AMR, as required by 10 CFR 54.21(a)(1).

2.3.3.16 Domestic Water System

2.3.3.16.1 Summary of Technical Information in the Application

In LRA Section 2.3.3.16, the applicant described the domestic water system. The domestic water system is a source of general purpose filtered well and/or city water to buildings of the plant. The major components are: a storage tank, level control system, accumulator, water heaters, pumps, piping, valves, and instrumentation. The accumulator is pressurized from the plant instrument air system. The system is NSR.

The failure of NSR SSCs in the domestic water system could potentially prevent the satisfactory accomplishment of an SR function.

The intended function, within the scope of license renewal, is to provide fluid pressure boundary.

In LRA Table 2.3.3-16, the applicant identified the following domestic water system component types that are within the scope of license renewal and subject to an AMR:

- accumulators
- fasteners
- heat exchangers
- pipe and fittings
- pumps
- valves

2.3.3.16.2 Staff Evaluation

The staff reviewed LRA Section 2.3.3.16 and the FSAR using the evaluation methodology described in SER Section 2.3. The staff conducted its review in accordance with the guidance described in SRP-LR Section 2.3.

In conducting its review, the staff evaluated the system functions described in the LRA and FSAR in accordance with the requirements of 10 CFR 54.4(a) to verify that the applicant had not omitted from the scope of license renewal any components with intended functions delineated under 10 CFR 54.4(a). The staff then reviewed those components that the applicant had identified as being within the scope of license renewal to verify that the applicant had not omitted any passive and long-lived components that should be subject to an AMR in accordance with the requirements of 10 CFR 54.21(a)(1).

The staff's review of LRA Section 2.3.3.16 identified an area in which additional information was necessary to complete the review of the applicant's scoping and screening results. The applicant responded to the staff's RAI as discussed below.

In RAI 2.3.3.16-3, dated September 21, 2005, the staff stated that drawing LR-220, sheet 2, shows a T-36 permanganate filter, at location G-3, as within the scope of license renewal in accordance with 10 CFR 54.4(a); however, LRA Table 2.3.3-16 does not list filters as a component type subject to an AMR. LRA Section 2.1.3.2 states that oil, grease, and filters (both system and component filters) have been treated as consumables because either a program for periodic replacement exists or a monitoring program exists that replaces these consumables, based on established performance criteria, when their condition begins to degrade, but before any loss of intended function. Therefore, if this filter is excluded as being subject to an AMR because it is subject to replacement as defined in 10 CFR 54.21(a)(1)(ii), the staff requested that the applicant describe the schedule for periodic replacement or the monitoring program and replacement criteria if they are replaced on condition. If the filter is not subject to an AMR, the staff requested that the applicant justify the exclusion of this filter from subject to an AMR, in accordance with the requirements of 10 CFR 54.21(a)(1).

In its response, by letter dated October 21, 2005, the applicant stated:

> Permanganate filter,T-36, is in scope of LR due to 10CFR54.4(a)(2) spatial orientation (spray) with an intended function of 'fluid pressure boundary,' and is subject to AMR. T-36 is included in Component Group 'Accumulator' in LRA Table 2.3.3.16.

Based on its review, the staff finds the applicant's response to RAI 2.3.3.16-3 acceptable because it adequately clarified that the permanganate filter is within the scope of license renewal in that the failure of its fluid pressure boundary intended function due to spatial orientation (i.e., spray) could prevent the satisfactory accomplishment of the functions of SR SSCs in accordance with 10 CFR 54.4(a)(2). The applicant further clarified that the permanganate filter is included in the LRA table under the component group "accumulator" as subject to an AMR in accordance with 10 CFR 54.21(a)(1). Therefore, the staff's concerns described in RAI 2.3.3.16-3 are resolved.

2.3.3.16.3 Conclusion

The staff reviewed the LRA, accompanying scoping boundary drawings, and RAI response to determine whether any SSCs that should be within the scope of license renewal had not been identified by the applicant. No omissions were identified. In addition, the staff performed a review to determine whether any components that should be subject to an AMR had not been identified by the applicant. No omissions were identified. On the basis of its review, the staff concludes that the applicant adequately identified the domestic water system components that are within the scope of license renewal, as required by 10 CFR 54.4(a), and the domestic water system components that are subject to an AMR, as required by 10 CFR 54.21(a)(1).

2.3.3.17 Chemical Addition System

2.3.3.17.1 Summary of Technical Information in the Application

In LRA Section 2.3.3.17, the applicant described the chemical addition system. The system provides chemical addition to: primary coolant, main feedwater, condensate, circulating water, and service water systems. These additions are to reduce radiation levels, preserve piping integrity, maintain heat transfer, and prevent biological growths.

The failure of NSR SSCs in the chemical addition system could potentially prevent the satisfactory accomplishment of an SR function.

The intended function, within the scope of license renewal, is to provide fluid pressure boundary.

In LRA Table 2.3.3-17, the applicant identified the following chemical addition system component types that are within the scope of license renewal and subject to an AMR:

- accumulator
- fasteners
- pipe and fittings
- pump
- valves and dampers

2.3.3.17.2 Staff Evaluation

The staff reviewed LRA Section 2.3.3.17 and the FSAR using the evaluation methodology described in SER Section 2.3. The staff conducted its review in accordance with the guidance described in SRP-LR Section 2.3.

In conducting its review, the staff evaluated the system functions described in the LRA and FSAR in accordance with the requirements of 10 CFR 54.4(a) to verify that the applicant had not omitted from the scope of license renewal any components with intended functions delineated under 10 CFR 54.4(a). The staff then reviewed those components that the applicant had identified as being within the scope of license renewal to verify that the applicant had not omitted any passive and long-lived components that should be subject to an AMR in accordance with the requirements of 10 CFR 54.21(a)(1).

The staff's review of LRA Section 2.3.3.17 identified areas in which additional information was necessary to complete the review of the applicant's scoping and screening results. The applicant responded to the staff's RAIs as discussed below.

In RAI 2.3.3.17-1, dated September 21, 2005, the staff stated that drawing LR-220, sheet 2, shows chemical addition tanks (T-19B/C) as within the scope of license renewal; however, neither the associated level gauges, tank drains, nor piping to the suction of the chemical addition pumps (P-15E/F/G) were included within the scope of license renewal at locations A-3 and B-4, respectively. Similarly, neither the level gauge nor the drain on the hydrazine addition tank (T-16) were shown as within the scope of license renewal on the same drawing at location C-8. In contrast, level gauges on the morpholine and boric acid tanks (T-15 and T-19A), as well as the lines to the suction of associated pumps (P-15A/B/C/D), are shown as within the scope of license renewal. Level gauges, tank drains, or piping to the suction of the chemical addition pumps apparently are located in an area where their failure could cause failure of SR components; therefore, they should be included within the scope of license renewal in accordance with 10 CFR 54.4(a)(2). Therefore, the staff requested that the applicant clarify the basis for the differences.

In its response, by letter dated October 21, 2005, the applicant stated:

> The non-pressurized chemical addition tanks (T-16, T-19B and T-19C) and piping are within spill retention dikes. The tanks are in scope because they are the major anchoring component for (a)(2), seismic II attached to I, piping. The other piping, etc., associated with these tanks (except as shown) were excluded from scope as there are no spatial interaction or seismic II attached to I concerns for (a)(2). In the cases of T-15 and T-19A, the tanks were included because they are the major anchoring component for (a)(2), seismic II attached to I, piping. Similar to the tanks discussed above, only the highlighted attached piping is in scope.

Based on its review, the staff finds the applicant's response to RAI 2.3.3.17-1 acceptable. The applicant has adequately clarified that the associated level gauges, tank drains, piping to the suction of the chemical addition pumps (P-15E/F/G), and level gauge or drain on the hydrazine addition tank (T-16) are not located in areas with SR equipment and therefore do not meet the requirements of 10 CFR 54.4(a)(2) for either seismic category II attached to seismic category I, or for spatial orientation (i.e., spray). Since the components cannot prevent an SR component from performing its intended function and are not within the scope of license renewal, the staff's concern described in RAI 2.3.3.17-1 is resolved.

In RAI 2.3.3.17-2, dated September 21, 2005, the staff stated that drawing LR-655, sheet 2, shows pumps P-101, P-100A, and P-100B, at locations E-4, D-4 and C-4, respectively, as within the scope of license renewal in accordance with 10 CFR 54.4(a)(2); however, the suction of these pumps is shown as not within the scope of license renewal. Therefore, the staff requested that the applicant justify the exclusion of the suction of these pumps from within the scope of license renewal in accordance with 10 CFR 54.4(a).

In its response, by letter dated October 21, 2005, the applicant stated:

> The non-pressurized chemical addition suction piping is within spill retention dikes and excluded from scope as there are no spatial interaction or seismic II attached to I concerns for (a)(2). The pump and discharge piping is pressurized with potential spatial interaction consequences, and is, therefore, in scope of license renewal.

Based on its review, the staff finds the applicant's response to RAI 2.3.3.17-2 acceptable because it adequately clarified that the nonpressurized chemical addition suction piping is not located in areas with SR equipment. The applicant further clarified that this portion of piping does not meet the requirements of 10 CFR 54.4(a)(2) for seismic category II attached to seismic category I or for spatial orientation (i.e., spray). Since the piping cannot prevent an SR component from performing its intended function and is not within the scope of license renewal, the staff's concern described in RAI 2.3.3.17-2 is resolved.

In RAI 2.3.3.17-6, dated September 21, 2005, the staff stated that drawing LR-653, sheet 1, shows pumps P-47A and P-47B as within the scope of license renewal in accordance with 10 CFR 54.4(a)(2); however, the piping and valves from these pumps, including basket strainers BS-5393 and BS-5394 up to tank T-44, are shown as not within the scope of license renewal. Therefore, the staff requested that the applicant justify the exclusion of this section of piping from the scope of license renewal in accordance with 10 CFR 54.4(a).

In its response, by letter dated October 21, 2005, the applicant stated:

> The non-pressurized chemical addition suction piping is within spill retention dikes and excluded from scope as there are no spatial interaction or seismic II attached to I concerns for (a)(2). The pump and discharge piping is pressurized with potential spatial interaction consequences, and is, therefore, in scope of license renewal.

Based on its review, the staff finds the applicant's response to RAI 2.3.3.17-6 acceptable because it adequately clarified that the piping and valves from pumps P-47A and P-47B, including basket strainers BS-5393 and BS-5394 up to tank T-44, are not located in areas with SR equipment. The applicant further clarified that these components do not meet 10 CFR 54.4(a)(2) for seismic category II attached to seismic category I or for spatial orientation (i.e., spray). Since the components cannot prevent an SR component from performing its intended function and are not within the scope of license renewal, the staff's concern described in RAI 2.3.3.17-6 is resolved.

2.3.3.17.3 Conclusion

The staff reviewed the LRA, accompanying scoping boundary drawings, and RAI responses to determine whether any SSCs that should be within the scope of license renewal had not been identified by the applicant. No omissions were identified. In addition, the staff performed a review to determine whether any components that should be subject to an AMR had not been identified by the applicant. No omissions were identified. On the basis of its review, the staff concludes that the applicant adequately identified the chemical addition system components that are within the scope of license renewal, as required by 10 CFR 54.4(a), and the chemical addition system components that are subject to an AMR, as required by 10 CFR 54.21(a)(1).

2.3.4 Steam and Power Conversion Systems

In LRA Section 2.3.4, the applicant identified the structures and components of the steam and power conversion systems that are subject to an AMR for license renewal.

The applicant described the supporting structures and components of the steam and power conversion systems in the following sections of the LRA:

- 2.3.4.1   condensate and condenser system (CDS)
- 2.3.4.2   demineralized makeup water system
- 2.3.4.3   feedwater system
- 2.3.4.4   heater extraction and drain system
- 2.3.4.5   main air ejection and gland seal system
- 2.3.4.6   main steam system
- 2.3.4.7   turbine generator system

The staff's review findings regarding LRA Sections 2.3.4.1 – 2.3.4.7 are presented in SER Sections 2.3.4.1 – 2.3.4.7, respectively.

2.3.4.1 Condensate and Condenser System (CDS)

2.3.4.1.1 Summary of Technical Information in the Application

In LRA Section 2.3.4.1, the applicant described the CDS. Within the CDS, the main condenser condenses the steam from the main turbine and main feed pump driver exhausts, as well as the flashed steam from the feedwater heater drains, and provides the design vacuum which establishes the steam-water cycle efficiency. Noncondensible gases are removed from the main condenser by air ejectors or vacuum pump which are evaluated in the air ejector system (AES). The condensate system provides the means for transferring the deaerated condensate from the condenser hotwell through the heat transfer surfaces of the air ejector condensers, the gland steam condenser, and the low pressure feedwater heaters to the suction of the main feed water pumps. Also included in this system are the condensate storage tank (CST) and the associated piping system for makeup to and reject of the condenser hot well. The CST and the primary makeup tank, which is covered in the demineralized make-up water system, provide a passive flow of water, by gravity, to the auxiliary feedwater (AFW) system. Piping from the CST to the AFW pump suctions is included in the feedwater system. The combined CST and primary make-up tank contain a sufficient amount of cooling water to remove decay heat following a reactor trip to allow for cool down of the primary coolant system to shutdown cooling entry conditions.

The CDS contains SR components that are relied upon to remain functional during and following DBEs. The failure of NSR SSCs in the CDS could potentially prevent the satisfactory accomplishment of an SR function. In addition, the CDS performs functions that support fire protection and SBO.

The intended function, within the scope of license renewal, is to provide fluid pressure boundary.

In LRA Table 2.3.4-1, the applicant identified the following CDS component types that are within the scope of license renewal and subject to an AMR:

- accumulators
- CST heater shell
- CST heater tubes
- filters/strainers
- feedwater (FW) heater shell and channel head
- fasteners
- heat exchangers
- pipe & fittings
- pumps
- valves & dampers

2.3.4.1.2 Staff Evaluation

The staff reviewed LRA Section 2.3.4.1 and the FSAR using the evaluation methodology described in SER Section 2.3. The staff conducted its review in accordance with the guidance described in SRP-LR Section 2.3.

In conducting its review, the staff evaluated the system functions described in the LRA and FSAR in accordance with the requirements of 10 CFR 54.4(a) to verify that the applicant had not omitted from the scope of license renewal any components with intended functions delineated under 10 CFR 54.4(a). The staff then reviewed those components that the applicant had identified as being within the scope of license renewal to verify that the applicant had not omitted any passive and long-lived components that should be subject to an AMR in accordance with the requirements of 10 CFR 54.21(a)(1).

The staff's review of LRA Section 2.3.4.1 identified areas in which additional information was necessary to complete the review of the applicant's scoping and screening results. The applicant responded to the staff's RAIs as discussed below.

In RAI 2.3.4.1-2, dated September 21, 2005, the staff stated that LRA Table 2.3.4-1 separately lists "CST Heater Shell," "CST Heater Tubes," "FW Heater Shell and Channel Head," and "Heat Exchangers" as component groups subject to an AMR. LRA Section 2.1.2.2 states that "heat exchangers were divided into subcomponents as necessary to identify all applicable material/environment/intended function combinations." Therefore, the staff requested that the applicant provide information in four areas:

(a) Clarify whether the component group "Heat Exchangers" in LRA Table 2.3.4-1, includes all subcomponents of the heat exchangers (shell, tubes, tube sheets, channel heads, etc.) as subject to an AMR.

(b) Specifically identify which heat exchangers (other than the CST heat exchanger and feedwater heaters) are included in the "Heat Exchangers" group in LRA Table 2.3.4-1.

(c) LRA Section 2.3.4.1 specifies feedwater heaters E-2A/B through E-5A/B as portions of the condensate and condenser system within the scope of license renewal. The staff requested that the applicant clarify why feedwater heaters E-1A/B, which are shown on drawing LR-207, sheet 1C were not specified in this LRA section and to explain how feedwater heaters E-1A/B differ from the other heaters.

(d) Clarify whether the subcomponents, except those listed above, of the CST heaters and FW heaters had been excluded from the scope of license renewal and from an AMR in accordance with 10 CFR 54.4(a) and 10 CFR 54.21(a)(1), respectively, and, if so, justify their exclusions.

In its response, by letter dated October 21, 2005, the applicant addressed each area of the staff's RAI:

(a) Component Group 'Heat Exchanger' listed in LRA Table 2.3.4-1 includes Condensate System heat exchanger components (channel head and shell) required for the heat exchangers to perform their 10CFR54.4(a)(2) spatial interaction intended function of 'provide fluid pressure boundary', in addition to the Condensate Storage Tank (CST) heat exchanger channel head which is safety related. The CST heat exchanger shell, CST heat exchanger tubes and the feedwater heater shell and channel head are listed as separate line items on LRA Table 2.3.4-1.

(b) Component Group 'Heat Exchanger' listed in LRA Table 2.3.4-1 includes the following Condensate System components as being subject to AMR due to 10CFR54.4(a)(2) spatial orientation: Feedwater Heater Drain Cooler Channel head (E-7A/B), Air Injector and Inner & After Condenser Channel head (E-8), Turbine Gland Seal Condenser Channel head (E-19), and Main Condenser Shell (E-10). This Component Group also includes the CST Heat Exchanger Channel head (E-27) as subject to AMR due to 10CFR54.4(a)(1). The E-7A/B Shells are evaluated in the Heater Extraction and Drain System, and the E-8 Shell and E-19 Shell are evaluated in the Main Air Ejection and Gland Seal System.

(c) E-1A/B should also be included with this group. LRA Section 2.3.4.1, page 2-165, 2nd paragraph, Item 3) is hereby changed to read, 'The CDS piping and components from the Main Condenser Hotwell, including the Hotwell, through the Condensate Pumps, Air Ejector Inter & After Condenser, Main Turbine Gland Seal Condenser, and up to the point where the piping exits the Turbine Building.' A new Item 4) is hereby added as follows: 'The CDS piping and components from the point where the piping enters the Turbine Building, through the Drain Coolers, Feedwater Heaters E-1A/B thru E-5A/B, to the Steam Generator Feed Pumps.'

(d) LRA Table 2.3.4-1, Component Group 'FW Heater Shell and Channel Head' includes FW Heaters E-1A/B thru E-5A/B Channel heads. The FW Heater Shell are included in the Heater Extraction and Drain System. The FW Heater Channel heads and Shells are in scope of license renewal due to 10CFR54.4(a)(2) spatial orientation (spray) and require an AMR. The FW Heater tubes and tube sheets are not in scope of license renewal due to not meeting the requirements of 10CFR54.4(a)(1), (2), or (3). LRA Table 2.3.4.-1, Component Groups 'CST Heater Shell, CST Heater Tubes, Heat Exchanger' (which includes CST Heater Channel head) are in scope of license renewal due to 10CFR54.4(a)(1) safety-related, and require an AMR. For completeness, the CST Heater Tubes component type listed in LRA Table 2.3.4-1 on page 2-168, and Table 3.4.2-1 on page 3-220, is hereby revised to 'CST Heater Tubes and Tube Sheet.' Component Group 'CST Heater Tubes and Tube Sheet' is in scope of license renewal due to 10CFR54.4(a)(1) safety-related, and requires an AMR.

The staff's review found the applicant's response to RAI 2.3.4.1-2 acceptable because:

(a)   The applicant adequately clarified that, in addition to the CST heat exchanger channel head, the component group "heat exchanger" in LRA Table 2.3.4-1 includes the condensate system heat exchanger components (channel head and shell) required for the heat exchangers to perform their 10 CFR 54.4(a)(2) fluid pressure boundary spatial interaction intended function. The applicant further clarified that the CST heat exchanger shell, CST heat exchanger tubes, and the feedwater heater shell and channel head are listed as separate line items in LRA Table 2.3.4-1.

(b)   The applicant specifically identified the condensate system components (other than the condensate storage tank heat exchanger and feedwater heaters), included under the "heat exchangers" component group in LRA Table 2.3.4-1 as subject to an AMR in accordance with 10 CFR 54.21(a)(1). The staff concurred with this specific list of components.

(c)   The applicant adequately clarified that the feedwater heaters (E-1A/B) shown on drawing LR-207, sheet 1C, are within the scope of license renewal in accordance with 10 CFR 54.4(a). LRA Section 2.3.4.1 was revised to include these feedwater heaters (E-1A/B) as portions of the condensate and condenser system.

(d)   The applicant adequately clarified that the only subcomponents of the CST heaters and feedwater heaters not subject to an AMR are the feedwater heater tubes and tube sheets. The applicant explained that the feedwater heater tubes and tube sheets do not meet any license renewal scoping criteria and, therefore, are not within the scope of license renewal in accordance with 10 CFR 54.4(a).

Therefore, the staff's concerns described in RAI 2.3.4.1-2 are resolved.

In RAI 2.3.4.1-3, dated September 21, 2005, the staff stated that drawing LR-207, sheet 1B, shows a flexible connection at location E-6 as within the scope of license renewal; however, the flexible connection at location E-3 is shown as not within the scope of license renewal. Therefore, the staff requested that the applicant justify the exclusion of the latter flexible connection from within the scope of license renewal in accordance with 10 CFR 54.4(a). In addition, LRA Table 2.3.4-1 does not list flexible connections as a component group subject to an AMR. These flexible connections serve a pressure boundary intended function and are passive and long-lived. Therefore, the staff requested that the applicant clarify if flexible connections already had been included in LRA Table 2.3.4-1 as part of any other component group and, if not, justify the exclusion of the flexible connections from being subject to an AMR, in accordance with 10 CFR 54.21(a)(1).

In its response, by letter dated October 21, 2005, the applicant stated:

> The flexible connections on drawing LR-M-207, Sheet 1B, at both locations E-3 and E-6 are included in scope of license renewal. The connections are not subject to AMR due to being replaced every 3 years as preventive maintenance activities. Drawing LR-M- 207, Sheet 1B, should highlight the flex connection, at location E-3, as being in scope of License Renewal.

Based on its review, the staff finds the applicant's response to RAI 2.3.4.1-3 acceptable because it adequately clarified that the flexible connections on drawing LR-M-207, sheet 1B, at locations E-3 and E-6, are within the scope of license renewal in accordance with 10 CFR 54.4(a). The applicant further clarified that flexible connections are replaced every three years as preventive maintenance activities. Since the flexible connections are periodically replaced, the applicant found that they are not subject to an AMR in accordance with 10 CFR 54.21(a)(1). Therefore, the staff's concerns described in RAI 2.3.4.1-3 are resolved.

2.3.4.1.3 Conclusion

The staff reviewed the LRA, accompanying scoping boundary drawings, and RAI responses to determine whether any SSCs that should be within the scope of license renewal had not been identified by the applicant. No omissions were identified. In addition, the staff performed a review to determine whether any components that should be subject to an AMR had not been identified by the applicant. No omissions were identified. On the basis of its review, the staff concludes that the applicant adequately identified the CDS components that are within the scope of license renewal, as required by 10 CFR 54.4(a), and the CDS components that are subject to an AMR, as required by 10 CFR 54.21(a)(1).

2.3.4.2 Demineralized Makeup Water System

2.3.4.2.1 Summary of Technical Information in the Application

In LRA Section 2.3.4.2, the applicant described the demineralized makeup water (DMW) system. The DMW system is the source of high quality filtered and demineralized water throughout the plant. The major components of this system are pumps, storage tanks, a reverse osmosis system, filters, tank heating heat exchangers, extensive piping, valves, and instrumentation.

The DMW system contains SR components that are relied upon to remain functional during and following DBEs. The failure of NSR SSCs in the DMW system could potentially prevent the satisfactory accomplishment of an SR function. In addition, the DMW system performs functions that support fire protection and SBO.

The intended function, within the scope of license renewal, is to provide fluid pressure boundary.

In LRA Table 2.3.4-2, the applicant identified the following DMW system component types that are within the scope of license renewal and subject to an AMR:

- accumulators
- fasteners
- filters
- heat exchangers
- pipe and fittings
- pumps
- transmitter/element
- valves & dampers

2.3.4.2.2 Staff Evaluation

The staff reviewed LRA Section 2.3.4.2 and the FSAR using the evaluation methodology described in SER Section 2.3. The staff conducted its review in accordance with the guidance described in SRP-LR Section 2.3.

In conducting its review, the staff evaluated the system functions described in the LRA and FSAR in accordance with the requirements of 10 CFR 54.4(a) to verify that the applicant had not omitted from the scope of license renewal any components with intended functions delineated under 10 CFR 54.4(a). The staff then reviewed those components that the applicant had identified as being within the scope of license renewal to verify that the applicant had not omitted any passive and long-lived components that should be subject to an AMR in accordance with the requirements of 10 CFR 54.21(a)(1).

The staff's review of LRA Section 2.3.4.2 identified an area in which additional information was necessary to complete the review of the applicant's scoping and screening results. The applicant responded to the staff's RAI as discussed below.

In RAI 2.3.4.2-1, dated September 21, 2005, the staff stated that LRA Section 2.3.4.2 identifies a system boundary within the scope of license renewal at valves MV-PMU100 (locked open) and MV-PMU109 (normally open) downstream of the primary system makeup storage tank (T-81). Also, drawing LR-220, sheet 1 (locations G4 and G-5), shows the piping upstream of and including these valves as within the scope of license renewal; however, the piping downstream from these valves, including the buried pipe, is shown as not within the scope of license renewal. Therefore, the staff requested that the applicant clarify the basis for excluding the piping downstream of valves MV-PMU100 and MV-PMU109 from within the scope of license renewal in accordance with 10 CFR 54.4(a). This LRA section states that T-81 provides condensate/feedwater supply to the condensate storage tank. Similarly, the staff requested that the applicant clarify the basis for excluding the following lines associated with the primary system makeup storage tank (T-90) from within the scope of license renewal, in accordance with the requirements of 10 CFR 54.4(a): lines HBD-14-3", HBD-15-2", and HBD-16-3" on drawing LR-652, sheet 1, at locations G-4 and G-5, and associated components from the primary system makeup storage tank (T-90) through and including the buried HCC piping downstream. FSAR Section 9.4.2.1 states that T-90 is one of the sources which supplies fuel pool makeup water.

In its response, by letter dated October 21, 2005, the applicant stated:

> The section of the piping through MV-PMU100 and MV-PMU109 up to the point where it enters the ground is in scope for (a)(2) due to spatial interaction with the adjacent Demineralized Water System piping. Thus, where it goes below grade there is no longer any concern for spray and that portion of the piping is not in scope. Drawing LR-M-220-1 should show the piping above ground (piping HC-11-1' and HC-11-3') as highlighted. LRA Section 2.3.4.2, third paragraph, Item 4, on Page 2-169, is hereby revised to read, '4) piping located downstream of T-81 through valves MV-PMU100 and MV-PMU109 to the point where the lines enter the ground.'

> T-90 is in scope for seismic reasons only. The piping from the tank does not need to be in scope until it exits the ground (for (a)(2) spatial interaction only) in the auxiliary building. FSAR Section 9.4.2.1 states, 'Fuel pool makeup water is supplied from the Primary Water Makeup Tank T-90, the Recycled Boric Acid Storage Tank T-96, the Safety Injection & Refueling Water Tank T-58 and Utility Water Tank T-91. In the event of a considerable loss of pool water, the fire system can be used to replenish the pool water content.' For License Renewal, Palisades credits the Safety Injection & Refueling Water Tank (T-58) for makeup to the fuel pool.

Based on its review, the staff finds the applicant's response to RAI 2.3.4.2-1 acceptable because it adequately clarified that the piping downstream of valves MV-PMU100 and MV-PMU109, including the buried piping, is not located in an area with SR components. Therefore, this portion of piping that goes below grade is not within the scope of license renewal because it cannot prevent an SR component from performing its intended function due to spatial interactions (i.e., spray) and does not meet 10 CFR 54.4(a)(2) requirements. The applicant further clarified that the primary system makeup storage tank T-90 is within the scope of license renewal for seismic reasons only and that the piping and associated components from T-90 are within the scope of license renewal only until the piping exits the ground in the auxiliary building. Piping buried or outside the auxiliary building cannot prevent any SR component from performing its intended function due to spatial interactions (i.e., spray). Therefore, lines HBD-14-3", HBD-15-2", and HBD-16-3" do not meet 10 CFR 54.4(a)(2) requirements and are not within the scope of license renewal. Finally, the applicant clarified that it credits the safety injection and refueling water tank (T-58) for makeup to the fuel pool; therefore, the primary system makeup storage tank (T-90) does not perform this license renewal intended function. Therefore, the staff's concerns described in RAI 2.3.4.2-1 are resolved.

2.3.4.2.3 Conclusion

The staff reviewed the LRA, accompanying scoping boundary drawings, and RAI response to determine whether any SSCs that should be within the scope of license renewal had not been identified by the applicant. No omissions were identified. In addition, the staff performed a review to determine whether any components that should be subject to an AMR had not been identified by the applicant. No omissions were identified. On the basis of its review, the staff concludes that the applicant adequately identified the DMW system components that are within the scope of license renewal, as required by 10 CFR 54.4(a), and the DMW system components that are subject to an AMR, as required by 10 CFR 54.21(a)(1).

2.3.4.3 Feedwater System

2.3.4.3.1 Summary of Technical Information in the Application

In LRA Section 2.3.4.3, the applicant described the feedwater system (FWS). The FWS consists of main feedwater (MFW) and auxiliary feedwater (AFW) systems. The MFW system provides a reliable source of water to the steam generators during normal operations. This water is preheated in the feedwater heaters to increase overall thermal efficiency. Flow rate is manually and/or automatically controlled to maintain steam generator level. Condensate from two parallel streams of low pressure feedwater heaters and from the discharge of two heater drain pumps is supplied to the suction of the two variable speed, turbine driven feedwater pumps. These pumps deliver the feedwater through parallel high pressure feedwater heaters to the steam generators. The AFW system supplies water to the secondary side of the steam generators for reactor decay heat removal when normal feedwater sources are unavailable during accident conditions. AFW is also designed to provide a supply of feedwater to the steam generators during start-up operations and to remove primary system sensible and decay heat during initial stages of shutdown operations. AFW can provide feedwater to any combination of steam generators from any one or combination of three pumps; two are motor-driven, and the third is steam-driven.

The FWS contains SR components that are relied upon to remain functional during and following DBEs. The failure of NSR SSCs in the FWS could potentially prevent the satisfactory accomplishment of an SR function. In addition, the FWS performs functions that support fire protection, EQ, ATWS, and SBO.

The intended functions within the scope of license renewal include the following:

- provides flow restriction (throttle)
- provides fluid pressure boundary

In LRA Table 2.3.4-3, the applicant identified the following FWS component types that are within the scope of license renewal and subject to an AMR:

- accumulators
- fasteners
- filters/strainers
- heat exchangers
- indicators/recorders
- pipe & fittings
- pumps
- traps (steam)
- turbines
- valves & dampers

2.3.4.3.2 Staff Evaluation

The staff reviewed LRA Section 2.3.4.3 and the FSAR using the evaluation methodology described in SER Section 2.3. The staff conducted its review in accordance with the guidance described in SRP-LR Section 2.3.

In conducting its review, the staff evaluated the system functions described in the LRA and FSAR in accordance with the requirements of 10 CFR 54.4(a) to verify that the applicant had not omitted from the scope of license renewal any components with intended functions delineated under 10 CFR 54.4(a). The staff then reviewed those components that the applicant had identified as being within the scope of license renewal to verify that the applicant had not omitted any passive and long-lived components that should be subject to an AMR in accordance with the requirements of 10 CFR 54.21(a)(1).

The staff's review of LRA Section 2.3.4.3 identified areas in which additional information was necessary to complete the review of the applicant's scoping and screening results. The applicant responded to the staff's RAIs as discussed below.

In RAI 2.3.4.3-2, dated September 21, 2005, the staff stated that drawing LR-207, sheet 2, shows restrictive orifices RO-0783A/B, at locations F-6 and G-6, as within the scope of license renewal. The function of these restrictive orifices is to provide minimum flow recirculation at maximum pressure for the AFW pump. In addition, other restrictive orifices are shown on the license renewal drawings for the feedwater system as within the scope of license renewal. LRA Table 2.3.4-3 does not list restrictive orifices as a component group subject to an AMR; however, the "flow restriction" intended function is listed in this table for the "pipe and fitting" component group. Therefore, the staff requested that the applicant clarify if the flow restrictive orifices were already included in LRA Table 2.3.4-3 in the component group "piping" and, if so, identify the intended function that the feedwater restrictive orifices serve or, if not, justify their exclusion from an AMR in accordance with 10 CFR 54.21(a)(1).

Additionally, flow nozzles and metering orifices (FE-0783A, FE-0783B, FE-0783C, FE-0737, FE-0736, FE-0727, FE-0749) used for flow measurement were shown as within the scope of license renewal on drawings for the feedwater system (locations F-6, G-6, D-6, A-5, C-5, G-4, and E-4, respectively); however, LRA Table 2.3.4-3 does not list flow nozzles or metering orifices as component groups subject to an AMR. Therefore, the staff requested that the applicant clarify if these flow nozzles and metering orifices are already included in LRA Table 2.3.4-3 as part of any other component group. If so, the staff requested that the applicant identify their intended function that should be maintained during the period of extended operation or, if not, justify the exclusion of flow nozzles and metering orifices from an AMR in accordance with 10 CFR 54.21(a)(1).

In its response, by letter dated October 21, 2005, the applicant stated:

> RO-0783A/B/C are subject to AMR and have intended functions of 'fluid pressure boundary' and 'flow restriction'. The other feedwater system restrictive orifices are subject to AMR and have an intended function of 'fluid pressure boundary.' The restrictive orifices are included in Component Group 'Pipe and Fittings' on LRA Table 2.3.4-3 on page 2-180.
>
> Flow elements (flow nozzles and metering orifices, referred to above) FE-0783A/B/C, 0737, 0736, 0727 and 0729 are subject to AMR and have an intended function of 'fluid pressure boundary'. The flow elements are included in Component Group 'Pipe and Fittings' on LRA Table 2.3.4-3 on page 2-180.

Based on its review, the staff finds the applicant's response to RAI 2.3.4.3-2 acceptable because it adequately clarified that the restrictive orifices RO-0783A/B/C perform the intended function of "flow restriction" and "fluid pressure boundary." The applicant further clarified that other restrictive orifices shown on the license renewal drawings for the feedwater system, and flow elements addressed in the RAI perform the intended function of "fluid pressure boundary" and are included under the component group "piping and fittings" in LRA Table 2.3.4-3 as subject to an AMR in accordance with 10 CFR 54.21(a)(1). Therefore, the staff's concerns described in RAI 2.3.4.3-2 are resolved.

In RAI 2.3.4.3-4, dated September 21, 2005, the staff stated that LRA Table 2.3.4-3 lists "turbines" as a component group subject to an AMR. The auxiliary feedwater pump turbine K-8, at location C-7, and its components were shown within the scope of license renewal on drawing LR-205; however, the turbine governor and mechanical speed sensor associated with this turbine were not listed in LRA Table 2.3.4-3 as component groups subject to an AMR. Therefore, the staff requested that the applicant clarify if these components are already included in LRA Table 2.3.4-3 under the component group "turbine" and, if not, to justify their exclusion from an AMR in accordance with 10 CFR 54.21(a)(1).

In its response, by letter dated October 21, 2005, the applicant stated, "The Auxiliary Feedwater Pump Turbine (K-8) governor and Mechanical Speed Sensor are in scope of license renewal, but are not subject to AMR due to being 'Active' components."

Based on its review, the staff finds the applicant's response to RAI 2.3.4.3-4 acceptable because it adequately clarified that the turbine governor and mechanical speed sensor of the auxiliary feedwater pump turbine are active per NEI 95-10 Appendix B guidance and not subject to an AMR in accordance with 10 CFR 54.21(a)(1). Therefore, the staff's concern described in RAI 2.3.4.3-4 is resolved.

2.3.4.3.3 Conclusion

The staff reviewed the LRA, accompanying scoping boundary drawings, and RAI responses to determine whether any SSCs that should be within the scope of license renewal had not been identified by the applicant. No omissions were identified. In addition, the staff performed a review to determine whether any components that should be subject to an AMR had not been identified by the applicant. No omissions were identified. On the basis of its review, the staff concludes that the applicant adequately identified the FWS components that are within the scope of license renewal, as required by 10 CFR 54.4(a), and the FWS components that are subject to an AMR, as required by 10 CFR 54.21(a)(1).

2.3.4.4 Heater Extraction and Drain System

2.3.4.4.1 Summary of Technical Information in the Application

In LRA Section 2.3.4.4, the applicant described the heater extraction and drain (HED) system. In the HED system, extraction steam is supplied to the feedwater heaters from the high pressure (HP) and low pressure (LP) turbines, as required by the plant thermal cycle. During normal operation, feedwater heating stages 1 through 4 are supplied by the LP turbines extraction steam, and heating stage 5 is supplied from the HP turbine exhaust crossunder to the moisture separator reheater. Feedwater heating stage 6, high pressure, is supplied by HP turbine extraction and the reheater drain tanks. All feedwater heaters with sufficient energy to overspeed the turbine have extraction nonreturn valves or turbine reheat stop and intercept valves to stop reverse steam flow following a turbine trip. The HED system collects drains from various heaters and returns them to the secondary water cycle. Feedwater heater 6 drains cascade to heater 5 and then to the moisture separator drain tank which supplies the heater drain pumps. Feedwater heaters 4, 3, 2, and 1 are also arranged in a cascaded drain system, but ultimately discharge to the main condenser. The system design is to inject the higher temperature drains to the shell side of the feedwater heaters closer to the steam generators. This allows the feedwater to be progressively heated until it is just a few degrees below main steam temperature when the feedwater is injected into the steam generators. This also increases the thermal efficiency of the steam/water cycle.

The failure of NSR SSCs in the HED system could potentially prevent the satisfactory accomplishment of an SR function. The HED system also performs functions that support fire protection.

The intended function, within the scope of license renewal, is to provide fluid pressure boundary.

In LRA Table 2.3.4-4, the applicant identified the following HED system component types that are within the scope of license renewal and subject to an AMR:

- accumulators
- fasteners
- heat exchangers
- pipe & fittings
- pumps
- transmitter/element
- traps (steam)
- valves & dampers

2.3.4.4.2 Staff Evaluation

The staff reviewed LRA Section 2.3.4.4 and the FSAR using the evaluation methodology described in SER Section 2.3. The staff conducted its review in accordance with the guidance described in SRP-LR Section 2.3.

In conducting its review, the staff evaluated the system functions described in the LRA and FSAR in accordance with the requirements of 10 CFR 54.4(a) to verify that the applicant had not omitted from the scope of license renewal any components with intended functions delineated under 10 CFR 54.4(a). The staff then reviewed those components that the applicant had identified as being within the scope of license renewal to verify that the applicant had not omitted any passive and long-lived components that should be subject to an AMR in accordance with the requirements of 10 CFR 54.21(a)(1).

The staff's review of LRA Section 2.3.4.4 identified areas in which additional information was necessary to complete the review of the applicant's scoping and screening results. The applicant responded to the staff's RAIs as discussed below.

In RAI 2.3.4.4-1, dated September 21, 2005, the staff stated that a portion of heaters E-2A/B is shown inside of the condenser on drawing LR-206, sheet 1A, at location C-4/5, as within the scope of license renewal (as is the associated boundary of the condenser). Heaters E-1A/B appeared to be similarly situated (same drawing, location B-4/5), yet the portion inside the condenser was not shown as within the scope of license renewal. Therefore, the staff requested that the applicant explain the rationale for the distinction and clarify the physical meaning of the condenser boundary that transects the heaters' symbol.

In its response, by letter dated October 21, 2005, the applicant stated:

> LR-M-206, Sheet 1A, is correct to show the portions of Heaters E-1A/B and E-2A/B that extend into the condenser as out of scope. As there is no safety related equipment inside the condenser, these portions of the heaters do not meet any of the scoping criteria.

Based on its review, the staff finds the applicant's response to RAI 2.3.4.4-1 acceptable because it adequately clarified that the portions of the heaters E-1A/B and E-2A/B extending into the condenser do not meet any of the 10 CFR 54.4(a) license renewal scoping criteria. As there is no SR equipment inside the condenser, they are not within the scope of license renewal. Therefore, the staff's concern described in RAI 2.3.4.4-1 is resolved.

In RAI 2.3.4.4-2, dated September 21, 2005, the staff stated that the September 14, 2005, drawing LR-207, sheet 1C, shows flexible connections, at locations E-3 and G-3, as within the scope of license renewal; however, LRA Table 2.3.4-1 does not list flexible connections as a component group subject to an AMR. These flexible connections serve a pressure boundary intended function and are passive and long-lived. Therefore, the staff requested that the applicant clarify if flexible connections are already included in LRA Table 2.3.4-4 as part of any other component group and, if so, identify their intended function or, if not, justify the exclusion of flexible connections from an AMR in accordance with 10 CFR 54.21(a)(1).

In its response, by letter dated October 21, 2005, the applicant stated:

> The Heater Drain Tank Pump seal cooling flexible connections are in scope of license renewal due to 10CFR54.4(a)(2) spatial orientation (spray), and are subject to AMR since they are replaced on condition rather than a set frequency. Therefore, the flexible connections are incorporated into Component Group 'Pipe & Fittings' in LRA Tables 2.3.4-4 and 3.4.2-4.

Based on its review, the staff finds the applicant's response to RAI 2.3.4.4-2 acceptable because it adequately clarified that the heater drain tank pump seal cooling flexible connections are included under the component group "piping and fittings" in LRA Table 2.3.4-4 as subject to an AMR, in accordance with 10 CFR 54.21(a)(1). Therefore, the staff's concern described in RAI 2.3.4.4-2 is resolved.

2.3.4.4.3 Conclusion

The staff reviewed the LRA, accompanying scoping boundary drawings, and RAI responses to determine whether any SSCs that should be within the scope of license renewal had not been identified by the applicant. No omissions were identified. In addition, the staff performed a review to determine whether any components that should be subject to an AMR had not been identified by the applicant. No omissions were identified. On the basis of its review, the staff concludes that the applicant adequately identified the HED system components that are within the scope of license renewal, as required by 10 CFR 54.4(a), and the HED system components that are subject to an AMR, as required by 10 CFR 54.21(a)(1).

2.3.4.5 Main Air Ejection and Gland Seal System

2.3.4.5.1 Summary of Technical Information in the Application

In LRA Section 2.3.4.5, the applicant described the main air ejection and gland seal (AES) system. In the AES system, noncondensible gases are removed from the main condenser during operation by the steam jet air ejectors, and during start-up by the condenser vacuum pump, hogging air ejector and the steam jet air ejectors. The condenser vacuum pump is used to establish a partial condenser vacuum during start-up, and to allow testing of the main condenser for leakage while the plant is shut down. The gland seal system provides steam seal on main turbine shaft to prevent air inleakage and maintain condenser vacuum. Steam supplies for sealing the main turbines pass through pressure reducing stations off the main steam supply line. The gland seal exhaust is condensed through the gland seal condenser and returned to steam-condensate cycle at the condenser hotwell.

The failure of NSR SSCs in the AES system could potentially prevent the satisfactory accomplishment of an SR function. The AES system also performs functions that support fire protection.

The intended function, within the scope of license renewal, is to provide fluid pressure boundary.

In LRA Table 2.3.4-5, the applicant identified the following AES system component types that are within the scope of license renewal and subject to an AMR:

- blowers fans compressor vacuum
- filters/strainers
- heat exchangers
- fasteners
- pipe & fittings
- traps (steam)
- valves & dampers

2.3.4.5.2 Staff Evaluation

The staff reviewed LRA Section 2.3.4.5 and the FSAR using the evaluation methodology described in SER Section 2.3. The staff conducted its review in accordance with the guidance described in SRP-LR Section 2.3.

In conducting its review, the staff evaluated the system functions described in the LRA and FSAR in accordance with the requirements of 10 CFR 54.4(a) to verify that the applicant had not omitted from the scope of license renewal any components with intended functions delineated under 10 CFR 54.4(a). The staff then reviewed those components that the applicant

had identified as being within the scope of license renewal to verify that the applicant had not omitted any passive and long-lived components that should be subject to an AMR in accordance with the requirements of 10 CFR 54.21(a)(1).

The staff's review of LRA Section 2.3.4.5 identified an area in which additional information was necessary to complete the review of the applicant's scoping and screening results. The applicant responded to the staff's RAI as discussed below.

In RAI 2.3.4.5-1, dated September 21, 2005, the staff stated that drawing LR-206, sheet 1C shows the following components within the scope of license renewal:

(a)   ejectors at several locations
(b)   drain traps at several locations
(c)   flexible connections at locations C-6 and D-6
(d)   vacuum pump (P-213) at location E-5

However, LRA Table 2.3.4-5 does not list these components as subject to an AMR. They serve a pressure boundary intended function and are passive and long-lived. Therefore, the staff requested that the applicant clarify if these components already had been included in LRA Table 2.3.4-5 as part of any other component group and, if not, justify their exclusion from an AMR in accordance with 10 CFR 54.21(a)(1). In the case of (b) above, the staff requested that the applicant clarify if drain traps are part of the component group "steam traps" in LRA Table 2.3.4-5.

In its response, by letter dated October 21, 2005, the applicant addressed each area of the staff's RAI:

(a)   Steam Jet Air Ejector (SJAE) Primary and Secondary are included in LRA Table 2.3.4-5 as Component Group - 'Blowers Fans Compressor Vacuum'

(b)   Drain traps are included in LRA Table 2.3.4-5 as Component Group - Traps (Steam).

(c)   Further evaluation has determined that the Gland Seal Condenser Exhauster (C-1A/B) and its piping and components do not perform an intended function per 10CFR54.4(a)(2) spatial orientation (spray). The exhauster provides a vacuum in the Gland Seal Condenser and discharges air/gas to outside the Turbine Building. Should the Exhauster inlet piping and components develop a throughwall flaw, in-leakage would occur. Should the exhauster discharge piping and components develop a through-wall flaw, air/gas would be discharged into the Turbine Building. However, no spray would occur and safety-related equipment located in the area would not be affected. Therefore, the Gland Seal Condenser Exhauster and its piping and components are hereby deleted from scope of License Renewal. The associated highlighting on drawing LR-M-206 Sheet 1C should be disregarded.

(d)   Vacuum pump (P-213) is included in LRA Table 2.3.4-5 as Component Group - Blowers Fans Compressor Vacuum.

Based on its review, the staff finds the applicant's response to RAI 2.3.4.5-1 acceptable for each area of the RAI because:

(a) The applicant adequately clarified that the steam jet air ejector primary and secondary, are included in LRA Table 2.3.4-5 under the "Blowers Fans Compressor Vacuum" component group as subject to an AMR, in accordance with 10 CFR 54.21(a)(1).

(b) The applicant adequately clarified that the drain traps are included in LRA Table 2.3.4-5 under the "traps (steam)" component group as subject to an AMR, in accordance with 10 CFR 54.21(a)(1).

(c) The applicant adequately clarified that the gland seal condenser exhausters and their associated piping and other components, including the flexible connections, do not meet the requirements of 10 CFR 54.4(a)(2) for spatial orientation (i.e., spray) and, therefore, cannot prevent an SR component from performing its intended function.

(d) The applicant adequately clarified that the vacuum pump P-213 is included in LRA Table 2.3.4-5 under the "Blowers Fans Compressor Vacuum" component group as subject to an AMR, in accordance with 10 CFR 54.21(a)(1).

Therefore, the staff's concerns described in RAI 2.3.4.5-1 are resolved.

2.3.4.5.3 Conclusion

The staff reviewed the LRA, accompanying scoping boundary drawings, and RAI response to determine whether any SSCs that should be within the scope of license renewal had not been identified by the applicant. No omissions were identified. In addition, the staff performed a review to determine whether any components that should be subject to an AMR had not been identified by the applicant. No omissions were identified. On the basis of its review, the staff concludes that the applicant adequately identified the AES system components that are within the scope of license renewal, as required by 10 CFR 54.4(a), and the AES system components that are subject to an AMR, as required by 10 CFR 54.21(a)(1).

2.3.4.6 Main Steam System

2.3.4.6.1 Summary of Technical Information in the Application

In LRA Section 2.3.4.6, the applicant described the main steam system (MSS). In the MSS, steam generated in the steam generators passes through two 36-inch headers and main steam isolation valves (MSIVs) to the turbine stop valves. Each main steam header is provided with 12 spring-loaded safety valves (MSSVs) and 2 atmospheric dump valves (ADVs) upstream of the MSIVs. In addition, there is a steam bypass to condenser downstream of the MSIVs. The main steam line also supplies steam for the steam jet air ejectors, the heating steam for the reheaters, the steam supply to the steam generator feed pump turbine drivers, and steam supply to the turbine-driven auxiliary feed pump turbine driver which is connected upstream of an MSIV. The MSIVs are provided to isolate the steam generators to: (1) prevent the uncontrolled release of radioactivity in the unlikely event of a steam generator tube failure, (2) lessen a rapid uncontrolled cool down of the PCS in the event of a main steam line break by limiting the blowdown from a single steam generator, and (3) limit the steam discharge to the containment from the affected steam generator in the event of the rupture of one main steam line inside containment.

The design-basis of the ADVs is to prevent lifting of the MSSVs following a turbine and reactor trip and to provide the capability to cool the plant to SDC system entry conditions when condenser vacuum is lost. The turbine bypass to the main condenser provides for removal of reactor decay heat following normal reactor shutdown. Overpressure protection for the shell side of the steam generators and the main steam line piping up to the inlet of the turbine stop valve is provided by the MSSVs. The MSSVs also provide protection against overpressurizing the primary coolant pressure boundary by providing a heat sink for the removal of energy from the PCS if the preferred heat sink, provided by the condenser and circulating water system, is not available.

A subsystem of main steam is steam generator blowdown. The steam generator blowdown subsystem is designed to process steam generator blowdown water. A minimum continuous blowdown is normally maintained for effective steam generator chemistry control. Other functions of the subsystem include the capability to clean up the condenser hotwell prior to start up by recirculating the water through the blowdown demineralizers, and the capability to recirculate steam generator secondary side water, for treatment purposes, during cold shutdown conditions.

The MSS contains SR components that are relied upon to remain functional during and following DBEs. The failure of NSR SSCs in the MSS could potentially prevent the satisfactory accomplishment of an SR function. In addition, the MSS performs functions that support fire protection, EQ, ATWS, and SBO.

The intended function, within the scope of license renewal, is to provide fluid pressure boundary.

In LRA Table 2.3.4-6, the applicant identified the following MSS component types that are within the scope of license renewal and subject to an AMR:

- accumulators
- ejectors
- fasteners
- filters/strainers
- heat exchangers
- indicators/recorders (level glasses)
- pipe & fittings
- pumps
- traps (steam)
- valves & dampers

2.3.4.6.2 Staff Evaluation

The staff reviewed LRA Section 2.3.4.6 and the FSAR using the evaluation methodology described in SER Section 2.3. The staff conducted its review in accordance with the guidance described in SRP-LR Section 2.3.

In conducting its review, the staff evaluated the system functions described in the LRA and FSAR in accordance with the requirements of 10 CFR 54.4(a) to verify that the applicant had not omitted from the scope of license renewal any components with intended functions delineated under 10 CFR 54.4(a). The staff then reviewed those components that the applicant had identified as being within the scope of license renewal to verify that the applicant had not omitted any passive and long-lived components that should be subject to an AMR in accordance with the requirements of 10 CFR 54.21(a)(1).

The staff's review of LRA Section 2.3.4.6 identified areas in which additional information was necessary to complete the review of the applicant's scoping and screening results. The applicant responded to the staff's RAIs as discussed below.

In RAI 2.3.4.6-1, dated September 21, 2005, the staff stated that one function of the ATWS mitigation system is to provide a diverse means of initiating a main turbine trip following an ATWS event. The first stage (impulse chamber) pressure on the high-pressure turbine provides an actuating signal to this mitigation system. Therefore, the in-line pressure transmitters, which sense this pressure, along with their associated piping and components, are within the scope of license renewal in accordance with 10 CFR 54.4(a)(3). However, drawing LR-205, sheet 1, shows these pressure transmitters, at location D-3, as not within the scope of license renewal. In addition, the component group "indicators/recorders" in LRA Table 2.3.4-6, is limited to "level glasses" only. This limitation implicitly excludes the pressure transmitters as subject to an AMR. Therefore, the staff requested that the applicant justify the exclusion of the above-mentioned pressure transmitters and associated piping and components from within the scope of license renewal in accordance with 10 CFR 54.4(a).

In its response, by letter dated October 21, 2005, the applicant stated, "The Palisades ATWS mitigating design does not use turbine first stage pressure as an input. It, rather, uses pressurizer pressure inputs. The associated instruments are in scope of license renewal, and are not subject to AMR due to being 'Active' per NEI 95-10."

Based on its review, the staff finds the applicant's response to RAI 2.3.4.6-1 acceptable because it adequately clarified that the turbine first stage pressure transmitters and associated piping and components on drawing LR-M-205, sheet 1, are not within the scope of license renewal because they are not used in the ATWS mitigation function. The ATWS mitigation function at PNP uses inputs from pressurizer pressure transmitters which are within the scope of license renewal in accordance with 10 CFR 54.4(a), but are "Active" per NEI 95-10 guidance. Therefore, the pressurizer pressure transmitters are not subject to an AMR in accordance with 10 CFR 54.21(a)(1). The staff concluded that the turbine first stage pressure transmitters are not within the scope of license renewal in accordance with 10 CFR 54.4(a)(3) for the ATWS regulated event. Therefore, the staff's concern described in RAI 2.3.4.6-1 is resolved.

In RAI 2.3.4.6-4, dated September 21, 2005, the staff stated that drawing LR-205, sheet 1, shows main steam piping runs that serve as main turbine control valves stem leakoff piping to the gland seal condenser, at locations C-5 through C-7 and E-5 through E-7; however, the only portion shown within the scope of license renewal was at location C-4/5, which continues to drawing LR-206, sheet 1C, for the gland seal condenser. The remainder of the piping was shown as not within the scope of license renewal with no clear indication of the location of the license renewal boundary. Therefore, the staff requested that the applicant explain why the identified piping was not within the scope of license renewal in accordance with 10 CFR 54.4(a).

In its response, by letter dated October 21, 2005, the applicant stated:

> The main turbine control valves' stem leakoff piping located on the Turbine Deck is not in scope of License Renewal due to not meeting 10CFR54.4(a)(1), (2), or (3) Criteria. This piping is not located in an area containing safety-related equipment. However, the portion of the piping below the turbine deck is in an area with safety related equipment, and requires an AMR due to 10CFR54.4(a)(2). This portion of the piping is highlighted on drawing LR-M-205 Sheet 1.

Based on its review, the staff finds the applicant's response to RAI 2.3.4.6-4 acceptable because it adequately clarified that the main turbine control valves' stem leakoff piping to the gland seal condenser, located on the turbine deck, is not in an area containing SR equipment. This portion of the main steam piping, located on the turbine deck, cannot prevent an SR component from performing its intended function and does not meet the requirements of 10 CFR 54.4(a)(2). As such, it is not within the scope of license renewal. Therefore, the staff's concern described in RAI 2.3.4.6-4 is resolved.

2.3.4.6.3 Conclusion

The staff reviewed the LRA, accompanying scoping boundary drawings, and RAI responses to determine whether any SSCs that should be within the scope of license renewal had not been identified by the applicant. No omissions were identified. In addition, the staff performed a review to determine whether any components that should be subject to an AMR had not been identified by the applicant. No omissions were identified. On the basis of its review, the staff concludes that the applicant adequately identified the MSS components that are within the scope of license renewal, as required by 10 CFR 54.4(a), and the MSS components that are subject to an AMR, as required by 10 CFR 54.21(a)(1).

2.3.4.7 Turbine Generator System

2.3.4.7.1 Summary of Technical Information in the Application

In LRA Section 2.3.4.7, the applicant described the turbine generator system (TGS). The TGS consists of the main turbines, main electrical generator and various supporting subsystems. The turbine is an 1,800 r/min tandem compound, three cylinder, quadruple flow, indoor unit. Saturated steam is supplied to the turbine throttle from the steam generators through four stop valves and four governing control valves. The steam flows through a double-flow, high-pressure turbine and four combination moisture separator-reheaters in parallel, and two double-flow, low-pressure turbines that exhaust to the main condenser. Turbine control is accomplished with a rapid response electrohydraulic control system. The turbine lubricating oil system supplies oil for lubricating the bearings. A bypass stream of turbine lubricating oil flows continuously through coalescing filters to remove water and other impurities. The generator is a hydrogen inner cooled unit connected directly to the turbine. It is rated for 955 MVA and has the capability to accept the gross output of the turbine at rated steam conditions. The generator is made up of a housing, stator, exciter, rotor and shaft with sleeve bearings and ventilation blower. Generator operation is supported by a hydrogen gas system, a seal oil system and a signal system. The turbine generator controls are the means by which the turbine generator is made to meet the electrical load demand placed upon it. The controls consist of the following five parts: (1)

operators interface panel, (2) digital controller & engineers console, (3) steam valve servo actuator assemblies, (4) high-pressure fluid supply, and (5) emergency trip.

The failure of NSR SSCs in the TGS could potentially prevent the satisfactory accomplishment of an SR function. The TGS also performs functions that support ATWS.

The intended function, within the scope of license renewal, is to provide fluid pressure boundary.

In LRA Table 2.3.4-7, the applicant identified the following TGS component types that are within the scope of license renewal and subject to an AMR:

- accumulators
- fasteners
- filters/strainers
- heat exchangers
- pipe & fittings
- pumps
- valves & dampers

2.3.4.7.2 Staff Evaluation

The staff reviewed LRA Section 2.3.4.7 and the FSAR using the evaluation methodology described in SER Section 2.3. The staff conducted its review in accordance with the guidance described in SRP-LR Section 2.3.

In conducting its review, the staff evaluated the system functions described in the LRA and FSAR in accordance with the requirements of 10 CFR 54.4(a) to verify that the applicant had not omitted from the scope of license renewal any components with intended functions delineated under 10 CFR 54.4(a). The staff then reviewed those components that the applicant had identified as being within the scope of license renewal to verify that the applicant had not omitted any passive and long-lived components that should be subject to an AMR in accordance with the requirements of 10 CFR 54.21(a)(1).

The staff's review of LRA Section 2.3.4.7 identified an area in which additional information was necessary to complete the review of the applicant's scoping and screening results. The applicant responded to the staff's RAI as discussed below.

In RAI 2.3.4.7-3, dated September 21, 2005, the staff noted that LRA Section 2.3.4.7 states that the system contains structures and/or components required by the CLB for ATWS. It further states that the turbine trip signal is an input to ATWS. The signal circuitry is in the TGS. The turbine stop valves are in the main steam system. FSAR Section 7.5.2.6 identifies the ATWS trip as an input to the "emergency trip 20 ET solenoid" and the emergency trip 20 ET solenoid directly releases electrohyrdraulic control (EHC) fluid to the drain; however, based on the information provided in the FSAR and the LRA, it is not clear which mechanical components (e.g., EHC components) of the turbine generator system are within the scope of license renewal in accordance with 10 CFR 54.4(a)(3) for ATWS. Therefore, the staff requested that the applicant provide information so that the staff can identify the mechanical components of the turbine generator system that are within the scope of license renewal in accordance with the requirements of the 10 CFR 54.4(a)(3) ATWS regulated event.

In its response, by letter dated October 21, 2005, the applicant stated:

> No mechanical components are credited for ATWS. Pressurizer High Pressure is the input to ATWS which results in Turbine Trip, and the referenced function note is incorrect. System Function TGS-AT in LRA Section 2.3.4.7, Page 2-198, is hereby revised to read, 'Pressurizer High Pressure is the input to ATWS which results in Turbine Trip. The signal circuitry for the ATWS turbine trip is in the Turbine Generator System. The Turbine stop valve is in the Main Steam System.'

Based on its review, the staff finds the applicant's response to RAI 2.3.4.7-3, acceptable because it adequately clarified that the "TGS-AT" system function note in LRA Section 2.3.4.7 is incorrect and that no mechanical components of the turbine generator system are within the scope of license renewal in accordance with the requirements of the ATWS regulated event of 10 CFR 54.4(a)(3). The applicant revised LRA Section 2.3.4.7 to clarify that only the signal circuitry for the ATWS turbine trip is within the TGS. Therefore, the staff's concern described in RAI 2.3.4.7-3 is resolved.

2.3.4.7.3 Conclusion

The staff reviewed the LRA, accompanying scoping boundary drawings, and RAI response to determine whether any SSCs that should be within the scope of license renewal had not been identified by the applicant. No omissions were identified. In addition, the staff performed a review to determine whether any components that should be subject to an AMR had not been identified by the applicant. No omissions were identified. On the basis of its review, the staff concludes that the applicant adequately identified the TGS components that are within the scope of license renewal, as required by 10 CFR 54.4(a), and the TGS components that are subject to an AMR, as required by 10 CFR 54.21(a)(1).

2.4 Scoping and Screening Results: Containments, Structures, and Component Supports

This section documents the staff's review of the applicant's scoping and screening results for containments, structures, and component supports. Specifically, this section discusses the following containments, structures, and component supports:

- auxiliary building
- component supports
- containment
- containment interior structures
- discharge structure
- feedwater purity building
- intake structure
- miscellaneous structural and bulk commodities
- switchyard and yard structures
- turbine building

In accordance with the requirements of 10 CFR 54.21(a)(1), the applicant identified and listed passive, long-lived SCs that are within the scope of license renewal and subject to an AMR. To verify that the applicant properly implemented its methodology, the staff focused its review on the implementation results. This approach allowed the staff to confirm that there were no omissions of containments, structures, and component supports components that meet the scoping criteria and are subject to an AMR.

Staff Evaluation Methodology. The staff's evaluation of the information provided in the LRA was performed in the same manner for all containments, structures, and component supports. The objective of the review was to determine if the components and supporting structures for a specific containment, structure, or component support, that appeared to meet the scoping criteria specified in the Rule, were identified by the applicant as within the scope of license renewal, in accordance with 10 CFR 54.4. Similarly, the staff evaluated the applicant's screening results to verify that all long-lived, passive components were subject to an AMR in accordance with 10 CFR 54.21(a)(1).

Scoping. To perform its evaluation, the staff reviewed the applicable LRA section and associated component drawings, focusing its review on components that had not been identified as within the scope of license renewal. The staff reviewed relevant licensing basis documents, including the final safety analysis report (FSAR), for each containment, structure, and component support to determine if the applicant had omitted components with intended functions delineated under 10 CFR 54.4(a) from the scope of license renewal. The staff also reviewed the licensing basis documents to determine if all intended functions delineated under 10 CFR 54.4(a) were specified in the LRA. If omissions were identified, the staff requested additional information to resolve the discrepancies.

Screening. Once the staff completed its review of the scoping results, the staff evaluated the applicant's screening results. For those containments, structures, and component supports with intended functions, the staff sought to determine: (1) if the functions are performed with moving parts or a change in configuration or properties, or (2) if they are subject to replacement based on a qualified life or specified time period, as described in 10 CFR 54.21(a)(1). For those that did not meet either of these criteria, the staff sought to confirm that these containments, structures, and component supports and components were subject to an AMR as required by 10 CFR 54.21(a)(1). If discrepancies were identified, the staff requested additional information to resolve them.

2.4.1 Auxiliary Building

2.4.1.1 Summary of Technical Information in the Application

In LRA Section 2.4.1, the applicant described the auxiliary building. The auxiliary building, with the exception of the administration area and the access control area, is a Class 1 structure. The following facilities, systems, and equipment are among those located in the auxiliary building: (1) control room; (2) emergency diesel generators and related auxiliaries; (3) new and spent fuel handling, storage and shipment facilities; (4) radwaste, chemical and volume control equipment; (5) safety injection system (majority); (6) component cooling system (majority); and (7) containment spray system (majority). The reinforced concrete enclosure containing the engineered safeguards equipment is located below grade. This building also houses the access control area, which controls access to and exit from the various radiation controlled zones. The new and spent fuel pools are located adjacent to the three-story concrete enclosure, which houses the control room, switchgear and EDGs. At elevation 590' there are electrical manholes in the 2.4kV switchgear room associated with SR bus 1-C, used for routing cables to various plant areas. The Class 1 auxiliary building radwaste addition structure houses additional gaseous, liquid and solid radwaste equipment. The building is constructed on a mat foundation. The reinforced concrete slabs and walls were designed as two-way slabs and bearing walls. During 1983, a reinforced concrete addition was appended to the north side of the auxiliary building for a technical support center (TSC), an electrical equipment room (EER), and an HVAC area. The TSC was constructed pursuant to NUREG-0696, the HVAC area as a result of the control room habitability requirements of NUREG-0737, and the EER area as a result of loads placed on the electrical system by the addition of the TSC and HVAC areas.

The auxiliary building contains SR components that are relied upon to remain functional during and following DBEs. The failure of NSR SSCs in the auxiliary building could potentially prevent the satisfactory accomplishment of an SR function. In addition, the auxiliary building performs functions that support fire protection, EQ, ATWS, and SBO.

The intended functions within the scope of license renewal include the following:

- provides spray shield or curbs for directing flow
- provides for thermal expansion and/or seismic separation
- provides rated fire barrier
- provides flood protection barrier
- provides fluid pressure boundary
- provides shielding against high-energy line breaks
- provides missile barrier
- provides pipe whip restraint
- provides shielding against radiation
- provides shelter/protection to SR components
- provides structural support to NSR components
- provides structural and/or functional support to SR components
- provides structural and/or functional support to regulated events components
- provides structural and/or functional support to SR equipment

In LRA Table 2.4.1-1, the applicant identified the following auxiliary building component types that are within the scope of license renewal and subject to an AMR:

- building framing - carbon steel, protected (column, hanger, beam, truss, decking, floor grating or plate, catwalk, threaded fastener, concrete expansion bolt, column base plate, weld, etc.)

- building framing - concrete, below grade (wall footing, foundation slab, grout, reinforcement, duct banks, cable pits, tunnels, etc.)

- building framing - concrete, exposed (foundations, concrete & masonry wall, beam, roof slab, grout, reinforcements, concrete around expansion & grouted anchors)

- building framing - concrete, protected (foundations, concrete & masonry wall, column, pedestal, beam, floor slab, grout, reinforcements, concrete around expansion & grouted anchors, cable pits, tunnels, etc.)

- flood barrier - carbon steel, exposed (water tight doors)

- flood barrier - carbon steel, protected (water tight doors and gate)

- fuel related component - carbon steel, protected (anchor bolts for SFP gates, liners, transfer tube appurtenances)

- fuel related component - stainless, protected (anchor bolts for SFP gates, liners, transfer tube appurtenances)

- fuel related component - stainless, borated (liner plates, gates, transfer tube expansion bellows)

- HELB/medium energy line break (MELB) component - carbon Steel, protected (doors, scuttle, blowout panels, floor drains & screens, guard pipes, louvers, whip restraints, bellows, spray shields, etc.)

- HELB/MELB component - concrete, protected (curbs & pipe whip restraint grout, concrete at locations of expansion & grouted anchors)

- HVAC component - stainless, protected (control room vestibule doors)

- HVAC component - carbon steel, protected (control room vestibule door)

- HVAC component - concrete, protected (control room vestibules, concrete & masonry walls, floors, ceilings)

- HVAC component - galvanized, protected (damper & louver frames)

- missile shield - carbon steel, protected (steel doors and structural steel missile barrier)

- operator access component - carbon steel, protected (stairs, floors, platforms, etc.)

- operator access component - concrete, protected (stairs, floors, platforms, concrete at locations of expansion & grouted anchors)

- operator access component - galvanized, protected (stairs, walkways, removable platform, welds, bolted connections)

2.4.1.2 Staff Evaluation

The staff reviewed LRA Section 2.4.1 and the FSAR using the evaluation methodology described in SER Section 2.4. The staff conducted its review in accordance with the guidance described in SRP-LR Section 2.4, "Scoping and Screening Results: Structures."

In conducting its review, the staff evaluated the structural component functions described in the LRA and FSAR in accordance with the requirements of 10 CFR 54.4(a) to verify that the applicant had not omitted from the scope of license renewal any components with intended functions delineated under 10 CFR 54.4(a). The staff then reviewed those components that the applicant had identified as being within the scope of license renewal to verify that the applicant had not omitted any passive and long-lived components that should be subject to an AMR in accordance with the requirements of 10 CFR 54.21(a)(1).

The staff's review of LRA Section 2.4.1 identified areas in which additional information was necessary to complete the review of the applicant's scoping and screening results. The applicant responded to the staff's RAIs as discussed below.

In RAI 2.4.1-1, dated September 14, 2005, the staff stated that the last paragraph of LRA page 2-201 refers to, "The portions of the Auxiliary Building that are in-boundary and contain components..." The staff requested that the applicant clarify the meaning of the phrase "in-boundary," as used in LRA Section 2.4, and the use of the auxiliary building as an example to explain the criteria used in defining the three dimensional "in-boundary" lines for structures that are within the scope of license renewal.

In its response, by letter dated October 14, 2005, the applicant stated:

> 'In-boundary' means 'in-scope of license renewal'. The in-scope boundaries are defined in the LRA structure description, as necessary. For example, on page 2-200, Section 2.4.1 Auxiliary Building states, 'The Auxiliary Building, with the exception of the administration area and the access control area, is a Class 1 structure.' The sentence on page 2-201 quoted in the question goes on to say, 'The portions of the Auxiliary Building that are in-boundary and contain components subject to an AMR include the original Auxiliary Building (except administrative areas), the Auxiliary Building Radwaste Addition and the Auxiliary Building TSC/EER/HVAC Addition.' This is further clarified in the scoping function S0200B-NSAS on page 2-203 where it includes the other non-class 1 access control portion of the Auxiliary Building in-scope of license renewal due to nonsafety affecting safety. As described in the function comment: 'The Access Control area of the Auxiliary Building is Consumers Design Class 3. However, concrete walls below the roof slab of Access Control Area supports the Load Distribution System (LDS) that is used to transport spent fuel assemblies. They are in-scope of 10CFR 54.4. These walls are part of the Auxiliary Building framing system which supports safety related components.'

> Thus, the only portion of the Auxiliary Building and additions that is not in-scope (inboundary) of license renewal is the administrative support area.

Based on its review, the staff finds the applicant's response to RAI 2.4.1-1 acceptable because it adequately clarified that "in-boundary" means "in-scope of license renewal." The applicant also used the auxiliary building as an example of the criteria for defining the three-dimensional "in-boundary" lines for in-scope structures and explained that the only portion of the auxiliary building and additions not within the scope of license renewal is the administrative support area. Therefore, the staff's concern described in RAI 2.4.1-1 is resolved.

In RAI 2.4.1-2, dated September 14, 2005, the staff stated that LRA Table 2.4.1-1 provides three fuel-related component entries. It is was not clear whether PNP's spent fuel racks, including their neutron absorbing material (if applicable), had been designated as items that require an AMR. If so the staff requested that the applicant indicate the location in the LRA where the items are listed. In addition, the table did not list cranes and their supporting rails as requiring an AMR. Therefore, the staff requested that the applicant provide additional information related to the scoping and screening of these components.

In its response, by letter dated October 14, 2005, the applicant stated:

> The 'Description' of each LRA Section 2.4.x provides a description of what can be found in that section. LRA Section 2.4.1, 'Auxiliary Building,' and the other

building-specific LRA sections, contain component groups associated with GALL Sections IIA and IIIA; that is, concrete and steel. The spent fuel racks, because of their component (ie, fuel) support function, are addressed on pages 2-209 and 2-222 in Section 2.4.2, 'Component Supports.' This section addresses component and piping supports for all in-scope structures. The various building cranes are addressed on pages 2-240 & 244 in Section 2.4.8, 'Miscellaneous Structures and Bulk Commodities.'

Based on its review, the staff finds the applicant's response to RAI 2.4.1-2 acceptable because the applicant adequately identified the location in the LRA where the spent fuel racks are addressed. The applicant also identified where the various building cranes are discussed. Therefore, the staff's concerns described in RAI 2.4.1-2 are resolved.

2.4.1.3 Conclusion

The staff reviewed the LRA, RAI responses, and related structural components to determine whether any SCs that should be within the scope of license renewal had not been identified by the applicant. No omissions were identified. In addition, the staff performed a review to determine whether any components that should be subject to an AMR had not been identified by the applicant. No omissions were identified. On the basis of its review, the staff concludes that the applicant adequately identified the auxiliary building components that are within the scope of license renewal, as required by 10 CFR 54.4(a), and the auxiliary building components that are subject to an AMR, as required by 10 CFR 54.21(a)(1).

2.4.2 Component Supports

2.4.2.1 Summary of Technical Information in the Application

In LRA Section 2.4.2, the applicant described the component supports commodity group. The component supports commodity group includes ASME and non-ASME pipe supports, ASME Class 1 equipment supports, as well as general mechanical and electrical component supports (e.g., HVAC, cable trays, conduits, etc.). These assets consist primarily of steel and concrete assets included under GALL Section IIIB for supports and GALL Section VII A1 and A2 for new fuel and spent fuel storage, respectively. They are grouped together based on their similar "support" function, material, environments, and expected AMPs. This commodity group includes the following categories of support components: (1) ASME Class 1 piping & mechanical component supports, (2) ASME Class 2 & 3 piping & mechanical component supports, (3) non-ASME piping & mechanical component supports, (4) electrical component supports, (5) new fuel storage rack, and (6) spent fuel storage rack-auxiliary building.

The component supports commodity group contains SR components that are relied upon to remain functional during and following DBEs. The failure of NSR SSCs in the commodity group could potentially prevent the satisfactory accomplishment of an SR function. In addition, the component supports commodity group performs functions that support fire protection, EQ, ATWS, and SBO.

The intended functions within the scope of license renewal include the following:

- provides for thermal expansion and/or seismic separation

- provides shielding against radiation
- provides structural support to NSR components
- provides structural and/or functional support to SR components
- provides structural and/or functional support to regulated events components

In LRA Table 2.4.2-1, the applicant identified the following component supports commodity group component groups that are within the scope of license renewal and subject to an AMR:

- ASME 1 support - containment - carbon steel, protected

- ASME 1 support - containment - concrete, protected

- ASME 1 support - containment - sliding material, cont cavity

- ASME 1 support - containment - sliding material, protected

- ASME 2 & 3 support - turbine (water treatment area) - carbon steel, protected

- ASME 2 & 3 support - turbine (water treatment area) - concrete, protected

- ASME Class 1 tubing support - auxiliary bldg, carbon steel, protected

- ASME Class 2 & 3 piping & mechanical component support - auxiliary bldg, carbon steel, protected

- ASME Class 2 & 3 piping & mechanical component support - auxiliary bldg, cast iron, protected

- ASME Class 2 & 3 piping & mechanical component support - auxiliary bldg, concrete, protected

- ASME Class 2 & 3 piping & mechanical component support - auxiliary bldg, galvanized, protected

- ASME Class 2 & 3 piping & mechanical component support - auxiliary bldg, sliding material, protected

- ASME Class 2 & 3 piping & mechanical component support - containment bldg, carbon steel, cont cavity

- ASME Class 2 & 3 piping & mechanical component support - containment bldg, carbon steel, protected

- ASME Class 2 & 3 piping & mechanical component support - containment bldg, concrete, protected

- ASME Class 2 & 3 piping & mechanical component support - containment bldg, galvanized, cont cavity

- ASME Class 2 & 3 piping & mechanical component support - containment bldg, galvanized, protected

- ASME Class 2 & 3 piping & mechanical component support - containment bldg, sliding material, protected

- ASME Class 2 & 3 piping & mechanical component support - containment bldg, stainless, borated

- ASME Class 2 & 3 piping & mechanical component support - intake structure bldg, carbon steel, protected
- ASME Class 2 & 3 piping & mechanical component support - intake structure bldg, concrete, protected
- ASME Class 2 & 3 piping & mechanical component support - intake structure bldg, galvanized, protected
- ASME Class 2 & 3 piping & mechanical component support - turbine bldg, carbon steel, protected
- ASME Class 2 & 3 piping & mechanical component support - turbine bldg, concrete, protected
- ASME Class 2 & 3 piping & mechanical component support - turbine bldg, galvanized, protected
- ASME Class 2 & 3 piping & mechanical component support - turbine bldg, sliding, protected
- ASME Class 2 & 3 piping & mechanical component support - yard, aluminum, exposed
- ASME Class 2 & 3 piping & mechanical component support - yard, carbon steel, below grade
- ASME Class 2 & 3 piping & mechanical component support - yard, carbon steel, exposed
- ASME Class 2 & 3 piping & mechanical component support - yard, concrete, exposed
- elec component support - auxiliary bldg, carbon steel, protected
- elec component support - auxiliary bldg, carbon steel, raw water
- elec component support - auxiliary bldg, concrete, protected
- elec component support - auxiliary bldg, concrete, raw water
- elec component support - auxiliary bldg, galvanized, protected
- elec component support - auxiliary bldg, galvanized, raw water
- elec component support - containment bldg, carbon steel, containment cavity
- elec component support - containment bldg, carbon steel, protected
- elec component support - containment bldg, concrete, protected
- elec component support - containment bldg, galvanized, containment cavity
- elec component support - containment bldg, galvanized, protected
- elec component support - discharge structure, carbon steel, protected
- elec component support - discharge structure, galvanized, protected
- elec component support - intake structure bldg, carbon steel, protected
- elec component support - intake structure bldg, concrete, protected
- elec component support - intake structure bldg, galvanized, protected

- elec component support - switch yard relay house group bldg, carbon steel, protected
- elec component support - switch yard relay house group bldg, concrete, protected
- elec component support - switch yard relay house group bldg, galvanized, protected
- elec component support - turbine bldg, carbon steel, protected
- elec component support - turbine bldg, concrete, protected
- elec component support - turbine bldg, galvanized, protected
- elec component support - yard, carbon steel, exposed
- elec component support - yard, concrete, exposed
- elec component support - yard, galvanized, exposed
- elec component support - yard, galvanized, raw water
- high strength bolting - containment building, carbon steel, protected
- non-ASME component support - auxiliary bldg, aluminum, protected
- non-ASME piping & mechanical component support - auxiliary bldg, carbon steel, protected
- non-ASME piping & mechanical component support - auxiliary bldg, concrete, protected
- non-ASME piping & mechanical component support - auxiliary bldg, galvanized, protected
- non-ASME piping & mechanical component support - auxiliary bldg, stainless, borated
- non-ASME piping & mechanical component support - boiler building, carbon steel, protected
- non-ASME piping & mechanical component support - boiler building, concrete, protected
- non-ASME piping & mechanical component support - containment bldg, aluminum, protected
- non-ASME piping & mechanical component support - containment bldg, carbon steel, containment cavity
- non-ASME piping & mechanical component support - containment bldg, carbon steel, protected
- non-ASME piping & mechanical component support - containment bldg, concrete, protected
- non-ASME piping & mechanical component support - containment bldg, galvanized, containment cavity
- non-ASME piping & mechanical component support - containment bldg, galvanized, protected
- non-ASME piping & mechanical component support - containment bldg, stainless, borated

- non-ASME piping & mechanical component support - discharge structure, carbon steel, protected
- non-ASME piping & mechanical component support - discharge structure, cast iron, protected
- non-ASME piping & mechanical component support - discharge structure, concrete, protected
- non-ASME piping & mechanical component support - discharge structure, galvanized, protected
- non-ASME piping & mechanical component support - feedwater purity bldg, carbon steel, protected
- non-ASME piping & mechanical component support - feedwater purity bldg, concrete, protected
- non-ASME piping & mechanical component support - intake structure bldg, carbon steel, protected
- non-ASME piping & mechanical component support - intake structure bldg, concrete, protected
- non-ASME piping & mechanical component support - intake structure bldg, galvanized, protected
- non-ASME piping & mechanical component support - turbine bldg, carbon steel, protected
- non-ASME piping & mechanical component support - turbine bldg, concrete, protected
- non-ASME piping & mechanical component support - turbine bldg, galvanized, protected
- non-ASME piping & mechanical component support - water treatment bldg, carbon steel, protected
- non-ASME piping & mechanical component support - water treatment bldg, concrete, protected
- non-ASME piping & mechanical component support - yard, carbon steel, exposed
- non-ASME piping & mechanical component support - yard, concrete, exposed
- non-ASME piping & mechanical component support - yard, galvanized, exposed
- spent fuel storage rack - auxiliary building, boron carbide, borated water
- spent fuel storage rack - auxiliary building, stainless steel, borated water

2.4.2.2 Staff Evaluation

The staff reviewed LRA Section 2.4.2 and the FSAR using the evaluation methodology described in SER Section 2.4. The staff conducted its review in accordance with the guidance described in SRP-LR Section 2.4.

In conducting its review, the staff evaluated the structural component functions described in the LRA and FSAR in accordance with the requirements of 10 CFR 54.4(a) to verify that the applicant had not omitted from the scope of license renewal any components with intended functions delineated under 10 CFR 54.4(a). The staff then reviewed those components that the applicant had identified as being within the scope of license renewal to verify that the applicant had not omitted any passive and long-lived components that should be subject to an AMR in accordance with the requirements of 10 CFR 54.21(a)(1).

The staff's review of LRA Section 2.4.2 identified an area in which additional information was necessary to complete the review of the applicant's scoping and screening results. The applicant responded to the staff's RAI as discussed below.

In RAI 2.4.2-1, dated September 14, 2005, the staff stated that a review of LRA Table 2.4.2-1 indicated that the applicant had not identified any pipe (penetration) passing through the containment as Class MC piping, as allowed by Figure NE-1120-1 of the ASME Code, Section III, Subsection NE. Therefore, the staff requested that the applicant verify that all Class MC pipe supports are within the scope of license renewal.

In its response, by letter dated October 14, 2005, the applicant stated:

> ASME Code Section III Subsection NE Article NE-1000 establishes the rules for material, design, etc. for metal containment vessels. Palisades' containment is steel lined, prestressed concrete, and, therefore, does not have Class MC piping or Class MC piping supports. See LRA Section 2.4.3 Containment description.
>
> Unlike Figure NE 1120-1 - 'Typical Containment Penetrations', that shows penetration sleeves for piping as Class MC, since they are welded to and supported by the steel liner, the Palisades penetration sleeves are anchored into the containment concrete. Refer to FSAR Figure 5.8-2 - for 'Containment Structure Typical Penetration Piping' and LRA Section 3.5.2.2.1.7, page 3-280. Piping, HVAC, and electrical penetrations are designed, fabricated, inspected, and installed in accordance with the ASME B&PV Code, Section III, Subsection B (see LRA page 4-48 and FSAR 5.8.6.3.2).

Based on its review, the staff finds the applicant's response to RAI 2.4.2-1 acceptable because the applicant explained that PNP does not have Class MC piping that penetrates through the containment. Therefore, the staff's concern described in RAI 2.4.2-1 is resolved.

2.4.2.3 Conclusion

The staff reviewed the LRA, RAI response, and related structural components to determine whether any SSCs that should be within the scope of license renewal had not been identified by the applicant. No omissions were identified. In addition, the staff performed a review to determine whether any components that should be subject to an AMR had not been identified by the applicant. No omissions were identified. On the basis of its review, the staff concludes that the applicant adequately identified the component supports commodity group components that are within the scope of license renewal, as required by 10 CFR 54.4(a), and the component supports commodity group components that are subject to an AMR, as required by 10 CFR 54.21(a)(1).

2.4.3 Containment

2.4.3.1 Summary of Technical Information in the Application

In LRA Section 2.4.3, the applicant described the containment structure. The containment structure consists of a post tensioned, reinforced concrete cylinder and dome connected to and supported by a massive reinforced concrete foundation slab. The containment structure is a Class 1 structure. The majority of the interior surface of the containment pressure boundary structure is lined with 1/4" thick welded, steel plate to ensure a high degree of leak tightness. Numerous mechanical and electrical systems penetrate the containment wall through steel penetrations that are welded to the containment liner plate. None of the mechanical or electrical penetrations require expansion joints. The post tensioning system consists of: (1) three groups of 55 dome tendons oriented at 120 degrees to each other for a total of 165 tendons anchored at the vertical face of the dome ring girder; (2) 178 vertical tendons anchored at the top surface of the ring girder and at the bottom of the base slab; and (3) six groups of hoop tendons enclosing 120 degrees of arc for a total of 502 tendons anchored at the 6 vertical buttresses.

The containment structure contains SR components that are relied upon to remain functional during and following DBEs. The failure of NSR SSCs in the containment structure could potentially prevent the satisfactory accomplishment of an SR function. In addition, the containment structure performs functions that support fire protection, EQ, ATWS, and SBO.

The intended functions within the scope of license renewal include the following:

- provides heat sink during SBO or design-basis accidents
- provides missile barrier
- provides pressure boundary or fission product retention barrier
- provides shielding against radiation
- provides shelter/protection to SR components
- provides structural and/or functional support to SR components
- provides structural and/or functional support to regulated events components

In LRA Table 2.4.3-1, the applicant identified the following containment structure component types that are within the scope of license renewal and subject to an AMR:

- containment shell & base slab - containment bldg - carbon steel, protected (air locks, equipment hatch, liner plate, penetrations)

- containment shell & base slab - containment bldg - concrete, below grade (base mat, foundation, wall, embedded steel, etc.)

- containment shell & base slab - containment bldg - concrete, exposed (dome, wall, base mat, embedded steel, etc.)

- containment shell & base slab - containment bldg - elastomer, protected (seals, gaskets, moisture barriers)

- containment shell & base slab - containment bldg - stainless steel, protected (fuel transfer tube, closure flange)

- containment shell prestressing system - containment bldg - carbon steel, exposed (includes tendons and anchorage components)

2.4.3.2 Staff Evaluation

The staff reviewed LRA Section 2.4.3 and the FSAR using the evaluation methodology described in SER Section 2.4. The staff conducted its review in accordance with the guidance described in SRP-LR Section 2.4.

In conducting its review, the staff evaluated the structural component functions described in the LRA and FSAR in accordance with the requirements of 10 CFR 54.4(a) to verify that the applicant had not omitted from the scope of license renewal any components with intended functions delineated under 10 CFR 54.4(a). The staff then reviewed those components that the applicant had identified as being within the scope of license renewal to verify that the applicant had not omitted any passive and long-lived components that should be subject to an AMR in accordance with the requirements of 10 CFR 54.21(a)(1).

2.4.3.3 Conclusion

The staff reviewed the LRA and related structural components to determine whether any SSCs that should be within the scope of license renewal had not been identified by the applicant. No omissions were identified. In addition, the staff performed a review to determine whether any components that should be subject to an AMR had not been identified by the applicant. No omissions were identified. On the basis of its review, the staff concludes that the applicant adequately identified the containment structure components that are within the scope of license renewal, as required by 10 CFR 54.4(a), and the containment structure components that are subject to an AMR, as required by 10 CFR 54.21(a)(1).

2.4.4 Containment Interior Structures

2.4.4.1 Summary of Technical Information in the Application

In LRA Section 2.4.4, the applicant described the containment interior structures. The containment interior structures, which consist of all structural elements within the containment shell, are Class 1 structures. The principal interior concrete structures are: (1) the primary shield wall, which forms the reactor cavity; (2) two steam generator compartments; (3) a refueling pool which is located between the steam generator compartments and above the reactor cavity; (4) an enclosed sump under the reactor cavity; and (5) major equipment supports including the steam generator pedestals. The primary shield wall (bioshield) is essentially a circular cylinder, lined with 0.25" steel plate, with concrete ranging in thickness from 7' to 8', with the inner 10" thickness acting as a sacrificial shield. The sacrificial shield is not reinforced, except for two horizontal "bands/hoops" of reinforcing steel, and is considered non-structural, non-load bearing concrete. The steam generator compartment walls form the secondary shield walls around the primary coolant loops. Pipe whip restraints are provided for the primary coolant pump suction, main steam, feedwater, and other high-pressure piping. Pipe whip barriers are also provided.

The containment interior structures contain SR components that are relied upon to remain functional during and following DBEs. The failure of NSR SSCs in the containment interior structures could potentially prevent the satisfactory accomplishment of an SR function. In addition, the containment interior structures perform functions that support, EQ, ATWS, and SBO.

The intended functions within the scope of license renewal include the following:

- provides spray shield or curbs for directing flow
- provides flood protection barrier
- provides heat sink during SBO or design-basis accidents
- provides shielding against high-energy line breaks
- provides missile barrier
- provides pipe whip restraint
- provides shielding against radiation
- provides structural support to NSR components
- provides structural and/or functional support to SR components
- provides structural and/or functional support to regulated events components

In LRA Table 2.4.4-1, the applicant identified the following containment interior structures component types that are within the scope of license renewal and subject to an AMR:

- building framing - carbon steel, protected (column, beam, truss, platform, floor grating or plate, catwalk, bracing, threaded fastener, concrete expansion bolt, column base plate, welds, etc.)

- building framing - concrete, containment cavity (reactor shield walls including reinforcements, inserts, grouted anchors)

- building framing - concrete, protected (concrete & masonry wall, column, pedestal, beam, slab, grout, reinforcements, concrete around expansion & grouted anchors, etc.)

- fuel related component - stainless, protected (refueling cavity liner, containment sump liner and screen, transfer tube)

- HELB/MELB component - carbon steel, protected (steel curbs, pipe whip restraints, spray shields, etc.)

- HELB/MELB component - concrete, protected (curbs, sump, pipe whip restraint, grout, concrete at locations of expansion & grouted anchors, etc.)

- HVAC component - carbon steel, protected (damper & louver mounting frames)

- HVAC component - galvanized, protected (damper & louver mounting frames)

- missile shield - concrete, containment cavity (removable missile shield above reactor vessel)

2.4.4.2 Staff Evaluation

The staff reviewed LRA Section 2.4.4 and the FSAR using the evaluation methodology described in SER Section 2.4. The staff conducted its review in accordance with the guidance described in SRP-LR Section 2.4.

In conducting its review, the staff evaluated the structural component functions described in the LRA and FSAR in accordance with the requirements of 10 CFR 54.4(a) to verify that the applicant had not omitted from the scope of license renewal any components with intended functions delineated under 10 CFR 54.4(a). The staff then reviewed those components that the applicant had identified as being within the scope of license renewal to verify that the applicant had not omitted any passive and long-lived components that should be subject to an AMR in accordance with the requirements of 10 CFR 54.21(a)(1).

2.4.4.3 Conclusion

The staff reviewed the LRA and related structural components to determine whether any SSCs that should be within the scope of license renewal had not been identified by the applicant. No omissions were identified. In addition, the staff performed a review to determine whether any components that should be subject to an AMR had not been identified by the applicant. No omissions were identified. On the basis of its review, the staff concludes that the applicant adequately identified the containment interior structures components that are within the scope of license renewal, as required by 10 CFR 54.4(a), and the containment interior structures components that are subject to an AMR, as required by 10 CFR 54.21(a)(1).

2.4.5 Discharge Structure

2.4.5.1 Summary of Technical Information in the Application

In LRA Section 2.4.5, the applicant described the discharge structure. The discharge structure is an NSR concrete structure with retaining walls that includes the mixing basin and make-up basin. The SR and NSR SWS piping combine into a common 24" discharge header. This discharge header runs underground, splits into two 24" headers, and discharges into the discharge structure through the north and south concrete walls. The discharge structure can also be used as a backup source of service water supply. In the event that water is not available to the intake structure due to a collapse of the intake crib or a similar type failure, approximately 17,000 gpm (gallons per minute) of flow to the intake structure may be supplied from the mixing basin or makeup basin via the warm water recirculation pump P-5, if available. The warm water recirculation pump's capability to provide water to the intake structure is an original design feature installed to mitigate circulating water system icing, and is not intended or required to provide an SR SWS makeup capability. Due to similarities in materials and environment, the intake crib is included with the discharge structure scoping. The intake crib is an NSR structure that functionally supports the ultimate heat sink function of maintaining adequate level in the Intake structure.

The failure of NSR SSCs in the discharge structure could potentially prevent the satisfactory accomplishment of an SR function.

The intended function, within the scope of license renewal, is to provide structural support to NSR components.

In LRA Table 2.4.5-1, the applicant identified the following discharge structure component types that are within the scope of license renewal and subject to an AMR:

- building framing - cast iron, raw water (sluice gates)

- building framing - discharge/intake crib - carbon steel, raw water (sluice gate, column, bracket, beam, bracing, threaded fastener, connector, weld, etc.)

- building framing - concrete, below grade (wall, footing, foundation, slab, grout, reinforcement, trenches, cable pits, tunnels, etc.)

- building framing - concrete, exposed (foundations, masonry & concrete wall, floor/roof slab, grout, reinforcements, steel shapes, concrete around expansion & grouted anchors)

- building framing - concrete, raw water (wall, footing, foundation, slab, grout, reinforcements)

2.4.5.2 Staff Evaluation

The staff reviewed LRA Section 2.4.5 and the FSAR using the evaluation methodology described in SER Section 2.4. The staff conducted its review in accordance with the guidance described in SRP-LR Section 2.4.

In conducting its review, the staff evaluated the structural component functions described in the LRA and FSAR in accordance with the requirements of 10 CFR 54.4(a) to verify that the applicant had not omitted from the scope of license renewal any components with intended functions delineated under 10 CFR 54.4(a). The staff then reviewed those components that the applicant had identified as being within the scope of license renewal to verify that the applicant had not omitted any passive and long-lived components that should be subject to an AMR in accordance with the requirements of 10 CFR 54.21(a)(1).

2.4.5.3 Conclusion

The staff reviewed the LRA and related structural components to determine whether any SSCs that should be within the scope of license renewal had not been identified by the applicant. No omissions were identified. In addition, the staff performed a review to determine whether any components that should be subject to an AMR had not been identified by the applicant. No omissions were identified. On the basis of its review, the staff concludes that the applicant adequately identified the discharge structure components that are within the scope of license renewal, as required by 10 CFR 54.4(a), and the discharge structure components that are subject to an AMR, as required by 10 CFR 54.21(a)(1).

2.4.6 Feedwater Purity Building

2.4.6.1 Summary of Technical Information in the Application

In LRA Section 2.4.6, the applicant described the feedwater purity building. The feedwater purity building (condensate and makeup demineralizer building) was constructed during the feedwater purity modification. It houses the raw water filtration system, the reverse osmosis pretreatment system, the makeup demineralizer system, various components of the condensate demineralizer system, regeneration chemicals handling system, feedwater purity service and instrument air, chemical storage, and a boiler room. Due to a concern for steam generator contamination from resin leakage and sodium release, the condensate demineralizer system has been rendered inoperable and retired in place; however, the water purity building is "in-scope" based on fire protection requirements to achieve safe shutdown. The feed water purity boiler room in the south end of the building houses a diesel fuel oil transfer pump and piping - this is the only area in this structure that is in-scope for license renewal.

The feedwater purity building performs functions that support fire protection.

The intended function, within the scope of license renewal, is to provide structural and/or functional support to regulated events components.

In LRA Table 2.4.6-1, the applicant identified the following feedwater purity building component types that are within the scope of license renewal and subject to an AMR:

- building framing - carbon steel, protected (column, beam, bracing, threaded fastener, concrete expansion bolt, column base plate, welds, etc.)

- building framing - concrete, below grade (pedestal, footing, foundation, slab, grout, reinforcement, trenches, cable pits, etc.)

- building framing - concrete, exposed (foundations, masonry/concrete wall, grout, reinforcement, concrete around expansion & grouted anchors, etc.)

- building framing - concrete, protected (foundations, masonry/concrete wall, column, pedestal, beam, slab, grout, reinforcement, concrete around expansion & grouted anchors, etc.)

2.4.6.2 Staff Evaluation

The staff reviewed LRA Section 2.4.6 and the FSAR using the evaluation methodology described in SER Section 2.4. The staff conducted its review in accordance with the guidance described in SRP-LR Section 2.4.

In conducting its review, the staff evaluated the structural component functions described in the LRA and FSAR in accordance with the requirements of 10 CFR 54.4(a) to verify that the applicant had not omitted from the scope of license renewal any components with intended functions delineated under 10 CFR 54.4(a). The staff then reviewed those components that the applicant had identified as being within the scope of license renewal to verify that the applicant had not omitted any passive and long-lived components that should be subject to an AMR in accordance with the requirements of 10 CFR 54.21(a)(1).

2.4.6.3 Conclusion

The staff reviewed the LRA and related structural components to determine whether any SSCs that should be within the scope of license renewal had not been identified by the applicant. No omissions were identified. In addition, the staff performed a review to determine whether any components that should be subject to an AMR had not been identified by the applicant. No omissions were identified. On the basis of its review, the staff concludes that the applicant adequately identified the feedwater purity building components that are within the scope of license renewal, as required by 10 CFR 54.4(a), and the feedwater purity building components that are subject to an AMR, as required by 10 CFR 54.21(a)(1).

2.4.7 Intake Structure

2.4.7.1 Summary of Technical Information in the Application

In LRA Section 2.4.7, the applicant described the intake structure. The portion of the intake structure above elevation 590 feet was designed to Consumer Powers Company (CP Co) Design Class I standards. Major items of equipment housed in the area include the three SR service water pumps, two dilution pumps, two diesel engine-driven fire pumps, and two 480V motor control centers providing electrical power to miscellaneous NSR equipment, including a motor-driven fire pump. The pump room east wall adjacent to the turbine building has a three-hour fire rating with a single three-hour fire door installed. All other walls and access have outdoor exposure with a small section common with the diesel engine-driven fire pump day tank room. Fire ratings are in excess of three hours.

The intake structure contains SR components that are relied upon to remain functional during and following DBEs. The failure of NSR SSCs in the intake structure could potentially prevent the satisfactory accomplishment of an SR function. In addition, the intake structure performs functions that support fire protection.

The intended functions within the scope of license renewal include the following:

- provides spray shield or curbs for directing flow
- provides rated fire barrier
- provides flood protection barrier
- provides missile barrier
- provides shelter/protection to SR components
- provides structural and/or functional support to SR components
- provides structural and/or functional support to regulated events components

In LRA Table 2.4.7-1, the applicant identified the following intake structure component types that are within the scope of license renewal and subject to an AMR:

- building framing - carbon steel, raw water (gates, guides, and trash racks)

- building framing - cast iron, raw water (sluice gates)

- building framing - galvanized, raw water (fasteners and anchor bolts)

- building framing - concrete, below grade (wall, foundation, slab, grout, reinforcement, trenches, cable pits, tunnels, etc.)

- building framing - concrete, exposed (wall, beam, floor slab, roof slab, grout, reinforcements, grouted anchors, etc.)

- building framing - concrete, protected (wall, beam, floor slab, grout, reinforcements, grouted anchors, etc.)

- building framing - concrete, raw water (wall, column, beam, footing, foundation slab, floor slab, grout, reinforcements, etc.)

- flood barrier - concrete, protected (concrete interior wall in southeast corner)

- HELB/MELB component - carbon steel, protected (steel curbs, floor drains, shields, etc.)

- HELB/MELB component - concrete, protected (concrete/masonry walls, curbs, etc.)

2.4.7.2 Staff Evaluation

The staff reviewed LRA Section 2.4.7 and the FSAR using the evaluation methodology described in SER Section 2.4. The staff conducted its review in accordance with the guidance described in SRP-LR Section 2.4.

In conducting its review, the staff evaluated the structural component functions described in the LRA and FSAR in accordance with the requirements of 10 CFR 54.4(a) to verify that the applicant had not omitted from the scope of license renewal any components with intended functions delineated under 10 CFR 54.4(a). The staff then reviewed those components that the applicant had identified as being within the scope of license renewal to verify that the applicant had not omitted any passive and long-lived components that should be subject to an AMR in accordance with the requirements of 10 CFR 54.21(a)(1).

2.4.7.3 Conclusion

The staff reviewed the LRA and related structural components to determine whether any SSCs that should be within the scope of license renewal had not been identified by the applicant. No omissions were identified. In addition, the staff performed a review to determine whether any components that should be subject to an AMR had not been identified by the applicant. No omissions were identified. On the basis of its review, the staff concludes that the applicant adequately identified the intake structure components that are within the scope of license renewal, as required by 10 CFR 54.4(a), and the intake structure components that are subject to an AMR, as required by 10 CFR 54.21(a)(1).

2.4.8 Miscellaneous Structural and Bulk Commodities

2.4.8.1 Summary of Technical Information in the Application

In LRA Section 2.4.8, the applicant described the miscellaneous structural and bulk commodities. This group includes miscellaneous component types (assets) of various materials that were identified as requiring an AMR, but which did not fit into the other structural commodity groups. Assets in this group include: (1) building crane bridges, trolleys, girders, and rails; (2) elastomers for watertight door seals/gaskets, emergency diesel generator vibration isolator elements, flood seals, seismic/expansion gap filler, waterstop, spray shield gasket, etc; (3) fire protection related commodities (seals, fire wrap, fire doors, concrete and concrete block fire barrier floors, ceilings and walls, etc.); (4) architectural commodities (e.g., roofing systems, siding, control room ceiling, etc.); (5) soil and rip rap related to water control structures - intake crib; and (6) insulation between the main steam and feedwater piping and containment concrete at the piping penetration. For the most part, these assets consist of component types that are included under GALL Report Section IIIA6.4-a and under various portions of GALL Report Section VII, as well as non-GALL items.

The miscellaneous structural and bulk commodities contain SR components that are relied upon to remain functional during and following DBEs. The failure of NSR SSCs in the miscellaneous structural and bulk commodities could potentially prevent the satisfactory accomplishment of an

SR function. In addition, the miscellaneous structural and bulk commodities perform functions that support fire protection, EQ, and SBO.

The intended functions within the scope of license renewal include the following:

- provides spray shield or curbs for directing flow
- provides for thermal expansion and/or seismic separation
- provides rated fire barrier
- provides flood protection barrier
- provides shelter/protection to SR components
- provides structural support to NSR components
- provides structural and/or functional support to SR components
- provides structural and/or functional support to regulated events components

In LRA Table 2.4.8-1, the applicant identified the following miscellaneous structural and bulk commodities component types that are within the scope of license renewal and subject to an AMR:

- built-up roofing - auxiliary bldg - tarred, exposed
- built-up roofing - discharge structure - tarred, exposed
- built-up roofing - intake structure bldg - tarred, exposed
- built-up roofing - switch yard relay house bldg - tarred, exposed
- built-up roofing - water treatment bldg - tarred, exposed
- crane - auxiliary bldg - carbon steel, protected
- crane - containment bldg - carbon steel, protected
- crane lift device - containment bldg - carbon steel, protected
- crane support - auxiliary bldg - carbon steel, protected
- crane support - containment bldg - carbon steel, protected
- fire barrier - auxiliary bldg - carbon steel, protected
- fire barrier - auxiliary bldg - concrete, exposed
- fire barrier - auxiliary bldg - concrete, protected
- fire barrier - auxiliary bldg - fire stop, protected
- fire barrier - auxiliary bldg - fire wrap, protected
- fire barrier - containment bldg - carbon steel, protected
- fire barrier - containment bldg - concrete, exposed
- fire barrier - intake structure bldg - carbon steel, protected
- fire barrier - intake structure bldg - concrete, exposed
- fire barrier - intake structure bldg - fire stop, protected
- fire barrier - intake structure bldg - fire wrap, protected

2-144

- fire barrier - turbine bldg - carbon steel, protected
- fire barrier - turbine bldg - concrete, exposed
- fire barrier - turbine bldg - concrete, protected
- fire barrier - turbine bldg - fire stop, protected
- fire barrier - turbine bldg - fire wrap, protected
- fire barrier - turbine bldg - water treatment bldg - concrete, exposed
- fire barrier - turbine bldg - water treatment bldg - concrete, protected
- fire barrier - turbine bldg - water treatment bldg - fire stop, protected
- fire barrier - turbine bldg - water treatment bldg - fire wrap, protected
- flood barrier - auxiliary bldg - elastomer, protected
- flood barrier - turbine building - elastomer, protected
- HELB/MELB civil/ structural component - auxiliary bldg - elastomer, protected
- HELB/MELB civil/ structural component - intake structure bldg - elastomer, protected
- mechanical general component support - containment bldg - elastomer, protected
- riprap - yard - soil, submerged
- roof flashing - auxiliary bldg - galvanized, exposed
- roof flashing - intake structure bldg - galvanized, exposed
- roof flashing - switchyard relay house - galvanized, exposed
- seal, gasket or filler - auxiliary bldg - elastomer, exposed
- seal, gasket or filler - auxiliary bldg - elastomer, protected
- seal, gasket or filler - containment bldg - elastomer, protected
- seal, gasket or filler - discharge structure - elastomer, exposed
- seal, gasket or filler - discharge structure - elastomer, protected
- seal, gasket or filler - intake structure bldg - elastomer, exposed
- seal, gasket or filler - switchyard relay house/ switchgear/ safeguard bldg - elastomer, exposed
- seal, gasket or filler - turbine bldg - elastomer, exposed
- seal, gasket or filler - turbine bldg - elastomer, protected

2.4.8.2 Staff Evaluation

The staff reviewed LRA Section 2.4.8 and the FSAR using the evaluation methodology described in SER Section 2.4. The staff conducted its review in accordance with the guidance described in SRP-LR Section 2.4.

In conducting its review, the staff evaluated the structural component functions described in the LRA and FSAR in accordance with the requirements of 10 CFR 54.4(a) to verify that the applicant had not omitted from the scope of license renewal any components with intended functions delineated under 10 CFR 54.4(a). The staff then reviewed those components that the applicant had identified as being within the scope of license renewal to verify that the applicant had not omitted any passive and long-lived components that should be subject to an AMR in accordance with the requirements of 10 CFR 54.21(a)(1).

The staff's review of LRA Section 2.4.8 identified an area in which additional information was necessary to complete the review of the applicant's scoping and screening results. The applicant responded to the staff's RAI as discussed below.

In RAI 2.4.8-1, dated September 14, 2005, the staff stated that LRA Table 2.4.8-1 lists water treatment building and riprap-yard-soil components with their intended functions. The staff requested that the applicant discuss the basis for not including these components in the LRA Section 2.4 listing of structures and structural commodities. The staff also requested that the applicant explain why miscellaneous commodities (i.e., fire barrier, elastomer, seal, gasket or filler), that are used in (1) containment interior structures, (2) feed water purity building (3) component supports, and (4) other in-scope SR/NSR structures) are not specifically listed in LRA Table 2.4.8-1 as requiring an AMR.

In its response, by letter dated October 14, 2005, the applicant stated:

> The water treatment building is part of and included with the turbine building, as discussed in section 2.4.10 on page 2-251. The rip-rap is not a stand alone structure. It is a component type that protects the intake crib. The rip-rap component type is included in the Miscellaneous and Bulk Commodities group due to its uniqueness, consisting of stone, sand, and concrete filled sacks, as described on page 2-240.
>
> Miscellaneous commodities are addressed as follows:
>
> Containment Interior Structures
> The containment interior structure elastomers identified in-scope of license renewal for panels and junction boxes are gaskets; and for other structural uses are caulk, thermal expansion/seismic separation joint filler, gap and crack seal, and gaskets. These are shown in Table 2.4.8-1 on page 2-246.
> (Note : Containment pressure boundary related elastomers - hatch gaskets and base slab to containment shell gap filler are covered in Section 2.4.3) There are no fire barriers for containment interior structures since the containment is one fire area. The only containment fire barrier is the containment shell.
>
> Feedwater Purity building
> Only a small portion of the Feedwater Purity building, the boiler room, is in scope of license renewal to support piping used as a backup emergency diesel generator fuel oil source. There are no elastomers identified as in scope. Similarly, there are no fire barriers in scope because the in-scope piping is only required in the event of a fire in the Intake Structure which houses the primary fuel oil supply transfer pumps that normally supply the emergency diesel

generators. It is not necessary to postulate multiple fires in separate areas. See system function S0800-FP on LRA page 2-235.

Component Supports
No elastomers, seals, gaskets or fire barriers have been identified in scope for component supports

Other In-Scope Safety or Non-Safety Structures
The listing of structures/components in Table 2.4.8-1 includes all other in scope elastomers, seals, gaskets and fire barriers requiring AMR.

Based on its review, the staff finds the applicant's response to RAI 2.4.8-1 acceptable because it adequately clarified that the water treatment building is considered as part of the turbine building. The applicant also clarified that the rip-rap component type was covered under the miscellaneous and bulk commodities group. The applicant adequately identified miscellaneous commodities used in containment interior structures, the feed water purify building, component supports, and other in-scope SR/NSR structures that require an AMR. In addition, the applicant provided the location in the LRA where each of the miscellaneous commodities are described. Therefore, the staff's concerns described in RAI 2.4.8-1 are resolved.

2.4.8.3 Conclusion

The staff reviewed the LRA, RAI response, and related structural components to determine whether any SSCs that should be within the scope of license renewal had not been identified by the applicant. No omissions were identified. In addition, the staff performed a review to determine whether any components that should be subject to an AMR had not been identified by the applicant. No omissions were identified. On the basis of its review, the staff concludes that the applicant adequately identified the miscellaneous structural and bulk commodities components that are within the scope of license renewal, as required by 10 CFR 54.4(a), and the miscellaneous structural and bulk commodities components that are subject to an AMR, as required by 10 CFR 54.21(a)(1).

2.4.9 Switchyard and Yard Structures

2.4.9.1 Summary of Technical Information in the Application

In LRA Section 2.4.9, the applicant described the switchyard and yard structures. The switch yard relay house group includes: (1) the switch yard relay house, (2) the switchgear 1F & 1G building, and (3) the safeguards bus building. These buildings are Consumers Design Class 3 structures. The switch yard foundations group includes reinforced concrete foundations for the switch yard relay house, safeguards building, and bus 1F/1G building. It also includes foundations for startup and safeguards transformers; foundation and framing for the high voltage towers between the plant and switchyard, foundation and framing for the overhead lines from the plant to switchyard disconnect 24R2 (including the subject disconnect and takeoff towers), foundation and framing for the underground line transitioning to the overhead bus, safeguards transformer, and disconnect 24 F1. Transformer foundations are reinforced concrete slab ongrade. Take-off towers have substantial, reinforced concrete spread footings, well below grade, and piers extending to 6" above grade. The switchgear housing and bus supports also are on spread footings with piers extending through backfill to above grade.

Also included are cable trench commodities, consisting of precast reinforced concrete "Trenwa" underground utility trench system comprised of trench walls and cover, which are set on a bed of gravel, precast reinforced concrete cable pits/manholes, and underground, reinforced concrete duct bank, rigid steel or PVC conduit encased in concrete. The tank foundations include reinforced concrete foundations for primary makeup storage tank, safety injection and refueling (SIRW) tank, condensate storage tank including valve pit, demineralized water storage tank, primary system makeup storage tank, utility water storage tank, diesel generator oil storage tank, and fuel oil storage tank. The foundations for primary makeup storage tank, condensate storage tank including valve pit, demineralized water storage tank, primary system makeup storage tank, and utility water storage tank are reinforced concrete rings, with backfill compacted to 95 percent to support the tank bottoms. The SIRW tank is supported on a concrete pad on top of the reinforced concrete roof of the auxiliary building. The Class 1 diesel generator fuel oil tank is housed in a below-grade vault, constructed of reinforced concrete (2' thick floor and walls, with an 18" thick roof) for missile protection.

The switchyard and yard structures perform functions that support SBO.

The intended functions within the scope of license renewal include the following:

- provides missile barrier
- provides structural and/or functional support to SR components
- provides structural and/or functional support to regulated events components
- provides structural and/or functional support to SR equipment

In LRA Table 2.4.9-1, the applicant identified the following switchyard and yard structures component types that are within the scope of license renewal and subject to an AMR:

- building framing - safeguard bus/switchgear - carbon steel, protected (floor beam, panels, welds, threaded fasteners, concrete expansion bolt, etc.)

- building framing - switchyard - concrete, below grade (grade beam, footing, trenches, slab, grout, reinforcement, trenches, cable pits, tunnels, etc.)

- building framing - switchyard - concrete, exposed (masonry/concrete wall, grout, reinforcements, foundations, concrete around expansion & grouted anchors, bus supports, etc.)

- building framing - switchyard - concrete, protected (masonry roof bearing walls, reinforcements, concrete around expansion & grouted anchors, etc.)

- metal roofing & siding - switchgear & safeguard bus - carbon steel, exposed (flashing)

- missile shield - yard - concrete, exposed (tank vault roof, pavement over buried piping)

- tank foundations - building & yard - concrete, below grade (concrete foundation)

- tank foundations - building & yard - concrete, exposed (concrete foundation)

2.4.9.2 Staff Evaluation

The staff reviewed LRA Section 2.4.9 and the FSAR using the evaluation methodology described in SER Section 2.4. The staff conducted its review in accordance with the guidance described in SRP-LR Section 2.4.

In conducting its review, the staff evaluated the structural component functions described in the LRA and FSAR in accordance with the requirements of 10 CFR 54.4(a) to verify that the applicant had not omitted from the scope of license renewal any components with intended functions delineated under 10 CFR 54.4(a). The staff then reviewed those components that the applicant had identified as being within the scope of license renewal to verify that the applicant had not omitted any passive and long-lived components that should be subject to an AMR in accordance with the requirements of 10 CFR 54.21(a)(1).

The staff's review of LRA Section 2.4.9 identified an area in which additional information was necessary to complete the review of the applicant's scoping and screening results. The applicant responded to the staff's RAI as discussed below.

In RAI 2.4.9-1, dated September 14, 2005, the staff stated that LRA Section 2.4.9 generally describes the switchyard and yard structure components that require an AMR. The staff requested that the applicant confirm that precast reinforced concrete elements, such as manholes, rigid steel or PVC conduits, and reinforced concrete duct banks, are items that require an AMR and to list them in LRA Table 2.4.9-1. In addition, the discussion related to "Tank Foundations" on LRA page 2-248 states, "...with backfill compacted to 95% to support the tank bottoms." The staff requested that the applicant clarify the meaning of the 95 percent number in the above phrase.

In its response, by letter dated October 14, 2005, the applicant stated:

> Precast reinforced concrete elements such as manholes, rigid steel, PVC conduits, and reinforced concrete duct banks are included in Table 2.4.9-1, Component Group 'Building Framing - Switchyard - Concrete, Below Grade' on page 2-249 of the LRA. They are discussed further on page 2-248 in two places that discuss cable trench commodities (including manholes, duct bank, conduit, etc.).

> The backfill inside the tank foundation was compacted to greater than 95% maximum density as determined by the Modified Proctor Method (ASTM D1557).

Based on its review, the staff finds the applicant's response to RAI 2.4.9-1 acceptable because it adequately identified the location in the LRA where the precast reinforced concrete elements are described. The applicant also explained that the tank foundation backfill was compacted to greater than 95 percent maximum density in accordance with ASTM Standard ASTM D 1557, "Test for Moisture-Density Relations of Soils and Soil Aggregate Mixtures using 10 pound Rammer and 18-inch Drop." Therefore, the staff's concerns described in RAI 2.4.9-1 are resolved.

2.4.9.3 Conclusion

The staff reviewed the LRA, RAI response, and related structural components to determine whether any SSCs that should be within the scope of license renewal had not been identified by the applicant. No omissions were identified. In addition, the staff performed a review to determine whether any components that should be subject to an AMR had not been identified by the applicant. No omissions were identified. On the basis of its review, the staff concludes that the applicant adequately identified the switchyard and yard structures components that are within the scope of license renewal, as required by 10 CFR 54.4(a), and the switchyard and yard structures components that are subject to an AMR, as required by 10 CFR 54.21(a)(1).

2.4.10 Turbine Building

2.4.10.1 Summary of Technical Information in the Application

In LRA Section 2.4.10, the applicant described the turbine building. The turbine building houses the turbine generator, condenser, feedwater heaters, condensate and feed water pumps, turbine auxiliaries and certain of the switchgear assemblies. The north end of the turbine building provides additional shop, laboratory and office space. The following areas of the turbine building were designed to CP Co Design Class 1 standards: (1) portion of the turbine building basement forming the auxiliary feedwater pump room and (2) portion of the turbine building known as the south electrical penetration room. The remainder of the turbine building is CP Co Design Class 3. Areas include the following: control room door enclosure, other tornado missile enclosures, containment escape hatch enclosure, boiler buildings, and water treatment building. The turbine building foundation is a series of concrete spread footings and piers for each steel building column or heavy equipment location with an interlocking grade beam arrangement. A grade slab fill is then provided to complete the turbine building floor. The turbine building has a steel frame superstructure with siding and a concrete curb wall at the foundation line. In addition, the turbine building has concrete and masonry wall construction in the areas adjoining the auxiliary and containment buildings. The boiler building and boiler building addition are CP Co Class 3 structures. They house the evaporator heating boiler and heating boiler. The boiler buildings support and protect diesel fuel oil components and associated piping. The water treatment area, like the turbine building, has a steel frame superstructure with siding and a concrete curb wall at the foundation line. The foundation and floor slab are reinforced concrete.

The turbine building contains SR components that are relied upon to remain functional during and following DBEs. The failure of NSR SSCs in the turbine building could potentially prevent the satisfactory accomplishment of an SR function. In addition, the turbine building performs functions that support fire protection and SBO.

The intended functions within the scope of license renewal include the following:

- provides flood protection barrier
- provides fluid pressure boundary
- provides shielding against high-energy line breaks
- provides missile barrier
- provides pipe whip restraint
- provides shelter/protection to SR components
- provides structural support to NSR components

- provides structural and/or functional support to SR components
- provides structural and/or functional support to regulated events components

In LRA Table 2.4.10-1, the applicant identified the following turbine building component types that are within the scope of license renewal and subject to an AMR:

- building framing - boiler buildings area - carbon steel, protected (column, hanger, beam, truss, decking, platform, floor grating or plate, catwalk, threaded fastener, concrete expansion bolt, column base plate, welds, etc.)

- building framing - boiler buildings area - concrete, below grade (wall, pedestal, grade beam, footing, foundation, slab, grout, reinforcement, cable pits, tunnels, etc.)

- building framing - boiler buildings area - concrete, exposed (masonry/concrete wall, grout, reinforcements, concrete around expansion & grouted anchors, etc.)

- building framing - boiler buildings area - concrete, protected (foundations, concrete & masonry wall, column, beam, floor roof slab, grout, reinforcements, concrete around expansion & grouted anchors, etc.)

- building framing - water treatment area - concrete, protected (concrete/ masonry wall, beam, floor slab, grout, reinforcements, etc.)

- building framing - carbon steel, protected (column, beam, truss, decking, platform, floor grating or plate, catwalk, bracing, threaded fastener, concrete expansion bolt, column base plate, welds, etc.)

- building framing - concrete, below grade (wall, pedestal, grade beam, footing, foundation, slab, grout, reinforcement, cable pits, tunnels, etc.)

- building framing - concrete, exposed (foundations, masonry/concrete wall, grout, reinforcements, concrete around expansion & grouted anchors, etc.)

- building framing - concrete, protected (foundations, concrete & masonry wall, column, beam, floor/roof slab, grout, reinforcements, concrete around expansion & grouted anchors, etc.)

- building framing - water treatment area - carbon steel, protected (column, beam, bracing, threaded fastener, weld, etc.)

- building framing - water treatment area - concrete, below grade (wall, pedestal, grade beam, footing, foundation, slab, grout, reinforcement, trenches, cable pits, tunnels, etc.)

- building framing - water treatment area - concrete, exposed (foundations, wall, grout, reinforcements, concrete around expansion & grouted anchors, etc.)

- building framing - water treatment area - concrete, protected (foundations, concrete/masonry wall, beam, floor slab, grout, reinforcements, concrete around expansion & grouted anchors, etc.)

- flood barrier - carbon steel, protected (flood doors, hatch, standpipe)

- HELB/MELB component - carbon steel, protected (curbs, floor drains, pipe whip restraints, spray shields, etc.)

- HELB/MELB component - concrete, protected (concrete/masonry wall, whip restraint grout, concrete around expansion & grouted anchors)

- HVAC component - carbon steel, protected (control room vestibule door)

- HVAC component - concrete, protected (control room vestibules, concrete & masonry walls, floors, ceilings)

- missile shield - concrete, exposed (concrete and/or masonry walls protecting component cooling water room door and containment escape hatch)

- missile shield - concrete, protected (concrete/masonry walls, floor & roof protecting control room door)

- operator access component - carbon steel, protected (stairs, floors, platforms)

- operator access component - concrete, protected (stairs, floors, platforms, concrete at locations of expansion & grouted anchors, etc.)

- operator access component - galvanized, protected (stairs, floors, platforms)

2.4.10.2 Staff Evaluation

The staff reviewed LRA Section 2.4.10 and the FSAR using the evaluation methodology described in SER Section 2.4. The staff conducted its review in accordance with the guidance described in SRP-LR Section 2.4.

In conducting its review, the staff evaluated the structural component functions described in the LRA and FSAR in accordance with the requirements of 10 CFR 54.4(a) to verify that the applicant had not omitted from the scope of license renewal any components with intended functions delineated under 10 CFR 54.4(a). The staff then reviewed those components that the applicant had identified as being within the scope of license renewal to verify that the applicant had not omitted any passive and long-lived components that should be subject to an AMR in accordance with the requirements of 10 CFR 54.21(a)(1).

2.4.10.3 Conclusion

The staff reviewed the LRA and related structural components to determine whether any SSCs that should be within the scope of license renewal had not been identified by the applicant. No omissions were identified. In addition, the staff performed a review to determine whether any components that should be subject to an AMR had not been identified by the applicant. No omissions were identified. On the basis of its review, the staff concludes that the applicant adequately identified the turbine building components that are within the scope of license renewal, as required by 10 CFR 54.4(a), and the turbine building components that are subject to an AMR, as required by 10 CFR 54.21(a)(1).

2.5 Scoping and Screening Results: Electrical and Instrumentation and Controls (I&C)

This section documents the staff's review of the applicant's scoping and screening results for electrical and I&C systems. Specifically, this section discusses the following electrical and I&C systems, which are within the scope of license renewal:

- cables and terminations commodity
- containment isolation and penetration system
- control rod drive system
- neutron monitoring system
- radiation monitoring system
- reactor protective system
- station power system
- switchgear system

In accordance with the requirements of 10 CFR 54.21(a)(1), the applicant identified and listed passive, long-lived SCs that are within the scope of license renewal and subject to an AMR. To verify that the applicant properly implemented its methodology, the staff focused its review on the implementation results. This approach allowed the staff to confirm that there were no omissions of electrical and I&C system components that meet the scoping criteria and are subject to an AMR.

Staff Evaluation Methodology. The staff's evaluation of the information provided in the LRA was performed in the same manner for all electrical and I&C systems. The objective of the review was to determine if the components and supporting structures for a specific electrical and I&C system, that appeared to meet the scoping criteria specified in the Rule, were identified by the applicant as within the scope of license renewal, in accordance with 10 CFR 54.4. Similarly, the staff evaluated the applicant's screening results to verify that all long-lived, passive components were subject to an AMR in accordance with 10 CFR 54.21(a)(1).

Scoping. To perform its evaluation, the staff reviewed the applicable LRA section and associated component drawings, focusing its review on components that had not been identified as within the scope of license renewal. The staff reviewed relevant licensing basis documents, including the final safety analysis report (FSAR), for each electrical and I&C system component to determine if the applicant had omitted components with intended functions delineated under 10 CFR 54.4(a) from the scope of license renewal. The staff also reviewed the licensing basis documents to determine if all intended functions delineated under 10 CFR 54.4(a) were specified in the LRA. If omissions were identified, the staff requested additional information to resolve the discrepancies.

Screening. Once the staff completed its review of the scoping results, the staff evaluated the applicant's screening results. For those systems and components with intended functions, the staff sought to determine: (1) if the functions are performed with moving parts or a change in configuration or properties, or (2) if they are subject to replacement based on a qualified life or specified time period, as described in 10 CFR 54.21(a)(1). For those that did not meet either of these criteria, the staff sought to confirm that these electrical and I&C systems and components were subject to an AMR as required by 10 CFR 54.21(a)(1). If discrepancies were identified, the staff requested additional information to resolve them.

2.5.1 Commodity Group Descriptions

After applying the scoping and screening methodology, the applicant categorized the SCs that are subject to an AMR into passive commodity groups. In LRA Section 2.5.1, the applicant identified the SCs of the electrical and I&C systems that are subject to an AMR for license renewal. SER Sections 2.5.1.1 - 2.5.1.8 present the staff's related review findings.

In LRA Section 2.5.1, the applicant described the commodity groups. In LRA Table 2.5-1, the applicant identified the following commodity group component types that are within the scope of license renewal and subject to an AMR:

- electrical cables and connections not subject to 10 CFR 50.49 EQ requirements

- electrical cables and connections used in instrumentation circuits not subject to 10 CFR 50.49 requirements that are sensitive to reduction in conductor IR (nuclear instrumentation and radiation monitoring systems)

- electrical portions of the non-EQ electrical and I&C penetration assemblies (cables and connections)

- fuse holders

- non-segregated phase bus and connections

- high-voltage transmission conductors

- high-voltage switchyard bus and connections

- insulation for inaccessible medium-voltage (2kV to 15kV) cables and connections not subject to 10 CFR 50.49 requirements

- high-voltage insulators

The commodity groups contain SR components that are relied upon to remain functional during and following DBEs. The failure of NSR SSCs in the commodity groups could potentially prevent the satisfactory accomplishment of an SR function. In addition, the commodity groups perform functions that support fire protection, EQ, ATWS, and SBO.

The intended functions within the scope of license renewal include the following:

- provides electrical connections
- insulates and supports an electrical conductor

2.5.1.1 Electrical Cables and Connections

2.5.1.1.1 Summary of Technical Information in the Application

In LRA Section 2.5.1, the applicant stated that the same type of cables and connections are included in the following two commodity groups:

- electrical cables and connections not subject to 10 CFR 50.49 EQ requirements

- electrical cables and connections used in instrumentation circuits not subject to 10 CFR 50.49 requirements that are sensitive to reduction in insulation resistance

Primarily, the component groups differ in the AMP selected to manage component aging. The component types that comprise these two commodity groups are described below.

Insulated Cables. Insulated cables encompass all in-scope cable types used in the plant. Cable insulation material groups are assessed on the basis of material and their respective material aging characteristics. The insulated cable materials subject to aging are metal and insulation; the metals used are copper and tinned copper; and the insulation materials are various elastomers and thermoplastics.

Electrical Connections - Splices. The applicant utilizes Raychem WCSF-N heat shrink tubing for all of its field splices. Raychem heat-shrinkable tubing consists of a proprietary material stated by Raychem to be made of a "modified polyolefin" with properties similar to those of cross-linked polyethylene (XLPE) and cross-linked polyolefin (XLPO).

Electrical Connections - Terminations (Terminal Blocks and Fuse Blocks). Termination blocks and fuse blocks that are part of an active component, were excluded from further AMR. Terminal blocks used at PNP, that are not part of an active component, were identified as Buchanan (Model NQB), States (ZWM and NT types), Cinch Jones (140 Series) and Westinghouse (542-245). The LR drawings were reviewed to capture the prominent terminal block brands and models used at PNP. All terminal blocks identified and reviewed as within the scope of license renewal had phenolic insulation material.

Electrical Connections - Connectors. Electrical connectors are used to connect the cable conductors to other cables or electrical devices. The three main types of connectors are compression, fusion, and plug-in (mated) connectors. Insulated plug-in (mated) connectors were the main focus of the connector aging review. The applicant stated that detailed drawings and vendor information may not be available to identify all insulating materials used in non-EQ connectors; therefore, the identification of insulation materials on all potentially in-scope plug-in connectors (both EQ and non-EQ) began with a review of other previous AMRs with references to NUREG/CR-6412. This review provided a listing of the predominantly-used connector insulation materials in the nuclear industry (i.e., ethylene propylene diene monomer, ethylene propylene rubber (EPR), Kapton, and XLPE). The plant EQ MEL and the EQ file reports were then used to validate completeness of the generic list and identify any additional connector insulation materials used at PNP. The EQ files identified additional materials, which were added to the generic materials list.

2.5.1.1.2 Staff Evaluation

The staff reviewed LRA Section 2.5.1 and the FSAR using the evaluation methodology described in SER Section 2.5. The staff conducted its review in accordance with the guidance described in SRP-LR Section 2.5, "Scoping and Screening Results: Electrical and Instrumentation and Controls Systems."

In conducting its review, the staff evaluated the system functions described in the LRA and FSAR in accordance with the requirements of 10 CFR 54.4(a) to verify that the applicant had not omitted from the scope of license renewal any components with intended functions delineated under 10 CFR 54.4(a). The staff then reviewed those components that the applicant had identified as being within the scope of license renewal to verify that the applicant had not omitted any passive and long-lived components that should be subject to an AMR in accordance with the requirements of 10 CFR 54.21(a)(1).

The applicant evaluated the cables and connectors as commodities across system boundaries on a plant-wide basis. In LRA Section 2.5.1, the applicant stated that because of similarities across system in material, environments, and aging management strategies, the various system components that required an AMR were assigned to a commodity group. AMRs were then performed on the electrical components as commodities, regardless of system association. The staff concludes that the passive function of the cables and connectors is to conduct electricity and they are subject to an AMR.

2.5.1.1.3 Conclusion

The staff reviewed the LRA and the FSAR to determine whether any SSCs that should be within the scope of license renewal had not been identified by the applicant. No omissions were identified. In addition, the staff performed a review to determine whether any components that should be subject to an AMR had not been identified by the applicant. No omissions were identified. On the basis of its review, the staff concludes that the applicant adequately identified the electrical cables and connections components that are within the scope of license renewal, as required by 10 CFR 54.4(a), and the components that are subject to an AMR, as required by 10 CFR 54.21(a)(1).

2.5.1.2 *Electrical Portions of the Non-EQ Electrical and I&C Penetration Assemblies*

2.5.1.2.1 Summary of Technical Information in the Application

In LRA Section 2.5.1, the applicant stated that there are 37 electrical containment penetrations (canisters) at PNP. Twenty-two (22) of the electrical penetrations fall within the EQ Program and 15 electrical penetrations are non-EQ. PNP uses two types of electrical containment penetrations (canisters) manufactured by two vendors; thirty-five (35) penetrations were manufactured by Viking Industries and two were manufactured by Conax. The two manufactured by Conax are EQ electrical penetrations and 15 of the 35 electrical penetrations manufactured by Viking Industries are non-EQ electrical penetrations.

The nonelectrical portions of the Viking Industries electrical penetration assemblies support the license renewal pressure boundary or structural function and are addressed in the civil/structural area. The electrical portions of the electrical penetration assemblies that electrically support the license renewal functions are the internal cable insulation materials and the connector insulation materials. The internal cables used in the Viking electrical penetrations are:

- coaxial cable (PE)
- Rockbestos firewall cable (SR)
- Anaconda Cable (EPR)
- ceramic bushings
- fiberglass

The cable insulation materials are included in the non-EQ insulated cable review.

2.5.1.2.2 Staff Evaluation

The staff reviewed LRA Section 2.5.1 and the FSAR using the evaluation methodology described in SER Section 2.5. The staff conducted its review in accordance with the guidance described in SRP-LR Section 2.5.

In conducting its review, the staff evaluated the system functions described in the LRA and FSAR in accordance with the requirements of 10 CFR 54.4(a) to verify that the applicant had not omitted from the scope of license renewal any components with intended functions delineated under 10 CFR 54.4(a). The staff then reviewed those components that the applicant had identified as being within the scope of license renewal to verify that the applicant had not omitted any passive and long-lived components that should be subject to an AMR in accordance with the requirements of 10 CFR 54.21(a)(1).

The electrical penetrations identified by the applicant as subject to an AMR are NSR, non-EQ, and used plant-wide to conduct electrical power (voltage and current), either continuously or intermittently between two sections of the electrical/I&C circuits supplying power to various equipment in the containment. The staff reviewed these component categories against the requirements in 10 CFR 54.4(a)(1) and 10 CFR 54.4(b) and found that those categories are encompassed by the requirements. The staff reviewed information in the FSAR and finds that the applicant has correctly identified the electrical portions of the non-EQ electrical and I&C penetration assemblies that are within the scope of license renewal.

2.5.1.2.3 Conclusion

The staff reviewed the LRA and the FSAR to determine whether any SSCs that should be within the scope of license renewal had not been identified by the applicant. No omissions were identified. In addition, the staff performed a review to determine whether any components that should be subject to an AMR had not been identified by the applicant. No omissions were identified. On the basis of its review, the staff concludes that the applicant adequately identified the electrical portions of the non-EQ electrical and I&C penetration assemblies components that are within the scope of license renewal, as required by 10 CFR 54.4(a), and the components that are subject to an AMR, as required by 10 CFR 54.21(a)(1).

2.5.1.3 Fuse Holders

2.5.1.3.1 Summary of Technical Information in the Application

In LRA Section 2.5.1, the applicant stated that all of the fuse blocks used at PNP, that are not part of an active component or larger active assembly. Those identified are located inside junction boxes in a controlled environment and are not subject to any aging mechanism, including cycling more than once on refueling outage.

2.5.1.3.2 Staff Evaluation

The staff reviewed LRA Section 2.5.1 and the FSAR using the evaluation methodology described in SER Section 2.5. The staff conducted its review in accordance with the guidance described in SRP-LR Section 2.5.

In conducting its review, the staff evaluated the system functions described in the LRA and FSAR in accordance with the requirements of 10 CFR 54.4(a) to verify that the applicant had not omitted from the scope of license renewal any components with intended functions delineated under 10 CFR 54.4(a). The staff then reviewed those components that the applicant had identified as being within the scope of license renewal to verify that the applicant had not omitted any passive and long-lived components that should be subject to an AMR in accordance with the requirements of 10 CFR 54.21(a)(1).

The applicant evaluated the fuse holder commodity across system boundaries on a plant-wide basis. AMRs were then performed on the electrical components as commodities, regardless of system association. The applicant correctly indicated that the passive function of the fuse holders is to conduct electricity and that the fuse holders are subject to an AMR. The staff reviewed these component categories against the requirements in 10 CFR 54.4(a)(1) and 10 CFR 54.4(b) and found that those categories are encompassed by the requirements. The staff reviewed information in the FSAR and finds that the applicant has correctly identified the fuse holders that are within the scope of license renewal.

2.5.1.3.3 Conclusion

The staff reviewed the LRA and the FSAR to determine whether any SSCs that should be within the scope of license renewal had not been identified by the applicant. No omissions were identified. In addition, the staff performed a review to determine whether any components that should be subject to an AMR had not been identified by the applicant. No omissions were identified. On the basis of its review, the staff concludes that the applicant adequately identified the fuse holders components that are within the scope of license renewal, as required by 10 CFR 54.4(a), and the components that are subject to an AMR, as required by 10 CFR 54.21(a)(1).

2.5.1.4 Non-Segregated Phase Bus and Connections

2.5.1.4.1 Summary of Technical Information in the Application

In LRA Section 2.5.1, the applicant stated that the non-segregated phase bus, supporting the SBO restoration path, is within the scope of license renewal. The structural hardware and enclosure housing components are addressed in the civil/structural area. The Unibus Inc. metal-enclosed bus contains the following vital electrical components, which provide the primary electrical conductivity and insulation functions:

- copper bus
- glass-reinforced polyester blocks (track-resistant)
- porcelain bus sleeve, or ceramic and fiberglass
- bus mounting boots

The non-segregated bus is provided with uniform heating and temperature control to prevent condensation. The contact surfaces of all electrical joints (connections) are either silver or tin-plated. Per design, no electrical joint compounds are used or required and no re-torquing of properly installed conductor connecting hardware is required on a routine basis. Flexible ground continuity connections are provided at bus expansion joints. The Unibus, Inc. metal enclosed bus system has been designed to eliminate components or consumables that might require periodic or routine replacement.

2.5.1.4.2 Staff Evaluation

The staff reviewed LRA Section 2.5.1 and the FSAR using the evaluation methodology described in SER Section 2.5. The staff conducted its review in accordance with the guidance described in SRP-LR Section 2.5.

In conducting its review, the staff evaluated the system functions described in the LRA and FSAR in accordance with the requirements of 10 CFR 54.4(a) to verify that the applicant had not omitted from the scope of license renewal any components with intended functions delineated under 10 CFR 54.4(a). The staff then reviewed those components that the applicant had identified as being within the scope of license renewal to verify that the applicant had not omitted any passive and long-lived components that should be subject to an AMR in accordance with the requirements of 10 CFR 54.21(a)(1).

The non-segregated phase buses were correctly identified by the applicant and are used to conduct electrical power (voltage and current), either continuously or intermittently between various equipment and components. The staff reviewed these component categories against the requirements in 10 CFR 54.4(a)(1) and 10 CFR 54.4(b) and found that those categories are encompassed by the requirements. The staff reviewed information in the FSAR and finds that the applicant has correctly identified the non-segregated phase bus and connections that are within the scope of license renewal.

2.5.1.4.3 Conclusion

The staff reviewed the LRA and the FSAR to determine whether any SSCs that should be within the scope of license renewal had not been identified by the applicant. No omissions were identified. In addition, the staff performed a review to determine whether any components that should be subject to an AMR had not been identified by the applicant. No omissions were identified. On the basis of its review, the staff concludes that the applicant adequately identified the non-segregated phase bus and connections components that are within the scope of license renewal, as required by 10 CFR 54.4(a), and the components that are subject to an AMR, as required by 10 CFR 54.21(a)(1).

2.5.1.5 High-Voltage Transmission Conductors

2.5.1.5.1 Summary of Technical Information in the Application

In LRA Section 2.5.1, the applicant stated that the switchyard is not included in the plant equipment defined in the CLB for mitigation of an SBO event; however, certain equipment, which provides offsite power for restoration from an SBO, is to be included within the scope of license renewal. Therefore, to be consistent with the guidance in ISG-2, the applicant included selected PNP equipment associated with the switchyard and the two qualified offsite power circuits, that can provide offsite power to the SR buses 1C and 1D following an SBO event, as within the scope of license renewal. The equipment brought within the scope of license renewal is addressed as a commodity, which consists of the following component types:

- all in-scope 345 kV station transmission cables are "Bluebird" 2, 156 MCM, 84/19 84/19 aluminum conductor steel reinforced (ACSR) and are constructed of aluminum and steel

- the transmission conductor between the switchyard bus and transformers or tie breakers

2.5.1.5.2 Staff Evaluation

The staff reviewed LRA Section 2.5.1 and the FSAR using the evaluation methodology described in SER Section 2.5. The staff conducted its review in accordance with the guidance described in SRP-LR Section 2.5.

In conducting its review, the staff evaluated the system functions described in the LRA and FSAR in accordance with the requirements of 10 CFR 54.4(a) to verify that the applicant had not omitted from the scope of license renewal any components with intended functions delineated under 10 CFR 54.4(a). The staff then reviewed those components that the applicant had identified as being within the scope of license renewal to verify that the applicant had not omitted any passive and long-lived components that should be subject to an AMR in accordance with the requirements of 10 CFR 54.21(a)(1).

The applicant correctly identified that the intended function of high-voltage transmission conductors is to provide electrical connections to specified sections of an electrical circuit to deliver voltage and current. The staff reviewed these component categories against the requirements in 10 CFR 54.4(a)(1) and 10 CFR 54.4(b) and found that those categories are encompassed by the requirements. The staff reviewed information in the FSAR and finds that the applicant has correctly identified the high-voltage transmission conductors that are within the scope of license renewal.

2.5.1.5.3 Conclusion

The staff reviewed the LRA and the FSAR to determine whether any SSCs that should be within the scope of license renewal had not been identified by the applicant. No omissions were identified. In addition, the staff performed a review to determine whether any components that should be subject to an AMR had not been identified by the applicant. No omissions were identified. On the basis of its review, the staff concludes that the applicant adequately identified the high-voltage transmission conductors components that are within the scope of license renewal, as required by 10 CFR 54.4(a), and the components that are subject to an AMR, as required by 10 CFR 54.21(a)(1).

2.5.1.6 High-Voltage Switchyard Bus and Connections

2.5.1.6.1 Summary of Technical Information in the Application

In LRA Section 2.5.1, the applicant described the high-voltage bus and connections. The 345 kV switchyard bus is 5" and 3 ½" aluminum alloy 6061-T6, schedule 40, seamless pipe. The aluminum bus bolted connection hardware is aluminum or stainless steel.

2.5.1.6.2 Staff Evaluation

The staff reviewed LRA Section 2.5.1 and the FSAR using the evaluation methodology described in SER Section 2.5. The staff conducted its review in accordance with the guidance described in SRP-LR Section 2.5.

In conducting its review, the staff evaluated the system functions described in the LRA and FSAR in accordance with the requirements of 10 CFR 54.4(a) to verify that the applicant had not omitted from the scope of license renewal any components with intended functions delineated under 10 CFR 54.4(a). The staff then reviewed those components that the applicant had identified as being within the scope of license renewal to verify that the applicant had not omitted any passive and long-lived components that should be subject to an AMR in accordance with the requirements of 10 CFR 54.21(a)(1).

The applicant correctly identified that the intended function of the high-voltage switchyard bus and connections is to provide electrical connections to specified sections of an electrical circuit to deliver voltage and current. The staff reviewed these component categories against the requirements in 10 CFR 54.4(a)(1) and 10 CFR 54.4(b) and found that those categories are encompassed by the requirements. The staff reviewed information in the FSAR and finds that the applicant has correctly identified the high-voltage switchyard bus and connections that are within the scope of license renewal.

2.5.1.6.3 Conclusion

The staff reviewed the LRA and the FSAR to determine whether any SSCs that should be within the scope of license renewal had not been identified by the applicant. No omissions were identified. In addition, the staff performed a review to determine whether any components that should be subject to an AMR had not been identified by the applicant. No omissions were identified. On the basis of its review, the staff concludes that the applicant adequately identified the high-voltage switchyard bus and connections components that are within the scope of license renewal, as required by 10 CFR 54.4(a), and the components that are subject to an AMR, as required by 10 CFR 54.21(a)(1).

2.5.1.7 Insulation for Inaccessible Medium-Voltage (2kV to 15kV) Cables and Connections

2.5.1.7.1 Summary of Technical Information in the Application

In LRA Section 2.5.1, the applicant described the inaccessible medium-voltage cables, including those in the SBO restoration path, that are within the scope of license renewal. These inaccessible medium-voltage cables were included with the non-EQ insulated cable commodity for review. Insulation materials are various elastomers and polymers.

2.5.1.7.2 Staff Evaluation

The staffs review of this commodity is included in SER Section 2.5.1.1.

2.5.1.7.3 Conclusion

The staff reviewed the LRA and the FSAR to determine whether any SSCs that should be within the scope of license renewal had not been identified by the applicant. No omissions were identified. In addition, the staff performed a review to determine whether any components that should be subject to an AMR had not been identified by the applicant. No omissions were identified. On the basis of its review, the staff concludes that the applicant adequately identified the insulation for inaccessible medium-voltage (2kV to 15kV) cables and connections components that are within the scope of license renewal, as required by 10 CFR 54.4(a), and the components that are subject to an AMR, as required by 10 CFR 54.21(a)(1).

2.5.1.8 High-Voltage Insulators

2.5.1.8.1 Summary of Technical Information in the Application

LRA Section 2.5.1, the applicant described high-voltage insulators as components that are used to insulate and support an electrical conductor. The high-voltage cable insulators are 5-3/4" x 10", 25,000 lbs. M&E strength, ASA 52-5 suspension insulators with a porcelain electrical insulation material, Portland cement, on a steel post. The switchyard bus post insulators are made of porcelain and cement.

2.5.1.8.2 Staff Evaluation

The staff reviewed LRA Section 2.5.1 and the FSAR using the evaluation methodology described in SER Section 2.5. The staff conducted its review in accordance with the guidance described in SRP-LR Section 2.5.

In conducting its review, the staff evaluated the system functions described in the LRA and FSAR in accordance with the requirements of 10 CFR 54.4(a) to verify that the applicant had not omitted from the scope of license renewal any components with intended functions delineated under 10 CFR 54.4(a). The staff then reviewed those components that the applicant had identified as being within the scope of license renewal to verify that the applicant had not omitted any passive and long-lived components that should be subject to an AMR in accordance with the requirements of 10 CFR 54.21(a)(1).

The applicant correctly identified that the intended function of the high-voltage insulators is to insulate and support an electrical conductor. The staff reviewed these component categories against the requirements in 10 CFR 54.4(a)(1) and 10 CFR 54.4(b) and found that those categories are encompassed by the requirements. The staff reviewed information in the FSAR and finds that the applicant has correctly identified the high-voltage insulators that are within the scope of license renewal.

2.5.1.8.3 Conclusion

The staff reviewed the LRA and the FSAR to determine whether any SSCs that should be within the scope of license renewal had not been identified by the applicant. No omissions were identified. In addition, the staff performed a review to determine whether any components that should be subject to an AMR had not been identified by the applicant. No omissions were identified. On the basis of its review, the staff concludes that the applicant adequately identified the high-voltage insulators components that are within the scope of license renewal, as required by 10 CFR 54.4(a), and the components that are subject to an AMR, as required by 10 CFR 54.21(a)(1).

2.6 Conclusion for Scoping and Screening

The staff reviewed the information in LRA Section 2, "Scoping and Screening Methodology for Identifying Structures and Components Subject to Aging Management Review, and Implementation Results." The staff determines that the applicant's scoping and screening methodology is consistent with the requirements of 10 CFR 54.21(a)(1) and the staff's position on the treatment of SR and NSR SSCs within the scope of license renewal and the SCs requiring an AMR is consistent with the requirements of 10 CFR 54.4 and 10 CFR 54.21(a)(1).

On the basis of its review, the staff concludes that the applicant has adequately identified those systems and components that are within the scope of license renewal, as required by 10 CFR 54.4(a), and those systems and components that are subject to an AMR, as required by 10 CFR 54.21(a)(1).

With regard to these matters, the staff concludes that there is reasonable assurance that the activities authorized by the renewed license will continue to be conducted in accordance with the CLB, and any changes made to the CLB, in order to comply with 10 CFR 54.21(a)(1), are in accordance with the NRC's regulations.

SECTION 3

AGING MANAGEMENT REVIEW RESULTS

This section of the safety evaluation report (SER) contains the evaluation of aging management programs (AMPs) and aging management reviews (AMRs) for Palisades Nuclear Plant (PNP), by the staff of the U.S. Nuclear Regulatory Commission (NRC or the staff). In Appendix B of its license renewal application (LRA), Nuclear Management Company, LLC (NMC or the applicant) described the 24 AMPs it relies on to manage or monitor the aging of passive and long-lived structures and components (SCs).

In LRA Section 3, the applicant provided the results of the AMRs for those SCs that were identified in LRA Section 2 as within the scope of license renewal and subject to an AMR.

3.0 Applicant's Use of the Generic Aging Lessons Learned Report

In preparing its LRA, the applicant credited U.S. Nuclear Regulatory Commission Regulatory Guide (NUREG)-1801, "Generic Aging Lessons Learned (GALL) Report," dated July 2001. The GALL Report contains the staff's generic evaluation of the existing plant programs and documents the technical basis for determining where existing programs are adequate without modification, and where existing programs should be augmented for the period of extended operation. The evaluation results documented in the GALL Report indicate that many of the existing programs are adequate to manage the aging effects for particular structures or components for license renewal without change. The GALL Report also contains recommendations on specific areas for which existing programs should be augmented for license renewal. An applicant may reference the GALL Report in its LRA to demonstrate that the programs at its facility correspond to those reviewed and approved in the GALL Report.

The purpose of the GALL Report is to provide the staff with a summary of staff-approved AMPs to manage or monitor the aging of SCs subject to an AMR. If an applicant commits to implementing these staff-approved AMPs, the time, effort, and resources used to review an applicant's LRA will be greatly reduced, thereby improving the efficiency and effectiveness of the license renewal review process. The GALL Report also serves as a reference for applicants and staff reviewers to quickly identify those AMPs and activities that the staff has determined will adequately manage or monitor aging during the period of extended operation.

The GALL Report identifies: (1) systems, structures, and components (SSCs); (2) SC materials; (3) the environments to which the SCs are exposed; (4) the aging effects associated with the materials and environments; (5) the AMPs credited with managing or monitoring the aging effects; and (6) recommendations for further applicant evaluations of aging management for certain component types.

The staff performed its review in accordance with the requirements of Title 10, Part 54, of the *Code of Federal Regulations* (10 CFR Part 54), "Requirements for Renewal of Operating Licenses for Nuclear Power Plants," the guidance provided in NUREG-1800, "Standard Review Plan for Review of License Renewal Applications for Nuclear Power Plant," (SRP-LR) dated July 2001, and the guidance provided in the GALL Report.

In addition to its review of the LRA, the staff conducted an onsite audit of selected AMRs and associated AMPs during the weeks of August 1 and June 20, 2005, respectively, as described in the "Audit and Review Report for Plant Aging Management Reviews and Programs for the Palisades Nuclear Plant License Renewal Application" (Audit and Review Report), dated October 20, 2005. The onsite audits and reviews are designed to maximize the efficiency of the staff's review of the LRA. The need for formal correspondence between the staff and the applicant is reduced and the result is an improvement in the review's efficiency. Also, the applicant could respond to questions and the staff could readily evaluate the applicant's responses.

3.0.1 Format of the License Renewal Application

The applicant submitted an application that followed the standard LRA format, as agreed to between the NRC and the Nuclear Energy Institute (NEI) (see letter dated April 7, 2003). This revised LRA format incorporates lessons learned from the staff's reviews of the previous five LRAs. These previous applications used a format developed from information gained during a staff and NEI demonstration project conducted to evaluate the use of the GALL Report in the staff's review process.

The organization of LRA Section 3 parallels Chapter 3 of the SRP-LR. The AMR results information in LRA Section 3 is presented in the following two table types:

- Table 1: Table 3.x.1 – where "3" indicates the LRA section number, "x" indicates the sub-section number from the GALL Report, and "1" indicates that this is the first table type in LRA Section 3.

- Table 2: Table 3.x.2-y – where "3" indicates the LRA section number, "x" indicates the sub-section number from the GALL Report, "2" indicates that this is the second table type in LRA Section 3, and "y" indicates the system table number.

The content of the previous applications and the PNP application is essentially the same. The intent of the revised format used for the PNP LRA was to modify the tables in Chapter 3 to provide additional information that would assist the staff in its review. In each Table 1, the applicant summarized the portions of the application that it considered to be consistent with the GALL Report. In each Table 2, the applicant identified the linkage between the scoping and screening results in Chapter 2 and the AMRs in Chapter 3.

3.0.1.1 Overview of Table 1

Each Table 3.x.1 (Table 1) provides a summary comparison of how the facility aligns with the corresponding tables of the GALL Report, Volume 1. The table is essentially the same as Tables 1 through 6 provided in the GALL Report, Volume 1, except that the "Type" column has been replaced by an "Item Number" column and the "Item Number in GALL" column has been replaced by a "Discussion" column. The "Item Number" column provides the reviewer with a means to cross-reference from Table 2 to Table 1. The "Discussion" column is used by the applicant to provide clarifying and amplifying information. The following are examples of information that might be contained within this column:

- further evaluation recommended – information or reference to where that information is located

- the name of a plant-specific program used

- exceptions to the GALL Report assumptions

- a discussion of how the line is consistent with the corresponding line item in the GALL Report when this consistency may not be intuitively obvious

- a discussion of how the item is different from the corresponding line item in the GALL Report (e.g., when there is exception taken to a GALL AMP)

The format of Table 1 allows the staff to align a specific Table 1 row with the corresponding GALL Report, Volume 1, table row so that the consistency can be easily checked.

3.0.1.2 Overview of Table 2

Each Table 3.x.2-y (Table 2) provides the detailed results of the AMRs for those components identified in LRA Section 2 as subject to an AMR. The LRA contains a Table 2 for each of the systems or components within a system grouping (e.g., reactor coolant systems, engineered safety features, auxiliary systems, etc.). For example, the engineered safety features group contains tables specific to the containment spray system, containment isolation system, and emergency core cooling system. Each Table 2 consists of the following nine columns:

(1) Component Type – The first column identifies the component types from LRA Section 2 subject to an AMR. The component types are listed in alphabetical order.

(2) Intended Function – The second column contains the license renewal intended functions for the listed component types. Definitions of intended functions are contained in LRA Table 2.1-1.

(3) Material – The third column lists the particular materials of construction for the component type.

(4) Environment – The fourth column lists the environment to which the component types are exposed. Internal and external service environments are indicated and a list of these environments is provided in LRA Table 3.0-1.

(5) Aging Effect Requiring Management – The fifth column lists aging effects requiring management (AERMs). As part of the AMR process, the applicant determined any AERMs for each combination of material and environment.

(6) Aging Management Programs – The sixth column lists the AMPs that the applicant used to manage the identified aging effects.

(7) GALL Report Volume 2 Line Item – The seventh column lists the GALL Report item(s) that the applicant identified as similar to the AMR results in the LRA. The applicant compared each combination of component type, material, environment, AERM, and AMP in Table 2 of the LRA to the items in the GALL Report. If there were no corresponding items in the GALL Report, the applicant left the column blank. In this way, the applicant identified the AMR results in the LRA tables that corresponded to the items in the GALL Report tables.

(8) Table 1 Item – The eighth column lists the corresponding summary item number from Table 1. If the applicant identifies AMR results in Table 2 that are consistent with the GALL Report, then the associated Table 3.x.1 line summary item number should be listed in Table 2. If there is no corresponding item in the GALL Report, then column eight is left blank. That way, the information from the two tables can be correlated.

(9) Notes – The ninth column lists the corresponding notes that the applicant used to identify how the information in Table 2 aligns with the information in the GALL Report. The notes identified by letters were developed by an NEI working group and will be used in future LRAs. Any plant-specific notes are identified by a number and provide additional information concerning the consistency of the line item with the GALL Report.

3.0.2 Staff's Review Process

The staff conducted the following three types of evaluations of the AMRs and associated AMPs:

(1) For items that the applicant stated were consistent with the GALL Report, the staff conducted either an audit or a technical review to determine consistency with the GALL Report.

(2) For items that the applicant stated were consistent with the GALL Report with exceptions and/or enhancements, the staff conducted either an audit or a technical review of the item to determine consistency with the GALL Report. In addition, the staff conducted either an audit or a technical review of the applicant's technical justification for the exceptions and the adequacy of the enhancements.

(3) For other items, the staff conducted a technical review pursuant to 10 CFR 54.21(a)(3).

The staff performed audits and technical reviews of the applicant's AMPs and AMRs. These audits and technical reviews determine whether the effects of aging on SCs can be adequately managed so that the intended functions can be maintained consistent with the plant's current licensing basis (CLB) for the period of extended operation, as required by 10 CFR Part 54.

3.0.2.1 Review of AMPs

For those AMPs for which the applicant had claimed consistency with the GALL Report AMPs, the staff conducted either an audit or a technical review to verify that the applicant's AMPs were consistent with the AMPs in the GALL Report. For each AMP that had one or more deviations, the staff evaluated each deviation to determine whether the deviation was acceptable and whether the AMP, as modified, would adequately manage the aging effect(s) for which it was credited. For AMPs that were not addressed in the GALL Report, the staff performed a full review to determine the adequacy of the AMPs. The staff evaluated the AMPs against the following 10 program elements defined in SRP-LR Appendix A.

(1) Scope of Program - The scope of program should include the specific SCs subject to an AMR for license renewal.

(2) Preventive Actions - Preventive actions should prevent or mitigate aging degradation.

(3) Parameters Monitored or Inspected - Parameters monitored or inspected should be linked to the degradation of the particular structure or component intended function(s).

(4) Detection of Aging Effects - Detection of aging effects should occur before there is a loss of structure or component intended function(s). This includes aspects such as method or technique (i.e., visual, volumetric, surface inspection), frequency, sample size, data collection, and timing of new/one-time inspections to ensure a timely detection of aging effects.

(5) Monitoring and Trending - Monitoring and trending should provide predictability of the extent of degradation, as well as timely corrective or mitigative actions.

(6) Acceptance Criteria - Acceptance criteria, against which the need for corrective action will be evaluated, should ensure that the structure or component intended function(s) are maintained under all CLB design conditions during the period of extended operation.

(7) Corrective Actions - Corrective actions, including root cause determination and prevention of recurrence, should be timely.

(8) Confirmation Process - Confirmation process should ensure that preventive actions are adequate and that appropriate corrective actions have been completed and are effective.

(9) Administrative Controls - Administrative controls should provide a formal review and approval process.

(10) Operating Experience - Operating experience of the AMP, including past corrective actions resulting in program enhancements or additional programs, should provide objective evidence to support the conclusion that the effects of aging will be managed adequately so that the SC intended functions will be maintained during the period of extended operation.

Details of the staff's audit evaluation of program elements (1) through (6) are documented in the Audit and Review Report and summarized in SER Section 3.0.3.

The staff reviewed the applicant's corrective action program and documented its evaluations in SER Section 3.0.4. The staff's evaluation of the corrective actions program included assessment of the following program elements: (7) "corrective actions," (8) "confirmation process," and (9) "administrative controls."

The staff reviewed the information concerning the (10) "operating experience" program element and documented its evaluation in the Audit and Review Report. The staff also included a summary of the program in SER Section 3.0.3.

The staff reviewed the final safety analysis report (FSAR) supplement for each AMP to determine if it provided an adequate description of the program or activity, as required by 10 CFR 54.21(d).

In some cases, the staff used draft GALL Report, Revision 1, dated January 2005, as a basis for accepting the applicant's AMRs that were not consistent with the GALL Report. By letter dated January 26, 2006, the applicant provided an assessment of the impact on the AMRs due to changes introduced when the final version of GALL Report, Revision 1, was issued in September 2005. The staff reviewed the applicant's assessment and concurred with its conclusions.

3.0.2.2 Review of AMR Results

Table 2 contains information concerning whether the AMRs align with the AMRs identified in the GALL Report. For a given AMR in Table 2, the staff reviewed the intended function, material, environment, AERM, and AMP combination for a particular component type within a system. The AMRs that correlate between a combination in Table 2 and a combination in the GALL Report were identified by a referenced item number in column seven, "NUREG-1801 Volume 2 Line Item." The staff also conducted onsite audits to verify the correlation. A blank column seven indicates that the applicant was unable to locate an appropriate corresponding combination in the GALL Report. The staff conducted a technical review of these combinations not consistent with the GALL Report. The next column, "Table 1 Item," provides a reference number that indicates the corresponding row in Table 1.

3.0.2.3 FSAR Supplement

Consistent with the SRP-LR, for the AMRs and associated AMPs that it reviewed, the staff also reviewed the FSAR supplement that summarizes the applicant's programs and activities for managing the effects of aging for the period of extended operation, as required by 10 CFR 54.21(d).

3.0.2.4 Documentation and Documents Reviewed

In performing its review, the staff used the LRA, LRA supplements, SRP-LR, and GALL Report. Also, during the onsite audit, the staff examined the applicant's justifications, as documented in the Audit and Review Report, to verify that the applicant's activities and programs will adequately manage the effects of aging on SCs. The staff also conducted detailed discussions and interviews with the applicant's license renewal project personnel and others with technical expertise relevant to aging management.

3.0.3 Aging Management Programs

SER Table 3.0.3-1 below presents the AMPs credited by the applicant and described in LRA Appendix B. The table also indicates the GALL Report AMP that the applicant claimed its AMP was consistent with, if applicable, and the SSCs for managing or monitoring aging. The section of the SER, in which the staff's evaluation of the program is documented, is also provided.

Table 3.0.3-1 PNP Aging Management Programs

PNP's AMP (LRA Section)	GALL Comparison	GALL AMP(s)	LRA Systems or Structures That Credit the AMP	Staff's SER Section
Existing AMPs				
Alloy 600 Program (B2.1.1)	Consistent	XI.M11	reactor coolant system	3.0.3.1.1
ASME Section XI IWB, IWC, IWD, IWF Inservice Inspection Program (B2.1.2)	Consistent	XI.M1 XI.M3 XI.S3	reactor coolant system; engineered safety features; auxiliary systems; containments, structures, and component supports	3.0.3.1.2

PNP's AMP (LRA Section)	GALL Comparison	GALL AMP(s)	LRA Systems or Structures That Credit the AMP	Staff's SER Section
Bolting Integrity Program (B2.1.3)	Consistent with enhancements	XI.M18	reactor coolant system; engineered safety features; auxiliary systems; steam and power conversion system; containments, structures, and component supports	3.0.3.2.1
Boric Acid Corrosion Program (B2.1.4)	Consistent with enhancements	XI.M10	reactor coolant system; engineered safety features; auxiliary systems; steam and power conversion system; containments, structures, and component supports; electrical and instrumentation and controls	3.0.3.2.2
Closed Cycle Cooling Water Program (B2.1.6)	Consistent with exceptions	XI.M21	reactor coolant system, engineered safety features, auxiliary systems	3.0.3.2.3
Containment Inservice Inspection Program (B2.1.7)	Consistent	X.S1 XI.S1 XI.S2	containments, structures, and component supports	3.0.3.1.4
Containment Leakage Testing Program (B2.1.8)	Consistent	XI.S4	containments, structures, and component supports	3.0.3.1.5
Diesel Fuel Monitoring and Storage Program (B2.1.9)	Consistent with exception and enhancements	XI.M30	auxiliary systems	3.0.3.2.4
Fire Protection Program (B2.1.10)	Consistent with exceptions and enhancements	XI.M26 XI.M27	auxiliary systems; containments, structures, and component supports	3.0.3.2.5
Flow Accelerated Corrosion Program (B2.1.11)	Consistent	XI.M17	reactor coolant system, steam and power conversion system	3.0.3.1.6
Open Cycle Cooling Water Program (B2.1.14)	Consistent	XI.M20	auxiliary systems	3.0.3.1.8
Overhead Load Handling Systems Inspection Program (B2.1.15)	Consistent with exception and enhancement	XI.M23	containments, structures, and component supports	3.0.3.2.7
Reactor Vessel Integrity Surveillance Program (B2.1.16)	Consistent with enhancements	XI.M31 Branch Technical Position RLSB-1	reactor coolant system	3.0.3.2.8

PNP's AMP (LRA Section)	GALL Comparison	GALL AMP(s)	LRA Systems or Structures That Credit the AMP	Staff's SER Section
Reactor Vessel Internals Inspection Program (B2.1.17)	Consistent with enhancements	XI.M16	reactor coolant system	3.0.3.2.9
Steam Generator Tube Integrity Program (B2.1.18)	Consistent	XI.M19	reactor coolant system	3.0.3.1.9
Structural Monitoring Program (B2.1.19)	Consistent with enhancement	XI.S5 XI.S6 XI.S7	containments, structures, and component supports	3.0.3.2.10
System Monitoring Program (B2.1.20)	Plant-specific	XI.M29	reactor coolant system, engineered safety features, auxiliary systems, steam and power conversion system	3.0.3.3.1
Water Chemistry Program (B2.1.21)	Consistent	XI.M2	reactor coolant system; engineered safety features; auxiliary systems; steam and power conversion system; containments, structures, and component supports	3.0.3.1.10
Compressed Air Monitoring Program (B2.1.22)	Consistent	XI.M24	auxiliary systems	3.0.3.1.11
Electrical Equipment Qualification Program (B3.1)	Consistent	X.E1	environmental qualification of electrical equipment TLAA	3.0.3.1.12
New AMPs				
Buried Services Corrosion Monitoring Program (B2.1.5)	Consistent	XI.M34	auxiliary systems, steam and power conversion system	3.0.3.1.3
Non-EQ Electrical Commodities Condition Monitoring Program (B2.1.12)	Consistent	XI.E1 XI.E2 XI.E3	electrical and instrumentation and controls	3.0.3.1.7
One-Time Inspection Program (B2.1.13)	Consistent with exception	XI.M29 XI.M32 XI.M33	reactor coolant system, engineered safety features, auxiliary systems, steam and power conversion system	3.0.3.2.6
Fatigue Monitoring Program (B3.2)	Consistent	X.M1	metal fatigue TLAA	3.0.3.1.13

3.0.3.1 AMPs That Are Consistent with the GALL Report

In LRA Appendix B, the applicant identified that the following AMPs were consistent with the GALL Report:

- Alloy 600 Program
- ASME Section XI IWB, IWC, IWD, IWF Inservice Inspection Program
- Buried Services Corrosion Monitoring Program
- Containment Inservice Inspection Program
- Containment Leakage Testing Program
- Flow Accelerated Corrosion Program
- Non-EQ Electrical Commodities Condition Monitoring Program
- Open Cycle Cooling Water Program
- Steam Generator Tube Integrity Program
- Water Chemistry Program
- Compressed Air Monitoring (added to LRA Appendix B)
- Electrical Equipment Qualification Program
- Fatigue Monitoring Program

3.0.3.1.1 Alloy 600 Program

Summary of Technical Information in the Application. In LRA Section B2.1.1, the applicant described the Alloy 600 Program, stating that this existing program is consistent with GALL AMP XI.M11, "Nickel-Alloy Nozzles and Penetrations."

The Alloy 600 Program manages aging due to primary water stress-corrosion cracking (PWSCC) of the primary coolant system (PCS) pressure boundary Alloy 600 components, including Inconel 82/182 weld joints, reactor vessel head penetrations, etc. The program includes:

(a) PWSCC susceptibility assessment using industry models to identify susceptible components

(b) monitoring and control of primary coolant chemistry to mitigate PWSCC

(c) in-service inspections (ISIs) of pressurizer penetrations, reactor vessel head penetrations, and Alloy 82/182 PCS pressure boundary welds in accordance with American Society of Mechanical Engineers (ASME) Section XI, Subsection IWB, Table IWB-2500-1

(d) augmented inspections or preemptive repair/replacement of susceptible components or welds

Staff Evaluation. During the audit and review, the staff confirmed the applicant's claim of consistency with the GALL Report. Details of the staff's evaluation of this AMP are documented in the Audit and Review Report. The staff determined that this AMP is consistent with the AMP described in the GALL Report, including the associated "operating experience" program element.

In LRA Section B2.1.1, the applicant stated that PNP has Alloy 600 penetrations, all of which are contained within the PCS. The staff asked for clarification of the Alloy 600 penetrations. By letter dated August 25, 2005, the applicant provided clarification of the nickel-alloy component locations. These nickel-alloy component locations within the scope of the Alloy 600 Program include the following:

- 45 reactor vessel upper head control rod drive (CRD) penetration nozzles
- 8 reactor vessel upper head incore instrument penetration nozzles
- 1 reactor vent line penetration nozzle
- 2 reactor flange leak detector taps
- 1 pressurizer power-operated relief valve (PORV) nozzle safe end weld
- 1 pressurizer spray nozzle safe end
- 1 pressurizer surge line nozzle safe end
- 3 pressurizer safety and relief valve nozzle flanges
- 8 pressurizer level nozzles
- 2 pressurizer temperature element nozzle penetrations
- 120 pressurizer heater sleeves
- 4 primary coolant piping safety injection and shutdown cooling inlet nozzles safe ends
- 1 primary coolant piping shutdown cooling outlet nozzle safe end
- 1 primary coolant piping surge line nozzle safe end
- 22 temperature measurement nozzles
- 1 hot leg drain nozzle
- 4 cold leg drain nozzles
- 18 pressure measurement and sampling nozzles
- 2 primary coolant piping spray nozzles
- 2 primary coolant piping charging inlet nozzles
- 4 steam generator (SG) primary bowl plugs

In the LRA, the applicant stated that inspections are based on the requirements prescribed by revised NRC Order EA-03-009 and Bulletin 2004-01, "Inspection of Alloy 82/182/600 Materials Used in the Fabrication of Pressurizer Penetrations and Steam Space Piping Connections at Pressurized-Water Reactors."

The staff reviewed the applicant's compliance with the revised NRC Order EA-03-009 and determines that its reactor pressure vessel (RPV) head aging management per revised NRC Order EA-03-009 was acceptable.

The staff noted that Bulletin 2004-01 addresses inspection of alloy for pressurizer penetrations and steam space connections. The staff asked the applicant how other PWSCC locations were addressed. The applicant indicated that it inspects all PCS nickel-alloy components. The staff reviewed the applicant's response to Bulletin 2004-01 and found it applied to all PCS components.

The staff also noted that Bulletin 2004-01 inspection commitments for all nickel-alloy primary system pressure boundary locations normally operated at greater than or equal to 350 °F is for the next two refueling outages only. Therefore, the staff determined that inspection commitments prescribed by the applicant's response to Bulletin 2004-01 do not provide adequate aging management.

By letter dated April 26, 2006, the applicant revised Commitment 6 and stated that it will revise its Alloy 600 Program to update the PWSCC corrosion rate assessments and inspection programs consistent with NRC requirements and industry commitments. A revised Alloy 600 Program description will be submitted for the staff's review and approval by March 24, 2008.

The staff finds the applicant's commitment to submit a revised Alloy 600 Program for review and approval by March 24, 2008, acceptable.

Operating Experience. In LRA Section B2.1.1, the applicant explained that a review of the industry operating experience related to the Alloy 600 Program revealed instances where degradation of material has occurred as a result of PWSCC. The review also considered related issues which included degradation of PCS hot leg piping and nozzles, thermal sleeves and instrument nozzles, reactor vessel head nozzles, CRD mechanism, thermocouple nozzles, and intrusion of demineralizer resins.

A review of the plant-specific operating experience revealed four instances where the Alloy 600 Program had been instrumental in discovering material degradation. Degradation was discovered in the pressurizer temperature element penetration, pressurizer safe end, and CRD nozzle penetration indications (2).

Two of the safe-ends on the primary coolant piping (12-inch schedule 140 surge nozzle and 12-inch schedule 140 shutdown cooling outlet nozzle) had the mechanical stress improvement process applied to them in 1995. The mechanical stress improvement process changes the residual stress patterns at these locations from tensile to compressive by plastically deforming the piping near the welds. The compressive residual stress is desired to help mitigate PWSCC.

The first bare metal visual examination of the reactor vessel head to meet the requirements of NRC Order EA-09-003 was completed during the spring 2003 refueling outage with results reported to the NRC. Qualified visual examination (VT)-2 examiners using direct visual techniques performed the examinations. Visual inspection was also performed to identify potential boric acid leaks from pressure-retaining components from above the reactor vessel. Results found that there were no accumulations of boric acid in the vicinity of, and no leakage of boric acid through, any of the reactor vessel head penetrations. All visual examinations of the reactor vessel head penetrations had acceptable results.

The first volumetric examinations performed in accordance with the revised NRC Order EA-03-009 were completed during the fall 2004 refueling outage. Results of the inspections identified two CRD nozzle penetrations indicating required repair. These repairs were completed during the outage.

In response to commitments made by the applicant in response to NRC Bulletin 2004-01, a bare metal visual examination of 100 percent of the pressurizer heater sleeve locations, including 360 degrees around each sleeve, and 36 Alloy 82/182/600 primary system pressure boundary locations, normally operated at greater than or equal to 350 °F, was performed during the fall 2004 refueling outage. Results of the bare metal visual examination of each of the penetrations reported to the staff were acceptable, with no accumulation of boric acid in the vicinity of any of the penetrations.

The staff reviewed the industry operating experience, which includes Bulletins 2001-01, 2002-01, 2002-02, and 2004-01, and the applicant's associated responses. The staff also reviewed the applicant's response to revised NRC Order EA-03-009. The staff finds that the applicant's inspection and responses to regulatory recommendations/requirements provide timely detection of degradation for nickel-alloy components.

The staff reviewed the operating experience provided in the LRA and program basis documents. The staff also interviewed the applicant's technical personnel and confirmed that the plant-specific operating experience did not reveal any degradation not bounded by industry experience.

FSAR Supplement. The applicant provided its FSAR supplement for the Alloy 600 Program in LRA Section A2.1. The staff reviewed this section and determines that the information in the FSAR supplement provides an adequate summary description of the program, as required by 10 CFR 54.21(d).

Conclusion. On the basis of its review and audit of the applicant's Alloy 600 Program, the staff determines that those program elements for which the applicant claimed consistency with the GALL Report are consistent. The staff concludes that the applicant has demonstrated that the effects of aging will be adequately managed so that the intended functions will be maintained consistent with the CLB for the period of extended operation, as required by 10 CFR 54.21(a)(3). The staff also reviewed the FSAR supplement for this AMP and concludes that it provides an adequate summary description of the program, as required by 10 CFR 54.21(d).

3.0.3.1.2 ASME Section XI IWB, IWC, IWD, IWF Inservice Inspection Program

Summary of Technical Information in the Application. In LRA Section B2.1.2, the applicant described the ASME Section XI IWB, IWC, IWD, IWF Inservice Inspection Program, stating that this existing program is consistent with GALL AMPs XI.M1, "ASME Section XI Inservice Inspection, Subsections IWB, IWC, and IWD," XI.M3, "Reactor Head Closure Studs," and XI.S3, "ASME Section XI, Subsection IWF."

The ASME Section XI IWB, IWC, IWD, IWF Inservice Inspection Program facilitates inspections to identify and correct degradation in Classes 1, 2, and 3 piping, components, their supports, and integral attachments. The program includes periodic visual, surface and/or volumetric examinations and leakage tests of all Classes 1, 2, and 3 pressure-retaining components, their supports, and integral attachments including welds, pump casings, valve bodies, pressure-retaining bolting, piping/component supports, and reactor head closure studs. These are identified in ASME Section XI, "Rules for Inservice Inspection of Nuclear Power Plant Components," as commitments requiring augmented ISIs and are within the scope of license renewal.

Staff Evaluation. During the audit and review, the staff confirmed the applicant's claim of consistency with the GALL Report. Details of the staff's evaluation of this AMP are documented in the Audit and Review Report. The staff determined that this AMP is consistent with the AMP described in the GALL Report, including the associated "operating experience" program element.

The staff's review of LRA Section B2.1.2 identified an area in which additional information was necessary to complete the review of the applicant's AMP. The applicant responded to the staff's RAI as discussed below.

In RAI-B2.1.2-1, dated June 3, 2005, the staff requested the applicant to specify the version of the ASME Code for the ASME Section XI IWB, IWC, IWD, IWF Inservice Inspection Program.

In its response, by letter dated July 1, 2005, the applicant stated:

> The Code of Federal Regulations, 10 CFR 50.55a, requires that inservice inspection of Class 1, 2, and 3 pressure retaining components, their integral attachments and supports, be conducted in accordance with the latest edition of ASME Section XI approved by the NRC twelve months prior to the start of a ten-year interval. The Palisades Inservice Inspection Program for the current (3rd) ten-year interval, which began on May 12, 1995, implements the 1989 edition, no addenda, of ASME Section XI as modified by 10 CFR 50.55a and approved Relief Requests and Code Cases. One of the NRC-approved relief requests authorizes implementation of a risk-informed inservice inspection program. This program will be in effect until the end of the current ten-year interval, which ends December 12, 2006.
>
> In 2006 NMC expects to update the inservice inspection code edition and addenda to those required for the fourth ten-year interval. It is anticipated that the Section XI version which will be incorporated into the Palisades program will be the 2001 edition through the 2003 addendum. This edition and addendum are the latest currently incorporated by reference in 10 CFR 50.55a(b).

By letter dated August 25, 2005, the applicant stated that it would revise its ASME Section XI IWB, IWC, IWD, IWF Inservice Inspection Program descriptions in LRA Appendices A and B to reflect the use of the 2001 Edition, including the 2002 and 2003 Addenda, of ASME Code Section XI. The revised program descriptions will identify exceptions to this code taken by the applicant, if any, that impact aging management effectiveness. Appropriate justification will also be provided to show that the exceptions, if any, still provide an acceptable level of aging management.

By letter dated October 31, 2005, the applicant provided its revised program descriptions for the staff's review and identified the following seven exceptions:

(1) risk-informed inspection program
(2) relief from regenerative heat exchanger (HX) examinations
(3) relief from pressurizer examinations
(4) relief from shutdown cooling HX examinations
(5) relief from SG examinations
(6) relief for single sided examinations
(7) relief for inside diameter surface examinations

In addition, the applicant stated that it had concluded that no changes were necessary to the ASME Section XI, Subsections IWB, IWC, IWD, IWF Inservice Inspection Program description in LRA Section A2.2.

The relief exceptions allow examination of the accessible volume of the weld as opposed to the requirement to examine 100 percent of the weld length due to design limitations. The staff reviewed the exceptions and determines that the applicant's bases for relief from examinations of the ASME Section XI, Subsections IWB, IWC, IWD, IWF Inservice Inspection Program are not adequate. If the applicant intends to use relief requests, it must follow the process as described in 10 CFR 50.55a. The relief request process is applicable for a 10-year inspection interval. The license renewal process under 10 CFR Part 54 does not address relief requests. In a teleconference dated November 15, 2005, the staff requested the applicant to provide additional information demonstrating the validity of exceptions 2 through 7, as identified in its October 31, 2005, letter.

By letter dated December 16, 2006, the applicant revised its program description to remove exceptions 2 through 7 of the October 31, 2005, letter. The applicant stated that the ASME Section XI, IWB, IWC, IWD, IWF Inservice Inspection Program is based on ASME Code Section XI, 2001 Edition, including 2002 and 2003 Addenda. Also, during the period of extended operation, the program will be maintained consistent with 10 CFR 50.55a. The applicant also stated that the risk-informed inservice inspection (RI-ISI) program would be applied as the exception to the GALL Report.

GALL AMP XI.M1 does not recognize RI-ISI programs as an alternative to the current ASME Section XI ISI requirement. The applicant implemented an RI-ISI program for Classes 1, 2, and 3, and non-class piping and obtained staff approval of the relief request to use RI-ISI in a letter dated May 19, 2003. Although the number of examinations is reduced, the risk from implementation of RI-ISI is expected to decrease slightly when compared to that estimated from the current requirements. The primary basis for the risk reduction is that examinations will be required for safety significant piping segments that may not be inspected per the current ASME Section XI IWB, IWC, IWD, IWF Inservice Inspection Program. In addition, RI-ISI is a living program that requires updating and expansion based on industry and site-specific inspection findings. However, the applicant will have to request approval to use the RI-ISI program during the period of extended operation in accordance with 10 CFR 50.55a 12 months prior to each specific interval. Therefore, the staff found that ASME Section XI, as referenced in 10 CFR 50.55a, as modified by a staff-approved or authorized RI-ISI program 12 months prior to each inspection interval, is acceptable for the period of extended operation.

In its program basis document, the applicant stated that its reactor vessel closure studs have a minimum tensile strength of 145 ksi, well below 170 ksi specified in GALL AMP XI.M3. The staff noted that 170 ksi specified in the GALL Report is the actual strength instead of the minimum strength identified in the design specification. As indicated in the Audit and Review Report, the applicant's basis document showed that the post-tempered maximum tensile strength would not exceed 170 ksi. The staff concludes that this strength is acceptable upon finding the material properties controlled by the specification to be less than 170 ksi and the material test reports filed with the purchase order demonstrating material properties less than 170 ksi.

Operating Experience. In LRA Section B2.1.2, the applicant explained that a review of the industry operating experience related to the ASME Section XI IWB, IWC, IWD, IWF Inservice Inspection Program revealed numerous instances where degradation of components, component supports, and bolting has occurred. The applicant looked at related issues including stress corrosion cracking (SCC) and crack initiation and growth due to thermal loading.

A review of plant-specific operating experience revealed 13 instances where the ASME Section XI IWB, IWC, IWD, IWF Inservice Inspection Program had been instrumental in discovering degradation in the following items: CRD housings, piping welds, component supports, bolting, temperature element penetration, reactor coolant pressurizer safe end, and engineered safeguards systems check valve.

In the LRA, the applicant stated that on several occasions, the ASME Section XI IWB, IWC, IWD, IWF Inservice Inspection Program had demonstrated that it provides reasonable assurance that aging effects are adequately managed for Classes 1, 2, and 3 components, component supports, and bolting. This management had been demonstrated through NRC inspection reports, Institute of Nuclear Power Operations (INPO) evaluations, audits, self-assessments, and the corrective action program.

The staff reviewed the operating experience provided in the LRA, and interviewed the applicant's technical personnel to confirm that the plant-specific operating experience revealed no degradation not bounded by industry experience.

FSAR Supplement. In LRA Section A2.2, the applicant provided the FSAR supplement for the ASME Section XI IWB, IWC, IWD, IWF Inservice Inspection Program. The staff reviewed this section and determines that the information in the FSAR supplement provides an adequate summary description of the program, as required by 10 CFR 54.21(d).

Conclusion. On the basis of its review and audit of the applicant's ASME Section XI IWB, IWC, IWD, IWF Inservice Inspection Program, the staff determines that those program elements for which the applicant claimed consistency with the GALL Report are consistent. In addition, the staff reviewed the exception and the associated justifications and determines that the AMP, with the exception, is adequate to manage the aging effects for which it is credited. The staff concludes that the applicant has demonstrated that the effects of aging will be adequately managed so that the intended functions will be maintained consistent with the CLB for the period of extended operation, as required by 10 CFR 54.21(a)(3). The staff also reviewed the FSAR supplement for this AMP and concludes that it provides an adequate summary description of the program, as required by 10 CFR 54.21(d).

3.0.3.1.3 Buried Services Corrosion Monitoring Program

Summary of Technical Information in the Application. In LRA Section B2.1.5, the applicant described the Buried Services Corrosion Monitoring Program, stating that this new program is consistent with GALL AMP XI.M34, "Buried Piping and Tanks Inspection."

The Buried Services Corrosion Monitoring Program manages aging effects on the external surfaces of carbon, low-alloy, and stainless steel components buried in soil or sand. This program includes (a) visual inspections of external surfaces of buried components for evidence of coating damage and substrate degradation to manage the effects of aging and (b) visual inspection of the external surfaces of buried stainless steel components for evidence of crevice corrosion, pitting, and microbiologically induced corrosion (MIC). The frequency of these inspections for carbon, low-alloy, and stainless steel will be based on opportunities for inspection, such as scheduled maintenance work. Age-related degradation of buried components susceptible to selective leaching is managed by the One-Time Inspection Program.

Staff Evaluation. During the audit and review, the staff confirmed the applicant's claim of consistency with the GALL Report. Details of the staff's audit evaluation of this AMP are documented in the Audit and Review Report. The staff determined that this AMP is consistent with the AMP described in the GALL Report, including the associated "operating experience" program element.

GALL AMP XI.M34 states that the program includes preventive measures to mitigate corrosion and periodic inspection to manage the effects of corrosion on the pressure-retaining capacity of buried carbon steel piping and tanks. Preventive measures are in accordance with standard industry practice for maintaining external coatings and wrappings.

As documented in the Audit and Review Report, the staff noted that the applicant uses the Buried Services Corrosion Monitoring Program to manage aging effects of buried stainless steel components. However, GALL AMP XI.M34 monitors and inspects coating and wrapping integrity of the external surfaces of buried carbon steel and tanks. Buried stainless steel components are not coated, so it was not clear from the information in the LRA and the program basis documents how the Buried Services Corrosion Monitoring Program manages the aging effect of stainless components in soil. By letter dated August 25, 2005, the applicant stated that its Buried Services Corrosion Monitoring Program requires a visual inspection of external surfaces of buried stainless steel components for evidence of MIC, crevice, and pitting corrosion. The applicant further stated that this requirement is not an exception to the GALL AMP. The staff finds the applicant's Buried Services Corrosion Monitoring Program acceptable because it manages the stainless piping aging effects beyond what is recommended in GALL AMP XI.M34.

As documented in the Audit and Review Report, the staff noted that in the "scope of program" program element of the Buried Services Corrosion Monitoring Program, the applicant stated that it had not identified any buried tanks in sand or soil. The staff requested the applicant to clarify whether the tanks, specifically those shown on scoping boundary drawing LR-C3, are located above ground and not in contact with soil/sand. The applicant responded that there is only one below-grade tank within the scope of license renewal. The diesel fuel oil storage tank is below grade, but contained in a vault and not exposed to a soil environment. The diesel fuel oil storage tank is managed by the Diesel Fuel Storage and Monitoring Program. The staff finds this response acceptable.

In LRA Section B2.1.5, the applicant stated that the frequency of the inspections for carbon, low-alloy, and stainless steel will be based on opportunities for inspection such as scheduled maintenance work. Also, in the "detection of aging effects" program element, the applicant stated that visual inspections of buried carbon steel, low-alloy, and stainless steel components will be based on plant operating experience, excavation of components for maintenance, or any other reason. GALL AMP XI.M34, under the "detection of aging effects" program element states that because the inspection frequency is plant-specific and depends on the plant operating experience, the applicant's proposed inspection frequency is to be further evaluated for the period of extended operation. As documented in the Audit and Review Report, the staff asked the applicant to specify the inspection frequency for the buried piping using its Buried Services Corrosion Monitoring Program. By letter dated July 1, 2005, the applicant stated that it would perform inspections within ten years after entering the period of extended operation.

The staff's review of LRA Section B2.1.5 identified an area in which additional information was necessary to complete the review of the applicant's AMP. The applicant responded to the staff's RAI as discussed below.

In RAI-B2.1.5-1, dated June 3, 2005, the staff requested the applicant to state the inspection frequency for the Buried Services Corrosion Monitoring Program.

In its response, by letter dated July 1, 2005, the applicant stated that its Buried Services Corrosion Monitoring Program includes inspection activities designed to detect degradation due to aging effects prior to loss of intended function. In Commitment 39 of this letter, the applicant stated that visual inspections of a sample of buried carbon, low-alloy, and stainless steel components would be performed within ten years after entering the period of extended operation unless opportunistic inspections had occurred within this ten-year period. Prior to the tenth year, the applicant would evaluate available data to determine whether inspections have assessed the condition of the components sufficiently. If data are insufficient, focused inspection(s) would be performed as needed.

As documented in the Audit and Review Report, the staff noted that Commitment 39 is not consistent with the level of inspection the staff determined appropriate for managing this aging effect. The staff determined that inspections to confirm that coating and wrapping are intact for steel components are effective to ensure that corrosion of external surfaces has not occurred and that intended function is maintained. For stainless steel components, visual inspections of the external surfaces are sufficient to assure that the intended function is maintained. Before the period of extended operation, the applicant shall confirm that there has been at least one opportunistic or focused inspection within the past ten years. After entering the period of extended operation, the applicant shall perform a focused inspection within ten years, unless an opportunistic inspection occurs within this ten-year period. Any credited inspection should be in areas with the highest likelihood and history of corrosion problems.

By letter dated September 16, 2005, the applicant revised Commitment 39 as follows:

> Visual inspections of a sample of buried carbon, low-alloy, and stainless steel components will be performed within ten years prior to entering, and within ten years after entering, the period of extended operation. Prior to the tenth year of each period, NMC will perform an evaluation of available data to determine sufficient opportunistic inspections have been performed within that period to assess the condition of the components. If insufficient data exists, focused inspection(s) will be performed as needed.

Based on its review, the staff finds the applicant's revised response to RAI-B2.1.5-1 acceptable because the applicant committed to visual inspections of samples of buried carbon, low-alloy, and stainless steel components within ten years before entering, and within ten years after entering, the period of extended operation. The staff determines that the inspection frequency of buried components is appropriate for managing this aging effect. Therefore, the staff's concern described in RAI-B2.1.5-1 is resolved.

In the LRA, the applicant stated that its Buried Services Corrosion Monitoring Program is consistent with GALL AMP XI.M34 and will be developed and implemented. Features of the program will include development and implementation of procedures for inspection of selected buried SSCs for corrosion, pitting, and MIC.

In addition, the applicant stated that the program elements describe the program to be implemented. The staff finds this program acceptable because in Commitment 15 the applicant agreed to implement the Buried Services Corrosion Monitoring Program with procedures for inspection of the buried components of corrosion, pitting, and MIC consistent with GALL AMP XI.M34.

Operating Experience. In LRA Section B2.1.5, the applicant explained that review of industry operating experience applicable to aging of buried services revealed diesel fuel line leakage from corrosion due to the absence of required coating. None of the industry operating instances revealed any new program issues attributed to exterior corrosion of buried services component materials.

A review of plant-specific operating experience related to the Buried Services Corrosion Monitoring Program and aging revealed that the following issues have been addressed:

- through wall leak in buried steam line
- generic program deficiencies from internal engineering programs audit
- operating experience related to buried fire main ruptures (see Fire Protection Program)

None of the plant operating issues or instances resulted from normal aging or reflected significant program deficiencies.

The staff reviewed the operating experience provided in the LRA and interviewed the applicant's technical personnel to confirm that the plant-specific operating experience revealed no degradation not bounded by industry experience.

On the basis of its review of industry and plant-specific operating experience and discussions with the applicant's technical personnel, the staff concludes that the applicant's Buried Services Corrosion Monitoring Program would adequately manage the aging effects identified in the LRA for which this AMP is credited.

The staff concludes that the corrective action program, based on internal and external plant operating experience, would capture operating experience in the future to support the conclusion that the effects of aging are adequately managed.

FSAR Supplement. In LRA Section A2.5, the applicant provided the FSAR supplement for the Buried Services Corrosion Monitoring Program. The staff reviewed this section and determines that the information in the FSAR supplement provides an adequate summary description of the program, as required by 10 CFR 54.21(d).

Conclusion. On the basis of its review and audit of the applicant's Buried Services Corrosion Monitoring Program and RAI response, the staff determines that those program elements for which the applicant claimed consistency with the GALL Report are consistent. The staff concludes that the applicant has demonstrated that the effects of aging will be adequately managed so that the intended functions will be maintained consistent with the CLB for the period of extended operation, as required by 10 CFR 54.21(a)(3). The staff also reviewed the

FSAR supplement for this AMP and concludes that it provides an adequate summary description of the program, as required by 10 CFR 54.21(d).

3.0.3.1.4 Containment Inservice Inspection Program

Summary of Technical Information in the Application. In LRA Section B2.1.7, the applicant described the Containment Inservice Inspection Program, stating that this existing program is consistent with GALL AMPs X.S1, "Concrete Containment Tendon Prestress," XI.S1, "ASME Section XI, Subsection IWE," and XI.S2, "ASME Section XI, Subsection IWL."

The Containment Inservice Inspection Program is designed to ensure that containment shell concrete, the post-tensioning system, and steel pressure-retaining elements continue to provide an acceptable level of structural integrity. In addition, it is designed to ensure that the liner (with associated moisture barriers), other leakage limiting steel barriers, and pressure-retaining bolted connections have not degraded. This program does not demonstrate actual containment leak tightness; that is done under the Containment Leakage Testing Program.

Staff Evaluation. During the audit and review, the staff confirmed the applicant's claim of consistency with the GALL Report. Details of the staff's evaluation of this AMP are documented in the Audit and Review Report. The staff determined that this AMP is consistent with the AMP described in the GALL Report, including the associated "operating experience" program element.

GALL AMPs XI.S1 and XI.S2 identify both the 1992 Edition with 1992 Addenda and the 1995 Edition with 1996 Addenda, as approved in 10 CFR 50.55a, as the applicable editions of the ASME Code for these programs. In the LRA, the applicant stated that its Containment Inservice Inspection Program follows ASME Code Section XI, 1998 Edition with no addenda.

By letter dated August 25, 2005, the applicant stated that the NRC had approved the use of the 1998 Edition in a letter dated September 27, 2002, "Palisades Plant - Evaluation of Containment In-service Inspection Relief Requests (TAC Nos. MB4216 and MB4218)." In this docketed letter, the first paragraph of the cover letter states:

> The licensee is seeking relief from the requirements of the 1992 edition with the 1992 addenda of the American Society of Mechanical Engineers (ASME) *Boiler and Pressure Vessel Code* (the Code), Section XI, Subsections IWE and IWL. As an alternative, the licensee proposes to use the provisions of Subsections IWE and IWL of the 1998 edition of the Code, which has not yet been incorporated by reference into 10 CFR 50.55a.

The final paragraph of the cover letter states:

> The licensee's proposed alternative to use the 1998 Edition of Subsections IWE and IWL, as supplemented by commitments in the licensee's February 21, 2002 letter, will provide an acceptable level of quality and safety for ensuring the integrity of the pressure boundary of the containment at PNP. Therefore, the proposed alternative is authorized pursuant to 10 CFR 50.55a(a)(3)(i). As requested, this relief is authorized for the balance of the current 10-year inspection interval.

The staff emphasized that, although the applicant had been granted a relief request to use the 1998 Edition instead of the 1992 or 1995 Edition of the ASME Code for IWE and IWL inspections, under license renewal this inconsistency with GALL AMPs XI.S1 and XI.S2 is an exception. After further discussions with the staff, the applicant agreed to change the Containment Inservice Inspection Program to reflect the use of the 1998 Edition of the ASME Code instead of the GALL Report specified 1992 or 1995 Edition as an exception and to provide written discussion on the code edition differences in ASME Code Section XI, Subsections IWE and IWL.

By letter dated August 25, 2005, the applicant stated that the Containment Inservice Inspection Program references the 1998 Edition with no addenda, for ASME Code Section XI, Subsections IWE and IWL, except that the personnel qualification process is based on the 1992 Edition through 1992 Addenda. The applicant stated that by another letter, it will revise the Containment Inservice Inspection Program description in the LRA to identify use of the 1998 Edition as an exception to the GALL Report. Exceptions taken to the 1998 Edition, if any, will be parts of the program description.

By letter dated October 31, 2005, the applicant stated that LRA Section B2.1.7 had been revised to show the Containment Inservice Inspection Program consistent with GALL AMPs XI.S1 and XI.S2 with exceptions. The Containment Inservice Inspection Program is still consistent with GALL AMP X.S1.

The applicant stated that the exceptions are as follow:

> NUREG Section XI.S1, references the 1992 Edition with the 1992 Addenda and the 1995 Edition with the 1996 Addenda, of ASME Code -Section XI, Subsection IWE, as approved in 10 CFR 50.55a, as providing acceptable aging management programs. It states that these codes and the additional requirements specified in 10 CFR 50.55a(b)(2) constitute an existing mandated program applicable to managing aging of steel containments, steel liners of concrete containments, and other containment components for license renewal. The Palisades Containment Inspection program implements ASME Section XI, Subsection IWE, 1998 edition...

> NUREG Section XI.S2, references both the 1992 Edition with the 1992 Addenda and the 1995 Edition with the 1996 Addenda, of ASME Code Section XI, Subsection IWL, as approved in 10 CFR 50.55a, as providing an acceptable aging management program. It states that these codes and the additional requirements specified in 10 CFR 50.55a(b)(2) constitute an existing mandated program applicable to managing aging of containment reinforced concrete and unbonded post-tensioning systems for license renewal. The Palisades Containment Inspection uses the 1998 edition of Subsection IWL as an alternative to the requirements of the 1992 edition and addenda for inspection of Class CC components...

The staff finds the exceptions acceptable because the applicant's 1998 ASME Code of record for ASME Code Section XI Inservice Inspection (Subsections IWE and IWL) is valid for a 10-year inspection interval under the CLB. The applicant will have to request approval to use the ASME Code Section XI Inservice Inspection (Subsections IWE and IWL) 12 months prior to each specific interval during the period of extended operation in accordance with 10 CFR 50.55a. Therefore, the staff finds that the ASME Code Section XI Edition in effect, as referenced in 10 CFR 50.55a, 12 months prior to each inspection interval, is acceptable for the period of extended operation.

GALL AMP XI.S2, under the "monitoring and trending" program element, has the following statement: "In addition to the random sampling used for tendon examination, one tendon of each type is selected from the first-year inspection sample and designated as a common tendon. Each common tendon is then examined during each inspection." The applicant made the following statement in LRA Section B2.1.7, under the "scope of program" program element, for concrete containment tendon prestress: "During the scheduled surveillances, the tendon prestress force is measured for a random sample of each tendon group. One tendon in each group is designated as the common tendon."

As documented in the Audit and Review Report, the staff requested the applicant to provide all the lift-off test results for the common tendon in each tendon group. The applicant stated that there was not really a common tendon for each group going back to the first-year inspection sample. However, going forward with the current operating license and anticipated extended operating license, the applicant had designated and established a common tendon for each group during the last surveillance inspection. After further discussions with the staff, the applicant agreed to change the Containment Inservice Inspection Program throughout where there is discussion of a common tendon to clarify that a common tendon in each group has only recently been established and that there are no common tendon test results from the first-year inspection sample forward. As there are no historic common tendons as specified in the GALL Report, the applicant designated the difference an exception to the GALL Report after revising the LRA.

By letter dated August 25, 2005, the applicant stated:

> The Palisades tendon surveillance program was directed by Plant Technical Specifications until 10 CFR 50.55a invoked testing in accordance with ASME Section XI, Subsection IWL, in 1996. Palisades Technical Specifications did not require the selection of common tendons. As a result, common tendons were not defined at Palisades until the 30-Year tendon surveillance conducted in 2002. The selected tendons did not meet the desired criteria in that they had been detensioned during the first tendon surveillance in the early 1970s. Therefore, LRA Section B2.1.7, Containment Inservice Inspection Program, Exceptions to NUREG 1801, is hereby revised to read, 'The generally accepted definition of common tendon does not completely correspond with the XI.S2, ASME Section XI Sub-Section IWL portion of the Containment Inservice Inspection Program, and is considered to be an exception to NUREG 1801.'

In this letter, the applicant further stated that this exception does not degrade the effectiveness of the program to assure an acceptable level of containment structural integrity at all times. The tendon surveillance program is designed to maintain the tendon force above minimum analysis requirements continuously from surveillance to surveillance. This maintenance is accomplished each surveillance by measuring tendon force, comparing the results against expected levels, and assuring that any expected relaxation will not reduce tendon forces below minimum requirements beyond at least the next surveillance. Structural integrity does not rely solely on the projection of forces in a designated common tendon to the end of the plant's life.

Based on its review, the staff finds the applicant's response acceptable because although an historic common tendon does not exist, common tendons have been designated going forward for the remaining current operating license period and potential period of extended operation. The program is still effective because during each tendon surveillance, tendon force measurements are taken and compared against expected levels for tendon force, assuring that the expected tendon relaxation will not reduce the tendon forces below minimum requirements until the next surveillance.

Operating Experience. In LRA Section B2.1.7, the applicant explained that a review of the industry operating experience related to the Containment Inservice Inspection Program revealed instances where degradation had occurred within containments. The applicant looked at related issues including degradation of containment liner plates, concrete, coatings, moisture barriers, bellows, tendons and tendon wires, tendon anchor heads, and penetrations.

A review of the plant-specific operating experience revealed some instances where the Containment Inservice Inspection Program had been instrumental in discovering material degradation. Containment degradation included liner plate corrosion, unacceptable tendon liftoff value, tendon gallery corrosion, tendon grease leakage, moisture barrier not in place, and tendon sheath water intrusion.

In addition, the applicant stated that the Containment Inservice Inspection Program had been effective in timely identifying material degradation, thus ensuring that age-related degradation of the containment would be effectively managed throughout the period of extended operation.

Furthermore, in the LRA, the applicant stated that the Containment Inservice Inspection Program adequately managed aging effects for the containment. This adequacy has been demonstrated through staff inspection reports, INPO evaluations, audits, self-assessments, and the corrective action program.

The staff's review of the operating experience related to the Containment Inservice Inspection Program revealed no significant failures of the containment shell concrete, post-tensioning system, and steel pressure-retaining elements due to degradation. The operating experience is typical of the types of minor material degradation found on the interior, the exterior, and with the tendon system for a prestressed concrete containment. The staff did not identify any operating experience that would require any modification to the program.

The staff reviewed the operating experience provided in the LRA and interviewed the applicant's technical personnel to confirm that the plant-specific operating experience revealed no degradation not bounded by industry experience.

FSAR Supplement. In LRA Section A2.7, the applicant provided the FSAR supplement for the Containment Inservice Inspection Program. The staff reviewed this section and determines that the information in the FSAR supplement provides an adequate summary description of the program, as required by 10 CFR 54.21(d).

Conclusion. On the basis of its review and audit of the applicant's Containment Inservice Inspection Program, the staff determines that those program elements for which the applicant claimed consistency with the GALL Report are consistent. In addition, the staff reviewed the exceptions and the associated justifications and determines that the AMP, with the exceptions, is adequate to manage the aging effects for which it is credited. The staff concludes that the applicant has demonstrated that the effects of aging will be adequately managed so that the intended functions will be maintained consistent with the CLB for the period of extended operation, as required by 10 CFR 54.21(a)(3). The staff also reviewed the FSAR supplement for this AMP and concludes that it provides an adequate summary description of the program, as required by 10 CFR 54.21(d).

3.0.3.1.5 Containment Leakage Testing Program

Summary of Technical Information in the Application. In LRA Section B2.1.8, the applicant described the Containment Leakage Testing Program, stating that this existing program is consistent with GALL AMP XI.S4, "10 CFR 50, Appendix J."

The Containment Leakage Testing Program ensures that containment leakage is maintained below the upper acceptance limit (La = 0.1% / day). This testing program, in conjunction with the Containment Inservice Inspection Program, provides assurance that age-related (and other) deterioration of the containment leakage limiting boundary is appropriately managed to ensure that postulated post-accident releases are limited to an acceptable level. The program is implemented through the following testing and examination activities: (a) overall containment leakage (integrated leakage rate or Type A) test to assess the leak-tight integrity of the entire pressure boundary, (b) visual examinations of the containment exterior and interior, and (c) local (Types B & C) tests to assess the leak-tight integrity of individual penetrations.

Staff Evaluation. During the audit and review, the staff confirmed the applicant's claim of consistency with the GALL Report. Details of the staff's evaluation of this AMP are documented in the Audit and Review Report. The staff determined that this AMP is consistent with the AMP described in the GALL Report, including the associated "operating experience" program element.

In implementing Option B for 10 CFR Part 50, Appendix J, testing, the applicant requested CLB exemptions from requirements of Appendix J and exceptions from guidance of NRC Regulatory Guide (RG) 1.163, "Performance-Based Containment Leak Test Program," and NEI 94-01, "Industry Guideline for Implementing Performance-Based Option of 10 CFR 50 Appendix J." RG 1.163 stipulates that containment purge valves are to be tested at least every 30 months. By letter dated March 30, 2001, the staff approved application of performance-based test interval extension criteria to the containment purge valves, "Palisades Plant Issuance of Amendment RE: Option B Containment Leak Rate Testing,"(ML010930230). However, the applicant retained the improved technical specification (TS) requirement to perform a leakage test on these valves every 184 days.

In Appendix J of 10 CFR Part 50, Option B states that Type B tests are to be performed at accident pressure. By letter dated June 1, 1989, the staff approved testing of the personnel air lock door seals at 10 rather than 55 psig.

NEI 94-01 specifically requires testing of air lock door seals following air lock door use when containment integrity is required, "Amendment No. 126 to Provisional Operating License No. DPR-20,"(ML020810123). By letter dated September 30, 1987, the staff approved the applicant's request to verify seal contact in lieu of performing a leakage test on emergency escape air lock door seals, "Palisades Plant-Issuance of Amendment RE: Containment Emergency Escape Air Lock Testing, and Exemption From Certain Requirements of 10 CFR Part 50, Appendix J," (ML020840256). These plant-specific alternatives were approved by the staff during implementation of Option B at PNP.

The staff reviewed the NRC-issued TS amendments, which approved the applicant's implementation of Option B with the exemptions. The applicant stated that these exemptions are not exceptions to GALL AMP XI.S4 elements because they are part of the CLB. The staff agreed and found these Containment Leakage Testing Program changes acceptable because NRC approval was obtained during implementation of Option B for 10 CFR Part 50, Appendix J, testing and they are contained in the TS.

Operating Experience. In LRA Section B2.1.8, the applicant explained that NUREG-1493, which provides the technical justification for the performance-based leakage testing program defined in Appendix J, Option B, includes a summary of industry testing experience. This summary demonstrates that performance-based leakage testing programs will almost always detect problems at an early stage and are, therefore, acceptable for managing containment leak-tight integrity. NUREG-1493 includes a few contrary examples; however, in most cases, these examples illustrate lack of administrative control rather than any technical deficiency in performance-based programs. While NUREG-1493 does not specifically address license renewal, it shows, by inference, that performance-based leakage testing programs are effective aging management tools.

A review of additional industry operating experience associated with the Containment Leakage Testing Program and aging reveals issues and instances related to Type B local leak rate test performed on containment penetration bellows and later invalidated by a subsequent containment integrated leak rate test. Site operating experience shows that no significant problems have been found during periodic Type A tests. This confirms that the local leakage rate testing program (in conjunction with periodic containment examinations) has always detected developing deterioration before it could result in a loss of containment leak-tight integrity (defined by overall leakage exceeding La).

Instances of excessive (in excess of the assigned administrative limit) component leakage have been detected by Types B & C tests over the operating lifetime of the plant. Most instances of excessive leakage are the results of active isolation valve seat deterioration; some are the results of air lock door seal misalignment or damage. Active isolation valves and air lock door seals (which are replaced at least once every three refueling outages), however, are not passive and long-lived components subject to aging management under this program.

Instances of problems with passive components are relatively rare. Two reported instances were found in a search through records going back through the mid 1980s. These are summarized below:

- In 1983, a leak of approximately 0.1 La was found at one conductor seal in electrical penetration EZ-104. The leak affected only one barrier so that minimum pathway leakage through the penetration was still essentially nil. The entire penetration was replaced during the 1985 outage.

- During a September 2001 Type C test on penetration MZ-66, a measured leak of approximately 0.15 La was identified primarily due to leakage through a manual isolation gate valve. The cause of the leakage was determined to be debris on the seat. The problem was corrected and leakage reduced to an acceptable level. Since the measured leakage was identified primarily to a single barrier, minimum pathway leakage through the penetration remained at a relatively low level.

In the LRA, the applicant stated that the Containment Leakage Testing Program has demonstrated through inspection reports, program health reports, and the corrective action program that it provides reasonable assurance that aging effects are managed for SSCs within the scope of license renewal.

The staff's review of the applicant's operating experience related to the Containment Leakage Testing Program revealed no significant failures of containment leakage barriers due to degradation. The cause of any significant leakage during recent containment leakage testing was debris on the seat of a manual isolation gate valve, not degradation from aging. The staff identified no Containment Leakage Testing Program operating experience that would require any program modification.

The staff reviewed the operating experience provided in the LRA and interviewed the applicant's technical personnel to confirm that the plant-specific operating experience revealed no degradation not bounded by industry experience.

FSAR Supplement. In LRA Section A2.8, the applicant provided the FSAR supplement for the Containment Leakage Testing Program. The staff reviewed this section and determines that the information in the FSAR supplement provides an adequate summary description of the program, as required by 10 CFR 54.21(d).

Conclusion. On the basis of its review and audit of the applicant's Containment Leakage Testing Program, the staff determines that those program elements for which the applicant claimed consistency with the GALL Report are consistent. The staff concludes that the applicant has demonstrated that the effects of aging will be adequately managed so that the intended functions will be maintained consistent with the CLB for the period of extended operation, as required by 10 CFR 54.21(a)(3). The staff also reviewed the FSAR supplement for this AMP and concludes that it provides an adequate summary description of the program, as required by 10 CFR 54.21(d).

3.0.3.1.6 Flow Accelerated Corrosion Program

Summary of Technical Information in the Application. In LRA Section B2.1.11, the applicant described the Flow Accelerated Corrosion Program, stating that this existing program is consistent with GALL AMP XI.M17, "Flow-Accelerated Corrosion."

The Flow Accelerated Corrosion Program manages aging effects due to flow-accelerated corrosion (FAC) on the internal surfaces of carbon or low-alloy steel piping, elbows, reducers, expanders, and valve bodies which contain high-energy fluids (both single phase and two phase). The program implements the Electric Power Research Institute (EPRI) guidelines in NSAC-202L-R2 for an effective FAC program and includes (a) an analysis using a predictive code such as the CHECWORKS computer code to determine critical locations, (b) baseline inspections to determine the extent of thinning at these locations, (c) follow-up inspections to confirm the predictions, and (d) repair or replacement of components, as necessary.

Staff Evaluation. During the audit and review, the staff confirmed the applicant's claim of consistency with the GALL Report. Details of the staff's evaluation of this AMP are documented in the Audit and Review Report. The staff determined that this AMP is consistent with the AMP described in the GALL Report, including the associated "operating experience" program element.

As documented in the Audit and Review Report, the staff noted that the "monitoring and trending" program element, as presented in the LRA, does not expressly commit to examinations in areas adjacent to locations where wall thickness was less than predicted. In the LRA, the applicant stated that the Flow Accelerated Program implements the EPRI guidelines in NSAC-202L-R2, which specifies such examinations. The staff found this program element consistent with the GALL Report and, therefore, acceptable.

The staff also noted that the "acceptance criteria" program element, as presented in the LRA, does not confirm that the number of refueling or operating cycles remaining would be computed. In the LRA, the applicant described the use of CHECWORKS, which generates this information in standard reports. As documented in the Audit and Review Report, the staff noted that the applicant also identified a minimum wall thickness acceptance criterion for SR piping at 87.5 percent of nominal wall thickness. By letter dated August 25, 2005, the applicant agreed to use the same value to trigger engineering analysis for NSR piping. The staff found this agreement consistent with the GALL Report and, therefore, acceptable.

Operating Experience. In LRA Section B2.1.11, the applicant explained that a review of the industry operating experience associated with the Flow Accelerated Corrosion Program and aging reveals issues and instances related to:

- feedwater heater shell degradation and ruptures
- feedwater and condensate line ruptures
- pipe wall thinning downstream of control valves and flow restricting devices
- valve body erosion
- extraction steam line ruptures
- moisture separator reheater drain tank drain line ruptures
- SG feedwater distribution piping and J-tube damage
- erosion of carbon steel ribs and tube supports in SGs

Various related NRC and/or industry generic communications have been issued and, in turn, have been incorporated into the program as applicable.

The applicant's review of plant-specific operating experience related to the Flow Accelerated Corrosion Program and aging revealed that the following issues had been addressed:

- FAC on 2-inch main steam line elbows

- higher than expected wear rates on 8-inch steam pipes and elbows on the outlet of moisture separator reheater

- main condenser tube leaks caused by FAC

- higher than expected wear rates on high pressure extraction steam piping to high pressure feedwater heater

- FAC on end-bell of low pressure feedwater heater

- valve body FAC on control valves and check valves

- FAC of feedwater heater shell side capped drains

- FAC damage to low-pressure turbine extraction sleeves

- FAC damage to extraction steam lines to high-pressure feedwater heaters

- FAC damage to moisture separator reheater vent line

- FAC of feedwater piping

- FAC of reducer downstream of control valve

- through-wall steam leak on steam generator flash tank

In addition, staff inspection reports, audits, self-assessments, and the corrective action program were reviewed for relevant information. There were no significant findings that indicated that the program is ineffective. Some weaknesses have been identified that resulted in appropriate corrective actions and enhancements.

The staff reviewed the applicant's FAC master plan to confirm that operating experience documented in the applicant's operating experience program and corrective action program reports had been incorporated into implementing procedures. The applicant's technical personnel were also interviewed to confirm that the plant-specific operating experience was bounded by industry experience. The staff found that the existing program has been effective in identifying, monitoring, and correcting FAC effects and can be expected to maintain piping wall thickness above the minimum design requirements.

FSAR Supplement. In LRA Section A2.11, the applicant provided the FSAR supplement for the Flow Accelerated Corrosion Program. The staff reviewed this section and determines that the information in the FSAR supplement provides an adequate summary description of the program, as required by 10 CFR 54.21(d).

Conclusion. On the basis of its review and audit of the applicant's Flow Accelerated Corrosion Program, the staff determines that those program elements for which the applicant claimed consistency with the GALL Report are consistent. The staff concludes that the applicant has demonstrated that the effects of aging will be adequately managed so that the intended functions will be maintained consistent with the CLB for the period of extended operation, as required by 10 CFR 54.21(a)(3). The staff also reviewed the FSAR supplement for this AMP and concludes that it provides an adequate summary description of the program, as required by 10 CFR 54.21(d).

3.0.3.1.7 Non-EQ Electrical Commodities Condition Monitoring Program

<u>Summary of Technical Information in the Application</u>. In LRA Section B2.1.12, the applicant described the Non-EQ Electrical Commodities Condition Monitoring Program, stating that this new program is consistent with GALL AMPs XI.E1, "Electrical Cables Not Subject to 10 CFR 50.49 Environmental Qualification Requirements," XI.E2, "Electrical Cables and Connections Not Subject to 10 CFR 50.49 Environmental Qualification Requirements Used in Instrumentation Circuits," and XI.E3, "Inaccessible Medium-Voltage Cables Not Subject to 10 CFR 50.49 Environmental Qualification Requirements."

The Non-EQ Electrical Commodities Condition Monitoring Program manages aging in selected non-EQ commodity groups within the scope of license renewal. Program activities are responsive to the staff guidance provided in the GALL Report and industry standards. The applicant identified each electrical commodity group requiring aging management for the three applicable sections of the GALL Report with the additional guidance provided in interim staff guidance (ISG)-2 and draft ISGs-5, 15, and 17, as follows:

- GALL AMP XI.E1 requires a periodic inspection program that visually inspects accessible cables and connections in adverse localized environments with any identified degradation evaluated and, as appropriate per plant procedures, entered into the plant corrective action process. The Non-EQ Electrical Commodities Condition Monitoring Program predominantly inspects for adverse aging from temperature, radiation, or moisture in the presence of oxygen.

- GALL AMP XI.E2 requires routine calibration tests to identify potential aging degradation of cables and connections used in low-level signal applications sensitive to reduction in insulation resistance (IR) such as radiation monitoring and nuclear instrumentation. This testing is revised as discussed in draft ISG-15, which allows testing every 10 years in lieu of TS surveillance test trending. The Non-EQ Electrical Commodities Condition Monitoring Program subjects sensitive instrumentation circuits, identified as requiring aging management, to periodic testing.

- GALL AMP XI.E3 requires a periodic test to indicate the condition of the conductor insulation for those cables within the scope of license renewal exposed to long periods of high moisture (greater than a few days at a time) and subjected to voltage stress (energized greater than 25% of the time). Periodic testing on these medium-voltage cables will indicate the insulation condition. The Non-EQ Electrical Commodities Condition Monitoring Program includes periodic inspections of underground raceway manholes for the accumulation of water over the medium-voltage cables. Periodic inspections of underground manholes for the accumulation of water in the medium-voltage cable manholes will minimize the effects of water inside the underground manholes.

<u>Staff Evaluation</u>. During the audit and review, the staff confirmed the applicant's claim of consistency with the GALL Report. Details of the staff's audit evaluation of this AMP are documented in the Audit and Review Report. The staff reviewed the enhancements and the associated justifications to determine whether the AMP, with the enhancements, remains adequate to manage the aging effects for which it is credited.

The scope of GALL AMP XI.E1 includes accessible electrical cables and connections within the scope of license renewal installed in localized adverse environments made adverse by heat or radiation in the presence of oxygen. As documented in the Audit and Review Report, the staff noted that the scope of the Non-EQ Electrical Commodities Condition Monitoring Program is different in that it includes only low-voltage cables and connections including pin connectors. It does not include medium-voltage cables or other cables and connections. As documented in the Audit and Review Report, the staff asked the applicant to explain how aging of accessible medium-voltage cables would be managed and why low-voltage cable electrical pinned connectors are identified as separate items. By letter dated August 25, 2005, the applicant responded that all cables are part of a spatial cable commodity. In Commitment 26, the applicant informed the staff that it would expand the scope of the Non-EQ Electrical Commodities Condition Monitoring Program to include all cables and connections within the scope of license renewal installed in localized adverse environments. The staff found the applicant's response acceptable because the scope of program is now consistent with that of GALL AMP XI.E1. By letter dated August 25, 2005, the applicant revised the LRA to expand the scope of the Non-EQ Electrical Commodities Condition Monitoring Program to include all cables and connections.

GALL AMP XI.E1 defines a localized adverse environment as a condition in a limited plant area that is significantly more severe than the specified service environment for the cable. The Non-EQ Electrical Commodities Condition Monitoring Program defines a localized adverse environment as exposure of any electrical insulation material to an aging environment significantly greater than the bounding design parameter value. The definition of adverse localized is inconsistent with the GALL AMP definition. As documented in the Audit and Review Report, the staff requested the applicant to explain the difference between significantly greater than the bounding design parameter value and significantly more severe than the specified service environment for the cable. By letter dated August 25, 2005, the applicant stated that the bounding design parameter has the same meaning as the specified service environment, that the terms are used interchangeably. The staff found the response acceptable because it clarified the definitions of the GALL AMP and the Non-EQ Electrical Commodities Condition Monitoring Program.

The applicant credited its 10 CFR Part 50, Appendix B, quality assurance (QA) program for the "corrective actions" program element; however, this program does not address the special requirements of GALL AMP XI.E1. This GALL AMP specifically requires that all unacceptable visual inspection results of cable and connection jacket surface anomalies be subject to engineering evaluations to consider component age and operating environment as well as the severity of the anomaly and whether such an anomaly has previously been correlated to degradation of conductor insulation or connections. Corrective actions may include, but not be limited to, testing, shielding, or otherwise changing the environment, or relocation or replacement of the affected cable or connection. When an unacceptable condition or situation is identified, a determination is made as to whether the same condition or situation is applicable to other accessible or inaccessible cables or connections. As documented in the Audit and Review Report, the staff requested the applicant to address these specific requirements. By letter dated August 25, 2005, the applicant informed the staff that it had committed to revise the Non-EQ Electrical Commodities Condition Monitoring Program to address the specific requirements (Commitment 26). The staff found the applicant's response acceptable because the "corrective action" program elements will be consistent with the GALL Report. By letter dated August 25,

2005, the applicant added these specific requirements to the Non-EQ Electrical Commodities Condition Monitoring Program, under the "corrective actions" program element.

The testing frequency in GALL AMP XI.E2, under the "detection of aging effects" program element, is different from the test frequency proposed in the Non-EQ Electrical Commodities Condition Monitoring Program. GALL AMP XI.E2 states that the test frequency of these cables shall be determined by the applicant based on an engineering evaluation, but shall not exceed ten years. The Non-EQ Electrical Commodities Condition Monitoring Program states that a ten-year frequency is an adequate period to identify cables and connection degradation to preclude excessive leakage currents as experience has shown that aging degradation is a slow process and test frequency should be determined by an engineering evaluation based on the test results. As documented in the Audit and Review Report, the staff questioned the applicant on the basis of the ten-year test frequency. In response, the applicant informed the staff that it had committed to revise the Non-EQ Electrical Commodities Condition Monitoring Program to state that test frequency of these cables shall be determined based on an engineering evaluation but shall not exceed ten years. The staff found this response acceptable because the proposed test frequency is consistent with GALL AMP XI.E2. By letter dated August 25, 2005, the applicant revised LRA Section B2.1.12 to include this change.

GALL AMP XI.E2 states that all cables within the scope of this program will be tested. The Non-EQ Electrical Commodities Condition Monitoring Program states that cables used in nuclear instrumentation circuits would be tested and if an unacceptable condition or situation is identified, a determination shall be made as to the applicability of the conditions on other cables in the nuclear instrumentation circuits. It was not clear to the staff whether only some or all nuclear instrumentation cables within the scope of the program would be tested. As documented in the Audit and Review Report, the staff requested the applicant to clarify this inconsistency. In its response, the applicant informed the staff that, to avoid confusion, it had revised the Non-EQ Electrical Commodities Condition Monitoring Program to test all sensitive cables and connections within the scope of license renewal and deleted the following text: "...and if an unacceptable condition or situation is identified, a determination shall be made as to applicability of the condition on other cables used in the nuclear instrumentation circuits." By letter dated August 25, 2005, the applicant revised LRA Section B2.1.12 to include this change.

In LRA Section B1.2, "Quality Assurance Program and Administrative Controls," the applicant credited the QA program for the "corrective actions" program element. However, this program does not address the special requirements of GALL AMP XI.E2, which specifically recommends an engineering evaluation when test acceptance criteria are not met to ensure that the intended functions of the electrical cable system can be maintained consistent with the CLB. Such an evaluation considers the significance of the test results, the operability of the component, reportable events, extent of the concern, likely root cause for not meeting test acceptance criteria, and likelihood of recurrence. As documented in the Audit and Review Report, the staff requested the applicant to address these specific requirements in the Non-EQ Electrical Commodities Condition Monitoring Program. The applicant clarified that it had addressed them in the AMP under the "corrective actions" program element. The staff found the applicant's response acceptable because the "corrective actions" program element is consistent with GALL AMP XI.E2. By letter dated August 25, 2005, the applicant revised LRA Section B2.1.12 to include this change.

The typical tests in the Non-EQ Electrical Commodities Condition Monitoring Program are different from the tests in the GALL Report. GALL AMP XI.E3, under the "parameters monitored or inspected" program element, states that a specific type of test shall be determined prior to the initial test and shall be a proven test like power factor, partial discharge, or polarization index, as described in EPRI TR-103834-P1-2, for detecting deterioration of the insulation system due to wetting or other state-of-the-art testing at the time the test is performed. The Non-EQ Electrical Commodities Condition Monitoring Program states that inaccessible medium-voltage insulated cables within the scope of license renewal subject to long periods of high-moisture conditions and voltage stress are tested by insulation resistance tests, time domain reflectometry tests, or other tests effective in determining cable insulation conditions. The staff noted that insulation resistance and time domain reflectometry are types of tests in GALL AMP XI.E2 for low-voltage sensitive instrumentation cables. As documented in the Audit and Review Report, the staff requested the applicant to provide the basis for insulation resistance or time domain reflectometry tests for inaccessible medium-voltage cables. The applicant responded that GALL AMP XI.E3 references EPRI TR-103834-P1-2, which discusses insulation resistance and time domain reflectometry tests. However, the applicant informed the staff that it had committed to revise the test methods to be consistent with GALL AMP XI.E3. The staff found the applicant's response acceptable because the test methods for detecting deterioration of insulated system for inaccessible medium-voltage would be the same as those of GALL AMP XI.E3. By letter dated August 25, 2005, the applicant revised LRA Section B2.1.12 to include this change.

GALL AMP XI.E3, under the "detection of aging effects" program element, states that inspection intervals for water collection should be based on actual field data, but should not exceed two years. The Non-EQ Electrical Commodities Condition Monitoring Program states that periodic inspections (periodicity based on inspection results) of underground manholes for the accumulation of water levels would be conducted. The applicant's AMP does not specify the maximum inspection duration for water collection. As documented in the Audit and Review Report, the staff requested the applicant to provide manhole inspection frequency. The applicant informed the staff that it had committed to revise the LRA to state that periodicity would be based on inspection results but not exceed two years. The staff found the applicant's response acceptable because the inspection interval for water collection in manholes is consistent with the GALL Report. By letter dated August 25, 2005, the applicant revised LRA Section B2.1.12 to include this change.

GALL AMP XI.E3 requires testing of all medium-voltage cables within the scope of the program exposed to significant voltage and moisture. The Non-EQ Electrical Commodities Condition Monitoring Program states that if an unacceptable condition or situation is detected, a determination would be made as to whether it is applicable to other inaccessible medium-voltage cables within the scope of the program. It was not clear to the staff whether all cables within the scope of the program or just a sample would be tested. As documented in the Audit and Review Report, the staff identified this confusion and asked the applicant to confirm that all cables within the scope of the program would be tested. In its response, the applicant informed the staff that it would revise LRA Section B2.1.12 to clearly state that all cables within the scope of the program would be tested. The staff found the applicant's response acceptable because it eliminated the confusion. By letter dated August 25, 2005, the applicant revised LRA Section B2.1.12 to include this change.

In LRA Section B1.2, the applicant credited the QA program for the "corrective actions" program element. However, the GALL AMP XI.E3 "corrective actions" program element requires such specifics as engineering evaluations when test acceptance criteria are not met in order to ensure that the intended function of the electrical system can be maintained consistent with the CLB. Such an evaluation is to consider the significance of the test results, operability of the component, reportable events, extent of the concern, potential root cause for not meeting the test acceptance criteria, corrective actions required, and likelihood of recurrence. As documented in the Audit and Review Report, the staff requested the applicant to address these specific requirements under the "correction actions" program element. The applicant informed the staff that it had committed to revise LRA Section B2.1.12 under the "corrective actions" program element to add such specific requirements. By letter dated August 25, 2005, the applicant revised LRA Section B2.1.12 to include these changes.

GALL AMP XI.E4, "Metal-Enclosed Bus" (previously ISG-17), states that a sample of metal-enclosed bus bolted connections shall be checked for a loose connection by using thermography or by measuring connection resistance with low-range ohmmeter. This program also inspects the internal portion of the metal-enclosed buses for cracks, corrosion, foreign debris, excessive dust buildup, and evidence of water intrusion. The bus insulating system would be inspected for signs of embrittlement, cracking, melting, swelling, or discoloration which may indicate overheating or aging degradation. The (internal) bus support would be inspected for structural integrity and signs of cracks. This program would be completed before the end of the initial 40-year license term and every 10 years thereafter. LRA Section B2.1.12 states that the periodic inspection shall also include signs of water leakage or contamination through the housing seals into the non-segregated bus and signs of localized heating from loose internal electrical connections that may lead to electrical failure. It fails to address inspection of all of metal-enclosed bus SCs like the external bus enclosure, internal bus enclosure, internal bus support, and bus insulating systems. As documented in the Audit and Review Report, the staff requested the applicant to explain how cracking, corrosion, foreign debris, etc. would be identified without internal inspections, how aging effects of internal bus support and metal-enclosed bus insulating systems would be managed, and how loose connections would be identified. The applicant informed the staff that it had committed to revise LRA Section B2.1.12 to state that periodic inspections also would include visual inspections for signs of water leakage or contamination through the housing seals into the non-segregated bus and signs of localized heating from loose internal electrical connections that may lead to electrical failure. This program would also provide for the inspection of the internal portion of the bus ducts for cracks, corrosion, foreign debris, excessive dust buildup, and evidence of water intrusion. The bus insulating system would be inspected for signs of cracking, melting, swelling, or discoloration which may indicate overheating or aging degradation. The (internal) bus supports would be inspected for structural integrity and signs of cracks. A representative sample of the bus connections would be inspected by thermography. The program would be completed before the end of the 40-year license term and every 10 years thereafter. The staff found the applicant's response acceptable because it conformed to GALL AMP XI.E4. By letter dated August 25, 2005, the applicant revised LRA Section B2.1.12 to include these changes.

In ISG-17 under the "preventive action" program element states that no actions are taken in this inspection program to prevent or mitigate aging degradation. No information was provided for this program element in LRA Section B2.1.12. As documented in the Audit and Review Report, the staff requested the applicant to include this program element. The applicant informed the staff that it had committed to revise LRA Section B2.1.12, under the "preventive actions" program element, to be consistent with ISG-17. The staff found the applicant's response acceptable and consistent with the program element described in ISG-17. By letter dated August 25, 2005, the applicant revised LRA Section B2.1.12 to include this program element.

In the LRA, the applicant stated that its Non-EQ Electrical Commodities Condition Monitoring Program is consistent with GALL AMPs XI.E1, XI.E2, and XI.E3 with enhancements. The applicant also stated that a Non-EQ Electrical Commodities Condition Monitoring Program would be developed and implemented. Features of the program would include development and implementation of procedures to inspect insulated cables and connectors, test sensitive instrumentation circuits, test medium-voltage cables, and inspect manhole water levels.

In addition, the applicant stated that the program elements describe the program to be implemented prior to the period of extended operation.

The staff found this statement acceptable because the applicant committed to develop the Non-EQ Electrical Commodities Condition Monitoring Program to include procedures for conducting periodic inspection of insulated cables and connectors, testing sensitive instrumentation circuits, testing medium-voltage cables, and inspecting manhole water levels consistent with GALL AMPs XI.E1, XI.E2, and XI.E3, and the guidance in ISG-2 and draft ISGs-5, 15, and 17.

Operating Experience. In LRA Section B2.1.12, the applicant explained that industry experience, as documented in SAND96-0344, "Aging Management Guideline for Commercial Nuclear Plants - Electrical Cables and Terminations," has shown three main causes of cable and connection failures well before a nominal 40- or 60-year service life:

(1)   Cables routed/installed in abnormal configurations, outside the prescribed or normal design guidelines and installation design criteria, may fail due to exposure to temperatures well above the expected normal ambient temperature. Polyvinyl chloride insulated cable insulation failures are the most common cable insulation failures to occur due to high temperature and/or radiation environments.

(2)   Sensitive instrumentation cable insulations (nuclear instrumentation and radiation monitoring) have less tolerance for "loss of-material properties" that adversely affect the circuit signals.

(3)   Medium-voltage power cable failures occur because of water-treeing (moisture and voltage stress).

The applicant further explained that site-specific experience has shown that existing routine switchyard inspections detect loose connections in the switchyard. Existing periodic and routine switchyard inspections preclude failures of connections in the switchyard.

Abnormal plant configurations were found to produce localized adverse environments in some specific cases. A corrective action document identified signs of cable jacket damage from improper design/installation that led to a localized adverse environment for the cables. In addition, LER 84-10 resulted from improper design and installation outside expected normal cable configurations. The corrective action program corrected both plant configurations to eliminate the identified localized adverse temperature environments.

A medium-voltage cable failure had occurred from the possible effects of water-treeing. LER 96-002 demonstrates that this commodity group warrants periodic testing to preclude or minimize future failures. PNP has also experienced that the underground manholes for the medium-voltage cables within the scope of license renewal have experienced moisture for periods greater than a few days at a time.

The applicant also explained that one cable commodity-related assessment was conducted to address over-loaded cable trays. This analysis calculated power cable ohmic heating temperatures in those overloaded tray sections and compared it against the respective cable temperature rating to ensure that proper operating conditions exist and are maintained. The results of this analysis were considered when reviewing the plant electrical cables and connections and were addressed when assessing and identifying those cables requiring aging management during the period of extended operation.

The applicant indicated that since the Non-EQ Electrical Commodities Condition Monitoring Program is a new program, no NRC inspection reports, audits, self assessments, or program-specific corrective actions are available.

The staff reviewed the operating experience provided in the LRA and interviewed the applicant's technical personnel to confirm that the plant-specific operating experience revealed no degradation not bounded by industry experience.

On the basis of its review of the industry and plant-specific operating experience and discussions with the applicant's technical personnel, the staff concludes that the applicant's Non-EQ Electrical Commodities Condition Monitoring Program would adequately manage the aging effects identified in the LRA for which this AMP is credited.

The staff recognized that the applicant's corrective action program, which documents internal and external plant operating experience issues, ensures that future operating experience will be reviewed and captured to provide objective evidence that the effects of aging are adequately managed.

FSAR Supplement. In LRA Section A2.12, the applicant provided the FSAR supplement for the Non-EQ Electrical Commodities Condition Monitoring Program. The staff reviewed this section and determines that the information in the FSAR supplement provides an adequate summary description of the program, as required by 10 CFR 54.21(d).

Conclusion. On the basis of its review and audit of the applicant's Non-EQ Electrical Commodities Condition Monitoring Program, the staff determines that those program elements for which the applicant claimed consistency with the GALL Report are consistent. The staff concludes that the applicant has demonstrated that the effects of aging will be adequately managed so that the intended functions will be maintained consistent with the CLB for the period of extended operation, as required by 10 CFR 54.21(a)(3). The staff also reviewed the FSAR supplement for this AMP and concludes that it provides an adequate summary description of the program, as required by 10 CFR 54.21(d).

3.0.3.1.8 Open Cycle Cooling Water Program

Summary of Technical Information in the Application. In LRA Section B2.1.14, the applicant described the Open Cycle Cooling Water Program, stating that this existing program is consistent with GALL AMP XI.M20, "Open-Cycle Cooling Water System."

The Open Cycle Cooling Water Program manages aging effects such as loss of material due to general, pitting, and crevice corrosion, erosion, MIC, and loss of heat transfer due to biological/corrosion product fouling (e.g., sedimentation, silting) caused by exposure of internal surfaces of metallic components to raw, untreated (e.g., service) water. The program's scope includes activities to manage aging in the service water system (SWS) and circulating water system (CWS). The aging effects are managed through (a) monitoring and control of biofouling, (b) flow balancing and flushing, (c) HX testing, and (d) routine inspection and maintenance program activities to ensure that aging effects do not impair component intended function. Inspection methods include visual, ultrasonic, radiographic, and eddy current tests. This program is responsive to NRC Generic Letter (GL) 89-13. PNP established a routine inspection and maintenance monitoring program for service water piping and components to ensure that corrosion, erosion, silting, and biofouling cannot degrade the performance of the safety-related (SR) systems supplied by service water to where they are unable to perform their intended functions.

Staff Evaluation. During the audit and review, the staff confirmed the applicant's claim of consistency with the GALL Report. Details of the staff's evaluation of this AMP are documented in the Audit and Review Report. The staff determined that this AMP is consistent with the AMP described in the GALL Report, including the associated "operating experience" program element.

As documented in the Audit and Review Report, the staff noted that the current "scope of program" program element, as described in the LRA, does not include commitments for two GL 89-13 guidelines incorporated in GALL AMP XI.M20. The specific components of the GL 89-13 program missing from the Open Cycle Cooling Water Program are a system walkdown inspection to ensure compliance with the CLB and a review of maintenance, operating, and training practices and procedures. As documented in the Audit and Review Report, in technical discussions with the staff, the applicant indicated that maintenance activities in fact meet these two components of GALL AMP XI.M20. The applicant stated that it would revise the LRA to include them. By letter dated August 25, 2005, the applicant stated that the original statements in the "scope of program" program element for the Open Cycle Cooling Water Program concerning implementation of recommendations (d) and (e) of GALL AMP XI.M20 were not correct, that they are included in existing plant procedures in the ongoing management of aging, and are not exceptions to the GALL Report. With this revision, the Open Cycle Cooling Water Program is consistent with the GALL AMP and, therefore, acceptable.

Operating Experience. In LRA Section B2.1.14, the applicant explained that a review of the industry operating experience associated with the Open Cycle Cooling Water Program and aging reveals issues and instances related to:

- accumulations of silt and corrosion products in service water piping, valves, and HXs

- accumulation of biological growth (mussels, clams, and shells) in service water piping, valves, and HXs
- MIC causing pitting attack of carbon steel and stainless steel service water piping, pump casings, and 90/10 copper/nickel heat exchanger tubes

A review of plant-specific operating experience related to the Open Cycle Cooling Water Program and aging revealed that the following issues had been addressed:

- defective tubes in the main condenser that required plugging due to MIC
- control room condensing unit condenser drain plug severely corroded due to MIC
- large zebra mussel accumulation near traveling screens and inside intake piping
- blockage of heat exchanger and cooler tubing
- corroded service water piping at threaded connections
- pinhole leaks in service water piping due to MIC
- switch failure due to sediment and corrosion (galvanic) blocking sensing line
- tubercles growing in carbon steel service water piping
- erosion of pipes, cooling coils, and HX tubes causing service water leaks

The applicant stated, in the LRA, that using the operating experience and corrective action programs to focus on industry and plant operating experience ensured that Open Cycle Cooling Water Program issues are addressed promptly and that age-related deterioration of SSCs within the scope of the Open Cycle Cooling Water Program would be effectively managed throughout the period of extended operation.

Furthermore, the applicant stated that the Open Cycle Cooling Water Program had demonstrated on several occasions that it provides reasonable assurance that aging effects are being managed for SSCs within the program. This demonstration is through NRC inspection reports, audits, self-assessments, and the corrective action program.

The staff reviewed a sample of the operating experience documents and noted that, as part of the corrective actions, program and system upgrades have been implemented. These included new system and component flushes, procedure changes, and new non-destructive examination (NDE) testing. There was no indication in the documents reviewed that the intended function of any system components had been lost before detection and corrective action. The staff's review found that the program had been effective in managing the aging effects. This finding was augmented by a review of recent system health reports for the service water, circulating water, and cooling towers systems. The system health reports indicated no issues and supported the conclusion that the applicant's program is effective in managing the aging effects for which this AMP is credited.

The staff reviewed the operating experience provided in the LRA and interviewed the applicant's technical personnel to confirm that the plant-specific operating experience revealed no degradation not bounded by industry experience.

FSAR Supplement. The applicant provided the FSAR Supplement for the Open Cycle Cooling Water Program in LRA Section A2.14. The current program, as described in the LRA, does not contain commitments for two of the GL 89-13 guidelines contained in the FSAR supplement table for the AMP. The specific components of the GL 89-13 program that are missing are a system walkdown inspection to ensure compliance with the licensing basis and a review of maintenance, operating, and training practices and procedures. By letter dated August 25, 2005, the applicant revised the LRA to add these statements to the "scope of program" description for the AMP. The staff reviewed this section and determines that the information in the FSAR supplement, with revision, provides an adequate summary description of the program, as required by 10 CFR 54.21(d).

Conclusion. On the basis of its review and audit of the applicant's Open Cycle Cooling Water Program, the staff determines that those program elements for which the applicant claimed consistency with the GALL Report are consistent. The staff concludes that the applicant has demonstrated that the effects of aging will be adequately managed so that the intended functions will be maintained consistent with the CLB for the period of extended operation, as required by 10 CFR 54.21(a)(3). The staff also reviewed the revised FSAR supplement for this AMP and concludes that it provides an adequate summary description of the program, as required by 10 CFR 54.21(d).

3.0.3.1.9 Steam Generator Tube Integrity Program

Summary of Technical Information in the Application. In LRA Section B2.1.18, the applicant described the Steam Generator Tube Integrity Program, stating that this existing program is consistent with GALL AMP XI.M19, "Steam Generator Tube Integrity."

The Steam Generator Tube Integrity Program manages the aging effects of SG tubes and tube repairs. The program also manages the aging effects of accessible SG secondary side internal components and incorporates the guidance of NEI 97-06. The program manages aging effects through a balance of mitigation, inspection, evaluation, repair, and leakage monitoring measures. Component degradation is mitigated by controlling primary and secondary water chemistry. Eddy current tests are used to detect SG tube flaws and degradation. Visual examinations identify degradation of accessible SG secondary side internal components. Primary to secondary leakage is monitored during plant operation. The program credits the Water Chemistry Program for primary and secondary water chemistry control. The program also satisfies ASME Code Section XI IWB-2500 Category B-Q requirements for performing volumetric examinations of SG tubes in the ASME Section XI IWB, IWC, IWD, IWF Inservice Inspection Program.

Staff Evaluation. During the audit and review, the staff confirmed the applicant's claim of consistency with the GALL Report. Details of the staff's evaluation of this AMP are documented in the Audit and Review Report. The staff determined that this AMP is consistent with the AMP described in the GALL Report, including the associated "operating experience" program element.

The staff concludes that the applicant's Steam Generator Tube Integrity Program provides reasonable assurance that the aging effects of SG tubes and tubes repairs would be adequately managed. The program also manages the aging effects of accessible SG secondary side internal components with the guidance of NEI 97-06. The staff found the applicant's Steam Generator Tube Integrity Program acceptable because it conforms to the recommended GALL AMP XI.M19.

Operating Experience. In LRA Section B2.1.18, the applicant explained that review of the industry operating experience related to the Steam Generator Tube Integrity Program revealed instances where degradation had occurred within the SGs. The applicant looked at related issues which included degradation of SG tubes, tube sheet, mechanical plugs, tube support plates, girth welds, anti-vibration bars, etc., plus degradation associated with loose parts, foreign objects, sludge, water chemistry, and wear.

A review of the plant-specific operating experience revealed several instances where the Steam Generator Tube Integrity Program had been instrumental in discovering material degradation. Steam Generator tube degradation was discovered in the following four areas: (1) top of tubesheet, (2) within the tubesheet, (3) U-bends, and (4) mechanical wear at eggcrate supports, vertical straps, and diagonal bars.

The SGs were replaced in late 1990. The new SGs are improved in design, material selection, and construction. Included in the new design was a change in the tube support from solid plate to egg crate dividers along with other features to minimize corrosion crevices and denting.

The staff reviewed the applicant's SG tube integrity inspection reports and an assessment of the 2003 refueling outage. As documented in the Audit and Review Report, the applicant explained that the SGs are Combustion Engineering (CE) Model 2530. Each SG has 8219 tubes. The tube material is mill annealed Alloy 600. Prior to installation of the replacement SGs, 308 tubes in SG E-50A and 309 tubes in SG E-50B potentially susceptible to fretting wear were plugged as a preventive measure. After nine cycles of operation, 72 additional tubes in SG E-50A and 54 additional tubes in SG E-50B have been plugged. Therefore, SG E-50A has 7839 active tubes with 4.62 percent of the tubes plugged. SG E-50B has 7856 active tubes with 4.42 percent of the tubes plugged. Active degradation mechanisms are structural wear in SG E-50B and axial outer diameter stress corrosion cracking (ODSCC) in SGs E-50A/B.

The staff reviewed the operating experience provided in the LRA, and interviewed the applicant's technical personnel to confirm that the plant-specific operating experience revealed no degradation not bounded by industry experience.

FSAR Supplement. In LRA Section A2.18, the applicant provided the FSAR supplement for the Steam Generator Tube Integrity Program. The staff reviewed this section and determines that the information in the FSAR supplement provides an adequate summary description of the program, as required by 10 CFR 54.21(d).

Conclusion. On the basis of its review and audit of the applicant's Steam Generator Tube Integrity Program, the staff determines that those program elements for which the applicant claimed consistency with the GALL Report are consistent. The staff concludes that the applicant has demonstrated that the effects of aging will be adequately managed so that the intended functions will be maintained consistent with the CLB for the period of extended operation, as required by 10 CFR 54.21(a)(3). The staff also reviewed the FSAR supplement for this AMP and concludes that it provides an adequate summary description of the program, as required by 10 CFR 54.21(d).

3.0.3.1.10 Water Chemistry Program

Summary of Technical Information in the Application. In LRA Section B2.1.21, the applicant described the Water Chemistry Program, stating that this existing program is consistent with GALL AMP XI.M2, "Water Chemistry."

The Water Chemistry Program is credited for managing aging effects such as loss of material due to general, pitting, and crevice corrosion, cracking due to SCC, and SG tube degradation caused by denting, intergranular attack (IGA), and ODSCC by controlling the environment to which internal surfaces of systems and components are exposed. The aging effects are minimized by controlling the chemical species that cause the underlying mechanisms. The program provides assurance that an elevated level of contaminants and, where applicable, oxygen does not exist in the systems and components covered by the program, thus minimizing aging effects, and maintaining each component's ability to perform the intended functions. The program is based on the guidelines in EPRI TR-105714, Revision 5, and TR-102134, Revision 5. The One-Time Inspection Program verifies that the Water Chemistry Program manages the effects of aging of selected components in low-flow or stagnant areas.

Staff Evaluation. During the audit and review, the staff confirmed the applicant's claim of consistency with the GALL Report. Details of the staff's evaluation of this AMP are documented in the Audit and Review Report. The staff determined that this AMP is consistent with the AMP described in the GALL Report, including the associated "operating experience" program element.

The GALL AMP XI.M2 "scope of program" program element states that water chemistry control is in accordance with the guidelines in EPRI TR-105714, Revision 3, for primary water chemistry in pressurized-water reactors (PWRs), EPRI TR-102134, Revision 3, for secondary water chemistry in PWRs, or later revisions or updates of these reports as approved by the staff. The Water Chemistry Program "scope of program" program element states that the program accomplishes this task by monitoring and controlling known detrimental contaminants like chlorides, fluorides, dissolved oxygen, and sulfate concentrations for water chemistry based on guidelines in EPRI TR-105714, Revision 5, for primary water chemistry and TR-102134, Revision 5, for secondary water chemistry. The program includes sampling activities for primary, borated, secondary, and makeup water systems. The applicant claimed consistency with the GALL Report and explained that use of a later revision of the EPRI guidelines is not considered an exception. As documented in the Audit and Review Report, the staff found this use different from the GALL Report recommendation and asked the applicant to compare the monitored parameters for Revision 3 and Revision 5 of the EPRI guidelines and explain why use of a later version is acceptable by verifying that none of the controlled parameters are relaxed in the later version.

In its response, by letter dated August 25, 2005, the applicant stated:

> NMC understands the NRC position that revisions of codes or standards used in AMPs that are not the same as those referenced in the GALL, are to be identified as exceptions to GALL. Therefore, code or standard revisions that are used in Palisades' programs, but are not referenced by either the 2001 or 2005 GALL descriptions, will be treated as exceptions to the GALL, and justification will be

provided as required. Revisions or supplements to the affected program descriptions will be submitted to the NRC.

XI.M2 - Primary Chemistry - The 2001 and 2005 GALL revisions reference EPRI TR-105714 Revision 3, and the Palisades AMP is based on Revision 5. NMC will prepare and submit a comparison of TR-105714 Revision 5 with Revision 3 to identify the material changes that impact aging management and justify their acceptability by October 31, 2005. If necessary, the Water Chemistry Program description will be revised to identify and justify use of TR-105714, Revision 5, as an exception to the GALL program description.

XI.M2 - Secondary Chemistry - 2001 and 2005 GALL revisions reference TR-102134 Revision 3, and the Palisades AMP is based on Revision 6. NMC will prepare and submit a comparison of TR-102134 Revision 6 with Revision 3 to identify the material changes that impact aging management and justify their acceptability by October 31, 2005. If necessary, the Water Chemistry Program description will be revised to identify and justify use of TR-102134, Revision 6, as an exception to the GALL program description.

Based on its review of the information provided in the applicant's response, the staff finds the applicant's use of the new EPRI document revisions acceptable.

By letter dated October 31, 2005, the applicant submitted a comparison of TR-105714, Revision 5 (issued as TR-102884, "Pressurized Water Reactor Primary Water Chemistry Guidelines"), with TR-105714, Revision 3, to identify the material changes that impact aging management and justify their acceptability. The applicant stated that after comparison of the two revisions of the EPRI primary water chemistry guidelines, it had concluded that Revision 5 provides an acceptable level of control to effectively manage aging of the associated systems. The applicant also stated that based on this comparison and to indicate the use of Revision 5 as an exception to the GALL Report, the LRA description of the primary chemistry portions of the Water Chemistry Program would be changed. The changes described below are relevant to the primary chemistry portions of the LRA description of the Water Chemistry Program.

By letter dated October 31, 2005, the applicant revised LRA Section B2.1.21 as follows:

On page B-154, under NUREG 1801 Consistency, replace the entire section with the following statement: 'The Water Chemistry program is an existing program that is consistent with, but includes exceptions to, NUREG 1801-Section XI.M2, 'Water Chemistry'.

On page B-154 under the heading of Exceptions to NUREG 1801, replace the entire section with the following statement: 'An exception is taken to the selected NUREG 1801 program elements listed below. The specific exceptions being taken are also discussed in the corresponding element discussions. They are repeated here for ease of review.

    1.   The Palisades Plant Water Chemistry program implements the PWR Primary Water Chemistry Guidelines, revision 5, which is issued as EPRI 102884, 'Pressurized Water Reactor Primary

Water Chemistry Guidelines, Revision 5' dated September 2003 (revised October 2003). Justification has been developed to show that this revision to the guidance has no detrimental affects [sic] on aging management of the water chemistry program.

On page B-159, under heading Conclusion, first paragraph, replace the last sentence with, 'The program is consistent with, but includes exceptions to, NUREG-1801, Section XI.M2, 'Water Chemistry."

The staff reviewed the applicant's comparison of Revision 3 with Revision 5 and found that the parameters of operation in Revision 5 were generally the same as those of Revision 3, or more conservative. In the few instances where the Revision 5 parameters were less conservative or new, the staff concurred with the applicant that the higher limit or new parameter was acceptable for aging management. On this basis, the staff finds the applicant's use of TR-105714, Revision 5, acceptable.

By letter dated October 31, 2005, the applicant submitted a comparison of TR-102134, Revision 6 (issued as TR-108224, "Pressurized Water Reactor Secondary Water Chemistry Guidelines"), with TR-102134, Revision 3, to identify and justify the acceptability of the material changes that impact aging management. The applicant stated that after comparison of Revision 6 with Revision 3, it concluded that Revision 6 provides an acceptable level of control to effectively manage aging of the systems. The applicant also stated that based on this comparison and to indicate the use of Revision 6 as an exception to the GALL Report, the LRA description of the secondary chemistry portions of the Water Chemistry Program had been changed. The changes described below are relevant to the secondary chemistry portions of the LRA description of the Water Chemistry Program and are based on the changes made to the primary chemistry portions.

By letter dated October 31, 2005, the applicant revised LRA Section B2.1.21 as follows:

On Page B-153, under heading Program Description, replace 'TR-102134, Rev. 5' in the first paragraphs, with 'TR-108224.' In the third paragraph, replace the last sentence to read, Palisades has adopted TR-102134, Rev. 5 (issued as TR-102884) and TR-102134 Rev. 6 (issued as TR-108224) which are later revisions of the same documents. To the end of the fifth paragraph, add the following, 'Revision 6 provides further details regarding how to best integrate these guidelines into this plant-specific optimization process.'

On page B-154, under the heading Exceptions to NUREG 1801, add the following exception number two:

2.   The Palisades Water Chemistry Program implements the guidelines of revision 6 of the EPRI PWR secondary water chemistry Guidelines, issued as EPRI TR-108224, 'Pressurized Water Reactor Secondary Water Chemistry Guidelines - Revision 6,' dated October 2003. Justification provided shows that this later revision to the guidance ensures that the Water Chemistry Program will continue to provide effective management of aging.

On Page B-1 54, under the heading Scope of Program, second sentence, replace TR-102134 with TR-108224.

On Page B-159, under the heading Conclusion, first paragraph, replace the last sentence with, 'This program is consistent with, but includes exceptions to, NUREG 1801 Section XI.M2, 'Water Chemistry.''

Based on its review of the applicant's comparison of Revision 6 with Revision 3, the staff finds that the parameters of operation in Revision 6 were generally the same as those of Revision 3, or more conservative. In the few instances where the Revision 6 parameters were less conservative, no longer measured, or renamed, the staff concurred with the applicant that the higher limit, elimination of the parameter, or renaming of the parameter was acceptable for aging management. On this basis the staff finds the applicant's use of TR-102134, Revision 6, acceptable.

GALL AMP XI.M2 states that in certain cases verification of the effectiveness of the chemistry control program is undertaken to ensure that significant degradation does not occur and that the component intended function will be maintained during the period of extended operation. An acceptable verification program is a one-time inspection of selected components at susceptible locations in the system. The applicant stated that it uses the One-Time Inspection Program to verify Water Chemistry Program management of aging effects of selected components in low-flow or stagnant areas. As documented in the Audit and Review Report, the staff noted that the LRA did not clarify that only the susceptible low-flow and stagnant locations require inspection by the One-Time Inspection Program.

By letter dated August 25, 2005, the applicant responded as follows:

While the most susceptible locations may be the low flow or stagnant potions of a particular system, it was not intended to limit the selection of susceptible locations to low flow or stagnant portions of a system. Upon implementation of the One-Time Inspection Program, NMC plans to group all identified components within the system with the same material, same environment, and same aging mechanism. From this group, the most susceptible locations will be selected for inspection. When determining the most susceptible locations, all portions of the system(s) will be considered, not just the low flow or stagnant sections. Therefore, the following changes are made to the Water Chemistry and One-Time Inspection Program descriptions in Appendix B of the LRA:

On Page B-97, under Program Description, revise the first bullet of the third paragraph to read, 'To verify the effectiveness of water chemistry control for managing the effects of aging in portions of piping exposed to a treated water environment.'

On Page B-103, under Detection of Aging Affects [sic] Related to XI.M32, One Time Inspection, revise the second paragraph in its entirety to read, 'To verify that the Water Chemistry Program and the Closed Cycle Cooling Water Program are mitigating the applicable aging effects, visual examinations or other appropriate NDE methodology will be used when components are inspected.'

On Page B-156, under Detection of Aging Affects [sic], revise the last sentence of the first paragraph to read, 'In addition, inspections of selected components at susceptible locations of a system, performed under the One-Time Inspection Program, provide verification of the effectiveness of the Water Chemistry Program.'

The staff reviewed the applicant's response and found that it provides adequate information about the selection of susceptible locations not limited to low-flow or stagnant portions of a system. Also, the applicant revised its One-Time Inspection Program to verify the effectiveness of the Water Chemistry Program. On these bases, the staff finds the applicant's response acceptable.

Operating Experience. In LRA Section B2.1.21, the applicant explained that a review of industry operating experience associated with the Water Chemistry Program and aging reveals issues and instances related to:

- cracking in SG welds
- cracking and pitting of SG tubes and components
- Alloy 600 cracking
- thinning of pipe and components due to erosion/corrosion
- cracking in safety injection accumulator nozzles
- high wear of reactor coolant pump aluminum oxide coated seals
- cracking of CRD housings
- cracking of pressurizer instrument tap nozzles
- cracking of safety injection piping
- cracking in feedwater piping
- chemical impurity intrusions into primary and secondary systems
- resin intrusions into the primary coolant systems

Various related NRC and/or industry generic communications have been issued and, in turn, incorporated into the program as applicable.

A review of plant-specific operating experience related to the Water Chemistry Program and aging revealed that the following two issues had been addressed: (1) defective tubes in the main condenser due to steam impingement wear and MIC pitting and (2) exceeding Action Level 3 limits for SG cation conductivity.

The applicant further explained that the second item involved exceeding Action Level 3 limits for SG cation conductivity, resulting in a plant shutdown. The cause of the high conductivity was traced to intrusion of glass-blasting material left in the turbine following a major overhaul/replacement of the turbine. Although project oversight weaknesses were identified in the events leading up to this chemistry excursion, proper chemistry monitoring quickly identified the rising cation conductivity levels, and subsequent actions prevented long term age-related degradation of components as Action Level 3 limits were exceeded for less than six hours. Compensatory actions were taken over the next cycle to ensure a high degree of contaminant removal or neutralization.

The staff reviewed the operating experience related to the applicant's Water Chemistry Program and found that there had been no significant increases of the controlled and diagnostic parameters above the EPRI guidelines action levels. The operating experience is typical of issues found in corrective actions where a parameter has exceeded an EPRI action level. The staff identified no operating experience issues related to the applicant's Water Chemistry Program that would require program modification.

The staff reviewed the operating experience provided in the LRA and interviewed the applicant's technical personnel to confirm that the plant-specific operating experience revealed no degradation not bounded by industry experience.

FSAR Supplement. In LRA Section A2.21, the applicant provided the FSAR supplement for the Water Chemistry Program. The staff reviewed this section and determines that the information in the FSAR supplement provides an adequate summary description of the program, as required by 10 CFR 54.21(d).

Conclusion. On the basis of its review and audit of the applicant's Water Chemistry Program, the staff determines that those program elements for which the applicant claimed consistency with the GALL Report are consistent. In addition, the staff reviewed the exceptions and the associated justifications and determines that the AMP, with the exceptions, is adequate to manage the aging effects for which it is credited. The staff concludes that the applicant has demonstrated that the effects of aging will be adequately managed so that the intended functions will be maintained consistent with the CLB for the period of extended operation, as required by 10 CFR 54.21(a)(3). The staff also reviewed the FSAR supplement for this AMP and concludes that it provides an adequate summary description of the program, as required by 10 CFR 54.21(d).

3.0.3.1.11 Compressed Air Monitoring Program

Summary of Technical Information in the Application. By letter dated October 31, 2005, the applicant submitted an LRA supplement providing the Compressed Air Program as new LRA Section B2.1.22, "Compressed Air Monitoring Program." The supplement also provides the FSAR description of the Compressed Air Program as new LRA Section A2.22, "Compressed Air Program." In LRA Section B2.1.22, the applicant described the Compressed Air Monitoring Program, stating that this existing program is consistent with GALL AMP XI.M24, "Compressed Air Monitoring."

The Compressed Air Monitoring Program manages aging effects on the internal surfaces of carbon steel, low-alloy steel, copper alloy, and stainless steel components within the scope of license renewal exposed to a compressed air environment. These include such components as piping, traps, HXs, filter housings, dryer housings, accumulators, and valve bodies made of materials like carbon steel, low alloy steel, copper alloys, and stainless steel. The program manages the aging effects of general, crevice, and pitting corrosion, and SCC. The program includes maintenance of the compressors, dryers, and filters associated with the plant instrument air system, high-pressure air system, feedwater purity air system, and associated back-up systems.

Staff Evaluation. During the audit and review, the staff confirmed the applicant's claim of consistency with the GALL Report. The staff determined that this AMP is consistent with the AMP described in the GALL Report, including the associated "operating experience" program element.

The staff concludes that the applicant's Compressed Air Monitoring Program provides reasonable assurance that the aging effects of internal surfaces of carbon steel, low-alloy steel, copper alloy, and stainless steel components within the scope of license renewal exposed to a compressed air environment are adequately managed. The applicant's program is consistent with the requirements and recommendations of ASME OM-S/G-1998, Part 17, ISA-7.0.01-1996, EPRI NP-7079, and EPRI TR-108147, as discussed in the GALL Report. The staff finds the applicant's Compressed Air Monitoring Program acceptable because it conforms to the recommended GALL AMP XI.M24.

Operating Experience. In LRA Section B2.1.22, the applicant explained that a review of industry operating experience associated with the Compressed Air Monitoring Program and aging reveals issues and instances related to:

- IN 81-38 - A review of problems related to contamination of air systems in operating nuclear power plants indicated that air-operated systems and components will occasionally become inoperable because they are contaminated with oil, water, desiccant, and rust or other corrosion products.

- IN 87-28, IN 87-28 S1 - A case study report entitled "Air Systems Problems at U.S. Light Water Reactors" discusses degradations of air systems and plant responses to air system losses. The report also highlights more than two dozen events in which, contrary to licensing assumptions, an SR system failed as a result of an air system degradation or failure.

- (LER) 50-237/94-005-3 - On April 30, 1994, Dresden Unit 2 was manually scrammed from 99 percent power due to rapid depressurization of the instrument air 1A header. The loss of instrument air was due to a mechanical failure of the threaded portion of the carbon steel inlet air supply piping to the 2A/1A receiver tank. Pipe failure was attributed to uniform pipe wall thinning from moisture induced corrosion compounded by the original construction threaded pipe installation, which was contrary to the original plant (butt weld) design specification. The 2A receiver tank and inlet pipe were replaced and the compressor returned to service. The safety significance of this event was minimal. The ADS system was available for reactor pressure relief and the LPCI and core spray systems were available for core cooling. A previous reactor scram on loss-of-instrument-air, as reported on January 16, 1993, was caused by dryer and backup air supply system failures. Corrective actions from that event would not have prevented this event.

In the LRA, the applicant further explained that it had incorporated the recommendations and requirements of various GLs, bulletins, and information notices (INs) into the program as applicable. None of the industry operating issues or instances reflect any new program issues. Furthermore, the applicant stated that plant-specific operating experience related to the Compressed Air Monitoring Program revealed no issues and instances related to aging.

The staff's review of the operating experience related to the applicant's Compressed Air Monitoring Program revealed no failures of the compressed air system components due to degradation. The staff identified no operating experience related to the Compressed Air Monitoring Program that would require any program modification.

The staff reviewed the operating experience provided in the LRA to confirm that the plant-specific operating experience revealed no degradation not bounded by industry experience.

FSAR Supplement. In LRA Section A2.22, the applicant provided the FSAR supplement for the Compressed Air Monitoring Program. The staff reviewed this section and determines that the information in the FSAR supplement provides an adequate summary description of the program, as required by 10 CFR 54.21(d).

Conclusion. On the basis of its review and audit of the applicant's Compressed Air Monitoring Program, the staff determines that those program elements for which the applicant claimed consistency with the GALL Report are consistent. The staff concludes that the applicant has demonstrated that the effects of aging will be adequately managed so that the intended functions will be maintained consistent with the CLB for the period of extended operation, as required by 10 CFR 54.21(a)(3). The staff also reviewed the FSAR supplement for this AMP and concludes that it provides an adequate summary description of the program, as required by 10 CFR 54.21(d).

3.0.3.1.12 Electrical Equipment Qualification Program

Summary of Technical Information in the Application. In LRA Section B3.1, the applicant described the Electrical Equipment Qualification Program, stating that this existing program is consistent with GALL AMP X.E1, "Environmental Qualification (EQ) of Electric Components."

The Electrical Equipment Qualification Program implements the requirements of 10 CFR 50.49, "Environmental Qualification of Electric Equipment Important to Safety for Nuclear Power Plants." Section 50.49 of 10 CFR defines the scope of components to be included, requires the preparation and maintenance of a list of in-scope components, and requires the preparation and maintenance of a qualification file that includes component performance specifications, electrical characteristics, the environmental conditions to which the components could be subjected, and the basis for qualification. Section 50.49(e)(5) of 10 CFR contains provisions for aging that require, in part, consideration of all significant types of aging degradation that can affect component functional capability. The section also requires replacement or refurbishment of qualified components prior to the end of their designated life, unless additional life is established through ongoing qualification. EQ programs manage component thermal, radiation, and cyclical aging through the use of aging evaluations based on 10 CFR 50.49(f) qualification methods.

Staff Evaluation. During the audit and review, the staff confirmed the applicant's claim of consistency with the GALL Report. Details of the staff's evaluation of this AMP are documented in the Audit and Review Report. The staff determined that this AMP is consistent with the AMP described in the GALL Report, including the associated "operating experience" program element.

As documented in the Audit and Review Report, the staff noted a difference between GALL AMP X.E1 and the Electrical Equipment Qualification Program under the program description. The results of the EQ of electrical equipment in LRA Section 4.4 indicate that the aging effects of the EQ of electrical equipment identified in the time-limited aging analysis (TLAA) will be managed during the period of extended operation, in accordance with 10 CFR 54.21(c)(1)(iii).

However, no information was provided in the Electrical Equipment Qualification Program on the attribute of a re-analysis of an aging evaluation to extend the qualification life of electrical equipment identified in the TLAA. The important attributes of a re-analysis are the analytical methods, data collection and reduction methods, underlying assumptions, acceptance criteria, and corrective actions. GALL AMP X.E1, under EQ component reanalysis attributes, clearly describes each attribute of a re-analysis under the program description. The applicant's Electrical Equipment Qualification Program does not include this information. The staff asked the applicant to provide these attributes under the program description. The applicant agreed to revise the Electrical Equipment Qualification Program to include EQ component re-analysis attributes as described in the program description for GALL AMP X.E1. The staff reviewed the program element and found it consistent with the GALL Report. The staff finds the applicant's response acceptable as consistent with the GALL AMP description. By letter dated August 25, 2005, the applicant revised LRA Section B3.1 to incorporate this change.

Operating Experience. In LRA Section B3.1, the applicant explained that a review of applicable operating experience was performed to determine whether there were deficiencies or recurring failures that would raise questions about the effectiveness of the Electrical Equipment Qualification Program. No significant items were found. Similar reviews were performed of inspection, assessment, and audit reports. No significant findings were identified. Some issues had been identified that resulted in appropriate corrective actions and enhancements.

The staff reviewed the operating experience provided in the LRA and interviewed the applicant's technical personnel to confirm that the plant-specific operating experience revealed no degradation not bounded by industry experience.

FSAR Supplement. In LRA Section A3.1, the applicant provided the FSAR supplement for the Electrical Equipment Qualification Program. The staff reviewed this section and determines that the information in the FSAR supplement provides an adequate summary description of the program, as required by 10 CFR 54.21(d).

Conclusion. On the basis of its review and audit of the applicant's Electrical Equipment Qualification Program, the staff determines that those program elements for which the applicant claimed consistency with the GALL Report are consistent. The staff concludes that the applicant has demonstrated that the effects of aging will be adequately managed so that the intended functions will be maintained consistent with the CLB for the period of extended operation, as required by 10 CFR 54.21(a)(3). The staff also reviewed the FSAR supplement for this AMP and concludes that it provides an adequate summary description of the program, as required by 10 CFR 54.21(d).

3.0.3.1.13 Fatigue Monitoring Program

Summary of Technical Information in the Application. In LRA Section B3.2, the applicant described the Fatigue Monitoring Program, stating that this new program is consistent with GALL AMP X.M1, "Metal Fatigue of Reactor Coolant Pressure Boundary."

The Fatigue Monitoring Program ensures that limits on fatigue usage are not exceeded during the period of extended operation. The program monitors and tracks selected cyclic loading transients (cycle counting) and their effects on susceptible components. The applicant selected this option under 10 CFR 54.21 to manage cracking due to metal fatigue of the reactor coolant pressure boundary (RCPB) during the period of extended operation. The Fatigue Monitoring Program provides the cycle counting activities credited in LRA Section 4.3 for confirming analytically derived cumulative usage values for applicable locations. Specific locations that may be subject to cyclic loading that could cause fatigue cracking are monitored using a computer-based monitoring program provided by EPRI called FatiguePro. If warranted, other monitoring methods in addition to cycle counting may also be employed under this program to monitor specific locations.

Staff Evaluation. During the audit and review, the staff confirmed the applicant's claim of consistency with the GALL Report. Details of the staff's audit evaluation of this AMP are documented in the Audit and Review Report. The staff reviewed the enhancements and the associated justifications to determine whether the AMP, with the enhancements, remains adequate to manage the aging effects for which it is credited.

In the LRA, the applicant discussed for the "scope of program" program element that components monitored would include those identified in NUREG/CR-6260, "Application of NUREG/CR-5999 Interim Fatigue Curves to Selected Nuclear Power Plant Components." As documented in the Audit and Review Report, the staff asked the applicant to clarify the selection of locations to be monitored. By letter dated August 25, 2005, the applicant elaborated on the selection of locations and stated that any locations more limiting than those identified in NUREG/CR-6260 would also be addressed. The staff finds this clarification acceptable.

In the LRA, the applicant also discussed the "preventive actions" and "parameters monitored or inspected" program elements. By letter dated August 25, 2005, the applicant confirmed that nondesign-basis transients will be addressed prior to the period of extended operation. As appropriate, allowable cycle counts will be determined or the cumulative usage factor (CUF) calculations will be updated if necessary to maintain a CUF < 1.0, as required by the ASME Code. The staff finds this proposal acceptable.

In the LRA discussion of the "monitoring and trending" program element, the applicant stated that running totals of cumulative fatigue usage at each monitored location are available on demand. By letter dated August 25, 2005, the applicant stated that the minimum frequency for determination of the CUF will be once per refueling cycle. The staff finds this frequency acceptable.

In the LRA, the applicant stated that its Fatigue Monitoring Program would be developed and implemented consistent with GALL AMP X.M1. Features of the program would include monitoring and tracking of selected cyclic loading transients (cycle counting) and their effects on critical RCPB components and other selected components. In addition, the applicant stated that the program elements describe the program to be implemented prior to the period of extended operation.

The staff finds this program acceptable because the applicant had committed to develop the Fatigue Monitoring Program to include monitoring and tracking of selected cyclic loading transients (cycle counting) and their effects on critical reactor pressure boundary components and other selected components consistently with GALL AMP X.M1 (Commitment 37).

Operating Experience. In LRA Section B3.2, the applicant explained that industry operating experience has shown that significant thermal stresses in piping connected to reactor coolant systems have caused piping failures and subsequent leakage at some plants. NRC Bulletin No. 88-08 and its supplements brought wide visibility to these issues beginning in June 1988. Later that same year, NRC Bulletin No. 88-11 covered issues pertaining specifically to thermal stratification and unexpected thermal cyclic loadings in the pressurizer surge line. NRC Bulletins 88-08 and 88-11 highlighted the need for evaluation and monitoring of cyclic loading events not considered in the original design of Class 1 piping and components. Subsequent to the identification of thermal stratification issues, significant resources have been expended to monitor cumulative damage due to thermal fatigue. Actual failures attributable to cumulative fatigue damage/thermal fatigue have been relatively rare but have occurred. EPRI has documented the significant operating experience in MRP-85, "Materials Reliability Program: Operating Experience Regarding Thermal Fatigue of Piping Connected to PWR Reactor Coolant Systems." This experience was reviewed for development of the Fatigue Monitoring Program, but is not repeated here.

The applicant further explained that the Fatigue Monitoring Program is a new program and, as such, there is no significant plant-specific operating experience with which to make a determination of its effectiveness; however, the applicant has been tracking and logging plant-specific transient cycles. A review of plant-specific analytical results and transient logs for all tracked transients, as well as industry operating experience related to metal fatigue, determined the need for more rigorous cycle counting methodology. In addition, review of the tracked transients revealed the following issues:

- Feedwater Flow Cycling – The use of auxiliary feedwater to slug feed feedwater into the SG feedwater nozzles during hot standby conditions has been shown to be a significant thermal stratification stress cycling event, not included in original design bases. The replacement SGs, installed in 1990, incorporated separate auxiliary feedwater nozzles which minimize the potential for water hammer and thermal stratification. Use of these separate auxiliary feedwater nozzles during hot standby has been shown to minimize the occurrence of thermally stratified feedwater flow into the SGs. Therefore, the applicant had determined that this transient need not be counted.

- Charging and Letdown Transients – Charging nozzles generally show a high design-basis fatigue usage due to the relatively large number of rapid temperature transients that can occur when either charging or letdown flow, or both, are terminated and reinitiated. These transients can occur during normal plant operation, including during surveillance testing of the control and isolation valves. Unlike many of the design-basis transients, the charging and letdown transients occur fairly often in normal plant operation and can be nearly as severe as the design-basis transient. Therefore, these plant transients will be added to the cycle tracking list. If warranted, the charging nozzles may be included as stress-based fatigue-monitored locations, as these nozzles tend to be the highest fatigue locations in the system and could serve to bound the other system locations.

- Pressurizer Spray Line Cycling – The pressurizer spray nozzle can be anticipated to show a relatively high design fatigue usage due to the large temperature shocks that regularly occur during heatup and cooldown conditions. Stratified flow in the pressurizer spray line was not considered significant because PNP operates with continuous pressurizer spray flow of 60 to 80 gpm. The applicant assumed that one spray actuation will occur during each plant cooldown. The fatigue damage to the spray nozzle could conceivably be tracked just by counting the number of plant cooldowns; however, main spray actuations of varying severities and duration also could occur during other plant conditions, such as plant heatup. Also, under certain conditions, the temperature difference between the source of the auxiliary spray and that of the main spray could be significant, thereby having the potential to cause significant thermal shocking at the auxiliary spray-to-main spray tee. Due to uncertainty in precisely monitoring the fatigue damage of the spray nozzle, track spray cycles will not be tracked, but may include the spray nozzle as a stress-based fatigue-monitored location.

- Pressurizer Surge Line Cycling – The issue of thermal stratification cycling in the surge line was addressed extensively in the industry's response to NRC Bulletin 88-11. CE evaluated the surge line and concluded that the fatigue usage of the pressurizer surge nozzle and the hot leg piping surge nozzle will be relatively high, but less than the ASME 1.0 CUF limit. It is a complex undertaking to identify the occurrence and magnitude of insurge or outsurge flow in the surge line. Due to the complexity and uncertainty in the ability to adequately monitor the fatigue damage of the surge lines and the connected nozzles, the applicant does not intend to track surge cycles, but may monitor both surge line nozzles using stress-based fatigue-monitoring.

- Failures in Unisolable Piping – Intermittent valve leakage, causing thermally stratified conditions in unisolable sections of piping, have caused failures in safety injection nozzles and in socket welds near the reactor coolant piping in several plants. Such conditions were the subject of NRC Bulletin 88-08. Based upon that experience, the high-pressure safety injection nozzles may be included in the Fatigue Monitoring Program as either cycle-based or stress-based fatigue locations.

The applicant stated that if necessary to maintain the projected CUF < 1.0 for these or other locations, stress-based fatigue monitoring would be implemented as part of the Fatigue Monitoring Program.

The staff reviewed the operating experience provided in the LRA and interviewed the applicant's technical personnel to confirm that the plant-specific operating experience revealed no degradation not bounded by industry experience.

On the basis of its review of the industry and plant-specific operating experience and discussions with the applicant's technical personnel, the staff concludes that the applicant's Fatigue Monitoring Program will adequately manage the aging effects identified in the LRA for which this AMP is credited.

FSAR Supplement. In LRA Section A3.2, the applicant provided the FSAR supplement for the Fatigue Monitoring Program. The staff reviewed this section and determines that the information in the FSAR supplement provides an adequate summary description of the program, as required by 10 CFR 54.21(d).

Conclusion. On the basis of its review and audit of the applicant's Fatigue Monitoring Program, the staff determines that those program elements for which the applicant claimed consistency with the GALL Report are consistent. The staff concludes that the applicant has demonstrated that the effects of aging will be adequately managed so that the intended functions will be maintained consistent with the CLB for the period of extended operation, as required by 10 CFR 54.21(a)(3). The staff also reviewed the FSAR supplement for this AMP and concludes that it provides an adequate summary description of the program, as required by 10 CFR 54.21(d).

3.0.3.2 AMPS That Are Consistent with the GALL Report with Exceptions or Enhancements

In LRA Appendix B, the applicant identified the following AMPs that were, or will be, consistent with the GALL Report, with exceptions or enhancements:

- Bolting Integrity Program
- Boric Acid Corrosion Program
- Closed Cycle Cooling Water Program
- Diesel Fuel Monitoring and Storage Program
- Fire Protection Program
- One-Time Inspection Program
- Overhead Load Handling Systems Inspection Program
- Reactor Vessel Integrity Surveillance Program
- Reactor Vessel Internals Inspection Program
- Structural Monitoring Program

For AMPs that the applicant claimed are consistent with the GALL Report, with exceptions or enhancements, the staff performed an audit to confirm that those attributes or features of the program for which the applicant claimed consistency with the GALL Report were indeed consistent. The staff also reviewed the exceptions and enhancements to the GALL Report to determine whether they were acceptable and adequate. The results of the staff's audit and reviews are documented in the following sections.

3.0.3.2.1 Bolting Integrity Program

Summary of Technical Information in the Application. In LRA Section B2.1.3, the applicant described the Bolting Integrity Program, stating that this existing program is consistent, with enhancements, with GALL AMP XI.M18, "Bolting Integrity."

The Bolting Integrity Program manages the aging effects associated with bolting through periodic inspections. The program also includes repair/replacement controls for ASME Code Section XI related bolting and generic guidance regarding material selection, thread lubrication, and assembly of bolted joints. The program considers the guidelines delineated in NUREG-1339 for a bolting integrity program, EPRI NP-5769 (with the exceptions noted in NUREG-1339) for SR bolting, and EPRI TR-104213 for nonsafety-related (NSR) bolting. The Bolting Integrity Program was created to permit direct comparison with the GALL Report. The program is considered as an existing program since most of the activities addressed by the program are already being performed. The program credits activities performed under the following three separate AMPs for the inspection of bolting: (1) ASME Section XI IWB, IWC, IWD, IWF Inservice Inspection Program, (2) Structural Monitoring Program, and (3) System Monitoring Program.

Staff Evaluation. The staff confirmed the applicant's claim of consistency with the GALL Report. The staff reviewed the enhancements and the associated justifications to determine whether the AMP, with the enhancements, remains adequate to manage the aging effects for which it is credited.

Two enhancements are scheduled prior to the period of extended operation to bring the Bolting Integrity Program into conformance with GALL AMP XI.M18. These enhancements consist of (1) reviewing and revising the ASME ISI master plan and plant maintenance procedures to reflect GALL Report guidance and (2) evaluating high-strength bolting used in component supports for cracking.

The staff reviewed the Bolting Integrity Program against the AMP elements found in the GALL Report, SRP-LR Section A.1.2.3, and SRP-LR Table A.1-1 and focused on how the program manages aging effects through the effective incorporation of ten program elements (i.e., "scope of program," "preventive actions," "parameters monitored or inspected," "detection of aging effects," "monitoring and trending," "acceptance criteria," "corrective actions," "confirmation process," "administrative controls," and "operating experience").

The applicant indicated that "corrective actions," "confirmation process," and "administrative controls" program elements are parts of the site-controlled QA program. The staff's evaluation of the QA program is discussed in SER Section 3.0.4. The remaining seven elements are discussed as follows.

(1) Scope of Program - LRA Section B2.1.3 states that the program covers all bolting and fasteners within the scope of license renewal, including SR bolting, bolting for nuclear steam supply system (NSSS) supports, bolting for other pressure-retaining bolting, and structural bolting.

The staff's review of LRA Section B2.1.3 identified areas in which additional information was necessary to complete the review of the applicant's program elements. The applicant responded to the staff's RAIs as discussed below.

In LRA Section B2.1.19 the Structural Monitoring Program is credited for the Bolting Integrity Program. However, the staff noted that bolting is not discussed in LRA Section B2.1.19.

In RAI B2.1.3-1, dated June 28, 2005, the staff requested that the applicant clarify why the Structural Monitoring Program, without any reference to bolting, can be credited for the Bolting Integrity Program.

In its response, by letter dated July 28, 2005, the applicant stated that while the summary description of LRA Section B2.1.19 does not explicitly mention bolting, bolting is included in the program scope. The program basis documentation for the Structural Monitoring Program specifically states that structural bolting is inspected for indications of potential problems including loss of coating integrity and obvious signs of corrosion and rust. Based on its review, the staff finds the applicant's response to RAI B2.1.3-1 acceptable because the Structural Monitoring Program does provide guidance on bolting inspection and is acceptable to be credited for the Bolting Integrity Program. Therefore, the staff's concerns in RAI B2.1.3-1 are resolved.

The staff found that bolting is also not mentioned in LRA Section B2.1.20, System Monitoring Program.

In RAI B2.1.3-2, dated June 28, 2005, the staff requested that the applicant (1) clarify why the System Monitoring Program, without having mentioned bolting, is credited for the Bolting Integrity Program; and (2) discuss how bolting integrity (e.g., cracking, loss of preload, and loss of material due to corrosion) would be determined during system walkdowns as required by the System Monitoring Program when bolts are covered with insulation.

In its response, by letter dated July 28, 2005, the applicant stated that although bolting is not specifically discussed in the System Monitoring Program, bolting is a component that will be inspected by the program. The intent of the System Monitoring Program is to inspect all accessible external surfaces of various component types (e.g., pump casings, valve bodies, piping, expansion joints) which would include bolted connections. The applicant stated that it does not remove insulation solely for the inspections performed by the System Monitoring Program because removal of insulation is not considered necessary to identify likely locations of potential degradation. The condition of the insulation (e.g., discoloration, evidence of wetting), in itself, provides an indirect indicator of the conditions beneath. If the insulation condition indicates that a potential problem exists beneath the insulation, this condition would be documented and corrective action (e.g., isolation, insulation removal, further inspection, repairs) would be initiated.

The applicant stated that the system walkdowns will look for evidence of degradation where pipe insulation is not removed in a manner similar to the ASME Code for SR piping and components. The inspection performed under the System Monitoring Program of insulated non-Class 1 pipe and closure joints will be similar to the visual examinations, VT-2, prescribed by the ASME Code for insulated Class 1 piping and pressure retaining bolted connections. ASME Section XI, Paragraph IWA-5242, Insulated Components, states that:

> (a) For systems borated for the purpose of controlling reactivity, insulation shall be removed from pressure retaining bolted connections for visual examination VT-2. For other components, visual examination VT-2 may be conducted without the removal of insulation by examining the accessible and exposed surfaces and joints of the insulation. Essentially vertical surfaces of insulation need only be examined at the lowest elevation where leakage may be detected. Essentially horizontal surfaces of insulation shall be examined at each insulation joint.

> (b) When examining insulated components, the examination of surrounding areas (including floor areas or equipment surfaces located underneath the

components) for evidence of leakage, or other areas to which such leakage may be channeled, shall be required.

(c) Discoloration or residue on surfaces examined shall be given particular attention to detect evidence of boric acid accumulation from borated reactor coolant leakage.

The applicant further stated that management of loss of bolting preload is addressed by maintenance and installation procedures. Loss of material due to corrosion will be identified by discovering the evidence of leakage prior to a loss of intended function. A corrosive environment is precluded through use of proper lubricants, and proper bolt torquing practices. Should leakage occur, the evidence of the leak would be discovered, investigated and resolved prior to a loss of intended function.

Based on its review, the staff agreed with the applicant's response to RAI 2.1.3-2 because the System Monitoring Program is acceptable to be credited for the Bolting Integrity Program since bolting will be inspected under the System Monitoring Program. With respect to inspecting bolting covered by insulation, the staff found that system walkdowns performed under the System Monitoring Program follow the visual inspection requirements in the ASME Code Section XI. Therefore, the staff's concerns in RAI B2.1.3-2 are resolved.

As stated above, LRA Section B2.1.2, the ASME Section XI IWB, IWC, IWD, IWF Inservice Inspection (ISI) Program is credited for the Bolting Integrity Program. The staff noted that the ASME Code Section XI specifies periodic inservice examination on a sampling basis, which means that some bolts may not be inspected.

In RAI B2.1.3-3, dated June 28, 2005, the staff requested that the applicant clarify whether all bolts at PNP covered under the LRA will be examined before entering the extended period of operation.

In its response, by letter dated July 28, 2005, the applicant described the scope of the credited programs for bolting as:

• The ASME Section XI IWB, IWC, IWD, IWF Inservice Inspection Program provides the requirements for inservice inspection of ASME Class 1, 2, and 3 piping, supports, and their integral attachments, which includes pressure retaining and support bolting. This program specifically discusses the inspection and lubrication of the reactor vessel head closure studs. The program supplements the ASME Section XI (Code Case N–491-2), Subsection IWF requirements, by applying the inspection requirements of Subsection IWB, Category B-G-1 to high yield strength (>150 ksi) bolting used in Nuclear Steam Supply System (NSSS) component supports.

• The System Monitoring Program provides the requirements for the inspection of non-safety related bolting within the scope of license renewal.

• The Structural Monitoring Program provides the requirements for the inspection of all structural bolting within the scope of license renewal. Other bolting and fasteners are also included within the scope of this program, such as those used in supports for cable trays, conduits and cabinet supports.

The applicant stated further that all ASME Class 1, 2, and 3 bolting is inspected by the ASME Section XI IWB, IWC, IWD, IWF Inservice Inspection Program a minimum of once each ten year inspection interval. ASME Section XI, Table IWB-2500-1 requires inspection of bolts, studs, and nuts for various Class 1 components. In all cases, the inspection requirement encompasses all bolts, studs, and nuts each inspection interval. There is no provision for a sampling size. ASME Section XI, Table IWC-2500-1 requires inspection of 100 percent of bolts and studs of size two-inch and greater at each bolted connection in Class 2 components.

The applicant stated that IWC-2500-1 also requires a visual VT-2 examination of all pressure retaining components during a system leakage test. The leakage test is required each inspection period. ASME Section XI, Table IWD-2500-1 requires a visual VT-2 examination of the pressure retaining boundary of Class 3 systems during a system leakage test each inspection period.

The applicant stated that non-ASME class bolting within the scope of license renewal that is reasonably accessible for external visual inspection will be periodically inspected under either the System Monitoring Program (system walkdowns) or the Structural Monitoring Program (general area inspections). The LRA includes commitments for certain enhancements to make each program an effective aging management program. Enhancements to the Bolting Integrity Program will be implemented prior to the period of extended operation. Upon completion of the identified enhancements, all bolting requiring aging management for license renewal will be included in an appropriate aging management program, and a sufficient percentage of the total population will be accessible for inspection to provide reasonable assurance that the bolting condition is adequate.

Based on its review, the staff finds that the applicant's response to RAI B.2.1.3-3 is acceptable because all bolts in Palisades will be examined at least once during the 10-year inspection interval per ASME Code Section XI and that the applicant will follow the ASME Code Section XI requirements. Therefore, the staff's concerns in RAI B2.1.3-3 are resolved.

In LRA Section B2.1.3, the applicant stated that the Bolting Integrity Program is adequate to manage loss of preload without hot torquing; therefore, hot torquing to establish a preload will not be credited for aging management of bolting.

In RAI B2.1.3-4, dated June 28, 2005, the staff requested that the applicant explain the adequacy of the Bolting Integrity Program to manage loss of preload without hot torquing.

In its response, by letter dated July 28, 2005, the applicant stated that, consistent with the recommendations in EPRI TR-104213, it has assessed the various aspects of proper preloading practices, including the potential benefits versus risks of hot torquing. Fastener preload procedures include instructions for preparation of joints and predetermination of fastener size, material, thread lubricant and design temperature. Yield strength, thread stress areas, and nut factors are determined. Final torque, maximum torque, and torque for each pass is calculated. Torque is generally applied in a minimum of four passes and measured during the sequence. Compression thicknesses are determined as part of the preload procedure. Gaskets are crushed progressively, in stages, and standard bolt patterns are used to crush gaskets uniformly.

Based on its review, the staff finds the applicant's response to RAI B2.1.3-4 acceptable because the Bolting Integrity Program is adequate to manage loss of preload because it has the proper technique and procedures for bolting torque to mitigate the likelihood of loss of bolting preload. Therefore, the staff's concerns in RAI B2.1.3-4 are resolved.

In LRA Section B2.1.3, the applicant stated that normal maintenance practices and quality verification procedures for pressure retaining bolting includes a check of bolt torque and uniformity of gasket compression. The staff noted that structural bolting was not included in that discussion.

In RAI B2.1.3-7, dated June 28, 2005, the staff requested that the applicant discuss whether there are similar procedures to verify the torque of structural bolting.

In its response, by letter dated July 28, 2005, the applicant stated that there are similar procedures to verify the torque of structural bolting. MSM-M-45, "Removal, Installation, and Repair of Pipe Supports" provides criteria for proper torquing or retorquing of bolted connections related to pipe supports. MSM-M-44,"HILTI Bolt Installation and Inspection," provides instructions for installation and or retorquing of the HILTI brand Drop-In anchors, and installation and or retorquing of the HILTI brand Kwik Bolt II Anchors. HILTI Drop-in Anchors are used for mounting small items. HILTI Kwik Bolt II Anchors may be used for piping; cable tray; heating, ventilation, and air conditioning; and small equipment supports as well as electrical components and instrumentation. Torquing requirements for new construction or modifications would not necessarily be covered by permanent plant maintenance procedures, but would be specified in new construction specifications or drawings. The applicant stated that Procedure MSM-M-48, "Standard Torque Tables," is applied to fasteners in standard bolting applications on pressure retaining components, and it is available for use in structural applications in the absence of other guidance.

Based on its review, the staff finds the applicant's response to RAI B2.1.3-7 acceptable because torquing procedures for structural bolting in the Bolting Integrity Program follow industry procedures. Therefore, the staff's concerns in RAI B2.1.3-7 are resolved.

In LRA Section B2.1.3, the applicant stated that MC component support bolting is evaluated in accordance with the acceptance standards of ASME Code Case N-491-2, Section 3410. The staff noted that Code Case N-491-2 provides guidance on examination and acceptance criteria, but it does not provide guidance on replacement or repair. GALL AMP XI.M18 recommends immediate replacement of the cracked bolt when indications of cracking is found in support bolting.

In RAI B2.1.3-8, dated June 28, 2005, the staff requested that the applicant discuss whether the Bolting Integrity Program has a process to replace a cracked bolt immediately.

In its response, by letter dated July 28, 2005, the applicant stated that if a degraded bolt is identified, the condition would be entered into the plant Corrective Action Program. The program, in turn, would require a prompt assessment of the effect of the condition on equipment operability and plant safety. Corrective actions would be assigned and completed commensurate with the safety and operational significance of the condition. The corrective action for a degraded bolt would likely include eventual replacement of that bolt, but the timing of the replacement would have to consider the safety-significance of the condition, the plant conditions needed to safely complete the work, the status of other plant equipment, the extent and complexity of repair, radiological conditions, and availability of replacement parts.

The applicant stated that the most prudent action, therefore, could include immediate replacement as emergency maintenance. However, the applicant could also consider actions such as isolation of the affected bolted joint from operating systems, analysis to verify adequate pressure boundary or structural integrity, and later replacement using normal maintenance planning and scheduling procedures. The timing and nature of any actions to be taken would be planned and managed under the corrective action process to assure the safety of the public, plant workers, and plant equipment.

Based on its review, the staff finds the applicant's response acceptable because the Bolting Integrity Program has appropriate procedures to replace cracked bolts. Therefore, the staff's concern in RAI B2.1.3-8 is resolved.

In LRA Section B2.1.3, the applicant stated that "...Palisades Bolting Integrity Program makes no distinction regarding "immediate" repairs, but instead relies upon the plant inservice inspection program and corrective action process to evaluate, prioritize and schedule repairs..."

In RAI B2.1.3-9, dated June 28, 2005, the staff noted that the applicant's approach is not consistent with GALL AMP XI.M18 which recommends that immediate repairs be performed for major leaks that may cause corrosion or contamination. Therefore, the staff requested that the applicant explain the inconsistency.

In its response, by letter dated July 28, 2005, the applicant stated that if a major leak were to occur that had a significant impact on plant equipment or plant safety, operators would take prompt action in accordance with procedures to bring the plant to a safe, stable condition. The leaking joint would be isolated, if practical, or the plant could even be shut down and depressurized, if warranted. Actions for such a severe condition would not be delayed for administrative processing of corrective action documents. Longer term repairs and recovery actions would be managed under the corrective action program. The applicant stated that whether or not immediate operational actions are taken, degraded conditions would be entered promptly into the plant corrective action program. Corrective actions regarding repair or replacement will be taken commensurate with the safety and operational significance of the degradation. The applicant's decision process to repair major leaks from degraded bolting will be the same as the decision process to repair cracked bolting as discussed in its response to RAI B2.1.3-8.

Based on its review, the staff finds the applicant's response to RAI B.2.1.3-9 acceptable because the Bolting Integrity Program has appropriate programs and procedures and manages immediate repairs for major leaks caused by degraded bolting. Therefore, the staff's concern in RAI B.2.1.3-9 is resolved.

In RAI B.2.1.3-1(b), dated July 12, 2005, the staff noted that the LRA AMR tables do not indicate whether any bolting is exposed to soil. Therefore, the staff requested the applicant to indicate whether any bolting is used in buried applications and to clarify whether the Bolting Integrity Program or the Buried Services Corrosion Monitoring Program manages such bolting.

In its response, by letter dated August 12, 2005, the applicant clarified that the fire protection system contains bolting both buried in soil and submerged in raw water. LRA Table 3.3.2-7 identifies these as fasteners and credits the Bolting Integrity Program for managing loss of preload and the Fire Protection Program for managing loss of material. The applicant further clarified that no other buried bolting has been identified in portions of other systems within the scope of license renewal.

Based on its review, the staff finds the applicant's response to RAI B.2.1.3-1(b) acceptable because, other than in the fire protection system, no other buried bolting or fasteners had been identified as within scope of license renewal. The Fire Protection Program is evaluated in SER Section 3.0.3.2.6. Therefore, the staff's concern described in RAI B.2.1.3-1(b) is resolved.

In RAI B.2.1.3-1(c), dated July 12, 2005, the staff noted that LRA Section B2.1.3 references EPRI NP-5769, which states that loss of preload is caused by a number of factors including stress relaxation (both at room temperature and elevated temperature), thermal cycling (particularly for gaskets), creep and flow of gasket material during initial compression, vibration and shock, and elastic interactions between separately-tightened bolts. Notes 206, 324, and 406 in LRA AMR tables state that loss of preload is included here in response to recent RAIs on non-primary system, high-temperature bolting that may experience loss of preload. Therefore, the staff requested the applicant to define "high temperature" and clarify whether loss of preload is managed for all closure bolting within the scope of license renewal, considering EPRI NP-5769 guidance that loss of preload may occur at room temperature.

In its response, by letter dated August 12, 2005, the applicant clarified that loss of preload is managed for all closure bolting within the scope of license renewal regardless of temperature. As stated in LRA Section B2.1.3, the Bolting Integrity Program covers all bolting within the scope of license renewal, including SR bolting, bolting for NSSS component supports, bolting for other pressure-retaining components, and structural bolting. The response further clarified that the applicant had not limited the management of loss of preload by the Bolting Integrity Program to only high-temperature applications. Based on its review, the staff finds the applicant's response to RAI B.2.1.3-1(c) acceptable because the applicant clarified that the management of loss of preload by the Bolting Integrity Program is not limited to high-temperature applications. Therefore, the staff's concern described in RAI B.2.1.3-1(c) is resolved.

In RAI B.2.1.3-1(d), dated July 12, 2005, the staff requested the applicant to clarify whether there were any in-scope, high-strength (actual yield strength greater than 150 ksi) structural bolting or fasteners used for NSSS component supports or other non-Class 1 applications.

In its response, by letter dated August 12, 2005, the applicant stated that structural and component support bolting drawings and specifications identify A490 bolting as the only bolting type in structural applications with the potential for actual yield strength greater than 150 ksi. Certified material test report documentation is not readily available to validate actual yield strength values. Therefore, A490 bolting is conservatively treated as high-strength bolting with yield strength potentially greater than 150 ksi.

The LRA and the applicant's response do not specify the nominal diameters applicable to the A490 high-strength bolts or whether these bolts are used in supporting ASME Class 1 components. If their diameters are greater than one inch and they are used in supporting ASME Class 1 components, volumetric examination of some of the bolts to address the SCC concern may be needed (refer to GALL AMP XI.M18).

By letter dated October 3, 2005, the staff asked the applicant to confirm that none of the A490 high-strength bolting with nominal diameter greater than one inch (if applicable) was used for Class 1 component supports. If some of the A490 high-strength bolting with nominal diameter greater that one inch was used for Class 1 component supports, the staff asked the applicant for the technical basis (e.g., operating experience) for not performing volumetric examination of the bolting to manage potential SCC.

In its response, by letter dated October 14, 2005, the applicant stated that Class 1 component supports do include A490 bolting of diameter greater than one inch. Specifically, A490 bolting is used in the following Class 1 component support applications: (a) SG upper guide lateral supports and associated horizontal snubbers, (b) pressurizer frame support members, and (c) primary coolant pump support structure base plates. In addition, the U-bolt lateral supports of the regenerative HXs were evaluated as high-strength material potentially susceptible to SCC. These U-bolt supports are not A490 bolting material, but rather SA-193-B7 material, which is a low-alloy quenched and tempered material. In certain configurations, this material potentially could be susceptible to SCC. For SCC to occur in high-strength, low-alloy quenched and tempered materials, the following three conditions must exist: (1) high tensile stress, (2) a corrosive environment, and (3) a susceptible material. Additionally, as SCC-caused failures have not been observed in bolting less than 1-1/4 inch, bolting one inch and less is not considered susceptible to SCC. A discussion of these factors for each of the applications of high-strength bolting materials follows.

For the SG, the applicant stated that A490 bolts are used in two applications for the upper guide lateral supports and snubbers. The "A-frame" upper guide base plates are through-bolted with 2-1/2 inch A490 bolts that are double nutted for reuse, indicating they are not heavily preloaded nor exposed to continuous high-tensile stress. The design configuration of the A490 bolts used for the snubber installations would permit removal and reinstallation, indicating that they would not be fully tensioned to permit reuse of the bolts and consequently would not be exposed to continuous high-tensile stress. However, the applicant conservatively assumed that snubber supports not double-nutted could be fully preloaded with SCC as a potential aging effect requiring management (AERM).

For the pressurizer, the applicant stated that the pressurizer support frame members utilize 1-3/8 inch embedded A490 anchor bolts and 1-3/8 inch A490 through-bolted base plates. The design of the restraint system places the anchoring concrete in compression, indicating the anchor bolts are not exposed to high, sustained loading under normal operating conditions. However, the applicant conservatively assumed that the A490 support bolting could be fully preloaded with SCC as a potential AERM.

For the primary coolant pumps, the applicant stated that the primary coolant pump sliding frame support members have base plates bolted to an embedded bearing plate with 1-1/2 inch A490 bolts and with 2-1/2 inch A490 through bolts with self-locking nuts. Additionally, the drawing for the pump supports indicates that all connection and anchor bolts are A490 unless noted otherwise. However, the drawing for the configuration provides specific instructions that bolts are to be "wrench tight," which does not fully preload the A490 bolts; therefore, a high sustained tensile stress is not present. Additionally, the concrete under the supports carries the dead and live loads so the anchor bolts are not subject to sustained dead and live loads. Therefore, the primary coolant pump bolts are not susceptible to SCC.

For the regenerative HX, the applicant stated that it is supported vertically by steel "dishes" and is horizontally "U-bolted" near the top and bottom of the steel frame. The 3/4-inch U-bolts are SA-193-B7 and the nuts are SA-194-2H. An interfacing compressible asbestos material is placed between the carbon steel U-bolt and the stainless steel HX shell so no high, sustained tensile stress is present. Also, the U-bolts are less than one inch in diameter, below the threshold of SCC susceptibility. Therefore, the regenerative HX U-bolts are not susceptible to SCC.

In summary, the applicant stated that for the Class 1 component support A490 and SA-193-B7 bolting applications SCC had been conservatively assumed to be applicable to only the SG snubber supports and the pressurizer supports. For those instances where SCC had been identified as an AERM, the applicant stated that it would inspect visually to detect SCC potential by evidence of corrosion or a corrosive environment. If the potential for cracking were found, the extent of any degradation would be measured by removal of the bolting for further inspection, proof testing by tension or torquing, in-situ ultrasonic testing, or hammer testing.

The staff reviewed the applicant's responses to RAI B.2.1.3-1(d) and determined that additional information was necessary to complete the review of the Bolting Integrity Program. The applicant stated that for those instances where SCC had been assumed as an AERM, it would inspect visually to detect SCC potential by evidence of corrosion or a corrosive environment; however, the applicant did not propose an inspection program, frequency of proposed inspection, or applicable criteria for assuming SCC potential and the need for further inspection, including in-situ ultrasonic testing. Therefore, in a public meeting on February 14, 2005, the staff asked for such additional information.

In its response, by letter dated April 26, 2006, the applicant noted that LRA Section B2.1.3 states: "The program credits activities performed under three separate aging management programs for the inspection of bolting. The three aging management programs are: (1) ASME Section XI IWB, IWC, IWD, IWF Inservice Inspection Program, (2) Structural Monitoring Program, and (3) System Monitoring Program."

As to bolting within the scope of the ASME Section XI IWB, IWC, IWD, IWF Inservice Inspection Program, the applicant stated, "The inspection criteria are established by the rules of Section XI. For high strength bolting within the scope of Section XI, the following amplification was provided in the NMC Response to Follow up Question Concerning NRC RAI B2.1.3-1(d) provided by NMC Letter Dated October 14, 2005." The applicant further stated, "The inspection frequency for all bolting is established by the rules of Section XI. Inspections are scheduled within the normal ten year intervals as the code specifies."

Additionally, in its response by letter dated April 26, 2006, as to high-strength bolting within the scope of the Structural Monitoring Program, the applicant stated:

Inspection frequencies for all bolting within the scope of the Structural Monitoring Program are discussed in LRA Section B2.1.19. The LRA Section B2.1.19, on Page B-137 states:

Initial baseline inspections under the Structural Monitoring Program were performed, as required by 10 CFR 50.65, starting in late 1996. A second complete inspection was performed in 1999 to validate the initial inspection results. Subsequent inspections follow a 10 year interval schedule that is similar to Inspection Plan B defined in the ASME Boiler & Pressure Vessel Code, Section XI, Table IWE-2412-1.

The 10 year inspection interval is divided into three 40 month periods. Approximately one third of the items in the program scope are examined in each period and all items are examined at least once during the 10 year interval. The first interval, first period, inspections have been completed, and Palisades is currently in the first interval, second period inspection cycle. Other features may have greater inspection frequencies such as watertight/flood barrier inspections (at least once per 5 years) and below-the-waterline water-control structures (once every 5 years).

Augmented inspection is required for items that have been repaired or that exhibit significant damage or deterioration. It may also be required for items subject to aggressive environments. Items that are tagged for augmented inspection following repair or for reasons of damage / deterioration are examined, at a minimum, in the period immediately following the one during which the repair was performed or the deleterious condition was found. Augmented inspection may be performed on a 40 month period basis or at more closely spaced intervals as specified by the Structural Monitoring Coordinator or in plant procedures.

As to high-strength bolting within the scope of the Structural Monitoring Program, the applicant also stated:

Examinations of structures and structural elements are performed by qualified personnel using techniques appropriate for the item, its environment and its intended function. Each individual selected to perform these examinations must have the following qualifications:

- A civil engineering degree from an accredited university.

- Familiarity with the design and performance requirements applicable to nuclear power plant structures and experience in the in-service examination and evaluation of these structures.

- A minimum of 5 years experience in engineering design and / or analysis of nuclear power plant structures.

In the response the applicant also noted that LRA Section B2.1.19 states, "System engineers and plant operators augment the formal examinations by noting the conditions of structures / structural elements during periodic system walk downs, and reporting observed damage / degradation to the Structural Monitoring Coordinator."

Based on its review, the staff finds the additional information provided by the applicant in response to RAI B.2.1.3-1(d) adequate and reasonable in scope and in technical content. Therefore, the staff's concerns described in RAI B.2.1.3-1(d) are resolved.

The staff confirmed that the "scope of program" program element satisfies the criterion defined in the GALL Report and SRP-LR Section A.1.2.3.1 and concludes that this program attribute is acceptable.

(2) Preventive Actions - In RAI B.2.1.3-2(a), dated July 12, 2005, the staff noted that LRA Section B2.1.3 states that the use of lubricants generally meets the recommendations of EPRI NP-5769 and NUREG-1339 and, under enhancements, the applicant plans a revision to the procedures that implement the credited programs, including use of lubricants and sealants to make the Bolting Integrity Program consistent with the GALL AMP XI.M18 description. As to the use of lubricants, the term "generally meet" is not clear. Therefore, the staff requested the applicant to clarify whether any exceptions to the GALL AMP and its bases are necessary for the use of lubricants.

In its response, by letter dated August 12, 2005, the applicant stated that the Bolting Integrity Program takes no exceptions to GALL AMP XI.M18 and that no exceptions are taken for use of lubricants. The applicant explained that the term "generally meet" should not have been used in the program description. In Commitment 10, the applicant stated that it would enhance the Bolting Integrity Program to incorporate the recommendations of EPRI NP-5769 and NUREG-1339. As discussed in LRA Section B1.1, the programs are described as they would exist after identified enhancements are incorporated. As the Bolting Integrity Program would contain the recommendations of EPRI NP-5769 and NUREG-1339 prior to the period of extended operation, the word "generally" was deleted from the description.

Based on its review, the staff finds the applicant's response to RAI B.2.1.3-2 acceptable because the applicant revised the program description and clarified that the Bolting Integrity Program would include the recommendations of EPRI NP-5769 and NUREG-1339 for lubricants. Therefore, the staff's concern described in RAI B.2.1.3-2 is resolved.

The staff confirmed that the "preventive actions" program element satisfies the criterion defined in the GALL Report and SRP-LR Section A.1.2.3.2 and concludes that this program attribute is acceptable.

(3) Parameters Monitored or Inspected - In LRA Section B2.1.3, the applicant stated that the specific parameters monitored or inspected are discussed in the three credited programs and include inspection of high-yield strength (> 150 ksi) bolting for NSSS component supports for cracking, pressure-retaining bolted joints for signs of leakage, and structural bolting for signs of aging degradation. The LRA states that this element is consistent with the corresponding element of GALL AMP XI.M18.

The staff confirmed that the "parameters monitored or inspected" program element satisfies the criterion defined in the GALL Report and SRP-LR Section A.1.2.3.3 and concludes that this program attribute is acceptable.

(4) Detection of Aging Effects - In RAI B.2.1.3-3(a), dated July 12, 2005, the staff stated that the GALL AMP XI.M18 "detection of aging effects" program element recommends such inspection techniques as removing the bolt, proof test by tension or torquing, by situ ultrasonic tests, or hammer tests. The applicant credits both visual and volumetric inspections of bolting; however, it was not clear how crack initiation due to cyclic loading or SCC is detected by visual inspection where the thread surface is not readily visible. Therefore, the staff requested the applicant to clarify (a) how crack initiation due to cyclic loading or SCC is detected by visual inspection for applications where the thread surface

is not readily visible, (b) whether leakage detection is the only method applied, and (c) whether removal of the bolting and inspection during maintenance or augmented inspection is credited.

In its response, by letter dated August 12, 2004, the applicant stated:

> NUREG-1801, AMP XI.M18, element 4, states: 'Structural bolting both inside and outside containment is inspected by visual inspection. Degradation of this bolting may be detected and measured either by removing the bolt, proof test by tension or torquing, by in situ ultrasonic tests, or hammer test. If this bolting is found corroded, a closer inspection is performed to assess extent of corrosion.' Management of Palisades' structural bolting is consistent with this discussion. The Palisades Bolting Integrity Program Basis Document states that structural bolting and fasteners, both inside and outside containment, are inspected by visual examination in accordance with the Structural Monitoring Program. The Structural Monitoring Program facilitates visual inspection of structural bolting. If visual degradation is observed requiring further evaluation, degradation (e.g. crack initiation due to cyclic loading or SCC) may be identified and measured by removing the bolting, proof test by tension or torquing, by in situ ultrasonic tests, or hammer tests.

> Structural bolting is typically not subject to significant cyclic loading or thermal stress, so cracking due to fatigue is not an applicable aging effect. For stress corrosion cracking to occur, three conditions must exist: high stress, a corrosive environment, and susceptible material. The corrosive environment is initially precluded through use of proper lubricants and proper bolt installation practices. Visual inspection for degradation will identify the potential for SCC by detecting evidence of corrosion or a corrosive environment. If the potential for cracking is found, the extent of any degradation may be identified and measured by removing the bolting for further inspection, proof testing by tension or torquing, in situ ultrasonic testing, or hammer testing.

> For inspection of pressure-retaining bolting, the Palisades Bolting Integrity Program credits the inspection requirements of ASME Section XI for Class 1, 2, and 3 pressure retaining components. For bolting in non-ASME classed pressure retaining components, Palisades applies visual inspection methodology similar to the VT-2 methodology used for ASME Section XI Class 3 pressure retaining components. See the response to RAI B2.1.20-2(e) in NMC letter dated July 25, 2005 for additional information.

Based on its review, the staff finds the applicant's response to RAI B.2.1.3-3(a) acceptable because the applicant clarified that several techniques are available to detect crack initiation due to cyclic loading or SCC for susceptible applications where the thread surface is not readily visible. As to pressure-retaining bolting, the applicant credits the ASME Code Section XI inspection requirements for Classes 1, 2, and 3 components and, for non-ASME bolting, a similar visual inspection methodology. The staff considered the use of ASME Code Section XI required inspections combined with other appropriate inspections during walkdowns and maintenance capable of detecting degradation of bolting where the thread surface is not readily visible. High-strength bolting used in NSSS component supports is addressed in the resolution of RAI B.2.1.3-1(d). Therefore, the staff's concern described in RAI B.2.1.3-3(a) is resolved.

In RAI B.2.1.3-3, dated July 12, 2005, the staff stated that LRA Section B2.1.3 identifies the ASME Section XI IWB, IWC, IWD, IWF Inservice Inspection Program for managing non-Class 1 SR pressure-retaining and support bolting. LRA Section B2.1.2 references 10 CFR 50.55a requirements for the ASME Code Section XI edition. GALL AMP XI.M3 recommends visual, surface, and volumetric examinations in accordance with IWA-2000 of the 1995 Edition through 1996 Addenda of ASME Code Section XI. Therefore, the staff requested the applicant to clarify whether detection of aging effects in bolting using the ASME Code Section XI edition cited in 10 CFR 50.55a is consistent with or conservative compared to the 1995 Edition through 1996 Addenda cited in GALL AMP XI.M3. If the new code edition is less restrictive, the applicant was asked to submit an evaluation to demonstrate that aging effects are adequately managed.

In its response, by letter dated August 12, 2005, the applicant stated:

> For clarification, the Palisades Reactor Head Closure Studs NUREG 1801 Section XI.M3 program) are age managed under the ASME Section XI IWB, IWC, IWD, IWF Aging Management Program (AMP) (LRA Section B.2.1.2). All other bolting that is age managed under license renewal falls under the Bolting Integrity Program (LRA Section B2.1.3), which implements the requirements of NUREG 1801 (GALL XI.M18 Bolting Integrity). However, the Bolting Integrity Program also credits the ASME section XI IWB, IWC, IWD, IWF Aging Management Program (LRA Section B.2.1.2) for the inspection of ASME Class 1, 2, and 3 bolting.
>
> NMC has discussed with the NRC staff various options for defining an appropriate code edition and addenda for the ASME Section XI IWB, IWC, IWD, and IWF Aging Management Program, and decoupling the AMP from the current ASME Section XI Inservice Inspection Program under 10 CFR 50.55a. In the near future, following NRC publication of NUREG 1801, Revision 1, NMC plans to submit a revised description of the ASME Section XI IWB, IWC, IWD, and IWF Aging Management Program that defines a new basis code and addenda. This revised AMP description will also impact all the other Palisades AMPs that credit this program. It is anticipated that the revised program description will credit the 2001 edition through 2003 addenda. Since this edition and addenda are expected to be endorsed by NUREG 1801, Revision 1, no additional justification of the code edition and addenda should be required.
>
> Therefore, the future submittal of the revised program description will constitute the specific response to this question.

Based on its review, the staff finds the applicant's response to RAI B.2.1.3-3 acceptable because the applicant clarified that the revised program description to be submitted would credit later code editions and addenda endorsed by the staff (Commitment 49). Therefore, the staff's concern described in RAI B.2.1.3-3 is resolved.

In RAI B.2.1.3-3(c), dated July 12, 2005, the staff stated that the "detection of aging effects" program element of LRA Section B2.1.3 indicates that structural bolting and fasteners, both inside and outside containment, are inspected visually in the Structural Monitoring Program. Therefore, the staff requested the applicant to clarify how the visual examination of structural bolting and fasteners is implemented pursuant to the Structural Monitoring Program, including a list of plant-specific programs or procedures that would be used to timely dispose of the identified bolting or fastener degradation events.

In its response, by letter dated August 12, 2005 the applicant stated:

> Palisades' existing Structural Monitoring Program procedure includes guidance for inspection of structural bolted connections. The procedure identifies common deficiencies in bolted connections including loose or missing fasteners, insufficient coverage of slotted holes with plate washers, insufficient contact between connected parts, etc. It also provides guidance on evaluation and disposition of inspection results. Identified deficiencies are directed to be corrected in a timely manner commensurate with their safety significance, their complexity, and other regulatory requirements. Use of the corrective action process and work request/work order process is specified to evaluate the operability of the affected SSC, the immediacy of the need for repair or replacement and ensure the appropriate level of cause determination and corrective action is completed for the identified deficiency.
>
> Personnel performing formal structural inspections are required to have the following qualifications as a minimum: be appropriately degreed engineers from an accredited college or university; be knowledgeable in the design, evaluation and performance of nuclear structures; and have five years minimum experience in structural engineering for nuclear structures. This level of qualification eliminates any delay in providing inspection results to qualified engineering personnel for evaluation and disposition.

Based on its review, the staff finds the applicant's response to RAI 2.1.3-3(c) acceptable because the applicant asserted that its existing Structural Monitoring Program procedure (a) includes guidance for inspection of structural bolted connections; (b) identifies common deficiencies in bolted connections including loose or missing fasteners, insufficient coverage of slotted holes with plate washers, insufficient contact between connected parts, etc; (c) provides adequate guidance on evaluation and disposition of inspection results; (d) provides adequate guidance for the corrective action process to timely dispose of identified deficiencies; and (e) clearly stipulates requirements for qualification of personnel performing formal structural inspections. Therefore, the staff's concern described in RAI 2.1.3-3(c) is resolved.

The LRA states that this element is consistent with the corresponding element of GALL AMP XI.M18. The staff finds that the "detection of aging effects" program element is acceptable because methods used to detect aging effects, including techniques identified by ASME Code Section XI, are consistent with the GALL Report.

The staff confirmed that the "detection of aging effects" program element satisfies the criterion defined in the GALL Report and SRP-LR Section A.1.2.3.4 and concludes that this program attribute is acceptable.

(5) Monitoring and Trending - LRA Section B2.1.3 states that examination schedules in the ASME Section XI IWB, IWC, IWD, IWF Inservice Inspection Program meet the requirements of ASME Code Section XI, Subsections IWB, IWC, and IWE, for Classes 1, 2, and MC pressure-retaining bolting, as well as Code Case N491-2 for Classes 1, 2, and 3 component support bolting. Corrective actions may include daily, weekly, or biweekly monitoring and trending of leakage. This element is described as consistent with GALL AMP XI.M18. The staff finds the overall "monitoring and trending" program element proposed by the applicant acceptable because ASME Code Section XI inspection results and inspections for leakage combined with an effective corrective action program would effectively monitor and trend the applicable aging effects consistent with GALL AMP XI.M18.

The staff confirmed that the "monitoring and trending" program element satisfies the criteria defined in the GALL Report and SRP-LR Section A.1.2.3.5 and concludes that this program attribute is acceptable.

(6) Acceptance Criteria - LRA Section B2.1.3 states that the acceptance criteria are consistent with the GALL Report and indications are evaluated in accordance with IWB-3000, IWC-3000 and IWE-3000 for Classes 1, 2, and MC pressure-retaining bolting. Classes 1, 2, and 3 and MC component support bolting are evaluated in accordance with the acceptable standards of ASME Code Case N-491-2, Section 3410, and all other bolting is evaluated in accordance with the requirements of the applicable monitoring program identified in the program description.

In RAI B.2.1.3-4(a), dated July 12, 2005, the staff noted that the "acceptance criteria" program element of GALL AMP XI.M18 recommends immediate repairs for major leaks that may cause corrosion or contamination. LRA Section B2.1.3 states that the Bolting Integrity Program makes no distinction of "immediate" repairs, but instead relies upon the ASME Section XI IWB, IWC, IWD, IWF Inservice Inspection Program and corrective action process to evaluate, prioritize, and schedule repairs. The "operating experience" program element in LRA Section B2.1.3 states that the Bolting Integrity Program has been effective in timely identifying bolting degradation, thus ensuring that age-related degradation of bolting will be effectively managed throughout the period of extended operation. The applicant identified six instances where the ASME Section XI IWB, IWC, IWD, IWF Inservice Inspection Program had been instrumental in detecting bolting degradation. The applicant concluded that reviews of internal and external assessments report that the program has effectively identified and disposed of issues that could have led to degraded conditions. Therefore, the staff requested the applicant to specifically address whether these reports represent objective evidence that the action request and corrective action processes have been effective in initiating timely repairs for bolting, including structural bolting and fasteners. The applicant was also asked to explain plant-specific operating experience with aging management of any non-Class 1 high-strength bolting, including structural bolting.

In its response, by letter dated August 12, 2005, the applicant stated:

> The plant specific operating experience for the bolting integrity program was discussed in NMC's response to RAI B2.1.3-10 in a letter dated July 28, 2005. This response discussed how the bolting degradation was detected, how the degradation was dispositioned, and the degradation mechanism for each. Each of these situations serves as an example of how the corrective action system has been timely and effective in addressing degraded plant equipment, including bolting.
>
> A review of Palisades' operating experience identified no examples of failure of high strength bolting.
>
> As discussed in NMC's response to RAI B2.1.3-9 in the letter dated July 28, 2005, '... degraded conditions would be entered promptly into the plant Corrective Action Program. The program, in turn, would require a prompt assessment of the effect of the condition on plant safety and equipment operability. Corrective actions would be assigned and completed commensurate with the safety and operational significance of the condition. The corrective action for a degraded bolt would likely include eventual replacement of that bolt, but the timing of the replacement would have to consider the safety-significance of the condition, the plant conditions needed to safely complete the work, the status of other plant equipment, the extent and complexity of repair, radiological conditions, availability of replacement parts, etc.'

Based on its review, the staff finds the applicant's response to RAI B.2.1.3-4(a) acceptable because the applicant clarified that failure of high-strength bolting had not been experienced and that the applicant had sufficient operating experience with bolting degradation for reasonable assurance that the corrective action process had been timely and effective in addressing degraded bolting. Therefore, the staff's concern described in RAI B.2.1.3-4(a) is resolved.

The staff confirmed that the "acceptance criteria" program element satisfies the criteria defined in the GALL Report and SRP-LR Section A.1.2.3.6 and concludes that this program attribute is acceptable.

(10) Operating Experience - In LRA Section B2.1.3, the applicant explained that industry operating experience related to the Bolting Integrity Program revealed numerous instances where degradation of bolting had occurred. The applicant looked at related issues including degradation of threaded fasteners due to SCC and fatigue loading, plus SCC in high-strength bolts used for NSSS component supports.

Plant-specific operating experience revealed six instances where the ASME Section XI IWB, IWC, IWD, IWF Inservice Inspection Program had been instrumental in detecting bolting degradation in the following items:

- piping flange bolts (one instance)
- pump studs (two instances)
- tank flange bolts (one instance)
- pipe support bolting (one instance)
- engineered safeguards system equipment bolting (one instance)

In RAI B2.1.3-10, dated June 28, 2005, the staff requested that the applicant discuss (1) how the six instances of bolting degradation were dispositioned, (2) how the six instances of bolting degradation were detected (e.g., by routine maintenance, system walkdown, or leakage monitoring system), and (3) describe degradation mechanism(s) of each of the six instances in question.

In its response, by letter dated July 28, 2005, the applicant stated that there were two instances of corrosion of components in the valve pit of the condensate storage tank. A corrective action document was initiated after corrosion was identified on carbon steel flanges during a valve line-up. Another corrective action document was initiated due to corroded bolts in a flange that were discovered during a walkdown in the valve pit. In both instances, the components were replaced. Both of these instances were due to the unventilated damp environment within the pit.

The applicant also discovered two instances of boric acid wastage of primary coolant pump flange bolts during walkdown which were contained in corrective action documents. The applicant removed boric acid and evaluated the condition of the bolts. The evaluation confirmed that the primary coolant pumps remained operable because bolting integrity was not compromised by the limited degradation of the bolting material. The degraded bolts were subsequently replaced during the next refueling outage.

The remaining two instances were identified through the applicant's solicitation from plant equipment personnel. The equipment personnel identified the possibility of fatigue of pipe supports and the potential for boric acid wastage of engineered safeguards system bolting material. The applicant stated that both of these concerns will be addressed by a combination of TLAAs and AMPs. No new aging mechanisms nor failures of AMPs were identified.

The staff finds reasonable assurance that the applicant's use of the operating experience program combined with the corrective action program and other assessments should provide objective evidence that the enhanced System Monitoring Program will adequately manage the aging effects in the systems and components that credit this program during the period of extended operation. Therefore, the staff's concerns described in RAI B2.1.3-10 are resolved.

The staff confirmed that the "operating experience" program element satisfies the criteria defined in the GALL Report and SRP-LR Section A.1.2.3.10 and concludes that this program attribute is acceptable.

FSAR Supplement. In LRA Section A2.3, the applicant provided the FSAR supplement for the Bolting Integrity Program. In RAI B.2.1.3-5(a), dated July 12, 2005, the staff noted that LRA Section A2.3 states that the program considers the guidelines delineated in NUREG-1339 for a bolting integrity program, EPRI NP-5769 (with the exceptions noted in NUREG-1339) for SR bolting, and EPRI TR-104213 for NSR bolting. The term "considers" is not definitive. Therefore, the staff requested the applicant to clarify whether the program relies on or is consistent with these documents.

In its response, by letter dated August 12, 2005, the applicant clarified that the first sentence of LRA Section A2.3 should read, "Palisades Bolting Integrity Program relies on the guidelines delineated in NUREG-1339, EPRI NP-5769 (with the exceptions noted in NUREG-1339) for safety-related bolting, and EPRI TR-104213 for non-safety related bolting." The applicant also clarified that these documents were used as source documents for developing the Bolting Integrity Program.

Based on its review, the staff finds the applicant's response to RAI B2.1.3-5(a) acceptable because the applicant changed the FSAR supplement to clarify that the Bolting Integrity Program relies on the guidelines delineated in NUREG-1339, EPRI NP-5769 (with the exceptions noted in NUREG-1339) for SR bolting, and EPRI TR-104213 for NSR bolting. Therefore, the staff's concern described in RAI B.2.1.3-5(a) is resolved.

In RAI A2.3.1, dated June 28, 2005, the staff noted that the first two sentences in Section A2.3 is not consistent with the FSAR supplement in the SRP-LR and requested that the applicant provide clarification.

In its response, by letter dated July 28, 2005, the applicant stated that the summary program description is intended to describe the Bolting Integrity Program for all systems and components in the scope of license renewal. This description is intended to encompass the requirements for bolting integrity described in SRP-LR, pages 3.3-17, 3.4-11, 3.5-19, 3.2-12, and 3.1-23. Since there are slight differences in the various bolting program descriptions in SRP-LR, the Bolting Integrity Program was more generally described in a way that was intended to be consistent with all the SRP-LR descriptions. In addition, since the Bolting Integrity Program addresses structural bolting and pressure retaining bolting, the applicant did not limit the Bolting Integrity Program scope by including the words "pressure retaining bolting" that appear in each SRP-LR description. Finally, the reference to "enhanced inspection techniques" in the SRP-LR is not clearly coupled to GALL AMP XI.M18 "Bolting Integrity" program description, so this statement was not considered relevant to the Bolting Integrity Program description.

The staff determines that the above clarification is satisfactory and that the information in LRA Section A2.3 provides an adequate summary description of the program, as required by 10 CFR 54.21(d).

The staff reviewed this section and determines that the information in the FSAR supplement and RAI response provide an adequate summary description of the program, as required by 10 CFR 54.21(d).

Conclusion. On the basis of its review of the applicant's Bolting Integrity Program and RAI responses, the staff determines that those program elements for which the applicant claimed consistency with the GALL Report are consistent. Also, the staff reviewed the enhancements and confirmed that the implementation of the enhancements prior to the period of extended operation would make the existing AMP consistent with the GALL Report AMP to which it was compared. The staff concludes that the applicant has demonstrated that the effects of aging will be adequately managed so that the intended functions will be maintained consistent with the CLB for the period of extended operation, as required by 10 CFR 54.21(a)(3). The staff also reviewed the revised FSAR supplement for this AMP and concludes that it provides an adequate summary description of the program, as required by 10 CFR 54.21(d).

3.0.3.2.2 Boric Acid Corrosion Program

Summary of Technical Information in the Application. In LRA Section B2.1.4, the applicant described the Boric Acid Corrosion Program, stating that this existing program is consistent, with enhancements, with GALL AMP XI.M10, "Boric Acid Corrosion."

The Boric Acid Corrosion Program monitors component degradation due to boric acid leakage through periodic inspections. It implements the recommendations of GL 88-05. The program requires periodic visual inspection of all systems within the scope of license renewal that contain borated water for evidence of leakage, accumulations of dried boric acid, or boric acid wastage. The program also provides for visual inspections and early detection of borated water leaks so that structures and electrical and mechanical components, that may be contacted by leaking borated water, will not be adversely affected such that their intended functions are impaired. Specifically, the Boric Acid Corrosion Program includes provisions for:

- identification of components exhibiting boric acid accumulations or leakage

- evaluation of the acceptability for continued service of components exhibiting boric acid accumulations or leakage

- trending and tracking of previously identified leaks or boric acid accumulations

- corrective actions per plant TS, FSAR, and administrative procedures

Staff Evaluation. The staff confirmed the applicant's claim of consistency with the GALL Report. The staff reviewed the enhancements and the associated justifications to determine whether the AMP, with the enhancements, remains adequate to manage the aging effects for which it is credited.

In the LRA the applicant described the three program enhancements:

(1) Detection of aging effects for susceptible SSCs: revise applicable plant procedures to include criteria for observing susceptible SSCs for boric acid leakage and degradation during system walkdowns.

(2) Acceptance criteria: revise applicable plant procedures to include explicit acceptance criteria for boric acid inspections.

(3) Detection of aging effects for structural steel and non-ASME component supports: revise applicable plant procedures to include periodic inspection of structural steel and non-ASME component supports for evidence of boric acid residue and boric acid wastage/corrosion.

In accordance with 10 CFR 54.21(a)(3), the staff reviewed the information in LRA Section B2.1.4 demonstrating that the Boric Acid Corrosion Program will ensure that the aging effects will be adequately managed so that intended functions will be maintained consistent with the CLB throughout the period of extended operation.

LRA Section B2.1.4 states that the Boric Acid Corrosion Program is consistent with the GALL Report with enhancements. The applicant stated that procedures will be revised or developed to identify susceptible components upon which borated water may have leaked and to ensure they are inspected for degradation. The staff examined these "enhancements" and determined that they are actually administrative or procedural improvements addressing GALL AMP XI.M10 recommendations; therefore, the staff does not consider them as GALL AMP enhancements.

The staff reviewed the Boric Acid Corrosion Program against the AMP elements found in the GALL Report, SRP-LR Section A.1.2.3, and SRP-LR Table A.1-1 and focused on how the program manages aging effects through the effective incorporation of ten elements (i.e., "scope of program," "preventive actions," "parameters monitored or inspected," "detection of aging effects," "monitoring and trending," "acceptance criteria," "corrective actions," "confirmation process," "administrative controls," and "operating experience").

The applicant indicated that the "corrective actions," "confirmation process," and "administrative controls" are parts of the site-controlled QA program. The staff's evaluation of the QA program is discussed in SER Section 3.0.4. The remaining seven elements are discussed as follows.

(1) Scope of Program - In LRA Section B2.1.4, the applicant stated that the Boric Acid Corrosion Program covers SSCs, including electrical components, from or on which borated water may leak. LRA Sections 3.1 through 3.6 list these SSCs; however, LRA Section B2.1.4 does not mention any prioritization of components according to susceptibility to boric acid corrosion.

The staff's review of LRA Section B2.1.4 identified areas in which additional information was necessary to complete the review of the applicant's program elements. The applicant responded to the staff's RAIs as discussed below.

In RAI B2.1.4-1, dated November 30, 2005, the staff noted that LRA Section B2.1.4 states that the program identifies components exhibiting boric acid leakage, evaluates the acceptability of these components for continued service, performs trending and tracking, and recommends corrective actions. GALL AMP XI.M10 requires the determination of the principal location of leakage (prioritization). Therefore, the staff requested the applicant to discuss the program's classification of components based on their susceptibility to corrosion from boric acid leakage and the program's determination of the scope and frequency of visual and other NDE inspections of these components. The staff also requested the applicant to provide information regarding provisions for managing potential boric acid leakage in inaccessible locations and areas covered by external insulation surfaces.

In its response, by letter dated January 13, 2006, the applicant stated:

Classification of components based on their susceptibility to corrosion from boric acid leakage

The Palisades program philosophy is that all leakage from systems containing boric acid should be minimized. Once a leak occurs our program addresses each leak by the same process with the underlying assumption that any components that may be exposed to the leakage could be susceptible to degradation from boric acid. The plant Boric Acid Corrosion Control Program procedure identifies the Primary Coolant System, Engineered Safeguards Systems, Chemical and Volume Control System and Liquid Radioactive Waste Systems as those with the highest probability of experiencing leakage of boric acid. The portions of these systems classified as ASME class 1, 2 or 3 are inspected at the frequencies specified in the ASME Section XI Inservice Inspection and Boric Acid Corrosion Control Program (BACCP) commitments. The ASME Section XI Code establishes the standards for inspection of code components based on their ASME classification. This includes criteria for visual or volumetric inspections as well as hydrostatic and leakage testing. Hydrostatic and leakage testing specified by the ASME Section XI Code establishes requirements to look for and determine the principle locations of leakage.

Those systems containing boric acid that are non-ASME Section XI class 1, 2 or 3 are inspected during system walkdowns. As a general practice the plant operations staff also looks for leakage from all plant systems on a continuing basis, and initiates corrective actions when appropriate to drive resolution of the leak.

<u>Program determination of the scope and frequency for visual and other nondestructive examination (NDE) inspections</u>

All in-scope systems receive periodic inspection in accordance with In-Service Inspection Pressure Testing, the Boric Acid Corrosion Control Program, and/or the System Monitoring Program. Generally, components are visually examined once per fuel cycle, except Primary Coolant System components, which may receive a mid cycle inspection (not necessarily VT-2) if conditions allow. Otherwise, indicators of primary coolant system leakage, such as containment pressure, humidity, radiation, sump level, air cooler drainage, and visual observations during containment entries are monitored for leak indicating trends. Areas outside of containment are generally accessible by operations, maintenance and engineering personnel during routine activities. The Boric Acid Corrosion Program credits the System Monitoring Program walkdowns and inspections for signs of boric acid leakage, residue or degradation of mechanical systems/components. Electrical connectors are specifically inspected on a periodic frequency for boric acid residue and degradation.

<u>Provisions for managing potential boric acid leakage in inaccessible locations and areas covered by insulation surfaces</u>

Leaks from inaccessible locations are managed in accordance with the plant's corrective action program and in accordance with Technical Specifications and Inservice Inspection requirements. If evidence of leakage is indicated at an inaccessible location, that leakage will be located and corrective actions taken. If continued operation is allowed, justification will be provided based on approved industry standards.

A detailed discussion of insulated or inaccessible locations was provided in NMC's January 20, 2003, response to the NRC's November 18, 2002, request for additional information (RAI) concerning the 60 Day response to NRC Bulletin 2002-01,'Reactor Pressure Vessel Head Degradation and Reactor Coolant Pressure boundary Integrity.'

Based on its review, the staff finds the applicant's response to RAI B2.1.4-1 acceptable; therefore, the staff's concern described in RAI B2.1.4-1 is resolved.

The staff confirmed that the "scope of program" program element satisfies the criterion defined in the GALL Report and SRP-LR Section A.1.2.3.1 and concludes that this program attribute is acceptable.

(2)   Preventive Actions - The applicant stated that the preventive actions include improving such maintenance practices as revising the valve packing program to improve packing techniques, performing periodic walkdowns to identify components that may require corrective maintenance, and monitoring locations where potential leakage could occur. In addition, timely repair of detected leakage prevents or mitigates boric acid corrosion.

The staff confirmed that the "preventive actions" program element satisfies the criterion defined in the GALL Report and SRP-LR Section A.1.2.3.2 and concludes that this program attribute is acceptable.

(3)   Parameters Monitored or Inspected - The applicant stated that visual inspections monitor the effects of boric acid corrosion on the intended functions of an affected structure or component. Borated water leakage results in deposits of white boric acid crystals and the presence of moisture that can be observed by visual inspections during system walkdowns.

The staff confirmed that the "parameters monitored or inspected" program element satisfies the criterion defined in the GALL Report and SRP-LR Section A.1.2.3.3. Visual inspections are expected to ensure that borated water leaks are properly managed. The staff concludes that this program attribute is acceptable.

(4)   Detection of Aging Effects - The applicant stated that degradation of components due to boric acid corrosion cannot occur without borated water leakage. Visual inspections are frequently conducted to identify necessary repairs and minimize the potential larger development of an undetected leak. The applicant also stated that evaluations are conducted when leaks are detected.

This program also credits the System Monitoring Program for the visual inspection of other SSCs outside the primary coolant pressure boundary that may be subject to the degrading effects of any borated water leakage.

The staff confirmed that the "detection of aging effects" program element satisfies the criterion defined in the GALL Report and SRP-LR Section A.1.2.3.4 and concludes that this program attribute is acceptable.

(5)   Monitoring and Trending - The applicant stated that monitoring and trending rely on visual inspections during normal plant operation and during shutdown for refueling. The program follows the guidelines in GL 88-05 and provides for timely detection of leakage by observation of boric acid crystal deposits during plant maintenance and walkdowns.

The staff confirmed that the "monitoring and trending" program element satisfies the criteria defined in the GALL Report and SRP-LR Section A.1.2.3.5. Trending of inspection results will enhance the applicant's ability to detect aging effects before a loss of intended function. The staff concludes that this program attribute is acceptable.

(6)   Acceptance Criteria - The applicant stated that plant procedures establish acceptance criteria and require corrective action or further evaluation if any leakage of borated water is noted.

The staff confirmed that the "acceptance criteria" program element satisfies the criteria defined in the GALL Report and SRP-LR Section A.1.2.3.6. Any anomalous signs of degradation would be evaluated by an engineer to determine material degradation. If found unacceptable, corrective measures would be implemented. The staff concludes that this program attribute is acceptable.

(10)   Operating Experience - In LRA Section B2.1.4, the applicant explained that a review of industry operating experience associated with the Boric Acid Corrosion Program and aging revealed issues related to:

- boric acid wastage of RCS piping and nozzles

- boric acid corrosion of reactor vessel head and closure studs from leaking borated water

- failure of valve packing gland bolts due to boric acid wastage

- failure of valve body to bonnet studs/nuts due to boric acid wastage

- boric acid wastage of reactor coolant pump closure flange studs

- boric acid corrosion of SG manway closure studs

- boric acid corrosion of high-pressure safety injection pump casing

These issues were addressed in various NRC and industry communications, which have been incorporated into the program as applicable.

The applicant also explained that a review of plant-specific operating experience related to the Boric Acid Corrosion Program and aging revealed that the following issues had been addressed:

- boric acid leaks in the containment spray header in containment at flanges with carbon steel bolting and a threaded spray nozzle connection

- boric acid wastage of primary coolant pump studs

- boric acid wastage of manual valve body-to-bonnet bolts

- corrosion of flanges for primary coolant pump component cooling water (CCW) connections due to external boric acid leakage

In RAI B2.1.4-2, dated November 30, 2005, the staff noted that LRA Section B2.1.4 lists a number of aging related issues addressed in various NRC and industry communications and indicates that the operating experience related to these issues has been incorporated into the AMP as applicable. Therefore, the staff requested the applicant to provide information about the program improvements directly related to lessons learned from the Davis-Besse vessel head degradation and the CRD mechanism penetration cracking discussed in NRC Bulletin 2001-01, "Circumferential Cracking of Reactor Pressure Vessel Head Penetration Nozzles," NRC Bulletin 2002-01, "Reactor Pressure Vessel Head Degradation and Reactor Coolant Pressure Boundary Integrity," NRC Bulletin 2002-02, "Reactor Pressure Vessel Head and Vessel Head Penetration Nozzle Inspection Programs," and NRC Order EA-03-009. The staff also requested discussion of examples of implementation of corrective actions in the program to prevent recurrence of degradation caused by boric acid leakage, as required by GL 88-05, "Boric Acid Corrosion of Carbon Steel Reactor Pressure Boundary Components in PWR Plants."

In its response, by letter dated January 13, 2006, the applicant listed various docketed letters that replied to the requested bulletins and orders. The applicant also included a short description of the program improvements directly related to lessons learned from the Davis-Besse vessel head degradation or PNP-specific CRD mechanism penetration cracking. The applicant discussed degradation caused by boric acid leakage:

> When boric acid leakage is identified, the source is repaired to stop the leakage. Two examples are highlighted where the corrective actions included a design or material change to reduce the probability of future boric acid leakage. In both of these instances the identified leakage was from the primary coolant system, and the corrective actions included replacement of the leaking components with components less likely to degrade in the operating environment and cause recurring exposure of the surrounding area to boric acid. First, the plant's 45 stainless steel control rod drive seal housings were replaced with Alloy 600 seal housings as a better material choice in the operating environment. Second, the design of eccentric reducers in the 45 control rod drive upper housing assemblies was modified to move a weld farther from the reducer to eliminate sensitization from welding in an area where large stresses exist due to the configuration of the reducer.

Based on its review, the staff finds the applicant's response to RAI B2.1.4-2 acceptable and concludes that the Boric Acid Corrosion Program would adequately manage the aging effects observed at PNP. Therefore, the staff's concern described in RAI B2.1.4-2 is resolved.

The staff confirmed that the "operating experience" program element satisfies the criteria defined in the GALL Report and SRP-LR Section A.1.2.3.10 and concludes that this program attribute is acceptable.

FSAR Supplement. In LRA Section A2.4, the applicant provided the FSAR supplement for the Boric Acid Corrosion Program. The staff reviewed this section and determines that the information in the FSAR supplement provides an adequate summary description of the program, as required by 10 CFR 54.21(d).

Conclusion. On the basis of its review of the applicant's Boric Acid Corrosion Program and RAI responses, the staff determines that those program elements for which the applicant claimed consistency with the GALL Report are consistent. Also, the staff reviewed the improvements and confirmed that the implementation of the improvements prior to the period of extended operation would make the existing AMP consistent with the GALL Report AMP to which it was compared. The staff concludes that the applicant has demonstrated that the effects of aging will be adequately managed so that the intended functions will be maintained consistent with the CLB for the period of extended operation, as required by 10 CFR 54.21(a)(3). The staff also reviewed the FSAR supplement for this AMP and concludes that it provides an adequate summary description of the program, as required by 10 CFR 54.21(d).

3.0.3.2.3 Closed Cycle Cooling Water Program

Summary of Technical Information in the Application. In LRA Section B2.1.6, the applicant described the Closed Cycle Cooling Water Program, stating that this existing program is consistent, with exceptions, with GALL AMP XI.M21, "Closed-Cycle Cooling Water System."

The Closed Cycle Cooling Water Program manages aging effects in closed cycle cooling water systems not subject to significant sources of contamination, in which water chemistry is controlled and heat is not directly rejected to the ultimate heat sink. The program includes maintenance of system corrosion inhibitor concentrations to minimize degradation and periodic or one-time testing and inspections to assess SSC aging. The program scope includes activities to manage aging in the CCW system, emergency diesel generator (EDG) jacket cooling water (emergency power system), and shield cooling system (SCS). The program credits the One-Time Inspection Program for the inspection of selected SCS and EDG system HXs and a representative sample of stagnant portions of the system piping. The inspections will check for fouling and evidence of corrosion or cracking. NDEs will be used, if practical and warranted, to verify pipe wall thickness at selected locations where loss of material has been experienced.

Staff Evaluation. During the audit and review, the staff confirmed the applicant's claim of consistency with the GALL Report. Details of the staff's audit evaluation of this AMP are documented in the Audit and Review Report. The staff reviewed the exceptions and the associated justifications to determine whether the AMP, with the exceptions, remains adequate to manage the aging effects for which it is credited.

As documented in the Audit and Review Report, the applicant stated that it used the TR-107396, Revision 1, version for the Closed Cycle Cooling Water Program, which is not consistent with GALL AMP XI.M21. By letter dated August 25, 2005, the applicant stated that by October 31, 2005, it would submit, for the staff's review and approval, a comparison of TR-107396, Revision 1, with Revision 0 to identify the material changes that impact aging management and justify their acceptability.

By letter dated October 31, 2005, the applicant submitted a comparison of TR-107396, Revision 1 (issued as TR-107820, "Closed Cycle Cooling Water Chemistry"), with Revision 0 to identify the material changes that impact aging management and justify their acceptability.

The applicant stated that the purpose of Revision 0 was to assist plants in developing water treatment strategies to protect carbon-steel and copper-containing systems from corrosion. This revision does not provide precise direction, but instead provides broad direction for plants to develop their own closed cooling water chemistry control programs by utilizing the guidance in the report to tailor specific station programs. Revision 0 does not have the now-typical operating table of "Control Parameters" and "Diagnostic Parameter" with respective sampling frequency and expected values; however, it does identify throughout parameters that should be monitored as "Control Parameters" or "Diagnostic Parameters." In general, Revision 0 allows plants a great deal of flexibility in developing their closed cooling water chemistry programs.

The applicant further stated that Revision 1 is significantly more directive. This revision incorporates the term "Action Level" and establishes values for which specific actions are "required" to be taken to correct conditions. Furthermore, it specifically establishes recommended monitoring frequencies and clearly identifies expected parameter values. For comparison of Revision 1 to Revision 0, the Revision 1 table of monitored parameters related to the applicant's water treatment regime was used. For each parameter in the Revision 1 table, the manner in which it was addressed by Revision 0 was discussed. As PNP uses nitrite corrosion inhibition for all of its closed cooling water systems, the differences between the other methods of corrosion inhibition were not discussed.

The applicant also stated that in Revision 0, there are parameters not specifically identified in Revision 1. Specifically, total organic carbon, dissolved oxygen, total alkalinity, calcium/magnesium, and refrigerants are all identified as "diagnostic" parameters in Revision 0. None of them (or monitoring of them) is considered to have any effect on the long-term health of closed cycle cooling water systems.

The applicant stated that after comparison of the two revisions of the EPRI Closed Cooling Water Chemistry Guidelines and concluded that while there were some minor changes from Revision 0 to Revision 1, they would not have any effect on the ability to properly manage the long-term aging of components within the closed cycle cooling water systems.

The applicant also stated that based on this comparison and to indicate the use of Revision 1 as an exception to the GALL Report, the LRA description of the Closed Cycle Cooling Water Program had been changed. These changes apply to the original LRA description of the program and supersede, in their entirety, the changes reported in response to a Closed Cycle Cooling Water Program question by letter dated August 25, 2005.

By letter dated October 31, 2005, the applicant revised LRA Section B2.1.6 as follows:

On Page B-41, under the heading **Program Description**, third paragraph, Revision 1 was added after TR-107396.

On Page B-42, under the heading of **Exceptions to NUREG-1801**, the entire section was replaced with the following statement: An exception is taken to the selected NUREG-1801 program elements listed below. The specific exceptions being taken are also discussed in the corresponding element discussions. They are repeated here for ease of review.

1. Preventive Actions: The PNP Plant Closed Cycle Cooling Water Program implements the guidelines of Revision 1 of the Closed Cycle Cooling Water Guidelines, issued as EPRI TR-1007820, 'Closed Cycle Cooling Water Chemistry,' dated April 2004. This guideline supersedes EPRI TR-107396, 'Closed Cycle Cooling Water Chemistry Guideline,' Revision 0, issued November 1997, referenced in NUREG-1801. Justification provided shows that this later revision to the guidelines has no detrimental affects on aging management of the closed cycle cooling water system.

On Page B-43, under the heading of **Preventive Actions**, last sentence of the first paragraph, Revision 1 was added after TR-107396.

On Page B-44, under the heading of **Parameters Monitored, Tested, and/or Inspected**, the last paragraph, including exception bullet, was replaced with the following statements. 'This program monitors the effects of corrosion by surveillance testing and inspection. For pumps, the parameters monitored include flow, discharge and suction pressures. For heat exchanges, the parameters include flow, inlet and outlet temperatures, and differential pressures as appropriate. This element is consistent with NUREG-1801, Section XI.M21, 'Closed-Cycle Cooling Water System."

On Pages B-44 and B-45, under **Detection of Aging Effects**, the last paragraph, including exception bullet, was replaced with the following statements. 'Performance and functional testing ensures acceptable functioning of system or components. For systems or components in continuous operation, performance adequacy is determined by monitoring data trends. Components not in operation are periodically tested to ensure operability. This element is consistent with NUREG-1801, Section XI.M21, 'Closed-Cycle Cooling Water System."

On Pages B-45 and B-46, under **Monitoring and Trending**, the last paragraph, including both exception bullets, was replaced with the following statements: 'Performance and functional testing are performed at least every 18 months to demonstrate system operability, and tests to evaluate heat removal capability of the system and degradation of system components are performed every five years. This element is consistent with NUREG-1801, Section XI.M21, 'Closed-Cycle Cooling Water System."

On Page B-46 and B-47, under **Acceptance Criteria**, the last paragraph, including both exception bullets, was replaced with the following statement: 'This element is consistent with NUREG-1801, Section XI.M21, 'Closed-Cycle Cooling Water System.''

The staff reviewed the applicant's comparison of Revision 0 with Revision 1 of the EPRI Closed Cooling Water Chemistry Guidelines. After review of the applicant's comparison, the staff found that the parameters of operation in Revision 1 were generally defined with specific values and sampling frequencies. Some of the parameters had been re-characterized as control or diagnostic parameters in Revision 1. For some parameters in Revision 1, there were no values specified as was also the case in Revision 0; thus, there was no difference between the revisions. There were parameters identified in Revision 0 not in Revision 1. Where some parameters had values specified in both revisions, in some instances the Revision 1 values were slightly less conservative. In the few instances where the Revision 1 parameters were less conservative, no longer measured, or re-characterized, the staff concurred with the applicant that the higher limit, elimination of the parameter, or re-characteriziation of the parameter was acceptable and the differences would have no effect on the applicant's ability to properly manage the long-term aging of components within the closed cycle cooling water systems. On this basis, the staff found the applicant's use of TR-107396, Revision 1, acceptable.

In the LRA, the applicant stated that its Closed Cycle Cooling Water Program is consistent with GALL AMP XI.M21 with exceptions. The Closed Cycle Cooling Water Program takes exception to the "parameters monitored or inspected," "monitoring and trending," and "acceptance criteria" program elements. The applicant stated that the exception is that it does not credit active flow testing for managing age-related degradation of CCW components.

The applicant stated that performance of selected HXs is monitored in accordance with the plant master HX testing plan. The performance and operation testing of selected pumps, including flow, suction, and discharge pressure, is monitored in accordance with the ASME Section XI, Subsection IWP, inservice testing program. In addition, as documented in the Audit and Review Report, the staff noted that PNP performs the required system and component performance testing as part of existing plant programs. By letter dated August 25, 2005, the applicant stated that the LRA had been revised to address these elements and to remove this exception. These elements, as revised, are consistent with the GALL Report. On the basis that this revision makes the AMP consistent with the GALL Report, the staff finds it acceptable.

Operating Experience. In LRA Section B2.1.6, the applicant explained that a review of industry operating experience associated with the Closed Cycle Cooling Water Program and aging reveals issues and instances related to:

- SCC in RCP oil cooler discharge piping
- corroded solder connections in diesel lube oil cooler due to inadequate corrosion inhibitor
- inoperable check valves (stuck open) due to corrosion product buildup
- cracks in CCW piping
- fouling of diesel cooling water HXs

Various related NRC and/or industry generic communications have been issued and, in turn, incorporated into the program as applicable.

The applicant further explained that a review of plant-specific operating experience related to the Closed Cycle Cooling Water Program and aging revealed that the following issues have been addressed:

- tube blockage and fouling in component cooling HX
- fuel pool HX tube breakage due to high CCW flow
- through-wall flaw in spent fuel pool cooling pipe

The staff reviewed selected condition reports as well as a recent system health and status report for the component cooling system. The condition reports reviewed revealed no instances in which the intended function of any system component had been lost. In addition, the applicant's "Health and Status Report for the Component Cooling System," dated June 23, 2005, indicated satisfactory system operation and no adverse chemistry trends. Based on its review of this information, the staff concludes that the applicant's Closed Cycle Cooling Water Program is effective in managing aging effects before any loss of component intended function.

The staff reviewed the operating experience provided in the LRA, and interviewed the applicant's technical personnel to confirm that plant-specific operating experience revealed no degradation not bounded by industry experience.

On the basis of its review of industry and plant-specific operating experience and discussions with the applicant's technical personnel, the staff concludes that the applicant's Closed Cycle Cooling Water Program would adequately manage the aging effects identified in the LRA for which this AMP is credited.

FSAR Supplement. In LRA Section A2.6, the applicant provided the FSAR supplement for the Closed Cycle Cooling Water Program. The staff reviewed this section and determines that the information in the FSAR supplement provides an adequate summary description of the program, as required by 10 CFR 54.21(d).

Conclusion. On the basis of its review and audit of the applicant's Closed Cycle Cooling Water Program, the staff determines that those program elements for which the applicant claimed consistency with the GALL Report are consistent. In addition, the staff reviewed the exceptions and associated justifications and determines that the AMP, with the exceptions, is adequate to manage the aging effects for which it is credited. The staff concludes that the applicant has demonstrated that the effects of aging will be adequately managed so that the intended functions will be maintained consistent with the CLB for the period of extended operation, as required by 10 CFR 54.21(a)(3). The staff also reviewed the FSAR supplement for this AMP and concludes that it provides an adequate summary description of the program, as required by 10 CFR 54.21(d).

3.0.3.2.4 Diesel Fuel Monitoring and Storage Program

Summary of Technical Information in the Application. In LRA Section B2.1.9, the applicant described the Diesel Fuel Monitoring and Storage Program, stating that this existing program is consistent, with exception and enhancements, with GALL AMP XI.M30, "Fuel Oil Chemistry."

The Diesel Fuel Monitoring and Storage Program assures the continued availability and quality of fuel oil to be used in diesel generators and diesel fire pumps. The program includes (a) monitoring and trending of fuel oil chemistry to maintain fuel oil quality and mitigate corrosion; (b) periodic draining, cleaning, and internal inspection of fuel oil storage tanks; and (c) verification of program effectiveness by a one-time measurement of fuel oil storage tank bottom thickness confirming the absence of an aging effect. Fuel oil quality is maintained by monitoring and controlling fuel oil contamination in accordance with the guidelines of the American Society for Testing Materials (ASTM) Standards D 1796, D 2276, D 2709, and D 4057. Exposure to fuel oil contaminants, such as water and microbiological organisms, is minimized by periodic draining and cleaning of tanks and by verifying the quality of new oil before its introduction into the storage tanks. However, corrosion may occur at locations in which contaminants may accumulate, as in tank bottoms. Accordingly, the effectiveness of the program is verified, through visual inspection and one-time ultrasonic thickness measurement of fuel oil storage tank bottom surface, to ensure that significant degradation is not occurring and that applicable component intended functions will be maintained during the period of extended operation.

Staff Evaluation. During the audit and review, the staff confirmed the applicant's claim of consistency with the GALL Report. Details of the staff's audit evaluation of this AMP are documented in the Audit and Review Report. The staff reviewed the exception and enhancements and the associated justifications to determine whether the AMP, with the exception and enhancements, remains adequate to manage the aging effects for which it is credited.

GALL AMP XI.M30 recommends, in the "acceptance criteria" program element, that modified ASTM D 2276 Method A be used in the determination of particulate size. The modification uses a filter with a pore size of 3.0 μm, instead of 0.8 μm. In the PNP testing procedure, which references this ASTM standard for particulate testing, the applicant uses a 0.8 μm pore filter instead of the GALL Report recommended 3.0 μm filter size. The staff found this use acceptable because the unmodified version is considered more conservative than the modified because of its smaller filter pore size. The smaller filter pore size will collect more particulate material from the fuel oil and lead to a conservative estimate of the amount of particulates present.

The staff reviewed those portions of the applicant's Diesel Fuel Monitoring and Storage Program for which the applicant claimed consistency with GALL AMP XI.M30 and found them consistent. Furthermore, the staff concludes that the applicant's Diesel Fuel Monitoring and Storage Program provides reasonable assurance of adequate management of the aging effect of loss of material due to general, pitting, and MIC by monitoring and controlling conditions that cause this aging effect and by monitoring the effectiveness of the program through surveillance and testing. The staff finds the applicant's Diesel Fuel Monitoring and Storage Program acceptable because it conforms to the recommended GALL AMP XI.M30, with the exception and enhancements.

In the LRA, the applicant stated that its Diesel Fuel Monitoring and Storage Program is consistent with the GALL Report with exception. The applicant stated that the exception is that PNP adds no biocides, stabilizers, or corrosion inhibitors to the fuel oil. As to not using biocides in the fuel oil, the applicant stated that no operational experience indicates a positive test for microbiological growth. In the event a test for microbiological growth would be positive, an evaluation per the corrective action program would determine whether biocides should be added.

As documented in the Audit and Review Report, the staff found this aspect of the exception acceptable because of the lack of physical evidence from the plant operating experience as well as the commitment through the corrective action program to investigate the use of biocides, if test results indicate the need (Commitment 16). The lack of physical evidence was confirmed during the interview of the applicant's technical personnel and through a review of selected test results.

As to fuel oil stabilizers, the applicant stated, in the LRA, that the fuel oil storage tanks have a relatively high fuel oil turnover rate. Tank T-10A has a typical operating volume of 38,000 gallons, with an average fuel consumption of 76,000 gallons per year. Tank T-926 has a typical operating volume of 15,000 gallons, with an average fuel consumption rate of 35,000 gallons per year. Due to the high fuel turnover rate in the storage tanks, the applicant had determined that there is no need to add fuel oil stabilizers to the diesel fuel. This determination is further strengthened by surveillance testing for the presence of particulates which, if significant, would result in filtering of tank contents. For example, the stored fuel in the T-10A fuel oil storage tank is filtered approximately every three years, or as needed. As documented in the Audit and Review Report, the staff found this aspect of the exception acceptable based on the relatively short resident time of the fuel oil in the storage tanks.

As to not adding corrosion inhibitors, the applicant stated, in the LRA, that the fuel oil is procured to meet ASTM D 975 standards, which include specifications and acceptance criteria for a copper strip corrosion test. Additionally, samples are periodically analyzed by an offsite facility for relative corrosiveness of the fuel oil by a copper strip corrosion test. All copper strip corrosion tests in the last five years have returned results that meet the ASTM standard. Consequently, the applicant adds no corrosion inhibitors to the diesel fuel.

During an interview of the plant's technical personnel, the staff confirmed that the applicant had specified the copper strip corrosion test as part of its fuel procurement specification and then performed a receipt acceptance test on a fuel oil sample with an outside laboratory. The applicant also indicated, as documented in the Audit and Review Report, that the fuel oil is periodically tested for relative corrosiveness by an outside laboratory. The staff finds this aspect of the exception of not using corrosion inhibitors acceptable based on the new fuel oil testing and the periodic testing of fuel oil in the tanks.

In the LRA, the applicant stated that its Diesel Fuel Monitoring and Storage Program is consistent with GALL AMP XI.M30 with an enhancement. The enhancement, in meeting the GALL Report "preventive actions" program element is to develop and implement procedures for periodic draining and cleaning of fuel oil storage tanks, EDG day tanks, and diesel fire pump day tanks, also procedures for periodic draining of water accumulated in the bottom of the fuel oil storage tanks, EDG day tanks, and diesel fire pump day tanks.

The applicant stated that it would develop a procedure for the periodic draining and cleaning of the fuel oil storage tanks to remove any water and sediment. As documented in the Audit and Review Report, the staff found these measures effective in mitigating corrosion inside diesel fuel oil tanks. On this basis, the staff finds this enhancement acceptable as such changes to the applicant's program would provide additional assurance that the aging effects will be adequately managed. This enhancement is consistent with the GALL Report recommendations.

In addition, the applicant stated that its Diesel Fuel Monitoring and Storage Program is consistent with the GALL Report with another enhancement. The enhancement in meeting the GALL Report "detection of aging effects" program element is to develop and implement procedures for periodic ultrasonic testing measurement of the thickness of the bottom of fuel oil storage tanks, EDG day tanks, and diesel fire pump day tanks. Also, the procedures for draining and cleaning of the tanks (preventive actions) include a visual inspection of interior tank surfaces for signs of degradation or corrosion with acceptance criteria, corrective actions, and documentation of inspection results.

In the LRA, the applicant stated that it would develop implementation procedures for non-destructive testing and visual inspections to ensure that significant degradation of the tank and tank bottoms does not occur without detection. As documented in the Audit and Review Report, the staff found that visual inspections and nondestructive testing to detect any signs of potential adverse degradation is consistent with the GALL Report recommendations for this program element. On this basis, the staff finds this enhancement acceptable as such changes to the applicant's program would provide additional assurance that the aging effects will be adequately managed. This enhancement is consistent with the GALL Report recommendations.

Operating Experience. In LRA Section B2.1.9, the applicant explained that a review of industry operating experience associated with the Diesel Fuel Monitoring and Storage Program and aging reveals issues and instances related to:

- fuel contamination leading to corrosion of fuel oil system components

- improper zinc coating curing and epoxy application by the manufacturer leads to zinc-fuel reaction creating adverse corrosion

- fuel oil leak caused by improper outer coating application

Various related NRC and/or industry generic communications have been issued and, in turn, have been incorporated into the program as applicable. Plant-specific operating experience related to the Diesel Fuel Monitoring and Storage Program identified no aging issues

The staff finds that most of the corrective action documents reviewed dealt with particulates in excess of administrative limits. Appropriate corrective actions like filtering of tanks as needed were implemented in response to these corrective action reports. On this basis, the staff concludes that there are no identified aging effects not effectively managed by the applicant's current program.

The staff reviewed the operating experience provided in the LRA and interviewed the applicant's technical personnel to confirm that plant-specific operating experience revealed no degradation not bounded by industry experience.

On the basis of its review of industry and plant-specific operating experience and discussions with the applicant's technical personnel, the staff concludes that the applicant's Diesel Fuel Monitoring and Storage Program would adequately manage the aging effects identified in the LRA for which this AMP is credited.

FSAR Supplement. In LRA Section A2.9, the applicant provided the FSAR supplement for the Diesel Fuel Monitoring and Storage Program. The staff reviewed this section and determines that the information in the FSAR supplement provides an adequate summary description of the program, as required by 10 CFR 54.21(d).

Conclusion. On the basis of its review and audit of the applicant's Diesel Fuel Monitoring and Storage Program, the staff determines that those program elements for which the applicant claimed consistency with the GALL Report are consistent. In addition, the staff reviewed the exception and the associated justifications and determines that the AMP, with the exception, is adequate to manage the aging effects for which it is credited. Also, the staff reviewed the enhancements and confirmed that the implementation of the enhancements prior to the period of extended operation would make the existing AMP consistent with the GALL Report AMP to which it was compared. The staff concludes that the applicant has demonstrated that the effects of aging will be adequately managed so that the intended functions will be maintained consistent with the CLB for the period of extended operation, as required by 10 CFR 54.21(a)(3). The staff also reviewed the FSAR supplement for this AMP and concludes that it provides an adequate summary description of the program, as required by 10 CFR 54.21(d).

3.0.3.2.5 Fire Protection Program

Summary of Technical Information in the Application. In LRA Section B2.1.10, the applicant described the Fire Protection Program, stating that this existing program is consistent, with exceptions and enhancements, with GALL AMPs XI.M26, "Fire Protection," and XI.M27, "Fire Water System."

The Fire Protection Program includes (a) fire barrier inspections, (b) electric and diesel-driven fire pump tests, and (c) periodic maintenance, testing, and inspection of water-based fire protection systems. Periodic visual inspections of fire barrier penetration seals, fire dampers, fire barrier walls, ceilings, and floors, and periodic visual inspections and functional tests of fire-rated doors are performed to ensure that function and operation are maintained. Periodic testing of the fire pumps ensures an adequate flow of firewater and no degradation of diesel fuel supply lines. Periodic maintenance, testing, and inspection activities of water-based fire protection systems provide reasonable assurance that fire water systems are capable of performing their intended function. Inspection and testing include periodic hydrant inspections, fire main flushing, sprinkler inspections, pipe wall thickness testing, and flow tests. Also included within the scope of the Fire Protection Program is aging management of spare cables for equipment required for Appendix R.

Staff Evaluation. During the audit and review, the staff confirmed the applicant's claim of consistency with the GALL Report. Details of the staff's audit evaluation of this AMP are documented in the Audit and Review Report. The staff reviewed the exceptions and enhancements and the associated justifications to determine whether the AMP, with the exceptions and enhancements, remains adequate to manage the aging effects for which it is credited.

ISG-4 and GALL AMP XI.M26 recommend periodic inspection and testing of the halon/carbon dioxide fire suppression systems. As documented in the Audit and Review Report, the staff noted that inspection of halon/carbon dioxide suppression systems is not included within the scope of the Fire Protection Program. By letter dated August 25, 2005, the applicant stated that there are no halon/carbon dioxide fire protection systems within the scope of license renewal under 10 CFR 54.4(a)(2). The applicant explained that PNP uses water spray to protect some areas (cable spreading room, etc.) protected by either carbon dioxide or halon at most plants. As there are no halon/carbon dioxide fire protection systems within the scope of license renewal, the Fire Protection Program is not required to manage them. The staff finds this omission acceptable because the applicant clarified that there are no halon or carbon dioxide systems within the scope of license renewal requiring aging management.

The applicant stated that its Fire Protection Program focuses on managing loss of material due to corrosion, MIC, or biofouling of carbon steel and cast-iron components exposed to water. As documented in the Audit and Review Report, the staff noted that LRA Table 3.3.2-7 lists bare copper, bronze, copper alloy, and stainless steel components in raw water with aging effects managed by the Fire Protection Program. By letter dated August 25, 2005, the applicant stated that the Fire Protection Program is intended to include all materials/components of LRA Table 3.3.2-7 that credit the program. The applicant revised the last sentence of the first paragraph of the "scope of program" description in LRA Section B2.1.10 to read, "The program focuses on managing loss of material due to corrosion, MIC, or biofouling of components; and aging management of fire barrier components." The staff finds this revision acceptable because it included all materials managed by the Fire Protection Program.

In the LRA, the applicant stated that its Fire Protection Program is consistent with GALL AMP XI.M26 with an exception to the "detection of aging effects" program element. The exception is that the applicant does not qualify the personnel performing the visual inspections of fire barrier walls, ceilings, floors, penetration seals, and fire doors to the ASME Code type of qualification per the GALL Report and ISG-04.

In the LRA, the applicant stated that per plant procedures, inspectors for fire barriers, doors, and fire seals are appropriately qualified to perform those inspections, but are not necessarily qualified to VT-1 or VT-3. There are no regulatory or other requirements specifying that these inspections be performed to VT-1 or VT-3 standards.

As documented in the Audit and Review Report, the staff asked the applicant to technically justify how the applicant's visual inspection method is equivalent and assures the same level of flaw identification and documentation as would be achieved by VT-1 and VT-3.

By letter dated August 25, 2005, the applicant stated:

> The ASME Section XI Code identifies inspection and acceptance criteria to apply to various systems/components (i.e., IWB-3520.1, IWB-3520.2) when using a VT-1 or VT-3 examination. IWA-2211, VT-1 Examination states: 'VT-1 examinations are conducted to detect discontinuities and imperfections on the surfaces of components, including such conditions as cracks, wear, corrosion, or

erosion.' IWA-2213, VT-3 Examination states: 'VT-3 examinations are conducted to determine the general mechanical and structural condition of components and their supports by verifying parameters such as clearances, settings, and physical displacements; and to detect discontinuities and imperfections, such as loss of integrity at bolted or welded connections, loose or missing parts, debris, corrosion, wear, or erosion. VT-3 includes examinations for conditions that could affect operation or functional adequacy of snubbers and constant load and spring type supports.'

Palisades' fire barrier penetration seal inspection surveillance procedure contains detailed inspection criteria, inspection methods, and acceptance criteria for each of the installed seal types. These requirements are equivalent to the level of detail required for a VT-1 inspection conducted under ASME Section XI. Completed inspection procedures are signed off as acceptable, or any unacceptable condition is documented in the Corrective Action System and repaired or replaced as required. NMC considers this as equivalent to VT-1 or VT-3 examinations as used in ASME Section XI and discussed in GALL.

It is also noted that the GALL Report, draft Revision 1, (as publicly released on August 12, 2005), Section XI.M26, removes reference to VT-1 and VT-3 from Detection of Aging Effects. It instead specifies that visual inspection by fire protection qualified inspectors of the fire barrier walls, ceilings, and floors, performed in walkdowns at least once every refueling outage, ensures timely detection of concrete cracking, spalling, and loss of material. Visual inspection by fire protection qualified inspectors detects any sign of degradation of the fire door such as wear and missing parts.

As the applicant clarified that inspections of fire barriers, doors, and fire seals are performed by qualified inspectors with procedures equivalent to VT-1 or VT-3 examinations, the staff finds this response acceptable.

In the LRA, the applicant also stated that its Fire Protection Program is consistent with GALL AMP XI.M27 with an exception to the "monitoring and trending" program element. The LRA states the exception that inspection and testing are as outlined in fire protection implementing procedures. National Fire Protection Association (NFPA) codes of record are identified in the Fire Protection Program Report or FSAR.

As documented in the Audit and Review Report, the staff asked the applicant to elaborate on the differences between code requirements NFPA-25 for monitoring and trending and the Fire Protection Program. By letter dated August 25, 2005, the applicant stated:

> Based on discussions during the interview and a detailed review of the applicable sections of NFPA-25, it has been concluded that Palisades complies with the specific monitoring and trending of results as specified in NFPA-25. This new understanding serves as the basis for not taking exception to the GALL on this issue. NMC's review of NFPA-25 for monitoring and trending of system performance testing identified two sections where requirements are stated.

NFPA-25 step 5-3.5.2 states that the pump test curve shall be compared to the unadjusted field acceptance test curve and the previous annual test curves.

This guidance is demonstrated in the fire suppression water system functional test and fire pump capacity test procedure which states, 'This procedure facilitates trending hydraulic performance of Fire Pumps P-9A, P-9B, and P-41, including comparison of current pump performance with original and historical pump performance.'

NFPA-25 step 4-3.1 states that underground and exposed piping shall be flow tested to determine the internal condition of the piping at minimum 5-year intervals. Flow test shall be made at flows representative of those expected during a fire for the purpose of comparing the friction loss characteristics of the pipe with those expected for the particular type of pipe involved, with due consideration given to the age of the pipe and to the results of previous flow tests. Any flow test results that indicate deterioration of available water flow and pressure shall be investigated to the complete satisfaction of the authority having jurisdiction to ensure that adequate flow and pressure are available for fire protection.

This testing guidance is demonstrated in the fire suppression water system flow test procedure, which states, 'To determine operability of fire suppression water system by performing a flow test to determine if there is any system degradation or obstruction. The procedure contains acceptance criteria and requirements to initiate a condition report if acceptance criteria are not met.' Flow testing is required to be performed every three (3) years.

Although Palisades is not committed to the requirements of NFPA-25, the Palisades program does meet the requirements as identified above. Continued implementation of this program provides reasonable assurance that the effects of aging of the applicable components will be adequately managed for the period of extended operation.

The staff finds this statement acceptable and agrees with removal of this exception because the applicant clarified that the Fire Protection Program meets NFPA-25 requirements.

In addition, in the LRA, the applicant stated that its Fire Protection Program is consistent with GALL AMP XI.M26 with another exception to the "acceptance criteria" program element: "Palisades inspection acceptance criteria states that no cracks of ¼" wide or greater are allowed." The applicant also indicated that acceptance criteria are derived from fire test reports.

As documented in the Audit and Review Report, the staff asked the applicant to justify acceptance of cracks no wider than ¼ inch. The applicant was also asked to identify where any cracks or defects in seals have been observed. By letter dated August 25, 2005, the applicant stated:

LRA Section B2.1.10, Fire Protection Program, Exceptions to NUREG 1801, is hereby revised to delete exception 3 on Page B-73. In addition, Acceptance Criteria on Page B-79 is hereby revised to read as follows:

Acceptance criteria are defined in the Palisades procedures used to perform tests and inspections of the Fire Protection System. Fire seal and conduit wrapping inspection results are acceptable if there are no visual indication of cracking, separation of seals from building structures and components, and no rupture or puncture of seals. Fire door inspection results are acceptable if there are no visual indications of wear, holes, damaged or missing parts, and clearances are within limits. Diesel-driven fire pump inspections are acceptable if there is no evidence of corrosion or leaks on the fuel oil supply line. Acceptance criteria for the diesel-driven fire pump capacity is contained within the test procedure.

The staff finds acceptable the revised Fire Protection Program "acceptance criteria" program element of no visual indication of cracking, no separation of seals from building structures and components, and no rupture or puncture of seals. Fire door inspection results are acceptable if there are no visual indications of wear, holes, or damaged or missing parts, and if clearances are within limits. Therefore, the staff agreed that the revised Fire Protection Program "acceptance criteria" program element is consistent with GALL AMP XI.M26, as clarified in ISG-4.

In the LRA, the applicant stated that its Fire Protection Program is consistent with GALL AMP XI.M26 with an enhancement in meeting the GALL Report "detection of aging effects" and "monitoring and trending" program elements: "The Structural Monitoring Program implementing procedures shall be revised to include specific inspection criteria and documentation requirements for verifying that walls; ceilings and floors that serve as Fire Protection Program fire barriers are verified to be free from aging related degradation that would impact the fire barrier's intended function."

In the LRA, the applicant further stated, under the "detection of aging effects" program element, that its Fire Protection Program credits the Structural Monitoring Program for aging management of fire barrier walls, ceilings, and floors. Fire doors are visually inspected and periodically tested by qualified inspectors for signs of corrosion, wear, or missing parts to ensure that function and operation are maintained. Also, under the "monitoring and trending" program element, the LRA states that the Fire Protection Program credits the Structural Monitoring Program for monitoring the condition of fire barrier walls, ceilings, and floors. At least 10 percent of the fire barrier penetration seals are visually inspected every 18 months for such signs of age-related degradation as seal separation from walls and components, cracking, rupture, and puncture of seals. Fire doors are tested and/or visually inspected by qualified inspectors semi-annually for signs of corrosion, wear, missing parts, and proper clearances to ensure that function and operation are maintained.

As documented in the Audit and Review Report, the staff verified that aging effects of fire walls, ceilings, and floors are managed by the applicant's Structural Monitoring Program. The staff found that revised Fire Protection Program procedures addressing more specifically aging-related degradation and documentation of fire door conditions would ensure detection of such degradation of all fire doors within the scope of license renewal before any loss of intended function. After implementation of this enhancement, this program element will be consistent with the GALL Report. On this basis, the staff finds this enhancement acceptable as such revision to the applicant's program will provide additional assurance that the effects of aging will be adequately managed.

In the LRA, the applicant also stated that its Fire Protection Program is consistent with GALL AMP XI.M26 with another enhancement in meeting the GALL Report "scope of program," "detection of aging effects," and "monitoring and trending" program elements: "Plant procedures shall be revised to more specifically address aging related degradation and expectations for documentation of fire door condition."

GALL AMP XI.M26 recommends at least bi-monthly visual inspection of hollow doors. The LRA does not specify inspection frequency. As documented in the Audit and Review Report, the staff asked the applicant to clarify whether it uses the same inspection frequency and, if not, to explain why and to add this explanation to the program exceptions. The applicant responded that the inspection interval for hollow metal fire doors is every six months. The inspection frequency has been every six months for many years, is considered satisfactory, and has not required a change without a trend of significant aging of hollow metal fire doors. PNP Procedure FPSP-SO-2 lists each of the fire doors, inspection requirements, and acceptance criteria. The applicant had committed in the LRA to revise inspection and acceptance criteria for fire door clearances. The staff finds this commitment acceptable as ISG-4, under "parameters monitored and inspected," states that hollow metal fire doors should be inspected visually on a plant-specific interval for integrity of door surfaces and for clearances. The plant-specific inspection intervals are determined by engineering evaluation to detect degradation of the fire doors prior to the loss of intended function. Therefore, this commitment is consistent with the ISG-4 recommendation and on this basis, the staff finds this enhancement acceptable as such changes to the applicant's program would provide additional assurance of adequate management of aging effects.

In addition, in LRA Section B2.1.10, the applicant stated that its Fire Protection Program is consistent with GALL AMPs XI.M26 and XI.M27 with another enhancement in meeting the GALL Report "detection of aging effects" and "monitoring and trending" program elements: "Develop and implement procedures to perform visual inspections for fire door clearances."

The applicant further stated that fire doors are tested or visually inspected by qualified inspectors semi-annually for signs of corrosion, wear, missing parts, and proper clearances to ensure that function and operation are maintained. The GALL Report recommends that fire door clearances are to be checked at least once bi-monthly; however, the LRA does not specify frequency of inspection. As documented in the Audit and Review Report, the staff asked the applicant to clarify that the same frequency is used and, if not, to provide justification. By letter dated August 25, 2005, the applicant stated that per ISG-4, a plant-specific interval is allowable. The applicant also stated that the inspection interval for hollow metal fire doors is every six months. This frequency has been every six months for many years, is considered satisfactory, and has not required a change without a trend of significant aging of hollow metal fire doors. As documented in the Audit and Review Report, a PNP procedure lists each of the fire doors, inspection requirements, and acceptance criteria. The staff finds this enhancement after implementation acceptable because the applicant revised the implementing procedures to be consistent with the GALL Report and ISG-4. In addition, the staff finds that such revision of the applicant's Fire Protection Program would provide additional assurance of adequate management of aging effects.

In the LRA, the applicant also stated that its Fire Protection Program is consistent with GALL AMP XI.M26 with another enhancement in meeting the GALL Report "detection of aging effects" and "monitoring and trending" program elements: "Revise diesel-driven fire pump performance test procedures to more specifically address the requirement to inspect and monitor the fuel oil supply line for aging related degradation, and to document inspection results."

In the LRA, the applicant further stated that fire pumps are tested every 18 months to ensure an adequate flow of fire water and that there has been no degradation of the fuel line to the diesel-driven fire pump. The fire protection system pressure is monitored continuously. Results of various surveillance tests are evaluated. Periodic full flow flushing of the main fire system underground piping assures that corrosion does not occur and that system function is maintained.

As documented in the Audit and Review Report, the staff asked the applicant for details on the diesel-driven fuel pump tests. By letter dated August 25, 2005, the applicant stated in Commitment 22 that it would, "Revise diesel-driven fire pump performance test procedures to more specifically address requirement to inspect and monitor fuel oil supply line for aging related degradation, and to document inspection results." As discussed in LRA Appendix B, the applicant described programs as if enhancements had been incorporated. The program, when enhancements are complete, would be consistent with the GALL Report. The applicant added that as its Fire Protection Program does not include a specific inspection of the diesel-driven fire pump fuel supply lines and no test results available for review.

As documented in the Audit and Review Report, the staff noted that the Fire Protection Program states that enhancement 4 "revises diesel-driven fire pump performance test procedures to more specifically address requirement to inspect and monitor fuel oil supply line for aging related degradation and to document inspection results." However, this enhancement does not indicate whether the revised procedures would be consistent with the GALL Report recommendations for the "detection of aging effects" and "monitoring and trending" program elements. By letter dated August 25, 2005, the applicant stated:

> LRA Section B2.1.10, Fire Protection Program, Detection of Aging Effects, second full paragraph on Page B-77, is hereby revised to read as follows:
>
>> Testing of the fire pumps (e.g., diesel-driven fire pump flow and discharge tests, sequential starting capability tests, and controller function tests) is performed every 18 months to ensure that an adequate flow of water is supplied and that there is no degradation of the fuel line to the diesel-driven fire pump.
>
> LRA Section B2.1.10, Fire Protection Program, Monitoring and Trending, third paragraph, is hereby revised to read as follows:
>
>> Testing of the fire pumps is performed every 18 months to ensure that an adequate supply of water is supplied and that there is no degradation of the fuel line to the diesel driven fire pump. The performance tests detect degradation of the fuel supply lines before loss of the component intended function, and provide data (e.g., pressure) necessary for trending. The applicant added that these revised statements are consistent with the GALL paragraphs quoted in the question.

After implementation of this enhancement, this program element would be consistent with the GALL Report and on this basis, the staff finds the enhancement acceptable.

In the LRA, the applicant also stated that its Fire Protection Program is consistent with GALL AMP XI.M27 with another enhancement. As stated in the LRA, the enhancement in meeting the GALL Report "detection of aging effects" program element is as follows:

> Develop and implement procedures for inspection of below grade fire protection system piping. Inspections shall occur when below grade piping is excavated for maintenance, and shall include pipe wall thickness (NDE or direct measurement) and documentation of aging related degradation of pipes. Procedures shall include acceptance criteria, and criteria for further corrective actions if acceptance criteria are not met.

In the LRA, the applicant further stated that below-grade fire protection system piping would be inspected for pipe wall thickness and age-related degradation during inspections of opportunity when the below-grade systems are excavated for maintenance.

As documented in the Audit and Review Report, the staff asked the applicant to clarify why LRA Table 3.3.2-7 lists "valves and dampers" and "piping and fittings" as component groups with a soil external environment. The applicant was asked to explain how the aging effects of loss of material and selective leaching (for buried cast iron piping) are managed by its Fire Protection Program for these components. In addition, as with the Buried Services Corrosion Monitoring Program, the applicant was asked for the inspection frequency of the buried components managed by its Fire Protection Program.

By letter dated August 25, 2005, the applicant stated that LRA Table 3.3.2-7 indicates that cast iron components in soil and raw water environments are managed for selective leaching by the One-Time Inspection Program. In LRA Section B2.1.13, the applicant summarized the application of the One-Time Inspection Program to selective leaching. The applicant added that the buried "valves and dampers" and "piping and fittings" of the fire protection system are managed for corrosion and MIC by the Fire Protection Program. LRA Section B2.1.10 summarizes the application of the program to below-grade fire protection system components.

By letter dated September 16, 2005, as to the inspection frequency of the buried components, the applicant stated:

> Visual inspections of a sample of buried carbon, low-alloy, and stainless steel components will be performed within ten years prior to entering, and within ten years after entering, the period of extended operation. Prior to the tenth year of each period, NMC will perform an evaluation of available data to determine sufficient opportunistic inspections have been performed within that period to assess the condition of the components. If insufficient data exists, focused inspection(s) will be performed as needed.

In reviewing the applicant's response, the staff found that after implementation of this enhancement, the program element would be consistent with the GALL Report. On this basis, the staff finds this enhancement acceptable.

In addition, in the LRA, the applicant stated that its Fire Protection Program is consistent with GALL AMPs XI.M26 and XI.M27 with another enhancement in meeting the GALL Report "parameters monitored or inspected" program element:

> Plant procedures shall be revised to more specifically address identification of aging related degradation and expectations for documentation of fire hydrant condition. Also, these revisions shall include provisions to perform flow testing for fire hydrants within the scope of license renewal that are credited for fire suppression in the Palisades current licensing basis.

In the LRA, the applicant stated that fire hydrants are flushed to test for flow restriction and proper hydrant operation and drainage. Hydrants are visually inspected for corrosion and damage, and proper thread and valve lubrication. Hydrants within the scope of license renewal credited for fire suppression in the CLB are flow tested.

GALL AMP XI.M27 lists nozzles, hydrant, hose stations, and standpipes as components with aging effects managed by this program. As documented in the Audit and Review Report, the staff noted that LRA Table 3.3.2-7 includes none of these as component groups as subject to an AMR. The applicant clarified that these components are included within "pipe and fittings" and "valves and dampers" in LRA Table 3.3.2-7. The applicant added that it has no exceptions to the surveillance frequencies in GALL AMP XI.M27, as clarified by ISG-04. Fire hydrant flushes and verification of unrestricted flow are conducted annually. After implementation of this enhancement, this program element would be consistent with the GALL Report. On this basis, the staff finds the enhancement acceptable.

In the LRA, the applicant also stated that its Fire Protection Program is consistent with GALL AMPs XI.M26 and XI.M27 with an additional enhancement in meeting the GALL Report "detection of aging effects" program element: "Develop and implement procedures to replace all sprinkler heads prior to the end of the 50 year service life, or for testing of a representative sample of sprinkler heads prior to the end of the 50 year service life and at 10 year intervals thereafter, per requirements of NFPA 25, Section 5.3."

The applicant further stated that sprinkler heads will be replaced or tested in accordance with NFPA 25 prior to exceeding their 50-year service life. The required testing would be repeated at ten-year intervals. The staff found that the enhancement to the applicant's Fire Protection Program would provide additional assurance of adequate management of aging effects because the program would be consistent with the ISG-04 recommendation for the sprinkler head inspections. After implementation of this enhancement, this program element would be consistent with the GALL Report. On this basis, the staff finds the enhancement acceptable.

Operating Experience. In LRA Section B2.1.10, the applicant explained that a review of industry operating experience associated with the Fire Protection Program and aging revealed issues and instances related to (a) fire water system piping corrosion and ruptures, (b) fire retardant coatings and materials, (c) fouling of components in contact with raw water, and (d) problems with fire barriers.

Various related NRC and/or industry generic communications had been issued and, in turn, incorporated into the program as applicable.

The applicant's review of plant-specific operating experience related to the Fire Protection Program and aging revealed that the following issues had been addressed: (a) blockage of fire protection piping with corrosion products, (b) deluge valve trim piping failures due to corrosion, (c) underground fire main rupture due to cyclic loadings, and (d) water tight fire door seal degradation.

The staff reviewed the operating experience provided in the LRA and interviewed the applicant's technical personnel to confirm that the plant-specific operating experience revealed no degradation not bounded by industry experience.

On the basis of its review of industry and plant-specific operating experience and discussions with the applicant's technical personnel, the staff concludes that the applicant's Fire Protection Program would adequately manage the aging effects identified in the LRA for which this AMP is credited.

FSAR Supplement. In LRA Section A2.10, the applicant provided the FSAR supplement for the Fire Protection Program. The staff reviewed this section and determines that the information in the FSAR supplement provides an adequate summary description of the program, as required by 10 CFR 54.21(d).

Conclusion. On the basis of its review and audit of the applicant's Fire Protection Program, the staff determines that those program elements for which the applicant claimed consistency with the GALL Report are consistent. In addition, the staff reviewed the exceptions and the associated justifications and determines that the AMP, with the exceptions, is adequate to manage the aging effects for which it is credited. Also, the staff reviewed the enhancements and confirmed that the implementation of the enhancements prior to the period of extended operation would make the existing AMP consistent with the GALL Report AMP to which it was compared. The staff concludes that the applicant has demonstrated that the effects of aging will be adequately managed so that the intended functions will be maintained consistent with the CLB for the period of extended operation, as required by 10 CFR 54.21(a)(3). The staff also reviewed the FSAR supplement for this AMP and concludes that it provides an adequate summary description of the program, as required by 10 CFR 54.21(d).

3.0.3.2.6 One-Time Inspection Program

Summary of Technical Information in the Application. In LRA Section B2.1.13, the applicant described the One-Time Inspection Program, stating that this new program is consistent, with exception and enhancements, with GALL AMPs XI.M29, "Aboveground Carbon Steel Tanks," XI.M32, "One-Time Inspection," and XI.M33, "Selective Leaching of Materials."

The One-Time Inspection Program addresses potentially long incubation periods for certain aging effects, including various corrosion mechanisms, cracking, and selective leaching, and provides a means of verifying that an aging effect either does not occur or progresses so slowly as to have a negligible effect on the intended function of the structure or component. Hence, the One-Time Inspection Program provides methods for verifying that an AMP is not needed, verifying the effectiveness of an existing program, or determining that degradation is occurring which will require evaluation and corrective action. The program includes (a) determination of appropriate inspection sample size; (b) identification of inspection locations; (c) selection of examination technique, with acceptance criteria; and (d) evaluation of results to determine the need for additional inspections or other corrective actions. The inspection sample includes locations where the most severe aging effect(s) would be expected to occur. Inspection methods may include visual (or remote visual), surface or volumetric examinations, or other established NDE techniques.

Staff Evaluation. During the audit and review, the staff confirmed the applicant's claim of consistency with the GALL Report. Details of the staff's audit evaluation of this AMP are documented in the Audit and Review Report. The staff reviewed the exception and enhancements and the associated justifications to determine whether the AMP, with the exception and enhancements, remains adequate to manage the aging effects for which it is credited.

As documented in the Audit and Review Report, the staff noted that the description of inspection and testing activities for the "detection of aging effects" program element of this AMP was insufficient. By letter dated July 1, 2005, the applicant provided a table clarifying which methods would be used to detect each of the aging effects managed by the One-Time Inspection Program, including detection of selective leaching and monitoring loss of material from the bottom of above-ground carbon steel tanks. The methods listed in the applicant's response, by letter dated July 1, 2005, monitor parameters directly related to the degradation of a component. As such, the staff finds the program consistent with the GALL Report and, therefore, acceptable.

The staff's review of the One-Time Inspection Program identified an area in which additional information was necessary to complete the review of the applicant's AMR results. The applicant responded to the staff's RAI as discussed below.

In RAI-3.1.2-1, dated June 3, 2005, the staff stated that AMPs that note "one-time inspections" should identify an acceptable inspection method for various situations. The LRA does not identify any specific inspection methods, but generally states that examination techniques would be visual, volumetric, or other appropriately established NDE methods. Therefore, the staff requested the applicant to identify the inspection methods for each one-time inspection listed.

In its response, by letter dated July 1, 2005, the applicant stated that specific inspection methods for individual components would be identified as part of implementation procedure development. The applicant stated that it would begin work on implementation later this year and plans to complete and implement procedures in 2006 (Commitment 27). The One-Time Inspection Program would be included in this effort. The one-time inspection methods are expected to be generally in accordance with the table provided in GALL Report, Volume 2, Revision 1, as follows:

Examples of Parameters Monitored or Inspected and Aging Effect for Specific Structure or Component

Aging Effect	Aging Mechanism	Parameter Monitored	Inspection Method
Loss of Material	Crevice Corrosion	Wall Thickness	Visual (VT-1) and/or Volumetric (RT or UT)
Loss of Material	Galvanic Corrosion	Wall Thickness	Visual (VT-3) and/or Volumetric (RT or UT)
Loss of Material	General Corrosion	Wall Thickness	Visual (VT-3) and/or Volumetric (RT or UT)
Loss of Material	MIC	Wall Thickness	Visual (VT-3) and/or Volumetric (RT or UT)
Loss of Material	Pitting corrosion	Wall Thickness	Visual (VT-1) and/or Volumetric (RT or UT)
Loss of Material	Selective Leaching	Wall Thickness	Hardness Test
Loss of Material	Erosion	Wall Thickness	Visual (VT-3) and/or Volumetric (RT or UT)
Loss of Heat Transfer	Fouling	Tube Fouling	Visual (VT-3) or remote visual
Cracking	SCC, Thermal Stratification and Turbulent Penetration	Cracks	Volumetric (RT or UT)
Loss of preload	Thermal Effects, Gasket Creep, and Self-Loosening	Loosening of Components	Visual (VT-3)

Based on its review, the staff finds the applicant's response to RAI-3.1.2-1 acceptable because the program would be consistent with the GALL Report; therefore, the staff's concern described in RAI-3.1.2-1 is resolved.

In the LRA, the applicant stated that its One-Time Inspection Program is consistent with GALL AMPs XI.M29, XI.M32, and XI.M33 with an exception. The One-Time Inspection Program takes exception to the "detection of aging effects" program element. As stated in the LRA, the exception to the program element is as follows:

> The current state of technology does not provide for an effective, reliable method of performing volumetric examinations of small bore socket welds. The combination of these one-time volumetric examinations of a 10% sample of Class 1 butt welds, 4" NPS and smaller, and the 100% VT-2 examinations of all Class 1 and 2 high safety significance (HSS) socket welds 2" NPS and under each refueling outage meets the intent of the SRP-LR and the GALL Report to provide aging management for small-bore Class 1 piping.

GALL AMP XI.M32 recommends plant-specific destructive examination of replaced piping due to plant modifications or NDE that will detect cracking on the inside surfaces of the small-bore piping. In the LRA, the applicant stated that the current state of technology provides no effective, reliable method of performing volumetric examinations of small-bore socket welds. The staff evaluated the combination of these one-time volumetric examinations of a 10 percent sample of Class 1 butt welds, nominal pipe size (NPS) 4-inch and smaller, and the 100 percent VT-2 examinations of all Classes 1 and 2 high safety significance socket welds NPS 2-inch and under during each refueling outage. The staff determines that this combination is a sufficient sample size for adequate aging management of small-bore Class 1 piping. On this basis, the staff finds the exception acceptable.

In the LRA, the applicant stated that its One-Time Inspection Program is consistent with GALL AMPs XI.M29, XI.M32, and XI.M33 and the One-Time Inspection Program will be developed and implemented. Features of the program would include:

- Controlling procedure and implementing documents for activities associated with the program. This procedure will include a listing of all SSCs that credit this program for aging management, the aging effects and mechanisms being managed, the materials and environments for the SSCs, grouping and inspection sampling techniques to be used, identification of inspection locations, acceptance criteria, inspection scope expansion criteria, and required actions for inspection results that fall outside acceptance criteria. Inspection results and evaluation of results should be documented, and records retrievable for the life of the plant.

- Controls to ensure that at least 10% of all Class 1 butt welds less than 4" NPS receive a volumetric examination prior to the end of, and within the last 5 years of, the current operating period, with the welds to be inspected chosen from the population of Class 1 HSS butt welds from the RI-ISI Program. In addition, ensure that 100% of all Class 1 and 2 HSS socket welds 2" NPS and under receive a VT-2 visual inspection each refueling outage.

In addition, the applicant explained that the program elements describe the program to be implemented prior to the period of extended operation.

The staff finds this program acceptable because the applicant had committed to develop its One-Time Inspection Program to include controlling procedure and implementing documents for activities associated with the program (Commitment 27). This procedure would list all SSCs that credit the program for aging management, the aging effects and mechanisms managed, the SSC materials and environments, grouping and inspection sampling techniques to be used, inspection locations, acceptance criteria, inspection scope expansion criteria, and required actions for inspection results outside acceptance criteria. Documented inspection results and evaluations would be retrievable for the life of the plant. The applicant had also committed to developing the One-Time Inspection Program to include controls to ensure that at least 10 percent of all Class 1 butt welds less than 4-inch NPS receive a volumetric examination prior to the end, and within the last five years, of the current operating period, with the inspected welds chosen from the population of Class 1 HSS butt welds from the RI-ISI program and that 100 percent of all Classes 1 and 2 HSS socket welds 2-inch NPS and under receive a VT-2 visual inspection each refueling outage. This commitment is consistent with GALL

AMPs XI.M32 and XI.M33 and the portions of XI.M29 associated with the thickness measurement of tank bottom surfaces.

Operating Experience. In LRA Section B2.1.13, the applicant explained that the One-Time Inspection Program is a new program to be implemented before the current operating license expires. The NDE inspection methods that will be used, such as visual (or remote visual), surface or volumetric, or other established techniques, are consistent with industry practice.

The staff reviewed the operating experience provided in the LRA and interviewed the applicant's technical personnel to confirm that plant-specific operating experience revealed no degradation not bounded by industry experience. On this basis, the staff finds that the one-time inspection is an acceptable method for confirming that existing programs adequately manage potential aging effects.

The staff recognized that the applicant's corrective action program, which records internal and external plant operating experience, ensures continued review of operating experience with incorporation of appropriate changes to AMPs. On the basis of its review of industry and plant-specific operating experience and discussions with the applicant's technical personnel, the staff concludes that the applicant's One-Time Inspection Program would adequately manage the aging effects identified in the LRA for which this AMP is credited.

FSAR Supplement. The applicant provided its FSAR supplement for the One-Time Inspection Program in LRA Section A2.13. The staff reviewed the FSAR supplement and determines that this AMP is credited with managing aging effects that may reduce the thickness of above-ground carbon steel tank bottoms. By letter dated August 25, 2005, the applicant stated that the FSAR summary for its One-Time Inspection Program would be modified to include reference to the aging management of above-ground carbon steel tank bottoms. The staff finds this modification consistent with the GALL Report and, therefore, acceptable. The staff reviewed this section and determines that the information in the FSAR supplement, with revision, provides an adequate summary description of the program, as required by 10 CFR 54.21(d).

Conclusion. On the basis of its review and audit of the applicant's One-Time Inspection Program, the staff determines that those program elements for which the applicant claimed consistency with the GALL Report are consistent. In addition, the staff reviewed the exception and the associated justifications and determines that the AMP, with the exception, is adequate to manage the aging effects for which it is credited. Also, the staff reviewed the enhancements and confirmed that the implementation of the enhancements prior to the period of extended operation would make the existing AMP consistent with the GALL Report AMP to which it was compared. The staff concludes that the applicant has demonstrated that the effects of aging will be adequately managed so that the intended functions will be maintained consistent with the CLB for the period of extended operation, as required by 10 CFR 54.21(a)(3). The staff also reviewed the revised FSAR supplement for this AMP and concludes that it provides an adequate summary description of the program, as required by 10 CFR 54.21(d).

3.0.3.2.7 Overhead Load Handling Systems Inspection Program

Summary of Technical Information in the Application. In LRA Section B2.1.15, the applicant described the Overhead Load Handling Systems Inspection Program, stating that this existing program is consistent, with exception and enhancement, with GALL AMP XI.M23, "Inspection of Overhead Heavy Load and Light Load (Related to Refueling) Handling Systems."

The Overhead Load Handling Systems Inspection Program provides for inspections of the structural components and rails of cranes and fuel handling machines associated with heavy load handling that are subject to the requirements of NUREG-0612 and within the scope of license renewal requiring aging management. These are the containment building polar crane, the spent fuel pool overhead crane, the containment building jib and boom cranes, and the reactor and spent fuel pool fuel handling machines. These cranes comply with the Maintenance Rule requirements provided in 10 CFR 50.65. The Overhead Load Handling Systems Inspections Program is primarily focused on structural components that make up the bridge and trolley of the overhead cranes within the scope of NUREG-0612.

Staff Evaluation. During the audit and review, the staff confirmed the applicant's claim of consistency with the GALL Report. Details of the staff's audit evaluation of this AMP are documented in the Audit and Review Report. The staff reviewed the exception and enhancement and the associated justifications to determine whether the AMP, with the exception and enhancement, remains adequate to manage the aging effects for which it is credited.

In the LRA, the applicant stated that its Overhead Load Handling Systems Inspection Program is consistent with GALL AMP XI.M23 with an exception to the GALL Report "parameters monitored or inspected" program element. As stated in the LRA, the exception to this program element is that, "Palisades does not track the number and magnitude of all lifts made by cranes. Administrative controls are implemented to ensure that only allowable loads are handled and fatigue failure of structural elements is not expected."

In the LRA, the applicant further stated that it does not track the number and magnitude of all lifts made by cranes. Administrative controls ensure that only allowable loads are handled and fatigue failure of structural elements is not expected. In LRA Section 4.7.1, the applicant summarized the crane load cycles TLAA and concluded that at the current service level, there are no fatigue concerns for the containment polar crane or the spent fuel pool crane as neither can realistically approach the 20,000 to 100,000 rated lifts assumed for its design evaluation during the period of extended operation. Frequent inspections of cranes for indications of functional failures are conducted and the applicant tracks the number and magnitude of lifts made that exceed the rated capacity of the cranes. These are called engineered lifts, follow the requirements of ANSI B30.2, and are generally used only for the polar crane lift of the reactor head with lead shielding. These lifts are numerically restricted and evaluated by engineering analysis.

The staff finds the applicant's explanation for the exception acceptable in that review of the number and magnitude of crane lifts is not necessary because the cranes within the scope of license renewal are used infrequently and the allowable limits are expected to provide adequate margin for the period of extended operation. In LRA Section 4.7.1, the applicant addressed the number of crane lifts in the crane load cycles TLAA. A qualitative review of the number and magnitude of lifts made by a crane is reasonable because recording the number and magnitude of every crane lift below design capacity would be an excessive documentation burden for cranes where their utilization is well below their design life. However, the applicant tracks the number and magnitude of crane lifts that exceed their rated capacity. These are designated engineered lifts with the number restricted and prior evaluation by engineering analysis performed. The staff reviewed the operating experience for the polar crane where the reactor vessel head with lead shielding was removed by engineered lifts during several refueling outages. The weight of the vessel head and shielding exceeded the rated capacity of the polar crane. The staff found the number and magnitude of the polar crane engineered lifts well documented. On the bases of the evaluation of the data in LRA Section 4.7.1, the infrequent service use of the cranes, and review of operating experience for the Overhead Load Handling Systems Inspection Program, the staff finds this exception acceptable.

In the LRA, the applicant also stated that the Overhead Load Handling Systems Inspection Program is consistent with GALL XI.M23 with an enhancement in meeting the GALL Report "scope of program," "detection of aging effects," and "acceptance criteria" program elements. The enhancement is as follows:

> Revise crane and fuel handling machine inspection procedures to specifically inspect for general corrosion on passive components making up the bridge, trolley, girders, etc., and to inspect rails of bridge cranes for wear. The revision should also include documentation of the results of these inspections, acceptance criteria, and qualification requirements for inspectors and crane supervisors.

> Note that the element descriptions describe the program as it will exist after the identified enhancements have been implemented. Enhancements are scheduled for completion prior to the period of extended operation.

The applicant further stated that this enhancement would add specific inspections for general corrosion and loss of material due to wear to the inspection procedures for cranes within the scope of license renewal. In addition, the inspection procedures would be revised to document the results of these inspections, define acceptance criteria in accordance with applicable industry standards and good industry practice, and provide qualification requirements for inspectors and crane supervisors.

The staff found that adding visual inspections for passive components making up the bridge, trolley, girders, etc., and requirements to inspect rails of bridge cranes for wear would be adequate to ensure that loss of material is detected before there is a loss of intended function. In addition, the staff found that revising the inspection procedures to evaluate significant indications against applicable industry standards and good practice is consistent with the recommendations in GALL AMP XI.M23. On the basis that implementing these specific inspection steps would bring the applicant's program into agreement with GALL AMP XI.M23, the staff finds this enhancement acceptable as such changes to the applicant's program would provide additional assurance of adequate management of aging effects.

Operating Experience. In LRA Section B2.1.15, the applicant explained that a review of industry operating experience associated with the Overhead Load Handling Systems Inspection Program revealed no issues and instances related to aging.

The applicant's review of plant-specific operating experience related to the Overhead Load Handling Systems Inspection Program and aging revealed that the following issues have been addressed:

- damage to spent fuel handling machine hoist cable
- movement identified on containment hatch crane's base structure
- containment crane rail attachment bolt grout pads cracked
- load limit of containment crane exceeded during head lift
- containment crane bridge rail splice weld cracks

The applicant further explained that the Overhead Load Handling Systems Inspection Program had demonstrated that it provides reasonable assurance that aging effects are managed for the program SSCs; however, no recent external or internal audits/assessments have been conducted on this program.

The staff reviewed the applicant's corrective action condition reports related to the Overhead Load Handling Systems Inspection Program. The staff found that there had been no failures from loss of material of structural components for cranes within the scope of license renewal. Any deficiencies in the cranes had been attributed to construction practices or general wear of active components. The staff also reviewed condition reports and engineering assistance requests associated with engineered lifts by the containment polar crane. The staff found the number and magnitude of the polar crane engineered lifts well documented and in accordance with applicable industry standards.

The staff reviewed the operating experience provided in the LRA and interviewed the applicant's technical personnel to confirm that the plant-specific operating experience revealed no degradation not bounded by industry experience.

On the basis of its review of industry and plant-specific operating experience and discussions with the applicant's technical personnel, the staff concludes that the applicant's Overhead Load Handling Systems Inspection Program would adequately manage the aging effects identified in the LRA for which this AMP is credited.

FSAR Supplement. In LRA Section A2.15, the applicant provided the FSAR supplement for the Overhead Load Handling Systems Inspection Program. The staff reviewed this section and determines that the information in the FSAR supplement provides an adequate summary description of the program, as required by 10 CFR 54.21(d).

Conclusion. On the basis of its review and audit of the applicant's Overhead Load Handling Systems Inspection Program, the staff determines that those program elements for which the applicant claimed consistency with the GALL Report are consistent. In addition, the staff reviewed the exception and the associated justifications and determines that the AMP, with the exception, is adequate to manage the aging effects for which it is credited. Also, the staff reviewed the enhancement and confirmed that the implementation of the enhancement prior to the period of extended operation, along with revising procedures (Commitment 28), would make the existing AMP consistent with the GALL Report AMP to which it was compared. The staff concludes that the applicant has demonstrated that the effects of aging will be adequately managed so that the intended functions will be maintained consistent with the CLB for the period of extended operation, as required by 10 CFR 54.21(a)(3). The staff also reviewed the

FSAR supplement for this AMP and concludes that it provides an adequate summary description of the program, as required by 10 CFR 54.21(d).

3.0.3.2.8 Reactor Vessel Integrity Surveillance Program

Summary of Technical Information in the Application. In LRA Section B2.1.16, the applicant described the Reactor Vessel Integrity Surveillance Program, stating that this existing program is consistent, with enhancements, with GALL AMP XI.M31, "Reactor Vessel Surveillance."

The Reactor Vessel Integrity Surveillance Program manages the aging effect reduction of fracture toughness due to neutron embrittlement of the low alloy steel reactor vessel. Monitoring methods will be in accordance with 10 CFR Part 50, Appendix H. The program includes:

- capsule insertion, withdrawal and materials testing/evaluation, (including USE and reference temperature nil ductility transition (RT_{NDT}) determinations)

- fluence and uncertainty calculations

- monitoring of effective full power years (EFPY)

- development of pressure-temperature limitations

- determination of low temperature overpressure protection (LTOP) set points

The program ensures the reactor vessel materials meet the fracture toughness requirements of 10 CFR Part 50, Appendix G, and have adequate margins against brittle fracture caused by pressurized thermal shock (PTS) in accordance with 10 CFR 50.61.

Staff Evaluation. The staff confirmed the applicant's claim of consistency with the GALL Report. The staff reviewed the enhancements and the associated justifications to determine whether the AMP, with the enhancements, remains adequate to manage the aging effects for which it is credited.

Appendix H of 10 CFR Part 50 specifies surveillance program criteria for 40 years of operation. GALL AMP XI.M31 specifies additional criteria for 60 years of operation. The staff determined that compliance with 10 CFR Part 50, Appendix H, criteria for capsule design, location, specimens, test procedures, and reporting remains appropriate for the applicant's Reactor Vessel Integrity Surveillance Program because these items, which satisfy 10 CFR Part 50, Appendix H, will now stay the same throughout the period of extended operation. To ensure that all capsules in the reactor vessel removed and tested during the period of extended operation still meet the test procedures and reporting requirements of ASTM E 185-82, the staff imposed conditions to address this specific concern. The 10 CFR Part 50, Appendix H, capsule withdrawal schedule during the period of extended operation is addressed according to the GALL Report's consideration of eight items for an acceptable Reactor Vessel Integrity Surveillance Program for 60 years of operation.

In LRA Section B2.1.16, the applicant described its Reactor Vessel Integrity Surveillance Program to manage aging in reactor vessel beltline materials. As stated in the LRA, this AMP is consistent with GALL AMP XI.M31 with the inclusion of the following four enhancements:

1. The Reactor Vessel Integrity Surveillance Program will ensure that pressure-temperature and LTOP curves are updated to reflect the additional neutron fluence accumulated during the extended operating period. Curves will be updated and submitted to NRC for approval prior to the period of extended operation.

2. Document and establish the requirement to save and store all pulled and tested reactor vessel surveillance capsules for future reconstitution use.

3. Evaluate and revise as necessary, the surveillance capsule withdrawal and testing schedule of FSAR Table 4-20 such that at least one capsule remains in the reactor vessel and is tested during the period of extended operation to monitor the effects of long-term exposure to neutron irradiation.

4. Develop a program level procedure to implement and control Technical Specification and FSAR activities associated with the Reactor Vessel Integrity Surveillance Program, including activities associated with surveillance capsules, pressure-temperature limit curves, LTOP setpoints, neutron embrittlement calculation methodology, neutron fluence calculations and control, and documentation requirements. The procedure title should be 'Reactor Vessel Integrity Surveillance Program.'

The staff examined these enhancements and determined that they are actually improvements addressing either GALL AMP XI.M31 recommendations or TS and FSAR activities associated with reactor vessel integrity which, until then, had been processed outside the scope of a typical Reactor Vessel Integrity Surveillance Program. For this AMP, the GALL Report recommends further evaluation of the proposed surveillance capsule withdrawal schedule. The staff reviewed LRA Section B2.1.16 to determine whether the GALL Report program had been properly applied to PNP. The staff also reviewed the program application to determine whether the AMP addressed the issue of an acceptable reactor vessel material surveillance program. It should be noted that the GALL Report does not provide an evaluation and technical basis of the ten attributes for the Reactor Vessel Integrity Surveillance Program. Rather, GALL AMP XI.M31 provides eight specific criteria for evaluating the acceptability of a reactor vessel integrity surveillance program. The staff's review of this AMP against the eight criteria or items is discussed below.

Items 1 to 3 and Item 8 relate to the monitoring of RPV embrittlement for upper shelf energy (USE) and P-T limits for 60 years in accordance with RG 1.99, Revision 2. The staff determined that the LRA satisfies these criteria with the evaluations and conclusions of SER Section 4.2.1 on Charpy USE and SER Section 4.2.3 on P-T limits.

According to FSAR Table 4-20, "Reactor Vessel Surveillance Coupon Removal Schedule," capsule W-280 will be withdrawn at the end of cycle 19. Assuming a 1.5-year cycle, the staff estimated that W-280 would be withdrawn in approximately spring of 2008. Improvement 1 to this AMP states that P-T limits and LTOP curves would be updated and submitted to the staff for review and approval prior to the period of extended operation to reflect the additional neutron fluence accumulated during the period of extended operation. The P-T limits and LTOP curves have been updated in the TS through the license amendment process. Therefore, use of the Reactor Vessel Integrity Surveillance Program to ensure the proper execution of this activity represents only an administrative improvement.

The staff's review of LRA Section B2.1.16 identified areas in which additional information was necessary to complete the review of the applicant's program elements. The applicant responded to the staff's RAIs as discussed below.

In RAI B2.1.16-1, dated November 30, 2005, the staff noted that LRA Section B2.1.16 states that enhancement 1 requires that P-T limits and LTOP curves be updated and submitted to the staff for review and approval prior to the period of extended operation to reflect additional neutron fluence accumulated during this period. Therefore, the staff requested the applicant to provide the approximate date (in calendar year) of withdrawal for capsule W-280 and confirm that the updated P-T limits and LTOP curves would incorporate information from the surveillance report on irradiated specimens from capsule W-280 when submitted to the staff prior to the period of extended operation. Further, the applicant stated that this program ensures the reactor vessel materials "have adequate margins against brittle fracture caused by pressurized thermal shock (PTS) in accordance with 10 CFR 50.61." The applicant was further asked to confirm that, like P-T limits and LTOP curves, PTS evaluation was also part of the Reactor Vessel Integrity Surveillance Program.

In its response, by letter dated January 13, 2006, the applicant stated:

> Palisades reactor vessel surveillance capsule W-280 is scheduled for removal following completion of the 19th operating cycle, which is estimated to occur in September 2007. Results of subsequent testing are required by 10 CFR 50 Appendix H to be reported within one year from the date of removal, so they would be available as input for revision of P-T limits and LTOP curves by September 2008. It is anticipated that the revised P-T limits and LTOP curves would be submitted to the NRC in 2010.... Therefore the test results from capsule W-280 would be used in the development of the new curves.
>
> As stated on page 4-16 and B-120 of the LRA, monitoring the reactor for compliance with pressurized thermal shock requirements is an integral part of the reactor vessel surveillance program.

Based on its review, the staff finds the applicant's response to RAI B2.1.16-1 acceptable because the applicant clarified that the updated P-T limits and LTOP curves to be submitted prior to the period of extended operation would reflect new information from irradiated specimens from capsule W-280. This submission also would report updated USE and PTS evaluations, including the predicted date for the RPV limiting material to reach the USE or the PTS screening criterion, and the dates for the appropriate submissions addressing the resolution of USE and PTS issues would be adjusted accordingly. Therefore, the staff's concern described in RAI B2.1.16-1 is resolved.

Item 4 relates to the disposition of pulled and tested capsules for possible future reconstitution use. Improvement 2 to this AMP is that the applicant would "document and establish the requirement to save and store all pulled and tested reactor vessel surveillance capsules for future reconstitution use." Therefore, this AMP is in accordance with the GALL AMP XI.M31 Item 4 recommendation.

Item 5 relates to the GALL Report recommendation that at least one capsule remain in the reactor vessel to be tested during the period of extended operation if the applicant's surveillance capsules' projected fluence (equivalent vessel fluence) at the end of 40 years is less than the 60-year fluence. Improvement 3 to this AMP is that the applicant's surveillance withdrawal and testing schedule in FSAR Table 4-20 would be evaluated and revised as necessary so that at least one capsule would remain in the reactor vessel for testing during the period of extended operation.

In RAI B2.1.16-2, dated November 30, 2005, the staff noted that LRA Section B2.1.16 states that enhancement 3 to this AMP would evaluate and revise the applicant's surveillance withdrawal and testing schedule of FSAR Table 4-20 as necessary so that at least one capsule would remain in the reactor vessel for testing during the period of extended operation. Item 5 of GALL XI.M31 provides this guideline for applicants whose surveillance capsules' projected fluence (equivalent vessel fluence) at the end of 40 years is less than the 60-year fluence. Therefore, the staff requested the applicant to confirm that the projected fluence at the end of 40 years for capsules W-280, W-260, and W-80 is less than the 60-year fluence. The applicant was also asked to provide the projected fluence and EFPYs for capsules W-280, W-260, and W-80 at the date of their scheduled withdrawals and to identify the capsules it intended to keep in the reactor vessel for testing during the period of extended operation.

In its response, by letter dated January 13, 2006, the applicant stated:

> Reactor vessel surveillance capsule W-280 is presently estimated to be removed in the fall of 2007. Capsules W-260 and W-80 are scheduled for removal during the extended operating period and are projected to have accumulated fluence of $2.57 \times 10^{19} \text{n/cm}^2$. The projected peak vessel fluence at 60 years (42.37 EFPY) is estimated to be $2.998 \times 10^{19} \text{n/cm}^2$.

Capsule	Capsule fluence at 40 years (24.63 EFPY)	Capsule fluence at removal	EFPY at removal	Refueling Outage	Removal Date
W-280	2.33×10^{19} n/cm^2	2.33×10^{19} n/cm^2	20.99	19	Fall 2007
W-260	2.57×10^{19} n/cm^2	2.88×10^{19} n/cm^2	29.23	25	Fall 2016
W-80	2.57×10^{19} n/cm^2	3.06×10^{19} n/cm^2	31.96	27	Fall 2019

Based on its review, the staff finds the applicant's response to RAI B2.1.16-2 acceptable because the applicant provided the projected fluence and withdrawal dates for the three capsules. Therefore, the staff's concern described in RAI B2.1.16-2 is resolved.

Consequently, the staff determined that with the inclusion of enhancement 3, the applicant's AMP meets Item 5 in GALL AMP XI.M31, which recommends that at least one capsule remain in the RPV for testing during the period of extended operation. The withdrawal schedule for this capsule shall be provided for the staff's review, as specified in the license condition for approving this AMP.

Item 6 recommends that all other standby capsules exceeding equivalent RPV fluence of 60 years be removed and placed in storage. LRA Section B2.1.16 does not provide this information; however, the applicant's response to RAI B2.1.16-2 confirmed that the planned surveillance capsule removal schedule is acceptable without the need to place any surveillance capsules into storage. All capsules in the applicant's program will receive meaningful levels of neutron exposure in accordance with the applicant's planned withdrawal schedule.

Item 7 does not apply to this AMP because PNP does not have in-vessel capsules and will not use alternative dosimetry to monitor neutron fluence during the period of extended operation.

Operating Experience. In LRA Section B2.1.16, the applicant explained that it had performed a search of industry and plant-specific operating experience related to the Reactor Vessel Integrity Surveillance Program. This search revealed the following two items relevant to this AMP: (1) GL 92-01, Revision 1, "Reactor Vessel Structural Integrity;" and (2) Supplement 1 to GL 92-01, Revision 1, "Reactor Vessel Structural Integrity." The applicant's response to these documents had been incorporated into the Reactor Vessel Integrity Surveillance Program. The applicant's review of inspection reports, QA audit/surveillance reports, and self-assessments since 1999 revealed no issues or findings that could impact the effectiveness of the Reactor Vessel Integrity Surveillance Program.

The staff reviewed the operating experience provided in the LRA and interviewed the applicant's technical personnel to confirm that the plant-specific operating experience revealed no degradation not bounded by industry experience.

FSAR Supplement. The applicant provided its FSAR supplement for the Reactor Vessel Integrity Surveillance Program in LRA Section A2.16. Appendix H of 10 CFR Part 50 requires licensees to submit proposed changes to their Reactor Vessel Integrity Surveillance Program withdrawal schedules to the staff for review and approval. To ensure that this reporting requirement will carry forward through the period of extended operation, the staff will impose the following license condition to the applicant's Reactor Vessel Integrity Surveillance Program:

> All capsules in the reactor vessel that are removed and tested must meet the test procedures and reporting requirements of ASTM E 185-82 to the extent practicable for the configuration of the specimens in the capsule. Any changes to the capsule withdrawal schedule, including spare capsules, must be approved by the NRC prior to implementation. All capsules placed in storage must be maintained for future insertion. Any changes to storage requirements must be approved by the NRC.

The staff reviewed this section and determines that the information in the FSAR supplement, with the license condition, provides an adequate summary description of the program, as required by 10 CFR 54.21(d).

Conclusion. On the basis of its review of the applicant's Reactor Vessel Integrity Surveillance Program and RAI responses, the staff determines that those program elements for which the applicant claimed consistency with the GALL Report are consistent. Also, the staff reviewed the improvements and confirmed that the implementation of the improvements prior to the period of extended operation would make the existing AMP consistent with the GALL Report AMP to which it was compared. The staff concludes that the applicant has demonstrated that the effects of aging will be adequately managed so that the intended functions will be maintained consistent with the CLB for the period of extended operation, as required by

10 CFR 54.21(a)(3). The staff also reviewed the FSAR supplement for this AMP and concludes that, with the license condition, it provides an adequate summary description of the program, as required by 10 CFR 54.21(d).

3.0.3.2.9 Reactor Vessel Internals Inspection Program

Summary of Technical Information in the Application. In LRA Section B2.1.17, the applicant described the Reactor Vessel Internals Inspection Program, stating that this existing program is consistent, with enhancements, with GALL AMP XI.M16, "PWR Vessel Internals."

The Reactor Vessel Internals Inspection Program manages the aging effects for reactor vessel internals. The program provides for:

- ISI in accordance with ASME Code Section XI requirements, including examinations performed during the 10-year ISI examination

- participation in industry initiatives to evaluate the significance of void swelling

- monitoring and control of reactor coolant water chemistry in accordance with the EPRI guidelines in TR-105714 (see Water Chemistry Program) to mitigate SCC or irradiation assisted stress corrosion cracking (IASCC)

- participation in industry initiatives that will generate additional data on aging mechanisms relevant to reactor vessel internals (RVI) and develop appropriate inspection techniques to permit detection and characterization of features of interest

Void swelling is an aging mechanism for RVI components that has the potential to cause reduction in fracture toughness and changes in dimensions aging effects.

Staff Evaluation. During the audit and review, the staff confirmed the applicant's claim of consistency with the GALL Report. Details of the staff's audit evaluation of this AMP are documented in the Audit and Review Report. The staff reviewed the enhancements and the associated justifications to determine whether the AMP, with the enhancements, remains adequate to manage the aging effects for which it is credited.

The staff noted that the applicant did not identify, as an exception, the difference between the edition of the ASME Code referenced in the LRA and that in the GALL AMP.

By letter dated August 25, 2005, the applicant stated that it would revise the ASME Section XI IWB, IWC, IWD, IWF Program descriptions in LRA Appendices A and B to reflect the ASME Code Section XI, 2001 Edition, including the 2002 and 2003 Addenda. By letter dated October 31, 2005, the applicant provided the revised program descriptions and identified seven exceptions. In addition, the applicant stated that no changes were necessary to the ASME Section XI, Subsections IWB, IWC, IWD, IWF Inservice Inspection Program descriptions in LRA Section A2.2. By letter dated December 16, 2006, the applicant revised its program description to remove exceptions 2 through 7 of the October 31, 2005, letter. The applicant stated that the ASME Section XI, IWB, IWC, IWD, IWF Inservice Inspection Program is based on the ASME Code Section XI, 2001 Edition, including 2002 and 2003 Addenda, and during the period of extended operation the program would be maintained consistent with the 10 CFR 50.55a requirement. The staff finds this revision an acceptable component of the program for RVI aging management.

As documented in the Audit and Review Report, the staff asked the applicant for additional information that would reflect all augmented examinations and enhanced inspection techniques. By letter dated August 25, 2005, the applicant revised Commitment 33, stating that it would participate in industry initiatives to generate additional data on RVI aging mechanisms, including void swelling, and to develop appropriate inspection techniques for detection and characterization of features of interest. The applicant further stated that recommendations for augmented inspections and techniques for this augmentation would be incorporated into its Reactor Vessel Internals Program as applicable. The applicant committed to submit its revised Reactor Vessel Internals Program for the staff's review and approval by March 24, 2009. On this basis, the staff finds Commitment 33 acceptable.

In the LRA, the applicant also stated that the ASME Section XI, IWB, IWC, IWD, IWF Inservice Inspection Program follows ASME Code Section XI, as required by 10 CFR 50.55a. However, GALL AMP XI.M16 references the 1995 Edition through 1996 Addenda. The ASME Section XI, IWB, IWC, IWD, IWF Inservice Inspection Program would be updated to later editions and addenda as required by 10 CFR 50.55a; the code edition and addenda would not necessarily be those referenced by this revision of the GALL Report. The staff finds this updating an acceptable component of the program for RVI aging management.

In addition, in the LRA, the applicant stated that its Reactor Vessel Internals Inspection Program is consistent with GALL AMP XI.M16 with enhancements in meeting the GALL Report "scope of program" and "detection of aging effects" program elements. Regarding the enhancements for these two program elements, the applicant stated that it will participate in the industry initiatives to evaluate the effect of changes in dimensions due to void swelling and report to the NRC, at least two years prior to the end of the current operating license, the results of the industry initiative and a schedule for augmented inspections that will be required, if any. The applicant also stated that it will participate in industry initiatives that will generate additional data on aging mechanisms relevant to RVI and develop appropriate inspection techniques to permit detection and characterization of features of interest. Any recommended augmented inspections would be incorporated as appropriate.

Based on its evaluation, the staff determined that the applicant had not considered void swelling a factor in the evaluation of reduction in fracture toughness; however, the applicant stated that the potential significance of void swelling would be assessed through monitoring industry operating experience and EPRI research. If assessed as significant, the need for augmented examinations for void swelling effects would be evaluated and the results reported to the NRC at least two years prior to the end of the current operating license.

As documented in the Audit and Review Report, the staff requested the applicant to provide a commitment to revise the Reactor Vessel Internals Inspection Program to reflect all appropriate augmented examinations and enhanced inspection techniques. The revised program would be submitted to the staff for review and approval prior to the period of extended operation.

By letter dated August 25, 2005, the applicant revised Commitment 33, stating that it would participate in industry initiatives to generate additional data on RVI aging mechanisms, including void swelling, and to develop appropriate inspection techniques to permit detection and characterization of features of interest. Recommendations for augmented inspection techniques would be incorporated into the Reactor Vessel Internals Inspection Program as applicable. The applicant committed to submit its revised Reactor Vessel Internals Program for the staff's review and approval by March 24, 2009. The staff accepted Commitment 33 to submit a revised Reactor Vessel Internals Inspection Program.

In the LRA, the applicant stated that on the basis of operating experience at CE PWRs and component-specific evaluation by the NSSS vendor, augmented inspections of the baffle-former bolting is not necessary because the most likely mechanism for the cracking of baffle-former bolts in foreign plants is IASCC of cold-worked 316 stainless steel. In addition, the preload and operating stresses imposed by the design of plants where this aging mechanism has been observed is significantly higher than at PNP.

The staff reviewed the program basis document for the Reactor Vessel Internals Inspection Program and vendor evaluations of the stresses imposed on the baffle-former bolts. These documents show that the baffle-former bolts are made of annealed 316 stainless steel that is not cold-worked. The documents also show that the bolt stress from preload is much less than in plants where failure has been observed, the differential pressure across the core shroud panels does not result in additional tensile loads on the bolts during normal operation, and the core shroud panel design transmits less load to the fasteners. On this basis, the staff concludes that the applicant's Reactor Vessel Internals Inspection Program would adequately manage cracking of baffle-former bolting without augmented inspections.

Operating Experience. In LRA Section B2.1.17, the applicant explained that it looked at operating experience issues, which included NRC INs 84-18, "Stress Corrosion Cracking in PWR Systems," and 98-11, "Cracking of Reactor Vessel Internal Baffle Former Bolts in Foreign Plants." A review of industry operating experience related to the Reactor Vessel Internals Inspection Program revealed several instances where degradation had occurred within the RVI. A variety of issues related to reactor internals included degradation of baffle former bolts, barrel former bolts, guide bar bolts, core support shield to core barrel bolts, guide funnels, guide tube support pins, and rod cluster control assemblies.

The applicant's review of plant-specific operating experience revealed two instances where the Reactor Vessel Internals Inspection Program had been instrumental in detecting material degradation. Degradation was detected in the core barrel and CRD mechanism seal housings.

The staff reviewed the operating experience provided in the LRA and interviewed the applicant's technical personnel to confirm that the plant-specific operating experience revealed no degradation not bounded by industry experience. The staff accepted Commitment 33 to submit a revised Reactor Vessel Internals Inspection Program for the staff's review and approval by March 24, 2009.

FSAR Supplement. In LRA Section A2.17, the applicant provided the FSAR supplement for the Reactor Vessel Internals Inspection Program:

> The Reactor Vessel Internals Inspection Program manages the aging effects for reactor vessel internals. The program provides for (a) Inservice Inspection (ISI) in accordance with ASME Section XI requirements, including examinations performed during the 10-year ISI examination; (b) Participation in industry

initiatives to evaluate the significance of void swelling; (c) Monitoring and control of reactor coolant water chemistry in accordance with the EPRI guidelines in TR-105714 (See Water Chemistry Program) to mitigate SCC or IASCC; and (d) Participation in industry initiatives that will generate additional data on aging mechanisms relevant to RVI and develop appropriate inspection techniques to permit detection and characterization of features of interest.

By letter dated August 25, 2005, the applicant revised Commitment 33 and agreed to submit the revised Reactor Vessel Internals Program for the staff's review and approval by March 24, 2009. The staff reviewed this section and determines that the information in the FSAR supplement and Commitment 33 provide an adequate summary description of the program, as required by 10 CFR 54.21(d).

Conclusion. On the basis of its review and audit of the applicant's Reactor Vessel Internals Inspection Program, the staff determines that those program elements for which the applicant claimed consistency with the GALL Report are consistent. Also, the staff reviewed the enhancements and confirmed that the implementation of the enhancements prior to the period of extended operation would make the existing AMP consistent with the GALL Report AMP to which it was compared. The staff concludes that the applicant has demonstrated that the effects of aging will be adequately managed so that the intended functions will be maintained consistent with the CLB for the period of extended operation, as required by 10 CFR 54.21(a)(3). The staff also reviewed the FSAR supplement and Commitment 33 for this AMP and concludes that they provide an adequate summary description of the program, as required by 10 CFR 54.21(d).

3.0.3.2.10 Structural Monitoring Program

Summary of Technical Information in the Application. In LRA Section B2.1.19, the applicant described the Structural Monitoring Program, stating that this existing program is consistent, with enhancement, with GALL AMPs XI.S5, "Masonry Wall Program," XI.S6, "Structures Monitoring Program," and XI.S7, "RG 1.127, Inspection of Water-Control Structures Associated with Nuclear Power Plants."

The Structural Monitoring Program is designed to ensure that age-related (as well as other) deterioration of plant structures (including masonry walls) and components within its scope is appropriately managed to ensure that each such structure or component retains the ability to perform its intended function. The program is implemented through visual examination of these structures, components, and other specified items. Damage or degradation found during visual examination may be further evaluated by measurements and testing techniques as appropriate. This program also implements provisions of the 10 CFR 50.65 Maintenance Rule that relate to masonry walls and water-control structures. It conforms to the guidance contained in RG 1.160 and Nuclear Utility Management and Resource Council (NUMARC) 93-01, as well as NEI 96-03. This NEI document, which supplements NUMARC 93-01, contains additional guidance specific to the monitoring of structures.

Staff Evaluation. During the audit and review, the staff confirmed the applicant's claim of consistency with the GALL Report. Details of the staff's audit evaluation of this AMP are documented in the Audit and Review Report. The staff reviewed the enhancement and the associated justifications to determine whether the AMP, with the enhancement, remains adequate to manage the aging effects for which it is credited.

GALL AMP XI.S5, under the "detection of aging effects" program element, has the following statement:

> The frequency of inspection is selected to ensure there is no loss of intended function between inspections. The inspection frequency may vary from wall to wall, depending on the significance of cracking in the evaluation basis. Unreinforced masonry walls that have not been contained by bracing warrant the most frequent inspection, because the development of cracks may invalidate the existing evaluation basis.

In LRA Section B2.1.19 under the "detection of aging effects" program element for masonry walls, the applicant stated that the program applies the same periodic visual examination techniques to masonry walls and to concrete structural elements as prescribed by this program element. Frequency of examinations may vary according to different reinforcement masonry configurations.

As documented in the Audit and Review Report, the applicant was asked to clarify whether unreinforced masonry walls without bracing were inspected more frequently than reinforced or braced masonry walls. The applicant responded that it did not differentiate block walls based on reinforcement, that all block walls were treated the same in inspections under the Structural Monitoring Program. By letter dated August 25, 2005, the applicant stated that a clarification to the Structural Monitoring Program had been provided to better demonstrate consistency with GALL AMP XI.S5. Specifically, LRA Section B2.1.19 was revised to read: "In addition, inspections for unreinforced block walls that are not contained by bracing will be performed on a more frequent basis than the periodicity of at least once every 10 year interval specified for reinforced or braced block walls." Also, in LRA Section A2.19, page A-8, the applicant added the following sentence to the end of the second paragraph: "In addition, the program specifies that inspections for unreinforced block walls that are not contained by bracing will be performed on a more frequent basis than the normal frequency of once each 10 year interval specified for reinforced or braced block walls." The staff finds the clarification acceptable. With the clarification statements, the applicant's Structural Monitoring Program is now consistent with the GALL AMP XI.S5 "detection of aging effects" program element.

GALL AMP XI.S6, under the "detection of aging effects" program element, states that, "For each structure/aging effect combination, the inspection methods, inspection schedule, and inspector qualifications are selected to ensure that aging degradation will be detected and quantified before there is loss of intended functions." As documented in the Audit and Review Report, the applicant stated that PNP has no aggressive below-grade environment nor flowing underground water and is not required to inspect inaccessible areas below the grade of structures within the scope of license renewal. The staff asked the applicant whether it would take advantage of inspection opportunities for structures within the scope of license renewal with inaccessible areas. In addition, as inaccessible areas become accessible by such means as excavation or other methods, the staff asked the applicant whether opportunistic inspections of those areas would be performed. In response, the applicant stated that when opportunities arise, the inspection of normally inaccessible areas of structures within the scope of license renewal would be made a required part of the program.

By letter dated August 25, 2005, the applicant stated:

> Due to the lack of aggressive groundwater at Palisades, a plant-specific program is not required to age manage inaccessible below grade concrete as discussed in ISG-3. However, to validate this determination, NMC will perform an inspection of opportunity on inaccessible concrete when excavation work uncovers a significant depth (i.e., several feet or more) of normally inaccessible concrete.

> Accordingly, the following sentence is added to PNP AMP B2.1.19, Structural Monitoring Program, after the first paragraph of the detection of aging effects program element section on Page B-141:

> > In addition, the program provides for inspections of opportunity of normally inaccessible below grade concrete when excavation work uncovers a significant depth (i.e., several feet or more) to provide access for inspection.

> In addition, the following sentence is added to LRA Section A2.19, Structural Monitoring Program, Page A-8, following the second sentence of the first paragraph:

> > In addition, the program provides for inspections of opportunity of normally inaccessible below grade concrete when excavation work uncovers a significant depth (i.e., several feet or more) to provide access for inspection.

The staff reviewed the applicant's response and found it acceptable because the applicant elected to take advantage of inspection opportunities for structures within the scope of license renewal with areas normally inaccessible. The inspection of normally inaccessible areas would verify that aging effects do not occur and that a plant-specific AMP is indeed not required.

In the LRA, the applicant stated that it has no aggressive below-grade environment. Although not required by GALL AMP XI.S6, the applicant's Structural Monitoring Program does not discuss the need or lack of need for periodic ground water monitoring to ensure that the below-grade water chemistry does not become aggressive. As documented in the Audit and Review Report, the staff asked the applicant to justify the lack of periodic ground water monitoring during the CLB and potential period of extended operation to check water chemistry for non-aggressiveness.

By letter dated September 2, 2005, the applicant stated that, as discussed in LRA Section 3.5.2.2.1.1, ground water chemistry records available for the current operating period show that water in contact with below-grade concrete is currently non-aggressive, and has been non-aggressive over at least the last 40 years. To ensure that ground water remains non-aggressive over the period of extended operation, ground water sampling for pH, chlorides, and sulfates will be part of the Structural Monitoring Program with periodicity not to exceed five years. Accordingly, the applicant made the following changes to LRA Section B2.1.19:

> On Page B-137, after the last paragraph of the section entitled program description, the following paragraph is added: 'For below grade inaccessible

concrete, Interim Staff Guidance #3 (ISG-3) discusses potential aging effects requiring management (AERMs) if the below grade environment is aggressive (pH <5.5, chlorides > 500 ppm, or sulfates > 1500 ppm. Historical groundwater sampling performed at Palisades shows that the below grade environment is and has been non-aggressive by a significant margin. As part of the Structural Monitoring Program, Palisades will continue to monitor groundwater on a periodic basis to ensure it remains non-aggressive such that the associated AERMs remain not applicable.'

On Page B-140, after the last paragraph of the section entitled XI.S6, Structures Monitoring, under parameters monitored, inspected, and/or tested, the following paragraph is added: 'Local groundwater will be sampled on a periodic basis to ensure pH values and concentrations of chlorides and sulfates remain below levels considered aggressive to concrete.'

On Page B-142, after the last paragraph of the section entitled XI.S6, Structures Monitoring, under monitoring and trending, the following paragraph is added: 'Groundwater sampling for pH, chlorides, and sulfates will be performed to ensure the below grade environment remains non-aggressive with a periodicity not to exceed every 5 years.'

On Page B-143, after the last paragraph of the section entitled XI.S6, Structures Monitoring, under acceptance criteria, the following paragraph is added: 'Groundwater sampling will verify a non-aggressive below grade environment exists, as described in ISG-3, by ensuring pH > 5.5, chlorides < 500 ppm and sulfates are < 1500 ppm.'

In addition, the applicant made conforming changes to various parts of LRA Section 3.5.2.2 to state that continued groundwater sampling is unnecessary. A conforming change was also made to the FSAR supplement summary description of the Structural Monitoring Program in LRA Section A2.19 by adding the following sentence at the end of the first paragraph: "As part of the Structural Monitoring Program, groundwater sampling for pH, chlorides, and sulfates will be performed, with a periodicity not to exceed every 5 years, to ensure the below grade environment remains non-aggressive."

The staff reviewed the applicant's response and found it acceptable. The applicant elected to perform periodic ground water monitoring during the CLB and potential period of extended operation to ensure that the below-grade water chemistry does not become aggressive. The monitoring of the ground water would verify that aging effects due to an aggressive environment will not occur.

In the LRA, the applicant stated that its Structural Monitoring Program is consistent with GALL AMPs XI.S5, XI.S6, and XI.S7 with an enhancement in meeting the GALL Report "scope of program" program element to incorporate all structural members listed in LRA Tables 3.5.2-1 through 3.5.2-10 within the scope of license renewal using the Structural Monitoring Program as an AMP. The enhancement would be incorporated into the Structural Monitoring Program prior to the period of extended operation.

The staff found that incorporation into the Structural Monitoring Program of all structural members listed in LRA Tables 3.5.2-1 through 3.5.2-10 would ensure detection of aging degradation of all passive and long-lived structural members within the scope of license renewal before loss of intended function. The staff found that by implementing this enhancement and specifying the structure/aging effect combinations managed by the applicant's Structural Monitoring Program would be consistent with the recommendation of GALL AMPs XI.S5, XI.S6, and XI.S7. On the basis that the scope of the Structural Monitoring Program would be well defined with the enhancement, the staff finds it acceptable as such changes to the program would provide additional assurance that the effects of aging would be adequately managed.

Operating Experience. In LRA Section B2.1.19, the applicant explained that a review of industry operating experience associated with the Structural Monitoring Program and aging reveals issues and instances related to (a) corrosion of steel ice condenser containment vessels caused by boric acid and condensation and (b) cracks in concrete floors caused by flexing and shrinkage.

Various related NRC and/or industry generic communications have been issued and, in turn, incorporated into the program as applicable.

The applicant also explained that a review of plant-specific operating experience related to the Structural Monitoring Program and aging revealed that the following issues had been addressed:

- settling of air compressor foundations
- watertight barrier degradation
- spalled concrete and exposed anchor bolts
- intake crib damage due to ice and to wave action
- cracking of concrete beams in the auxiliary building
- corrosion of condenser rock anchors caused by standing water and debris
- degradation of snubber anchor support structure concrete and grout
- deterioration of floor plugs due to leaking water
- moisture separator reheater foundation cracking
- cracks in concrete duct bank
- cracks in west ESS room west wall
- spalled concrete on wall of 1-2 diesel generator exhaust plenum
- groundwater leaks in auxiliary feedwater pump room floor

In the LRA, the applicant stated that its Structural Monitoring Program had demonstrated that it provides reasonable assurance that aging effects are managed for its SSCs. Additionally, this management has been demonstrated through inspection reports, program health reports, and the corrective action program.

The staff reviewed the operating experience for watertight barrier degradation, spalled concrete, and exposed anchor bolts, cracking of concrete beams in the auxiliary building, corrosion of condenser rock anchors caused by standing water and debris, and identification of floor plugs due to leaking water and found that the applicant's existing Structural Monitoring Program had been effective in detecting deterioration of plant SCs within the program's scope. Such deficiencies were placed in the corrective action program and dispositioned for repair, accept as is, or accept as is with additional monitoring. Operating experience demonstrated that the

Structural Monitoring Program is effective in ensuring adequate management of age-related deterioration of plant SCs within the scope of license renewal so that they maintain their ability to perform their intended functions.

The staff's review of operating experience with groundwater leaks in the auxiliary feedwater pump room floor revealed some Structural Monitoring Program areas which needed strengthening. The groundwater leaks in the auxiliary feedwater pump room were identified on a corrective action program report. From the written description of the groundwater leakage into the feedwater pump room, the staff believed the deficiency to have occurred since original plant construction. The written description identified water leakage into the room carrying sand at a rate sufficient to overflow installed catch barriers. The disposition of the deficiency was that it was an existing condition that had been occurring for years and evaluated by engineering as not a concern. The staff discussed with the applicant how severe the condition was and why it had not been repaired. The applicant's engineering personnel stated that the written description on the corrective action program report had exaggerated the severity of the condition, that the water ingress was seepage and not leakage, and that there was no sand. The staff agreed that if the condition was not as severe as stated on the corrective action program report, repairs might not be necessary, but stated that augmented inspections were needed of the Structural Monitoring Program for areas that exhibit deterioration.

In the LRA, the applicant stated that initial baseline structural inspections under its Structural Monitoring Program, as required by 10 CFR 50.65, were started in late 1996. A second complete inspection in 1999 validated the initial inspection results. As documented in the Audit and Review Report, the staff asked the applicant to review the structural inspections performed in 1996 and 1999 of the auxiliary feedwater pump room to determine whether the water ingress condition had been noted. The applicant stated that the condition had not been identified during either inspection. As a result of the staff's questions, the applicant initiated a new corrective action program report for documentation. With the strengthening of the applicant's Structural Monitoring Program through its corrective action process, the staff concludes that all deterioration of plant structures would be identified, documented, evaluated, and repaired as necessary with augmented inspections as required per the program.

The staff reviewed the operating experience provided in the LRA and interviewed the applicant's technical personnel to confirm that plant-specific operating experience revealed no degradation not bounded by industry experience.

On the basis of its review of industry and plant-specific operating experience and discussions with the applicant's technical personnel, the staff concludes that the applicant's Structural Monitoring Program would adequately manage the aging effects identified in the LRA for which this AMP is credited.

FSAR Supplement. In LRA Section A2.19, the applicant provided its FSAR supplement for the Structural Monitoring Program, which states that the program is designed to ensure that age-related (as well as other) deterioration of plant SCs (including masonry walls) within its scope is appropriately managed to ensure that each structure or component retains the ability to perform its intended function. The program is implemented through visual examination of these SCs and other specified items. Damage or degradation found during visual examination may be further evaluated by measurements and testing techniques as appropriate.

This program also implements provisions of the 10 CFR 50.65 Maintenance Rule relating to masonry walls and water-control structures. It conforms to the guidance contained in RG 1.160 and NUMARC 93-01, as well as NEI 96-03. This NEI document, which supplements NUMARC 93-01, contains additional guidance specific to monitoring structures. In Commitment 35, the applicant stated that it would enhance the program procedures prior to the period of extended operation.

The applicant made a conforming change to LRA Section A2.19 by adding the following sentence at the end of the first paragraph: "As part of the Structural Monitoring Program, groundwater sampling for pH, chlorides, and sulfates will be performed, with a periodicity not to exceed every five years, to ensure the below grade environment remains non-aggressive." Also, in LRA Section A2.19, page A-8, the applicant added the following sentence to the end of the second paragraph: "In addition, the program specifies that inspections for unreinforced block walls that are not contained by bracing will be performed on a more frequent basis than the normal frequency of once each 10 year interval specified for reinforced or braced block walls." In addition, the following sentence was added to LRA Section A2.19, page A-8, following the second sentence of the first paragraph: "In addition, the program provides for inspections of opportunity of normally inaccessible below grade concrete when excavation work uncovers a significant depth (i.e., several feet or more) to provide access for inspection."

The staff reviewed this section and determines that the information in the FSAR supplement, with revision, provides an adequate summary description of the program, as required by 10 CFR 54.21(d).

Conclusion. On the basis of its review and audit of the applicant's Structural Monitoring Program, the staff determines that those program elements for which the applicant claimed consistency with the GALL Report are consistent. Also, the staff reviewed the enhancement and confirmed that the implementation of the enhancement prior to the period of extended operation would make the existing AMP consistent with the GALL Report AMP to which it was compared. The staff concludes that the applicant has demonstrated that the effects of aging will be adequately managed so that the intended functions will be maintained consistent with the CLB for the period of extended operation, as required by 10 CFR 54.21(a)(3). The staff also reviewed the revised FSAR supplement for this AMP and concludes that it provides an adequate summary description of the program, as required by 10 CFR 54.21(d).

3.0.3.3 AMPs That Are Not Consistent with or Not Addressed in the GALL Report

In LRA Appendix B, the applicant identified that the System Monitoring Program is a plant-specific AMP. For the AMP that is not consistent with or not addressed by the GALL Report, the staff performed a complete review of the AMP to determine whether it was adequate to monitor or manage aging. The staff's review of this plant-specific AMPs is documented in the following section of this SER.

3.0.3.3.1 System Monitoring Program

Summary of Technical Information in the Application. In LRA Section B2.1.20, the applicant described the System Monitoring Program, stating that this is an existing, plant-specific program.

The System Monitoring Program is a condition monitoring program to identify degraded conditions on external surfaces of piping, tanks and other components and equipment prior to the loss of the systems' and components' intended function. The program is identified as consistent with GALL AMP XI.M29, "Aboveground Carbon Steel Tanks," that are applicable to the accessible external surfaces of the applicable tanks. The program includes an enhancement scheduled prior to the period of extended operation to enhance system walkdown procedures. . The applicant stated that implementation of the System Monitoring Program provides reasonable assurance that aging effects will be managed such that the SSCs within the scope of this program will continue to perform their intended functions consistent with the current licensing basis for the period of extended operation.

Staff Evaluation. In accordance with 10 CFR 54.21(a)(3), the staff reviewed the information included in LRA Section B2.1.20, regarding the applicant's demonstration of the System Monitoring Program to ensure that the effects of aging, as discussed above, will be adequately managed so that the intended functions will be maintained consistent with the CLB for the period of extended operation.

In LRA Section B.2.1.20 the applicant described its program to manage aging of the various systems and components within the scope of license renewal. The program is plant-specific; therefore, the staff reviewed the program using the guidance in Branch Technical Position RLSB-1 in SRP-LR, Appendix A. The staff's evaluation focused on management of aging effects through incorporation of the following 10 elements from RLSB-1: "scope of program," "preventive actions," "parameters monitored or inspected," "detection of aging effects," "monitoring and trending," "acceptance criteria," "corrective actions," "confirmation process," "administrative controls," and "operating experience." The applicant indicated that the "corrective actions," "confirmation process," and "administrative controls" for license renewal are consistent with the corresponding SRP-LR and the GALL Report AMP elements and are addressed by the quality assurance program addressed in LRA Section B1.2 and evaluated in SER Section 3.0.4. The evaluation of the remaining seven elements is provided below. The staff also reviewed the FSAR supplement to determine whether it provides an adequate description of the program. During its review, the staff determined that additional information was required.

(1) Scope of Program - The LRA states that the System Monitoring Program is credited for managing the aging effects of normally accessible surfaces of piping, tanks, and other components and equipment within the scope of license renewal. This element is identified as consistent with portions of GALL AMP XI.M29 that are applicable to the accessible external surfaces of aboveground carbon steel tanks.

The staff's review of LRA Section B2.1.20 identified areas in which additional information was necessary to complete the review of the applicant's program elements. The applicant responded to the staff's RAIs as discussed below.

In RAI B2.1.20-1(a), dated June 22, 2005, the staff requested that the applicant clarify if the entire exposed surface of the aboveground steel tanks are included within scope of the System Monitoring Program. In its response, by letter dated July 25, 2005, the applicant stated that the System Monitoring Program is designed to cover the entire accessible exposed surfaces of aboveground steel tanks down to their surface contact with soil or concrete. The applicant added that the program will not rely on sampling locations when completing inspections and the entire accessible exposed surface is that which can be accessed by an individual taking advantage of installed plant walkways, ladders and platforms. The applicant's response was not entirely acceptable because the staff was concerned that the entire exposed surface of the aboveground steel tanks may not be accessible for inspection from existing installed plant walkways, ladders, and platforms. The applicant was requested to clarify if all exposed surfaces of tanks are accessible for inspection from existing installed plant walkways, ladders and platforms or to clarify if temporary ladders and platforms are to be installed to support inspections. In a teleconference on September 14, 2005, the applicant explained that temporary movable platforms will be made available to provide access for inspections of all accessible external tank surfaces. The applicant agreed to provide written clarification that temporary access is to be made available to support tank inspections of all exposed external surfaces. By letter dated April 26, 2006, the applicant confirmed that portable ladders and platforms will be used as necessary to gain access to exposed tank surfaces for visual inspections under the System Monitoring Program. This response clarified that necessary access will be made available to support inspections and all issues related to RAI B.2.1.20-1(a) are resolved.

In RAI B2.1.20-1(b), dated June 22, 2005, the staff stated that the LRA AMR tables credit the System Monitoring Program for managing external surfaces of elastomers, but elastomers are not specifically addressed in the System Monitoring Program. Therefore, the staff requested that the applicant clarify if elastomers are within scope of the System Monitoring Program and, if so, consider the unique aging degradation characteristics of elastomers in the specified environment and include a discussion of elastomers within each element of the program. In its response, by letter dated July 25, 2005, the applicant clarified that there are no elastomers in mechanical systems that are required to be managed by an AMP. This response is evaluated below under RAI B2.1.20-1(c) and the applicant subsequently revised their position by including elastomers for certain systems within scope of the program. By letter dated November 18, 2005, the applicant modified the description of the System Monitoring Program to address elastomers within each element of the program.

Considering the limited shelf life and service life of elastomers for the specified environment, in RAI B.2.1.20-1(c), dated June 22, 2005, the staff requested that the applicant clarify if elastomers meet the definition of long-lived components within scope of license renewal. In its response, by letter dated July 25, 2005, the applicant responded that NMC has determined that the elastomers listed in the Heating Ventilation and Air Conditioning (HVAC) system and the Service Water System AMR Tables of the LRA are not long-lived components that require aging management. Therefore, the applicant concluded, there are no elastomers in mechanical systems that are required to be managed by an Aging Management Program and the line items for elastomers in Tables 2.3.3-9, 3.3.2-9 and 3.3.2-12 and Sections 3.3.2.1-9 and 3.3.2.1-12 should be deleted. The applicant clarified that, based on this determination, NMC will enhance the preventive maintenance program to periodically inspect and replace as necessary the expansion joints/flexible connections in the HVAC system and Service Water System that are within scope of license renewal. The staff was concerned that, although these passive components do not require an aging management program if they are not long-lived components, the Part 54.21 rule requires an IPA for passive components unless they are subject to replacement based on a qualified life or specified time period. Therefore, the applicant must either, (1) clarify that these components are not in scope because they will be replaced based on a qualified life or specified time period, or (2) clarify that these passive components are considered in scope because they are long-lived and will be periodically inspected by an aging management program such as the system monitoring program or preventive maintenance program. The program that manages the elastomers should include appropriate criteria for inspecting the elastomers and this criteria is addressed below in RAI B2.1.20-3 under acceptance criteria. In a conference call with the applicant on September 14, 2005, it was agreed that this concern was considered unresolved and deferred. By letter dated November 18, 2005, the applicant revised their response to RAI B2.1-20-1(c) in its entirety to read:

> Mechanical elastomers, including rubber, are included the LRA for the Heating, Ventilation and Air Conditioning (HVAC) System (LRA Tables 2.3.3-9, 3.3.2-9 and section 3.3.2.1.9) and the Service Water System (Table 3.3.2-12 and Section 3.3.2.1.12).

> The elastomers listed in the Heating, Ventilation and Air Conditioning System are long lived and require aging management as described in LRA Table 2.3.3-9, Table 3.3.2-9, Section 3.3.2.1.9, and Section 3.3.2.2.2. This is consistent with GALL sections VIIF1, F2, F3, and F4.

> NMC has determined that the elastomers in the Service Water System (SWS) AMR tables of the LRA are not long lived components that require aging management because they are replaced on a periodic basis. Therefore, there are no elastomers in the SWS system that are required to be managed by an Aging Management Program, and the 'Pipe & Fittings' entry on page 3-183 for the material 'Rubber' is deleted along with the associated Aging Management Program, NUREG-1801 volume 2 Line Item, Table 1 Item, and Notes. In addition, in Section 3.3.2.1.12 on page 3-96, the bullet 'Rubber' under 'Materials' is deleted. Additionally, in Section 3.3.2.2.2 on page 3-103, the last two sentences of the second paragraph, which refer to management of elastomers in systems other than HVAC, are hereby deleted.

The staff finds the scope of program to be comprehensive and acceptable because it includes the components that credit this program, as identified in the AMR tables. All items related to RAIs B2.1.20-1(b) and B2.1.20-1(c) are resolved.

The staff confirmed that the "scope of program" program element satisfies the criterion defined in SRP-LR Section A.1.2.3.1 and concludes that this program attribute is acceptable.

(2) Preventive Actions - The LRA states that there are no preventive actions associated with this program. The applicant does not credit protective coatings or sealants and the System Monitoring Program is a condition monitoring program and, thus, there are no preventative actions. The staff concurred with this assessment and does not identify the need for any preventative actions associated with this program.

(3)   Parameters Monitored or Inspected - The LRA states that the program utilizes periodic visual inspections to monitor for leakage and evidence of material degradation. Above ground carbon steel tank external coatings or paint are inspected to provide indication of the condition of the material underneath the coating or paint. The LRA states that this element is consistent with the corresponding element of NUREG-1801, Section XI.M29, Above Ground Carbon Steel Tanks. The staff finds that the parameters monitored or inspected will provide symptomatic evidence of potential degradation and, therefore, are acceptable.

The staff confirmed that the "parameters monitored or inspected" program element satisfies the criterion defined in SRP-LR Section A.1.2.3.3 and concludes that this program attribute is acceptable.

(4)   Detection of Aging Effects - The LRA states that the System Monitoring Program uses visual inspection activities to detect degradation due to aging effects prior to loss of intended function. The minimum walkdown frequency for those components that are accessible during normal plant operation is identified as annual. Systems and components that are only accessible during plant outages are inspected at least once per refueling outage. The inspection frequency may be increased based on the safety significance, production significance, discovery and/or operating experience of each system.

In RAI B.2.1.20-2(a), dated June 22, 2005, the staff requested that the applicant clarify what codes and standards or manufacturer's recommendations are applied to determine that the technique and frequency are adequate to detect degradation before the loss of intended component function.

In its response, by letter dated July 25, 2005, the applicant indicated that NMC will identify specific methods of inspection for individual components as part of the System Monitoring Program implementation development later in 2005 and 2006. The applicant stated that the System Monitoring Program will adopt recommendations for inspection techniques and frequencies from applicable codes, industry standards and/or manufacturer's recommendations. The applicant further stated that industry documents such as EPRI 1009743, GS-7086 and API 575 will be used as source documents to define the tank testing and inspection requirements.

After reviewing the applicant's RAI response, the staff was concerned that applicable industry codes and standards are not identified within the AMP as justification that the technique and frequency are justified and consistent with industry practice. For an existing program, this information should be available now for review. In a teleconference on September 14, 2005, the applicant was requested to either (1) identify specific codes and standards or manufacturer's recommendations that will be applied to the system monitoring program, or (2) as a condition of the license, clarify that this program will be updated at a later date and submitted to the staff for review prior to the period of extended operation. In a teleconference with the applicant on September 14, 2005, the applicant clarified that applicable industry codes and standards such as EPRI 1009743, EPRI GS-7086 and API 575 will be applied to manage external surfaces of system components and tanks in addition to any other applicable standards or manufacturer's recommendations identified in the future. The applicant agreed to provide a written clarification to identify applicable codes and standards. By letter dated April 26, 2006, the applicant confirmed that industry standards EPRI 1009743, GS-7086 and AP 575 would be used as source documents to define tank testing and inspection requirements performed under the System Monitoring Program. For other general system inspections, consistent with industry practice and GALL AMP XI.M36, other industry standards provide the bases for system walkdowns under the System Monitoring Program. These industry standards identified by the applicant are INPO Good Practice TS-413, EPRI Technical Report 1007933 and EPRI Technical Report 1009743. This response identified appropriate industry standards and all issues related to RAI B2.1.20-2(a) are resolved by this response.

In RAI B2.1.20-2(b), dated June 22, 2005, the staff stated that the LRA AMR Tables credit the System Monitoring Program for managing change in material properties and cracking for elastomers used inside/outside containment. Therefore, the staff requested that the applicant clarify how visual inspections are performed to detect changes in material properties and identify if other methods such as hardness testing or physical manipulation in combination with visual inspections are appropriate. In its response, by letter dated July 25, 2005, the applicant stated that there are no elastomers managed by the System Monitoring Program. By letter dated November 18, 2005, the applicant revised their response in its entirety to read:

> Visual inspections will detect cracking resulting from changes in material properties such as loss of flexibility and embrittlement. Physical manipulation during the visual inspection will verify flexibility of the elastomers.

The staff found that visual inspections to detect cracking resulting from changes in material properties combined with physical manipulation of the elastomer to verify flexibility are appropriate methods to detect elastomer degradation.

In RAI B2.1.20(c), dated June 22, 2005, the staff requested that the applicant clarify if sampling is applied to inspect a group of SCs or clarify if the entire surfaces of all accessible components are inspected. In its response by letter dated July 25, 2005, the applicant clarified that the System Monitoring Program is designed to inspect the entire accessible exposed surface of the components included within scope of the program. The applicant stated that the program will not rely on sampling of locations when completing inspections.

The staff finds the response acceptable because the applicant clarified that the program is designed to inspect the entire accessible exposed surface of the components included in the program and will not rely on sampling. The related staff scope concern regarding accessibility is addressed in RAI B2.1.20-1(a). The related staff concern regarding surfaces covered by insulation is addressed below under RAI B2.1.20-2(d).

In RAI B2.1.20-2(d), dated June 22, 2005, the staff requested that the applicant clarify if insulation will be removed to provide access for inspections and if insulation is not removed, to justify the basis that any potential degradation will be detected. In its response, by letter dated July 25, 2005, the applicant responded that, except for Class 1 borated water systems, NMC will not remove insulation because the condition of the insulation provides a good indirect indicator of the conditions beneath. Industry guidance, such as API-570, suggest that either insulation be removed or imaging techniques be applied to detect degradation under insulation. After reviewing the applicant's response, the staff was concerned that, if insulation is not removed or other imaging techniques are not applied, degradation occurring under insulation may not necessarily be detected. The applicant was requested to consider industry guidance on corrosion under insulation and clarify if removal of insulation at selected locations or imaging techniques may be prudent. For example, API-570 identifies common locations susceptible to corrosion under insulation including the extent of visual external and corrosion under insulation inspections at suspect locations. In a teleconference with the applicant on September 14, 2005, it was agreed that this concern is considered unresolved and deferred. By letter dated April 26, 2006, the applicant agreed to enhance the System Monitoring Program to verify that corrosion under insulation is not occurring. The applicant agreed to perform inspections of opportunity to assess the external surface condition when insulation is removed for maintenance or surveillance. To verify that there were a sufficient number of inspection opportunities to provide a representative indication of system condition, the System Monitoring program will require a periodic review of documented under insulation inspection results. If there were insufficient opportunities for inspection, insulation will be removed from additional sample locations to assess system condition under insulation. This program requirement will be implemented prior to March 24, 2011. The applicant's response to enhance the System Monitoring Program is acceptable and all issues related to RAI B2.1.20-2(d) are resolved.

In RAI B2.1.20-2(e), dated June 22, 2005, the staff requested that the applicant clarify how evidence of corrosion and wear will be detected where the threaded surfaces are not readily visible. In its response, by letter dated July 25, 2005, the applicant stated that bolting is a component that will be inspected during system walkdowns. As indicated in the response to RAI B2.1.20-2(d), insulation will not be removed to provide access for inspections, unless insulation conditions indicate that a potential problem exists under the insulation. Further, the applicant response indicated that the inspections of non-class 1 closure bolts will be similar to the visual examinations prescribed by the ASME Section XI Code. The staff was concerned that, if insulation is not removed or other imaging techniques are not applied, degradation occurring under insulation may not necessarily be detected. As identified in the response to RAI B2.1.20-2(d) discussed above, the applicant has agreed to remove insulation to perform representative inspections and all issues related to RAI B2.1.20-2(d) and RAI B2.1.20-2(e) are resolved.

The staff finds that the applicant's response is acceptable on the basis that visual inspections of accessible surfaces of systems and components with insulation removed on a periodic basis combined with physical manipulation of elastomers during visual inspections are capable of detecting the aging effects that are covered by this program. The use of visual inspections and physical manipulation of elastomers to detect external degradation is consistent with industry practice.

The staff confirmed that the "detection of aging effects" program element satisfies the criterion defined in SRP-LR Section A.1.2.3.4 and concludes that this program attribute is acceptable.

(5) Monitoring and Trending - The LRA indicates that the periodic system walkdown inspections provide for timely detection of aging effects. Walkdown results are documented to provide a historical record of items monitored during the walkdowns and the results are monitored and trended if significant material loss is detected. The staff finds that the overall monitoring and trending proposed by the applicant is acceptable because there is reasonable assurance that documentation of periodic walkdown inspection results combined with an effective Corrective Action Program will effectively monitor and trend the applicable aging effects.

The staff confirmed that the "monitoring and trending" program element satisfies the criteria defined in SRP-LR Section A.1.2.3.5 and concludes that this program attribute is acceptable.

(6) Acceptance Criteria - The LRA states that system walkdown procedures require that signs of significant degradation to paint, coatings or exposed external steel surfaces found that may affect a component's ability to perform its intended function be entered into the Corrective Action program for resolution.

In RAI B2.1.20-3, dated June 22, 2005, the staff stated that SRP-LR Section A.1.2.3.6 Item 4 states that qualitative inspections should be performed by the same predetermined criteria as quantitative inspections by personnel in accordance with ASME Code and through approved site specific programs. Therefore, the staff requested that the applicant identify the inspection criteria such as ASME Code VT-1 examination or other industry standards and identify the qualifications of personnel performing the inspections.

In its response, by letter dated July 25, 2005, the applicant stated that specific inspection criteria and individual inspector qualifications for the System Monitoring program will be determined as part of the implementation phase of the license renewal project. The applicant will begin working on implementation later in 2005 (Commitment 41) and plans to complete draft aging management programs and their associated implementing procedures in 2006. The staff was concerned that, for an existing program, such information should be available now for review. The applicant was requested to either (1) identify the inspection criteria and qualifications of inspectors that will be applied to the system monitoring program, or (2) as a commitment, clarify that this program will be updated at a later date and submitted to the staff for review prior to the period of extended operation. In a teleconference with the applicant on September 14, 2005, it was agreed that this concern is considered unresolved and deferred.

By letter dated March 30, 2006, in response to an unresolved item related to RAI B2.1.20-1(c), the applicant clarified the basis for visual and physical manipulation strategies to manage elastomer aging. The applicant clarified that procedures use a combination of visual inspection and manipulation to determine the condition of the elastomers. Visual inspections will detect cracking resulting from changes in material properties and physical manipulation during physical inspection will verify flexibility of the elastomer. The applicant credits that they have successfully utilized physical manipulation of elastomers to deterioration by feel for indications of being hard and brittle or soft and tacky. The applicant also credits other LRAs including DC Cook and Millstone that applied similar techniques for elastomers. In regard to elastomers, the staff finds that, on the basis of standard practices and operating experience, the applicant has demonstrated that appropriate inspection techniques will be applied to manage aging of elastomers within scope of the program. After reviewing the RAI response, the staff was concerned that the basis for inspection criteria and inspector qualifications to manage other materials is not identified. By letter dated April 26, 2006, the applicant identified guidance consistent with GALL AMP XI.M36, "External Surfaces Monitoring." The applicant identified that persons performing system walkdowns are qualified using a formal NMC fleet-wide standard process according to NMC Fleet Mentoring Guide FL-ESP-SYS-005M that defines and documents the knowledge and practical demonstration requirements for persons performing walkdowns. This standard also references plant procedure, EM-20 which provides plant specific guidance for walkdowns. The applicant's response that clarifies person's performing plant walkdowns are qualified according to plant procedures is satisfactory and all items related to RAI B2.1.20-3 are resolved.

The staff confirmed that the "monitoring and trending" program element satisfies the criteria defined in SRP-LR Section A.1.2.3.6 and concludes that this program attribute is acceptable.

(10)   Operating Experience - The LRA states that Palisades has a comprehensive operating experience program that monitors industry issues/events and assesses these for applicability to its own operations. The LRA concludes that, through NRC inspection reports, audits, self-assessments and the corrective action program, the Palisades system monitoring has been demonstrated that aging effects are being managed.

The staff finds that the applicant's use of the operating experience program combined with the corrective action program and other assessments should provide objective evidence to support the conclusion that the enhanced System Monitoring Program will adequately manage the aging effects in the systems and components that credit this program during the period of extended operation.

The staff confirmed that the "operating experience" program element satisfies the criteria defined in SRP-LR Section A.1.2.3.10. The staff concludes that this program attribute is acceptable.

FSAR Supplement. In LRA Section A2.20, the applicant provided the FSAR supplement for the System Monitoring Program. By letter dated March 30, 2006, the applicant revised the description of the A2.20 System Monitoring Program to include elastomers. The staff reviewed this section and determines that the information in the FSAR supplement provides an adequate summary description of the program, as required by 10 CFR 54.21(d).

Conclusion. On the basis of its review and audit of the applicant's System Monitoring Program and RAI responses, the staff concludes that the applicant demonstrated that the effects of aging will be adequately managed so that the intended functions will be maintained consistent with the CLB for the period of extended operation, as required by 10 CFR 54.21(a)(3). The staff also reviewed the FSAR supplement for this AMP and concludes that it provides an adequate summary description of the program, as required by 10 CFR 54.21(d).

3.0.4 Quality Assurance Program Attributes Integral to Aging Management Programs

Pursuant to 10 CFR 54.21(a)(3), the applicant is required to demonstrate that the effects of aging on SCs subject to an AMR will be adequately managed so that their intended functions will be maintained consistent with the CLB for the period of extended operation. SRP-LR Branch Technical Position RLSB-1, "Aging Management Review - Generic," describes ten attributes of an acceptable AMP. Three of these ten attributes are associated with the QA activities of "corrective action," "confirmation process," and "administrative control." SRP-LR Table A.1-1, "Elements of an Aging Management Program for License Renewal," provides the following description of these quality attributes:

- corrective actions, including root cause determination and prevention of recurrence, should be timely

- the confirmation process should ensure that preventive actions are adequate and that appropriate corrective actions have been completed and are effective

- administrative controls should provide a formal review and approval process

SRP-LR Branch Technical Position IQMB-1, "Quality Assurance For Aging Management Programs," notes that those aspects of the AMP that affect quality of SR SSCs are subject to the QA requirements of 10 CFR Part 50, Appendix B, "Quality Assurance Criteria for Nuclear Power Plants and Fuel Reprocessing Plants." Additionally, for NSR SCs subject to an AMR, the applicant's existing 10 CFR Part 50, Appendix B, QA program may be used to address the "corrective action," "confirmation process," and "administrative control" program elements. Branch Technical Position IQMB-1 provides the following guidance with regard to the QA attributes of AMPs:

- SR SCs are subject to 10 CFR Part 50, Appendix B, requirements, which are adequate to address all quality-related aspects of an AMP consistent with the facility's CLB for the period of extended operation.

- For NSR SCs subject to an AMR for license renewal, an applicant has the option to expand the scope of its 10 CFR Part 50, Appendix B, program to include these SCs addressing "corrective action," "confirmation process," and "administrative control" elements for aging management during the period of extended operation. In this case, the applicant should document such a commitment in the FSAR supplement in accordance with 10 CFR 54.21(d).

3.0.4.1 Summary of Technical Information in Application

LRA Section 3 provides an AMR summary for each unique structure, component, or commodity group determined to require aging management during the period of extended operation. This summary includes identification of AERMs and AMPs used to manage these aging effects. Topical Report, "Quality Program Description for Nuclear Power Plants (Part 2) - Palisades Nuclear Plant (CPC-2A)," and LRA Appendix B demonstrate how the programs manage aging effects using attributes consistent with industry and NRC guidance. In LRA Section B.1.2, "Quality Assurance Program and Administrative Controls," the applicant referenced Topical Report CPC-2A. This program includes the "corrective action," "confirmation process," and "administrative controls" program elements and is applied to both SR and NSR SSCs within the scope of license renewal.

In LRA Section B.1.2, the applicant discussed the implementation of 10 CFR Part 50, Appendix B, and its consistency with the summary in SRP-LR Section A.2. The QA program included the "corrective action," "confirmation process," and "administrative controls" program elements. The QA program implementation procedures would be expanded to apply these three program elements to both SR and NSR SSCs subject to an AMR for license renewal. The staff reviewed the LRA to verify that the three program elements had been addressed. The three elements are applicable as follows.

Corrective Action. Corrective actions are implemented in accordance with the requirements of 10 CFR Part 50, Appendix B, and the NRC-approved Topical Report CPC-2A. Controls are established to assure that conditions adversely affecting quality are identified and documented and that appropriate remedial action is taken. For significant conditions adversely affecting quality, necessary corrective action is promptly determined and recorded. Corrective action includes determining the cause and extent of the condition and taking appropriate action to prevent similar problems in the future. The controls also assure that corrective action is implemented promptly.

Corrective actions are implemented through the initiation of an action request in accordance with plant procedures. Equipment deficiencies may be initially documented by a work order, but the corrective action process specifies that an action request also be initiated if required. This approach ensures that identified problems are corrected promptly.

Confirmation Process. The confirmation process is part of the Corrective Action Program implemented in accordance with the requirements of 10 CFR Part 50, Appendix B, and the NRC-approved Topical Report CPC-2A. The aging management activities required by this program would also uncover any unsatisfactory condition due to ineffective corrective action. Administrative procedures include provisions for identification, evaluation, assignment, tracking, monitoring, reviewing, verifying, and approving corrective actions to ensure effectiveness. The corrective action process is also monitored for potentially adverse trends. An adverse trend due to recurring or repetitive adverse conditions must be documented in an action request. The post-maintenance testing procedure includes provisions for verifying the completion and effectiveness of corrective actions for equipment deficiencies, establishes criteria for the selection and documentation of post-maintenance tests, provides guidelines to ensure equipment will perform its intended function prior to return to service, and ensures that the original equipment deficiency is corrected and that a new deficiency has not been created.

Administrative Control. Plant AMPs are administered through various plant implementation documents subject to administrative controls, including a formal review and approval process in accordance with the requirements of 10 CFR Part 50 and, therefore, consistent with the SRP-LR.

Plant AMPs are implemented through a variety of procedures. Implementing documents are subject to administrative controls, including a formal review and approval process, in accordance with 10 CFR Part 50, Appendix B, requirements and Topical Report CPC-2A.

Administrative procedures provide guidance on procedures and administrative documents. Uniform guidelines and requirements are provided for preparing, revising, reviewing, and approving procedures. Usage and adherence requirements are also defined for plant procedures.

3.0.4.2 Staff Evaluation

The staff reviewed the applicant's AMPs described in LRA Section B.1.2. In this section, the applicant references Topical Report CPC-2A. The purpose of this review was to assure that the aging management activities were consistent with the staff's guidance described in SRP-LR Section A.2, "Quality Assurance for Aging Management Programs (Branch Technical Position IQMB-1)," regarding QA attributes of AMPs. The staff also sampled the AMRs for chemistry control and electrical equipment qualification and found that the attributes of corrective action, confirmation process, and administrative document control were adequately discussed. Based on its review, the descriptions and applicability of the plant-specific AMPs, and their associated quality attributes provided in Topical Report CPC-2A and SRP-LR Appendix B1.2, the staff concluded that the program descriptions are consistent with the staff's position and the branch technical position discussed in IQMB-1.

The staff's review identified an area in which additional information was necessary to complete the review of the applicant's QA program (and its attributes of "corrective action," "confirmation process," and "administrative controls"). The applicant responded to the staff's RAI as discussed below.

In RAI 3.0-1, dated September 8, 2005, the staff requested the applicant to supplement the description in LRA Appendix A to include a description of the QA program attributes, including references to pertinent implementing guidance, as necessary, credited for the programs described in LRA Appendix B.1.2. The descriptions in LRA Appendix A should provide the staff with sufficient information to determine whether the quality attributes for the AMPs are consistent with the review acceptance criteria contained in SRP-LR Section A.2.

In its response, by letter dated October 6, 2005, the applicant provided information about QA attributes applied in AMPs. The applicant included a new LRA Section A1.1 with implementation of the recently approved NMC Quality Assurance Topical Report NMC-1 and a description of the program elements of "corrective action," "confirmation process," and "administrative controls" that would be applied to both SR and NSR SSCs subject to an AMR. The applicant also updated references to reflect implementation of new NMC Quality Assurance Topical Report NMC-1.

Based on its review of the additional information, the staff finds the applicant's response to RAI 3.0-1 acceptable; therefore, the staff's concern described in RAI 3.0-1 is resolved.

3.0.4.3 Conclusion

The staff concludes that the QA attributes of the applicant's AMPs are consistent with 10 CFR 54.21(a)(3). Specifically, the applicant described the QA attributes of the programs and activities for managing the effects of aging for both SR and NSR SSCs within the scope of license renewal and stated that its QA program addresses the program elements of "corrective action," "confirmation process," and "administrative control." Therefore, the staff concludes that the applicant's QA description for its AMPs is acceptable.

3.1 Aging Management of Reactor Coolant System

This section of the SER documents the staff's review of the applicant's AMR results for the RCS components and component groups of the following:

- primary coolant system
- reactor vessel
- reactor vessel internals
- replacement steam generators

3.1.1 Summary of Technical Information in the Application

In LRA Section 3.1, the applicant provided AMR results for the RCS components and component groups. In LRA Table 3.1.1, "Summary of Aging Management Evaluations in Chapter IV of NUREG-1801 for Reactor Coolant System," the applicant provided a summary comparison of its AMRs to those evaluated in the GALL Report for RCS components and component groups.

The applicant's AMRs incorporated applicable operating experience in the determination of aging effects requiring management (AERMs). These reviews included evaluation of plant-specific and industry operating experience. The plant-specific evaluation included reviews of condition reports and discussions with appropriate site personnel to identify AERMs. The applicant's review of industry operating experience included a review of the GALL Report and issues identified since the issuance of the GALL Report.

3.1.2 Staff Evaluation

The staff reviewed LRA Section 3.1 to determine whether the applicant had provided sufficient information to demonstrate that the aging effects for the RCS components within the scope of license renewal and subject to an AMR will be adequately managed so that the intended function(s) will be maintained consistent with the CLB for the period of extended operation, as required by 10 CFR 54.21(a)(3).

The staff conducted an onsite audit of AMRs to confirm the applicant's claim that certain identified AMRs were consistent with the GALL Report. The staff did not repeat its review of the matters described in the GALL Report; however, the staff did verify that the material presented in the LRA was applicable and that the applicant had identified the appropriate GALL Report AMRs. The staff's evaluations of the AMPs are documented in SER Section 3.0.3. Details of the staff's audit evaluation are documented in the Audit and Review Report and summarized in SER Section 3.1.2.1.

In the onsite audit, the staff also selected AMRs consistent with the GALL Report for which further evaluation is recommended. The staff confirmed that the applicant's further evaluations were consistent with the acceptance criteria in SRP-LR Section 3.1.2.2. The staff's audit evaluations are documented in the Audit and Review Report and summarized in SER Section 3.1.2.2.

In the onsite audit, the staff also conducted a technical review of the those remaining AMRs not consistent with or not addressed in the GALL Report. The audit and technical review evaluated whether all plausible aging effects were identified and whether the aging effects listed were appropriate for the combination of materials and environments specified. The staff's audit evaluations documented in the Audit and Review Report are summarized in SER Section 3.1.2.3. The staff's evaluation of its technical review is also documented in SER Section 3.1.2.3.

Finally, the staff reviewed the AMP summary descriptions in the FSAR supplement to ensure that they adequately describe the programs credited with managing or monitoring RCS component aging.

SER Table 3.1-1 below summarizes the staff's evaluation of components, aging effects/mechanisms, and AMPs listed in LRA Section 3.1 and addressed in the GALL Report.

Table 3.1-1 Staff Evaluation for Reactor Coolant System Components in the GALL Report

Component Group	Aging Effect/ Mechanism	AMP in GALL Report	AMP in LRA	Staff Evaluation
Reactor coolant pressure boundary components (Item Number 3.1.1-01)	Cumulative fatigue damage	TLAA, evaluated in accordance with 10 CFR 54.21(c)	TLAA	This TLAA is evaluated in SER Section 4.3, "Metal Fatigue"
Steam generator shell assembly (Item Number 3.1.1-02)	Loss of material due to pitting and crevice corrosion	Inservice inspection; water chemistry	ASME Section XI IWB, IWC, IWD, IWF Inservice Inspection Program (B2.1.2); Steam Generator Tube Integrity Program (B2.1.18); Water Chemistry Program (B2.1.21)	Consistent with GALL, which recommends further evaluation (See SER Sections 3.1.2.1.1 and 3.1.2.2.2)
Pressure vessel ferritic materials that have a neutron fluence greater than 10^{17} n/cm^2 (E > 1 MeV) (Item Number 3.1.1-04)	Loss of fracture toughness due to neutron irradiation embrittlement	TLAA, evaluated in accordance with Appendix G of 10 CFR 50 and RG 1.99	TLAA	This TLAA is evaluated in SER Section 4.2, "Reactor Vessel Neutron Embrittlement"
Reactor vessel beltline shell and welds (Item Number 3.1.1-05)	Loss of fracture toughness due to neutron irradiation embrittlement	Reactor vessel surveillance	Reactor Vessel Integrity Surveillance Program (B2.1.16)	Consistent with GALL, which recommends further evaluation (See SER Section 3.1.2.2.3)
Westinghouse and B&W baffle/former bolts (Item Number 3.1.1-06)	Loss of fracture toughness due to neutron irradiation embrittlement and void swelling	Plant-specific	Reactor Vessel Internals Inspection Program (B2.1.17)	(See SER Section 3.1.2.2.3)

Component Group	Aging Effect/ Mechanism	AMP in GALL Report	AMP in LRA	Staff Evaluation
Small-bore reactor coolant system and connected systems piping (Item Number 3.1.1-07)	Crack initiation and growth due to SCC, intergranular SCC, and thermal and mechanical loading	Inservice inspection; water chemistry; one-time inspection	ASME Section XI IWB, IWC, IWD, IWF Inservice Inspection Program (B2.1.2); One-Time Inspection Program (B2.1.13); Water Chemistry Program (B2.1.21)	Consistent with GALL, which recommends further evaluation (See SER Sections 3.1.2.1.2 and 3.1.2.2.4)
Vessel shell (Item Number 3.1.1-10)	Crack growth due to cyclic loading	TLAA	TLAA	This TLAA is evaluated in SER Section 4.7.6
Reactor internals (Item Number 3.1.1-11)	Changes in dimension due to void swelling	Plant-specific	Reactor Vessel Internals Inspection Program (B2.1.17)	Consistent with GALL, which recommends further evaluation (See SER Section 3.1.2.2.6)
PWR core support pads, instrument tubes (bottom head penetrations), pressurizer spray heads, and nozzles for the steam generator instruments and drains (Item Number 3.1.1-12)	Crack initiation and growth due to SCC and/or primary water stress corrosion cracking (PWSCC)	Plant-specific	Alloy 600 Program (B2.1.1); Reactor Vessel Internals Inspection Program (B2.1.17); Water Chemistry Program (B2.1.21)	Consistent with GALL, which recommends further evaluation (See SER Section 3.1.2.2.7)
Cast austenitic stainless steel (CASS) reactor coolant system piping (Item.Number 3.1.1-13)	Crack initiation and growth due to SCC	Plant-specific	None	Not applicable (See SER Section 3.1.2.2.7)
Pressurizer instrumentation penetrations and heater sheaths and sleeves made of Ni-alloys (Item Number 3.1.1-14)	Crack initiation and growth due to PWSCC	Inservice inspection; water chemistry	Alloy 600 Program (B2.1.1); Water Chemistry Program (B2.1.21)	Consistent with GALL, which recommends further evaluation (See SER Section 3.1.2.2.7)
Westinghouse and B&W baffle former bolts (Item Number 3.1.1-15)	Crack initiation and growth due to SCC and IASCC	Plant-specific	Reactor Vessel Internals Inspection Program (B2.1.17)	(See SER Section 3.1.2.2.8)

Component Group	Aging Effect/ Mechanism	AMP in GALL Report	AMP in LRA	Staff Evaluation
Westinghouse and B&W baffle former bolts (Item Number 3.1.1-16)	Loss of preload due to stress relaxation	Plant-specific	Reactor Vessel Internals Inspection Program (B2.1.17)	(See SER Section 3.1.2.2.9)
Steam generator feedwater impingement plate and support (Item Number 3.1.1-17)	Loss of section thickness due to erosion	Plant-specific	None	Not applicable (See SER Section 3.1.2.2.10)
(Alloy 600) Steam generator tubes, repair sleeves, and plugs (Item Number 3.1.1-18)	Crack initiation and growth due to PWSCC, outside diameter stress corrosion cracking (ODSCC), and/or intergranular attack (IGA) or loss of material due to wastage and pitting corrosion, and fretting and wear; or deformation due to corrosion at tube support plate intersections	Steam generator tubing integrity; water chemistry	Steam Generator Tube Integrity Program (B2.1.18), Water Chemistry Program (B2.1.21)	Consistent with GALL, which recommends further evaluation (See SER Section 3.1.2.2.11)
Tube support lattice bars made of carbon steel (Item Number 3.1.1-19)	Loss of section thickness due to FAC	Plant-specific	None	Not applicable (See SER Section 3.1.2.2.12)
Carbon steel tube support plate (Item Number 3.1.1-20)	Ligament cracking due to corrosion	Plant-specific	Steam Generator Tube Integrity Program (B2.1.18), Water Chemistry Program (B2.1.21)	Consistent with GALL, which recommends further evaluation (See SER Section 3.1.2.2.13)
Steam generator feedwater inlet ring and supports (Item Number 3.1.1-21)	Loss of material due to flow accelerated corrosion	Combustion engineering (CE) steam generator feedwater ring inspection	Flow Accelerated Corrosion Program (B2.1.11)	(See Section 3.1.2.2.14)

Component Group	Aging Effect/ Mechanism	AMP in GALL Report	AMP in LRA	Staff Evaluation
Reactor vessel closure studs and stud assembly (Item Number 3.1.1-22)	Crack initiation and growth due to SCC and/or IGSCC	Reactor head closure studs	ASME Section XI IWB, IWC, IWD, IWF Inservice Inspection Program (B2.1.2)	Consistent with GALL, which recommends no further evaluation (See SER Section 3.1.2.1) (Reactor head closure studs included in program)
CASS pump casing and valve body (Item Number 3.1.1-23)	Loss of fracture toughness due to thermal aging embrittlement	Inservice inspection	ASME Section XI IWB, IWC, IWD, IWF Inservice Inspection Program (B2.1.2)	Consistent with GALL, which recommends no further evaluation (See SER Section 3.1.2.1)
CASS piping (Item Number 3.1.1-24)	Loss of fracture toughness due to thermal aging embrittlement	Thermal aging embrittlement of CASS	None	Not applicable (No CASS piping in primary coolant system)
BWR piping and fittings; steam generator components (Item Number 3.1.1-25)	Wall thinning due to flow-accelerated corrosion	Flow-accelerated corrosion	Flow Accelerated Corrosion Program (B2.1.11)	Consistent with GALL, which recommends no further evaluation (See SER Section 3.1.2.1.3)
Reactor coolant pressure boundary (RCPB) valve closure bolting, manway and holding bolting, and closure bolting in high pressure and high temperature systems (Item Number 3.1.1-26)	Loss of material due to wear; loss of preload due to stress relaxation; crack initiation and growth due to cyclic loading and/or SCC	Bolting integrity	Bolting Integrity Program (B2.1.3)	Consistent with GALL, which recommends no further evaluation (See SER Section 3.1.2.1)
CRD nozzle (Item Number 3.1.1-35)	Crack initiation and growth due to PWSCC	Ni-alloy nozzles and penetrations; water chemistry	Alloy 600 Program (B2.1.1), Water Chemistry Program (B2.1.21)	Consistent with GALL, which recommends no further evaluation (See SER Section 3.1.2.1)

Component Group	Aging Effect/ Mechanism	AMP in GALL Report	AMP in LRA	Staff Evaluation
Reactor vessel nozzles safe ends and CRD housing; reactor coolant system components (except CASS and bolting) (Item Number 3.1.1-36)	Crack initiation and growth due to cyclic loading, and/or SCC and PWSCC	Inservice inspection; water chemistry	Alloy 600 Program (B2.1.1); ASME Section XI IWB, IWC, IWD, IWF Inservice Inspection Program (B2.1.2); Water Chemistry Program (B2.1.21)	Consistent with GALL, which recommends no further evaluation (See SER Section 3.1.2.1.4)
Reactor vessel internals CASS components (Item Number 3.1.1-37)	Loss of fracture toughness due to thermal aging, neutron irradiation embrittlement, and void swelling	Thermal aging and neutron irradiation embrittlement	None	Not applicable (No CASS reactor vessel internals)
External surfaces of carbon steel components in reactor coolant system pressure boundary (Item Number 3.1.1-38)	Loss of material due to boric acid corrosion	Boric acid corrosion	Boric Acid Corrosion Program (B2.1.4); System Monitoring Program (B2.1.20)	Consistent with GALL, which recommends no further evaluation (See SER Section 3.1.2.1.5)
Steam generator secondary manways and handholds (CS) (Item Number 3.1.1-39)	Loss of material due to erosion	Inservice inspection	None	Not applicable (Steam generators are not once through)
Reactor internals, reactor vessel closure studs, and core support pads (Item Number 3.1.1-40)	Loss of material due to wear	Inservice inspection	ASME Section XI IWB, IWC, IWD, IWF Inservice Inspection Program (B2.1.2)	Consistent with GALL, which recommends no further evaluation (See SER Section 3.1.2.1)
Pressurizer integral support (Item Number 3.1.1-41)	Crack initiation and growth due to cyclic loading	Inservice inspection	ASME Section XI IWB, IWC, IWD, IWF Inservice Inspection Program (B2.1.2)	Consistent with GALL, which recommends no further evaluation (See SER Section 3.1.2.1)
Upper and lower internals assembly (Westinghouse) (Item Number 3.1.1-42)	Loss of preload due to stress relaxation	Inservice inspection; loose part and/or neutron noise monitoring	None	Not applicable (CE plant)

Component Group	Aging Effect/ Mechanism	AMP in GALL Report	AMP in LRA	Staff Evaluation
Reactor vessel internals in fuel zone region [except Westinghouse and Babcock & Wilcox (B&W) baffle bolts] (Item Number 3.1.1-43)	Loss of fracture toughness due to neutron irradiation embrittlement, and void swelling	PWR vessel internals; water chemistry	Reactor Vessel Internals Inspection Program (B2.1.17)	Consistent with GALL, which recommends no further evaluation (See SER Section 3.1.2.1)
Steam generator upper and lower heads; tubesheets; primary nozzles and safe ends (Item Number 3.1.1-44)	Crack initiation and growth due to SCC, PWSCC and IASCC	Inservice inspection; water chemistry	Alloy 600 Program (B2.1.1); ASME Section XI IWB, IWC, IWD, IWF Inservice Inspection Program (B2.1.2); Water Chemistry Program (B2.1.21)	Consistent with GALL, which recommends no further evaluation (See SER Section 3.1.2.1.6)
Vessel internals (except Westinghouse and B&W baffle former bolts) (Item Number 3.1.1-45)	Crack initiation and growth due to SCC and IASCC	PWR vessel internals; water chemistry	Reactor Vessel Internals Inspection Program (B2.1.17), Water Chemistry Program (B2.1.21)	Consistent with GALL, which recommends no further evaluation (See SER Section 3.1.2.1.7)
Reactor internals (B&W screws and bolts) (Item Number 3.1.1-46)	Loss of preload due to stress relaxation	Inservice inspection; loose part monitoring	None	Not applicable (CE plant)
Reactor vessel closure studs and stud assembly (Item Number 3.1.1-47)	Loss of material due to wear	Reactor head closure studs	ASME Section XI IWB, IWC, IWD, IWF Inservice Inspection Program (B2.1.2)	Consistent with GALL, which recommends no further evaluation (See SER Section 3.1.2.1) (Reactor head closure studs included in program)
Reactor internals (Westinghouse upper and lower internal assemblies; CE bolts and tie rods) (Item Number 3.1.1-48)	Loss of preload due to stress relaxation	Inservice inspection; loose part monitoring	ASME Section XI IWB, IWC, IWD, IWF Inservice Inspection Program (B2.1.2)	Consistent with GALL, which recommends no further evaluation (See SER Section 3.1.2.1.8)

The staff's review of the RCS component groups followed several approaches. One approach, documented in SER Section 3.1.2.1, discusses the staff's review of AMR results for components the applicant indicated are consistent with the GALL Report and require no further evaluation. Another approach, documented in SER Section 3.1.2.2, discusses the staff's review of AMR results for components the applicant indicated are consistent with the GALL Report and for which further evaluation is recommended. A third approach, documented in SER Section 3.1.2.3, discusses the staff's review of AMR results for components the applicant indicated are not consistent with or not addressed in the GALL Report. The staff's review of AMPs credited to manage or monitor aging effects of the RCS components is documented in SER Section 3.0.3.

3.1.2.1 AMR Results That Are Consistent with the GALL Report

Summary of Technical Information in the Application. In LRA Section 3.1.2.1, the applicant identified the materials, environments, and AERMs. The applicant identified the following programs that manage the aging effects of the RCS components:

- Alloy 600 Program
- ASME Section XI IWB, IWC, IWD, IWF Inservice Inspection Program
- Bolting Integrity Program
- Boric Acid Corrosion Program
- Closed Cycle Cooling Water Program
- Flow Accelerated Corrosion Program
- One-Time Inspection Program
- Reactor Vessel Integrity Surveillance Program
- Reactor Vessel Internals Inspection Program
- Steam Generator Tube Integrity Program
- System Monitoring Program
- Water Chemistry Program

Staff Evaluation. In LRA Tables 3.1.2-1 through 3.1.2-4, the applicant summarized AMRs for the RCS components and identified which AMRs it considered consistent with the GALL Report.

For component groups evaluated in the GALL Report for which the applicant had claimed consistency and for which the GALL Report does not recommend further evaluation, the staff performed an audit and review to determine whether the plant-specific components in these GALL Report component groups were bounded by the GALL Report evaluation.

The applicant provided a note for each AMR line item. The notes describe how the information in the tables aligns with the information in the GALL Report. The staff audited those AMRs with Notes A through E, which indicate how the AMR was consistent with the GALL Report.

Note A indicates that the AMR line item is consistent with the GALL Report for component, material, environment, and aging effect. In addition, the AMP is consistent with the GALL Report AMP. The staff audited these line items to verify consistency with the GALL Report and the validity of the AMR for the site-specific conditions.

Note B indicates that the AMR line item is consistent with the GALL Report for component, material, environment, and aging effect. In addition, the AMP takes some exceptions to the AMP identified in the GALL Report. The staff audited these line items to verify consistency with the GALL Report. The staff verified that the identified exceptions to the GALL Report AMPs had been reviewed and accepted by the staff. The staff also determined whether the AMP identified

3-135

by the applicant was consistent with the AMP identified in the GALL Report and whether the AMR was valid for the site-specific conditions.

Note C indicates that the component for the AMR line item, although different from, is consistent with the GALL Report for material, environment, and aging effect. In addition, the AMP is consistent with the AMP identified by the GALL Report. This note indicates that the applicant was unable to find a listing of some system components in the GALL Report; however, the applicant identified a different component in the GALL Report that had the same material, environment, aging effect, and AMP as the component under review. The staff audited these line items to verify consistency with the GALL Report. The staff also determined whether the AMR line item of the different component applied to the component under review and whether the AMR was valid for the site-specific conditions.

Note D indicates that the component for the AMR line item, although different from, is consistent with the GALL Report for material, environment, and aging effect. In addition, the AMP takes some exceptions to the AMP identified in the GALL Report. The staff audited these line items to verify consistency with the GALL Report. The staff verified whether the AMR line item of the different component was applicable to the component under review. The staff verified whether the exceptions to the GALL Report AMPs had been reviewed and accepted by the staff. The staff also determined whether the AMP identified by the applicant was consistent with the AMP identified in the GALL Report and whether the AMR was valid for the site-specific conditions.

Note E indicates that the AMR line item is consistent with the GALL Report for material, environment, and aging effect, but a different AMP is credited. The staff audited these line items to verify consistency with the GALL Report. The staff also determined whether the identified AMP would manage the aging effect consistent with the AMP identified in the GALL Report and whether the AMR was valid for the site-specific conditions.

The staff did not repeat its review of the matters described in the GALL Report; however, the staff did verify that the material presented in the LRA was applicable and that the applicant had identified the appropriate GALL Report AMRs. The staff's evaluation is discussed below.

3.1.2.1.1 Loss of Material Due to Pitting and Crevice Corrosion

In the discussion column of LRA Table 3.1.1, Item 3.1.1-02, the applicant stated that the ASME Section XI IWB, IWC, IWD, IWF Inservice Inspection and Water Chemistry Programs manage loss of material due to pitting and crevice corrosion. In LRA Table 3.1.2-4, the applicant stated that these programs would be applied to the feedwater inlet nozzles and thermal sleeves.

The GALL Report recommends the use of programs consistent with GALL AMPs XI.M1, "ASME Section XI Inservice Inspection, Subsections IWB, IWC, and IWD," and XI.M2, "Water Chemistry," to manage this aging effect.

The staff's evaluations of the applicant's ASME Section XI IWB, IWC, IWD, IWF Inservice Inspection and Water Chemistry Programs are documented in SER Sections 3.0.3.1.2 and 3.0.3.1.10, respectively.

The staff's review of LRA Table 3.1.1, Item 3.1.1-02, identified an area in which additional information was necessary to complete the review of the applicant's AMR results. The applicant responded to the staff's RAI as discussed below.

In RAI 3.1-1, dated June 3, 2005, the staff stated that during its audit and review, it was not apparent how the applicant intended to manage thermal sleeve aging effects. Therefore, the staff requested the applicant to provide information on how it planned to manage aging of thermal sleeves.

In its response dated November 18, 2005, the applicant stated that loss of material from the carbon steel steam generator feedwater inlet nozzles and thermal sleeves would be managed by the Water Chemistry Program and the ASME Section XI Program. The applicant also stated that, although the thermal sleeves are not accessible for direct inspection, the results of the Inservice Inspection Program inspections of the feedwater inlet nozzles are representative of the condition of the thermal sleeves and sufficient to manage any potential aging effects.

By letter dated March 30, 2006, the applicant stated that, although thermal sleeves are not pressure boundary components, residual stresses from the expansion process used for securing thermal sleeves cannot be quantified. Therefore, the applicant included cracking due to SCC and PWSCC as an AERM for thermal sleeves. The applicant further stated that:

> 'Stainless Steel Thermal Sleeves' are age managed by the ASME Section XI ISI Program and the Water Chemistry Program. The 'Alloy 600 Thermal Sleeves' are being age managed by the Alloy 600 Program and the Water Chemistry Program. Note that the thermal sleeves are inaccessible for direct inspection; however, the results of the ASME Section XI ISI and Alloy 600 program inspections on the adjacent nozzle and safe end welds are considered conservatively representative (given the higher applied stress conditions) of the condition of the thermal sleeves.

The staff finds the applicant's proposed use of the ASME Section XI IWB, IWC, IWD, IWF Inservice Inspection and Water Chemistry to manage cracking and the use of the Water Chemistry Program to manage loss of material due to crevice and pitting corrosion in the SG consistent with the GALL Report and, therefore, acceptable.

3.1.2.1.2 Crack Initiation and Growth Due to Thermal and Mechanical Loading or SCC of Small-Bore Piping < 4" NPS)

In the discussion column of LRA Table 3.1.1, Item 3.1.1-07, the applicant stated that crack initiation and growth due to thermal and mechanical loading or SCC of small-bore piping would be managed by the ASME Section XI IWB, IWC, IWD, IWF Inservice Inspection and the Water Chemistry Programs. The GALL Report recommends the use of programs consistent with GALL AMPs XI.M1, "ASME Section XI Inservice Inspection, Subsections IWB, IWC, and IWD," XI.M2, "Water Chemistry," and XI.M32, "One-Time Inspection," to manage these aging effects.

By letter dated August 27, 2005, the applicant stated that it would also use the One-Time Inspection Program to manage aging of all PCS components handled according to LRA Table 3.1.1, Item 3.1.1-07. The staff finds this management consistent with the GALL Report and, therefore, acceptable.

In LRA Table 3.1.2-1, the applicant also stated that stainless steel flow elements, pipe, and pressure test fittings of the chemical and volume control (CVC) system are managed by the Water Chemistry, ASME Section XI IWB, IWC, IWD, IWF Inservice Inspection, and One-Time Inspection Programs. The staff finds this management consistent with the GALL Report and, therefore, acceptable.

In the LRA, the applicant also associated LRA Table 3.1.1, Item 3.1.1-07, with the stainless steel reactor head vent and flow elements in the CVC system. For these components, loss of material was identified as an aging effect to be managed by the Water Chemistry and ASME Section XI IWB, IWC, IWD, IWF Inservice Inspection Programs. By letter dated August 27, 2005, the applicant agreed that the AERM should be cracking instead of loss of material. This correction is consistent with the GALL Report and, therefore, acceptable to the staff.

3.1.2.1.3 Wall Thinning Due to Flow-Accelerated Corrosion

In the discussion column of LRA Table 3.1.1, Item 3.1.1-25, the applicant stated that the Flow Accelerated Corrosion Program manages wall thinning due to FAC acting on feedwater inlet nozzles and thermal sleeves.

The GALL Report recommends a program consistent with GALL AMP XI.M17, "Flow-Accelerated Corrosion," to manage aging of those SG components within the scope of license renewal also subject to FAC.

The staff's evaluation of the applicant's Flow-Accelerated Corrosion Program is documented in SER Section 3.0.3.1.6. The staff confirmed that the thermal sleeves in question are addressed in the implementation of this program and, therefore, this aging effect would be adequately managed.

The staff's review of LRA Table 3.1.1, Item 3.1.1-02, identified an area in which additional information was necessary to complete the review of the applicant's AMR results. The applicant responded to the staff's RAI as discussed below.

In RAI 3.1-1, dated June 3, 2005, the staff stated that from its audit and review it was not apparent how the applicant intended to manage thermal sleeve aging effects. Therefore, the staff requested information on how it planned to manage such aging.

By letter dated November 18, 2005, the applicant stated that loss of material due to FAC of the feedwater inlet nozzles and thermal sleeves would be managed by the Flow Accelerated Corrosion Program. The applicant further stated for thermal sleeves inaccessible for direct inspection, that the Flow Accelerated Corrosion Program feedwater inlet nozzle inspection results represent the condition of the thermal sleeves and are sufficient to manage any potential aging effects.

The staff finds that the applicant's proposed use of the Flow-Accelerated Corrosion Program is consistent with the GALL Report and, therefore, acceptable.

3.1.2.1.4 Crack Initiation and Growth Due to Cyclic Loading, and/or SCC, and PWSCC

In the discussion column of LRA Table 3.1.1, Item 3.1.1-36, the applicant stated that the Alloy 600, ASME Section XI IWB, IWC, IWD, IWF Inservice Inspection, and Water Chemistry Programs manage crack initiation and growth due to cyclic loading, SCC, and PWSCC. However, in no instance did the applicant apply all three of these AMPs to manage this aging effect. In LRA Table 3.1.2-1, only the Alloy 600 and Water Chemistry Programs were applied to manage these aging mechanisms for Alloy 600/690 safe ends. To manage CASS valve bodies and pump casings, only the ASME Section XI IWB, IWC, IWD, IWF Inservice Inspection and Water Chemistry Programs were applied.

The GALL Report recommends management of this aging mechanism for safe ends by programs consistent with GALL AMPs XI.M1, "ASME Section XI Inservice Inspection, Subsections IWB, IWC, and IWD," and XI.M2, "Water Chemistry." However, for nozzles of nickel-alloy (LRA Table 3.1.1, Item 3.1.1-35), the GALL Report states that a program consistent with GALL AMP XI.M11, "Nickel-Alloy Nozzles and Penetrations," is acceptable and does not identify SCC as an AERM.

The staff determines that the Alloy 600 and Water Chemistry Programs are appropriate for managing crack initiation and growth in the nickel-alloy safe ends, consistent with the GALL Report, and, therefore, acceptable.

Similarly, for CASS components, the GALL Report recommends an AMP consistent with GALL AMP XI.M12, "Thermal Aging Embrittlement of Cast Austenitic Stainless Steel (CASS)," which explicitly states that for pump casings and valve bodies a program consistent with GALL AMP XI.M1 is sufficient.

On these bases, the staff determines that the use of the ASME Section XI IWB, IWC, IWD, IWF Inservice Inspection Program and the Water Chemistry Program for managing crack initiation and growth in the CASS pumps and valves is consistent with the GALL Report and, therefore, acceptable.

Finally, management of crack initiation and growth in the PCS, reactor vessel system, and CVC system stainless steel components exposed to treated (borated) water by both the ASME Section XI IWB, IWC, IWD, IWF Inservice Inspection Program and the Water Chemistry Program is consistent with the GALL Report and, therefore, acceptable.

By letter dated March 30, 2006, the applicant stated that although thermal sleeves are not pressure boundary components, residual stresses from the expansion process used for securing thermal sleeves cannot be quantified. Therefore, the applicant included cracking due to SCC and PWSCC as an AERM for thermal sleeves. The applicant further stated that:

> 'Stainless Steel Thermal Sleeves' are age managed by the ASME Section XI ISI Program and the Water Chemistry Program. The 'Alloy 600 Thermal Sleeves' are being age managed by the Alloy 600 Program and the Water Chemistry Program. Note that the thermal sleeves are inaccessible for direct inspection; however, the results of the ASME Section XI ISI and Alloy 600 program inspections on the adjacent nozzle and safe end welds are considered conservatively representative

(given the higher applied stress conditions) of the condition of the thermal sleeves.

The staff finds the applicant's proposed use of the ASME Section XI IWB, IWC, IWD, IWF Inservice Inspection, Water Chemistry, and Alloy 600 Programs to manage pitting and crevice corrosion in the SG consistent with the GALL Report and, therefore, acceptable.

The staff's review finds that the applicant has appropriately addressed crack initiation and growth due to cyclic loading, and/or SCC, and PWSCC, as recommended by the GALL Report.

3.1.2.1.5 Loss of Material Due to Boric Acid Corrosion

In the discussion column of LRA Table 3.1.1, Item 3.1.1-38, the applicant stated that the Boric Acid Corrosion Program manages loss of material due to boric acid corrosion. In LRA Tables 3.1.2-1 and 3.1.2-2, the Boric Acid Corrosion Program is augmented with the Bolting Integrity Program for low-alloy and stainless steel bolting and fasteners exposed to air.

The GALL Report recommends management of this aging effect by a program consistent with GALL AMP XI.M10, "Boric Acid Corrosion."

The staff finds that augmentation of the Boric Acid Corrosion Program with the Bolting Integrity Program (for fasteners) exceeds the recommendations of the GALL Report and is, therefore, acceptable.

In LRA Table 3.1.2-1, the applicant stated that the epoxy-coated carbon steel of the pressurizer quench tank is exposed to containment air and treated water and proposed to manage loss of material from this surface with the System Monitoring Program, augmented with the One-Time Inspection Program. The FSAR states that the tank normally contains nitrogen and demineralized water. In the course of normal operation, borated water is introduced as well.

The staff's review of LRA Table 3.1.2-1 identified an area in which additional information was necessary to complete the review of the applicant's AMR results. The applicant responded to the staff's RAI as discussed below.

In RAI 3.1.2-1(a), dated October 24, 2005, the staff requested the applicant to clarify the nature of the pressurizer quench tank interior environment and justify the application of its System Monitoring and One-Time Inspection Programs to manage loss of material in this environment.

In its response, by letter dated November 18, 2005, the applicant stated:

> The environments for the internal surfaces of the pressurizer quench tank are Treated Water (borated water), and Gas (nitrogen) for the space above the borated water level in the tank. The quench tank is normally less than 212 degrees F.

The intent of the Epoxy Coated Carbon Steel line items in LRA Table 3.1.2-1 was as follows:

Containment Air (Ext) is for the external surface of the pressurizer quench tank, carbon steel, and the System Monitoring Program was assigned to manage loss of material due to general corrosion.

Treated Water (Int) is for the internal surfaces of the pressurizer quench tank, epoxy coated carbon steel, and the One-Time Inspection Program was assigned to verify the condition of the epoxy coating for the period of extended operation.

Based on its review, the staff finds the applicant's response to RAI 3.1.2-1(a) acceptable. The staff determines that the System Monitoring Program is capable of detecting the loss of material aging effect on the external surface of the pressurizer quench tank before loss of intended function. In addition, the staff concludes that the One-Time Inspection Program is capable of detecting degradation of the epoxy coating on the internal surfaces of the pressurizer quench tank before loss of intended function. Therefore, the staff's concern described in RAI 3.1.2-1(a) is resolved.

3.1.2.1.6 Crack Initiation and Growth Due to SCC, PWSCC, and/or IASCC

In the discussion column of LRA Table 3.1.1, Item 3.1.1-44, the applicant stated that the ASME Section XI IWB, IWC, IWD, IWF Inservice Inspection and Water Chemistry Programs are used to manage crack initiation and growth due to SCC, PWSCC, and IASCC.

In LRA Table 3.1.2-1, the applicant proposed to use the Alloy 600 and Water Chemistry Programs to manage this aging effect for Alloy 600 cladding of the PCS. This proposed use is inconsistent with LRA Table 3.1.1, Item 3.1.1-44, which states that steam generator upper and lower heads; tubesheets; primary nozzles and safe ends are managed by the ASME Section XI IWB, IWC, IWD, IWF Inservice Inspection and Water Chemistry Programs for crack initiation and growth due to SCC, PWSCC, and IASCC.

The GALL Report recommends management of crack initiation and growth for safe ends by programs consistent with GALL AMPs XI.M1, "ASME Section XI Inservice Inspection, Subsections IWB, IWC, and IWD" and XI.M2, "Water Chemistry." However, for nozzles of nickel-alloy (LRA Table 3.1.1, Item 3.1.1-35), the GALL Report states that a program consistent with GALL AMP XI.M11, "Nickel-Alloy Nozzles and Penetrations," is acceptable for management of cracking due to PWSCC.

On the basis that the programs listed in LRA Table 3.1.2-1 are consistent with the GALL Report, which does not suggest that SCC or IASCC require aging management in Alloy 600 cladding, the staff determines that aging management of crack initiation and growth of this component type by the Alloy 600 and Water Chemistry Programs is consistent with the GALL Report and, therefore, acceptable.

In LRA Table 3.1.2-4, the applicant proposed to use only the Water Chemistry Program to manage crack initiation and growth in the stainless steel manway cover diaphragm and the primary divider plate components of the replacement SG.

On the basis that the primary divider plate is subject only to the differential pressure across the SG and is not intended to perform a system fluid pressure boundary function, the staff finds acceptable the applicant's use of only the Water Chemistry Program to manage crack initiation and growth in the divider plate.

As documented in the Audit and Review Report, the staff asked the applicant to justify the use of only its Water Chemistry Program to manage this aging affect for the steel manway cover. By letter dated August 27, 2005, the applicant stated that the stainless steel manway cover component type would be managed by the ASME Section XI IWB, IWC, IWD, IWF Inservice Inspection Program in addition to the Water Chemistry Program. The staff finds this management consistent with the GALL Report and, therefore, acceptable.

The staff's review finds that the applicant has addressed crack initiation and growth due to SCC, PWSCC, and IASCC, as recommended by the GALL Report.

3.1.2.1.7 Crack Initiation and Growth Due to SCC and IASCC

In the discussion column of LRA Table 3.1.1, Item 3.1.1-45, the applicant stated that the Reactor Vessel Internals Inspection and Water Chemistry Programs manage crack initiation and growth due to SCC and IASCC. In LRA Table 3.1.2-3, the applicant proposed only the Reactor Vessel Internals Inspection Program to manage this aging effect in the stainless steel spacer shim and instrument sleeve. As documented in the Audit and Review Report, the staff asked the applicant to justify this proposal.

By letter dated August 27, 2005, the applicant stated that the stainless steel spacer shim and instrument sleeve would be managed by the Water Chemistry Program in addition to the Reactor Vessel Internals Inspection Program. The staff finds this management consistent with the GALL Report and, therefore, acceptable.

The staff's review found management of crack initiation and growth due to SCC and IASCC consistent with the GALL Report and, therefore. acceptable.

3.1.2.1.8 Loss of Preload

In the discussion column of LRA Table 3.1.1, Item 3.1.1-48, the applicant stated that the ASME Section XI IWB, IWC, IWD, IWF Inservice Inspection Program manages loss of preload due to stress relaxation.

The GALL Report recommends management of loss of preload by programs consistent with GALL AMPs XI.M1, "ASME Section XI Inservice Inspection, Subsections IWB, IWC, and IWD," and XI.M14, "Loose Part Monitoring."

As documented in the Audit and Review Report, the staff asked the applicant to clarify how the GALL Report recommendation was satisfied. The applicant pointed out that GALL AMP XI.M14 would not be effective until component failure had occurred. By letter dated August 27, 2005, the applicant revised the statement of aging management for this mechanism to note that the program is not identical to the combination of programs recommended in the GALL Report.

Inspection in accordance with examination category B-3 of ASME Code Section XI, Subsection IWB, includes VT-3 examinations of bolted connections to detect such gross loss of preload as looseness or improper fit prior to failure of a connection. Therefore, the applicant does not rely on a loose parts monitoring program because this approach would require failure of the bolting before becoming effective. The staff finds this non-reliance acceptable because the applicant uses inspections to prevent failure of bolted connections.

Conclusion. The staff evaluated the applicant's claim of consistency with the GALL Report. The staff also reviewed information pertaining to the applicant's consideration of recent operating experience and proposals for managing the associated aging effects. On the basis of its review, the staff concludes that the AMR results, which the applicant claimed to be consistent with the GALL Report, are consistent with the GALL Report AMRs. Therefore, the staff concludes that the applicant has demonstrated that the aging effects for these components will be adequately managed so that their intended function(s) will be maintained consistent with the CLB for the period of extended operation, as required by 10 CFR 54.21(a)(3).

3.1.2.2 AMR Results That Are Consistent with the GALL Report, for Which Further Evaluation is Recommended

Summary of Technical Information in the Application. In LRA Section 3.1.2.2, the applicant provided further evaluation of aging management as recommended by the GALL Report for the RCS components. The applicant provided information concerning how it will manage the following aging effects:

- cumulative fatigue damage

- loss of material due to pitting and crevice corrosion

- loss of fracture toughness due to neutron irradiation embrittlement

- crack initiation and growth due to thermal and mechanical loading or SCC

- crack growth due to cyclic loading

- changes in dimension due to void swelling

- crack initiation and growth due to SCC or PWSCC

- crack initiation and growth due to SCC or IASCC

- loss of preload due to stress relaxation

- loss of section thickness due to erosion

- crack initiation and growth due to PWSCC, ODSCC, or intergranular attack or loss of material due to wastage and pitting corrosion or loss of section thickness due to fretting and wear or denting due to corrosion of carbon steel tube support plate

- loss of section thickness due to FAC

- ligament cracking due to corrosion

- loss of material due to FAC

<u>Staff Evaluation</u>. For component groups evaluated in the GALL Report for which the applicant had claimed consistency with the GALL Report and for which further evaluation is recommended, the staff audited and reviewed the applicant's evaluations to determine whether they adequately address those issues. In addition, the staff reviewed the applicant's further evaluations against the criteria in SRP-LR Section 3.1.2.2. Details of the staff's audit are documented in the Audit and Review Report. The staff's evaluation of the aging effects is discussed in the following sections.

3.1.2.2.1 Cumulative Fatigue Damage

In LRA Section 3.1.2.2.1, the applicant stated that fatigue is a TLAA, as defined in 10 CFR 54.3. Applicants must evaluate TLAAs in accordance with 10 CFR 54.21(c)(1). SER Section 4.3 documents the staff's review of the applicant's evaluation of this TLAA.

3.1.2.2.2 Loss of Material Due to Pitting and Crevice Corrosion

The staff reviewed LRA Section 3.1.2.2.2 against the criteria in SRP-LR Section 3.1.2.2.2.

In LRA Section 3.1.2.2.2, the applicant addressed loss of material in the SG shell assembly due to pitting and crevice corrosion. The applicant stated that the concerns of IN 90-04 are not applicable since the SGs were replaced in 1990 and pitting corrosion of the SG shell is not known to exist.

SRP-LR Section 3.1.2.2.2 states that loss of material due to pitting and crevice corrosion could occur in the SG shell assembly. The existing program relies on control of chemistry to mitigate corrosion and ISI to detect loss of material. The extent and schedule of the existing SG inspections are designed to ensure that flaws cannot attain a depth sufficient to threaten the integrity of the welds; however, according to IN 90-04, if general pitting corrosion of the shell exists, the program may not be sufficient to detect it.

The applicant credited its Water Chemistry Control and ASME Section XI IWB, IWC, IWD, IWF Inservice Inspection Programs for managing loss of material due to pitting and crevice corrosion on the internal surfaces of the SG shell and recommended the Steam Generator Tube Integrity Program to manage pitting and crevice corrosion. The Steam Generator Tube Integrity Program incorporates the guidance of NEI 97-06, "Steam Generator Program Guidelines," dated January 2001, to verify the integrity of the secondary-side internal surfaces of the SGs. A combination of the Water Chemistry, ASME Section XI IWB, IWC, IWD, IWF Inservice Inspection, and Steam Generator Tube Integrity Programs manages this aging effect.

The staff's review of LRA Section 3.1.2.2.2 identified an area in which additional information was necessary to complete the review of the applicant's AMR results. The applicant responded to the staff's RAI as discussed below.

In RAI-3.1.2.2.2-1, dated June 3, 2005, the staff noted that the LRA states that an augmented inspection of SG shell assemblies for loss of material due to pitting/crevice corrosion was not applicable. The GALL Report recommends such augmented inspection based on industry experience and extended exposure of the shell material to the water environment. Therefore, the staff requested the applicant to provide technical justification for the statement of non-applicability.

In its response, by letter dated July 1, 2005, the applicant stated:

> Section 3.1.2.2.2 of the SRP-LR states that the loss of material due to pitting and crevice corrosion could occur in the steam generator shell assembly. The existing program relies on control of water chemistry to mitigate corrosion, and inservice inspection to detect the loss of material. The extent and schedule of the existing steam generator inspections ensure that flaws cannot attain a depth sufficient to threaten the integrity of the welds. However, the NRC states in NRC IN 90-04, 'Cracking of the Upper Shell-to-Transition Cone Girth Welds in Steam Generators,' dated January 26, 1990, if pitting and crevice corrosion of the shell exists, the program may not be sufficient to detect pitting and corrosion. The GALL Report recommends augmented inspections to manage this aging effect.

> In Section 3.1.2.2.2.1 of the LRA, Palisades states that pitting/crevice corrosion is not known to exist in the steam generator shells, and, therefore, augmented inspections are not necessary. This statement was based upon the following operating experience: (1) In February of 2000 the steam generator program engineer, who was also a certified welding inspector, completed a 360 degree walk down of the secondary side internal wall, at the elevation of the main feedwater ring. During that walk down no evidence of ID pitting was identified; and (2) During the 2003 refueling outage, 2 complete steam generator shell circumferential welds were examined from the OD using volumetric inspection techniques. These weld inspections did not identify evidence of internal pitting in the associated steam generator shell area.

> The original Palisades steam generators were replaced in 1990 with Combustion Engineering Model 2530 steam generators. Since then, Palisades has maintained secondary water chemistry in accordance with EPRI guidelines. The combination of these factors, coupled with continued water chemistry maintenance and ISI inspections provides reasonable assurance that pitting/crevice corrosion will not threaten the steam generator shell pressure boundary function during the period of extended operation. This is consistent with the staffs conclusions in past SERs, including the SER for Plant Farley (NUREG 1825).

Based on its review, the staff finds the applicant's response to RAI-3.1.2.2.2-1 acceptable because it provided a case supported by operating experience for no augmented inspections. Therefore, the staff's concern described in RAI-3.1.2.2.2-1 is resolved.

As documented in the Audit and Review Report, the staff interviewed the applicant's technical personnel and reviewed the most recent SG inspection records. The staff confirmed that these records document the absence of pitting and crevice corrosion.

Based on the programs identified above, the staff concludes that the applicant has met the criteria of SRP-LR Section 3.1.2.2.2. For those line items that apply to LRA Section 3.1.2.2.2, the staff determines that the application is consistent with the GALL Report and the applicant has demonstrated that the aging effects will be adequately managed so that the intended function(s) will be maintained consistent with the CLB during the period of extended operation, as required by 10 CFR 54.21(a)(3).

3.1.2.2.3 Loss of Fracture Toughness Due to Neutron Irradiation Embrittlement

The staff reviewed LRA Sections 3.1.2.2.3.1 through 3.1.2.2.3.3 against the criteria in SRP-LR Section 3.1.2.2.3.

In LRA Sections 3.1.2.2.3.1 through 3.1.2.2.3.3, the applicant addressed loss of fracture toughness due to neutron irradiation embrittlement.

LRA Sections 3.1.2.2.3.1 and 3.1.2.2.3.2 evaluate TLAAs, as defined in 10 CFR 54.3. Applicants must evaluate TLAAs in accordance with 10 CFR 54.21(c)(1). SER Section 4.2 documents the staff's review of the applicant's evaluation of these TLAAs.

SRP-LR Section 3.1.2.2.3 states that certain aspects of neutron irradiation embrittlement are TLAAs and that loss of fracture toughness due to neutron irradiation embrittlement and void swelling could occur in Westinghouse and Babcock and Wilcox (B&W) baffle/former bolts. The GALL Report recommends further evaluation to ensure adequate management of this aging effect.

In LRA Section 3.1.2.2.3.3, the applicant stated that this issue is not applicable, that PNP is a CE PWR with no baffle/former bolt configuration. By letter dated August 27, 2005, the applicant clarified the basis for the conclusion that embrittlement of the bolts is not a concern. Although PNP uses baffle/former bolts, the material in these bolts is annealed 316 stainless steel. The applicant stated that the stresses on the bolts are significantly lower in a CE PWR because the bolt stress from preload is a smaller percentage of the yield strength, the differential pressure across the core shroud panels does not produce large tensile loads on the baffle/former bolts during normal operation, and the core shroud panel design allows for some flexing of the former plate relative to the core barrel.

The staff reviewed evaluations by the reactor vendor and confirmed that the baffle-former bolts are of annealed 316 stainless steel. This material is significantly less brittle than the cold-worked 316 stainless steel or other materials used in designs where this aging mechanism has been observed. In addition, the loading of these fasteners is low enough to obviate the concern. On these bases, the staff concludes that the applicant's Reactor Vessel Internals Inspection Program would adequately manage loss of fracture toughness for baffle-former bolting without augmented inspections.

Based on the programs identified above, the staff concludes that the applicant has met the criteria of SRP-LR Section 3.1.2.2.3. For those line items that apply to LRA Section 3.1.2.2.3, the staff determines that the application is consistent with the GALL Report and the applicant has demonstrated that the aging effects will be adequately managed so that the intended function(s) will be maintained consistent with the CLB during the period of extended operation, as required by 10 CFR 54.21(a)(3).

3.1.2.2.4 Crack Initiation and Growth Due to Thermal and Mechanical Loading or Stress Corrosion Cracking

The staff reviewed LRA Section 3.1.2.2.4 against the criteria in SRP-LR Section 3.1.2.2.4.

In LRA Section 3.1.2.2.4, the applicant addressed crack initiation and growth in the small-bore RCS and connected system piping less than NPS 4 due to SCC, intergranular stress corrosion cracking (IGSCC), and thermal and mechanical loading.

SRP-LR Section 3.1.2.2.4 states that crack initiation and growth due to thermal and mechanical loading or SCC (including IGSCC) could occur in small-bore RCS and connected system piping smaller than NPS 4. The existing program relies on control of water chemistry to mitigate SCC and ASME Code Section XI ISI to detect it. The GALL Report recommends a plant-specific destructive examination or NDE that permits inspection of the inside surfaces of the piping to ensure that cracking has not occurred and that the component intended function would be maintained during the period of extended operation. The AMPs should be augmented by verifying that service-induced weld cracking does not occur in the small-bore piping less than NPS 4, including pipe, fittings, and branch connections. A one-time inspection of a sample of locations is an acceptable method to ensure that the aging effect does not occur and that the component's intended function will be maintained during the period of extended operation.

In the LRA, the applicant stated that crack initiation and growth due to SCC had been identified as an AERM in small-bore (less than NPS 4) PCS piping and branch lines. Aging management of service-induced cracking would be accomplished by a combination of the Water Chemistry and ASME Section XI, Subsections IWB, IWC, and IWD Inservice Inspection Programs. In addition, inspections of a sample of small-bore PCS piping would be performed.

Based on the programs identified above, the staff concludes that the applicant has met the criteria of SRP-LR Section 3.1.2.2.4. For those line items that apply to LRA Section 3.1.2.2.4, the staff determines that the application is consistent with the GALL Report and the applicant has demonstrated that the aging effects will be adequately managed so that the intended function(s) will be maintained consistent with the CLB during the period of extended operation, as required by 10 CFR 54.21(a)(3).

3.1.2.2.5 Crack Growth Due to Cyclic Loading

The staff reviewed LRA Section 3.1.2.2.5 against the criteria in SRP-LR Section 3.1.2.2.5.

In LRA Section 3.1.2.2.5, the applicant addressed crack growth due to cyclic loading for reactor vessel components fabricated from SA-508 Class 2 material.

SRP-LR Section 3.1.2.2.5 states that crack growth due to cyclic loading could occur in reactor vessel shell and RCS piping and fittings. Growth of intergranular separations (underclad cracks) in a low-alloy or carbon steel heat-affected zone under austenitic stainless steel cladding is a TLAA to be evaluated for the period of extended operation for all the SA-508 Class 2 forgings where the cladding was deposited with a high heat input welding process. The methodology for evaluating the underclad flaw should be consistent with the current well-established flaw evaluation procedure and criterion in ASME Code Section XI.

By letter dated April 26, 2006, the applicant revised LRA Section 3.1.2.2.5 by replacing the existing information with the following:

> NUREG-1800 states that crack growth due cyclic loading could occur in reactor
> vessel shell and reactor coolant system piping and fittings. Growth of

intergranular separations (underclad cracks) in low-alloy or carbon steel heat affected zone under austenitic stainless steel cladding is a time-limited aging analysis (TLAA) to be evaluated for the period of extended operation for all the SA 508-Cl 2 forgings where the cladding was deposited with a high heat input welding process.

Underclad cracking in carbon/low-alloy steel, which has been clad with austenitic stainless steel using weld-overlay processes, has been identified as an aging effect requiring management and is addressed as a TLAA. An evaluation of the TLAA for underclad cracking is contained in Section 4.7.6.

On the basis of its review, the staff determines that the applicant had correctly identified, consistent with the SRP-LR, crack growth due to cyclic loading as a TLAA. Applicants must evaluate TLAAs in accordance with 10 CFR 54.21(c)(1). SER Section 4 documents the staff's review of the applicant's evaluation of this TLAA.

3.1.2.2.6 Changes in Dimension Due to Void Swelling

The staff reviewed LRA Section 3.1.2.2.6 against the criteria in SRP-LR Section 3.1.2.2.6.

In LRA Section 3.1.2.2.6, the applicant addressed changes in dimension of reactor internals due to void swelling.

SRP-LR Section 3.1.2.2.6 states that changes in dimension due to void swelling could occur in reactor internal components. The GALL Report recommends further evaluation to ensure adequate management of this aging effect. The RVI receives a VT-3 according to category B-3 of ASME Code Section XI, Subsection IXB. This inspection is not sufficient to detect the effects of changes in dimension due to void swelling. The GALL Report recommends evaluation of a plant-specific AMP. The applicant shall participate in industry programs to investigate aging effects and determine the appropriate AMP, provide a plant-specific AMP, provide the basis for concluding that void swelling is not an issue, or provide a program to manage the effects of changes in dimension due to void swelling and the loss of ductility associated with swelling.

In the LRA, the applicant stated that industry activities are underway to determine whether changes in dimension due to void swelling are aging effects requiring management for license renewal. The applicant stated that it continues to participate in industry investigations of aging effects applicable to the RVI, as well as initiatives to develop and qualify methods for detection and management. The applicant stated that it would incorporate applicable results of industry initiatives related to void swelling in its Reactor Vessel Internals Inspection Program.

Based on the programs identified above, the staff concludes that the applicant has met the criteria of SRP-LR Section 3.1.2.2.6. For those line items that apply to LRA Section 3.1.2.2.6, the staff determines that the application is consistent with the GALL Report and the applicant has demonstrated that the aging effects will be adequately managed so that the intended function(s) will be maintained consistent with the CLB during the period of extended operation, as required by 10 CFR 54.21(a)(3).

3.1.2.2.7 Crack Initiation and Growth Due to Stress Corrosion Cracking or Primary Water Stress Corrosion Cracking

The staff reviewed LRA Sections 3.1.2.2.7.1 through 3.1.2.2.7.3 against the criteria in SRP-LR Section 3.1.2.2.7, which addresses several areas discussed below.

Crack Initiation and Growth Due to Stress Corrosion Cracking or Primary Water Stress Corrosion Cracking (PWR Components). The staff reviewed LRA Section 3.1.2.2.7.1 against the criteria in SRP-LR Section 3.1.2.2.7.

In LRA Section 3.1.2.2.7.1, the applicant addressed crack initiation and growth due to SCC or PWSCC in core support pads, instrument tubes (bottom head penetrations), pressurizer spray heads, and nozzles for the SG instruments and drains.

SRP-LR Section 3.1.2.2.7 states that crack initiation and growth due to SCC and PWSCC could occur in PWR core support pads (or core guide lugs), instrument tubes (bottom head penetrations), pressurizer spray heads, and nozzles for the SG instruments and drains. The GALL Report recommends further evaluation to ensure adequate management of these aging effects. The GALL Report recommends evaluation of a plant-specific AMP because existing programs may not be capable of mitigating or detecting crack initiation and growth due to SCC.

In the LRA, the applicant stated that this grouping includes the surge nozzle thermal sleeve, safety injection nozzle thermal sleeve, charging inlet nozzle thermal sleeve, resistance temperature detectors (RTD) nozzles, pressure measurement nozzle, sampling nozzle, and partial nozzle replacement. Reactor vessel items included in this grouping are the lower shell and bottom head cladding, surveillance capsule holders, core stabilizing lugs, core stop and support lugs, and the flow baffle and skirt, the reactor head O-ring, leakoff tubing, and valves. SG items included in this grouping are the tube plate cladding, channel head divider plate, and primary nozzle closure rings. The EPRI material reliability program, in conjunction with the PWR owners group, is developing a strategic plan to manage and mitigate cracking of nickel-based alloy items. The guidance developed by the EPRI material reliability program will identify critical locations for inspection and for augmentation of existing ISI inspections, as appropriate. The results of the strategic plan will be incorporated into the applicant's Alloy 600 Program and, as applicable, its Water Chemistry and ASME Section XI IWB, IWC, IWD, IWF Inservice Inspection Programs.

Based on the programs identified above, the staff concludes that the applicant has met the criteria of SRP-LR Section 3.1.2.2.7. For those line items that apply to LRA Section 3.1.2.2.7.1, the staff determines that the application is consistent with the GALL Report and the applicant has demonstrated that the aging effects will be adequately managed so that the intended function(s) will be maintained consistent with the CLB during the period of extended operation, as required by 10 CFR 54.21(a)(3).

Crack Initiation and Growth Due to SCC or PWSCC (CASS Components). The staff reviewed LRA Section 3.1.2.2.7.2 against the criteria in SRP-LR Section 3.1.2.2.7.

In LRA Section 3.1.2.2.7.2, the applicant addressed crack initiation and growth due to SCC in CASS RCS piping.

SRP-LR Section 3.1.2.2.7 states that crack initiation and growth due to SCC could occur in PWR CASS RCS piping and fittings and pressurizer surge line nozzles. The GALL Report recommends further evaluation of piping that does not meet either TR-105714 reactor water chemistry guidelines or the material guidelines of NUREG-0313, "BWR Coolant Pressure Boundary," dated January 1988.

The staff's review of LRA Section 3.1.2.2.7.2 identified an area in which additional information was necessary to complete the review of the applicant's AMR results. The applicant responded to the staff's RAI as discussed below.

In RAI-3.1.2.2.7.2-1, dated June 3, 2005, the staff stated that the applicant's AMP for CASS thermal embrittlement includes neither a flaw tolerance evaluation nor an enhanced volumetric inspection, as recommended in the GALL Report. Therefore, the staff requested the applicant to clarify and discuss these omissions.

By letter dated July 1, 2005, the applicant responded that as stated in LRA Section B2.0, PNP has no AMP for CASS and no PCS CASS material other than valve bodies and pump casings managed by the applicant's ASME Section XI IWB, IWC, IWD, IWF Inservice Inspection Program.

The staff finds the applicant's response to RAI-3.1.2.2.7.2-1 acceptable because PNP has no PCS CASS material other than valve bodies and pump casings effectively managed by its ASME Section XI IWB, IWC, IWD, IWF Inservice Inspection Program. Therefore, the staff's concern described in RAI-3.1.2.2.7.2-1 is resolved.

In the LRA, the applicant stated that there is no CASS piping and that only the primary coolant pump casing and valve bodies of PORV isolation valves are fabricated of CASS. On this basis, the staff determines that for CASS piping, crack initiation and growth due to SCC or PWSCC requires no aging management.

Based on the programs identified above, the staff concludes that the applicant has met the criteria of SRP-LR Section 3.1.2.2.7. For those line items that apply to LRA Section 3.1.2.2.7.2, the staff determines that the application is consistent with the GALL Report and the applicant has demonstrated that the aging effects will be adequately managed so that the intended function(s) will be maintained consistent with the CLB during the period of extended operation, as required by 10 CFR 54.21(a)(3).

Crack Initiation and Growth Due to SCC or PWSCC (Nickel Alloys). The staff reviewed LRA Section 3.1.2.2.7.3 against the criteria in SRP-LR Section 3.1.2.2.7.

In LRA Section 3.1.2.2.7.3, the applicant addressed crack initiation and growth due to PWSCC in pressurizer instrumentation penetrations and heater sheaths and sleeves made of nickel alloys.

SRP-LR Section 3.1.2.2.7 states that crack initiation and growth due to PWSCC could occur in pressurizer instrumentation penetrations and heater sheaths and sleeves made of nickel alloys. The existing program relies on ASME Code Section XI ISI and on control of water chemistry to mitigate PWSCC. However, the existing program should be augmented to manage the effects of SCC on the intended function of nickel-alloy components. The GALL Report recommends that the applicant provide a plant-specific AMP or participate in industry programs to determine an appropriate AMP for PWSCC of Inconel 182 welds.

In the LRA, the applicant stated that nickel-based alloy material is identified for the pressurizer instrumentation nozzles, heater sheaths and sleeves, and thermal sleeves. Pressurizer components included in this grouping are the instrument nozzles and electric heaters (penetration nozzles and plugs, original heater sheath, heater sleeve, and end plugs). The programs credited for the management of PWSCC of these nickel-based alloy items are the Alloy 600 and Water Chemistry Programs, supplemented by the ASME Section XI IWB, IWC, IWD, IWF Inservice Inspection Program.

As described in LRA Section 3.1.2.2.7.1, the Alloy 600 Program includes participation in industry programs to identify critical locations for inspection and augmentation of existing ISI inspections where appropriate. PNP has had several instances of SCC in Alloy 600 welds and heat-affected zones, some of which resulted in through-wall leaks.

Based on the programs identified above, the staff concludes that the applicant has met the criteria of SRP-LR Section 3.1.2.2.7. For those line items that apply to LRA Section 3.1.2.2.7.3, the staff determines that the application is consistent with the GALL Report and the applicant has demonstrated that the aging effects will be adequately managed so that the intended function(s) will be maintained consistent with the CLB during the period of extended operation, as required by 10 CFR 54.21(a)(3).

3.1.2.2.8 Crack Initiation and Growth Due to Stress Corrosion Cracking or Irradiation-Assisted Stress Corrosion Cracking

The staff reviewed LRA Section 3.1.2.2.8 against the criteria in SRP-LR Section 3.1.2.2.8.

In LRA Section 3.1.2.2.8, the applicant addressed crack initiation and growth due to SCC and IASCC of Westinghouse and B&W baffle former bolts.

SRP-LR Section 3.1.2.2.8 states that crack initiation and growth due to SCC or IASCC could occur in baffle/former bolts in Westinghouse and B&W reactors.

In the LRA, the applicant stated that the RVI does not include baffle/former bolts. By letter dated August 27, 2005, the applicant clarified the basis for the conclusion that embrittlement of the bolts is not a concern. Although PNP uses baffle/former bolts, the material used in these bolts is annealed 316 stainless steel. The applicant stated that the stresses on the bolts are significantly lower in a CE PWR because the bolt stress from preload is a smaller percentage of the yield strength, the differential pressure across the core shroud panels does not produce large tensile loads on the baffle/former bolts during normal operation, and the core shroud panel design allows for some flexing of the former plate relative to the core barrel.

The staff reviewed evaluations by the reactor vendor and confirmed that the baffle-former bolts are of annealed 316 stainless steel. This material is significantly less brittle than the cold-worked 316 stainless steel or other materials used in designs where this aging mechanism has been observed. In addition, the loading of these fasteners is low enough to obviate the concern. On these bases, the staff concludes that the applicant's Reactor Vessel Internals Inspection

Program would adequately manage cracking of baffle-former bolting without augmented inspections.

Based on the programs identified above, the staff concludes that the applicant has met the criteria of SRP-LR Section 3.1.2.2.8. For those line items that apply to LRA Section 3.1.2.2.8, the staff determines that the application is consistent with the GALL Report and the applicant has demonstrated that the aging effects will be adequately managed so that the intended function(s) will be maintained consistent with the CLB during the period of extended operation, as required by 10 CFR 54.21(a)(3).

3.1.2.2.9 Loss of Preload Due to Stress Relaxation

The staff reviewed LRA Section 3.1.2.2.9 against the criteria in SRP-LR Section 3.1.2.2.9.

In LRA Section 3.1.2.2.9, the applicant addressed loss of preload due to stress relaxation.

SRP-LR Section 3.1.2.2.9 states that loss of preload due to stress relaxation could occur in baffle/former bolts in Westinghouse and B&W reactors.

In the LRA, the applicant stated that the RVI does not include baffle/former bolts. By letter dated August 27, 2005, the applicant clarified the basis for the conclusion that embrittlement of the bolts is not a concern. Although PNP uses baffle/former bolts, the material used in these bolts is annealed 316 stainless steel. The applicant stated that the stresses on the bolts are significantly lower in a CE PWR because the bolt stress from preload is a smaller percentage of the yield strength, the differential pressure across the core shroud panels does not produce large tensile loads on the baffle/former bolts during normal operation, and the core shroud panel design allows for some flexing of the former plate relative to the core barrel.

The staff reviewed evaluations by the reactor vendor and confirmed that the baffle-former bolts are of annealed 316 stainless steel. The loading of these fasteners is low enough to obviate the concern. On this basis, the staff concludes that the applicant's Reactor Vessel Internals Inspection Program would adequately manage loss of preload of baffle-former bolting without augmented inspections.

Based on the programs identified above, the staff concludes that the applicant has met the criteria of SRP-LR Section 3.1.2.2.9. For those line items that apply to LRA Section 3.1.2.2.9, the staff determines that the application is consistent with the GALL Report and the applicant has demonstrated that the aging effects will be adequately managed so that the intended function(s) will be maintained consistent with the CLB during the period of extended operation, as required by 10 CFR 54.21(a)(3).

3.1.2.2.10 Loss of Section Thickness Due to Erosion

The staff reviewed LRA Section 3.1.2.2.10 against the criteria in SRP-LR Section 3.1.2.2.10.

In LRA Section 3.1.2.2.10, the applicant addressed loss of section thickness of SG feedwater impingement plate and support due to erosion.

SRP-LR Section 3.1.2.2.10 states that loss of section thickness due to erosion could occur in SG feedwater impingement plates and supports. The GALL Report recommends further evaluation of a plant-specific AMP to ensure adequate management of this aging effect.

In the LRA, the applicant stated that the SGs do not include impingement plates. Therefore, the staff determines that loss of section thickness due to erosion of impingement plates requires no aging management.

3.1.2.2.11 Crack Initiation and Growth Due to PWSCC, ODSCC, or Intergranular Attack or Loss of Material Due to Wastage and Pitting Corrosion or Loss of Section Thickness Due to Fretting and Wear or Denting Due to Corrosion of Carbon Steel Tube Support Plate

The staff reviewed LRA Section 3.1.2.2.11 against the criteria in SRP-LR Section 3.1.2.2.11.

In LRA Section 3.1.2.2.11, the applicant addressed crack initiation and growth due to PWSCC, ODSCC, or IGA in (Alloy 600) SG tubes, repair sleeves, and plugs; loss of material in (Alloy 600) SG tubes, repair sleeves, and plugs due to wastage and pitting corrosion; and fretting and wear or deformation due to corrosion at tube support plate intersections.

SRP-LR Section 3.1.2.2.11 states that crack initiation and growth due to PWSCC, ODSCC, or IGA or loss of material due to wastage and pitting corrosion or deformation due to corrosion could occur in Alloy 600 components of the SG tubes, repair sleeves, and plugs. All PWR licensees have committed voluntarily to an SG degradation management program described in NEI 97-06, "Steam Generator Program Guidelines," dated January 2001; these guidelines are currently under staff review. The GALL Report recommends development of an AMP based on the recommendations of staff-approved NEI 97-06 guidelines, or other alternate regulatory basis for SG degradation management, to ensure adequate management of this aging effect.

In the LRA, the applicant stated that crack initiation and growth due to PWSCC, SCC, or IGA or loss of material due to wastage and pitting corrosion or deformation due to corrosion could occur in nickel-based alloy components of the SG tube plugs. To manage these aging effects, the applicant credited its Steam Generator Tube Integrity Program supplemented by its Water Chemistry and ASME Section XI IWB, IWC, IWD, IWF Inservice Inspection Programs. The Steam Generator Tube Integrity Program assessment of tube integrity and plugging or repair criteria of flawed tubes are in accordance with plant TS and NEI 97-06 guidelines. For general and pitting corrosion, the acceptance criteria are in accordance with NEI 97-06 guidelines.

The applicant also stated that new, replacement recirculating SGs were installed in 1990. These new SGs incorporate many enhancements in design and materials of construction to minimize aging effects; however, cracking due to PWSCC and IGA/IGSCC and loss of material due to pitting and wear could occur in the SG tubes and plugs.

In the LRA, the applicant also stated that its Water Chemistry Program conforms to the guidelines in EPRI TR-105714 and TR-102134. The Water Chemistry Program mitigates such aging effects as cracking due to PWSCC and IGA/IGSCC and loss of material due to pitting and wear by controlling the environment to which the SG tubes and plugs are exposed. These aging effects are minimized by controlling the chemical species that cause the underlying mechanisms for these aging effects. The program provides assurance that no elevated level of contaminants and oxygen exists in either the primary or secondary sides of the SGs, and thus minimizes aging effects. The Water Chemistry Program has maintained the desired primary and secondary water chemistry and has detected abnormal conditions since initial plant operation. The applicant suggested that the Water Chemistry Program mitigates cracking due to PWSCC and IGA/IGSCC and loss of material due to pitting and wear in the SG tubes and plugs. Finally, the applicant stated that verification of the effectiveness of the program would ensure that these aging effects do not occur.

In addition, in the LRA, the applicant stated that its Steam Generator Tube Integrity Program manages these aging effects for the SG tubes and plugs in order to confirm the effectiveness of the Water Chemistry Program. The Steam Generator Tube Integrity Program was developed to meet the guidelines of NEI 97-06 and manages these aging effects through prevention, inspection, evaluation, repair, and leakage monitoring measures. ECT is used to detect SG tube flaws and degradation. SG tubes not meeting TS limits for continued operation are removed from service by installation of tube plugs fabricated from heat-treated Alloy 690 material. Although these plugs have a high resistance to PWSCC, they are routinely inspected as a part of the program. A tube integrity assessment following each SG tube inspection ensures that performance criteria have been met for the previous operating period and will continue to be met for the next period.

Based on the programs identified above, the staff concludes that the applicant has met the criteria of SRP-LR Section 3.1.2.2.11. For those line items that apply to LRA Section 3.1.2.2.11, the staff determines that the application is consistent with the GALL Report and the applicant has demonstrated that the aging effects will be adequately managed so that the intended function(s) will be maintained consistent with the CLB during the period of extended operation, as required by 10 CFR 54.21(a)(3).

3.1.2.2.12 Loss of Section Thickness Due to Flow-Accelerated Corrosion (FAC)

The staff reviewed LRA Section 3.1.2.2.12 against the criteria in SRP-LR Section 3.1.2.2.12.

In LRA Section 3.1.2.2.12, the applicant addressed loss of section thickness due to FAC of tube support lattice bars made of carbon steel.

SRP-LR Section 3.1.2.2.12 states that loss of section thickness due to FAC could occur in tube support lattice bars made of carbon steel. The GALL Report recommends evaluation of a plant-specific AMP and, on the basis of the guidelines of GL 97-06, development of an inspection program for SG internals to ensure adequate management of this aging effect.

In the LRA, the applicant stated that the SGs include no carbon steel tube support lattice bars. Therefore, loss of section thickness of such bars is not an applicable aging effect. On this basis, the staff determines that loss of section thickness of tube support lattice bars made of carbon steel requires no aging management.

3.1.2.2.13 Ligament Cracking Due to Corrosion

The staff reviewed LRA Section 3.1.2.2.13 against the criteria in SRP-LR Section 3.1.2.2.13.

In LRA Section 3.1.2.2.13, the applicant addressed ligament cracking due to corrosion of the carbon steel tube support plate.

SRP-LR Section 3.1.2.2.13 states that ligament cracking due to corrosion could occur in carbon steel components in the SG tube support plate. All PWR licensees have committed voluntarily to an SG degradation management program described in NEI 97-06; these guidelines are currently under staff review. The GALL Report recommends development of an AMP based on the recommendations of NEI 97-06 guidelines, or other alternate regulatory basis for SG degradation management, to ensure adequate management of this aging effect.

In the LRA, the applicant stated that the SGs have a carbon steel tube bundle support assembly, but stainless steel eggcrate tube lattice support rings.

The staff determines that ligament cracking due to corrosion of the support plate requires no aging management because the replacement SGs incorporate an eggcrate tube lattice support.

The applicant also stated that its Water Chemistry Program conforms to EPRI TR-105714 and TR-102134 guidelines and mitigates aging effects by controlling the environment to which the SG stainless steel tube support plates are exposed. In the LRA, the applicant stated that verification of the effectiveness of the program would ensure that this aging effect does not occur.

In addition, in the LRA, the applicant stated that its Steam Generator Tube Integrity Program manages this aging effect for the SG stainless steel tube support plates in order to confirm the effectiveness of the Water Chemistry Program. The Steam Generator Tube Integrity Program was developed to meet the guidelines of NEI 97-06. The program manages this aging effect through prevention, inspection, evaluation, repair, and leakage monitoring measures. Periodic visual inspections of accessible areas verify the integrity of secondary-side components, including the SG stainless steel tube support plates. A combination of the Water Chemistry and Steam Generator Tube Integrity Programs manages this aging effect.

Based on the programs identified above, the staff concludes that the applicant has met the criteria of SRP-LR Section 3.1.2.2.13. For those line items that apply to LRA Section 3.1.2.2.13, the staff determines that the application is consistent with the GALL Report and the applicant has demonstrated that the aging effects will be adequately managed so that the intended function(s) will be maintained consistent with the CLB during the period of extended operation, as required by 10 CFR 54.21(a)(3).

3.1.2.2.14 Loss of Material Due to Flow-Accelerated Corrosion (FAC)

The staff reviewed LRA Section 3.1.2.2.14 against the criteria in SRP-LR Section 3.1.2.2.14.

In LRA Section 3.1.2.2.14, the applicant addressed loss of material due to FAC of SG feedwater inlet ring and supports.

SRP-LR Section 3.1.2.2.14 states that loss of material due to FAC could occur in feedwater inlet ring and supports. As noted in CE IN 90-04, NRC IN 91-19 and LER 50-362/90-05-01, this form of degradation has been detected only in certain CE System 80 SGs. The GALL Report recommends further evaluation to ensure adequate management of this aging effect. The GALL Report recommends evaluation of a plant-specific AMP because existing programs may not be capable of mitigating or detecting loss of material due to FAC.

By letter dated November 18, 2005, the applicant stated that the steam generator tube integrity program is credited for managing wall thinning/FAC of the carbon steel feedwater ring component. The FAC is for pipe inside surface wall thinning. The staff is not aware that SG tube integrity program can manage the inside surface wall thinning due to FAC.

By letter, dated March 30, 2006, the applicant stated:

> For completeness, therefore, the following changes are made to the new line item added for the feedwater ring in the November 18, 2005 letter:
>
> 1. In Table 3.1.2-4 on page 3-61, to the new line item 'Feedwater Ring', a third Aging Management Program, 'Flow Accelerated Corrosion' is hereby added. The associated NUREG 1801 Volume 2, Table 1, and Notes entries are 'IV.D1.3-a', '3.1.1-21', and 'A, 126', respectively.
>
> 2. On page 3-63, the new plant-specific note 126 is revised to read as follows, 'The Flow Accelerated Corrosion Program is credited for managing wall thinning/flow accelerated corrosion. The Steam Generator Tube Integrity Program and the Water Chemistry Program are credited for other loss of material mechanisms.'

Based on the addition of the Flow Accelerated Corrosion Program, the staff concludes that the applicant has met the criteria of SRP-LR Section 3.1.2.2.14. For those line items that apply to LRA Section 3.1.2.2.14, the staff determines that the application is consistent with the GALL Report and the applicant has demonstrated that the aging effects will be adequately managed so that the intended function(s) will be maintained consistent with the CLB during the period of extended operation, as required by 10 CFR 54.21(a)(3).

3.1.2.2.15 Quality Assurance for Aging Management of Nonsafety-Related Components

SER Section 3.0.4 provides the staff's evaluation of the applicant's QA program.

Conclusion. On the basis of its review, for component groups evaluated in the GALL Report for which the applicant claimed consistency with the GALL Report and for which further evaluation is recommended, the staff determines that the applicant has adequately addressed those issues. The staff finds that the applicant has demonstrated that the aging effects will be adequately managed so that the intended function(s) will be maintained consistent with the CLB for the period of extended operation, as required by 10 CFR 54.21(a)(3).

3.1.2.3 AMR Results That Are Not Consistent with or Not Addressed in the GALL Report

Summary of Technical Information in the Application. In LRA Tables 3.1.2-1 through 3.1.2-4, the staff reviewed additional details of AMR results for material, environment, AERM, and AMP combinations not consistent with or not addressed in the GALL Report.

In LRA Tables 3.1.2-1 through 3.1.2-4, the applicant indicated, via Notes F through J, that the combination of component type, material, environment, and AERM does not correspond to a line item in the GALL Report. The applicant provided further information concerning how the aging effects will be managed. Specifically, Note F indicates that the material for the AMR line item component is not evaluated in the GALL Report. Note G indicates that the environment for the AMR line item component and material is not evaluated in the GALL Report. Note H indicates that the aging effect for the AMR line item component, material, and environment combination is not evaluated in the GALL Report. Note I indicates that the aging effect identified in the GALL Report for the line item component, material, and environment combination is not applicable. Note J indicates that neither the component nor the material and environment combination for the line item is evaluated in the GALL Report.

Staff Evaluation. For component type, material, and environment combinations not evaluated in the GALL Report, the staff reviewed the applicant's evaluation to determine whether the applicant had demonstrated that the aging effects will be adequately managed so that the intended function(s) will be maintained consistent with the CLB during the period of extended operation. The staff's evaluation is discussed in the following sections.

3.1.2.3.1 Reactor Coolant System – Primary Coolant System – Summary of Aging Management Evaluation – LRA Table 3.1.2-1

The staff reviewed LRA Table 3.1.2-1, which summarizes the results of AMR evaluations for the PCS component groups.

The applicant proposed to manage loss of material from low-alloy steel bolting and fasteners exposed to plant indoor air environment with the Bolting Integrity Program.

The Bolting Integrity Program was reviewed by the staff and is addressed separately in SER Section 3.0.3.2.1. The staff determined that Bolting Integrity Program management of loss of material from low-alloy steel bolting and fasteners exposed to the PCS plant indoor air environment is acceptable.

In LRA Table 3.1.2-1, the applicant proposed the One-Time Inspection Program to manage loss of material from the epoxy-coated, carbon-steel pressurizer quench tank exposed to treated water. The staff's evaluation of this AMR, as documented in SER Section 3.1.2.1.5, is applicable here.

On the basis of its review, as discussed above, the staff concludes that the applicant has demonstrated that the aging effects associated with the components will be adequately managed so that the intended function(s) will be maintained consistent with the CLB for the period of extended operation, as required by 10 CFR 54.21(a)(3).

3.1.2.3.2 Reactor Coolant System – Reactor Vessel – Summary of Aging Management Evaluation – LRA Table 3.1.2-2

The staff reviewed LRA Table 3.1.2-2, which summarizes the results of AMR evaluations for the reactor vessel component groups.

The applicant proposed the ASME Section XI IWB, IWC, IWD, IWF Inservice Inspection Program to manage loss of nickel-based alloy due to fretting of the reactor vessel core stabilizer lugs in a treated water environment

The staff's review of the ASME Section XI IWB, IWC, IWD, IWF Inservice Inspection Program is documented in SER Section 3.0.3.1.2. The reactor vessel interior and core support structure are visually examined (VT-3) in accordance with ASME Code Section XI ISI. As identified in IWA-2213 of ASME Code Section XI, VT-3 examinations detect such discontinuities and imperfections as corrosion, wear, or erosion. The staff concludes that the applicant had appropriately addressed the aging effect of loss of material due to fretting for the reactor vessel core stabilizer lugs by crediting ASME Section XI IWB.

On the basis of its review, as discussed above, the staff concludes that the applicant has demonstrated that the aging effects associated with the components will be adequately managed so that the intended function(s) will be maintained consistent with the CLB for the period of extended operation, as required by 10 CFR 54.21(a)(3).

3.1.2.3.3 Reactor Coolant System – Reactor Vessel Internals – Summary of Aging Management Evaluation – LRA Table 3.1.2-3

The staff reviewed LRA Table 3.1.2-3, which summarizes the results of AMR evaluations for the RVI component groups.

The applicant proposed the Reactor Vessel Internals Inspection Program to manage reduction in fracture toughness of the stainless steel instrument sleeve of the upper guide structure.

The GALL Report recommends programs consistent with GALL AMPs XI.M16, "PWR Vessel Internals," and XI.M2, "Water Chemistry," to manage RVI in the fuel zone region.

By letter dated August 27, 2005, the applicant stated that its Water Chemistry Program would also manage aging of the instrument sleeve.

On the basis of its review of the applicant's programs, aging effects, plant-specific, and industry operating experience, as discussed above, the staff concludes that, with this change, the applicant has demonstrated that the aging effects associated with the components will be adequately managed so that the intended function(s) will be maintained consistent with the CLB for the period of extended operation, as required by 10 CFR 54.21(a)(3).

3.1.2.3.4 Reactor Coolant System – Replacement Steam Generators – Summary of Aging Management Evaluation – LRA Table 3.1.2-4

The staff reviewed LRA Table 3.1.2-4, which summarizes the results of AMR evaluations for the replacement SGs component groups.

The applicant proposed only the Water Chemistry Program to manage loss of material from the low-alloy steel tube bundle wrapper exposed to treated water.

By letter dated September 2, 2005, the applicant stated that the effectiveness of its Water Chemistry Program would be verified by the Steam Generator Tube Integrity Program.

The staff's reviews of the Water Chemistry and Steam Generator Tube Integrity Programs are documented in SER Sections 3.0.3.1.10 and 3.0.3.1.7, respectively. The Water Chemistry Program is credited for managing such aging effects as loss of material due to general, pitting and crevice corrosion, cracking due to SCC, and SG tube degradation caused by denting, IGA, and ODSCC by controlling the environment to which internal surfaces of systems and components are exposed. The Steam Generator Tube Integrity Program provides opportunities to visually examine SG internals to demonstrate adequate management of aging effects.

The staff concludes that the applicant had appropriately addressed management of loss of material from the low-alloy steel tube bundle wrapper exposed to treated water in the replacement SG system.

On the basis of its review, as discussed above, the staff concludes that the applicant has demonstrated that the aging effects associated with the components will be adequately managed so that the intended function(s) will be maintained consistent with the CLB for the period of extended operation, as required by 10 CFR 54.21(a)(3).

Conclusion. On the basis of its review, the staff finds that the applicant has appropriately evaluated AMR results involving material, environment, AERMs, and AMP combinations not evaluated in the GALL Report. The staff finds that the applicant has demonstrated that the aging effects will be adequately managed so that the intended functions will be maintained consistent with the CLB for the period of extended operation, as required by 10 CFR 54.21(a)(3).

3.1.3 Conclusion

The staff concludes that the applicant has demonstrated that the aging effects associated with the RCS components will be adequately managed so that the intended function(s) will be maintained consistent with the CLB for the period of extended operation, as required by 10 CFR 54.21(a)(3).

The staff also reviewed the applicable FSAR supplement program summaries and concludes that they adequately describe the AMPs credited for managing aging of the RCS, as required by 10 CFR 54.21(d).

3.2 Aging Management of Engineered Safety Features (ESF)

This section of the SER documents the staff's review of the applicant's AMR results for the ESF components and component groups of the ESS.

3.2.1 Summary of Technical Information in the Application

In LRA Section 3.2, the applicant provided AMR results for the ESF components and component groups. In LRA Table 3.2.1, "Summary of Aging Management Evaluations in Chapter V of NUREG-1801 for Engineered Safety Features," the applicant provided a summary comparison of its AMRs to those evaluated in the GALL Report for ESF components and component groups.

The applicant's AMRs incorporated applicable operating experience in the determination of AERMs. These reviews included evaluation of plant-specific and industry operating experience. The plant-specific evaluation included reviews of condition reports and discussions with appropriate site personnel to identify AERMs. The applicant's review of industry operating experience included a review of the GALL Report and issues identified since the issuance of the GALL Report.

3.2.2 Staff Evaluation

The staff reviewed LRA Section 3.2 to determine whether the applicant had provided sufficient information to demonstrate that the aging effects for the ESF components within the scope of license renewal and subject to an AMR will be adequately managed so that the intended function(s) will be maintained consistent with the CLB for the period of extended operation, as required by 10 CFR 54.21(a)(3).

The staff conducted an onsite audit of AMRs to confirm the applicant's claim that certain identified AMRs were consistent with the GALL Report. The staff did not repeat its review of the matters described in the GALL Report; however, the staff did verify that the material presented in the LRA was applicable and that the applicant had identified the appropriate GALL Report AMRs. The staff's evaluations of the AMPs are documented in SER Section 3.0.3. Details of the staff's audit evaluation are documented in the Audit and Review Report and summarized in SER Section 3.2.2.1.

In the onsite audit, the staff also selected AMRs consistent with the GALL Report for which further evaluation is recommended. The staff confirmed that the applicant's further evaluations were consistent with the acceptance criteria in SRP-LR Section 3.2.2.2. The staff's audit evaluations are documented in the Audit and Review Report and summarized in SER Section 3.2.2.2.

In the onsite audit, the staff also conducted a technical review of those remaining AMRs not consistent with or not addressed in the GALL Report. The audit and technical review evaluated whether all plausible aging effects were identified and whether the aging effects listed were appropriate for the combination of materials and environments specified. The staff's audit evaluations documented in the Audit and Review Report are summarized in SER Section 3.2.2.3. The staff's evaluation of its technical review is also documented in SER Section 3.2.2.3.

Finally, the staff reviewed the AMP summary descriptions in the FSAR supplement to ensure that they adequately describe the programs credited with managing or monitoring ESF component aging.

SER Table 3.2-1 below summarizes the staff's evaluation of components, aging effects/mechanisms, and AMPs listed in LRA Section 3.2 and addressed in the GALL Report.

Table 3.2-1 Staff Evaluation for Engineered Safety Features Systems Components in the GALL Report

Component Group	Aging Effect/ Mechanism	AMP in GALL Report	AMP in LRA	Staff Evaluation
Piping, fittings, and valves in emergency core cooling system (Item Number 3.2.1-01)	Cumulative fatigue damage	TLAA, evaluated in accordance with 10 CFR 54.21(c)	TLAA	This TLAA is evaluated in SER Section 4.3, "Metal Fatigue"

Component Group	Aging Effect/ Mechanism	AMP in GALL Report	AMP in LRA	Staff Evaluation
Components in containment spray (PWR only), standby gas treatment (BWR only), containment isolation, and emergency core cooling systems (Item Number 3.2.1-03)	Loss of material due to general corrosion	Plant-specific	Containment Leakage Testing Program (B2.1.8), One-Time Inspection Program (B2.1.13), System Monitoring Program (B2.1.20), Water Chemistry Program (B2.1.21)	Consistent with GALL, which recommends further evaluation (See SER Section 3.2.2.2.2)
Components in containment spray (PWR only), standby gas treatment (BWR only), containment isolation, and emergency core cooling systems (Item Number 3.2.1-05)	Loss of material due to pitting and crevice corrosion	Plant-specific	Containment Leakage Testing Program (B2.1.8), One-Time Inspection Program (B2.1.13), Water Chemistry Program (B2.1.21)	Consistent with GALL, which recommends further evaluation (See SER Section 3.2.2.2.3)
Containment isolation valves and associated piping (Item Number 3.2.1-06)	Loss of material due to microbiologically influenced corrosion	Plant-specific	Containment Leakage Testing Program (B2.1.8), One-Time Inspection Program (B2.1.13), Water Chemistry Program (B2.1.21)	Consistent with GALL, which recommends further evaluation (See SER Section 3.2.2.2.4)
High pressure safety injection (charging) pump miniflow orifice (Item Number 3.2.1-08)	Loss of material due to erosion	Plant-specific	None	Not applicable (See SER Section 3.2.2.2.6)
External surface of carbon steel components (Item Number 3.2.1-10)	Loss of material due to general corrosion	Plant-specific	System Monitoring Program (B2.1.20)	Consistent with GALL, which recommends further evaluation (See SER Section 3.2.2.2.2)
Piping and fittings of CASS in emergency core cooling system (Item Number 3.2.1-11)	Loss of fracture toughness due to thermal aging embrittlement	Thermal aging embrittlement of CASS	ASME Section XI IWB, IWC, IWD, IWF Inservice Inspection Program (B2.1.2)	Consistent with GALL, which recommends no further evaluation (See SER Section 3.2.2.1.1)
Components serviced by open-cycle cooling system (Item Number 3.2.1-12)	Local loss of material due to corrosion and/or buildup of deposit due to biofouling	Open-cycle cooling water system	None	Not applicable (no ESF system components serviced by open-cycle cooling water system)

Component Group	Aging Effect/ Mechanism	AMP in GALL Report	AMP in LRA	Staff Evaluation
Components serviced by closed-cycle cooling system (Item Number 3.2.1-13)	Loss of material due to general, pitting, and crevice corrosion	Closed-cycle cooling water system	Closed Cycle Cooling Water Program (B2.1.6), One-Time Inspection Program (B2.1.13), Water Chemistry Program (B2.1.21)	Consistent with GALL, which recommends no further evaluation (See SER Section 3.2.2.1.2)
Pumps, valves, piping, and fittings in containment spray and emergency core cooling systems (Item Number 3.2.1-15)	Crack initiation and growth due to SCC	Water chemistry	Water Chemistry Program (B2.1.21), One-Time Inspection Program (B2.1.13)	Consistent with GALL, which recommends no further evaluation (See SER Section 3.2.2.1.3)
Carbon steel components (Item Number 3.2.1-17)	Loss of material due to boric acid corrosion	Boric acid corrosion	Boric Acid Corrosion Program (B2.1.4)	Consistent with GALL, which recommends no further evaluation (See SER Section 3.2.2.1)
Closure bolting in high pressure or high temperature systems (Item Number 3.2.1-18)	Loss of material due to general corrosion, loss of preload due to stress relaxation, and crack initiation and growth due to cyclic loading or SCC	Bolting integrity	Bolting Integrity Program (B2.1.3)	Consistent with GALL, which recommends no further evaluation (See SER Section 3.2.2.1.4)

The staff's review of the ESF component groups followed several approaches. One approach, documented in SER Section 3.2.2.1, discusses the staff's review of AMR results for components the applicant indicated are consistent with the GALL Report and require no further evaluation. Another approach, documented in SER Section 3.2.2.2, discusses the staff's review of AMR results for components the applicant indicated are consistent with the GALL Report and for which further evaluation is recommended. A third approach, documented in SER Section 3.2.2.3, discusses the staff's review of AMR results for components the applicant indicated are not consistent with or not addressed in the GALL Report. The staff's review of AMPs credited to manage or monitor aging effects of the ESF components is documented in SER Section 3.0.3.

3.2.2.1 AMR Results That Are Consistent with the GALL Report

Summary of Technical Information in the Application. In LRA Section 3.2.2.1, the applicant identified the materials, environments, and AERMs. The applicant identified the following programs that manage the aging effects of ESF components:

- ASME Section XI IWB, IWC, IWD, IWF Inservice Inspection Program
- Bolting Integrity Program
- Boric Acid Corrosion Program

- Closed Cycle Cooling Water Program
- One-Time Inspection Program
- System Monitoring Program
- Water Chemistry Program

Staff Evaluation. In LRA Table 3.2.2-1, the applicant summarized AMRs for the ESF components and identified which AMRs it considered consistent with the GALL Report.

For component groups evaluated in the GALL Report for which the applicant had claimed consistency and for which the GALL Report does not recommend further evaluation, the staff performed an audit and review to determine whether the plant-specific components in these GALL Report component groups were bounded by the GALL Report evaluation.

The applicant provided a note for each AMR line item. The notes describe how the information in the tables aligns with the information in the GALL Report. The staff audited those AMRs with Notes A through E, which indicate how the AMR was consistent with the GALL Report.

Note A indicates that the AMR line item is consistent with the GALL Report for component, material, environment, and aging effect. In addition, the AMP is consistent with the GALL Report AMP. The staff audited these line items to verify consistency with the GALL Report and the validity of the AMR for the site-specific conditions.

Note B indicates that the AMR line item is consistent with the GALL Report for component, material, environment, and aging effect. In addition, the AMP takes some exceptions to the AMP identified in the GALL Report. The staff audited these line items to verify consistency with the GALL Report. The staff verified that the identified exceptions to the GALL Report AMPs had been reviewed and accepted by the staff. The staff also determined whether the AMP identified by the applicant was consistent with the AMP identified in the GALL Report and whether the AMR was valid for the site-specific conditions.

Note C indicates that the component for the AMR line item, although different from, is consistent with the GALL Report for material, environment, and aging effect. In addition, the AMP is consistent with the AMP identified by the GALL Report. This note indicates that the applicant was unable to find a listing of some system components in the GALL Report; however, the applicant identified a different component in the GALL Report that had the same material, environment, aging effect, and AMP as the component under review. The staff audited these line items to verify consistency with the GALL Report. The staff also determined whether the AMR line item of the different component applied to the component under review and whether the AMR was valid for the site-specific conditions.

Note D indicates that the component for the AMR line item, although different from, is consistent with the GALL Report for material, environment, and aging effect. In addition, the AMP takes some exceptions to the AMP identified in the GALL Report. The staff audited these line items to verify consistency with the GALL Report. The staff verified whether the AMR line item of the different component was applicable to the component under review. The staff verified whether the exceptions to the GALL Report AMPs had been reviewed and accepted by the staff. The staff also determined whether the AMP identified by the applicant was consistent with the AMP identified in the GALL Report and whether the AMR was valid for the site-specific conditions.

Note E indicates that the AMR line item is consistent with the GALL Report for material, environment, and aging effect, but a different AMP is credited. The staff audited these line items to verify consistency with the GALL Report. The staff also determined whether the identified AMP would manage the aging effect consistent with the AMP identified in the GALL Report and whether the AMR was valid for the site-specific conditions.

In LRA Table 3.2.2-1, the applicant provided a summary of the AMR results for component types associated with the ESS. The summary information for each component type included intended function, material, environment, AERM, AMPs, GALL Report, Volume 2, item, cross reference to LRA Table 3.2.1, and generic and plant-specific notes related to consistency with the GALL Report.

By letter dated May 5, 2005, the applicant provided additions to LRA Table 3.2.2-1. As documented in the Audit and Review Report, numerous questions were developed by the staff, prompted by the manner in which component types are grouped in the LRA table rather than the manner in which aging effects are managed. To facilitate the staff's further review, the applicant developed a new LRA Table 3.2.2-1; the review of AMRs on the revised table is described in this SER. By letter dated August 27, 2005, the applicant submitted the revised table. The staff used the revised table for its audit and review of the ESF systems.

The staff's review of LRA Section 3.2 identified an area in which additional information was necessary to complete the review of the applicant's AMR results. The applicant responded to the staff's RAI as discussed below.

In RAI-3.2.1-1, dated June 3, 2005, the staff stated that the Flow Accelerated Corrosion Program is not listed as an AMP in LRA Sections 3.2.1 and 3.2.3. For license renewal, such a program is typically necessary to manage FAC effects. Therefore, the staff requested the applicant to verify whether it intended to credit this AMP in these LRA sections.

In its response, by letter dated July 1, 2005, the applicant stated that it does not credit the Flow Accelerated Corrosion Program in LRA Sections 3.2.2.1 and 3.2.2.2.6 because there are no FAC-susceptible material and environment combinations in those ESF components. Most ESF components in contact with fluid are stainless steel, which is FAC-resistant material. Further, except for some CASS valves, the ESF components were evaluated in their normal standby static condition of 120 °F.

Based on its review, the staff finds the applicant's response to RAI-3.2.1-1 acceptable because the ESS lacks FAC-susceptible material and environment combinations. Therefore, the staff's concern described in RAI-3.2.1-1 is resolved.

The staff did not repeat its review of the matters described in the GALL Report; however, the staff did verify that the material presented in the LRA was applicable and that the applicant had identified the appropriate GALL Report AMRs. The staff's evaluation is discussed below.

3.2.2.1.1 Loss of Fracture Toughness Due to Thermal Aging Embrittlement of CASS

In the discussion column of LRA Table 3.2.1, Item 3.2.1-11, the applicant stated that the ASME Section XI IWB, IWC, IWD, IWF Inservice Inspection Program manages reduction in fracture toughness due to thermal aging embrittlement of CASS components in the ESS.

The GALL Report recommends the use of a program consistent with GALL AMP XI.M12, "Thermal Aging Embrittlement of Cast Austenitic Stainless Steel (CASS)," which states that for pump casings and valve bodies, the recommendations of GALL AMP XI.M1, "ASME Section XI Inservice Inspection, Subsections IWB, IWC, and IWD," are sufficient to manage this aging effect.

As documented in the Audit and Review Report, the staff noted that, although the applicant was not using GALL AMP XI.M12, it was applying an AMP recommended by the GALL Report for pump casings and valve bodies.

The staff's evaluation of the applicant's ASME Section XI IWB, IWC, IWD, IWF Inservice Inspection Program is documented in SER Section 3.0.3.1.2. The staff finds management by this program of reduction in fracture toughness of valve bodies due to thermal aging embrittlement consistent with the GALL Report and, therefore, acceptable.

3.2.2.1.2 Loss of Material Due to General, Pitting, and Crevice Corrosion

In the discussion column of LRA Table 3.2.1, Item 3.2.1-13, the applicant stated that the Closed-Cycle Cooling Water and Water Chemistry Programs are used to manage loss of material due to general, pitting, and crevice corrosion from components served by the closed-cycle cooling system. In LRA Table 3.2.2-1, the applicant's Water Chemistry Program manages loss of material from ESF components exposed to treated water other than closed-cycle cooling water. For the cast iron containment spray pump HX shell, this program is augmented with the One-Time Inspection Program.

The GALL Report recommends the use of a program consistent with GALL AMP XI.M21, "Closed-Cycle Cooling Water System."

The staff's reviews of the applicant's Closed-Cycle Cooling Water, Water Chemistry, and One-Time Inspection Programs are documented in SER Sections 3.0.3.2.3, 3.0.3.1.10, and 3.0.3.2.6, respectively.

The staff finds management of this aging effect by the Closed-Cycle Cooling Water Program consistent with the GALL Report and, therefore, acceptable. The staff also concludes that the Water Chemistry Program augmented with the One-Time Inspection Program (for carbon steel and aluminum components as well as such stainless steel components as the applicant chooses to manage with this combination of AMPs) to confirm the effectiveness of the Water Chemistry Program is recommended by the GALL Report as an acceptable method for managing loss of material exposed to treated water other than closed-cycle cooling water. Cast iron subject to loss of material due to selective leaching was managed through the One-Time Inspection Program, as recommended by the GALL Report. On these bases, the staff concludes that the applicant had appropriately addressed the aging effect of loss of material for those components.

3.2.2.1.3 Crack Initiation and Growth Due to Stress Corrosion Cracking

In the discussion column of LRA Table 3.2.1, Item 3.2.1-15, the applicant stated that the Water Chemistry Program manages crack initiation and growth due to SCC. In LRA Table 3.2.2-1, the applicant supplemented this program with the One-Time Inspection Program for the aluminum safety injection and refueling water tank, stainless steel safety injection and refueling water tank, HX tubes, piping, fittings, and valves of the ESS.

The GALL Report recommends the use of a program consistent with GALL AMP XI.M2, "Water Chemistry."

The staff finds that the applicant's Water Chemistry Program augmented with its One-Time Inspection Program (for aluminum as well as stainless steel components the applicant chooses to manage with this combination of AMPs) exceeds the GALL Report recommendation and, therefore, forms an acceptable method for management of crack initiation and growth in the ESS.

3.2.2.1.4 Loss of Preload Due to Stress Relaxation

In the discussion column of LRA Table 3.2.1, Item 3.2.1-18, the applicant stated that the Bolting Integrity Program manages loss of material due to general corrosion, loss of preload due to stress relaxation, and crack initiation and growth due to cyclic loading and/or SCC for closure bolting in high-pressure or high-temperature systems.

The GALL Report recommends the use of a program consistent with GALL AMP XI.M18, "Bolting Integrity," to manage loss of material due to general corrosion and crack initiation and growth due to cyclic loading and/or SCC. The application of GALL AMP XI.M18 to loss of preload is not mentioned in GALL Report Table 2, "Summary of Aging Management Programs for the Engineered Safety Features Evaluated in Chapter V of the GALL Report." However, in LRA Table 3.2.2-1, the applicant applied the Bolting Integrity Program to the management of all these aging effects for all closure bolting in the ESS.

The staff finds that loss of material, loss of preload, and crack initiation and growth will be adequately managed so that the intended functions will be maintained during the period of extended operation, as required by 10 CFR 54.21(a)(3).

Conclusion. The staff evaluated the applicant's claim of consistency with the GALL Report. The staff also reviewed information pertaining to the applicant's consideration of recent operating experience and proposals for managing the associated aging effects. On the basis of its review, the staff concludes that the AMR results, which the applicant claimed to be consistent with the GALL Report, are consistent with the GALL Report AMRs. Therefore, the staff concludes that the applicant has demonstrated that the aging effects for these components will be adequately managed so that their intended function(s) will be maintained consistent with the CLB for the period of extended operation, as required by 10 CFR 54.21(a)(3).

3.2.2.2 AMR Results That Are Consistent with the GALL Report, for Which Further Evaluation Is Recommended

Summary of Technical Information in the Application. In LRA Section 3.2.2.2, the applicant provided further evaluation of aging management as recommended by the GALL Report for the ESF components. The applicant provided information concerning how it will manage the following aging effects:

- cumulative fatigue damage

- loss of material due to general corrosion
- local loss of material due to pitting and crevice corrosion
- local loss of material due to MIC
- changes in properties due to elastomer degradation
- local loss of material due to erosion
- buildup of deposits due to corrosion

Staff Evaluation. For component groups evaluated in the GALL Report for which the applicant had claimed consistency with the GALL Report and for which further evaluation is recommended, the staff audited and reviewed the applicant's evaluations to determine whether they adequately address those issues. In addition, the staff reviewed the applicant's further evaluations against the criteria in SRP-LR Section 3.2.2.2. Details of the staff's audit are documented in the Audit and Review Report. The staff's evaluation of the aging effects is discussed in the following sections.

3.2.2.2.1 Cumulative Fatigue Damage

In LRA Section 3.2.2.2.1, the applicant stated that fatigue is a TLAA, as defined in 10 CFR 54.3. Applicants must evaluate TLAAs in accordance with 10 CFR 54.21(c)(1). SER Section 4.3 documents the staff's review of the applicant's evaluation of this TLAA.

3.2.2.2.2 Loss of Material Due to General Corrosion

The staff reviewed LRA Sections 3.2.2.2.2.1 and 3.2.2.2.2.2 against the criteria in SRP-LR Section 3.2.2.2.2.

In LRA Section 3.2.2.2.2.1, the applicant stated that this issue applies to BWRs only.

The GALL Report states that this item applies only to BWRs. For this reason, the staff found that this item need not be addressed.

In LRA Section 3.2.2.2.2.2, the applicant addressed loss of material of components in the containment spray, containment isolation, and emergency core cooling systems due to general corrosion. The LRA states that this aging effect on carbon steel or cast iron in air does not affect stainless steel components in containment spray and emergency core cooling systems. The external surfaces of carbon steel and cast iron components in the containment spray and emergency core cooling systems are susceptible to general corrosion in an air environment. The System Monitoring Program is credited with managing this aging effect.

The applicant also stated that containment isolation components are addressed with their individual systems. Programs credited for aging management of general corrosion are identified in each Table 2 for the system containing the penetration. The aging effect/mechanism for containment isolation in aqueous systems is managed by the Water Chemistry Program and supplemented by the One-Time Inspection Program in gaseous systems, containment sump level instrumentation, and radwaste systems.

SRP-LR Section 3.2.2.2.2.2 states that loss of material due to general corrosion could occur in the containment spray and spray nozzle components, containment isolation valves and associated piping, and on the external surfaces of carbon steel components. The GALL Report recommends further evaluation on a plant-specific basis to ensure adequate management of the aging effect.

The staff reviewed the applicant's further evaluation and finds that the applicant has selected appropriate plant-specific AMPs to manage the loss of material due to general corrosion.

Based on the programs identified above, the staff concludes that the applicant has met the criteria of SRP-LR Section 3.2.2.2.2. For those line items that apply to LRA Section 3.2.2.2.2, the staff determines that the application is consistent with the GALL Report and the applicant has demonstrated that the aging effects will be adequately managed so that the intended function(s) will be maintained consistent with the CLB during the period of extended operation, as required by 10 CFR 54.21(a)(3).

3.2.2.2.3 Local Loss of Material Due to Pitting and Crevice Corrosion

The staff reviewed LRA Sections 3.2.2.2.3.1 and 3.2.2.2.3.2 against the criteria in SRP-LR Section 3.2.2.2.3.

In LRA Section 3.2.2.2.3.1, the applicant stated that this issue applies to BWRs only.

The GALL Report states that this item applies only to BWRs. For this reason, staff found that this item need not be addressed.

In LRA Section 3.2.2.2.3.2, the applicant addressed loss of material of components in containment spray, containment isolation, and emergency core cooling systems due to pitting and crevice corrosion. The applicant stated that the containment spray and emergency core cooling system components are stainless steel, susceptible to pitting and crevice corrosion, and managed by the Water Chemistry Program supplemented by the One-Time Inspection Program. The stainless steel containment isolation components are exposed to borated water, susceptible to pitting and crevice corrosion, and managed by the Water Chemistry Program supplemented by the One-Time Inspection Program. In addition, the applicant stated that all containment isolation valves and associated piping are currently tested on a set frequency by the Containment Leakage Testing Program and the testing will be continued throughout the period of extended operation. The safety injection and refueling water tank bottom is located on the auxiliary building roof, not buried. The bottom edge is sealed around its circumference from exposure to the weather. Supply piping enters and exits the bottom of the tank through the roof of the auxiliary building. This arrangement exposes the bottom surface of the tank to a plant indoor air environment. Loss of material due to crevice and pitting is not a potential aging effect/mechanism for aluminum in air.

SRP-LR Section 3.2.2.2.3 states that local loss of material from pitting and crevice corrosion could occur in the PWR containment spray components, containment isolation valves and associated piping, and the buried portion of the refueling water tank external surface.

Based on the programs identified above, the staff concludes that the applicant has met the criteria of SRP-LR Section 3.2.2.2.3. For those line items that apply to LRA Section 3.2.2.2.3, the staff determines that the application is consistent with the GALL Report and the applicant has demonstrated that the aging effects will be adequately managed so that the intended function(s) will be maintained consistent with the CLB during the period of extended operation, as required by 10 CFR 54.21(a)(3).

3.2.2.2.4 Local Loss of Material Due to Microbiologically Influenced Corrosion

The staff reviewed LRA Section 3.2.2.2.4 against the criteria in SRP-LR Section 3.2.2.2.4.

In LRA Section 3.2.2.2.4, the applicant addressed loss of material of containment isolation valves and associated piping due to MIC. Loss of material due to MIC is considered a potential aging effect/mechanism that requires management, even though the component has no potential source of MIC contamination in treated water. The treated water environment is effectively controlled by the existing Water Chemistry Program.

In the LRA, the applicant stated that containment isolation valve bodies and connecting piping are addressed with their individual systems. Programs credited for aging management are identified in each Table 2 for the system containing the penetration. All containment isolation valves and associated piping are currently tested on a set frequency by the Containment Leakage Testing Program and the testing will be continued in the period of extended operation.

The applicant also stated that the Water Chemistry Program is credited for managing this aging effect/mechanism on containment isolation valves and associated piping in aqueous systems. This management is supplemented by the One-Time Inspection Program to determine whether the aging effect/mechanism exists and, if so, how rapidly it progresses.

In addition, in the LRA, the applicant stated that the high temperature (> 210 °F) of the main steam, main feedwater, and steam generator blowdown containment isolation components precludes MIC. Also, the One-Time Inspection Program is credited for managing this aging effect/mechanism in gaseous systems, containment sump level instrumentation, and radwaste system containment isolation components.

SRP-LR Section 3.2.2.2.4 states that local loss of material due to MIC could occur in containment isolation valves and associated piping in systems not addressed in other chapters of the GALL Report. The GALL Report recommends further evaluation to ensure adequate management of the aging effect. Acceptance criteria are described in SRP-LR Appendix A.1.

The staff reviewed the applicant's further evaluation and finds that the applicant has selected appropriate plant-specific AMPs to manage the loss of material due to MIC.

Based on the programs identified above, the staff concludes that the applicant has met the criteria of SRP-LR Section 3.2.2.2.4. For those line items that apply to LRA Section 3.2.2.2.4, the staff determines that the application is consistent with the GALL Report and the applicant has demonstrated that the aging effects will be adequately managed so that the intended function(s) will be maintained consistent with the CLB during the period of extended operation, as required by 10 CFR 54.21(a)(3).

3.2.2.2.5 Changes in Properties Due to Elastomer Degradation

The staff reviewed LRA Section 3.2.2.2.5 against the criteria specified in SRP-LR Section 3.2.2.2.5.

In LRA Section 3.2.2.2.5, the applicant stated that this aging effect applies to BWRs only.

SRP-LR Section 3.2.2.2.5 states that changes in properties due to elastomer degradation could occur in seals associated with the standby gas treatment system ductwork and filters. SRP-LR Table 3.2-1 states that further evaluation of this aging effect applies to BWR plants only.

The staff finds this aging effect not applicable to PNP.

3.2.2.2.6 Local Loss of Material Due to Erosion

The staff reviewed LRA Section 3.2.2.2.6 against the criteria in SRP-LR Section 3.2.2.2.6.

In LRA Section 3.2.2.2.6, the applicant addressed loss of material of the HPSI (charging) pump miniflow orifice due to erosion. In the LRA, the applicant stated that the HPSI pumps are not used for normal charging and loss of material due to erosion of miniflow orifices is not applicable.

SRP-LR Section 3.2.2.2.6 states that local loss of material due to erosion could occur in the HPSI pump miniflow orifice. This aging effect/mechanism applies only to charging pumps normally used in the chemical and volume control systems. The GALL Report recommends further evaluation to ensure adequate management of the aging effect. Acceptance criteria are described in SRP-LR Appendix A.1.

On the basis that the HPSI pump miniflow orifice is only subject to infrequent flow, the staff found that loss of material due to erosion of the miniflow orifice need not be managed.

3.2.2.2.7 Buildup of Deposits Due to Corrosion

The staff reviewed LRA Section 3.2.2.2.7 against the criteria specified in SRP-LR Section 3.2.2.2.7.

In LRA Section 3.2.2.2.7, the applicant stated that this aging effect applies to BWRs only.

SRP-LR Section 3.2.2.2.7 states that plugging of components due to general corrosion could occur in the spray nozzles and flow orifices of the drywell and suppression chamber spray system. SRP-LR Table 3.2-1 states that further evaluation for this aging effect is applicable to BWR plants only.

The staff finds this aging effect not applicable to PNP.

3.2.2.2.8 Quality Assurance for Aging Management of Nonsafety-Related Components

SER Section 3.0.4 provides the staff's evaluation of the applicant's QA program.

Conclusion. On the basis of its review, for component groups evaluated in the GALL Report for which the applicant claimed consistency with the GALL Report and for which further evaluation is recommended, the staff determines that the applicant has adequately addressed the issues that were further evaluated. The staff finds that the applicant has demonstrated that the aging effects will be adequately managed so that the intended function(s) will be maintained consistent with the CLB for the period of extended operation, as required by 10 CFR 54.21(a)(3).

3.2.2.3 AMR Results That Are Not Consistent with or Not Addressed in the GALL Report

Summary of Technical Information in the Application. In LRA Table 3.2.2-1, the staff reviewed additional details concerning AMR results for material, environment, AERM, and AMP combinations not consistent with or not addressed in the GALL Report.

In LRA Table 3.2.2-1, the applicant indicated, via Notes F through J, that the combination of component type, material, environment, and AERM does not correspond to a line item in the GALL Report. The applicant provided further information concerning how the aging effects will be managed. Specifically, Note F indicates that the material for the AMR line item component is not evaluated in the GALL Report. Note G indicates that the environment for the AMR line item component and material is not evaluated in the GALL Report. Note H indicates that the aging effect for the AMR line item component, material, and environment combination is not evaluated in the GALL Report. Note I indicates that the aging effect identified in the GALL Report for the line item component, material, and environment combination is not applicable. Note J indicates that neither the component nor the material and environment combination for the line item is evaluated in the GALL Report.

Staff Evaluation. For component type, material, and environment combinations not evaluated in the GALL Report, the staff reviewed the applicant's evaluation to determine whether the applicant had demonstrated that the aging effects will be adequately managed so that the intended function(s) will be maintained consistent with the CLB during the period of extended operation. The staff's evaluation is discussed in the following sections.

3.2.2.3.1 Engineered Safety Features – Engineering Safeguards System (ESS) – Summary of Aging Management Evaluation – LRA Table 3.2.2-1

The staff reviewed LRA Table 3.2.2-1, which summarizes the results of AMR evaluations for the ESS component groups.

In LRA Table 3.2.2-1, the applicant proposed the Closed-Cycle Cooling Water Program, or a combination of the Water Chemistry and One-Time Inspection Programs, to manage heat transfer degradation of stainless steel HX tubes and cooling coils of the ESS exposed to treated water.

The staff's reviews of the applicant's Closed-Cycle Cooling Water, Water Chemistry, and One-Time Inspection Programs are documented in SER Sections 3.0.3.2.3, 3.0.3.1.10, and 3.0.3.2.6, respectively. The Closed Cycle Cooling Water Program manages aging effects in closed cycle cooling water systems and includes maintenance of system corrosion inhibitor concentrations to minimize degradation and periodic or one-time testing and inspections to assess aging. The existing Water Chemistry Program minimizes aging effects by controlling the chemical species that cause or allow them. The program provides assurance of no elevated level of contaminants and, where applicable, that oxygen does not exist in the systems and components covered by the program, maintaining each component's ability to perform intended functions. The One-Time Inspection Program verifies the Water Chemistry Program's management of the aging effects of selected components. On these bases, the staff finds management of heat transfer degradation in the ESS acceptable.

In LRA Table 3.2.2-1, the applicant proposed the Water Chemistry Program to manage cracking of stainless steel primary coolant pump seal cooler coils exposed to treated water.

The staff's review of the applicant's Water Chemistry Program is documented in SER Section 3.0.3.1.10. The program manages such aging effects as cracking due to SCC by controlling the environment to which internal surfaces of systems and components are exposed. The staff found no evidence of the aging effect of cracking due to SCC of stainless steel components exposed to primary treated water. On this basis, the staff finds that the applicant has appropriately addressed this aging effect by crediting the Water Chemistry Program.

In LRA Table 3.2.2-1, the applicant proposed the Closed Cycle Cooling Water Program to manage cracking of stainless steel low-pressure safety injection pump coils and containment spray pump coils exposed to treated water.

The staff's review of the applicant's Closed Cycle Cooling Water Program is documented in SER Section 3.0.3.2.3. The program manages aging effects in closed cycle cooling water systems and includes (a) maintenance of system corrosion inhibitor concentrations to minimize degradation and (b) periodic or one-time testing and inspections to assess aging. On these bases, the staff finds management of crack initiation and growth in the ESS acceptable.

On the basis of its review, as discussed above, the staff concludes that the applicant has demonstrated that the aging effects associated with the components will be adequately managed so that the intended function(s) will be maintained consistent with the CLB for the period of extended operation, as required by 10 CFR 54.21(a)(3).

Conclusion. On the basis of its review, the staff finds that the applicant has appropriately evaluated AMR results involving material, environment, AERMs, and AMP combinations not evaluated in the GALL Report. The staff finds that the applicant has demonstrated that the aging effects will be adequately managed so that the intended functions will be maintained consistent with the CLB for the period of extended operation, as required by 10 CFR 54.21(a)(3).

3.2.3 Conclusion

The staff concludes that the applicant has demonstrated that the aging effects associated with the ESF components will be adequately managed so that the intended function(s) will be maintained consistent with the CLB for the period of extended operation, as required by 10 CFR 54.21(a)(3).

The staff also reviewed the applicable FSAR supplement program summaries and concludes that they adequately describe the AMPs credited for managing aging of the ESF, as required by 10 CFR 54.21(d).

3.3 Aging Management of Auxiliary Systems

This section of the SER documents the staff's review of the applicant's AMR results for the auxiliary systems components and component groups of the following:

- chemical addition system
- chemical and volume control system
- circulating water system
- component cooling water system
- compressed air system
- containment air recirculation and cooling system
- domestic water system
- emergency power system
- fire protection system
- fuel oil system
- heating, ventilation, and air conditioning system
- miscellaneous gas system
- radwaste system
- service water system
- shield cooling system
- spent fuel pool cooling system
- waste gas system

3.3.1 Summary of Technical Information in the Application

In LRA Section 3.3, the applicant provided AMR results for the auxiliary systems components and component groups. In LRA Table 3.3.1, "Summary of Aging Management Evaluations in Chapter VII of NUREG-1801 for Auxiliary Systems," the applicant provided a summary comparison of its AMRs to those evaluated in the GALL Report for auxiliary systems components and component groups.

The applicant's AMRs incorporated applicable operating experience in the determination of AERMs. These reviews included evaluation of plant-specific and industry operating experience. The plant-specific evaluation included reviews of condition reports and discussions with appropriate site personnel to identify AERMs. The applicant's review of industry operating experience included a review of the GALL Report and issues identified since the issuance of the GALL Report.

3.3.2 Staff Evaluation

The staff reviewed LRA Section 3.3 to determine whether the applicant had provided sufficient information to demonstrate that the aging effects for the auxiliary systems components within the scope of license renewal and subject to an AMR will be adequately managed so that the intended function(s) will be maintained consistent with the CLB for the period of extended operation, as required by 10 CFR 54.21(a)(3).

The staff conducted an onsite audit of AMRs to confirm the applicant's claim that certain identified AMRs were consistent with the GALL Report. The staff did not repeat its review of the matters described in the GALL Report; however, the staff did verify that the material presented in the LRA was applicable and that the applicant had identified the appropriate GALL Report AMRs. The staff's evaluations of the AMPs are documented in SER Section 3.0.3. Details of the staff's audit evaluation are documented in the Audit and Review Report and summarized in SER Section 3.3.2.1.

In the onsite audit, the staff also selected AMRs consistent with the GALL Report for which further evaluation is recommended. The staff confirmed that the applicant's further evaluations were consistent with the acceptance criteria in SRP-LR Section 3.3.2.2. The staff's audit evaluations are documented in the Audit and Review Report and summarized in SER Section 3.3.2.2.

In the onsite audit, the staff also conducted a technical review of the remaining AMRs not consistent with or not addressed in the GALL Report. The audit and technical review evaluated whether all plausible aging effects were identified and whether the aging effects listed were appropriate for the combination of materials and environments specified. The staff's audit evaluations documented in the Audit and Review Report are summarized in SER Section 3.3.2.3. The staff's evaluation of its technical review is also documented in SER Section 3.3.2.3.

Finally, the staff reviewed the AMP summary descriptions in the FSAR supplement to ensure that they adequately describe the programs credited with managing or monitoring auxiliary systems component aging.

SER Table 3.3-1 below summarizes the staff's evaluation of components, aging effects/mechanisms, and AMPs listed in LRA Section 3.3 and addressed in the GALL Report.

Table 3.3-1 Staff Evaluation for Auxiliary Systems Components in the GALL Report

Component Group	Aging Effect/ Mechanism	AMP in GALL Report	AMP in LRA	Staff Evaluation
Components in spent fuel pool cooling and cleanup (Item Number 3.3.1-01)	Loss of material due to general, pitting, and crevice corrosion	Water chemistry and one-time inspection	One-Time Inspection Program (B2.1.13), Water Chemistry Program (B2.1.21)	Consistent with GALL, which recommends further evaluation (See SER Section 3.3.2.2.1)
Linings in spent fuel pool cooling and cleanup system; seals and collars in ventilation systems (Item Number 3.3.1-02)	Hardening, cracking and loss of strength due to elastomer degradation; loss of material due to wear	Plant-specific	System Monitoring Program (B2.1.20), One-Time Inspection Program (B2.1.13)	Consistent with GALL, which recommends further evaluation (See SER Section 3.3.2.2.2)

Component Group	Aging Effect/ Mechanism	AMP in GALL Report	AMP in LRA	Staff Evaluation
Components in load handling, chemical and volume control system (PWR), and reactor water cleanup and shutdown cooling systems (older BWR) (Item Number 3.3.1-03)	Cumulative fatigue damage	TLAA, evaluated in accordance with 10 CFR 54.21(c)	TLAA	This TLAA is evaluated in SER Section 4.3, "Metal Fatigue"
Heat exchangers in reactor water cleanup system (BWR); high pressure pumps in chemical and volume control system (PWR) (Item Number 3.3.1-04)	Crack initiation and growth due to SCC or cracking	Plant-specific	None	Not applicable (See SER Section 3.3.2.2.4)
Components in ventilation systems, diesel fuel oil system, and emergency diesel generator systems; external surfaces of carbon steel components (Item Number 3.3.1-05)	Loss of material due to general, pitting, and crevice corrosion, and MIC	Plant-specific	Bolting Integrity Program (B2.1.3), Fire Protection Program (B2.1.10), One-Time Inspection Program (B2.1.13), System Monitoring Program (B2.1.20)	Consistent with GALL, which recommends further evaluation (See SER Section 3.3.2.2.5)
Components in reactor coolant pump oil collect system of fire protection (Item Number 3.3.1-06)	Loss of material due to galvanic, general, pitting, and crevice corrosion	One-time inspection	One-Time Inspection Program (B2.1.13)	Consistent with GALL, which recommends further evaluation (See SER Section 3.3.2.2.6)
Diesel fuel oil tanks in diesel fuel oil system and emergency diesel generator system (Item Number 3.3.1-07)	Loss of material due to general, pitting, and crevice corrosion, MIC, and biofouling	Fuel oil chemistry and one-time inspection	Diesel Fuel Monitoring and Storage Program (B2.1.9), One-Time Inspection Program (B2.1.13)	Consistent with GALL, which recommends further evaluation (See SER Section 3.3.2.2.7)

Component Group	Aging Effect/ Mechanism	AMP in GALL Report	AMP in LRA	Staff Evaluation
Heat exchangers in chemical and volume control system (Item Number 3.3.1-09)	Crack initiation and growth due to SCC and cyclic loading	Water chemistry and a plant-specific verification program	ASME Section XI IWB, IWC, IWD, IWF Inservice Inspection Program (B2.1.2), One-Time Inspection Program (B2.1.13), Water Chemistry Program (B2.1.21)	Consistent with GALL, which recommends further evaluation (See SER Section 3.3.2.2.9)
Neutron absorbing sheets in spent fuel storage racks (Item Number 3.3.1-10)	Reduction of neutron absorbing capacity and loss of material due to general corrosion (Boral, boron steel)	Plant-specific	None	(See SER Section 3.3.2.2.10)
New fuel rack assembly (Item Number 3.3.1-11)	Loss of material due to general, pitting, and crevice corrosion	Structures monitoring	None	Not applicable (New fuel rack aluminum with different AMP)
Neutron absorbing sheets in spent fuel storage racks (Item Number 3.3.1-12)	Reduction of neutron absorbing capacity due to Boraflex degradation	Boraflex monitoring	None	Not applicable (Neutron absorbing sheets not used)
Spent fuel storage racks and valves in spent fuel pool cooling and cleanup (Item Number 3.3.1-13)	Crack initiation and growth due to stress corrosion cracking	Water chemistry	Water Chemistry Program (B2.1.21)	Consistent with GALL, which recommends no further evaluation (See SER Section 3.3.2.1)
Closure bolting and external surfaces of carbon steel and low-alloy steel components (Item Number 3.3.1-14)	Loss of material due to boric acid corrosion	Boric acid corrosion	Boric Acid Corrosion Program (B2.1.4)	Consistent with GALL, which recommends no further evaluation (See SER Section 3.3.2.1)
Components in or serviced by closed-cycle cooling water system (Item Number 3.3.1-15)	Loss of material due to general, pitting, and crevice corrosion, and MIC	Closed-cycle cooling water system	Closed Cycle Cooling Water Program (B2.1.6), One-Time Inspection Program (B2.1.13), Water Chemistry Program (B2.1.21)	Consistent with GALL, which recommends no further evaluation (See SER Section 3.3.2.1.3)

Component Group	Aging Effect/ Mechanism	AMP in GALL Report	AMP in LRA	Staff Evaluation
Cranes including bridge and trolleys and rail system in load handling system (Item Number 3.3.1-16)	Loss of material due to general corrosion and wear	Overhead heavy load and light load handling systems	Overhead Load Handling Systems Inspection Program (B2.1.15)	Consistent with GALL, which recommends no further evaluation (See SER Section 3.3.2.1)
Components in or serviced by open-cycle cooling water systems (Item Number 3.3.1-17)	Loss of material due to general, pitting, crevice, and galvanic corrosion, MIC, and biofouling; buildup of deposit due to biofouling	Open-cycle cooling water system	Closed Cycle Cooling Water Program (B2.1.6), One-Time Inspection Program (B2.1.13), Open Cycle Cooling Water Program (B2.1.14)	Consistent with GALL, which recommends no further evaluation (See SER Section 3.3.2.1.4)
Buried piping and fittings (Item Number 3.3.1-18)	Loss of material due to general, pitting, and crevice corrosion, and MIC	Buried piping and tanks surveillance or Buried piping and tanks inspection	Buried Services Corrosion Monitoring Program (B2.1.5)	Consistent with GALL, which recommends further evaluation (See SER Section 3.3.2.2.11)
Components in compressed air system (Item Number 3.3.1-19)	Loss of material due to general and pitting corrosion	Compressed air monitoring	Compressed Air Monitoring Program (B2.1.22)	Consistent with GALL, which recommends no further evaluation (See SER Section 3.3.2.1.6)
Components (doors and barrier penetration seals) and concrete structures in fire protection (Item Number 3.3.1-20)	Loss of material due to wear; hardening and shrinkage due to weathering	Fire protection	Fire Protection Program (B2.1.10)	Consistent with GALL, which recommends no further evaluation (See SER Section 3.3.2.1)
Components in water-based fire protection (Item Number 3.3.1-21)	Loss of material due to general, pitting, crevice, and galvanic corrosion, MIC, and biofouling	Fire water system	Fire Protection Program (B2.1.10)	Consistent with GALL, which recommends no further evaluation (See SER Section 3.3.2.1)
Components in diesel fire system (Item Number 3.3.1-22)	Loss of material due to galvanic, general, pitting, and crevice corrosion	Fire protection and fuel oil chemistry	Diesel Fuel Monitoring and Storage Program (B2.1.9), Fire Protection Program (B2.1.10)	Consistent with GALL, which recommends no further evaluation (See SER Section 3.3.2.1)

Component Group	Aging Effect/ Mechanism	AMP in GALL Report	AMP in LRA	Staff Evaluation
Tanks in diesel fuel oil system (Item Number 3.3.1-23)	Loss of material due to general, pitting, and crevice corrosion	Aboveground carbon steel tanks	Diesel Fuel Monitoring and Storage Program (B2.1.9), One-Time Inspection Program (B2.1.13), System Monitoring Program (B2.1.20)	Consistent with GALL, which recommends no further evaluation (See SER Section 3.3.2.1.7)
Closure bolting (Item Number 3.3.1-24)	Loss of material due to general corrosion; crack initiation and growth due to cyclic loading and SCC	Bolting integrity	Bolting Integrity Program (B2.1.3), Boric Acid Corrosion Program (B2.1.4), System Monitoring Program (B2.1.20)	Consistent with GALL, which recommends no further evaluation (See SER Section 3.3.2.1.8)
Components (aluminum bronze, brass, cast iron, cast steel) in open-cycle and closed-cycle cooling water systems, and ultimate heat sink (Item Number 3.3.1-29)	Loss of material due to selective leaching	Selective leaching of materials	One-Time Inspection Program (B2.1.13)	Consistent with GALL, which recommends no further evaluation (See SER Section 3.3.2.1.9)
Fire barriers, walls, ceilings, and floors in fire protection (Item Number 3.3.1-30)	Concrete cracking and spalling due to freeze-thaw, aggressive chemical attack, and reaction with aggregates; loss of material due to corrosion of embedded steel	Fire protection and structures monitoring	Fire Protection Program (B2.1.10), Structural Monitoring Program (B2.1.19)	Consistent with GALL, which recommends no further evaluation (See SER Section 3.3.2.1)

The staff's review of the auxiliary systems component groups followed several approaches. One approach, documented in SER Section 3.3.2.1, discusses the staff's review of AMR results for components the applicant indicated are consistent with the GALL Report and require no further evaluation. Another approach, documented in SER Section 3.3.2.2, discusses the staff's review of AMR results for components the applicant indicated are consistent with the GALL Report and for which further evaluation is recommended. A third approach, documented in SER Section 3.3.2.3, discusses the staff's review of AMR results for components the applicant indicated are not consistent with or not addressed in the GALL Report. The staff's review of AMPs credited to manage or monitor aging effects of the auxiliary systems components is documented in SER Section 3.0.3.

3.3.2.1 AMR Results That Are Consistent with the GALL Report

Summary of Technical Information in the Application. In LRA Section 3.3.2.1, the applicant identified the materials, environments, and AERMs. The applicant identified the following programs that manage the aging effects of the auxiliary systems components:

- ASME Section XI IWB, IWC, IWD, IWF Inservice Inspection Program
- Bolting Integrity Program
- Boric Acid Corrosion Program
- Buried Services Corrosion Monitoring Program
- Closed Cycle Cooling Water Program
- Diesel Fuel Monitoring and Storage Program
- Fire Protection Program
- One-Time Inspection Program
- Open Cycle Cooling Water Program
- System Monitoring Program
- Water Chemistry Program

Staff Evaluation. In LRA Tables 3.3.2-1 through 3.3.2-17, the applicant summarized AMRs for the auxiliary systems components and identified which AMRs it considered consistent with the GALL Report.

For component groups evaluated in the GALL Report for which the applicant had claimed consistency and for which the GALL Report does not recommend further evaluation, the staff performed an audit and review to determine whether the plant-specific components in these GALL Report component groups were bounded by the GALL Report evaluation.

The applicant provided a note for each AMR line item. The notes describe how the information in the tables aligns with the information in the GALL Report. The staff audited those AMRs with Notes A through E, which indicate how the AMR was consistent with the GALL Report.

Note A indicates that the AMR line item is consistent with the GALL Report for component, material, environment, and aging effect. In addition, the AMP is consistent with the GALL Report AMP. The staff audited these line items to verify consistency with the GALL Report and the validity of the AMR for the site-specific conditions.

Note B indicates that the AMR line item is consistent with the GALL Report for component, material, environment, and aging effect. In addition, the AMP takes some exceptions to the AMP identified in the GALL Report. The staff audited these line items to verify consistency with the GALL Report. The staff verified that the identified exceptions to the GALL Report AMPs had been reviewed and accepted by the staff. The staff also determined whether the AMP identified by the applicant was consistent with the AMP identified in the GALL Report and whether the AMR was valid for the site-specific conditions.

Note C indicates that the component for the AMR line item, although different from, is consistent with the GALL Report for material, environment, and aging effect. In addition, the AMP is consistent with the AMP identified by the GALL Report. This note indicates that the applicant was unable to find a listing of some system components in the GALL Report; however, the applicant identified a different component in the GALL Report that had the same material, environment, aging effect, and AMP as the component under review. The staff audited these line items to verify consistency with the GALL Report. The staff also determined whether the AMR line item of the different component applied to the component under review and whether the AMR was valid for the site-specific conditions.

Note D indicates that the component for the AMR line item, although different from, is consistent with the GALL Report for material, environment, and aging effect. In addition, the AMP takes some exceptions to the AMP identified in the GALL Report. The staff audited these line items to verify consistency with the GALL Report. The staff verified whether the AMR line item of the different component was applicable to the component under review. The staff verified whether the exceptions to the GALL Report AMPs had been reviewed and accepted by the staff. The staff also determined whether the AMP identified by the applicant was consistent with the AMP identified in the GALL Report and whether the AMR was valid for the site-specific conditions.

Note E indicates that the AMR line item is consistent with the GALL Report for material, environment, and aging effect, but a different AMP is credited. The staff audited these line items to verify consistency with the GALL Report. The staff also determined whether the identified AMP would manage the aging effect consistent with the AMP identified in the GALL Report and whether the AMR was valid for the site-specific conditions.

The staff did not repeat its review of the matters described in the GALL Report; however, the staff did verify that the material presented in the LRA was applicable and that the applicant identified the appropriate GALL Report AMRs. The staff's evaluation is discussed below.

3.3.2.1.1 Cumulative Fatigue Damage

In LRA Table 3.3.1, Item 3.3.1-03, the applicant stated that cumulative fatigue damage is managed using a TLAA.

As documented in the Audit and Review Report, the staff asked the applicant to clarify the assignment in LRA Table 3.3.2-11 of this Table 1 item to the management of loss of carbon steel HX material exposed to air. By letter dated August 27, 2005, the applicant stated that the Table 1 item had been changed from 3.3.1-03 to 3.3.1-05 which changed the aging effect from cumulative fatigue to loss of material due to general, pitting, and crevice corrosion, and MIC. The staff finds this change consistent with the GALL Report and, therefore, acceptable.

3.3.2.1.2 Crack Initiation and Growth Due to Cracking or Stress Corrosion Cracking (SCC)

In LRA Table 3.3.1, Item 3.3.1-04, the applicant stated that for high-pressure pumps in the chemical and volume control (CVC) system, crack initiation and growth due to SCC or cracking is managed by a plant-specific AMP. As documented in the Audit and Review Report, the staff noted that the applicant applied LRA Table 3.3.1, Item 3.3.1-04, in LRA Table 3.3.2-1 for cracking of stainless steel valves and instrument assemblies in treated water. By letter dated August 27, 2005, the applicant stated that stress corrosion cracking/intergranular attack (SCC/IGA) is an AERM for CVC system components constructed of stainless steel in an environment of treated water in containment where the temperatures of the CVC system are greater than 140 °F. Also, SCC/IGA is an AERM for the CVC system heat traced piping located in the auxiliary building with temperatures greater than the 140 °F threshold to sustain SCC/IGA. Not all CVC system components have temperatures greater than 140 °F. SCC/IGA, including crack initiation and growth, is managed where applicable by the applicant's ASME Section XI IWB, IWC, IWD, IWF Inservice Inspection, Closed Cycle Cooling Water, One-Time Inspection, and/or Water Chemistry Programs. The applicant also stated that Note 303, "Cracking is applicable for applications greater than 140 °F," was added to LRA Page 3-198. Furthermore, this note was added in LRA Table 3.3.2-1 to every line item that had an existing note associated with the cracking AERM. The staff reviewed the applicant's response and found it acceptable as consistent with the GALL Report.

.3.3.2.1.3 Loss of Material Due to General Corrosion, Pitting, Crevice Corrosion, and MIC

In the discussion column of LRA Table 3.3.1, Item 3.3.1-15, the applicant credited its Closed Cycle Cooling Water, One-Time Inspection, and Water Chemistry Programs for managing loss of material of the components in or serviced by the closed-cycle cooling water (CCCW) system. In LRA Table 3.3.2-3, the applicant identified Table 3.3.1, Item 3.3.1-15, as the corresponding Table 1 item for the copper alloy HX in the treated water environment. In LRA Table 3.3.2-3, the applicant credited its Closed Cycle Cooling Water Program for managing the loss of material and the One-Time Inspection Program for managing selective leaching of the copper alloys in HXs in treated water and stated that the combinations of these material, environment, aging effects, and AMPs are consistent with GALL Report Items VII.C2.3-a and VII.C2.4-a, respectively. However, GALL Report Items VII.C2.3-a and VII.C2.4-a materials are carbon steel and cast iron. As documented in the Audit and Review Report, the staff requested that the applicant provide clarification.

By letter dated August 27, 2005, the applicant stated:

> LRA Table 3.3.2-3, page 3-126, Component Type Heat Exchanger is revised as follows:
>
> The NUREG 1801 Volume 2, Table 1 and Note information for copper alloy Heat Exchanger components in an external treated water environment that are being managed by the Closed Cooling Water Program (existing GALL reconciliation VII.C2.4- a, 3.3.1-15 and notes 323, C) are hereby changed to VII.C1.3-a, 3.3.1-17, and 323, E, respectively.
>
> The NUREG 1801 Volume 2, Table 1 and Note information for copper alloy Heat Exchanger components in an external treated water environment that are being managed by the One-Time Inspection Program (existing GALL reconciliation VII.C2.3-a, 3.3.1-15 and notes 301, C) are hereby changed to VII.C1.3-a, 3.3.1-29 and notes 301, E, respectively.

The staff reviewed the applicant's revisions and found them acceptable as consistent with the GALL Report because the revision changes the Closed Cycle Cooling Water Program to the Open Cycle Cooling Water Program and adds selective leaching.

3.3.2.1.4 Loss of Material Due to General Corrosion, Pitting, Crevice Corrosion, MIC and Biofouling; Buildup of Deposit Due to Biofouling

In the discussion column of LRA Table 3.3.1, Item 3.3.1-17, the applicant credited its Closed Cycle Cooling Water, One-Time Inspection, and Open Cycle Cooling Water Programs for managing loss of material of the components in or serviced by the open-cycle cooling water system. LRA Table 3.3.2-12 shows the Open Cycle Cooling Water Program managing loss of cast iron and copper alloy material in raw water (internal) and references GALL Report Item VII.C1.2-a, which describes selective leaching as a mechanism. Also, in LRA Table 3.3.2-16, the applicant showed the One-Time Inspection Program managing loss of copper alloy material in raw water (internal) and referred to GALL Report Item VII.C1.1-a, which describes selective leaching as a mechanism. As documented in the Audit and Review Report, the applicant was requested to clarify how the Open Cycle Cooling Water Program manages loss of material due to selective leaching for cast iron and copper alloys in raw water.

By letter dated August 27, 2005, the applicant stated that as to copper alloy valves and dampers exposed to raw water, the remainder of the copper alloy components contains less than 15 percent zinc and, therefore, is not subject to selective leaching. The applicant further stated:

> Plant specific notes 301 and 399 in the LRA provide assistance in determining where selective leaching is included under a particular line item where the AERM column only indicates loss of material. However, to further facilitate review, LRA Table 3.3.2-12, page 3-185, valves & dampers, fluid pressure boundary, cast iron, raw water, loss of material, One-Time Inspection Program, Table 1 item 3.3.1-17 is hereby changed to read 3.3.1-29. In addition, LRA plant specific note #304 is added to this line item.

> Note 304 on page 3-198 is hereby changed to read, 'This component contains less than 15% zinc; therefore, selective leaching is not a potential aging mechanism.'

The staff reviewed the applicant's response and found it acceptable as consistent with the GALL Report because the One Time Inspection Program includes a Selective Leaching of Materials Program.

3.3.2.1.5 Loss of Material Due to General, Pitting, Crevice Corrosion, and MIC

As documented in the Audit and Review Report, the staff noted that LRA Table 3.3.2-8 refers to LRA Table 3.3.1, Item 3.3.1-18, for carbon steel pipes and fittings and to GALL Report Item VII.H1.1-b. The environments identified in GALL Report Item VII.H1.1-b are soil and groundwater. However, the LRA indicates that the environments are plant indoor air and raw water. Therefore, the staff requested the applicant to clarify this difference.

By letter dated August 27, 2005, the applicant stated that the diesel fuel oil system components in plant indoor air managed by its System Monitoring Program with GALL Report Item VII.H1,1-b, Table 1 Item 3.3.1-18, and Note A should be GALL Report Item VII.I.1-b with Table 1 Item 3.3.1-5 and Note A.

At PNP, components in soil or ground water are included in the raw water environment. The raw water is rough filtered to remove large particles. Biocides may be added to control micro- or macro-organisms. Another designation of raw water is water that leaks from any system. Damp soil (moist soil/earth) containing ground water is also included in this environment. This grouping includes structural carbon or low-alloy steel in raw water or in non-borated treated water. Buried components were evaluated as exposed to a raw water environment. With this clarification, the staff determines that management of carbon steel pipes and fittings in soil as well as diesel fuel oil components in plant indoor air is consistent with the GALL Report and, therefore, acceptable.

3.3.2.1.6 Loss of Material Due to General Corrosion, Pitting, and Crevice Corrosion

In the discussion column of LRA Table 3.3.1, Item 3.3.1-19, the applicant credited the One-Time Inspection Program for managing loss of material of compressed air system components.

In LRA Table 3.3.2-4 the applicant referred to Table 3.3.1, Item 3.3.1-19, and to GALL Report Items VII.D.1.1-a, 2-a, 3-a, 5-a, and 6-a. Also, In LRA Tables 3.3.2-4 and 3.3.2-10, the applicant referred to Item 3.3.1-19 and to GALL Report Item VII.D.1.5-a. The LRA indicates that the AMP is the One-Time Inspection Program. The AMP listed in the GALL Report refers to GALL AMP XI.M24, "Compressed Air Monitoring." GALL AMP XI.M24 incorporates air quality measuring and maintenance.

As documented in the Audit and Review Report, the applicant was asked to explain use of the One-Time Inspection Program in lieu of GALL AMP XI.M24 for managing loss of material of carbon steel, cast iron, and galvanized components in air. By letter dated August 27, 2005, the applicant stated that it would develop a Compressed Air Monitoring Program. The LRA was updated to describe this new program and the compressed air system components that had credited the One-Time Inspection Program to manage aging effects in an internal environment of air credited the applicant's new Compressed Air Monitoring Program. The staff reviewed the applicant's update and found it acceptable as consistent with the GALL Report.

In LRA Table 3.3.2-4, the applicant referred to Table 3.3.1, Item 3.3.1-19, and to GALL Report Item VII.D.5-a. The applicant's indication that the AMP is the Boric Acid Corrosion Program was not consistent with the GALL Report reference. As documented in the Audit and Review Report, the applicant was asked to explain how the GALL Report item and Table 1 item applied to the indicated material, environment, aging effect, and AMP. By letter dated August 27, 2005, the applicant stated that for the Boric Acid Corrosion Program, GALL Report Item VII.D.5-a had been revised to GALL Report Item VII.I.1-a, external steel surfaces exposed to air with borated water, with reference to LRA Table 3.3.1, Item 3.3.1-14. The staff reviewed the applicant's revision and found it acceptable as consistent with the GALL Report.

In LRA Table 3.3.2-10, the applicant referred to Table 3.3.1, Item 3.3.1-19, and to GALL Report Items VII.D.1.1-a, 2-a, 3-a, 5-a, and 6-a. The applicant indicated that its Bolting Integrity Program manages loss of carbon steel miscellaneous components material in containment/plant indoor air (external) environments. As documented in the Audit and Review Report, the applicant was asked to explain how the GALL Report item and Table 1 item applied the indicated material, environment, and AMP.

By letter dated August 27, 2005, the applicant stated that on Page 3-171 of the LRA, the component type of miscellaneous mechanical had been changed to fasteners as there were no AERMs for the copper alloy manifold or the stainless steel monitor. GALL Report, Items VII.D.1-a, .2-a, .3-a, .5-a, and .6-a were corrected to a single GALL Report Item VII.I.1-b managed by the System Monitoring Program with references to Table 3.3.1, Item 3.3.1-05, and Note A. This correction is consistent with the GALL Report and, therefore, acceptable to the staff.

For loss of material, the Bolting Integrity Program was changed to the System Monitoring Program. The staff reviewed the applicant's response and found it acceptable as consistent with the GALL Report. Also, loss of preload managed by the Bolting Integrity Program would not have a GALL Report, Volume 2, or LRA Table 1 reference and would have notes of 324 and H. On Page 3-172, the continuation would be fasteners with notes of 324 and H. The staff reviewed the applicant's changes and found them acceptable because loss of preload is not an AERM for miscellaneous mechanical and fastener component types in auxiliary systems. However, the applicant conservatively decided to manage this aging effect with its Bolting Integrity Program.

3.3.2.1.7 Loss of Material Due to General Corrosion, Pitting, and Crevice Corrosion

In the discussion column of LRA Table 3.3.1, Item 3.3.1-23, the applicant credited its Diesel Fuel Monitoring and Storage, One-Time Inspection, and System Monitoring Programs for managing loss of material due to general, pitting, and crevice corrosion for above-ground carbon steel tanks of the diesel fuel oil system.

As documented in the Audit and Review Report, the staff noted that in LRA Table 3.3.1, Item 3.3.1-23, the System Monitoring Program is credited to manage loss of material from the external surfaces of cast iron pumps in air. The staff requested the applicant to clarify the reference in LRA Table 3.3.1, Item 3.3.1-23 to cast iron pumps.

By letter dated August 27, 2005, the applicant stated that the cast iron diesel fuel oil pumps in plant indoor air would be managed by the System Monitoring Program and revised the LRA to indicate GALL Report, Volume 2, Item VII.I.1-b and Table 3.3.1, Item 3.3.1-5. This revision is consistent with the GALL Report and, therefore, acceptable to the staff.

3.3.2.1.8 Loss of Material Due to General Corrosion; Crack Initiation and Growth Due to Cyclic Loading and SCC.

In the discussion column of LRA Table 3.3.1, Item 3.3.1-24, the applicant referred to its Boric Acid Corrosion Monitoring, One-Time Inspection, and System Monitoring Programs for managing loss of material for closure bolting. The GALL Report recommends the Bolting Integrity Program for managing this aging effect. LRA Tables 3.3.2-1, 3.3.2-2, 3.3.2-3, 3.3.2-5, 3.3.2-7, 3.3.2-8, 3.3.2-9, 3.3.2-11, 3.3.2-14, and 3.3.2-15 credit the Bolting Integrity Program for managing loss of preload and loss of material for carbon steel and low-alloy steel fasteners in air and refers to GALL Report Item VII.I.2-a and Table 3.3.1, Item 3.3.1-24. However, GALL Report Item VII.I.2-a addresses the loss of material of carbon and low-alloy steel only. As documented in the Audit and Review Report, the applicant was asked to explain the applicability of this GALL Report item to loss of preload for these components.

By letter dated August 27, 2005, the applicant stated that the tables listing management by the Bolting Integrity Program and GALL Report Item VII.I.2-a refers to Note 324. This note states that loss of preload had been included in response to recent RAIs on non-primary system, high-temperature bolting that may experience loss of preload, that the Bolting Integrity Program manages potential bolting AERMs and event-driven degradation, and that GALL Report reconciliation is based on loss of material. Therefore, loss of preload was conservatively evaluated for NSR components. The staff reviewed the applicant's response and determines that it exceeded the recommendation of the GALL Report and was, therefore, acceptable.

For cast iron steam traps exposed to plant indoor air, LRA Table 3.3.2-6 references GALL Report Item VII.I.2-b, which is for carbon and low-alloy steel in moist air to manage crack initiation and growth through the Bolting Integrity Program. However, in the LRA, the item is for cast iron traps in air and manages loss of material using the System Monitoring Program. As documented in the Audit and Review Report, the applicant was asked to explain the LRA reference to this GALL Report item.

By letter dated August 27, 2005, the applicant stated that the emergency power system traps in plant indoor air managed by the System Monitoring Program with GALL Report Item VII.I.2-b, LRA Table 1, Item 3.3.1-24, and Notes 399 and A should be GALL Report Item VII.I.1-b, external steel surfaces exposed to condensation, with LRA Table 1, Item 3.3.1-5, and Notes 399 and A. The staff reviewed the applicant's statement and found it acceptable as consistent with the GALL Report.

3.3.2.1.9 Loss of Material Due to Selective Leaching

In the discussion column of LRA Table 3.3.1, Item 3.3.1-15, the applicant credited its Closed Cycle Cooling Water, One-Time Inspection, and Water Chemistry Programs for managing loss of material for the components in or serviced by the CCCW systems. LRA Tables 3.3.2-11, 3.3.2-12, and 3.3.2-15 refer to Table 1, Item 3.3.1-29 (selective leaching). The affected material is carbon steel, to which selective leaching does not apply. The staff requested the applicant to explain how the Table 1 item was used for carbon steel.

By letter dated August 27, 2005, the applicant stated that in LRA Tables 3.3.2-11, 3.3.2-12, and 3.3.2-15, GALL Report Item VII.C3.1-a, Table 1 Item 3.3.1-29 should be 3.3.1-15. The staff reviewed the applicant's response and found it acceptable as consistent with the GALL Report because Item 3.3.1-15 states the use of the Closed Cycle Cooling Water, One Time Inspection, and Water Chemistry Programs for loss of material.

Conclusion. The staff evaluated the applicant's claim of consistency with the GALL Report. The staff also reviewed information pertaining to the applicant's consideration of recent operating experience and proposals for managing the associated aging effects. On the basis of its review, the staff concludes that the AMR results, which the applicant claimed to be consistent with the GALL Report, are consistent with the GALL Report AMRs. Therefore, the staff concludes that the applicant has demonstrated that the aging effects for these components will be adequately managed so that their intended function(s) will be maintained consistent with the CLB for the period of extended operation, as required by 10 CFR 54.21(a)(3).

3.3.2.2 AMR Results That Are Consistent with the GALL Report, for Which Further Evaluation is Recommended

Summary of Technical Information in the Application. In LRA Section 3.3.2.2, the applicant provided further evaluation of aging management, as recommended by the GALL Report, for the auxiliary systems components. The applicant provided information concerning how it will manage the following aging effects:

- loss of material due to general, pitting, and crevice corrosion

- hardening and cracking or loss of strength due to elastomer degradation or loss of material due to wear

- cumulative fatigue damage

- crack initiation and growth due to cracking or SCC

- loss of material due to general, microbiologically influenced, pitting, and crevice corrosion

- loss of material due to general, galvanic, pitting, and crevice corrosion

- loss of material due to general, pitting, crevice, and microbiologically influenced corrosion and biofouling

- crack initiation and growth due to SCC and cyclic loading

- reduction of neutron-absorbing capacity and loss of material due to general corrosion

- loss of material due to general, pitting, crevice, and microbiologically influenced corrosion

Staff Evaluation. For component groups evaluated in the GALL Report for which the applicant had claimed consistency with the GALL Report and for which the GALL Report recommends further evaluation, the staff audited and reviewed the applicant's evaluations to determine whether they adequately address those issues. In addition, the staff reviewed the applicant's further evaluations against the criteria in SRP-LR Section 3.3.2.2. Details of the staff's audit are documented in the Audit and Review Report. The staff's evaluation of the aging effects is discussed in the following sections.

3.3.2.2.1 Loss of Material Due to General, Pitting, and Crevice Corrosion

The staff reviewed LRA Section 3.3.2.2.1 against the criteria in SRP-LR Section 3.3.2.2.1.

The Water Chemistry Program relies on monitoring and control of reactor water chemistry based on guidelines of EPRI TR-105714 for primary water chemistry of EPRI TR-102134 for secondary water chemistry to manage the effects of loss of material from pitting or crevice corrosion. However, high concentrations of impurities at crevices and locations of stagnant flow conditions could cause pitting or crevice corrosion. Therefore, the effectiveness of the Chemistry Control Program should be verified to ensure that corrosion does not occur. The GALL Report recommends further evaluation of programs to manage loss of material from pitting and crevice corrosion to verify the effectiveness of the Water Chemistry Program. A one-time inspection of select components at susceptible locations is an acceptable method to ensure that corrosion does not occur and that the component's intended function will be maintained during the period of extended operation.

In LRA Section 3.3.2.2.1, the applicant addressed loss of material of components in the spent fuel pool (SFP) cooling and cleanup system due to general, pitting, and crevice corrosion.

SRP-LR Section 3.3.2.2.1.1 states that loss of material due to general, pitting, and crevice corrosion could occur in the HX channel head and access cover, tubes, and tubesheets in the SFP cooling and cleanup system.

SRP-LR Section 3.3.2.2.1.2 states that loss of material due to pitting and crevice corrosion could occur in the filter housing, valve bodies, and nozzles of the ion exchanger in the SFP cooling and cleanup system.

In the LRA, the applicant stated that the materials of the SFP cooling components addressed in the GALL Report are carbon steel with elastomer linings, which differ from the stainless steel

SFP cooling system components. The applicant stated that this aging effect is not applicable to the SFP cooling system.

The staff reviewed the LRA Table 3.3.2-14 AMR for the SFP cooling system and verified that the system contains no carbon steel components with elastomer linings. This review determined that this Table 1 line item is not applicable for components in the SFP cooling system.

The applicant proposed to manage loss of material of stainless steel pipe, pump, and SFP HX and valve component types exposed to chemically treated borated water using the Water Chemistry Program, which is consistent with GALL AMP XI.M2, "Water Chemistry." The applicant also uses the One-Time Inspection Program to verify the effectiveness of the Water Chemistry Program. The staff reviewed the applicant's Water Chemistry and One-Time Inspection Programs as documented in SER Sections 3.0.3.1.10 and 3.0.3.2.6, respectively. The staff finds the applicant's Water Chemistry and One-Time Inspection Programs acceptable for managing this aging effect for the stainless steel components in treated water.

Based on the programs identified above, the staff concludes that the applicant has met the criteria of SRP-LR Section 3.3.2.2.1. For those line items that apply to LRA Section 3.3.2.2.1, the staff determines that the application is consistent with the GALL Report and the applicant has demonstrated that the aging effects will be adequately managed so that the intended function(s) will be maintained consistent with the CLB during the period of extended operation, as required by 10 CFR 54.21(a)(3).

3.3.2.2.2 Hardening and Cracking or Loss of Strength Due to Elastomer Degradation or Loss of Material Due to Wear

The staff reviewed LRA Section 3.3.2.2.2 against the criteria in SRP-LR Section 3.3.2.2.2.

In LRA Section 3.3.2.2.2, the applicant addressed hardening and cracking or loss of strength due to elastomer degradation or loss of material due to wear.

SRP-LR Section 3.3.2.2.2 states that hardening and cracking due to elastomer degradation could occur in elastomer linings of the filter, valve, and ion exchangers in SFP cooling and cleanup systems. Hardening and loss of strength due to elastomer degradation could occur in the collars and seals of the duct and in the elastomer seals of the filters in the control room area, auxiliary and radwaste area, and primary containment heating and ventilation systems, and in the collars and seals of the duct in the diesel generator building ventilation system. Loss of material due to wear could occur in the collars and seals of the duct in the ventilation systems. The GALL Report recommends further evaluation to ensure that these aging effects are adequately managed.

In LRA Section 3.3.2.2.2, the applicant stated that the materials of the SFP cooling system components subject to an AMR have no elastomers. For the ventilation systems, elastomers are evaluated for cracking and changes of material properties due to thermal and radiation exposure. The applicant's System Monitoring Program manages degradation of elastomers at the seals and flexible connections. Elastomers are used in other systems. For those systems, management of elastomer degradation is by the System Monitoring Program.

Based on the programs identified above, the staff concludes that the applicant has met the criteria of SRP-LR Section 3.3.2.2.2. For those line items that apply to LRA Section 3.3.2.2.2, the staff determines that the application is consistent with the GALL Report and the applicant has demonstrated that the aging effects will be adequately managed so that the intended function(s) will be maintained consistent with the CLB during the period of extended operation, as required by 10 CFR 54.21(a)(3).

3.3.2.2.3 Cumulative Fatigue Damage

In LRA Section 3.3.2.2.3, the applicant stated that fatigue is a TLAA, as defined in 10 CFR 54.3. Applicants must evaluate TLAAs in accordance with 10 CFR 54.21(c)(1). SER Section 4.3 documents the staff's review of the applicant's evaluation of this TLAA.

3.3.2.2.4 Crack Initiation and Growth Due to Cracking or Stress Corrosion Cracking

The staff reviewed LRA Section 3.3.2.2.4 against the criteria in SRP-LR Section 3.3.2.2.4.

In LRA Section 3.3.2.2.4, the applicant addressed crack initiation and growth of high-pressure pumps in CVC systems due to SCC or cracking.

SRP-LR Section 3.3.2.2.4 states that crack initiation and growth due to SCC could occur due to cracking in the CVC system high-pressure pump. The GALL Report recommends further evaluation to ensure that these aging effects are adequately managed.

In the LRA, the applicant stated that this paragraph of the SRP-LR, which discusses cracking in the high-pressure pump, is not applicable because the pump temperature is less than 140 °F, which is below the temperature threshold required to support cracking.

The staff reviewed the applicant's further evaluation discussion and concludes that cracking in the high-pressure pump in the CVC system is not applicable. On this basis, the staff determines that cracking for the high-pressure pumps in the CVC system requires no management.

3.3.2.2.5 Loss of Material Due to General, Microbiologically Influenced (MIC), Pitting, and Crevice Corrosion

The staff reviewed LRA Section 3.3.2.2.5 against the criteria in SRP-LR Section 3.3.2.2.5.

In LRA Section 3.3.2.2.5, the applicant addressed loss of material from components in ventilation systems, diesel fuel oil systems, and emergency diesel generator (EDG) systems and from external surfaces of carbon steel components due to general, pitting, and crevice corrosion, and MIC.

SRP-LR Section 3.3.2.2.5 states that loss of material due to general, pitting, and crevice corrosion could occur in: the piping or filter housing and/or supports in the control room area; the auxiliary and radwaste area; the primary containment heating and ventilation systems; the piping of the diesel generator building ventilation system; the above-ground piping and fittings, valves, and pumps in the diesel fuel oil system; and the diesel engine starting air, combustion air intake, and combustion air exhaust subsystems in the EDG system. Loss of material could occur due to: general, pitting, and crevice corrosion, and MIC in the duct fittings, access doors, closure bolts, equipment frames, and housing of the duct; pitting and crevice corrosion in the heating/cooling coils of the air handler heating/cooling; due to general corrosion on the external surfaces of all carbon steel structures and components, including bolting exposed to operating temperatures < 212 °F in the ventilation systems. The GALL Report recommends further evaluation to ensure that these aging effects are adequately managed.

In the LRA, the applicant stated that SRP-LR Section 3.3.2.2.5 discusses the loss of material from corrosion that could occur on internal and external surfaces of components exposed to plant indoor air as well as atmosphere and weather. Specifically included are the ventilation, diesel fuel oil, emergency diesel starting air, and combustion air intake and exhaust systems and the external carbon steel surfaces of auxiliary systems.

The applicant also stated that for the internal environments of applicable auxiliary systems, the Open Cycle Cooling Water Program, Diesel Fuel Monitoring and Storage Program, One-Time Inspection Program (including tank internal inspection), and Fire Protection Program are credited for managing the aging effect of loss of material.

The applicant further stated, in the LRA, that for the external surfaces of all carbon steel components in auxiliary systems, the System Monitoring Program is credited for managing the aging effect of loss of material. The Open Cycle Cooling Water and Fire Protection Programs are credited to augment the System Monitoring Program for managing external aging effects in the service water and fire protection systems, respectively. Closure bolting is managed by the Bolting Integrity Program.

As documented in the Audit and Review Report, the staff asked the applicant to clarify how the Open Cycle Cooling Water Program manages the external surfaces of carbon steel components in auxiliary systems. By letter dated September 2, 2005, the applicant stated that the Open Cycle Cooling Water Program had been erroneously included in LRA Section 3.3.2.2.5. The revised LRA Section 3.3.2.2.5 includes no reference to this program.

The LRA Table 2s address this Table 1 item management of loss of material due to general, pitting, and crevice corrosion, and MIC for internal and external surfaces of components exposed to plant indoor air as well as atmosphere and weather for the ventilation, diesel fuel oil, emergency diesel starting air, combustion air intake and exhaust systems and the external carbon steel surfaces of auxiliary systems.

The Fire Protection Program manages loss of material of the fuel oil system carbon steel valves and dampers in plant indoor (external) environment. LRA Section B2.1.10 states that for the detection of aging effects program element, the applicant credits its One-Time Inspection Program for aging management of the reactor coolant pump (RCP) oil collection tank, piping, and valve bodies for wall thickness and aging degradation.

The staff reviewed the applicant's One-Time Inspection, System Monitoring, Fire Protection, and Bolting Integrity Programs as documented in SER Sections 3.0.3.2.6, 3.0.3.3.1, 3.0.3.2.5, and 3.0.3.2.1, respectively.

Based on the programs identified above, the staff concludes that the applicant has met the criteria of SRP-LR Section 3.3.2.2.5. For those line items that apply to LRA Section 3.3.2.2.5, the staff determines that the application is consistent with the GALL Report and the applicant has demonstrated that the aging effects will be adequately managed so that the intended function(s) will be maintained consistent with the CLB during the period of extended operation, as required by 10 CFR 54.21(a)(3).

3.3.2.2.6 Loss of Material Due to General, Galvanic, Pitting, and Crevice Corrosion

The staff reviewed LRA Section 3.3.2.2.6 against the criteria in SRP-LR Section 3.3.2.2.6.

In LRA Section 3.3.2.2.6, the applicant addressed loss of material of tanks, piping, valve bodies, and tubing in the RCP oil collection system in fire protection due to galvanic, general, pitting, and crevice corrosion.

SRP-LR Section 3.3.2.2.6 states that loss of material due to general, galvanic, pitting, and crevice corrosion could occur in tanks, piping, valve bodies, and tubing in the RCP oil collection system in fire protection. The Fire Protection Program relies on a combination of visual and volumetric examinations in accordance with 10 CFR Part 50, Appendix R, guidelines and Branch Technical Position 9.5-1 to manage loss of material from corrosion. However, corrosion may occur at locations where water from wash downs may accumulate. Therefore, the effectiveness of the program should be verified to ensure that corrosion does not occur. The GALL Report recommends further evaluation of programs to manage loss of material due to general, galvanic, pitting, and crevice corrosion to verify the effectiveness of the program. A one-time inspection of the bottom half of the interior surface of the tank of the RCP oil collection system is an acceptable method to ensure that corrosion does not occur and that the component's intended function will be maintained during the period of extended operation.

In the LRA, the applicant stated that the Fire Protection Program has a separate oil collection system for each primary coolant pump (PCP). The GALL Report assumes that the RCP oil collection tanks and piping are constructed from carbon steel, copper, and stainless steel alloys. However, the oil collection tank, drip pans, and oil lift pump enclosures are stainless steel. Annealed copper tubing connects these components and directs the path of any oil leakage from the PCP motor. These materials located inside containment are normally exposed to containment air internal and external environments. Any oil collected will be lubricating oil without any water entrainment. Loss of material due to general and galvanic corrosion is not applicable to this oil collection system. If water condensed from air or spray from some water source entered the system, then crevice and pitting corrosion could be an aging effect. A one-time inspection of the tank and piping was credited to ensure that loss of material due to crevice and pitting corrosion does not occur.

The applicant credited the One-Time Inspection Program for inspections of carbon steel, cast iron, and copper alloy components in steam and power and auxiliary systems in lube oil (internal and external) environment. The One-Time Inspection Program provides methods for verifying an AMP is not needed, for verifying the effectiveness of an existing program, or for determining the occurrence of degradation requiring evaluation and corrective action. The staff reviewed the applicant's One-Time Inspection Program as documented in SER Section 3.0.3.2.6.

Based on the programs identified above, the staff concludes that the applicant has met the criteria of SRP-LR Section 3.3.2.2.6. For those line items that apply to LRA Section 3.3.2.2.6, the staff determines that the application is consistent with the GALL Report and the applicant has demonstrated that the aging effects will be adequately managed so that the intended function(s) will be maintained consistent with the CLB during the period of extended operation, as required by 10 CFR 54.21(a)(3).

3.3.2.2.7 Loss of Material Due to General, Pitting, Crevice, and Microbiologically Influenced Corrosion and Biofouling

The staff reviewed LRA Section 3.3.2.2.7 against the criteria in SRP-LR Section 3.3.2.2.7.

In LRA Section 3.3.2.2.7, the applicant addressed loss of diesel fuel oil tank material in the diesel fuel oil and EDG systems due to general, pitting, and crevice corrosion, MIC, and biofouling.

SRP-LR Section 3.3.2.2.7 states that loss of material due to: general, pitting, and crevice corrosion, MIC, and biofouling could occur in the internal surface of tanks in the diesel fuel oil system; and general, pitting, and crevice corrosion, and MIC in the diesel fuel oil tanks in the EDG system. The existing AMP relies on the Fuel Oil Chemistry Program to monitor and control fuel oil contamination in accordance with the guidelines of ASTM Standards D4057, D1796, D2709, and D2276 to manage loss of material due to corrosion or biofouling. Corrosion or biofouling may occur at locations where contaminants accumulate. The effectiveness of the Chemistry Control Program should be verified to ensure that corrosion does not occur. The GALL Report recommends further evaluation of programs to manage corrosion/biofouling to verify the effectiveness of the programs. A one-time inspection of selected components at susceptible locations is an acceptable method to ensure that corrosion does not occur and that the component's intended function will be maintained during the period of extended operation.

In the LRA, the applicant stated that the Diesel Fuel Monitoring and Storage Program manages components in the fuel oil system associated with EDGs and diesel fire pumps, including storage tanks, day tanks, piping, valve bodies, and other passive components. These components rely on the Diesel Fuel Monitoring and Storage Program to minimize the potential for degradation and loss of intended function. The program manages the conditions that would cause general and pitting corrosion, and MIC of the diesel fuel tank internal surfaces.

In the LRA, the applicant also stated that the One-Time Inspection Program provides reasonable assurance that aging effects will be managed such that SSCs within the scope of this program will continue to perform their intended functions consistent with the CLB for the period of extended operation.

Based on a review of the information in LRA Section 3.3.2.2.7, the staff concludes that the applicant had applied the appropriate AMPs, including its One-Time Inspection Program to assure management of aging effects.

Based on the programs identified above, the staff concludes that the applicant has met the criteria of SRP-LR Section 3.3.2.2.7. For those line items that apply to LRA Section 3.3.2.2.7, the staff determines that the application is consistent with the GALL Report and the applicant has demonstrated that the aging effects will be adequately managed so that the intended function(s) will be maintained consistent with the CLB during the period of extended operation, as required by 10 CFR 54.21(a)(3).

3.3.2.2.8 Quality Assurance for Aging Management of Nonsafety-Related Components

SER Section 3.0.4 provides the staff's evaluation of the applicant's QA program.

3.3.2.2.9 Crack Initiation and Growth Due to Stress Corrosion Cracking and Cyclic Loading

The staff reviewed LRA Section 3.3.2.2.9 against the criteria in SRP-LR Section 3.3.2.2.9.

In LRA Section 3.3.2.2.9, the applicant addressed crack initiation and growth of CVC system HXs due to SCC and cyclic loading.

SRP-LR Section 3.3.2.2.9 states that crack initiation and growth due to SCC and cyclic loading could occur in the channel head and access cover, tubesheet, tubes, shell and access cover, and closure bolting of the regenerative HX and in the channel head and access cover, tubesheet, and tubes of the letdown HX in the CVC system. The Water Chemistry Program relies on monitoring and control of water chemistry based on TR-105714 guidelines for primary water chemistry in PWRs to manage the effects of crack initiation and growth due to SCC and cyclic loading. The effectiveness of the Water Chemistry Program should be verified to ensure that crack initiation and growth do not occur. The GALL Report recommends further evaluation to manage crack initiation and growth from SCC and cyclic loading for these systems to verify the effectiveness of the Water Chemistry Program. A one-time inspection of select components and susceptible locations is an acceptable method to ensure that crack initiation and growth do not occur and that component intended functions will be maintained during the period of extended operation.

In the LRA, the applicant credited the Water Chemistry Program for managing such aging effects as loss of material due to general, pitting, and crevice corrosion, MIC, cracking due to SCC, and fouling due to corrosion product buildup in stagnant and low-flow regions by controlling the environment to which internal surfaces of systems and components are exposed. The aging effects are minimized by controlling the chemical species that cause the underlying mechanisms that result in these aging effects. The program provides assurance that no elevated level of contaminants and oxygen exists in the systems and components covered by the program, thus minimizing the occurrences of aging effects and maintaining component ability to perform intended functions. The program is based on the guidelines in EPRI TR-105714, Revision 5, and TR-102134, Revision 5.

In the LRA, the applicant credited the One-Time Inspection and Water Chemistry Programs, and its corresponding GALL AMP XI.M2 for managing the aging effects in stagnant or low-flow areas of components. The Water Chemistry Program may be credited for managing aging effects of components where water flow is low or stagnant conditions exist. However, water chemistry sampling points may not indicate conditions in stagnant or low-flow locations. Therefore, confirmatory inspections are appropriate. Accordingly, to ensure that significant degradation does not occur and that components will continue to perform intended functions during the period of extended operation, a one-time inspection of selected components will be performed.

Based on a review of the applicant's information in LRA Section 3.3.2.2.9, the staff concludes that the applicant had applied the appropriate AMPs including its One-Time Inspection Program to ensure management of aging effects.

Based on the programs identified above, the staff concludes that the applicant has met the criteria of SRP-LR Section 3.3.2.2.9. For those line items that apply to LRA Section 3.3.2.2.9, the staff determines that the application is consistent with the GALL Report and the applicant has demonstrated that the aging effects will be adequately managed so that the intended function(s) will be maintained consistent with the CLB during the period of extended operation, as required by 10 CFR 54.21(a)(3).

3.3.2.2.10 Reduction of Neutron-Absorbing Capacity and Loss of Material Due to General Corrosion

The staff reviewed LRA Section 3.3.2.2.10 against the criteria in SRP-LR Section 3.3.2.2.10.

In LRA Section 3.3.2.2.10, the applicant addressed reduction of neutron-absorbing capacity and loss of material due to general corrosion.

SRP-LR Section 3.3.2.2.10 states that reduction of neutron-absorbing capacity and loss of material due to general corrosion could occur in the neutron-absorbing sheets of the spent fuel storage rack in the spent fuel storage. The GALL Report recommends further evaluation to ensure adequate management of these aging effects.

In LRA Section 3.3.2.2.10, the applicant stated that the spent fuel storage racks contain boron carbide and Boraflex neutron-absorbing material. Due to industry concerns about degradation of Boraflex, soluble boron is maintained at 1720 ppm per TS limiting conditions of operation 3.7.15 to maintain k-effective less than or equal to 0.95. Criticality calculations take credit for the soluble boron in the SFP water and conclude that Boraflex material need not be credited. For the storage racks containing boron carbide, no credit is taken for soluble boron. In industry experience, no boron carbide sheathed in stainless steel has experienced loss of material due to corrosion. As to reduction of neutron-absorbing capability due to corrosion, the NRC memorandum, "Resolution of Spent Fuel Storage Pool Action Plan Issue," dated July 26, 1996, Section 3.3.1 states, "degradation in neutron absorption performance has not been observed in materials other than Boraflex."

In the LRA, the applicant stated:

> Therefore, no plant specific aging management program is required to manage reduction of neutron-absorbing capacity and loss of material due to general corrosion, as addressed in NUREG-1801 Item VIIA2.1-b, in VIIA2, Spent Fuel Storage. The stainless steel sheathing for the boron carbide will be age managed for loss of material due to pitting or crevice corrosion cracking via the Water Chemistry Program, consistent with NUREG-1801 item VIIA2.1c. Boraflex, as discussed in NUREG-1801 item VIIA2.1a is not in-scope of license renewal since it performs no intended function.

The staff's review of LRA Section 3.3.2.2.10 identified areas in which additional information was necessary to complete the review of the applicant's aging management results. The applicant responded to the staff's RAI as discussed below.

In RAI 3.5.2-1, dated August 23, 2005, staff requested that the applicant provide a detailed description of the neutron absorbing panels, to include, at a minimum, the B_4C content, existing areal density, BC matrix, and composition. In addition, the staff requested that the applicant clarify that this material is also known as Carborundum.

In its response, by letter dated September 23, 2005, the applicant stated:

> The boron carbide neutron absorber plate is B_4C powder bonded together in a carbonaceous matrix. The absorber is 50% B_4C by volume with the remainder being phenolic binder material and voids. The areal density is 0.0959 grams per cm squared. Specifications for the B_4C powder used for the absorber plates require that the median particle size be 125 microns by volume consistent with maintaining the criticality allowance for heterogeneity. Specifications also require that no more than 3% of the boron in the powder be in the oxide (B_2O_3) form. This B_2O_3 boron is not credited in the minimum boron loading considered in the criticality analysis.
>
> While the Carborundum Company is the manufacturer of Palisades' boron carbide (B_4C) plates, the B_4C material is not the material referred to by the trade name carborundum. NMC's research indicates that the material known by trade name carborundum is silicon carbide and not boron carbide.

In RAI 3.5.3-2, dated August 23, 2005, the staff requested that the applicant further discuss and justify that there is no aging effect applicable to carborundum given the discussion in the referenced documents during the site audit. The applicant was also requested to explain the possible cause for stuck fuel assemblies in the spent fuel pool discussed in the referenced documents and discuss why this component is not defined as a TLAA per 10 CFR 54.3.

In its response, by letter dated September 23, 2005, the applicant stated:

> 1) NMC concurs with the questioner's statement that having fuel assemblies sticking in spent fuel racks is not prevalent in the industry, even though racks of similar material and environments exist at other facilities. This is consistent with NMC's conclusion that Palisades' condition is more a problem of rack size as it relates to fuel assembly size (i.e. the fuel assembly is a tight fit in the cell), and/or the possible lack of adequate vent holes to relieve the gas generated during irradiation of the absorber material.
>
> 2) Palisades' documentation from when the racks were originally installed indicates that vent hole drilling was problematic, and there was uncertainty whether all cells were successfully vented. Correspondence indicates vent hole positions were altered because initial hole drilling did not provide satisfactory venting of the cells. It is probable, then, that not all cells were properly drilled and vented during original rack installation. Thus, improperly drilled vent holes, and not corrosion or other aging effects, is the most likely cause for the stuck assemblies documented in the CAPs referenced above.
>
> 3) The compatibility of stainless steel in treated borated water with proper chemistry controls is well established in the GALL and the Palisades LRA. Material compatibility between the boron carbide neutron absorber plates and the stainless steel spent fuel racks is excellent because the absorber plates do not

exhibit a galvanic potential with respect to the stainless steel storage cans. The gas generated by the neutron absorber binder material under irradiation is primarily hydrogen, and is reducing rather than oxidizing in nature, which would minimize corrosion. Therefore, NMC has concluded that the sticking of fuel assemblies in the rack is not due to rack degradation related to corrosion or aging.

The applicant further stated that the NUS fuel racks with the B_4C absorber should not be considered a TLAA for, based on the following:

1) The Carborundum Test Report does not specify 40 years or any specific period as the qualified service life. Rather, it subjects the test specimens to a fluence value of $10E^{11}$ rads gamma radiation to ensure the physical and mechanical properties are maintained "...for a period of time in excess of the designed original life of the reactor facility."

2) The Carborundum Test Report CB078-299 describes test conditions for the coupons that are more severe than actually encountered during use in the Palisades racks. For example, the test coupons were immersed in a circulating fluid during irradiation that does not represent actual use in the racks. In the Palisades racks the material is encapsulated with only a small vent hole providing a possible exposure point to spent fuel pool water. This configuration would substantially reduce any dissolution or washout of the B_4C material.

3) Since the year 2000, Palisades has sampled the spent fuel pool water periodically for Total Organic Carbon (TOC), with the results being very low, stable, TOC values (typically less than 0.2 ppm). These sample results indicate that the B_4C absorber is not degrading. This periodic confirmatory sampling adequately assures that neutron absorber degradation is not occurring.

4) Palisades has been in contact with another nuclear plant which has the same basic rack design with its B_4C absorbers. This plant has a test coupon surveillance program in progress that has results dating back to 1982. It shows that test coupon B^{10} loading has not degraded during the duration of the test. The most recent test results from early in 2005 are essentially the same as the results documented in the early 1980's. This plant's test program inspects the test coupons every three (3) years. After each test, the coupons are placed adjacent to fuel assemblies freshly discharged from the core, which would subject them to higher than average neutron and gamma doses. Thus the test coupons receive bounding high irradiation doses that exceed those likely to be received by the actual absorber material in the racks. It is expected that, should a degrading trend be noted at another plant in the future, industry operating experience would be reported that would alert Palisades to a potential developing rack problem.

In a supplemental letter, dated November 18, 2005, the applicant stated:

...margin for any minor degradation of neutron absorption capability exists since the criticality analysis for the Palisades spent fuel racks in question (manufactured by NUS) includes conservatisms. This margin comes from the fact that the purchase specification for Boron 10 (B10) areal density in the rack required a 0.0959 grams/cm2 value whereas the criticality analysis credits only 90% of that B10 density. Additionally, the analysis assumptions for each B4C absorber panel were slightly narrower and thinner than design (assumed width was 0.1" less and the panel thickness 0.02" thinner), thus providing additional boron areal density margin in the analysis. In addition, the criticality analysis currently assumes the spent fuel pool water is unborated.

The applicant further stated:

> ...to further validate that the NUS racks have no aging effects requiring management and will remain acceptable for the extended period of operation, NMC will perform a neutron absorption ("blackness") test of selected cells in the NUS spent fuel racks prior to March 24, 2011, to validate that there is no significant degradation of the neutron absorption capability. An additional test will be performed within the first 10 years following the start of the period of extended operation. If degradation is identified in either test, an evaluation of the condition will be performed under the NMC Corrective Action Program. This evaluation will consider the potential need for additional or more frequent testing if applicable.

On this basis of its review, the staff determines that, although there is no AMP in place, sufficient procedures are available to monitor degradation in the neutron absorbing panels. When the neutron absorbing panels were installed, a coupon program was not implemented to monitor panel degradation, therefore, coupons were not installed in the SFP.

Given the absence of a coupon program, the applicant has been using the operating experience of a sister plant as a method for monitoring neutron degradation. The sister plant removes a coupon for testing every three years. Test results have shown that no degradation has occurred up-to-date. Any degradation trend reported at this plant will alert the applicant of a potential problem and corrective actions will be taken. In addition, the applicant tests for the presence of organic carbon in the SFP. The absorbing panels currently in place are bonded in a carbonaceous matrix. Any increase in the amount of organic carbon present in the SFP could be coming from the neutron absorbing panels, which would alert the applicant of a possible neutron panel degradation. This test is currently performed on a monthly basis.

The neutron absorbing panels are used to maintain criticality in the SFP. The areal density of the panels, along with the thickness and width, are used in the criticality analysis of the SFP. If the absorbing panels have sustained degradation, their capacity to maintain criticality control could be reduced. Therefore, in anticipation of possible degradation, the applicant has introduced some conservatism in their criticality calculation by crediting only 90 percent of the minimum areal density of the absorbing panels.

Also, in the criticality analysis the values the applicant used for width and thickness of the panels are slightly less than in their original design. This provides another source of conservatism into the calculation. In addition to neutron absorbing panels, criticality is controlled by using borated water. The applicant's criticality analysis currently assumes that the SFP water is unborated.

During discussions with the applicant, the staff indicated that additional information is required to complete the staff's evaluation. Since no coupons exist for testing, the applicant must provide assurance that they will be capable of identifying and mitigating any degradation of the boron carbide over the period of extended operation. The applicant has committed to perform neutron absorption testing of some cells before the start of period of extended operation, followed by testing within the first 10 years after the start of the period of operation (Commitment 43). The test to be performed, identified as blackness testing, will provide information on the physical state of the panels, and it will help identify if any degradation is occurring. If any degradation is identified, the applicant committed to perform an evaluation of the condition of the panels under their corrective action program. The results of this evaluation will determine if more testing is required to be performed.

On the basis of its review, the staff concludes that the applicant has demonstrated that the aging effects associated with the neutron absorbing panels will be adequately managed so that the intended functions will be maintained consistent with the CLB for the period of extended operation, as required by 10 CFR 54.21(a)(3).

3.3.2.2.11 Loss of Material Due to General, Pitting, Crevice, and Microbiologically Influenced Corrosion

The staff reviewed LRA Section 3.3.2.2.11 against the criteria in SRP-LR Section 3.3.2.2.11.

In LRA Section 3.3.2.2.11, the applicant addressed loss of material of buried piping and fittings due to general, pitting, and crevice corrosion, and MIC.

SRP-LR Section 3.3.2.2.11 states that loss of material due to general, pitting, and crevice corrosion, and MIC could occur in the underground piping and fittings in the open-cycle cooling water system (SWS) and in the diesel fuel oil system. The Buried Piping and Tanks Inspection Program relies on industry practice, frequency of pipe excavation, and operating experience to manage the effects of loss of material from general, pitting, and crevice corrosion, and MIC. The effectiveness of the Buried Piping and Tanks Inspection Program should be verified to evaluate an applicant's inspection frequency and operating experience with buried components to ensure that loss of material does not occur.

In the LRA, the applicant stated that its Buried Services Corrosion Monitoring Program manages aging effects on the external surfaces of carbon, low-alloy, and stainless steel components (e.g., tanks, piping) buried in soil or sand. This program includes (a) visual inspections of external surfaces of buried components for evidence of coating damage and substrate degradation to manage the aging effects and (b) visual inspection of the external surfaces of buried stainless steel components for evidence of crevice corrosion, pitting, and MIC. The periodicity of these inspections for carbon, low-alloy, and stainless steel will be based on opportunities for inspection like scheduled maintenance work. In response to staff audit questions, by letter dated September 16, 2005, the applicant revised Commitment 39 as follows:

> Visual inspections of a sample of buried carbon, low-alloy, and stainless steel components will be performed within ten years prior to entering, and within ten years after entering, the period of extended operation. Prior to the tenth year of each period, NMC will perform an evaluation of available data to determine sufficient opportunistic inspections have been performed within that period to assess the condition of the components. If insufficient data exists, focused inspection(s) will be performed as needed.

The staff reviewed the applicant's Buried Services Corrosion Monitoring Program as documented in SER Section 3.0.3.1.3.

Based on the programs identified above, the staff concludes that the applicant has met the criteria of SRP-LR Section 3.3.2.2.11. For those line items that apply to LRA Section 3.3.2.2.11, the staff determines that the application is consistent with the GALL Report and the applicant has demonstrated that the aging effects will be adequately managed so that the intended function(s) will be maintained consistent with the CLB during the period of extended operation, as required by 10 CFR 54.21(a)(3).

Conclusion. On the basis of its review, for component groups evaluated in the GALL Report for which the applicant claimed consistency with the GALL Report and for which further evaluation is recommended, the staff determines that the applicant has adequately addressed the issues that were further evaluated. The staff finds that the applicant has demonstrated that the aging effects will be adequately managed so that the intended function(s) will be maintained consistent with the CLB for the period of extended operation, as required by 10 CFR 54.21(a)(3).

3.3.2.3 AMR Results That Are Not Consistent with or Not Addressed in the GALL Report

Summary of Technical Information in the Application. In LRA Tables 3.3.2-1 through 3.3.2-17, the staff reviewed additional details of AMR results for material, environment, AERM, and AMP combinations not consistent with or not addressed in the GALL Report.

In LRA Tables 3.3.2-1 through 3.3.2-17, the applicant indicated, via Notes F through J, that the combination of component type, material, environment, and AERM does not correspond to a line item in the GALL Report. The applicant provided further information concerning how the aging effects will be managed. Specifically, Note F indicates that the material for the AMR line item component is not evaluated in the GALL Report. Note G indicates that the environment for the AMR line item component and material is not evaluated in the GALL Report. Note H indicates that the aging effect for the AMR line item component, material, and environment combination is not evaluated in the GALL Report. Note I indicates that the aging effect identified in the GALL Report for the line item component, material, and environment combination is not applicable. Note J indicates that neither the component nor the material and environment combination for the line item is evaluated in the GALL Report.

Staff Evaluation. For component type, material, and environment combinations not evaluated in the GALL Report, the staff reviewed the applicant's evaluation to determine whether the applicant had demonstrated that the aging effects will be adequately managed so that the intended function(s) will be maintained consistent with the CLB during the period of extended operation. The staff's evaluation is discussed in the following sections.

The staff's review identified areas in which additional information was necessary to complete the review of the applicant's aging management results. The applicant responded to the staff's RAIs as discussed below.

In RAI 3.3-1, dated March 8, 2006, the staff stated that in LRA Sections 3.3 and 3.4, the applicant proposed to manage loss of material of carbon steel/cast iron components internally exposed to raw water/steam using only the One-Time Inspection Program. The One-Time

Inspection Program provides a means of verifying that an aging effect is either not occurring or progressing so slowly as to have a negligible effect on the intended function of the structure or component. Carbon steel/cast iron is subject to significant corrosion under moisture. Therefore, the staff requested that the applicant provide the justification for only using the One-Time Inspection Program.

In its response, by letter dated March 30, 2006, the applicant stated:

> For those cases where 'Loss of Material - Selective Leaching' is the Aging Effect Requiring Management, the One Time Inspection Program is the correct program, and it need not be coupled with another program. As indicated in the License Renewal Application Section B2.1.13 on pages B-97 - B-107, the Palisades One-Time Inspection Program contains the elements of the GALL program "Selective Leaching of Materials". As stated under Conclusion on page B-106, "This program is consistent with NUREG-1801, Section XI.M33, "Selective Leaching of Materials"."

> For other cases, however, NMC concurs that the One-Time Inspection Program should not be used by itself as an aging management program if aging is anticipated. While many of the items in Sections 3.3 and 3.4 have controlled internal environments that should preclude aging degradation, there are some items in these sections that are not subject to another aging management program. For completeness and clarity, therefore, the affected line items are revised to identify the applicable aging management program and/or credited activities, either as a replacement for, or in addition to, the One Time Inspection Program.

The individual responses to this general RAI for those carbon steel and cast iron line items which only cite the One-Time Inspection Program are provided in SER Sections 3.3.2.3.5, 3.3.2.3.9, 3.3.2.3.11, 3.3.2.3.15, and 3.3.2.3.16.

In RAI 3.3-2, dated March 8, 2006, the staff stated that in LRA Sections 3.3 and 3.4, the applicant proposed to manage loss of material of metal components exposed to oil environment using the One-Time Inspection Program. The GALL Report recommended that loss of material and reduction of heat transfer aging effects are to be managed by a plant specific or lubricating/fuel oil analysis AMP along with the One-Time Inspection Program to ensure the effectiveness of the AMP. Therefore, the staff requested that the applicant provide the following:

(1) The justification of methods used to ensure the oil remains free of contaminants, which might degrade the components.

(2) The description of preventive maintenance to ensure that the reduction of heat transfer function does not reach unacceptable levels for components performing heat transfer functions.

In its response, by letter dated March 30, 2006, the applicant stated:

> NMC concurs that the One-Time Inspection Program should not be used by itself as an aging management program for the various materials in oil environments if

aging is anticipated. In the context of this question, however, routine oil sampling and analysis activities will maintain the oil system contaminants (primarily water and particulates) within acceptable limits, thereby preserving an environment that is not conducive to loss of material, cracking, or reduction of heat transfer. For completeness, therefore, the affected line items are revised to indicate that oil sampling and analysis activities would maintain the environments such that aging would not be expected, and One Time Inspection is used for confirmation that significant aging is not occurring.

The individual responses to this general RAI for the changes to the affected LRA line items in Sections 3.3 and 3.4, which only cite One-Time Inspection for the oil environment, are provided in SER Sections 3.3.2.3.4, 3.3.2.3.6, 3.3.2.3.12, 3.4.2.3.3, and 3.4.2.3.7.

3.3.2.3.1 Auxiliary Systems – Chemical and Volume Control System – Summary of Aging Management Evaluation – LRA Table 3.3.2-1

The staff reviewed LRA Table 3.3.2-1, which summarizes the results of AMR evaluations for the CVC system component groups.

The applicant proposed the One-Time Inspection Program to manage loss of material of stainless steel boric acid storage tanks exposed to an internal air (potential for borated water exposure) environment.

The staff's review of the One-Time Inspection Program is documented in SER Section 3.0.3.2.6. The program addresses potentially long incubation periods for certain aging effects, including various corrosion mechanisms, cracking, and selective leaching, and provides a means of verifying that an aging effect either does not occur or progresses so slowly as to have negligible effect on the intended function of the structure or component. Current industry research and operating experience confirm that stainless steel in air is not subject to aging effects that could be of concern during the period of extended operation. Stainless steel is not susceptible to general corrosion when subjected to borated water environments, as cited in EPRI NP-7077, Revision 2, Project 2493, "PWR Primary Water Chemistry Guideline," dated November 1990. On this basis, the staff concludes that management by the One-Time Inspection Program of loss of stainless steel material exposed to an internal air (potential for borated water exposure) environment in the CVC system is acceptable.

In LRA Table 3.3.2-1, the applicant proposed the Boric Acid Corrosion Program to manage loss of material of brass oil cooler shell exposed to a plant indoor air environment (external).

The applicant's Boric Acid Corrosion Program was reviewed by the staff as documented in SER Section 3.0.3.2.2. The staff determines that management of loss of brass oil cooler shell material exposed to an external air environment in the CVC system is acceptable.

The applicant proposed the One-Time Inspection Program to manage loss of stainless CVC spray steel nozzle material exposed to an external gas environment.

The staff reviewed the applicant's One-Time Inspection Program as documented in SER Section 3.0.3.2.6. Current industry research and operating experience confirm that stainless steel in air/gas is not subject to aging effects that could be of concern during the period of extended operation. On this basis, the staff concludes that One-Time Inspection Program management of loss of the stainless steel CVC spray component type material exposed to an external gas environment in the CVC system is acceptable.

The applicant proposed the Closed Cycle Cooling Water Program to manage loss of copper-nickel oil cooler tubes material exposed to an internal treated water environment.

The staff reviewed the Closed Cycle Cooling Water Program as documented in SER Section 3.0.3.2.3. The program manages aging effects in CCCW systems not subject to significant sources of contamination in which water chemistry is controlled and heat is not directly rejected to the ultimate heat sink. On the basis of its review of the applicant's plant-specific and industry operating experience, the staff determines that the aging effect of loss of copper-nickel material exposed to an internal treated water environment is effectively managed by the Closed Cycle Cooling Water Program. On this basis, the staff concludes that management of loss of CVC system material is acceptable.

The applicant proposed the One-Time Inspection Program to manage loss of brass oil cooler shell and stainless steel CVC oil tubing material in the CVC system exposed to an internal oil environment.

The staff reviewed the applicant's One-Time Inspection Program as documented in SER Section 3.0.3.2.6. On the basis of its review, the staff concludes that One-Time Inspection Program management of loss of brass oil cooler shell and stainless steel CVC oil tubing material exposed to an internal oil environment in the CVC system is acceptable.

The applicant proposed the Bolting Integrity Program to manage loss of preload of stainless steel fasteners exposed to both a plant indoor air external environment and a containment air external environment.

The applicant's Bolting Integrity Program was reviewed by the staff as documented in SER Section 3.0.3.2.1. The staff concludes that management of loss of preload of stainless steel fasteners exposed to both a plant indoor air external environment and a containment air external environment in the CVC system is acceptable.

The applicant proposed the Closed Cycle Cooling Water Program augmented by the One-Time Inspection Program to manage loss of material from component types stainless steel letdown HX tubes tube sheets and letdown HX tubes exposed to a treated water external environment.

The staff reviewed the Closed Cycle Cooling Water and One-Time Inspection Programs as documented in SER Sections 3.0.3.2.3 and 3.0.3.2.6, respectively. On the basis of its review, the staff determines that the aging effect of loss of stainless steel material exposed to an external treated water environment is effectively managed by the Closed Cycle Cooling Water Program augmented by One-Time Inspection Program.

The applicant proposed the ASME Section XI IWB, IWC, IWD, IWF Inservice Inspection Program to manage the reduction of fracture toughness of CASS letdown stop valve CV-2001 exposed to a treated water internal environment.

As stated in GALL AMP XI.M12, "Thermal Aging Embrittlement of Cast Austenitic Stainless Steel (CASS)," the staff's conservative bounding integrity analysis shows that thermally-aged CASS valve bodies are resistant to failure and the adequacy of ISI according to ASME Code Section XI has been demonstrated by an NRC bounding integrity analysis. On this basis, the staff concludes that the applicant had appropriately addressed the aging effect of reduction of fracture toughness, as recommended by the GALL Report.

The applicant proposed the One-Time Inspection Program to manage loss of stainless steel flow elements material exposed to an internal treated water environment.

The staff reviewed the One-Time Inspection Program as documented in SER Section 3.0.3.2.6. The PWR treated water environment contains boron, which is a recognized corrosion inhibitor. On the basis of its review, the staff concludes that the One-Time Inspection Program effectively manages the aging effect of loss of stainless steel material exposed to an internal treated water environment.

On the basis of its review, as discussed above, the staff concludes that the applicant has demonstrated that the aging effects associated with the components will be adequately managed so that the intended function(s) will be maintained consistent with the CLB for the period of extended operation, as required by 10 CFR 54.21(a)(3).

3.3.2.3.2 Auxiliary Systems – Circulating Water System – Summary of Aging Management Evaluation – LRA Table 3.3.2-2

The staff reviewed LRA Table 3.3.2-2, which summarizes the results of AMR evaluations for the circulating water system component groups.

The results of these evaluations are all consistent with the GALL Report.

On the basis of its review, as discussed above, the staff concludes that the applicant has demonstrated that the aging effects associated with the components will be adequately managed so that the intended function(s) will be maintained consistent with the CLB for the period of extended operation, as required by 10 CFR 54.21(a)(3).

3.3.2.3.3 Auxiliary Systems – Component Cooling Water System – Summary of Aging Management Evaluation – LRA Table 3.3.2-3

The staff reviewed LRA Table 3.3.2-3, which summarizes the results of AMR evaluations for the component cooling water (CCW) system component groups.

The applicant proposed the Closed Cycle Cooling Water Program to manage loss of carbon steel accumulators material internally exposed to an air environment. As documented in the Audit and Review Report, the applicant clarified that the component type identified by this line item is the CCW surge tank and addressed the internal air in the tank (above the water line). The Closed Cycle Cooling Water Program takes credit for internal one-time inspections of CCW components. The area above the water line in the tank would be inspected for aging degradation.

The staff reviewed the Closed Cycle Cooling Water Program as documented in SER Section 3.0.3.2.3. On the basis of its review, the staff concludes that the applicant had appropriately addressed this aging effect using the Closed Cycle Cooling Water Program, which takes credit for one-time inspections.

In LRA Table 3.3.2-3, the applicant proposed the Closed Cycle Cooling Water and One-Time Inspection Programs to manage loss of material/pitting, crevice and galvanic corrosion of nickel-based alloy materials for coolers exposed to an external treated water (i.e., CCW) environment.

The staff reviewed the applicant's Closed Cycle Cooling Water and One-Time Inspection Programs as documented in SER Sections 3.0.3.2.3 and 3.0.3.2.6, respectively. On the basis of its review of the applicant's plant-specific and industry operating experience, the staff determines that the aging effects of loss of nickel-based alloy material exposed to a treated water (i.e., CCW) environment are effectively managed by the Closed Cycle Cooling Water and One-Time Inspection Programs and concluded that the management of loss of nickel-based alloy coolers material externally exposed to treated water in the CCW system is acceptable.

In LRA Table 3.3.2-3, the applicant proposed the One-Time Inspection and Water Chemistry Programs to manage loss of nickel-based alloy coolers material exposed to an internal treated water environment.

The staff reviewed the One-Time Inspection and Water Chemistry Programs as documented in SER Sections 3.0.3.2.6 and 3.0.3.1.10, respectively. The Water Chemistry Program is credited for managing such aging effects as loss of material due to general, pitting, and crevice corrosion, cracking due to SCC, and SG tube degradation caused by denting, IGA, and ODSCC by controlling the environment to which internal surfaces of systems and components are exposed. On the basis of its review of the applicant's plant-specific and industry operating experience, the staff determines that the aging effect of loss of nickel-based alloy material exposed to an internal treated water environment is effectively managed by the One-Time Inspection and Water Chemistry Programs and concludes that the management of loss of material for nickel-based alloy coolers exposed to an internal treated water environment in the CCW system is acceptable.

In LRA Table 3.3.2-3, the applicant proposed the Bolting Integrity Program to manage loss of preload of stainless steel fasteners exposed to both plant indoor air (external) and containment air (external) environments.

The staff reviewed the applicant's Bolting Integrity Program as documented in SER Section 3.0.3.2.1. The staff concludes that the management of loss of preload of stainless steel fasteners exposed to both plant indoor air external and containment air external environments in cooling water system components is acceptable.

In LRA Table 3.3.2-3, the applicant proposed the Boric Acid Corrosion Program to manage loss of copper alloy pipe and fittings and bronze valves and dampers materials exposed to an external containment air environment.

The staff reviewed the Boric Acid Corrosion Program as documented in SER Section 3.0.3.2.2. The staff concludes that the management of loss of copper alloy and bronze materials for pipe and fittings and valve and dampers component types exposed to an external containment air environment in the CCW system is acceptable.

As documented in the Audit and Review Report, the staff noted that in LRA Table 3.3.2-3, the applicant proposed the Closed Cycle Cooling Water Program to manage cracking and loss of material for stainless steel waste gas compressors in a gas environment. By letter dated August 27, 2005, the applicant stated that cracking due to SCC/IGA is not a potential aging mechanism for the interior of the waste gas compressor (C-50A/B) cooler tubes because temperatures are not greater than 140 °F. The applicant also stated that LRA Table 3.3.2-3 had been revised to delete cracking of the stainless steel waste gas compressor tubes as this aging mechanism does not occur. The compressor tubes are cooled by CCW and the tube temperature is less than 140 °F. As to the loss of stainless steel waste gas compressor material in a gas environment, the staff reviewed industry experience and found no aging effect identified for stainless steel in a dry gas environment. However, the applicant proposed its Closed Cycle Cooling Water Program to manage loss of stainless steel waste gas compressor material in gas. The staff finds this proposal acceptable because no aging effect is expected to occur.

On the basis of its review, as discussed above, the staff concludes that the applicant has demonstrated that the aging effects associated with the components will be adequately managed so that the intended function(s) will be maintained consistent with the CLB for the period of extended operation, as required by 10 CFR 54.21(a)(3).

3.3.2.3.4 Auxiliary Systems – Compressed Air System – Summary of Aging Management Evaluation – LRA Table 3.3.2-4

The staff reviewed LRA Table 3.3.2-4, which summarizes the results of AMR evaluations for the compressed air system component groups.

In LRA Table 3.3.2-4, the applicant proposed the Boric Acid Corrosion Program to manage loss of copper alloy, brass, and bronze materials for component types of fasteners, filters/strainers, HXs, pipe and fittings, and valves and dampers exposed to external containment air, external plant indoor air, and air environments.

The staff reviewed the Boric Acid Corrosion Program as documented in SER Section 3.0.3.2.2. The staff concludes that the management of loss of copper alloy, brass, and bronze materials for component types of fasteners, filters/strainers, HXs, pipe and fittings, and valves and dampers exposed to external containment air, external plant indoor air, and air environments in the compressed air system is acceptable.

In LRA Table 3.3.2-4, the applicant proposed the One-Time Inspection Program to manage loss and/or cracking of copper alloy, bronze, brass, aluminum, and stainless steel materials for component types of filters/strainers, HXs, pipe and fittings, traps (steam), valves and dampers exposed to an internal air environment.

As documented in the Audit and Review Report, the staff noted that the applicant used the One-Time Inspection Program in lieu of the Compressed Air Monitoring Program, which is recommended by the GALL Report, to manage loss of material and/or cracking for the components in the compressed air system. By letter dated August 27, 2005, the applicant stated that it would prepare a formal Compressed Air Monitoring Program. The staff reviewed the

Compressed Air Monitoring Program as documented in SER Section 3.0.3.1.11. The LRA was updated by letter dated October 31, 2005, to describe this new program. The compressed air system components that had credited the One-Time Inspection Program to manage aging effects in an internal environment of air now credit the new Compressed Air Monitoring Program. On this basis, the staff concludes that the management of loss of material and/or cracking of copper alloy, bronze, brass, aluminum, and stainless steel materials in the compressed air system is acceptable. The staff reviewed the Compressed Air Monitoring Program as documented in SER Section 3.0.3.1.11.

In LRA Table 3.3.2-4, the applicant proposed the Bolting Integrity Program to manage loss of preload of carbon steel and copper alloy fasteners exposed to both plant indoor air external and containment air external environments and stainless steel fasteners exposed to a plant indoor air external environment.

The staff reviewed the Bolting Integrity Program as documented in SER Section 3.0.3.2.1. The staff concludes that the management of loss of preload of carbon steel and copper alloy fasteners exposed to both plant indoor air external and containment air external environments and stainless steel fasteners exposed to a plant indoor air external environment in the compressed air system is acceptable.

In LRA Table 3.3.2-4, the applicant proposed the One-Time Inspection Program to manage loss of stainless steel filters and strainers and brass valves and dampers materials exposed to internal oil and oil environments.

The GALL Report recommends management of the loss of material aging effect by a plant-specific or lubricating/fuel oil analysis AMP with a One-Time Inspection Program to ensure AMP effectiveness.

In its response to general RAI 3.3-2, the applicant stated, "In Table 3.3.2-4 on page 3-132, for the line item 'Filters/Strainers' with an 'Oil (Int)' environment, note 311 is hereby added with the existing Notes."

The staff's review found the applicant's response acceptable because, with the addition of note 311, selective leaching of materials is included in the One-Time Inspection Program.

In LRA Table 3.3.2-4, the applicant proposed the Open Cycle Cooling Water Program to manage loss of bronze HX material exposed to an internal air environment.

The staff reviewed the Open Cycle Cooling Water Program as documented in SER Section 3.0.3.1.8. The program manages aging effects caused by exposure of internal surfaces of metallic components to raw, untreated (e.g., service) water. On the basis of its review, the staff concludes that the management of loss of bronze HX material exposed to an internal air environment in the compressed air system is acceptable.

In LRA Table 3.3.2-4, the applicant proposed the Open Cycle Cooling Water Program to manage loss of bronze and carbon steel HX material exposed to external and internal raw water environments. The staff concludes that the management of loss of carbon steel and bronze HX material exposed to raw water in the compressed air system is acceptable.

In LRA Table 3.3.2-4, the applicant proposed the One-Time Inspection Program to manage cracking of brass valves and dampers exposed to an oil environment.

The staff reviewed the One-Time Inspection Program as documented in SER Section 3.0.3.2.6. On the basis of its review of the applicant's plant-specific and industry operating experience, the staff determines that the aging effect of cracking of brass material exposed to an oil environment is effectively managed by the One-Time Inspection Program and concludes that the management of cracking of brass valves and dampers exposed to an oil environment in the compressed air system is acceptable.

On the basis of its review, as discussed above, the staff concludes that the applicant has demonstrated that the aging effects associated with the components will be adequately managed so that the intended function(s) will be maintained consistent with the CLB for the period of extended operation, as required by 10 CFR 54.21(a)(3).

3.3.2.3.5 Auxiliary Systems – Containment Air Recirculation and Cooling System – Summary of Aging Management Evaluation – LRA Table 3.3.2-5

The staff reviewed LRA Table 3.3.2-5, which summarizes the results of AMR evaluations for the containment air recirculation and cooling system component groups.

In LRA Table 3.3.2-5, the applicant proposed the System Monitoring Program to manage heat transfer degradation of copper alloy containment air cooler coils exposed to an air external environment.

In its response, to general RAI 3.3-1, the applicant stated that, "In Table 3.3.2-5, the line item for 'Drip Pans' of 'Carbon Steel' in 'Raw Water (Int),' the 'One-Time Inspection Program' is hereby replaced by the 'System Monitoring Program' to manage Loss of Material. The corresponding NUREG 1801 Volume 2, Table 1, and Notes entries are 'VII.CI.1-a' and '3.3.1-15' and 'C,' respectively."

The staff review found the applicant's use of the System Monitoring Program acceptable.

The applicant's System Monitoring Program was reviewed by the staff as addressed separately in SER Section 3.0.3.3.1. The staff concludes that the management of heat transfer degradation of copper alloy containment air cooler coils exposed to an air external environment in the containment air recirculation and cooling system is acceptable.

In LRA Table 3.3.2-5, the applicant proposed the Bolting Integrity Program to manage loss of preload of stainless steel fasteners exposed to external containment air.

The staff reviewed the applicant's Bolting Integrity Program as documented in SER Section 3.0.3.2.1. The staff concludes that the management of loss of preload of stainless steel fasteners exposed to external containment air in the containment air recirculation and cooling system is acceptable.

On the basis of its review, as discussed above, the staff concludes that the applicant has demonstrated that the aging effects associated with the components will be adequately managed so that the intended function(s) will be maintained consistent with the CLB for the period of extended operation, as required by 10 CFR 54.21(a)(3).

3.3.2.3.6 Auxiliary Systems – Emergency Power System – Summary of Aging Management Evaluation – LRA Table 3.3.2-6

The staff reviewed LRA Table 3.3.2-6, which summarizes the results of AMR evaluations for the emergency power system component groups.

In LRA Table 3.3.2-6, the applicant proposed the Closed Cycle Cooling Water Program to manage loss of material/pitting, crevice, and galvanic corrosion of copper alloy material for pipe and fittings component types exposed to an internal treated water environment.

The staff reviewed the applicant's Closed Cycle Cooling Water Program as documented in SER Section 3.0.3.2.3. On the basis of its review of the applicant's plant-specific and industry operating experience, the staff determines that the aging effect of loss of material/pitting, crevice, and galvanic corrosion of copper alloy material exposed to an internal treated water (i.e., CCW) environment is effectively managed by the Closed Cycle Cooling Water Program and concludes that the management of loss of material/pitting, crevice, and galvanic corrosion in the emergency power system is acceptable.

In LRA Table 3.3.2-6, the applicant proposed the Diesel Fuel Monitoring and Storage Program as augmented by the Fire Protection or One-Time Inspection Programs to manage cracking and loss of brass, bronze, and copper alloy materials for valves and dampers component types exposed to an internal oil environment.

The staff reviewed the applicant's Diesel Fuel Monitoring and Storage, Fire Protection, and One-Time Inspection Programs as documented in SER Sections 3.0.3.2.4, 3.0.3.2.5, and 3.0.3.2.6, respectively. The Diesel Fuel Monitoring and Storage Program assures the continued availability and quality of fuel oil used in diesel generators and diesel fire pumps. The Fire Protection Program includes (a) fire barrier inspections, (b) electric and diesel-driven fire pump tests, and (c) periodic maintenance, testing, and inspection of water-based fire protection systems. On the basis of its review of the applicant's plant-specific and industry operating experience, the staff determines that the aging effects of cracking and loss of brass, bronze, and copper alloy materials exposed to an internal oil environment are effectively managed by the Diesel Fuel Monitoring and Storage Program as augmented by the Fire Protection Program or the One-Time Inspection Program and concludes that the management of cracking and loss of material in the emergency power system is acceptable.

In its response to general RAI 3.3-2, the applicant stated, "Note 311 is hereby changed from 'Not Used' to 'The Oil environment for this component is managed by periodic oil sampling and analysis activities, and the effectiveness of these activities will be verified by a one-time inspection."

The applicant further stated that the Note 312 is hereby changed from "Not used" to the following:

> This is a component in the Primary Coolant Pump oil collection system which is designed to collect oil leakage from a Primary Coolant Pump motor. It is being

age managed for Fluid Pressure Boundary even though the system is vented to the containment atmosphere, and would only experience a small static head pressure. Any oil entering the tank from a Primary Coolant Pump oil leak would be oil that is subject to sampling and analysis. Therefore, Palisades does not expect significant age-related degradation of these components, and will perform a one-time inspection to verify that this is the case.

The applicant also stated that the Note 309 is changed from "Not Used" to the following:

This is a component in the Control Room Air Conditioning Refrigeration Unit. The environment is refrigerant with entrained oil. There is no moisture in circulation with the oil/refrigerant, and, therefore, no degradation is expected. A one-time inspection will be performed to verify that this is the case only if a refrigeration unit is required to be opened for maintenance.

The staff review found the applicant's response acceptable because the addition of note 311 adds the Selective Leaching of Materials Program to the One-Time Inspection Program.

In LRA Table 3.3.2-6, the applicant proposed the One-Time Inspection Program to manage loss and/or cracking of copper alloy, bronze, and brass materials for component types of HX instrument valve assemblies and tubing, pipe and fittings, and valves and dampers exposed to an internal or external air environment.

On the basis of its review of current industry research and operating experience, the staff finds that air on metal would not result in aging of concern during the period of extended operation. Therefore, the staff concludes that there are no significant AERMs for metal in an air environment. The One-Time Inspection Program can provide a means of verifying that aging effects either do not occur or progress so slowly as to have negligible effects. On this basis, the staff finds this management acceptable.

In LRA Table 3.3.2-6, the applicant proposed the One-Time Inspection Program to manage loss of stainless steel materials for component types of pipe and fittings and valves and dampers exposed to an internal air environment.

The staff reviewed the applicant's One-Time Inspection Program as documented in SER Section 3.0.3.2.6. On the basis of its review, the staff concludes that the management of loss of stainless steel material exposed to an internal air environment in the emergency power system by the One-Time Inspection Program is acceptable.

In LRA Table 3.3.2-6, the applicant proposed the Closed Cycle Cooling Water Program to manage loss of bronze valves and dampers and stainless steel instrument valve assemblies and tubing materials exposed to an internal treated water environment.

The staff reviewed the applicant's Closed Cycle Cooling Water Program as documented in SER Section 3.0.3.2.6. On the basis of its review, the staff concludes that the management of loss of bronze and stainless steel material in the emergency power system is acceptable.

In LRA Table 3.3.2-6, the applicant proposed the One-Time Inspection Program to manage loss of material due to selective leaching of bronze instrument valve assemblies and tubing, cast iron traps, and brass valves and dampers exposed to an internal air environment.

The staff reviewed the applicant's One-Time Inspection Program as documented in SER Section 3.0.3.2.6. On the basis of its review, the staff concludes that the management by the One-Time Inspection Program of loss of brass, cast iron, and bronze materials exposed to an internal air environment in the emergency power system is acceptable.

In LRA Table 3.3.2-6, the applicant proposed the One-Time Inspection Program to manage heat transfer degradation of carbon steel coolers and carbon steel and copper alloy HXs exposed to an external oil environment.

The GALL Report recommends management of loss of material and reduction of heat transfer aging effects by a plant-specific or lubricating/fuel oil analysis AMP with a One-Time Inspection Program to ensure AMP effectiveness.

In its response to general RAI 3.3-2, the applicant stated, "In Table 3.3.2-6, on page 3-144, for the line item 'Heat Exchangers' with an 'Oil (Ext)' environment, note 311 is hereby added with the existing Notes for 'Heat Transfer - Degradation,' with the existing Notes for 'Loss of Material - Selective Leaching,' and with the existing Notes for 'Loss of Material.'"

The staff review found the applicant's response acceptable because, with the addition of note 311, selective leaching of materials is included in the One-Time Inspection Program.

In LRA Table 3.3.2-6, the applicant proposed the Diesel Fuel Monitoring and Storage Program as augmented by the Fire Protection or One-Time Inspection Programs to manage loss of cast iron pumps, stainless steel instrument valve assemblies and tubing, and stainless steel valves and dampers materials exposed to a fuel oil environment.

The staff reviewed the applicant's Diesel Fuel Monitoring and Storage, Fire Protection, and One-Time Inspection Programs as documented in SER Sections 3.0.3.2.4, 3.0.3.2.5, and 3.0.3.2.6, respectively. On the basis of its review, the staff concludes that the management of loss of material for stainless steel and cast iron materials exposed to an internal fuel oil environment in the emergency power system is acceptable.

In LRA Table 3.3.2-6, the applicant proposed the One-Time Inspection Program to manage loss of material due to selective leaching of copper alloy HXs and cast iron pumps exposed to an internal or external oil environment.

The staff reviewed the applicant's One-Time Inspection Program as documented in SER Section 3.0.3.2.6. The applicant's program is consistent with GALL AMP XI.M33, "Selective Leaching of Materials." On this consistency basis, the staff concludes that the applicant had appropriately addressed the aging mechanism as recommended by the GALL Report.

In LRA Table 3.3.2-6, the applicant proposed the Open Cycle Cooling Water Program to manage the buildup of deposits on carbon steel coolers and HXs exposed to an internal raw water environment.

The staff reviewed the Open Cycle Cooling Water Program as documented in SER Section 3.0.3.1.8. The program manages aging effects caused by exposure of internal surfaces of metallic components to raw, untreated (e.g., service) water. On the basis of its review, the staff concludes that management of the buildup of deposits on carbon steel coolers and HXs exposed to an internal raw water environment in the emergency power system is acceptable.

In LRA Table 3.3.2-6, the applicant proposed the Open Cycle Cooling Water Program to manage heat transfer degradation of carbon steel HXs exposed to a raw water environment.

The staff reviewed the Open Cycle Cooling Water Program as documented in SER Section 3.0.3.1.8. On the basis of its review, the staff concludes that management of heat transfer degradation of carbon steel HXs exposed to a raw water environment in the emergency power system is acceptable.

In LRA Table 3.3.2-6, the applicant proposed the Bolting Integrity Program to manage loss of preload of stainless steel fasteners exposed to a plant indoor air external environment. The staff reviewed the applicant's Bolting Integrity Program as documented in SER Section 3.0.3.2.1. The staff concludes that management of loss of preload of stainless steel fasteners exposed to a plant indoor air external environment in the emergency power system is acceptable.

In LRA Table 3.3.2-6, the applicant proposed the System Monitoring Program to manage loss of galvanized filters/strainers material exposed to an atmosphere/weather external environment. The staff reviewed the applicant's System Monitoring Program as documented in SER Section 3.0.3.3.1. The staff concludes that management of loss of galvanized filters/strainers material exposed to an atmosphere/weather external environment in the emergency power system is acceptable.

In LRA Table 3.3.2-6, the applicant proposed the Closed Cycle Cooling Water Program to manage heat transfer degradation of carbon steel and copper alloy HXs exposed to an internal or external treated water environment.

The staff reviewed the applicant's Closed Cycle Cooling Water Program as documented in SER Section 3.0.3.2.3. On the basis of its review, the staff concludes that management of heat transfer degradation for carbon steel and copper alloy HXs exposed to an internal or external treated water environment in the emergency power system is acceptable.

In LRA Table 3.3.2-6, the applicant proposed the Diesel Fuel Oil Monitoring and Storage Program to manage loss of carbon steel valves and dampers material exposed to an internal air environment.

The staff reviewed the applicant's Diesel Fuel Oil Monitoring and Storage Program as documented in SER Section 3.0.3.2.4. On the basis of its review of the applicant's plant-specific and industry operating experience, the staff determined that the aging effect of loss of carbon steel material exposed to an internal air environment is effectively managed by the Diesel Fuel Oil Monitoring and Storage Program and concludes that management of the loss of material for carbon steel valves and dampers exposed to an internal air environment in the emergency power system is acceptable.

On the basis of its review, as discussed above, the staff concludes that the applicant has demonstrated that the aging effects associated with the components will be adequately managed so that the intended function(s) will be maintained consistent with the CLB for the period of extended operation, as required by 10 CFR 54.21(a)(3).

3.3.2.3.7 Auxiliary Systems – Fire Protection System – Summary of Aging Management Evaluation – LRA Table 3.3.2-7

The staff reviewed LRA Table 3.3.2-7, which summarizes the results of AMR evaluations for the fire protection system component groups.

In LRA Table 3.3.2-7, the applicant proposed the Boric Acid Corrosion Program to manage loss of copper alloy and brass materials for component types of sprinkler heads and pipe and fittings exposed to an external plant indoor air or containment air environment.

The staff reviewed the applicant's Boric Acid Corrosion Program as documented in SER Section 3.0.3.2.2. The staff concludes that management of loss of copper alloy and brass materials for component types of sprinkler heads and pipe and fittings exposed to an external plant indoor air or containment air environment in the fire protection system is acceptable.

In LRA Table 3.3.2-7, the applicant proposed the Fire Protection and System Monitoring Programs to manage loss of cast iron valves and dampers material exposed to an external atmosphere/weather environment.

The staff reviewed the applicant's Fire Protection Program as documented in SER Section 3.0.3.2.5. The program includes (a) fire barrier inspections, (b) electric and diesel-driven fire pump tests, and (c) periodic maintenance, testing, and inspection of water-based fire protection systems. The staff reviewed the applicant's System Monitoring Program as documented in SER Section 3.0.3.3.1. The staff determined that the aging effect of loss of cast iron valves and dampers material exposed to an external atmosphere/weather environment is effectively managed by the Fire Protection and System Monitoring Programs. On this basis, the staff concludes that management of loss of cast iron valves and dampers material exposed to an external atmosphere/weather environment in the fire protection system is acceptable.

In LRA Table 3.3.2-7, the applicant proposed the System Monitoring Program to manage loss of cast iron filters and strainers material exposed to an external plant indoor air environment.

The staff reviewed the applicant's System Monitoring Program as documented in SER Section 3.0.3.3.1. The staff concludes that management of loss of cast iron filters and strainers material exposed to an external plant indoor air environment in the fire protection system is acceptable.

In LRA Table 3.3.2-7, the applicant proposed the Fire Protection Program to manage loss of cast iron valves and dampers material exposed to an internal air environment.

The staff reviewed the applicant's Fire Protection Program as documented in SER Section 3.0.3.2.5. On the basis of its review, the staff concludes that management of loss of cast iron valves and dampers material exposed to an internal air environment in the fire protection system is acceptable.

In LRA Table 3.3.2-7, the applicant proposed the Bolting Integrity Program to manage loss of preload of carbon steel and stainless steel fasteners exposed to an external plant indoor air environment.

The staff reviewed the applicant's Bolting Integrity Program as documented in SER Section 3.0.3.2.1. The staff determined that management of loss of preload of carbon steel and stainless steel fasteners exposed to an external plant indoor air environment in the fire protection system is acceptable.

In LRA Table 3.3.2-7, the applicant proposed the Boric Acid Corrosion and System Monitoring Programs to manage loss of brass and cast iron materials of valves and dampers exposed to an external plant indoor air environment.

The staff reviewed the applicant's Boric Acid Corrosion and System Monitoring Programs as documented in SER Sections 3.0.3.2.2 and 3.0.3.3.1, respectively. The staff concludes that management of loss of brass and cast iron materials of valves and dampers exposed to an external plant indoor air environment in the fire protection system is acceptable.

In LRA Table 3.3.2-7, the applicant proposed the Fire Protection Program to manage the buildup of deposits on stainless steel and brass valves and dampers, and brass sprinkler heads exposed to an internal raw water environment.

The staff reviewed the applicant's Fire Protection Program as documented in SER Section 3.0.3.2.5. On the basis of its review, the staff concludes that management of buildup of deposits in the fire protection system is acceptable.

In LRA Table 3.3.2-7, the applicant proposed the One-Time Inspection Program to manage loss of stainless steel accumulators material exposed to an internal oil environment.

The staff reviewed the applicant's One-Time Inspection Program as documented in SER Section 3.0.3.2.6. On the basis of its review of the applicant's plant-specific and industry operating experience, the staff determined that the aging effect of loss of stainless steel material exposed to an internal oil environment is effectively managed by the One-Time Inspection Program and concludes that management of loss of material for stainless steel accumulators exposed to an internal oil environment in the fire protection system is acceptable.

In LRA Table 3.3.2-7, the applicant proposed the Bolting Integrity Program to manage loss of preload of carbon steel fasteners exposed to a raw water external environment.

The staff reviewed the applicant's Bolting Integrity Program as documented in SER Section 3.0.3.2.1. The staff concludes that management of loss of preload of carbon steel fasteners exposed to a raw water external environment in the fire protection system is acceptable.

In LRA Table 3.3.2-7, the applicant proposed the Bolting Integrity Program to manage loss of preload of carbon steel fasteners exposed to a soil external environment.

The staff concludes that management of loss of preload of carbon steel fasteners exposed to a soil external environment in the fire protection system is acceptable.

In LRA Table 3.3.2-7, the applicant proposed the One-Time Inspection Program to manage loss of material due to selective leaching of cast iron filters and strainers, pipe and fittings, and pumps exposed to an internal or external raw water environment.

The staff reviewed the applicant's One-Time Inspection Program as documented in SER Section 3.0.3.2.6. The program is consistent with GALL AMP XI.M33, "Selective Leaching of Materials." On this consistency basis, the staff concludes that the applicant had appropriately addressed the aging mechanism as recommended by the GALL Report.

In LRA Table 3.3.2-7, the applicant proposed the One-Time Inspection Program to manage loss of cast iron valves and dampers material exposed to an internal raw water environment.

The staff reviewed the applicant's One-Time Inspection Program as documented in SER Section 3.0.3.2.6. On the basis of its review, the staff concludes that management of loss of cast iron material of valves and dampers exposed to an internal raw water environment in the fire protection system is acceptable.

In LRA Table 3.3.2-7, the applicant proposed the One-Time Inspection Program to manage loss of material due to selective leaching of cast iron valves and dampers exposed to an external soil environment.

The staff reviewed the applicant's One-Time Inspection Program as documented in SER Section 3.0.3.2.6. The program is consistent with GALL AMP XI.M33. On this consistency basis, the staff concludes that the applicant had appropriately addressed the aging mechanism as recommended by the GALL Report.

In LRA Table 3.3.2-7, the applicant proposed the Fire Protection Program to manage the loss of brass valves and dampers material exposed to an internal raw water environment.

The staff reviewed the applicant's Fire Protection Program as documented in SER Section 3.0.3.2.5.

On the basis of its review, as discussed above, the staff concludes that the applicant has demonstrated that the aging effects associated with the components will be adequately managed so that the intended function(s) will be maintained consistent with the CLB for the period of extended operation, as required by 10 CFR 54.21(a)(3).

3.3.2.3.8 Auxiliary Systems – Fuel Oil System – Summary of Aging Management Evaluation – LRA Table 3.3.2-8

The staff reviewed LRA Table 3.3.2-8, which summarizes the results of AMR evaluations for the fuel oil system component groups.

In LRA Table 3.3.2-8, the applicant proposed the Diesel Fuel Monitoring and Storage Program as augmented by the One-Time Inspection Program to manage loss of material or cracking of copper alloy pipe and fittings and brass, bronze, and stainless steel valves and dampers exposed to an internal oil environment.

The staff reviewed the applicant's Diesel Fuel Monitoring and Storage and One-Time Inspection Programs as documented in SER Sections 3.0.3.2.4 and 3.0.3.2.6, respectively. The Diesel Fuel Monitoring and Storage Program assures the continued availability and quality of fuel oil used in diesel generators and diesel fire pumps. The One-Time Inspection Program addresses potentially long incubation periods for certain aging effects, including various corrosion mechanisms, cracking, and selective leaching, and provides a means of verifying that an aging effect either does not occur or progresses so slowly as to have negligible effect on the intended function of the structure or component. On the basis of its review, the staff concludes that management of loss of material or cracking of copper alloy, brass, bronze, and stainless steel components exposed to an internal oil environment in the fuel oil system is acceptable. In LRA Table 3.3.2-8, the applicant proposed the Diesel Fuel Monitoring and Storage Program as augmented by the Fire Protection and One-Time Inspection Programs to manage loss of stainless steel pipe and fittings material exposed to an internal fuel oil environment.

The staff reviewed the applicant's Diesel Fuel Monitoring and Storage, Fire Protection, and One-Time Inspection Programs as documented in SER Sections 3.0.3.2.4, 3.0.3.2.5, and 3.0.3.2.6, respectively. The Fire Protection Program includes (a) fire barrier inspections, (b) electric and diesel-driven fire pump tests, and (c) periodic maintenance, testing, and inspection of water-based fire protection systems. Periodic visual inspections of fire barrier penetration seals, fire dampers, fire barrier walls, ceilings, and floors, and periodic visual inspections and functional tests of fire-rated doors ensure that function and operation are maintained. On the basis of its review, the staff concludes that the management of loss of stainless steel pipe and fittings material exposed to an internal fuel oil environment in the fuel oil system is acceptable.

In LRA Table 3.3.2-8, the applicant proposed the Diesel Fuel Monitoring and Storage Program to manage loss of carbon steel accumulators and pipe and fittings material exposed to an internal air environment.

The staff reviewed the applicant's Diesel Fuel Monitoring and Storage Program as documented in SER Section 3.0.3.2.4. On the basis of its review, the staff concludes that management of loss of carbon steel accumulators and pipe and fittings material exposed to an internal air environment in the fuel oil system is acceptable.

On the basis of its review, as discussed above, the staff concludes that the applicant has demonstrated that the aging effects associated with the components will be adequately managed so that the intended function(s) will be maintained consistent with the CLB for the period of extended operation, as required by 10 CFR 54.21(a)(3).

3.3.2.3.9 Auxiliary Systems – Heating, Ventilation, and Air Conditioning System – Summary of Aging Management Evaluation – LRA Table 3.3.2-9

The staff reviewed LRA Table 3.3.2-9, which summarizes the results of AMR evaluations for the HVAC system component groups.

In LRA Table 3.3.2-9, the applicant proposed the One-Time Inspection Program to manage loss of copper alloy materials for component type of HXs exposed to an external gas environment.

On the basis of its review of current industry research and operating experience, the staff finds that gas on copper alloys would not cause aging of concern during the period of extended operation. Therefore, the staff concludes that there are no significant AERMs for metal in an air environment and that the One-Time Inspection Program can provide a means of verifying that aging effects either do not occur or progress so slowly as to have negligible effects. On this basis, the staff finds this proposal acceptable.

In LRA Table 3.3.2-9, the applicant proposed the One-Time Inspection Program to manage loss of stainless steel and carbon steel valves and dampers, cast iron traps, and carbon steel pipe and fittings materials exposed to an internal raw water environment. For the HVAC system, condensation has been designated as a raw water environment.

In its response to general RAI 3.3-1, the applicant stated that the Note 365 "Not Used" is hereby replaced with new Note 365 to read, "Palisades will perform internal visual inspections of opportunity when maintenance provides such an opportunity." To assure that these inspections are implemented, applicable Palisades procedures will be enhanced to inspect and document the internal condition of applicable in-scope components when maintenance provides an opportunity. Applicable components are those that have an internal environment of water, are constructed of materials that are potentially susceptible to internal aging degradation in a wetted environment, but are not subject to another AMP (e.g., Water Chemistry, Open Cycle Cooling Water Programs) that would manage the internal environment such that aging degradation of the internal surfaces would not be expected. The applicant stated that in LRA Table 3.3.2-9, to the line item for "Pipe and Fittings" of "Carbon Steel" in "Raw Water (Int)," an additional Note 365 is hereby added to Notes.

The staff review found the applicant's response acceptable because new note 365 states that the applicant will perform internal visual inspections of opportunity when maintenance provides such an opportunity.

The staff reviewed the applicant's One-Time Inspection Program as documented in SER Section 3.0.3.2.6. On the basis of its review, the staff concludes that the management of loss of stainless steel, carbon steel, and cast iron component materials exposed to an internal raw water (condensation) environment in the HVAC system is acceptable.

In LRA Table 3.3.2-9, the applicant proposed the One-Time Inspection Program to manage loss of carbon steel and copper alloy HXs, carbon steel pipe and fittings, and bronze valves and dampers materials exposed to an internal steam environment.

The staff reviewed the applicant's One-Time Inspection Program as documented in SER Section 3.0.3.2.6. On the basis of its review of the applicant's plant-specific and industry operating experience, the staff determined that the aging effect of loss of bronze, carbon steel, and copper alloy materials exposed to an internal steam environment is effectively managed by the One-Time Inspection Program. On this determination, the staff concludes that the management of loss of bronze valves and dampers, carbon steel and copper alloy HXs, and carbon steel pipe and fittings materials exposed to an internal steam environment in the HVAC system is acceptable.

In LRA Table 3.3.2-9, the applicant proposed the One-Time Inspection Program to manage loss of the component type cast iron traps material exposed to an internal steam environment.

The One-Time Inspection Program provides a means of verifying that an aging effect either does not occur or progresses so slowly as to have negligible effect on the intended function of the structure or component. Carbon steel/cast iron is subject to significant corrosion under moisture.

In its response to general RAI 3.3-1, the applicant stated:

> In Table 3.3.2-9, on page 3-169, to the line item for 'Traps (Steam)' of 'Cast Iron' in 'Steam (Int),' 'Water Chemistry Program' is hereby added in addition to the current 'One Time Inspection Program' to manage Loss of Material. The corresponding NUREG 1801 Volume 2 and Table 1 items are blank and the Notes entry is '390, 393, G.'

The staff review found the applicant's response acceptable because the applicant added the Water Chemistry Program to manage loss of material.

In LRA Table 3.3.2-9, the applicant proposed the One-Time Inspection Program to manage degradation of heat transfer of copper alloy HXs exposed to an external air environment.

The staff reviewed the applicant's One-Time Inspection Program as documented in SER Section 3.0.3.2.6. On the basis of its review, the staff concludes that the management of degradation of heat transfer for copper alloy HXs exposed to an external air environment in the HVAC system is acceptable.

In LRA Table 3.3.2-9, the applicant proposed the One-Time Inspection Program to manage loss of copper alloy HX material exposed to an internal raw water environment.

In LRA Table 3.3.2-9, the applicant proposed the One-Time Inspection Program for the internal exposure and the System Monitoring Program for the external exposure to manage loss of carbon steel valves and dampers material exposed to a steam environment.

In its response to general RAI 3.3-1, the applicant stated:

> In Table 3.3.2-9, on page 3-170, the line item for 'Valves and Dampers' of 'Carbon Steel' in 'Raw Water (Int),' the 'Open Cycle Cooling Water Program' is hereby added to replace the current 'One Time Inspection Program' to manage Loss of Material. The corresponding NUREG 1801 Volume 2 and Table 1 items are blank and the Notes entry is '395, G.'

The staff review found the applicant's response acceptable because the applicant added the Open Cycle Cooling Water Program to manage loss of material.

For the external surfaces, the applicant proposed its System Monitoring Program to manage this aging effect. The applicant's System Monitoring Program was reviewed by the staff as documented in SER Section 3.0.3.3.1. The staff concludes that the management of loss of carbon steel valves and dampers material exposed to an external steam environment in the HVAC system is acceptable.

On the basis of its review, as discussed above, the staff concludes that the applicant has demonstrated that the aging effects associated with the components will be adequately managed so that the intended function(s) will be maintained consistent with the CLB for the period of extended operation, as required by 10 CFR 54.21(a)(3).

3.3.2.3.10 Auxiliary Systems – Miscellaneous Gas System – Summary of Aging Management Evaluation – LRA Table 3.3.2-10

The staff reviewed LRA Table 3.3.2-10, which summarizes the results of AMR evaluations for the miscellaneous gas system component groups.

In LRA Table 3.3.2-10, the applicant proposed the Boric Acid Corrosion Program to manage loss of copper alloys, bronze, and brass materials for component types of pipe and fittings and valves and dampers exposed to a containment air or plant indoor air environment.

The staff reviewed the applicant's Boric Acid Corrosion Program as documented in SER Section 3.0.3.2.2. The staff concludes that the management of loss of copper alloys, bronze, and brass materials for component types of pipe and fittings and valves and dampers exposed to a containment air or plant indoor air environment in the miscellaneous gas system is acceptable.

In LRA Table 3.3.2-10, the applicant proposed the Bolting Integrity Program to manage loss of preload and material of stainless steel and copper alloy fasteners exposed to a containment air or plant air environment.

The staff reviewed the applicant's Bolting Integrity Program as documented in SER Section 3.0.3.2.1. The staff concludes that the management of loss of preload and material of stainless steel and copper alloy fasteners exposed to a containment air or plant air environment in the miscellaneous gas system is acceptable.

On the basis of its review, as discussed above, the staff concludes that the applicant has demonstrated that the aging effects associated with the components will be adequately managed so that the intended function(s) will be maintained consistent with the CLB for the period of extended operation, as required by 10 CFR 54.21(a)(3).

3.3.2.3.11 Auxiliary Systems – Radwaste System – Summary of Aging Management Evaluation – LRA Table 3.3.2-11

The staff reviewed LRA Table 3.3.2-11, which summarizes the results of AMR evaluations for the radwaste system component groups.

In LRA Table 3.3.2-11, the applicant proposed the Closed Cycle Cooling Water Program to manage loss of copper alloy and bronze materials for component type HXs exposed to a treated water environment.

The staff reviewed the applicant's Closed Cycle Cooling Water Program as documented in SER Section 3.0.3.2.3. On the basis of its review, the staff concludes that the management of loss of copper alloy and bronze materials in components exposed to a treated water environment in the radwaste system is acceptable.

In LRA Table 3.3.2-11, the applicant proposed the Boric Acid Corrosion Program to manage loss of bronze material for component types HXs and pumps exposed to an external plant indoor air environment. The staff concludes that management of loss of bronze material for component types HXs and pumps exposed to an external plant indoor air environment in the radwaste system is acceptable.

In LRA Table 3.3.2-11, the applicant proposed the One-Time Inspection Program to manage loss of carbon steel material for component type pipe and fittings exposed to a gas environment.

The staff reviewed the applicant's One-Time Inspection Program as documented in SER Section 3.0.3.2.6. On the basis of its review, the staff concludes that the management of loss of material for carbon steel components exposed to a gas environment in the radwaste system is acceptable.

In LRA Table 3.3.2-11, the applicant proposed the One-Time Inspection Program to manage loss of cast iron material for component type piping and fittings exposed to a treated water environment. In LRA plant-specific Note 381, the applicant stated that for the radwaste system, treated water is not controlled by water chemistry and can be considered warm moist air.

In its response to general RAI 3.3-1, the applicant stated, "that in Table 3.3.2-1 1, on page 3-1 76, to the line item for 'Pipe and Fittings' of 'Carbon Steel' in 'Raw Water (Int),' note 365 is hereby added with the existing Notes"

The staff review found the applicant's response acceptable because new note 365 states that the applicant will perform internal visual inspections of opportunity when maintenance provides such an opportunity.

In LRA Table 3.3.2-11, the applicant proposed the One-Time Inspection Program to manage loss of material from carbon steel HXs exposed to an internal steam environment.

In its response general to RAI 3.3-1, the applicant stated, "In Table 3.3.2-1 1, on page 3-175, to the line item for 'Heat Exchangers' of 'Carbon Steel' in 'Steam (Int),' the 'Water Chemistry Program' is hereby added in addition to the current 'One Time Inspection Program' to manage Loss of Material. The corresponding NUREG 1801 Volume 2 and Table 1 items are blank, and the Notes entry is '331, J.'"

In LRA Table 3.3.2-11, the applicant proposed the One-Time Inspection Program to manage loss of carbon steel materials for component type accumulators exposed to an internal air environment.

The staff reviewed the applicant's One-Time Inspection Program as documented in SER Section 3.0.3.2.6. On the basis of its review, the staff concludes that the management of loss of carbon steel components material exposed to an internal air environment in the radwaste system is acceptable.

In LRA Table 3.3.2-11, the applicant proposed the Bolting Integrity Program to manage loss of preload of stainless steel fasteners exposed to a plant indoor air environment.

The staff reviewed the applicant's Bolting Integrity Program as documented in SER Section 3.0.3.2.1. The staff concludes that the management of loss of preload of stainless steel fasteners exposed to a plant indoor air environment in the radwaste system is acceptable.

In LRA Table 3.3.2-11, the applicant proposed the One-Time Inspection Program to manage loss of carbon steel component type pump material exposed to an external raw water environment.

In its response to general RAI 3.3-1, the applicant stated, "In Table 3.3.2-1 1, on page 3-177, to the line items for "Pumps" of "Carbon Steel" in "Raw Water (Ext)" and "Raw Water (Int)," note 365 is hereby added with the existing Notes."

The staff review found the applicant's response acceptable because new note 365 states that the applicant will perform internal visual inspections of opportunity when maintenance provides such an opportunity.

On the basis of its review, as discussed above, the staff concludes that the applicant has demonstrated that the aging effects associated with the components will be adequately managed so that the intended function(s) will be maintained consistent with the CLB for the period of extended operation, as required by 10 CFR 54.21(a)(3).

3.3.2.3.12 Auxiliary Systems – Service Water System – Summary of Aging Management Evaluation – LRA Table 3.3.2-12

The staff reviewed LRA Table 3.3.2-12, which summarizes the results of AMR evaluations for the SWS component groups.

In LRA Table 3.3.2-12, the applicant proposed the One-Time Inspection Program to manage loss of carbon steel materials for component type HX exposed to a gas/air internal environment.

The staff reviewed the applicant's One-Time Inspection Program as documented in SER Section 3.0.3.2.6. On the basis of its review, the staff concludes that the management of loss of material for SWS carbon steel components exposed to a gas/air internal environment is acceptable.

In LRA Table 3.3.2-12, the applicant proposed the Bolting Integrity Program to manage loss of preload of low-alloy steel materials for component type fasteners exposed to a containment air and plant indoor air environment.

The Bolting Integrity Program was reviewed by the staff as documented in SER Section 3.0.3.2.1. The staff concludes that the management of loss of preload of SWS low-alloy steel fasteners exposed to containment air and plant indoor air environment is acceptable.

In LRA Table 3.3.2-12, the applicant proposed the One-Time Inspection Program to manage loss of carbon steel materials for component type HX exposed to an internal oil environment.

The GALL Report recommends management of loss of material and reduction of heat transfer aging effects by a plant-specific or lubricating/fuel oil analysis AMP with a One-Time Inspection Program to ensure AMP effectiveness.

In its response to general RAI 3.3-2, the applicant stated, "In Table 3.3.2-12, on page 3-180, for the line item 'Heat Exchanger' with an 'Oil (Int)' environment, note 311 is hereby added with the existing Notes."

The staff review found the applicant's response acceptable because, with the addition of note 311, selective leaching of materials is included in the One-Time Inspection Program.

In LRA Table 3.3.2-12, the applicant proposed the System Monitoring Program to manage cracking of rubber materials for component type pipe and fittings exposed to a plant indoor air environment.

The staff reviewed the applicant's System Monitoring Program as documented in SER Section 3.0.3.3.1. The staff concludes that the management of cracking of rubber materials for component type pipe and fittings exposed to a plant indoor air environment in the SWS is acceptable.

On the basis of its review, as discussed above, the staff concludes that the applicant has demonstrated that the aging effects associated with the components will be adequately managed so that the intended function(s) will be maintained consistent with the CLB for the period of extended operation, as required by 10 CFR 54.21(a)(3).

3.3.2.3.13 Auxiliary Systems – Shield Cooling System – Summary of Aging Management Evaluation – LRA Table 3.3.2-13

The staff reviewed LRA Table 3.3.2-13, which summarizes the results of AMR evaluations for the shield cooling system (SCS) component groups.

In LRA Table 3.3.2-13, the applicant proposed the Bolting Integrity Program to manage loss of preload of carbon steel material for component type fasteners exposed to containment air and plant indoor air environments.

The staff reviewed the applicant's Bolting Integrity Program as documented in SER Section 3.0.3.2.1. The staff concludes that the management of loss of preload of carbon steel material for SCS component type fasteners exposed to containment air and plant indoor air environments is acceptable.

On the basis of its review, as discussed above, the staff concludes that the applicant has demonstrated that the aging effects associated with the components will be adequately managed so that the intended function(s) will be maintained consistent with the CLB for the period of extended operation, as required by 10 CFR 54.21(a)(3).

3.3.2.3.14 Auxiliary Systems – Spent Fuel Pool Cooling System – Summary of Aging Management Evaluation – LRA Table 3.3.2-14

The staff reviewed LRA Table 3.3.2-14, which summarizes the results of AMR evaluations for the SFP cooling system component groups.

In LRA Table 3.3.2-14, the applicant proposed the Bolting Integrity Program to manage loss of preload of stainless steel material for component type fasteners exposed to a containment air and plant indoor air environment.

The staff reviewed the applicant's Bolting Integrity Program as documented in SER Section 3.0.3.2.1. The staff concludes that the management of loss of preload of stainless steel material for component type fasteners exposed to containment air and plant air environments in the SFP cooling system is acceptable.

In LRA Table 3.3.2-14, the applicant proposed the Closed Cycle Cooling Water Program to manage heat transfer degradation of stainless steel material of component type SFP HX tube and tubesheet exposed to an external treated water environment.

The staff reviewed the applicant's Closed Cycle Cooling Water Program as documented in SER Section 3.0.3.2.3. On the basis of its review, the staff concludes that the management of heat transfer degradation for stainless steel material exposed to an external treated water environment in the SFP cooling system is acceptable.

In LRA Table 3.3.2-14, the applicant proposed the Closed Cycle Cooling Water and Water Chemistry Programs to manage heat transfer degradation of stainless steel materials for component type SFP HX tube and tubesheet exposed to an internal treated water environment.

The staff reviewed the applicant's Closed Cycle Cooling Water and Water Chemistry Programs as documented in SER Sections 3.0.3.2.3 and 3.0.3.1.10, respectively. The Water Chemistry Program is credited for managing such aging effects as loss of material due to general, pitting, and crevice corrosion, cracking due to SCC, and SG tube degradation caused by denting, IGA, and ODSCC by controlling the environment to which internal surfaces of systems and components are exposed. On the basis of its review, the staff concludes that the management of heat transfer degradation for stainless steel materials exposed to an internal treated water environment in the SFP cooling system is acceptable.

On the basis of its review, as discussed above, the staff concludes that the applicant has demonstrated that the aging effects associated with the components will be adequately managed so that the intended function(s) will be maintained consistent with the CLB for the period of extended operation, as required by 10 CFR 54.21(a)(3).

3.3.2.3.15 Auxiliary Systems – Waste Gas System – Summary of Aging Management Evaluation – LRA Table 3.3.2-15

The staff reviewed LRA Table 3.3.2-15, which summarizes the results of AMR evaluations for the waste gas system component groups.

In LRA Table 3.3.2-15, the applicant proposed the One-Time Inspection Program to manage loss of carbon steel material of component types accumulator, filter/strainer, and valves and dampers exposed to an internal gas environment.

In its response to general RAI 3.3-1, the applicant stated, "In Table 3.3.2-15, on page 3-192, to the line item for 'Accumulator' of 'Carbon Steel' in 'Raw Water (Int),' note 365 is hereby added with the existing Notes"

The staff review found the applicant's response acceptable because new note 365 states that the applicant will perform internal visual inspections of opportunity when maintenance provides such an opportunity.

The staff reviewed the applicant's One-Time Inspection Program as documented in SER Section 3.0.3.2.6. The staff determined that steel in an inert gas environment exhibits no aging effect and that the structure or component, therefore, would remain capable of performing its intended functions consistent with the CLB for the period of extended operation. This determination is based on the fact that gaseous corrosion (dry corrosion) usually involves reaction with high-temperature gases. On the basis of its review, the staff concludes that the management of loss of carbon steel material for components exposed to an internal gas environment in the waste gas system is acceptable.

On the basis of its review, as discussed above, the staff concludes that the applicant has demonstrated that the aging effects associated with the components will be adequately managed so that the intended function(s) will be maintained consistent with the CLB for the period of extended operation, as required by 10 CFR 54.21(a)(3).

3.3.2.3.16 Auxiliary Systems – Domestic Water System – Summary of Aging Management Evaluation – LRA Table 3.3.2-16

The staff reviewed LRA Table 3.3.2-16, which summarizes the results of AMR evaluations for the domestic water system component groups.

In LRA Table 3.3.2-16, the applicant proposed the One-Time Inspection Program to manage loss of carbon steel and cast iron materials for component type accumulators exposed to an internal air environment.

The staff reviewed the applicant's One-Time Inspection Program as documented in SER Section 3.0.3.2.6. On the basis of its review, the staff concludes that the management of loss of material for carbon steel or cast iron components exposed to an internal air environment in the domestic water system is acceptable.

In LRA Table 3.3.2-16, the applicant proposed the One-Time Inspection Program to manage loss of carbon steel materials for component type accumulator exposed to an internal raw water environment.

In its response to general RAI 3.3-1, the applicant stated, "In Table 3.3.2-16, on page 3-194, to the line item for 'Accumulator' of 'Carbon Steel' in 'Raw Water (Int),' note 365 is hereby added with the existing Notes."

The staff review found the applicant's response acceptable because the applicant added the Water Chemistry Program to manage the loss of material.

In LRA Table 3.3.2-16, the applicant proposed the One-Time Inspection Program to manage loss of carbon steel materials for component type HX exposed to an internal raw water environment.

In its response to general RAI 3.3-1, the applicant stated, "In Table 3.3.2-16, on page 3-195, to the line item for 'Pipe and Fittings' of 'Carbon Steel' in 'Raw Water (Int),' note 365 is hereby added with the existing Notes."

The staff review found the applicant's response acceptable because new note 365 states that the applicant will perform internal visual inspections of opportunity when maintenance provides such an opportunity.

In LRA Table 3.3.2-16, the applicant proposed the One-Time Inspection Program to manage loss of carbon steel materials for component type HX exposed to a internal steam environment.

In its response to general RAI 3.3-1, the applicant stated, "that in Table 3.3.2-16, on page 3-194, to the line item for 'Heat Exchanger' of 'Carbon Steel' in 'Steam (Int),' the 'Water Chemistry Program' is hereby added in addition to the current 'One Time Inspection Program' to manage Loss of Material. The corresponding NUREG 1801 Volume 2 and Table 1 items are blank, and the Notes entry is J.'"

The staff review found the applicant's response acceptable because the applicant added the Water Chemistry Program to manage loss of material.

On the basis of its review, as discussed above, the staff concludes that the applicant has demonstrated that the aging effects associated with the components will be adequately managed so that the intended function(s) will be maintained consistent with the CLB for the period of extended operation, as required by 10 CFR 54.21(a)(3).

3.3.2.3.17 Auxiliary Systems – Chemical Addition System – Summary of Aging Management Evaluation – LRA Table 3.3.2-17

The staff reviewed LRA Table 3.3.2-17, which summarizes the results of AMR evaluations for the chemical addition system component groups.

In LRA Table 3.3.2-17, the applicant proposed the One-Time Inspection Program to manage loss of carbon steel material for component type accumulators exposed to an internal air environment.

The staff reviewed the applicant's One-Time Inspection Program as documented in SER Section 3.0.3.2.6. On the basis of its review, the staff concludes that the management of loss of material for carbon steel accumulators exposed to an internal air environment in the chemical addition system is acceptable.

On the basis of its review, as discussed above, the staff concludes that the applicant has demonstrated that the aging effects associated with the components will be adequately managed so that the intended function(s) will be maintained consistent with the CLB for the period of extended operation, as required by 10 CFR 54.21(a)(3).

Conclusion. On the basis of its review, the staff finds that the applicant has appropriately evaluated AMR results involving material, environment, AERMs, and AMP combinations not evaluated in the GALL Report. The staff finds that the applicant has demonstrated that the aging effects will be adequately managed so that the intended functions will be maintained consistent with the CLB for the period of extended operation, as required by 10 CFR 54.21(a)(3).

3.3.3 Conclusion

The staff concludes that the applicant has demonstrated that the aging effects associated with the auxiliary systems components will be adequately managed so that the intended function(s) will be maintained consistent with the CLB for the period of extended operation, as required by 10 CFR 54.21(a)(3).

The staff also reviewed the applicable FSAR supplement program summaries and concludes that they adequately describe the AMPs credited for managing aging of the auxiliary systems, as required by 10 CFR 54.21(d).

3.4 Aging Management of Steam and Power Conversion System

This section of the SER documents the staff's review of the applicant's AMR results for the steam and power conversion system components and component groups of the following:

- condensate and condenser system
- demineralized makeup water system
- feedwater system
- heater extraction and drain system
- main air ejection and gland seal system
- main steam system
- turbine generator and crane system

3.4.1 Summary of Technical Information in the Application

In LRA Section 3.4, the applicant provided AMR results for the steam and power conversion system components and component groups. In LRA Table 3.4.1, "Summary of Aging Management Evaluations in Chapter VII of NUREG-1801 for Steam and Power Conversion," the applicant provided a summary comparison of its AMRs to those evaluated in the GALL Report for steam and power conversion system components and component groups.

The applicant's AMRs evaluated and incorporated plant-specific and industry operating experience in the determination of AERMs from plant-specific condition reports and discussions with site personnel and from the GALL Report and issues identified since its publication.

3.4.2 Staff Evaluation

The staff reviewed LRA Section 3.4 to determine whether the applicant had provided sufficient information to demonstrate that the aging effects for the steam and power conversion system components within the scope of license renewal and subject to an AMR will be adequately managed so that the intended function(s) will be maintained consistent with the CLB for the period of extended operation, as required by 10 CFR 54.21(a)(3).

The staff conducted an onsite audit of AMRs to confirm the applicant's claim that certain identified AMRs were consistent with the GALL Report. The staff did not repeat its review of the matters described in the GALL Report; however, the staff did verify that the material presented in the LRA was applicable and that the applicant had identified the appropriate GALL Report AMRs. The staff's evaluations of the AMPs are documented in SER Section 3.0.3. Details of the staff's audit evaluation are documented in the Audit and Review Report and summarized in SER Section 3.4.2.1.

In the onsite audit, the staff also selected AMRs consistent with the GALL Report for which further evaluation is recommended. The staff confirmed that the applicant's further evaluations were consistent with the acceptance criteria in SRP-LR Section 3.4.2.2. The staff's audit evaluations are documented in the Audit and Review Report and summarized in SER Section 3.4.2.2.

In the onsite audit, the staff also conducted a technical review of the those remaining AMRs not consistent with or not addressed in the GALL Report. The audit and technical review evaluated whether all plausible aging effects were identified and whether the aging effects listed were appropriate for the combination of materials and environments specified. The staff's audit evaluations documented in the Audit and Review Report are summarized in SER Section 3.4.2.3. The staff's evaluation of its technical review is also documented in SER Section 3.4.2.3.

Finally, the staff reviewed the AMP summary descriptions in the FSAR supplement to ensure that they adequately describe the programs credited with managing or monitoring steam and power conversion system component aging.

SER Table 3.4-1 below summarizes the staff's evaluation of components, aging effects/mechanisms, and AMPs listed in LRA Section 3.4 and addressed in the GALL Report.

Table 3.4-1 Staff Evaluation for Steam and Power Conversion Systems Components in the GALL Report

Component Group	Aging Effect/ Mechanism	AMP in GALL Report	AMP in LRA	Staff Evaluation
Piping and fittings in main feedwater line, steam line and AFW piping (PWR only) (Item Number 3.4.1-01)	Cumulative fatigue damage	TLAA, evaluated in accordance with 10 CFR 54.21(c)	TLAA	This TLAA is evaluated in SER Section 4.3, "Metal Fatigue"
Piping and fittings, valve bodies and bonnets, pump casings, tanks, tubes, tubesheets, channel head and shell (except main steam system) (Item Number 3.4.1-02)	Loss of material due to general (carbon steel only), pitting, and crevice corrosion	Water chemistry and one-time inspection	One-Time Inspection Program (B2.1.13), Water Chemistry Program (B2.1.21)	Consistent with GALL, which recommends further evaluation (See SER Section 3.4.2.2.2)

Component Group	Aging Effect/ Mechanism	AMP in GALL Report	AMP in LRA	Staff Evaluation
Auxiliary feedwater (AFW) piping (Item Number 3.4.1-03)	Loss of material due to general, pitting, and crevice corrosion, MIC, and biofouling	Plant-specific	Fire Protection Program (B2.1.10), One-Time Inspection Program (B2.1.13), Open Cycle Cooling Water Program (B2.1.14)	Consistent with GALL, which recommends further evaluation (See SER Section 3.4.2.2.3)
Oil coolers in AFW system (lubricating oil side possibly contaminated with water (Item Number 3.4.1-04)	Loss of material due to general (carbon steel only), pitting, and crevice corrosion and MIC	Plant-specific	None	Not applicable (See SER Section 3.4.2.2.5) (PNP does not have bearing oil coolers)
External surface of carbon steel components (Item Number 3.4.1-05)	Loss of material due to general corrosion	Plant-specific	Bolting Integrity Program (B2.1.3), System Monitoring Program (B2.1.20)	Consistent with GALL, which recommends further evaluation (See SER Section 3.4.2.2.4)
Carbon steel piping and valve bodies (Item Number 3.4.1-06)	Wall thinning due to flow-accelerated corrosion	Flow-accelerated corrosion	Flow Accelerated Corrosion Program (B2.1.11)	Consistent with GALL, which recommends no further evaluation (See SER Section 3.4.2.1)
Carbon steel piping and valve bodies in main steam system (Item Number 3.4.1-07)	Loss of material due to pitting and crevice corrosion	Water chemistry	One-Time Inspection Program (B2.1.13), Water Chemistry Program (B2.1.21)	Consistent with GALL, which recommends no further evaluation (See SER Section 3.4.2.1.1)
Closure bolting in high-pressure or high-temperature systems (Item Number 3.4.1-08)	Loss of material due to general corrosion; crack initiation and growth due to cyclic loading and/or SCC	Bolting integrity	Bolting Integrity Program (B2.1.3), System Monitoring Program (B2.1.20)	Consistent with GALL, which recommends no further evaluation (See SER Section 3.4.2.1.2)
Heat exchangers and coolers/condensers serviced by open-cycle cooling water (Item Number 3.4.1-09)	Loss of material due to general (carbon steel only), pitting, and crevice corrosion, MIC, and biofouling; buildup of deposit due to biofouling	Open-cycle cooling water system	None	Not applicable (Applicant evaluated by heat exchanger subcomponents)

Component Group	Aging Effect/ Mechanism	AMP in GALL Report	AMP in LRA	Staff Evaluation
Heat exchangers and coolers/condensers serviced by closed-cycle cooling water (Item Number 3.4.1-10)	Loss of material due to general (carbon steel only), pitting, and crevice corrosion	Closed-cycle cooling water system	None	Not applicable (Applicant evaluated by heat exchanger subcomponents)
External surface of aboveground condensate storage tank (Item Number 3.4.1-11)	Loss of material due to general (carbon steel only), pitting, and crevice corrosion	Aboveground carbon steel tanks	One-Time Inspection Program (B2.1.13), System Monitoring Program (B2.1.20)	Consistent with GALL, which recommends no further evaluation (See SER Section 3.4.2.1.3)
External surface of buried condensate storage tank and AFW piping (Item Number 3.4.1-12)	Loss of material due to general, pitting, and crevice corrosion and MIC	Buried piping and tanks surveillance or Buried piping and tanks inspection	Buried Services Corrosion Monitoring Program (B2.1.5)	Consistent with GALL, which recommends further evaluation (See SER Section 3.4.2.2.5)

The staff's review of the steam and power conversion system component groups followed several approaches. One approach, documented in SER Section 3.4.2.1, discusses staff's review of AMR results for components the applicant indicated are consistent with the GALL Report and require no further evaluation. Another approach, documented in SER Section 3.4.2.2, discusses the staff's review of AMR results for components the applicant indicated are consistent with the GALL Report and for which further evaluation is recommended. A third approach, documented in SER Section 3.4.2.3, discusses the staff's review of AMR results for components the applicant indicated are not consistent with or not addressed in the GALL Report. The staff's review of AMPs credited to manage or monitor aging effects of the steam and power conversion system components is documented in SER Section 3.0.3.

3.4.2.1 AMR Results That Are Consistent with the GALL Report

Summary of Technical Information in the Application. In LRA Section 3.4.2.1, the applicant identified the materials, environments, and AERMs. The applicant identified the following programs that manage the aging effects of the steam and power conversion system components:

- Bolting Integrity Program
- Boric Acid Corrosion Program
- Buried Services Corrosion Monitoring Program
- Flow Accelerated Corrosion Program
- One-Time Inspection Program
- System Monitoring Program
- Water Chemistry Program

Staff Evaluation. In LRA Tables 3.4.2-1 through 3.4.2-7, the applicant summarized AMRs for the steam and power conversion system components and identified which AMRs it considered consistent with the GALL Report.

For component groups evaluated in the GALL Report for which the applicant had claimed consistency and for which the GALL Report does not recommend further evaluation, the staff performed an audit and review to determine whether the plant-specific components in these GALL Report component groups were bounded by the GALL Report evaluation.

The applicant provided a note for each AMR line item. The notes describe how the information in the tables aligns with the information in the GALL Report. The staff audited those AMRs with Notes A through E, which indicate how the AMR was consistent with the GALL Report.

Note A indicates that the AMR line item is consistent with the GALL Report for component, material, environment, and aging effect. In addition, the AMP is consistent with the GALL Report AMP. The staff audited these line items to verify consistency with the GALL Report and the validity of the AMR for the site-specific conditions.

Note B indicates that the AMR line item is consistent with the GALL Report for component, material, environment, and aging effect. In addition, the AMP takes some exceptions to the AMP identified in the GALL Report. The staff audited these line items to verify consistency with the GALL Report. The staff verified that the identified exceptions to the GALL Report AMPs had been reviewed and accepted by the staff. The staff also determined whether the AMP identified by the applicant was consistent with the AMP identified in the GALL Report and whether the AMR was valid for the site-specific conditions.

Note C indicates that the component for the AMR line item, although different from, is consistent with the GALL Report for material, environment, and aging effect. In addition, the AMP is consistent with the AMP identified by the GALL Report. This note indicates that the applicant was unable to find a listing of some system components in the GALL Report; however, the applicant identified a different component in the GALL Report that had the same material, environment, aging effect, and AMP as the component under review. The staff audited these line items to verify consistency with the GALL Report. The staff also determined whether the AMR line item of the different component applied to the component under review and whether the AMR was valid for the site-specific conditions.

Note D indicates that the component for the AMR line item, although different from, is consistent with the GALL Report for material, environment, and aging effect. In addition, the AMP takes some exceptions to the AMP identified in the GALL Report. The staff audited these line items to verify consistency with the GALL Report. The staff verified whether the AMR line item of the different component was applicable to the component under review. The staff verified whether the exceptions to the GALL Report AMPs had been reviewed and accepted by the staff. The staff also determined whether the AMP identified by the applicant was consistent with the AMP identified in the GALL Report and whether the AMR was valid for the site-specific conditions.

Note E indicates that the AMR line item is consistent with the GALL Report for material, environment, and aging effect, but a different AMP is credited. The staff audited these line items to verify consistency with the GALL Report. The staff also determined whether the identified

AMP would manage the aging effect consistent with the AMP identified in the GALL Report and whether the AMR was valid for the site-specific conditions.

The staff did not repeat its review of the matters described in the GALL Report; however, the staff did verify that the material presented in the LRA was applicable and that the applicant had identified the appropriate GALL Report AMRs. The staff's evaluation is discussed below.

3.4.2.1.1 Loss of Material Due to Pitting and Crevice Corrosion

In the discussion column of LRA Table 3.4.1, Item 3.4.1-07, the applicant stated that the loss of material due to pitting and crevice corrosion is managed by the Water Chemistry and One-Time Inspection Programs. As documented in the Audit and Review Report, the staff noted that for HXs in the main steam (MS) system, the applicant used only the One-Time Inspection Program to manage loss of material for carbon steel exposed to steam. By letter dated August 27, 2005, the applicant revised the LRA to add the Water Chemistry Program to manage this aging effect.

On the basis of its review, the staff finds that the applicant has appropriately addressed the aging mechanism, as recommended by the GALL Report.

3.4.2.1.2 Loss of Material Due to General Corrosion; Crack Initiation and Growth Due to Cyclic Loading and/or Stress Corrosion Cracking

In the discussion column of LRA Table 3.4.1, Item 3.4.1-08, the applicant stated that the loss of material due to general corrosion, crack initiation and growth due to cyclic loading, or SCC is managed by the Bolting Integrity and System Monitoring Programs. As documented in the Audit and Review Report, the staff noted that the applicant used the Bolting Integrity Program to manage the loss of pre-load for carbon steel fasteners in each of the steam and power conversion systems. For carbon steel fasteners in the steam and power conversion system, the GALL Report does not recommend managing loss of pre-load. However, in the LRA, the applicant conservatively included this aging effect and appropriately used the Bolting Integrity Program to manage it. By letter dated August 27, 2005, the applicant revised the LRA to change Note C to H and to delete the reference to Item 3.4.1-08.

On the basis of its review, the staff finds that the applicant has conservatively addressed the loss of preload for carbon steel fasteners in the steam and power conversion system.

3.4.2.1.3 Loss of Material Due to General (Carbon Steel Only), Pitting, and Crevice Corrosion

In the discussion column of LRA Table 3.4.1, Item 3.4.1-11, the applicant stated that the loss of material due to general (carbon steel only), pitting, and crevice corrosion is managed by the One-Time Inspection and System Monitoring Programs. As documented in the Audit and Review Report, the staff noted that for accumulators in the condensate and condenser system and the demineralized makeup water system, the applicant used only the One-Time Inspection Program to manage the loss of material for carbon steel exposed to weather. For carbon steel accumulators in sun, weather, humidity, and moisture, the GALL Report recommends the use of GALL AMP XI.M29, "Above Ground Carbon Steel Tanks Program." By letter dated August 27, 2005, the applicant clarified that for carbon steel accumulators, its System Monitoring Program had been selected to manage the external surface and its One-Time Inspection Program to manage the bottom thickness instead of GALL AMP XI.M29. Therefore, the note designation should have been E rather than A and B and for accumulators in the demineralized makeup water system, the note designation should have been E rather than C and D.

On the basis of its review, the staff finds that the applicant has selected appropriate AMPs (System Monitoring Program for external surfaces and One-Time Inspection Program for tank bottom thickness) to address loss of material due to general (carbon steel only), pitting, and crevice corrosion for accumulators exposed to weather in the condensate and condenser system and the demineralized makeup water system. The System Monitoring Program manages the loss of material of external surfaces of accumulators by visual inspection for material degradation. The One-Time Inspection Program manages the loss of material of the accumulator bottom surface by measuring the tank bottom surface thickness during the period of extended operation.

Conclusion. The staff evaluated the applicant's claim of consistency with the GALL Report. The staff also reviewed information pertaining to the applicant's consideration of recent operating experience and proposals for managing the associated aging effects. On the basis of its review, the staff concludes that the AMR results, which the applicant claimed to be consistent with the GALL Report, are consistent with the GALL Report AMRs. Therefore, the staff concludes that the applicant has demonstrated that the aging effects for these components will be adequately managed so that their intended function(s) will be maintained consistent with the CLB for the period of extended operation, as required by 10 CFR 54.21(a)(3).

3.4.2.2 AMR Results That Are Consistent with the GALL Report, for Which Further Evaluation is Recommended

Summary of Technical Information in the Application. In LRA Section 3.4.2.2, the applicant provided further evaluation of aging management, as recommended by the GALL Report for the steam and power conversion system components. The applicant provided information concerning how it will manage the following aging effects:

- cumulative fatigue damage
- loss of material due to general, pitting, and crevice corrosion
- loss of material due to general, pitting, and crevice corrosion, MIC, and biofouling
- general corrosion
- loss of material due to general, pitting, crevice, and microbiologically influenced corrosion

Staff Evaluation. For component groups evaluated in the GALL Report for which the applicant had claimed consistency with the GALL Report and for which the GALL Report recommends further evaluation, the staff audited and reviewed the applicant's evaluations to determine whether they adequately address those issues. In addition, the staff reviewed the applicant's further evaluations against the criteria in SRP-LR Section 3.4.2.2. Details of the staff's audit are documented in the Audit and Review Report. The staff's evaluation of the aging effects is discussed in the following sections.

3.4.2.2.1 Cumulative Fatigue Damage

In LRA Section 3.4.2.2.1, the applicant stated that fatigue is a TLAA, as defined in 10 CFR 54.3. Applicants must evaluate TLAAs in accordance with 10 CFR 54.21(c)(1). SER Section 4.3 documents the staff's review of the applicant's evaluation of this TLAA.

3.4.2.2.2 Loss of Material Due to General, Pitting, and Crevice Corrosion

The staff reviewed LRA Section 3.4.2.2.2 against the criteria in SRP-LR Section 3.4.2.2.2.

In LRA Section 3.4.2.2.2, the applicant addressed loss of material of piping and fittings, valve bodies and bonnets, pump casings, tanks, tubes, tubesheets, and channel head and shell (except for the MS system) due to general (carbon steel only), pitting, and crevice corrosion.

SRP-LR Section 3.4.2.2.2 states that the management of loss of material due to general, pitting, and crevice corrosion should be evaluated further for carbon steel piping and fittings, valve bodies and bonnets, pump casings, pump suction and discharge lines, tanks, tubesheets, and channel heads and shells (except for MS system) components and for loss of material due to pitting and crevice corrosion for stainless steel tanks and heat exchanger/cooler tubes. The Water Chemistry Program relies on monitoring and control of water chemistry based on the EPRI TR-102134 guidelines for secondary water chemistry to manage the effects of loss of material due to general, pitting, or crevice corrosion. However, corrosion may occur at locations of stagnant flow conditions. Therefore, the effectiveness of the chemistry control program should be verified to ensure that corrosion does not occur. The GALL Report recommends further evaluation of programs to manage loss of material due to general, pitting, and crevice corrosion to verify the effectiveness of the Water Chemistry Program. A one-time inspection of select components and susceptible locations is an acceptable method to ensure that corrosion is not occurring and that the component intended functions will be maintained during the period of extended operation.

In the LRA, the applicant stated that under the One-Time Inspection Program, a representative sample of the components would be chosen for inspection. The focus, when practical, would be placed on bounding or lead components. Factors considered when choosing components for inspection are time in service, severity of operating conditions, and operating experience. The examination techniques would be visual, volumetric, or other appropriately established NDE methods capable of management of the aging effect of loss of material due to galvanic and general corrosion, MIC, pitting and crevice corrosion, and selective leaching.

The AMPs recommended by the GALL Report for management of this aging effect are GALL AMPs XI.M2, "Water Chemistry" and XI.M32, "One-Time Inspection." As documented in the Audit and Review Report, the staff noted that the applicant did not use the Water Chemistry Program to manage this aging effect as recommended. By letter dated August 27, 2005, the applicant stated that the Water Chemistry Program would be added for those AMRs from which it had been omitted. With this change, the applicant made further evaluation by using the One-Time Inspection Program to verify the effectiveness of the Water Chemistry Program to manage this aging effect.

Based on the programs identified above, the staff concludes that the applicant has met the criteria of SRP-LR Section 3.4.2.2.2. For those line items that apply to LRA Section 3.4.2.2.2, the staff determines that the application is consistent with the GALL Report and the applicant has demonstrated that the aging effects will be adequately managed so that the intended function(s) will be maintained consistent with the CLB for the period of extended operation, as required by 10 CFR 54.21(a)(3).

3.4.2.2.3 Loss of Material Due to General, Pitting and Crevice Corrosion, Microbiologically Influenced Corrosion, and Biofouling

The staff reviewed LRA Section 3.4.2.2.3 against the criteria in SRP-LR Section 3.4.2.2.3.

In LRA Section 3.4.2.2.3, the applicant addressed loss of material of auxiliary feedwater (AFW) piping due to general, pitting, and crevice corrosion, MIC, and biofouling.

SRP-LR Section 3.4.2.2.3 states that loss of material due to general corrosion, pitting and crevice corrosion, MIC, and biofouling could occur in carbon steel piping and fittings for untreated water from the backup water supply in the AFW system. The GALL Report recommends further evaluation to ensure adequate management of these aging effects.

In the LRA, the applicant stated that the portion of the lines from the SWS to the AFW system is addressed as part of the SWS (SRP-LR Table 3.3.1, Item 17). The aging effect of loss of material is managed by the Open Cycle Cooling Water Program, which is consistent with GALL AMP XI.M20, "Open-Cycle Cooling Water System," and aging effects will be managed for SSCs within the scope of this program to continue to perform intended functions.

In the LRA, the applicant stated that the portion of the lines from the fire protection water lines to the AFW system is addressed as part of the fire protection system (SRP-LR Table 3.3.1, Item 21). The aging effect of loss of material is managed by the Fire Protection Program, which is consistent with, but includes exceptions to, GALL AMP XI.M27, "Fire Water System."

As documented in the Audit and Review Report, the staff noted that the applicant used only the One-Time Inspection Program to manage this aging effect for carbon steel heat exchangers exposed to raw water in the auxiliary feedwater system. By letter dated August 27, 2005, the applicant stated that it would add the Open Cycle Cooling Water Program for this AMR line item and change the LRA Table 3.4.1 reference to Item 3.4.1-09. The GALL Report recommends managing this aging effect with GALL AMP XI.M20. With this change, the applicant's further evaluation used the One-Time Inspection Program to verify the effectiveness of the Open Cycle Cooling Water Program to manage this aging effect.

Additionally, as documented in the Audit and Review Report, the staff noted that the applicant used only the One-Time Inspection Program to manage this aging effect for carbon steel valves and dampers exposed to raw water in the auxiliary feedwater system. By letter dated August 27, 2005, the applicant stated that it would add the Open Cycle Cooling Water Program as well as the Fire Protection Program for this AMR line item and would change the LRA Table 3.4.1 reference to Item 3.4.1-03. The GALL Report recommends managing this aging effect with a plant-specific program. With this change, the applicant's further evaluation used the One-Time Inspection Program to verify the effectiveness of the Open Cycle Cooling Water Program and the Fire Protection Program to manage this aging effect.

Based on the programs identified above, the staff concludes that the applicant has met the criteria of SRP-LR Section 3.4.2.2.3. For those line items that apply to LRA Section 3.4.2.2.3, the staff determines that the application is consistent with the GALL Report and the applicant has demonstrated that the aging effects will be adequately managed so that the intended function(s) will be maintained consistent with the CLB for the period of extended operation, as required by 10 CFR 54.21(a)(3).

3.4.2.2.4 General Corrosion

The staff reviewed LRA Section 3.4.2.2.4 against the criteria in SRP-LR Section 3.4.2.2.4.

In LRA Section 3.4.2.2.4, the applicant stated that the external aging effect of general corrosion is managed by the System Monitoring Program. The external surfaces of various component types (e.g., pump casings, valve bodies, piping, etc.) are visually inspected for leakage and evidence of such degradation as loss of material due to corrosion. The external aging effect of general corrosion of closure boltings is managed by the Bolting Integrity Program.

SRP-LR Section 3.4.2.2.4 states that loss of material due to general corrosion could occur on the external surfaces of all carbon steel structures and components (SCs), including closure boltings, exposed to an operating temperature of less than 212 °F. The GALL Report recommends further evaluation to ensure adequate management of this aging effect.

In the LRA, the applicant further stated that the attributes of the System Monitoring Program are consistent with SRP-LR Appendix A.1 criteria and that this aging effect would be managed for the subject SSCs to continue to perform intended functions. In the LRA, the applicant stated that its Bolting Integrity Program is consistent with GALL AMP XI.M18, "Bolting Integrity."

The staff finds that the applicant has met the criteria of SRP-LR Section 3.4.2.2.4 for further evaluation.

Based on the programs identified above, the staff concludes that the applicant has met the criteria of SRP-LR Section 3.4.2.2.4. For those line items that apply to LRA Section 3.4.2.2.4, the staff determines that the application is consistent with the GALL Report and the applicant has demonstrated that the aging effects will be adequately managed so that the intended function(s) will be maintained consistent with the CLB for the period of extended operation, as required by 10 CFR 54.21(a)(3).

3.4.2.2.5 Loss of Material Due to General, Pitting, Crevice, and Microbiologically Influenced Corrosion

The staff reviewed LRA Sections 3.4.2.2.5.1 and 3.4.2.2.5.2 against the criteria in SRP-LR Section 3.4.2.2.5.

In LRA Section 3.4.2.2.5.1, the applicant stated that this paragraph of the SRP-LR is not applicable. The applicant also stated that the bearing oil coolers for the AFW pump turbines addressed in GALL Report Items VIII.G.5-a through VIII.G.5-d are not applicable.

In LRA Section 3.4.2.2.5.2, the applicant addressed loss of material of the external surface of the buried condensate storage tank and AFW piping due to general, pitting and crevice corrosion, and MIC. The LRA states that PNP has no buried storage tanks. Loss of material in buried carbon steel and stainless steel piping is managed by the Buried Services Corrosion Monitoring Program, which is consistent with GALL AMP XI.M34, "Buried Piping and Tanks Inspection," and would be effective in maintaining the intended functions of these underground piping systems.

SRP-LR Section 3.4.2.2.5 states that loss of material due to general (carbon steel only), pitting and crevice corrosion, and MIC could occur in stainless steel and carbon steel shells, tubes, and tubesheets within the bearing oil coolers (for steam turbine pumps) in AFW system. The GALL Report recommends further evaluation to ensure adequate management of these aging effects. Acceptance criteria are described in SRP-LR Appendix A.1.

SRP-LR Section 3.4.2.2.5 also states that loss of material due to general, pitting and crevice corrosion, and MIC could occur in underground piping and fittings and emergency condensate storage tank in the AFW system and the underground condensate storage tank in the condensate system. The Buried Piping and Tanks Inspection Program relies on industry practice, frequency of pipe excavation, and operating experience to manage the effects of loss of material from general corrosion, pitting and crevice corrosion, and MIC. The effectiveness of the Buried Piping and Tanks Inspection Program should be verified to evaluate an applicant's inspection frequency and operating experience with buried components, ensuring that loss of material does not occur.

The AFW pump turbines are not equipped with bearing oil coolers; therefore, the staff found the aging effect discussed in SRP-LR Section 3.4.2.2.5.1 not applicable.

On the basis that there are no buried storage tanks in the steam and power conversion system, the staff finds this aging effect not applicable for this component type. For buried carbon steel and stainless steel AFW piping, the staff finds that the applicant's Buried Services Corrosion Monitoring Program has met the criteria of SRP-LR Section 3.4.2.2.5.2.

Based on the programs identified above, the staff concludes that the applicant has met the criteria of SRP-LR Section 3.4.2.2.5. For those line items that apply to LRA Section 3.4.2.2.5, the staff determines that the application is consistent with the GALL Report and the applicant has demonstrated that the aging effects will be adequately managed so that the intended function(s) will be maintained consistent with the CLB for the period of extended operation, as required by 10 CFR 54.21(a)(3).

3.4.2.2.6 Quality Assurance for Aging Management of Nonsafety-Related Components

SER Section 3.0.4 provides the staff's evaluation of the applicant's QA program.

Conclusion. On the basis of its review, for component groups evaluated in the GALL Report for which the applicant claimed consistency with the GALL Report and for which further evaluation is recommended, the staff determined that the applicant has adequately addressed the issues that were further evaluated. The staff finds that the applicant has demonstrated that the aging effects will be adequately managed so that the intended function(s) will be maintained consistent with the CLB for the period of extended operation, as required by 10 CFR 54.21(a)(3).

3.4.2.3 AMR Results That Are Not Consistent with or Not Addressed in the GALL Report

Summary of Technical Information in the Application. In LRA Tables 3.4.2-1 through 3.4.2-7, the staff reviewed additional details of AMR results for material, environment, AERM, and AMP combinations not consistent with or not addressed in the GALL Report.

In LRA Tables 3.4.2-1 through 3.4.2-7, the applicant indicated, via Notes F through J, that the combination of component type, material, environment, and AERM does not correspond to a line item in the GALL Report. The applicant provided further information concerning how the aging effects will be managed. Specifically, Note F indicates that the material for the AMR line item component is not evaluated in the GALL Report. Note G indicates that the environment for the AMR line item component and material is not evaluated in the GALL Report. Note H indicates that the aging effect for the AMR line item component, material, and environment combination is not evaluated in the GALL Report. Note I indicates that the aging effect identified in the GALL Report for the line item component, material, and environment combination is not applicable. Note J indicates that neither the component nor the material and environment combination for the line item is evaluated in the GALL Report.

Staff Evaluation. For component type, material, and environment combinations not evaluated in the GALL Report, the staff reviewed the applicant's evaluation to determine whether the applicant had demonstrated that the aging effects will be adequately managed so that the intended function(s) will be maintained consistent with the CLB during the period of extended operation. The staff's evaluation is discussed in the following sections.

3.4.2.3.1 Steam and Power Conversion System – Condensate and Condenser System – Summary of Aging Management – LRA Table 3.4.2-1

The staff reviewed LRA Table 3.4.2-1, which summarizes the results of AMR evaluations for the condensate and condenser system component groups.

The applicant proposed the Buried Services Corrosion Monitoring Program to manage loss of stainless steel materials for component type pipe and fittings exposed to an external soil environment.

The staff reviewed the applicant's Buried Services Corrosion Monitoring Program as documented in SER Section 3.0.3.1.3. The program manages aging effects on the external surfaces of carbon, low-alloy, and stainless steel components buried in soil or sand. On the basis of its review, the staff finds the management of loss of stainless steel in the condensate and condenser system exposed to soil acceptable.

In LRA Table 3.4.2-1, the applicant proposed the One-Time Inspection Program to manage loss of material (selective leaching) of cast iron materials for component type of pumps exposed to a treated water environment.

The staff reviewed the applicant's One-Time Inspection Program as documented in SER Section 3.0.3.2.6. The program addresses potentially long incubation periods for certain aging effects, including various corrosion mechanisms, cracking, and selective leaching, and verifies that an aging effect either does not occur or progresses so slowly as to have a negligible effect on the intended function of the structure or component. On the basis of its review, the staff finds the aging effect of loss material of (selective leaching) of cast iron exposed to a treated water environment is effectively managed by a program consistent with GALL AMP XI.M33, "Selective Leaching of Materials." In the LRA, selective leaching is managed by the One-Time Inspection Program. On this basis, the staff finds management of loss of material (selective leaching) in the condensate and condenser system acceptable.

On the basis of its review, as discussed above, the staff concludes that the applicant has demonstrated that the aging effects associated with the components will be adequately managed so that the intended function(s) will be maintained consistent with the CLB for the period of extended operation, as required by 10 CFR 54.21(a)(3).

3.4.2.3.2 Steam and Power Conversion System – Demineralized Makeup Water System – Summary of Aging Management – LRA Table 3.4.2-2

The staff reviewed LRA Table 3.4.2-2, which summarizes the results of AMR evaluations for the demineralized makeup water system component groups.

In LRA Table 3.4.2-2, the applicant proposed the Buried Services Corrosion Monitoring Program to manage loss of material from stainless steel pipe and fittings exposed to an external soil environment.

The staff reviewed the applicant's Buried Services Corrosion Monitoring Program as documented in SER Section 3.0.3.1.3. On the basis of its review, the staff finds management of loss of stainless steel exposed to soil in the demineralized makeup water system acceptable.

On the basis of its review, as discussed above, the staff concludes that the applicant has demonstrated that the aging effects associated with the components will be adequately managed so that the intended function(s) will be maintained consistent with the CLB for the period of extended operation, as required by 10 CFR 54.21(a)(3).

3.4.2.3.3 Steam and Power Conversion System – Feedwater System – Summary of Aging Management – LRA Table 3.4.2-3

The staff reviewed LRA Table 3.4.2-3, which summarizes the results of AMR evaluations for the feedwater system component groups.

In LRA Table 3.4.2-3, the applicant proposed the Water Chemistry Program augmented by the One-Time Inspection Program to manage cracking of stainless steel materials for component types valves and dampers and pipe and fittings exposed to an internal treated water environment.

The staff reviewed the applicant's Water Chemistry and One-Time Inspection Programs as documented in SER Sections 3.0.3.1.10 and 3.0.3.2.6, respectively. The Water Chemistry Program is credited for managing such aging effects as loss of material due to general, pitting, and crevice corrosion, cracking due to SCC, and steam generator tube degradation caused by denting, IGA, and ODSCC by controlling environments to which internal surfaces of systems and components are exposed. The One-Time Inspection Program addresses potentially long incubation periods for certain aging effects, including various corrosion mechanisms, cracking, and selective leaching, and verifies that an aging effect either does not occur or progresses so slowly as to have a negligible effect on the intended function of the structure or component. On the basis of its review, the staff finds management of cracking of stainless steel components exposed to treated water in the feedwater system acceptable.

In LRA Table 3.4.2-3, the applicant proposed the One-Time Inspection and Water Chemistry Programs to manage cracking of stainless steel materials for component types of valves and dampers exposed to an internal steam environment.

The staff reviewed the applicant's One-Time Inspection and Water Chemistry Programs as documented in SER Sections 3.0.3.2.6 and 3.0.3.1.10, respectively. On the basis of its review, the staff finds management of cracking of stainless steel components exposed to steam in the feedwater system acceptable.

In LRA Table 3.4.2-3, the applicant proposed the One-Time Inspection Program to manage loss of material of stainless steel for component type of pipe and fittings exposed to an internal oil environment.

The GALL Report recommends management of the loss of material aging effect by a plant-specific or lubricating/fuel oil analysis AMP with a one-time inspection to ensure AMP effectiveness.

In its response, to general RAI 3.3-2, the applicant stated that a new section (Section A2.23) that addresses future oil sampling and analysis activities is hereby added to LRA Appendix A and reads as follows:

A2.23 Oil Sampling and Analysis

For selected components, in-scope for license renewal, that have an internal environment of oil, and are constructed of materials that are potentially susceptible to internal aging degradation in that environment, the oil shall be subject to periodic sampling and analysis. The purpose of these activities is to ensure that oil system contaminants (primarily water and particulates) are maintained within acceptable limits, thereby preserving an environment that is not conducive to loss of material or reduction of heat transfer. Associated activities include (a) determination of appropriate analysis to be performed, (b) frequency of analysis, (c) acceptance criteria, (d) trending of results, and (e) corrective actions, if required. These activities ensure that the lubricating oil environment of these components is maintained such that water and contaminants are minimized.

The staff review found the applicant response acceptable since the applicant has added the Oil Sampling Analysis Program with a verification by the One Time Inspection Program. This position is consistent with the GALL recommendation, therefore, this RAI is resolved.

In LRA Table 3.4.2-3, the applicant proposed the Buried Services Corrosion Monitoring Program to manage loss of material of stainless steel for component type of pipe and fittings exposed to an external soil environment.

The staff reviewed the applicant's Buried Services Corrosion Monitoring Program as documented in SER Section 3.0.3.1.3. On the basis of its review, the staff finds management of loss of material of stainless steel components exposed to soil in the feedwater system acceptable.

In LRA Table 3.4.2-3, the applicant proposed the One-Time Inspection and Water Chemistry Programs to manage loss of material of stainless steel for component type of valves and dampers exposed to an internal steam environment.

The staff reviewed the applicant's One-Time Inspection and Water Chemistry Programs as documented in SER Sections 3.0.3.2.6 and 3.0.3.1.10, respectively. On the basis of its review, the staff finds management of loss of material of stainless steel components exposed to steam in the feedwater system acceptable.

In LRA Table 3.4.2-3, the applicant proposed the One-Time Inspection Program to manage loss of material of carbon steel for component types of traps (steam), turbines and valves and dampers exposed to internal air environments.

The staff reviewed the applicant's One-Time Inspection Program as documented in SER Section 3.0.3.2.6. On the basis of its review, the staff finds management of loss of material of carbon steel components exposed to air in the feedwater system with the One-Time Inspection Program acceptable because the applicant conservatively proposed to manage this aging effect with its One-Time Inspection Program.

In LRA Table 3.4.2-3, the applicant proposed the One-Time Inspection Program to manage loss (selective leaching) of cast iron materials for component types of valves and dampers exposed to a treated water environment.

The staff reviewed the applicant's One-Time Inspection Program as documented in SER Section 3.0.3.2.6. On the basis of its review of the applicant's plant-specific and industry operating experience, the staff finds the aging effect of loss of material (selective leaching) of cast iron materials exposed to a treated water environment effectively managed by a program consistent with GALL AMP XI.M33, "Selective Leaching of Materials." In the LRA, selective leaching is managed by the One-Time Inspection Program. On this basis, the staff finds management of loss of material (selective leaching) of cast iron valves and dampers exposed to treated water in the feedwater system acceptable.

In LRA Table 3.4.2-3, the applicant proposed the One-Time Inspection Program to manage loss of stainless steel materials for component types of pipe and fittings and valves and dampers exposed to a plant indoor air environment.

The staff reviewed the applicant's One-Time Inspection Program as documented in SER Section 3.0.3.2.6. On the basis of its review, the staff finds management by the One-Time Inspection Program of loss of stainless steel material exposed to air in the feedwater system acceptable.

In LRA Table 3.4.2-3, the applicant proposed the System Monitoring Program to manage loss of stainless steel materials for component type pipe and fittings exposed to an external weather environment.

The System Monitoring Program was reviewed by the staff as addressed in SER Section 3.0.3.3.1. This plant-specific program manages aging effects for normally accessible, external surfaces of piping, tanks, and other components and equipment within the scope of license renewal. The staff finds management of loss of stainless steel material exposed to weather in the feedwater system acceptable.

On the basis of its review, as discussed above, the staff concludes that the applicant has demonstrated that the aging effects associated with the components will be adequately managed so that the intended function(s) will be maintained consistent with the CLB for the period of extended operation, as required by 10 CFR 54.21(a)(3).

3.4.2.3.4 Steam and Power Conversion System – Heater Extraction and Drain System – Summary of Aging Management – LRA Table 3.4.2-4

The staff reviewed LRA Table 3.4.2-4, which summarizes the results of AMR evaluations for the heater extraction and drain system component groups.

In LRA Table 3.4.2-4, the applicant proposed the One-Time Inspection and Water Chemistry Programs to manage cracking of stainless steel materials for component types of pipe and fittings and valves and dampers exposed to an internal steam environment.

The staff reviewed the applicant's One-Time Inspection and Water Chemistry Programs as documented in SER Sections 3.0.3.2.6 and 3.0.3.1.10, respectively. On the basis of its review, the staff finds management of cracking of stainless steel exposed to steam in the heater extraction and drain system acceptable.

In LRA Table 3.4.2-4, the applicant proposed the Water Chemistry Program augmented by the One-Time Inspection Program to manage the cracking of stainless steel materials for component type of transmitter/element exposed to an internal treated water environment.

The staff reviewed the applicant's Water Chemistry and One-Time Inspection Programs as documented in SER Sections 3.0.3.1.10 and 3.0.3.2.6, respectively. On the basis of its review, the staff finds management of cracking of stainless steel exposed to treated water in the heater extraction and drain system acceptable.

In LRA Table 3.4.2-4, the applicant proposed the One-Time Inspection and Water Chemistry Programs to manage loss of stainless steel materials for component types of pipe and fittings and valves and dampers exposed to an internal steam environment.

The staff reviewed the applicant's One-Time Inspection and Water Chemistry Programs as documented in SER Sections 3.0.3.2.6 and 3.0.3.1.10, respectively. On the basis of its review, the staff finds management of loss of material in the heater extraction and drain system acceptable.

In LRA Table 3.4.2-4, the applicant proposed the Water Chemistry Program augmented by the One-Time Inspection Program to manage loss of stainless steel materials for component type transmitter/element exposed to an internal treated water environment.

The staff reviewed the applicant's Water Chemistry and One-Time Inspection Programs as documented in SER Sections 3.0.3.1.10 and 3.0.3.2.6, respectively. On the basis of its review, the staff finds management of loss of stainless steel material exposed to treated water in the heater extraction and drain system acceptable.

In LRA Table 3.4.2-4, the applicant does not manage loss of material due to general corrosion for cast iron valves and dampers exposed to a plant indoor air environment. On the basis of its review of the applicant's plant-specific and industry operating experience, the staff agreed that there was no aging effect to be managed for this material and environment combination in the heater extraction and drain system.

As documented in the Audit and Review Report, the staff noted that the applicant managed loss of material due to selective leaching for cast iron valves and dampers exposed to a steam environment. By letter dated August 27, 2005, the applicant clarified that aging effects of cast iron in a steam environment are considered the same as those of cast iron in condensate (treated water). Therefore, loss of material due to selective leaching is a valid aging effect for this component type.

The staff reviewed the applicant's One-Time Inspection Program as documented in SER Section 3.0.3.2.6. On the basis of its review of the applicant's plant-specific and industry operating experience, the staff finds the aging effect of loss of material (selective leaching) of cast iron materials exposed to a treated water environment effectively managed by a program consistent with GALL AMP XI.M33, "Selective Leaching of Materials." Selective leaching is managed by the One-Time Inspection Program. On this consistency basis, the staff finds management of loss of material (selective leaching) in the heater extraction and drain system acceptable.

On the basis of its review, as discussed above, the staff concludes that the applicant has demonstrated that the aging effects associated with the components will be adequately managed so that the intended function(s) will be maintained consistent with the CLB for the period of extended operation, as required by 10 CFR 54.21(a)(3).

3.4.2.3.5 Steam and Power Conversion System – Main Air Ejection and Gland Seal System – Summary of Aging Management – LRA Table 3.4.2-5

The staff reviewed LRA Table 3.4.2-5, which summarizes the results of AMR evaluations for the main air ejection and gland seal system component groups.

In LRA Table 3.4.2-5, the applicant proposed the One-Time Inspection Program to manage loss of carbon steel materials for component types of blowers/fans (compressor vacuum), filters/strainers, pipe and fittings, and valves and dampers exposed to internal air environment.

The staff reviewed the applicant's One-Time Inspection Program as documented in SER Section 3.0.3.2.6. On the basis of its review, the staff finds management by the One-Time Inspection Program of loss of material of carbon steel components exposed to air in the main air ejection and gland seal system acceptable.

As documented in the Audit and Review Report, the staff noted that the applicant managed loss of material due to selective leaching for cast iron blowers/fans (compressor vacuum) and valves and dampers exposed to a steam environment. By letter dated August 27, 2005, the applicant clarified that aging effects of cast iron in a steam environment are considered the same as those of cast iron in condensate (treated water). Therefore, loss of material due to selective leaching is a valid aging effect for this component type.

The staff reviewed the applicant's One-Time Inspection Program as documented in SER Section 3.0.3.2.6. On the basis of its review of the applicant's plant-specific and industry operating experience, the staff found the aging effect of loss of material (selective leaching) of cast iron materials exposed to a treated water environment effectively managed by a program consistent with GALL AMP XI.M33, "Selective Leaching of Materials." In the LRA, selective leaching is managed by the One-Time Inspection Program. On this basis, the staff finds management of loss of material (selective leaching) in the main air ejection and gland seal system acceptable.

On the basis of its review, as discussed above, the staff concludes that the applicant has demonstrated that the aging effects associated with the components will be adequately managed so that the intended function(s) will be maintained consistent with the CLB for the period of extended operation, as required by 10 CFR 54.21(a)(3).

3.4.2.3.6 Steam and Power Conversion System – Main Steam System – Summary of Aging Management – LRA Table 3.4.2-6

The staff reviewed LRA Table 3.4.2-6, which summarizes the results of AMR evaluations for the MS system component groups.

In LRA Table 3.4.2-6, the applicant proposed the Water Chemistry Program to manage cracking of stainless steel materials for component types of indicators/recorders, pipe and fittings, and valves and dampers exposed to an internal steam environment.

The staff reviewed the applicant's Water Chemistry Program as documented in SER Section 3.0.3.1.10. On the basis of its review, the staff finds management of cracking of stainless steel exposed to air in the MS system acceptable.

In LRA Table 3.4.2-6, the applicant proposed the Water Chemistry Program to manage loss of stainless steel materials for component types of indicators/recorders, pipe and fittings, and valves and dampers exposed to an internal steam environment.

The staff reviewed the applicant's Water Chemistry Program as documented in SER Section 3.0.3.1.10. On the basis of its review, the staff finds management of loss of material of stainless steel exposed to steam in the MS system acceptable.

In LRA Table 3.4.2-6, the applicant proposed the Boric Acid Corrosion Program to manage loss of copper alloy materials for component type of pipe and fittings exposed to a plant indoor air environment.

The Boric Acid Corrosion Program was reviewed by the staff as addressed in SER Section 3.0.3.2.2. The program monitors through periodic inspections component degradation due to boric acid leakage. The staff finds management of loss of copper alloy materials exposed to a plant indoor air environment in the feedwater system acceptable.

On the basis of its review, as discussed above, the staff concludes that the applicant has demonstrated that the aging effects associated with the components will be adequately managed so that the intended function(s) will be maintained consistent with the CLB for the period of extended operation, as required by 10 CFR 54.21(a)(3).

3.4.2.3.7 Steam and Power Conversion System – Turbine Generator System – Summary of Aging Management – LRA Table 3.4.2-7

The staff reviewed LRA Table 3.4.2-7, which summarizes the results of AMR evaluations for the turbine generator system component groups.

In LRA Table 3.4.2-7, the applicant proposed the One-Time Inspection Program to manage loss of stainless steel materials for component types of accumulators and pipe and fittings exposed to an internal oil environment.

The GALL Report recommends management of the loss of material aging effect by a plant-specific or lubricating/fuel oil analysis AMP in conjunction with the One-Time Inspection Program to ensure the effectiveness of the AMP.

In its response to general RAI 3.3-2, the applicant stated, as noted in SER Section 3.4.2.3.3, a new section (Section A2.23) that addresses future oil sampling and analysis activities is added to LRA Appendix A.

The staff review found the applicant response acceptable since the applicant has added the Oil Sampling Analysis Program with a verification by the One Time Inspection Program. This position is consistent with the GALL recommendation, therefore, this RAI is resolved.

On the basis of its review, as discussed above, the staff concludes that the applicant has demonstrated that the aging effects associated with the components will be adequately managed so that the intended function(s) will be maintained consistent with the CLB for the period of extended operation, as required by 10 CFR 54.21(a)(3).

Conclusion. On the basis of its review, the staff finds that the applicant has appropriately evaluated AMR results involving material, environment, AERMs, and AMP combinations not evaluated in the GALL Report. The staff finds that the applicant has demonstrated that the aging effects will be adequately managed so that the intended functions will be maintained consistent with the CLB for the period of extended operation, as required by 10 CFR 54.21(a)(3).

3.4.3 Conclusion

The staff concludes that the applicant has demonstrated that the aging effects associated with the steam and power conversion system components will be adequately managed so that the intended function(s) will be maintained consistent with the CLB for the period of extended operation, as required by 10 CFR 54.21(a)(3).

The staff also reviewed the applicable FSAR supplement program summaries and concludes that they adequately describe the AMPs credited for managing aging of the steam and power conversion system, as required by 10 CFR 54.21(d).

3.5 Aging Management of Containments, Structures, and Component Supports

This section of the SER documents the staff's review of the applicant's AMR results for the containments, structures, and component supports of the following:

- auxiliary building
- component supports
- containment
- containment interior structures
- discharge structure
- feedwater purity building
- intake structure
- miscellaneous structural and bulk commodities
- switchyard and yard structures
- turbine building

3.5.1 Summary of Technical Information in the Application

In LRA Section 3.5, the applicant provided AMR results for the containments, structures, and component supports, components and component groups. In LRA Table 3.5.1, "Summary of Aging Management Evaluations in Chapters II and III of NUREG-1801 for Structures and Component Supports," the applicant provided a summary comparison of its AMRs to those evaluated in the GALL Report for containments, structures, and component supports, components and component groups.

The applicant's AMRs incorporated applicable operating experience in the determination of AERMs. These reviews included evaluation of plant-specific and industry operating experience. The plant-specific evaluation included reviews of condition reports and discussions with appropriate site personnel to identify AERMs. The applicant's review of industry operating experience included a review of the GALL Report and issues identified since the issuance of the GALL Report.

3.5.2 Staff Evaluation

The staff reviewed LRA Section 3.5 to determine whether the applicant had provided sufficient information to demonstrate that the aging effects for the containments, structures, and component supports components within the scope of license renewal and subject to an AMR will be adequately managed so that the intended function(s) will be maintained consistent with the CLB for the period of extended operation, as required by 10 CFR 54.21(a)(3).

The staff conducted an onsite audit of AMRs to confirm the applicant's claim that certain identified AMRs were consistent with the GALL Report. The staff did not repeat its review of the matters described in the GALL Report; however, the staff did verify that the material presented in the LRA was applicable and that the applicant had identified the appropriate GALL Report AMRs. The staff's evaluations of the AMPs are documented in SER Section 3.0.3. Details of the staff's audit evaluation are documented in the Audit and Review Report and summarized in SER Section 3.5.2.1.

In the onsite audit, the staff also selected AMRs consistent with the GALL Report for which further evaluation is recommended. The staff confirmed that the applicant's further evaluations were consistent with the acceptance criteria in SRP-LR Section 3.5.2.2. The staff's audit evaluations are documented in the Audit and Review Report and summarized in SER Section 3.5.2.2.

In the onsite audit, the staff also conducted a technical review of the those remaining AMRs not consistent with, or not addressed in, the GALL Report. The audit and technical review evaluated whether all plausible aging effects were identified and whether the aging effects listed were appropriate for the combination of materials and environments specified. The staff's audit evaluations documented in the Audit and Review Report are summarized in SER Section 3.5.2.3. The staff's evaluation of its technical review is also documented in SER Section 3.5.2.3.

Finally, the staff reviewed the AMP summary descriptions in the FSAR supplement to ensure that they adequately describe the programs credited with managing or monitoring aging for the containments, structures, and component supports components.

SER Table 3.5-1 below summarizes the staff's evaluation of components, aging effects/mechanisms, and AMPs listed in LRA Section 3.5 and addressed in the GALL Report.

Table 3.5-1 Staff Evaluation for Containments, Structures, and Component Supports in the GALL Report

Component Group	Aging Effect/ Mechanism	AMP in GALL Report	AMP in LRA	Staff Evaluation
Common Components of All Types of PWR and BWR Containment				
Penetration sleeves, penetration bellows, and dissimilar metal welds (Item Number 3.5.1-01)	Cumulative fatigue damage (CLB fatigue analysis exists)	TLAA, evaluated in accordance with 10 CFR 54.21(c)	TLAA	This TLAA is evaluated in SER Section 4.6, "Containment Liner Plate and Penetrations Load Cycles"
Penetration sleeves, bellows, and dissimilar metal welds (Item Number 3.5.1-02)	Cracking due to cyclic loading, or crack initiation and growth due to SCC	Containment ISI and Containment leak rate test	None	Not applicable (See SER Section 3.5.2.2.1)
Penetration sleeves, penetration bellows, and dissimilar metal welds (Item Number 3.5.1-03)	Loss of material due to corrosion	Containment ISI and Containment leak rate test	Containment Inservice Inspection Program (B2.1.7), Containment Leakage Testing Program (B2.1.8)	Consistent with GALL, which recommends no further evaluation (See SER Section 3.5.2.1)

Component Group	Aging Effect/ Mechanism	AMP in GALL Report	AMP in LRA	Staff Evaluation
Personnel airlock and equipment hatch (Item Number 3.5.1-04)	Loss of material due to corrosion	Containment ISI and Containment leak rate test	Containment Inservice Inspection Program (B2.1.7), Containment Leakage Testing Program (B2.1.8)	Consistent with GALL, which recommends no further evaluation (See SER Section 3.5.2.1)
Personnel airlock and equipment hatch (Item Number 3.5.1-05)	Loss of leak tightness in closed position due to mechanical wear of locks, hinges and closure mechanism	Containment leak rate test and Plant Technical Specifications	Containment Leakage Testing Program (B2.1.8)	Consistent with GALL, which recommends no further evaluation (See SER Section 3.5.2.1.1)
Seals, gaskets, and moisture barriers (Item Number 3.5.1-06)	Loss of sealant and leakage through containment due to deterioration of joint seals, gaskets, and moisture barriers	Containment ISI and Containment leak rate test	Containment Inservice Inspection Program (B2.1.7), Containment Leakage Testing Program (B2.1.8)	Consistent with GALL, which recommends no further evaluation (See SER Section 3.5.2.1.2)
PWR Concrete (Reinforced and Prestressed) and Steel Containment				
Concrete elements: foundation, walls, dome (Item Number 3.5.1-07)	Aging of accessible and inaccessible concrete areas due to leaching of calcium hydroxide, aggressive chemical attack, and corrosion of embedded steel	Containment ISI	Containment Inservice Inspection Program (B2.1.7)	Consistent with GALL, which recommends further evaluation (See SER Section 3.5.2.2.1)
Concrete elements: foundation (Item Number 3.5.1-08)	Cracks, distortion, and increases in component stress level due to settlement	Structures Monitoring	Structural Monitoring Program (B2.1.19)	Consistent with GALL, which recommends further evaluation (See SER Section 3.5.2.2.1)
Concrete elements: foundation (Item Number 3.5.1-09)	Reduction in foundation strength due to erosion of porous concrete subfoundation	Structures Monitoring	None	Not applicable (See SER Section 3.5.2.2.1)
Concrete elements: foundation, dome, and wall (Item Number 3.5.1-10)	Reduction of strength and modulus due to elevated temperature	Plant specific	None	Not applicable (See SER Section 3.5.2.2.1)
Prestressed containment: tendons and anchorage components (Item Number 3.5.1-11)	Loss of prestress due to relaxation, shrinkage, creep, and elevated temperature	TLAA, evaluated in accordance with 10 CFR 54.21(c)	TLAA	This TLAA is evaluated in SER Section 4.5, "Concrete Containment Tendon Prestress Analysis"

Component Group	Aging Effect/ Mechanism	AMP in GALL Report	AMP in LRA	Staff Evaluation
Steel elements: liner plate, containment shell (Item Number 3.5.1-12)	Loss of material due to corrosion in accessible and inaccessible areas	Containment ISI and Containment leak rate test	Boric Acid Corrosion Program (B2.1.4), Containment Inservice Inspection Program (B2.1.7), Containment Leakage Testing Program (B2.1.8)	Consistent with GALL, which recommends further evaluation (See SER Sections 3.5.2.1.3 and 3.5.2.2.1)
Steel elements: protected by coating (Item Number 3.5.1-14)	Loss of material due to corrosion in accessible areas only	Protective coating monitoring and maintenance	None	Not applicable (No credit is taken for coatings)
Prestressed containment: tendons and anchorage components (Item Number 3.5.1-15)	Loss of material due to corrosion of prestressing tendons and anchorage components	Containment ISI	Containment Inservice Inspection Program (B2.1.7)	Consistent with GALL, which recommends no further evaluation (See SER Section 3.5.2.1)
Concrete elements: foundation, dome, and wall (Item Number 3.5.1-16)	Scaling, cracking, and spalling due to freeze-thaw; expansion and cracking due to reaction with aggregate	Containment ISI	Containment Inservice Inspection Program (B2.1.7)	Consistent with GALL, which recommends no further evaluation (See SER Sections 3.5.2.1.4 and 3.5.2.2.1)
Class I Structures				
All Groups except Group 6: accessible interior/exterior concrete & steel components (Item Number 3.5.1-20)	All types of aging effects	Structures Monitoring	Structural Monitoring Program (B2.1.19)	Consistent with GALL, which recommends further evaluation (See SER Sections 3.5.2.1.5 and 3.5.2.2.2)
Groups 1-3, 5, 7-9: inaccessible concrete components, such as exterior walls below grade and foundation (Item Number 3.5.1-21)	Aging of inaccessible concrete areas due to aggressive chemical attack, and corrosion of embedded steel	Plant specific	Structural Monitoring Program (B2.1.19)	Consistent with GALL, which recommends further evaluation (See SER Sections 3.5.2.1.6 and 3.5.2.2.2)

Component Group	Aging Effect/ Mechanism	AMP in GALL Report	AMP in LRA	Staff Evaluation
Group 6: all accessible/ inaccessible concrete, steel, and earthen components (Item Number 3.5.1-22)	All types of aging effects, including loss of material due to abrasion, cavitation, and corrosion	Inspection of Water-Control Structures or FERC/US Army Corps of Engineers dam inspections and maintenance	Structural Monitoring Program (B2.1.19)	Consistent with GALL, which recommends no further evaluation (See SER Section 3.5.2.1.7) (Structural Monitoring Program covers GALL AMP)
Group 5: liners (Item Number 3.5.1-23)	Crack initiation and growth from SCC and loss of material due to crevice corrosion	Water Chemistry Program and Monitoring of spent fuel pool water level	Structural Monitoring Program (B2.1.19), Water Chemistry Program (B2.1.21)	Consistent with GALL, which recommends no further evaluation (See SER Section 3.5.2.1.8)
Groups 1-3, 5, 6: all masonry block walls (Item Number 3.5.1-24)	Cracking due to restraint, shrinkage, creep, and aggressive environment	Masonry Wall	Structural Monitoring Program (B2.1.19)	Consistent with GALL, which recommends no further evaluation (See SER Section 3.5.2.1.9) (Masonry walls covered under applicant's Structural Monitoring Program)
Groups 1-3, 5, 7-9: foundation (Item Number 3.5.1-25)	Cracks, distortion, and increases in component stress level due to settlement	Structures Monitoring	Structural Monitoring Program (B2.1.19)	Consistent with GALL, which recommends further evaluation (See SER Sections 3.5.2.1.10 and 3.5.2.2.2)
Groups 1-3, 5-9: foundation (Item Number 3.5.1-26)	Reduction in foundation strength due to erosion of porous concrete subfoundation	Structures Monitoring	None	Not applicable (See SER Sections 3.5.2.1.11 and 3.5.2.2.2)
Groups 1-5: concrete (Item Number 3.5.1-27)	Reduction of strength and modulus due to elevated temperature	Plant-specific	None	Not applicable (See SER Sections 3.5.2.1.12 and 3.5.2.2.2)
Groups 7, 8: liners (Item Number 3.5.1-28)	Crack initiation and growth due to SCC; Loss of material due to crevice corrosion	Plant-specific	None	Not applicable (Component not within scope of license renewal)

Component Group	Aging Effect/ Mechanism	AMP in GALL Report	AMP in LRA	Staff Evaluation
Component Supports				
All Groups: support members: anchor bolts, concrete surrounding anchor bolts, welds, grout pad, bolted connections, etc. (Item Number 3.5.1-29)	Aging of component supports	Structures Monitoring	Structural Monitoring Program (B2.1.19)	Consistent with GALL, which recommends further evaluation (See SER Section 3.5.2.1.13 and 3.5.2.2.3)
Groups B1.1, B1.2, and B1.3: support members: anchor bolts, welds (Item Number 3.5.1-30)	Cumulative fatigue damage (CLB fatigue analysis exists)	TLAA, evaluated in accordance with 10 CFR 54.21(c)	None	Not applicable (See SER Section 3.5.2.2.3)
All Groups: support members: anchor bolts, welds (Item Number 3.5.1-31)	Loss of material due to boric acid corrosion	Boric acid corrosion	Boric Acid Corrosion Program (B2.1.4)	Consistent with GALL, which recommends no further evaluation (See SER Section 3.5.2.1.14)
Groups B1.1, B1.2, and B1.3: support members: anchor bolts, welds, spring hangers, guides, stops, and vibration isolators (Item Number 3.5.1-32)	Loss of material due to environmental corrosion; loss of mechanical function due to corrosion, distortion, dirt, overload, etc.	ISI	ASME Section XI IWB, IWC, IWD, IWF Inservice Inspection Program (B2.1.2)	Consistent with GALL, which recommends no further evaluation (See SER Section 3.5.2.1.15)
Group B1.1: high strength low-alloy bolts (Item Number 3.5.1-33)	Crack initiation and growth due to SCC	Bolting integrity	Bolting Integrity Program (B2.1.3)	Consistent with GALL, which recommends no further evaluation (See SER Section 3.5.2.1)

The staff's review of the containments, structures, and component supports groups followed several approaches. One approach, documented in SER Section 3.5.2.1, discusses the staff's review of AMR results for components the applicant indicated are consistent with the GALL Report and require no further evaluation. Another approach, documented in SER Section 3.5.2.2, discusses the staff's review of AMR results for components the applicant indicated are consistent with the GALL Report and for which further evaluation is recommended. A third approach, documented in SER Section 3.5.2.3, discusses the staff's review of AMR results for components the applicant indicated are not consistent with, or not addressed in, the GALL Report. The staff's review of AMPs credited to manage or monitor aging effects of the containments, structures, and component supports components is documented in SER Section 3.0.3.

3.5.2.1 AMR Results That Are Consistent with the GALL Report

Summary of Technical Information in the Application. In LRA Section 3.5.2.1, the applicant identified the materials, environments, and AERMs. The applicant identified the following programs that manage the aging effects of the containments, structures, and component supports components:

- ASME Section XI IWB, IWC, IWD, IWF Inservice Inspection Program
- Bolting Integrity Program
- Boric Acid Corrosion Program
- Containment Inservice Inspection Program
- Containment Leakage Testing Program
- Fire Protection Program
- Overhead Load Handling Systems Inspection Program
- Structural Monitoring Program
- Water Chemistry Program

Staff Evaluation. In LRA Tables 3.5.2-1 through 3.5.2-10, the applicant summarized AMRs for the containments, structures, and component supports components and identified which AMRs it considered consistent with the GALL Report.

For component groups evaluated in the GALL Report for which the applicant had claimed consistency and for which the GALL Report does not recommend further evaluation, the staff performed an audit and review to determine whether the plant-specific components in these GALL Report component groups were bounded by the GALL Report evaluation.

The applicant provided a note for each AMR line item. The notes describe how the information in the tables aligns with the information in the GALL Report. The staff audited those AMRs with Notes A through E, which indicate how the AMR was consistent with the GALL Report.

Note A indicates that the AMR line item is consistent with the GALL Report for component, material, environment, and aging effect. In addition, the AMP is consistent with the GALL Report AMP. The staff audited these line items to verify consistency with the GALL Report and the validity of the AMR for the site-specific conditions.

Note B indicates that the AMR line item is consistent with the GALL Report for component, material, environment, and aging effect. In addition, the AMP takes some exceptions to the AMP identified in the GALL Report. The staff audited these line items to verify consistency with the GALL Report. The staff verified that the identified exceptions to the GALL Report AMPs had been reviewed and accepted by the staff. The staff also determined whether the AMP identified by the applicant was consistent with the AMP identified in the GALL Report and whether the AMR was valid for the site-specific conditions.

Note C indicates that the component for the AMR line item, although different from, is consistent with the GALL Report for material, environment, and aging effect. In addition, the AMP is consistent with the AMP identified by the GALL Report. This note indicates that the applicant was unable to find a listing of some system components in the GALL Report; however, the applicant identified a different component in the GALL Report that had the same material, environment, aging effect, and AMP as the component under review. The staff audited these line items to verify consistency with the GALL Report. The staff also determined whether the AMR line item of the different component applied to the component under review and whether the AMR was valid for the site-specific conditions.

Note D indicates that the component for the AMR line item, although different from, is consistent with the GALL Report for material, environment, and aging effect. In addition, the AMP takes some exceptions to the AMP identified in the GALL Report. The staff audited these line items to verify consistency with the GALL Report. The staff verified whether the AMR line item of the different component was applicable to the component under review. The staff verified whether the exceptions to the GALL Report AMPs had been reviewed and accepted by the staff. The staff also determined whether the AMP identified by the applicant was consistent with the AMP identified in the GALL Report and whether the AMR was valid for the site-specific conditions.

Note E indicates that the AMR line item is consistent with the GALL Report for material, environment, and aging effect, but a different AMP is credited. The staff audited these line items to verify consistency with the GALL Report. The staff also determined whether the identified AMP would manage the aging effect consistent with the AMP identified in the GALL Report and whether the AMR was valid for the site-specific conditions.

The staff did not repeat its review of the matters described in the GALL Report; however, the staff did verify that the material presented in the LRA was applicable and that the applicant identified the appropriate GALL Report AMRs. The staff's evaluation is discussed below.

3.5.2.1.1 Loss of Leak Tightness in Closed Position Due to Mechanical Wear of Locks, Hinges and Closure Mechanism

As documented in the Audit and Review Report, the staff noted that in LRA Table 3.5.2-3 for component type "Containment Shell & Base Slab - Containment Bldg., Carbon Steel, Protected," the GALL Report item is II.A3.2-b, which cites GALL AMP XI.S4 and plant TS. The applicant listed the Containment Leakage Testing Program. As documented in the Audit and Review Report, the staff requested that the applicant explain why the plant TS were not listed as this line item was cited as consistent with the GALL Report.

By letter dated August 27, 2005, the applicant stated that TS Section 3.6 prescribes testing requirements for the containment pressure boundary, including air locks, that should have been included in LRA Table 3.5.2-3 for component type "Containment Shell & Base Slab - Containment Bldg., Carbon Steel, Protected" along with the Containment Leakage Testing Program. The AMR line item was changed to credit plant TS as stated in the GALL Report.

The staff's review found the applicant's response acceptable because the applicant had appropriately addressed the aging effect/mechanism as recommended by the GALL Report.

3.5.2.1.2 Loss of Sealant and Leakage Through Containment Due to Deterioration of Joint Seals, Gaskets, and Moisture Barriers

As documented in the Audit and Review Report, the staff noted that in LRA Table 3.5.2-3 for component type "Containment Shell & Base Slab - Containment Building, Elastomer, Protected" the GALL Report item is II.A3.3-a, which states that the aging effect of loss of leak tightness is monitored by 10 CFR Part 50, Appendix J, leak rate tests for pressure boundary, seals, and gaskets. However, LRA Table 3.5.2-3 shows the aging effects change in material properties and cracking (which are aging mechanisms, not aging effects, per the GALL Report line item) as managed by the Containment Leakage Testing Program. Therefore, the staff requested that the applicant explain why the Containment Leakage Testing Program was not shown as an AMP for loss of leak tightness as an aging effect and why it was shown as an AMP for change in material properties and cracking. Loss of leak tightness was shown as an aging effect in LRA Section 3.5.2.1.3.

By letter dated August 27, 2005, the applicant stated that: the aging effects/mechanisms of change in material property/irradiation, thermal exposure and cracking/irradiation, and thermal exposure, ultraviolet, had been evaluated based on industry guidance; the net effect of such aging effects, if not managed, is loss of pressure boundary as the intended function for this component type; and that the loss of pressure boundary intended function is considered equivalent to the GALL Report loss of leak tightness aging effect of GALL Report Item II.A3.3-a, so the alignment was made. The loss of leak tightness aging effect in LRA Section 3.5.2.1.3 is associated with component type "Containment Shell & Base Slab - Containment Building, Carbon Steel, Protected" shown in LRA Table 3.5.2-3. The AMR line item was changed to add a plant-specific note clarifying that the aging effect/mechanism nomenclature in the GALL Report is different, but the end result of the change in material properties and cracking aging effects is failure of the pressure boundary intended function, which is equivalent to loss of leak tightness.

The staff's review found the applicant's response acceptable as the applicant had appropriately addressed the aging effect/mechanism as recommended by the GALL Report.

3.5.2.1.3 Loss of Material Due to Corrosion in Accessible and Inaccessible Areas

As documented in the Audit and Review Report, the staff noted that in LRA Table 3.5.2-3 for component type "Containment Shell & Base Slab - Containment Bldg. Carbon Steel, Protected" one GALL Report line item for loss of material is II.A1.2-a, which discusses cleaning up borated water spills, but does not specify the Boric Acid Corrosion Program to manage this cleanup. Therefore, the staff requested that the applicant justify the notes for this Table 2 line item that the Boric Acid Corrosion Program management of loss of carbon steel liner material is consistent with the GALL Report.

The GALL Report recommends that boric acid spills be cleaned up promptly. It is not specific as to how to assure promptness. As documented in the Audit and Review Report, the applicant stated that the Boric Acid Corrosion Program is the process which implements the GALL Report recommendation of prompt cleanup of boric acid spills and, as such, meets the GALL Report recommendation.

The staff's review found the applicant's response acceptable as the applicant had appropriately addressed the aging effect/mechanism as recommended by the GALL Report.

As documented in the Audit and Review Report, the staff noted that in LRA Table 3.5.1, Item 3.5.1-12, the further evaluation of aging management recommended by SRP-LR Section 3.5.2.2.1.4 is documented in LRA Section 3.5.2.2.1.4, which refers to GALL Report Item II.A2.1-a for four conditions that must be satisfied to find no need for a plant-specific program. Therefore, the staff requested that the applicant explain why it had referred to GALL

Report Item II.A2.1-a for metal PWR containments instead of Item IIA1.2-a for prestressed concrete PWR containments.

By letter dated August 27, 2005, the applicant stated that the reference to GALL Report Item II.A2.1-a in LRA Section 3.5.2.2.1.4 was a typographical error that should have been Item II.A1.2-a. This statement was substantiated by alignment to Item II.A1.2-a for component type "Containment Shell & Base Slab - Containment Bldg. Carbon Steel, Protected" in LRA Table 3.5.2-3. Therefore, LRA Section 3.5.2.2.1.4 was revised to change Item II.A2.1-a to Item II.A1.2-a.

The staff's review found the applicant's response acceptable as the applicant had appropriately addressed the aging effect/mechanism as recommended by the GALL Report.

3.5.2.1.4 Scaling, Cracking, and Spalling Due to Freeze-thaw; Expansion and Cracking Due to Reaction with Aggregate

As documented in the Audit and Review Report, the staff noted that in LRA Table 3.5.2-3 for component type "Containment Shell & Base Slab - Containment Bldg, Concrete, Exposed" the AERMs are cracking, cracking and expansion, and loss of material corresponding to LRA Table 3.5.1, Item 3.5.1-16. However, in LRA Table 3.5.2-3 for component type "Containment Shell & Base Slab - Containment Bldg. Concrete, Below Grade," the only AERMs are cracking and expansion and loss of material corresponding to LRA Table 3.5.1, Item 3.5.1-16, with cracking not shown. The staff requested that the applicant explain why cracking was shown corresponding to LRA Table 3.5.1, Item 3.5.1-16, for above-grade containment concrete, but not for below-grade containment concrete.

As documented in the Audit and Review Report, the applicant explained that there are in fact no AERMs for below-grade concrete as discussed in the associated plant-specific notes and further evaluation sections of the LRA consistent with ISG-3. The GALL Report alignments for the below-grade concrete were included for completeness. However, it would have been more thorough, though unnecessary, for the below-grade concrete component type to refer to LRA Table 3.5.1, Item 3.5.1-16, for cracking as for the above-grade concrete. It was worth noting that the above-grade concrete is managed in accordance with ISG-3.

The applicant has shown GALL Report alignment for below-grade concrete even though there are no aging effects. Cracking was inadvertently not shown as this aging effect does not occur; therefore, the staff finds this explanation acceptable.

The staff's review found the applicant's response acceptable as the applicant had appropriately addressed the aging effect/mechanism as recommended by the GALL Report.

3.5.2.1.5 All Types of Aging Effects

As documented in the Audit and Review Report, the staff noted that in LRA Table 3.5.2-1 for component type "Building Framing - Concrete, Protected," one of the AERMs shown is loss of strength. The GALL Report line item shown is Item III.A3.1-b. In the GALL Report, Item III.A3.1-b is for concrete exterior above and below grade. Therefore, the staff requested that the applicant explain why this component type and the GALL Report line item were shown together. This question also applied to LRA Table 3.5.2-1 for component type "Operator Access

Component - Concrete, Protected;" to LRA Table 3.5.2-6 for component type "Building Framing - Concrete, Protected;" to LRA Table 3.5.2-9 for component type "Building Framing - Switchyard - Concrete, Protected;" to LRA Table 3.5.2-10 for component type "Building Framing-Boiler Buildings Area-Concrete, Protected;" to LRA Table 3.5.2-10 for component type "Building Framing-Concrete, Protected;" to LRA Table 3.5.2-10 for component type "Building Framing-Water Treatment Area - Concrete, Protected" and to LRA Table 3.5.2-10 for component type "Operator Access Component - Concrete, Protected."

By letter dated August 27, 2005, the applicant stated that the aging effect/mechanism in question is loss of strength/leaching of calcium hydroxide, the same aging effect/mechanism for GALL Report Item III.A3.1-b, that ISG-3 Item III.A1.1-b specifies the Structural Monitoring Program to inspect for evidence of calcium hydroxide leaching, and that although the aging effect is associated with a flowing water exterior environment, the applicant conservatively decided to utilize the Structural Monitoring Program to inspect the interior of exterior walls to ensure that no leaching of calcium hydroxide occurs due to ground water migration (flow) through the concrete. Accordingly, the alignment to GALL Report Item III.A3.1-b was made. The AMR line item for each of the component types was changed to add a plant-specific note to clarify applicability of GALL Report alignment when the GALL Report has "flowing water" and the applicant has "plant indoor air" as the environment.

The staff's review found the applicant's response acceptable as the applicant had appropriately addressed the aging effect/mechanism as recommended by the GALL Report.

As documented in the Audit and Review Report, the staff noted that in LRA Table 3.5.2-1 for component type "Building Framing - Carbon Steel, Protected," the material is carbon steel, the environment plant indoor air, and the aging effect loss of material corresponding to GALL Report Item III.A3.2-a. However, the AMP shown is the Boric Acid Corrosion Program instead of the Structural Monitoring Program. Therefore, the staff requested that the applicant explain why the note was H instead of E (consistent with the GALL Report for material, environment, aging effect but crediting a different AMP).

By letter dated August 27, 2005, the applicant stated that the aging effect/mechanism managed by the Boric Acid Corrosion Program is loss of material/boric acid corrosion and that Note H (aging effect not in the GALL Report for this component, material, and environment combination) was chosen because the GALL Report does not manage this aging mechanism for the component type. However, the loss of material aging effect is still consistent with the GALL Report line item, so use of Note E (consistent with the GALL Report for material, environment, and aging effect but a different AMP credited) could also be appropriate. The AMR line item was revised to change Note H to Note E, as discussed.

The staff's review found the applicant's response acceptable as the applicant had appropriately addressed the aging effect/mechanism as recommended by the GALL Report.

As documented in the Audit and Review Report, the staff noted that in LRA Table 3.5.2-4 for component type "Building Framing - Carbon Steel, Protected," the material is carbon steel, the environment plant indoor air, and the aging effect loss of material corresponding to GALL Report Item III.A4.2-a. However, the AMP shown is the Boric Acid Corrosion Program instead of the Structural Monitoring Program. Therefore, the staff requested that the applicant explain why the note was H instead of E (consistent with the GALL Report for material, environment, aging effect but crediting a different AMP).

By letter dated August 27, 2005, the applicant stated that the aging effect/mechanism managed by the Boric Acid Corrosion Program is loss of material/boric acid corrosion. Note H (aging effect not in the GALL Report for this component, material, and environment combination) was chosen because the GALL Report does not manage this aging mechanism for the component type. However, the aging effect of loss of material is still consistent with the GALL Report line item, so use of Note E (consistent with the GALL Report for material, environment, and aging effect but a different AMP credited) could also be appropriate. The AMR line item was revised to change Note H to Note E, as discussed.

The staff's review found the applicant's response acceptable as the applicant had appropriately addressed the aging effect/mechanism as recommended by the GALL Report.

As documented in the Audit and Review Report, the staff noted that in LRA Table 3.5.2-8 for component type "Roof Flashing-Auxiliary Bldg-Galvanized, Exposed," the note shown was A only. The staff requested that the applicant explain why Note 581 was not shown also to address galvanizing.

By letter dated August 27, 2005, the applicant stated that Note 581 would be appropriate here and for the two other galvanized component types that follow for the intake structure and switchyard relay house. The AMR line items for these three galvanized components was revised to add Note 581.

The staff's review found the applicant's response acceptable as the applicant had appropriately addressed the aging effect/mechanism as recommended by the GALL Report.

As documented in the Audit and Review Report, the staff noted that in LRA Table 3.5.2-9 for component type "Tank Foundations-Building and Yard-Concrete, Exposed," the environment is atmosphere/weather. For the aging effect change in material properties the GALL Report Item is III.A8.1-b, for which the environment is flowing water. Therefore, the staff requested that the applicant explain how an atmosphere/weather environment is the same as a flowing water environment.

By letter dated August 27, 2005, the applicant stated that the aging effect/mechanism in question is loss of strength/leaching of calcium hydroxide, the same aging effect/mechanism for GALL Report Item III.A8.1-b. Although flowing water is not permanent in the outdoor environment, surface runoff conservatively assumed to occur on occasion due to rainfall was evaluated and aligned to this line item. The AMR line item was changed to add a plant-specific note to clarify applicability of the GALL Report alignment when the GALL Report has "flowing water" and the applicant has "atmosphere/weather" as the environment.

The staff's review found the applicant's response acceptable as the applicant had appropriately addressed the aging effect/mechanism as recommended by the GALL Report.

As documented in the Audit and Review Report, the staff noted that in LRA Table 3.5.2-10 for component type "Building Framing-Concrete, Exposed," the environment is atmosphere/weather. For the aging effect change in material properties, the GALL Report Item is III.A3.1-b. The GALL Report environment for Item III.A3.1-b is flowing water. Therefore, the staff requested that the applicant explain how an atmosphere/weather environment is the same as a flowing water environment. The question also applied to LRA Table 3.5.2-10 for component type "Building Framing-Water Treatment Area - Concrete, Exposed."

By letter dated August 27, 2005, the applicant stated that the aging effect/mechanism in question is loss of strength/leaching of calcium hydroxide, the same aging effect/mechanism for GALL Report Item III.A3.1-b. Although flowing water is not permanent in the outdoor environment, surface runoff conservatively assumed to occur on occasion due to rainfall was evaluated and aligned to this line item. The two AMR line items were changed to add a plant-specific note to clarify applicability of GALL Report alignment when the GALL Report has "flowing water" and the applicant has "atmosphere/weather" as the environment.

The staff's review found the applicant's response acceptable as the applicant had appropriately addressed the aging effect/mechanism as recommended by the GALL Report.

3.5.2.1.6 Aging of Inaccessible Concrete Areas Due to Aggressive Chemical Attack, and Corrosion of Embedded Steel

As documented in the Audit and Review Report, the staff noted that in LRA Table 3.5.2-9 for component type "Tank Foundations - Building and Yard - Concrete, Exposed," the environment is atmosphere/weather. For the aging effect "cracking, loss of bond/material" the GALL Report Item is III.A8.1-d and the Table 3.5.1 Item is 3.5.1-21, which is for inaccessible concrete areas. The GALL Report environment for Item III.A8.1-d is exposure to aggressive environment and the component is foundation below-grade. For Table 3.5.1, Item 3.5.1-21, further evaluation is provided in LRA Section 3.5.2.2.2.2 concluding that aging management is not required for cracking, loss of bond, and loss of material due to corrosion of embedded steel for below-grade inaccessible concrete. Therefore, the staff requested that the applicant explain the rationale for the AMR association of this component with GALL Report Item III.A8.1-d and Table 3.5.1, Item 3.5.1-21. The staff also asked the applicant to explain Note A (consistent with the GALL Report).

By letter dated August 27, 2005, the applicant stated that the alignment to GALL Report Item III.A8.1-d was made for the same component type (tank foundation) and aging effect/mechanism. The applicant recognized that the environments are different, but the material, environment, aging effects, and AMP combination is consistent with GALL Report Item III.A1.1-d (with ISG-3 clarifications) such that the overall alignment was judged consistent. A clarifying note to that effect would have been helpful. A more appropriate alignment may have been with GALL Report Item III.A1.1-d with a standard Note C (component different but consistent with GALL Report item for material, environment and aging effect, AMP consistent with GALL Report AMP). The AMR line item was changed to add a plant-specific note to summarize this explanation.

The staff's review found the applicant's response acceptable as the applicant had appropriately addressed the aging effect/mechanism, as recommended by the GALL Report.

3.5.2.1.7 All Types of Aging Effects, Including Loss of Material Due to Abrasion, Cavitation, and Corrosion

As documented in the Audit and Review Report, the staff noted that in LRA Table 3.5.2-5 for component type "Building Framing - Cast Iron, Raw Water," there is Note 582, which states that cast iron is considered consistent with carbon steel and evaluated the same, but with the additional aging effect/mechanism of loss of material due to selective leaching also evaluated. Therefore, the staff requested that the applicant explain where in the Structural Monitoring Program selective leaching and the inspection for it are discussed.

The applicant explained that there was a related RAI on cast iron, selective leaching, and the Structural Monitoring Program. The response, submitted on July 28, 2005, provides the correction that for cast iron in raw water selective leaching is an AERM with the LRA showing the One-Time Inspection Program managing this aging effect.

The staff's review found the applicant's response acceptable as the applicant had appropriately addressed the aging effect/mechanism as recommended by the GALL Report.

As documented in the Audit and Review Report, the staff noted that in LRA Table 3.5.2-7 for component type "Building Framing - Concrete, Protected," for aging effect loss of material, GALL Report Item III.A6.1-a is referenced. Therefore, the staff requested that the applicant explain why Item III.A6.1-a for a weather environment is shown while the environment shown for this component type is plant indoor air.

By letter dated August 27, 2005, the applicant stated that GALL Report Item III.A6.1-a is for loss of material (spalling, scaling) and cracking/freeze-thaw. Review of the AMR shows that loss of material/freeze-thaw was evaluated as not an AERM. Thus, there should not be a GALL Report alignment for Item III.A6.1-a. The AMR line item aligned with GALL Report Item III.A6.1-a was removed from the subject component.

The staff's review found the applicant's response acceptable as the applicant had appropriately addressed the aging effect/mechanism as recommended by the GALL Report.

As documented in the Audit and Review Report, the staff noted that in LRA Table 3.5.2-7 for component type "Building Framing - Concrete, Protected," for the aging effect loss of strength GALL Report Item III.A6.1-b is referenced. Therefore, the staff requested that the applicant explain why Item III.A6.1-b for a flowing water environment was shown while the environment for this component type is plant indoor air.

By letter dated August 27, 2005, the applicant stated that the aging effect/mechanism in question was loss of strength/leaching of calcium hydroxide, the same aging effect/mechanism for GALL Report Item III.A6.1-b, that ISG-3 Item III.A1.1-b specifies that the Structural Monitoring Program be used to inspect for evidence of calcium hydroxide leaching and that although the aging effect is associated with a flowing water exterior environment, the applicant had conservatively decided to utilize the Structural Monitoring Program to inspect the interior of exterior walls to ensure that no leaching of calcium hydroxide occurs due to ground water migration (flow) through the concrete. Accordingly, the alignment to GALL Report Item III.A6.1-b was made. The AMR line item was changed to add a plant-specific note to clarify applicability of the GALL Report item when the GALL Report has "flowing water" and the applicant has "plant indoor air" as the environment.

The staff's review found the applicant's response acceptable as the applicant had appropriately addressed the aging effect/mechanism as recommended by the GALL Report.

As documented in the Audit and Review Report, the staff noted that in LRA Table 3.5.2-7 for component type "Flood Barrier - Concrete, Protected," for the aging effect loss of strength GALL Report Item III.A6.1-b, was referenced. Therefore, the staff requested that the applicant explain why Item III.A6.1-b for a flowing water environment was shown while the environment for this component type is plant indoor air.

By letter dated August 27, 2005, the applicant stated that the aging effect/mechanism in question is loss of strength/leaching of calcium hydroxide, the same aging effect/mechanism for GALL Report Item III.A6.1-b, that ISG-3 Item III.A1.1-b specifies that the Structural Monitoring Program be used to inspect for evidence of calcium hydroxide leaching, and that although the aging effect is associated with a flowing water exterior environment, the applicant conservatively decided to utilize the Structural Monitoring Program to inspect the interior of exterior walls to ensure that no leaching of calcium hydroxide occurs due to ground water migration (flow) through the concrete. Accordingly, the alignment to GALL Report Item III.A6.1-b was made. The AMR line item was changed to add a plant-specific note to clarify applicability of GALL Report alignment when the GALL Report has "flowing water" and the applicant has "plant indoor air" as the environment.

The staff's review found the applicant's response acceptable as the applicant had appropriately addressed the aging effect/mechanism as recommended by the GALL Report.

3.5.2.1.8 Crack Initiation and Growth from SCC and Loss of Material Due to Crevice Corrosion

As documented in the Audit and Review Report, the staff noted that in LRA Table 3.5.2-2 for component type "Spent Fuel Storage Rack - Auxiliary Building, Stainless Steel, Borated Water," that GALL Report Item III.A5.2-b and LRA Table 3.5.1, Item 3.5.1-23, were not associated with this component nor a Note C assigned. The staff requested that the applicant explain why GALL Report Item III.A5.2-b and LRA Table 3.5.1, Item 3.5.1-23, were not associated with this component and Note C not assigned.

By letter dated August 27, 2005, the applicant stated that the spent fuel storage rack, auxiliary building, stainless steel, borated water is aligned with GALL Report Item VII.A2.1-c as the appropriate component match (spent fuel storage racks), but that the associated GALL Report aging effect/mechanism (crack initiation and growth/SCC) is not applicable as SFP temperatures are below the temperature threshold for the effect. However, the applicant manages loss of material due to crevice/pitting corrosion. Thus, Note H (aging effect not in the GALL Report for this component) was utilized. Item III.A5.2-b with Note C would also be applicable, however, as the same material, environment, aging effects, and AMP combination, but a different component. The AMR line item was changed to align with GALL Report Item III.A5.2-b with Note C for the subject component type.

The staff's review found the applicant's response acceptable since the applicant had appropriately addressed the aging effect/mechanism as recommended by the GALL Report.

3.5.2.1.9 Cracking Due to Restraint, Shrinkage, Creep, and Aggressive Environment

As documented in the Audit and Review Report, the staff noted that in LRA Table 3.5.2-4 for the component type "HELB/MELB Component - Concrete, Protected," the GALL Report, Revision 1, item is III.A3-11, which is for masonry walls. Therefore, the staff requested that the applicant explain why if the subcomponent is masonry walls GALL Report, Revision 0, Item III.A3.3-a was not shown for this AMR with a note other than H.

By letter dated August 27, 2005, the applicant stated that the staff's observation that the cracking aging effect with a standard Note H for a masonry block wall was correct. Accordingly, the appropriate GALL Report alignment for this AERM should be to GALL Report, Revision 0, Item III.A3.3-a, with a standard Note A rather than to no GALL Report line item with Note H as indicated in LRA Table 3.5.2-4. The AMR line item was changed to add the GALL Report line item with Note A, as discussed. Also, block walls were added to the component description in LRA Table 3.5.2-4 for the subject component. The changes were also applied to component type "Building Framing - Concrete, Protected."

The staff's review found the applicant's response acceptable as the applicant had appropriately addressed the aging effect/mechanism as recommended by the GALL Report.

As documented in the Audit and Review Report, the staff noted that in LRA Table 3.5.2-1 for component type "HELB/MELB Component - Concrete, Protected," the aging effect is cracking with the referenced GALL Report Item III.A3.3-a and LRA Table 3.5.1, Item 3.5.1-24. These items are for masonry walls. Therefore, the staff requested that the applicant explain why the component type has no reference to masonry walls like other component types in the Table 2. This question also applied to LRA Table 3.5.2-1 for component type "Operator Access Component - Concrete, Protected;" to LRA Table 3.5.2-7 for component type "Building Framing - Concrete, Exposed;" to LRA Table 3.5.2-7 for component type "Building Framing - Concrete, Protected;" to LRA Table 3.5.2-10 for component type "Building Framing - Water Treatment Area - Concrete, Exposed," and to LRA Table 3.5.2-10 for component type "Operator Access Component - Concrete, Protected."

By letter dated August 27, 2005, the applicant stated that, as was evident from the component type naming scheme, the applicant scoped civil/structural components based on design attributes, building, material, and environment and that for the concrete elements concrete and masonry block walls were grouped together rather than separated. Thus, aging effects for both component types were evaluated where the component type included, or could include, masonry walls. The examples listed in parentheses in the component type title in the LRA were representative, but not necessarily fully inclusive, of all structural members in the component group. All components identified in the question included masonry walls.

The staff's review found the applicant's response acceptable as the applicant had appropriately addressed the aging effect/mechanism as recommended by the GALL Report.

3.5.2.1.10 Cracks, Distortion and Increases in Component Stress Level Due to Settlement

As documented in the Audit and Review Report, the staff noted that LRA Table 3.5.2-1 shows for component type "Building Framing - Concrete, Below Grade" GALL Report Item III.A3.1-h for cracking and no AMP required, but for "Building Framing - Concrete, Exposed," cracking monitored by the Structural Monitoring Program. Therefore, the staff requested that the applicant explain how concrete cracking can occur above grade from settlement, but not below-grade if settlement of the auxiliary building foundation does not occur at all. This question applied to the following::

- LRA Table 3.5.2-1 for component type "Building Framing - Concrete, Protected"

- LRA Table 3.5.2-6 for component types "Building Framing - Concrete, Exposed" and "Building Framing - Concrete, Protected"

- LRA Table 3.5.2-9 for component types "Building Framing - Switchyard - Concrete, Exposed," "Building Framing - Switchyard - Concrete, Protected," and "Missile Shield - Yard - Concrete, Exposed"

- LRA Table 3.5.2-10 for component types "Building Framing - Boiler Buildings Area - Concrete, Exposed"" "Building Framing - Boiler Buildings Area - Concrete, Protected,""Building Framing - Concrete, Exposed," "Building Framing - Concrete, Protected," "Building Framing - Water Treatment Area - Concrete, Exposed," "Building Framing - Water Treatment Area - Concrete, Protected," "HVAC Component - Concrete, Protected," "Missile Shield - Concrete, Exposed,""Missile Shield - Concrete, Protected," and "Operator Access Component - Concrete, Protected"

By letter dated August 27, 2005, the applicant stated that GALL Report Item III.A3.1-h is considered similar to GALL Report Item III.A1.1-a for the aging effect/mechanism loss of material (spalling, scaling) and cracking/freeze-thaw. Accordingly, guidance was taken from ISG-3 Item III.A1.1-a exempting inaccessible areas from inspections if air content requirements are met, subsequent inspections find no freeze-thaw degradations, and the Structures Monitoring Program inspects accessible concrete for cracking due to freeze-thaw. Therefore, the applicant included inspections for cracking/freeze-thaw of accessible concrete within the scope of the Structural Monitoring Program and, as PNP meets the other criteria, inspections of inaccessible concrete were not required. Each AMR line item for the subject components was changed to add plant-specific Note 547, which explains this exemption in the LRA.

The applicant added Note 547 to each AMR line item in the question. The staff finds that the note correctly explains why some inspections are not performed. The staff's review found the applicant's response acceptable as the applicant appropriately addressed the aging effect/mechanism, as recommended by the GALL Report.

As documented in the Audit and Review Report, the staff noted that the discussion column for LRA Table 3.5.1, Item 3.5.1-25, refers to the LRA Section 3.5.2.2.1.2, which is for containment, for further evaluation. Therefore, the staff requested that the applicant explain why there was no reference to LRA Section 3.5.2.2.2.1 for further evaluation of Groups 1-3, 5, and 7-9 (Class 1 structures).

By letter dated August 27, 2005, the applicant stated that LRA Table 3.5.1 line item 25 incorrectly references SRP-LR Section 3.5.2.2.1.2 for non-containment structures when in fact Section 3.5.2.2.2.1 is the appropriate SRP-LR (and LRA) section and that, therefore, LRA Table 3.5.1, Item 3.5.1-25 should refer instead to LRA Section 3.5.2.2.2.1 for the Class 1 structures. The LRA Table 3.5.1, Item 3.5.1-25, discussion column was revised to show the correct LRA section for further evaluation.

The staff's review found the applicant's response acceptable as the applicant had appropriately addressed the aging effect/mechanism as recommended by the GALL Report.

3.5.2.1.11 Reduction in Foundation Strength Due to Erosion of Porous Concrete Subfoundation

As documented in the Audit and Review Report, the staff noted that the LRA Table 3.5.1, Item 3.5.1-26, discussion column refers to LRA Section 3.5.2.2.1.2, which is for containment, for further evaluation. The staff requested that the applicant explain why there was no reference to LRA Section 3.5.2.2.2.1 for further evaluation of Groups 1-3, 5, and 7-9 (Class 1 structures).

By letter dated August 27, 2005, the applicant stated that SRP-LR Table 3.5.1 line item 26 refers incorrectly to SRP-LR Section 3.5.2.2.1.2 for noncontainment structures when in fact Section 3.5.2.2.2.1 is the appropriate SRP-LR (and LRA) section and that, therefore, LRA Table 3.5.1, Item 3.5.1-26, should refer instead to LRA Section 3.5.2.2.2.1 for the Class 1 structures. The LRA Table 3.5.1, Item 3.5.1-26, discussion column was revised to show the correct LRA section for further evaluation.

The staff's review found the applicant's response acceptable as the applicant had appropriately addressed the aging effect/mechanism as recommended by the GALL Report.

3.5.2.1.12 Reduction in Strength and Modulus Due to Elevated Temperature

As documented in the Audit and Review Report, the staff noted that the LRA Table 3.5.1, Item 3.5.1-27, discussion column refers to LRA Section 3.5.2.2.1.3, which is for containment, for further evaluation. The staff requested that the applicant explain why there was no reference to LRA Section 3.5.2.2.2.1 for further evaluation of Groups 1-5 (Class 1 structures).

By letter dated August 27, 2005, the applicant stated that SRP-LR Table 3.5.1 line item 27 refers incorrectly to SRP-LR Section 3.5.2.2.1.3 for noncontainment structures when in fact Section 3.5.2.2.2.1 is the appropriate SRP-LR (and LRA) section and that, therefore, LRA Table 3.5.1, Item 3.5.1-27, should refer instead to LRA Section 3.5.2.2.2.1 for the Class 1 structures. The LRA Table 3.5.1, Item 3.5.1-27, discussion column was revised to show the correct LRA section for further evaluation.

The staff's review found the applicant's response acceptable as the applicant had appropriately addressed the aging effect/mechanism as recommended by the GALL Report.

3.5.2.1.13 Aging of Component Supports

As documented in the Audit and Review Report, the staff noted that in LRA Table 3.5.2-2 for component type "Non-ASME Piping & Mechanical Component Support - Boiler Building, Concrete, Protected," the aging effect is loss of material with reference to GALL Report Items III.B2.2-a and III.B4.3-a and LRA Table 3.5.1, Item 3.5.1-29. GALL Report Items III.B2.2-a and III.B4.3-a are for concrete material and the aging effect of reduction in concrete anchor capacity. Therefore, the staff requested that the applicant explain how the LRA Table 3.5.2-2 line item for this component could show the material as carbon steel and the aging effect as loss of material and still show Note A (consistent with the GALL Report).

By letter dated August 27, 2005, the applicant stated that the carbon steel material shown for this component type was a typographical error that should have been shown as concrete as evaluated and summarized in the AMR. The AMR line item was revised to change carbon steel to concrete as discussed.

The staff's review found the applicant's response acceptable as the applicant had appropriately addressed the aging effect/mechanism as recommended by the GALL Report.

As documented in the Audit and Review Report, the staff noted that LRA Table 3.5.2-10 shows for component type "Missile Shield - Concrete, Exposed" the aging effect reduction in concrete anchor capacity, GALL Report Item III.B5.2-a, and Note A (consistent with the GALL Report). However, concrete at locations of expansion and grouted anchors, etc. was not shown in the component type as for other similar AMR line items. Therefore, the staff requested that the applicant explain why concrete locations of expansion and grouted anchors were not shown in the component type column if this AMR line item was consistent with the GALL Report. This question also applied to component type "Missile Shield - Concrete Protected;" to component type "Missile Shield - Yard - Concrete, Exposed" and to component type "HVAC Component - Concrete, Protected."

By letter dated August 27, 2005, the applicant stated that, as was evident from the component type naming scheme, it had scoped civil/structural components based on design attributes, building, material, and environment. For the concrete types, "concrete" was used generically to include concrete at locations of expansion and grouted anchors. Thus, aging effects for concrete and concrete at expansion and grouted anchors were both evaluated. The examples shown in parentheses in the component type column represented, but did not necessarily include all structural members in the component group. All components identified in the question included concrete at locations of expansion and grouted anchors.

The staff's review found the applicant's response acceptable as the applicant had appropriately addressed the aging effect/mechanism as recommended by the GALL Report.

As documented in the Audit and Review Report, the staff noted that in LRA Table 3.5.2-1 for component type "Building Framing - Carbon Steel, Protected," one of the AMPs to manage loss of material is the Structural Monitoring Program. GALL Report Item III.B5.1-a is shown here corresponding to LRA Table 3.5.1, Item 3.5.1-31. However, GALL Report, Volume 1, Table 5 relates Item III.B5.1-a to Item 3.5.1-29, not 3.5.1-31. The staff requested that the applicant explain why LRA Table 3.5.1, Item 3.5.1-31, was shown related to GALL Report Item III.B5.1-a.

By letter dated August 27, 2005, the applicant stated that LRA Table 3.5.2-1, Item 3.5.1-31, was a typographical error and that the appropriate relation should have been shown as Item 3.5.1-29 in the AMR. The AMR line item was revised to change 3.5.1-31 to 3.5.1-29 as discussed.

The staff's review found the applicant's response acceptable as the applicant had appropriately addressed the aging effect/mechanism as recommended by the GALL Report.

3.5.2.1.14 Loss of Material Due to Boric Acid Corrosion

As documented in the Audit and Review Report, the staff noted that in LRA Table 3.5.2-4 for component type "Building Framing - Carbon Steel, Protected," the aging effect is loss of material with GALL Report Item III.B5.1-b and Table 3.5.1, Item 3.5.1-29. GALL Report Item III.B5.1-b is associated with Table 3.5.1, Item 3.5.1-31, in GALL Report Table 5. The staff asked the applicant why Table 3.5.1, Item 3.5.1-29, was shown for this component AMR instead of Item 3.5.1-31.

By letter dated August 27, 2005, the applicant stated that Table 3.5.2-4, Item 3.5.1-29, was a typographical error and that the appropriate alignment should have been shown as Item 3.5.1-31 in the AMR. The AMR line item was revised to change 3.5.1-29 to 3.5.1-31, as discussed.

The staff's review found the applicant's response acceptable as the applicant had appropriately addressed the aging effect/mechanism as recommended by the GALL Report.

As documented in the Audit and Review Report, the staff noted that in LRA Table 3.5.2-1 for component type "Operator Access Component - Carbon Steel, Protected," one of the AMPs to manage loss of material is the Boric Acid Corrosion Program. The applicant showed Note A for this line item. However, in the same table for component type "Operator Access Component - Galvanized, Protected," the applicant showed Note C for the same aging effect and AMP combination. The staff requested that the applicant explain why the first line item shows Note A and the second Note C when the only difference in components appeared to be galvanizing.

By letter dated August 27, 2005, the applicant stated that the use of Note C was conservative because, as the reviewer had noted, the only difference was the galvanizing. As Note 581 for "Operator Access Component - Galvanized, Protected" states, "Galvanized material is treated the same as carbon steel. No credit is taken for the galvanized coating." Thus, Note A could have been used. The AMR line item was revised to change the note from C to A.

The staff's review found the applicant's response acceptable as the applicant had appropriately addressed the aging effect/mechanism as recommended by the GALL Report.

As documented in the Audit and Review Report, the staff noted that in LRA Table 3.5.2-2 for component type "ASME Class 1 Tubing Support - Auxiliary Bldg, Carbon Steel, Protected," one AMP to manage loss of material was the Boric Acid Corrosion Program. GALL Report Item III.B1.1.1-b was shown with LRA Table 3.5.1, Item 3.5.1-31. However, Note 583 for this line item relates not to Item III.B1.1.1-b, but to Item III.B5.1-b. Therefore, the staff requested that the applicant explain why Note 583 was shown for this line item.

By letter dated August 27, 2005, the applicant explained that Note 583 applies also to GALL Report Item B1.1.1-b, that it would have been more appropriate not to have included the specific GALL Report line item (i.e., GALL Report Item III.B5-1b) and to have kept the reference generic (i.e., GALL Report) to be useful for similar circumstances. Note 583 was revised to delete the specific GALL Report line item reference.

The staff's review found the applicant's response acceptable as the applicant had appropriately addressed the aging effect/mechanism as recommended by the GALL Report.

As documented in the Audit and Review Report, the staff noted that in LRA Table 3.5.2-2 for component type "ASME Class 2 and 3 Piping and Mechanical Component Support - Auxiliary Bldg, Carbon Steel, Protected," one AMP to manage loss of material is the Boric Acid Corrosion

Program. GALL Report Item III.B1.2.1-b was shown with LRA Table 3.5.1, Item 3.5.1-31. However, Note 583 for this line item relates not to Item III.B1.2.1-b, but to III.B5.1-b. The staff requested that the applicant explain why Note 583 is shown for this line item. This question also applied to the same galvanized component in LRA Table 3.5.2-2; to component type "Elec Component Support - Auxiliary Bldg, Carbon Steel, Protected;" to component type "Elec Component Support - Auxiliary Building, Galvanized, Protected;" to component type "Non-ASME Piping & Mechanical Component Support - Auxiliary Bldg, Carbon Steel, Protected," and to component type "Non-ASME Piping & Mechanical Component Support - Auxiliary Building, Galvanized, Protected."

By letter dated August 27, 2005, the applicant explained that Note 583 applies also to GALL Report Items III.B1.2.1-b, III.B2.1-b, III.B3.1-b, and III.B4.1-b aligned with the subject component types and that it would have been more appropriate not to have included the specific GALL Report line item (i.e., GALL Report Item III.B5-1b) and to have kept the reference generic (i.e., GALL Report) to be useful for similar circumstances. Note 583 was revised to delete the specific GALL Report line item reference.

The staff's review found the applicant's response acceptable as the had applicant appropriately addressed the aging effect/mechanism as recommended by the GALL Report.

As documented in the Audit and Review Report, the staff noted that in LRA Table 3.5.2-4 for component type "HVAC Component - Carbon Steel, Protected," one AMP to manage loss of material was the Boric Acid Corrosion Program. GALL Report Item III.B5.1-b was shown with LRA Table 3.5.1, Item 3.5.1-31. As documented in the Audit and Review Report, the staff asked the applicant why GALL Report Item III.B4.1-b is not shown for this line item and why Note A is shown instead of Note C. This question also applied to the HVAC galvanized component in LRA Table 3.5.2-4.

By letter dated August 27, 2005, the applicant stated that a more appropriate alignment would have been to GALL Report Item III.B4.1-b with a Note A rather than Item III.B5.1-b. The AMR line item was revised to change the GALL Report line item and note as discussed.

The staff's review found the applicant's response acceptable as the applicant had appropriately addressed the aging effect/mechanism as recommended by the GALL Report.

3.5.2.1.15 Loss of Material Due to Environmental Corrosion; Loss of Mechanical Function Due to Corrosion, Distortion, Dirt, Overload, Etc.

As documented in the Audit and Review Report, the staff noted that in LRA Table 3.5.2-2 for component type "ASME Class 1 Tubing Support - Auxiliary Bldg, Carbon Steel Protected," one of the GALL Report items shown is III.B1.1.1-a, which states the environment inside containment. The staff requested that the applicant justify Note A (consistent with the GALL Report) for this Table 3.5.2-2 line item when the component is in the auxiliary building and not in the containment.

By letter dated August 27, 2005, the applicant stated that both the auxiliary building and containment are indoor air environments with little difference between the two other than temperature and radiation exposure. The auxiliary building environment is equivalent to or slightly less harsh than that of the containment. As the material, environment, aging effect, and

AMP are consistent with GALL Report Item IIIB1.1.1-a, the note was deemed appropriate. LRA Table 3.5.2-2 was revised to include a plant-specific note for the line item to clarify why the environments are equivalent.

The staff's review found the applicant's response acceptable as the applicant had appropriately addressed the aging effect/mechanism as recommended by the GALL Report.

Conclusion. The staff evaluated the applicant's claim of consistency with the GALL Report. The staff also reviewed information pertaining to the applicant's consideration of recent operating experience and proposals for managing the associated aging effects. On the basis of its review, the staff concludes that the AMR results, which the applicant claimed to be consistent with the GALL Report, are consistent with the GALL Report AMRs. Therefore, the staff concludes that the applicant has demonstrated that the aging effects for these components will be adequately managed so that their intended function(s) will be maintained consistent with the CLB for the period of extended operation, as required by 10 CFR 54.21(a)(3).

3.5.2.2 AMR Results That Are Consistent with the GALL Report, for Which Further Evaluation Is Recommended

Summary of Technical Information in the Application. In LRA Section 3.5.2.2, the applicant provided further evaluation of aging management as recommended by the GALL Report for the containments, structures, and component supports components. The applicant provided information concerning how it will manage the following aging effects:

PWR Containment:

- aging of inaccessible concrete areas
- cracking, distortion, and increase in component stress level due to settlement; reduction of foundation strength due to erosion of porous concrete subfoundations, if not covered by structures monitoring program
- reduction of strength and modulus of concrete structures due to elevated temperature
- loss of material due to corrosion in inaccessible areas of steel containment shell or liner plate
- loss of prestress due to relaxation, shrinkage, creep, and elevated temperature
- cumulative fatigue damage
- cracking due to cyclic loading and SCC

Class 1 Structures:

- aging of structures not covered by structures monitoring program
- aging management of inaccessible areas

Component Supports:

- aging of supports not covered by structures monitoring program
- cumulative fatigue damage due to cyclic loading

Staff Evaluation. For component groups evaluated in the GALL Report for which the applicant had claimed consistency with the GALL Report and for which the GALL Report recommends further evaluation, the staff audited and reviewed the applicant's evaluations to determine whether they adequately address those issues. In addition, the staff reviewed the applicant's further evaluations against the criteria in SRP-LR Section 3.5.2.2. Details of the staff's audit are documented in the Audit and Review Report. The staff's evaluation of the aging effects is discussed in the following sections.

3.5.2.2.1 PWR Containment

The staff reviewed LRA Section 3.5.2.2.1 against the criteria in SRP-LR Section 3.5.2.2.1, which addresses several areas that are discussed below.

Aging of Inaccessible Concrete Areas. The staff reviewed LRA Section 3.5.2.2.1.1 against the criteria in SRP-LR Section 3.5.2.2.1.1.

In LRA Section 3.5.2.2.1.1, the applicant addressed aging of the containment concrete foundation, walls, and dome due to leaching of calcium hydroxide, aggressive chemical attack, and corrosion of embedded steel.

SRP-LR Section 3.5.2.2.1.1 states that cracking, spalling, and increases in porosity and permeability due to leaching of calcium hydroxide and aggressive chemical attack and cracking, spalling, loss of bond, and loss of material due to corrosion of embedded steel could occur in inaccessible areas of concrete and steel containments. The GALL Report recommends further evaluation of plant-specific programs to manage the aging effects for inaccessible areas if specific criteria defined in the GALL Report cannot be satisfied.

In LRA Section 3.5.2.2.1.1 the applicant provided a detailed description for the following containment aging effects:

- loss of material (spalling, scaling) and cracking due to freeze-thaw (GALL Report Item IIA1.1-a)

- increase in porosity, permeability, and loss of strength due to leaching (GALL Report Item IIA1.1-b)

- increase in porosity and permeability, cracking, loss of material (spalling, scaling) due to aggressive chemical attack (GALL Report Item IIA1.1-c)

- expansion and cracking due to reaction with aggregates (GALL Report Item IIA1.1-d)

- aggressive groundwater environment-embedded steel (GALL Report Item IIA1.1-e)

As documented in the Audit and Review Report, the staff noted that the applicant's Structural Monitoring Program does not discuss the need or lack of need for periodic ground water monitoring to ensure that below-grade water chemistry does not become aggressive. The applicant was asked to justify not monitoring ground water periodically during the CLB and for the potential period of extended operation to check water chemistry for non-aggressiveness.

By letter dated September 2, 2005, the applicant stated that, as discussed in LRA Section 3.5.2.2.1.1, ground water chemistry records available for the current operating period indicate that water in contact with below-grade concrete has been non-aggressive over at least the last 40 years. To ensure that ground water remains non-aggressive over the period of extended operation, ground water sampling for pH, chlorides, and sulfates would be performed as part of the Structural Monitoring Program at least every five years. Accordingly, changes were made to the Structural Monitoring Program.

In addition, conforming changes from the September 2, 2005, letter were made where LRA Section 3.5.2.2 states that continued groundwater sampling is unnecessary. The new paragraph that replaced the former paragraph reads as follows: "As part of the Structural Monitoring Program, Palisades will continue to monitor groundwater on a periodic basis to ensure it remains non-aggressive, such that the associated aging effects remain not applicable."

The new paragraph was added to the LRA to replace the former paragraph, as follows:

> On Page 3-271, replaced existing paragraph that begins, 'In addition it is concluded that additional groundwater monitoring'

> On Page 3-273, replaced existing paragraph that begins, 'In addition it is concluded that additional groundwater monitoring'

The staff's review found the applicant's response acceptable as the applicant had appropriately addressed the aging effect/mechanism as recommended by the GALL Report.

The staff's audit and review found that loss of material and cracking of accessible containment concrete due to freeze-thaw; increase in porosity, permeability, and loss of strength of accessible containment concrete due to leaching of calcium hydroxide; increase in porosity and permeability, cracking, loss of material of accessible containment concrete due to aggressive chemical attack; expansion and cracking of accessible containment concrete due to reaction with aggregates; and cracking, loss of bond, and loss of material of accessible containment concrete due to corrosion of embedded steel would be adequately managed by the Containment Inservice Inspection Program in accordance with ISG-3. As documented in the Audit and Review Report, the staff interviewed the applicant's technical personnel and reviewed relevant operating experience to confirm that such aging effects due to these aging mechanisms either had not been observed or that when observed corrective action had been taken.

In addition, the staff found that loss of material and cracking of inaccessible containment concrete due to freeze-thaw; increase in porosity, permeability, and loss of strength of inaccessible containment concrete due to leaching of calcium hydroxide; increase in porosity and permeability, cracking, loss of material of inaccessible containment concrete due to aggressive chemical attack; expansion and cracking of inaccessible containment concrete due to reaction with aggregates; and cracking, loss of bond, and loss of material of inaccessible containment concrete due to corrosion of embedded steel are not plausible aging effects of these aging mechanisms in accordance with ISG-3. Through interviews with the applicant's technical personnel and review of applicable documentation, the staff found that the concrete containment is designed in accordance with American Concrete Institute (ACI) 318-63 and constructed of ingredients conforming to ACI and ASTM standards in accordance with ISG-3. In addition, ground water sample testing and flow rate monitoring have demonstrated that no aggressive environment exists for inaccessible concrete. The applicant had demonstrated that aggregates used for containment concrete do not cause aggregate reactions. The staff found that the recommendations of ISG-3 have been satisfied and that no plant-specific AMP for inaccessible containment concrete is required.

Based on the programs identified above, the staff concludes that the applicant has met the criteria of SRP-LR Section 3.5.2.2.1.1. For those line items that apply to LRA Section 3.5.2.2.1.1, the staff determined that the applicant is consistent with the GALL Report and had demonstrated that the aging effects will be adequately managed so that the intended function(s) will be maintained consistent with the CLB during the period of extended operation, as required by 10 CFR 54.21(a)(3).

Cracking, Distortion, and Increase in Component Stress Level Due to Settlement; Reduction of Foundation Strength Due to Erosion of Porous Concrete Subfoundations, If Not Covered by Structures Monitoring Program. The staff reviewed LRA Section 3.5.2.2.1.2 against the criteria in SRP-LR Section 3.5.2.2.1.2.

In LRA Section 3.5.2.2.1.2, the applicant addressed cracks, distortion, and increases in concrete foundation stress level due to settlement. Also in this LRA section, the applicant addressed reduction of foundation strength due to erosion of porous concrete subfoundations.

SRP-LR Section 3.5.2.2.1.2 states that cracking, distortion, and increase in component stress level due to settlement could occur in concrete and steel containments and that reduction of foundation strength due to erosion of porous concrete subfoundations can occur in all types of containments. Some plants may rely on a de-watering system to lower the site ground water level. If the plant's CLB credits a de-watering system, the GALL Report recommends verification of the continued function of the de-watering system during the period of extended operation. The GALL Report recommends no further evaluation if this activity is included within the scope of the applicant's Structures Monitoring Program.

In LRA Section 3.5.2.2.1.1 the applicant provided a detailed description for the following containment aging effects:

* cracks; distortion; increase in component stress level due to settlement (GALL Report Item IIA1.1-f)

* reduction in foundation strength, cracking, differential settlement due to erosion of porous concrete subfoundation (GALL Report Item IIA1.1-g)

On the basis of its audit and review, the staff finds that cracking, distortion, and increase in component stress level due to containment settlement and reduction of containment foundation strength due to erosion of porous concrete subfoundations are not plausible aging effects of these nonexistent aging mechanisms. The applicant stated that the aging effects due to settlement are not expected for the containment structure because it is founded on highly dense, compacted sand that remained after removal of the sand dunes for site preparation. In addition, there is no porous concrete subfoundation below the containment base mat. The staff determined that no AMP is required because these aging mechanisms do not exist. However, the applicant conservatively elected to use the Structural Monitoring Program to monitor the above-grade exposed containment concrete for the aging effect of cracking due to settlement, which election the staff finds acceptable.

Based on the programs identified above, the staff concludes that the applicant has met the criteria of SRP-LR Section 3.5.2.2.1.2. For those line items that apply to LRA Section 3.5.2.2.1.2, the staff determined that the applicant is consistent with the GALL Report and had demonstrated that the aging effects will be adequately managed so that the intended function(s) will be maintained consistent with the CLB during the period of extended operation, as required by 10 CFR 54.21(a)(3).

Reduction of Strength and Modulus of Concrete Structures Due to Elevated Temperature. The staff reviewed LRA Section 3.5.2.2.1.3 against the criteria in SRP-LR Section 3.5.2.2.1.3.

In LRA Section 3.5.2.2.1.3, the applicant addressed reduction of strength and modulus of concrete structures due to elevated temperature.

SRP-LR Section 3.5.2.2.1.3 states that reduction of strength and modulus of elasticity due to elevated temperatures could occur in PWR concrete and steel containments. The GALL Report recommends a plant-specific AMP and further evaluation when any portion of the concrete containment components exceeds specified temperature limits (i.e., general area temperature 66 °C (150 °F) and local area temperature 93 °C (200 °F)).

In the LRA, the applicant stated that, for the containment concrete foundation, dome, and exterior wall, this aging effect is not applicable. In LRA Section 3.5.2.2.1.3, the applicant stated that during normal operation all areas within the containment building do not experience elevated temperatures greater than 150 °F general and 200 °F local. Therefore, change in material properties (reduction of strength and modulus of concrete) due to elevated temperature is not an AERM for the containment concrete. The staff determined through discussions with the applicant's technical personnel that in operating experience the containment concrete has experienced no aging effects due to elevated temperatures.

On the basis that PNP does not have a containment concrete elevated temperature aging mechanism, the staff finds that this aging effect and aging mechanism is not applicable.

Loss of Material due to Corrosion in Inaccessible Areas of Steel Containment Shell or Liner Plate. The staff reviewed LRA Section 3.5.2.2.1.4 against the criteria in SRP-LR Section 3.5.2.2.1.4.

In LRA Section 3.5.2.2.1.4, the applicant addressed loss of material of the steel liner plate due to corrosion in inaccessible areas.

SRP-LR Section 3.5.2.2.1.4 states that loss of material due to corrosion could occur in inaccessible areas of the steel containment shell or the steel liner plate for all types of containments. The GALL Report recommends further evaluation of plant-specific programs to manage this aging effect for inaccessible areas if specific criteria defined in the GALL Report cannot be satisfied.

In the LRA, the applicant stated:

> NUREG-1801, Item IIA2.1-a, states that loss of material due to corrosion is not significant if four conditions are satisfied. Each condition, and a Palisades discussion for that condition, is itemized below:
>
> 1. Concrete meeting the requirements of ACI 318 or 349 and the guidance of 201.2R was used for the containment concrete in contact with the embedded containment shell or liner.
>
> The Palisades containment structure was designed and constructed in accordance with ACI-318-63, ACI-301-72 (proposed) and the ASME Pressure Vessel Code, Sections III, VIII and IX, 1965. (reference FSAR Section 5.1.6.2). Palisades' concrete, meeting the requirements of ACI 318 (and is consistent with the guidance of 201.2R-77), was used for the containment concrete in contact with the embedded containment shell or liner, as discussed in FSAR Section 5.8.2 and Section 5.8.7.1. These materials produced an excellent high strength, dense, sound concrete.
>
> 2. The concrete is monitored to ensure that it is free of penetrating cracks that provide a path for water seepage to the surface of the containment shell or liner.
>
> The containment exterior concrete is monitored by the Palisades Containment Inservice Inspection Program to ensure that it is free of penetrating cracks that might provide a path for water seepage to the surface of the containment shell or liner. The containment interior 18" reinforced concrete floor placed over the containment bottom steel liner is monitored by the Palisades Structural Monitoring Program to ensure that penetrating cracks are not occurring.
>
> 3. The moisture barrier, at the junction where the shell or liner becomes embedded, is subject to aging management activities in accordance with IWE requirements.
>
> The moisture barrier, at the junction where the shell or liner becomes embedded, is subject to aging management activities in accordance with Palisades Containment Inservice Inspection Program requirements.
>
> 4. Borated water spills and water ponding on the containment concrete floor are not common and when detected are cleaned up in a timely manner.
>
> Borated water spills and water ponding on the containment concrete floor are not common, and when detected, are cleaned up in a timely manner, in accordance with the Palisades Boric Acid Corrosion Program.

In the LRA, the applicant stated that, based on satisfaction of the four conditions, corrosion is not significant for inaccessible areas of the containment liner.

The staff's audit and review found all of the GALL Report criteria satisfied. In the LRA, the applicant stated that the containment concrete in contact with the steel liner plate is designed in accordance with ACI 318-63, and meets the requirements of guideline ACI 201.2R-77. Accessible concrete of the containment structure is monitored for penetrating cracks by the applicant's Containment Inservice Inspection Program. In addition, the applicant stated, the accessible portions of the steel liner plate and moisture barrier where the liner becomes embedded are inspected by the same program. Spills (e.g., borated water spill) are cleaned up promptly in accordance with the Boric Acid Corrosion Program. Operating experience demonstrates that the aging effect of loss of material due to corrosion has not been significant for the liner plate. The staff finds no additional plant-specific AMP required to manage inaccessible areas of the containment steel liner plate.

Based on the programs identified above, the staff concludes that the applicant has met the criteria of SRP-LR Section 3.5.2.2.1.4. For those line items that apply to LRA Section 3.5.2.2.1.4, the staff determined that the applicant is consistent with the GALL Report and had demonstrated that the aging effects will be adequately managed so that the intended function(s) will be maintained consistent with the CLB during the period of extended operation, as required by 10 CFR 54.21(a)(3).

Loss of Prestress Due to Relaxation, Shrinkage, Creep, and Elevated Temperature. In LRA Section 3.5.2.2.1.5, the applicant stated that loss of prestress forces due to relaxation, shrinkage, creep, and elevated temperature for PWR prestressed concrete containments is a TLAA, as defined in 10 CFR 54.3. Applicants must evaluate TLAAs in accordance with 10 CFR 54.21(c)(1). SER Section 4.5 documents the staff's review of the applicant's evaluation of this TLAA.

Cumulative Fatigue Damage. In LRA Section 3.5.2.2.1.6, the applicant stated that fatigue is a TLAA, as defined in 10 CFR 54.3. Applicants must evaluate TLAAs in accordance with 10 CFR 54.21(c)(1). SER Sections 4.6.1 and 4.6.2 document the staff's review of the applicant's evaluation of this TLAA for the containment liner and penetrations, respectively.

Cracking Due to Cyclic Loading and SCC. The staff reviewed LRA Section 3.5.2.2.1.7 against the criteria in SRP-LR Section 3.5.2.2.1.7.

In LRA Section 3.5.2.2.1.7, the applicant addressed cracking in penetration sleeves, bellows, and dissimilar metal welds due to cyclic loading or crack initiation and growth in penetration sleeves, bellows, and dissimilar metal welds due to SCC. The applicant stated, in the LRA, that for penetration sleeves, bellows, and dissimilar metal welds, this aging effect is not applicable.

SRP-LR Section 3.5.2.2.1.7 states that cracking of containment penetrations (including penetration sleeves, penetration bellows, and dissimilar metal welds) due to cyclic loading or SCC could occur in all types of PWR and BWR containments. Cracking can also occur in vent line bellows, vent headers and downcomers due to SCC for BWR containments. A VT-3 examination would not detect such cracks. The GALL Report recommends further evaluation of the inspection methods implemented to detect these aging effects.

In the LRA, the applicant stated that no expansion bellows are required in the containment design and that all piping and ventilation penetrations are of the rigid welded type and solidly anchored to the containment shell, thus precluding any requirement for expansion bellows (Reference FSAR Section 5.8.6.2.2). Stress concentrations around openings in the liner plate were calculated from the theory of elasticity. These stress concentrations were then reduced by thickening the liner plate around each penetration in accordance with the ASME B&PV Code, Section III, 1965 Edition.

The applicant also stated in the LRA that anchor bolts are parts of each penetration assembly. When the penetration assembly has no significant external loads, the anchors maintain the strain compatibility between the liner plate and the concrete. When significant loads are present, the anchors control the inward displacement of the liner plate. The stress level in the anchor bolts from external loads is in accordance with the American Institute of Steel Construction (AISC) Code.

Therefore, in the LRA, the applicant stated that no further evaluation is required as these GALL Report items are not applicable to the design.

As documented in the Audit and Review Report, the staff determined through discussions with the applicant's technical personnel that due to the low stress design configuration and low number of cycles, fatigue is not an applicable aging effect for containment penetrations and that further evaluation of inspection methods that detect fatigue-related aging effects is not required. On the basis that PNP has no severe cyclic loading and SSC containment penetration aging mechanism, the staff found this aging effect (cracking) not applicable.

Based on the programs identified above, the staff concludes that the applicant has met the criteria of SRP-LR Section 3.5.2.2.1.7. For those line items that apply to LRA Section 3.5.2.2.1.7, the staff determined that the applicant is consistent with the GALL Report and had demonstrated that the aging effects will be adequately managed so that the intended function(s) will be maintained consistent with the CLB during the period of extended operation, as required by 10 CFR 54.21(a)(3).

3.5.2.2.2 Class 1 Structures

The staff reviewed LRA Section 3.5.2.2.2 against the criteria in SRP-LR Section 3.5.2.2.2, which addresses several areas discussed below.

Aging of Structures Not Covered by Structures Monitoring Program. The staff reviewed LRA Section 3.5.2.2.2.1 against the criteria in SRP-LR Section 3.5.2.2.2.1.

In LRA Section 3.5.2.2.2.1, the applicant addressed all types of aging effects for all groups (except Group 6) of accessible interior/exterior concrete and steel components.

SRP-LR Section 3.5.2.2.2.1 states that the GALL Report recommends further evaluation of certain structure/aging effect combinations not covered by the Structures Monitoring Program, including the following nine items:

(1) scaling, cracking, and spalling due to repeated freeze-thaw for Groups 1-3, 5, and 7-9 structures

(2) scaling, cracking, spalling, and increase in porosity and permeability due to leaching of calcium hydroxide and aggressive chemical attack for Groups 1-5 and 7-9 structures

(3) expansion and cracking due to reaction with aggregates for Groups 1-5 and 7-9 structures

(4) cracking, spalling, loss of bond, and loss of material due to corrosion of embedded steel for Groups 1-5 and 7-9 structures

(5) cracks, distortion, and increase in component stress level due to settlement for Groups 1-3, 5, 7-9 structures

(6) reduction of foundation strength due to erosion of porous concrete subfoundation for Groups 1-3 and 5-9 structures

(7) loss of material due to corrosion of structural steel components for Groups 1-5 and 7-8 structures

(8) loss of strength and modulus of concrete structures due to elevated temperatures for Groups 1-5 structures

(9) crack initiation and growth due to SCC and loss of material due to crevice corrosion of stainless steel liner for Groups 7 and 8 structures

Further evaluation is necessary only for structure/aging effect combinations not covered by the Structures Monitoring Program.

Technical details of the aging management issue are presented in SRP-LR Section 3.5.2.2.1.2 for items 5 and 6 and SRP-LR Section 3.5.2.2.1.3 for item 8.

In LRA Section 3.5.2.2.1.1 the applicant provided a detailed description for the following containment aging effects:

- loss of material (spalling, scaling) and cracking due to freeze-thaw - (GALL Report Items IIIA3.1-a, III.A6.1-a, IIIA8.1-a)

The staff's audit and review found that loss of material (spalling and scaling) and cracking of accessible concrete due to freeze-thaw for Groups 1-3, 5, and 7-9 structures would be adequately managed by the applicant's Structural Monitoring Program as clarified by ISG-3. As documented in the Audit and Review Report, the staff interviewed the applicant's technical personnel and reviewed relevant operating experience to confirm that these aging effects due to this aging mechanism either had not been observed or that when observed corrective action had been taken.

In addition, the staff found that loss of material and cracking of inaccessible concrete for Groups 1-3, 5, and 7-9 structures are not plausible aging effects due to the nonexistence of this aging mechanism. The applicant stated that Class 1 structures are designed in accordance with ACI 318-63 for adequate air entrainment to prevent cracking from freeze-thaw. The applicant stated that concrete for Class 1 structures also meets the ACI 201.2R-77 guidelines for low water to cement ratio, which results in a dense concrete mix resistant to water intrusion. Through interviews with the applicant's technical personnel and review of applicable documentation, the staff found the concrete of Class 1 structures designed in accordance with ACI 318-63 and constructed of ingredients conforming to ACI and ASTM standards. The staff determined that the recommendations of the GALL Report had been satisfied and that no plant-specific AMP for inaccessible concrete of Class 1 structures is required for this aging effect.

In LRA Section 3.5.2.2.1.1 the applicant provided a detailed description for the following containment aging effects:

- increase in porosity, permeability, and loss of strength due to leaching - (GALL Report Items IIIA3.1-b, III.A6.1-b, IIIA8.1-b)

The staff's audit and review found that increase in porosity, permeability, and loss of strength due to leaching of calcium hydroxide of accessible concrete for Groups 1-5 and 7-9 structures would be adequately managed by the applicant's Structural Monitoring Program. As documented in the Audit and Review Report, the staff interviewed the applicant's technical personnel and reviewed relevant operating experience to confirm that these aging effects due to this mechanism either had not been observed or that when observed, corrective action had been taken.

In addition, the staff found that increase in porosity, permeability, and loss of strength due to leaching of calcium hydroxide of inaccessible concrete for Groups 1-5 and 7-9 structures are not plausible aging effects due to the nonexistence of this aging mechanism. The applicant stated that Class 1 structures are designed in accordance with ACI 318-63 and meet ACI 201.2R-77 requirements for good quality, dense, well cured, and low permeability concrete which does not leach. In addition, the applicant stated that the rate of groundwater flow is 650 feet per year, not an aggressive rate. The staff found that the recommendations of the GALL Report had been satisfied and that no plant-specific AMP for inaccessible concrete of Class 1 structures is required for this aging effect.

In LRA Section 3.5.2.2.1.1 the applicant provided a detailed description for the following containment aging effects:

- increase in porosity and permeability, cracking, loss of material (spalling, scaling) due to aggressive chemical attack (GALL Report Items III.A3.1-f, III.A4, III.A6.1-e, III.A8.1-e)

As documented in the Audit and Review Report, the staff noted that the applicant's Structural Monitoring Program does not discuss the need or lack of need for periodic ground water monitoring to ensure that the below-grade water chemistry does not become aggressive. The applicant was asked to justify not monitoring ground water periodically during the CLB and potential period of extended operation to check water chemistry for non-aggressiveness.

By letter dated September 2, 2005, the applicant stated that, as discussed in LRA Section 3.5.2.2.1.1, ground water chemistry records available for the current operating period indicate that water in contact with below-grade concrete has been non-aggressive over at least the last 40 years. To ensure that ground water remains non-aggressive over the period of extended operation, ground water sampling for pH, chlorides, and sulfates would be performed as part of the Structural Monitoring Program at least every five years. Accordingly, changes were made to the Structural Monitoring Program.

In addition, conforming changes were made where LRA Section 3.5.2.2 states that continued groundwater sampling was unnecessary. The new paragraph that replaced the former paragraph reads: "As part of the Structural Monitoring Program, Palisades will continue to monitor groundwater on a periodic basis to ensure it remains non-aggressive, such that the associated aging effects remain not applicable."

The new paragraph was added to the LRA to replace the former paragraph, as follows:

> On Page 3-286, replace existing paragraph that began, 'It is also concluded that it
> is not necessary to monitor groundwater chemistry.'

The staff's review found the applicant's response acceptable as the applicant had appropriately addressed the aging effect/mechanism, as recommended by the GALL Report.

The staff's audit and review found that increase in porosity, permeability, cracking, and loss of material due to aggressive chemical attack of accessible concrete of Groups 1-5 and 7-9 structures would be adequately managed by the applicant's Structural Monitoring Program. As documented in the Audit and Review Report, the staff interviewed members the applicant's technical personnel and reviewed relevant operating experience to confirm that these aging effects due to this mechanism had not been observed or that when observed corrective action had been taken.

In addition, the staff found that increase in porosity, permeability, cracking, and loss of material of Groups 1-5 and 7-9 structures due to aggressive chemical attack of inaccessible concrete are not plausible aging effects due to the nonexistence of this aging mechanism. The applicant stated that Class 1 structures are designed in accordance with ACI 318-63 and meet ACI 201.2R-77 requirements for good quality, dense, well cured, and low-permeability concrete which provides an acceptable degree of protection against an aggressive environment. In addition, the applicant stated that the rate of groundwater flow is 650 feet per year, not an aggressive rate. The staff found that the recommendations of the GALL Report had been satisfied and that no plant-specific AMP for inaccessible concrete of Class 1 structures is required for this aging effect.

In LRA Section 3.5.2.2.1.1 the applicant provided a detailed description for the following containment aging effects:

- expansion and cracking due to reaction with aggregates - (GALL, IIIA3.1-c, IIIA4.1-b, IIIA6.1-c, IIIA8.1-c)

The staff's audit and review found that expansion and cracking of accessible concrete of Groups 1-5 and 7-9 structures due to reaction with aggregates would be adequately managed by the applicant's Structural Monitoring Program. As documented in the Audit and Review Report, the staff interviewed the applicant's technical personnel and reviewed relevant operating experience to confirm that these aging effects due to this mechanism had not been observed or that when observed corrective action had been taken.

In addition, the staff found that expansion and cracking of inaccessible concrete of Groups 1-5 and 7-9 structures due to reaction with aggregates are not plausible aging effects due to the nonexistence of this aging mechanism. The applicant stated that the aggregates used in the concrete of Class 1 structures were specifically investigated, tested, and examined in accordance with ASTM C33, "Standard Specification of Concrete Aggregates," for materials not susceptible to aggregate reactions. In addition, low alkali cement, which mitigates expansion, was used in the concrete mix. The staff found that the recommendations of the GALL Report had been satisfied and that no plant-specific AMP for inaccessible concrete of Class 1 structures is required for this aging effect.

3-274

In LRA Section 3.5.2.2.1.1 the applicant provided a detailed description for the following containment aging effects:

- cracking, spalling, loss of bond, and loss of material due to corrosion of embedded steel for Groups 1-5, 7-9 structures

As documented in the Audit and Review Report, the staff noted that the applicant's Structural Monitoring Program does not discuss the need or lack of need for periodic ground water monitoring to ensure that the below-grade water chemistry does not become aggressive. The applicant was asked to justify not monitoring ground water periodically during the CLB and potential period of extended operation to check water chemistry for non-aggressiveness.

By letter dated September 2, 2005, the applicant stated that, as discussed in LRA Section 3.5.2.2.1.1, ground water chemistry records available for the current operating period indicate that water in contact with below-grade concrete has been non-aggressive over at least the last 40 years. To ensure that ground water remains non-aggressive over the period of extended operation, ground water sampling for pH, chlorides, and sulfates would be performed as part of the Structural Monitoring Program at least every 5 years. Accordingly, changes were made to the Structural Monitoring Program.

In addition, conforming changes were made where LRA Section 3.5.2.2 states that continued groundwater sampling was unnecessary. The new paragraph that replaced the former paragraph reads: "As part of the Structural Monitoring Program, Palisades will continue to monitor groundwater on a periodic basis to ensure it remains non-aggressive, such that the associated aging effects remain not applicable."

The new paragraph was added to the LRA to replace the former paragraph, as follows:

On Page 3-289, replace existing paragraph that begins, 'It is also concluded that it is not necessary to monitor groundwater chemistry'

The staff's review found the applicant's response acceptable as the applicant appropriately addressed the aging effect/mechanism, as recommended by the GALL Report.

The staff's audit and review found that cracking, spalling, loss of bond, and loss of material due to corrosion of embedded steel of accessible concrete for Groups 1-5 and 7-9 structures would be adequately managed by the applicant's Structural Monitoring Program. As documented in the Audit and Review Report, the staff interviewed the applicant's technical personnel and reviewed relevant operating experience to confirm that these aging effects due to this mechanism either had not been observed or that when observed corrective action had been taken.

In addition, the staff found that cracking, spalling, loss of bond, and loss of material due to corrosion of embedded steel of Group 1-5 and 7-9 structures inaccessible concrete are not plausible aging effects due to the nonexistence of this aging mechanism. The applicant stated that the design and construction for dense, well cured, and low-permeability reinforced concrete protects against exposure to aggressive environments. In addition, the applicant stated that groundwater sample measurements have confirmed that the rate of groundwater flow is 650 feet per year, not an aggressive rate. The staff found that the recommendations of the GALL Report had been satisfied and that no plant-specific AMP for inaccessible Class 1 structures concrete is required for this aging effect.

In LRA Section 3.5.2.2.1.1 the applicant provided a detailed description for the following containment aging effects:

- cracks, distortion, and increase in component stress level due to settlement for Groups 1-3, 5, 7-9 structures

The staff's audit and review found that cracking, distortion, and increase in component stress level due to settlement of Groups 1-3, 5, and 7-9 structures are not plausible aging effects due to the nonexistence of this aging mechanism. The applicant stated that aging effects due to settlement are not anticipated for Group 1-3, 5, and 7-9 structures because they are founded on highly dense, compacted sand that remained after removal of the sand dunes for site preparation. The staff determined that no AMP is required as these aging effects do not occur. However, the applicant conservatively elected to use its Structural Monitoring Program to monitor the above-grade exposed concrete of Groups 1-3, 5, and 7-9 structures for the aging effect of cracking due to settlement, which election the staff found acceptable.

In LRA Section 3.5.2.2.1.1 the applicant provided a detailed description for the following containment aging effects:

- reduction of foundation strength due to erosion of porous concrete subfoundation for Groups 1-3, 5-9 structures

The staff's audit and review found that reduction of foundation strength of Groups 1-3 and 5-9 structures due to erosion of porous concrete subfoundations is not a plausible aging effect due to the nonexistence of the aging mechanism. The applicant stated that there are no porous concrete subfoundations below building foundations of Groups 1-3 and 5-9 structures. The staff determined that no AMP is not required because this aging effect does not occur.

In LRA Section 3.5.2.2.1.1 the applicant provided a detailed description for the following containment aging effects:

- loss of material due to corrosion of structural steel components for Groups 1-5, 7-8 structures

The applicant stated that the aging effect of loss of material of Groups 1-5 and 7-8 structures due to corrosion of steel components is managed by the Structural Monitoring Program.

The staff reviewed the AMR results for the aging effect of loss of material of structural steel components from corrosion and confirmed that the applicant's Structural Monitoring Program manages each of the affected structures. The staff's audit and review found that the applicant had appropriately evaluated AMR results for this aging effect, that corrosion of structural steel components would be adequately managed by the applicant's Structures Monitoring Program, and that no detailed further evaluation was required.

In LRA Section 3.5.2.2.1.1 the applicant provided a detailed description for the following containment aging effects:

- loss of strength and modulus of concrete structures due to elevated temperatures for Groups 1-5

The applicant stated that during normal operation, temperatures for all general concrete areas in Groups 1-5 Class 1 structures remain below 150 °F and local area temperatures remain below 200 °F, except for the containment primary shield wall, which is exposed to a temperature of 200 °F at the cavity liner and needs further evaluation. The applicant stated that the primary shield wall has a SCS designed to remove heat from the shield wall and limit thermal stresses in the structural concrete. The system assures that the concrete in the reactor cavity does not overheat from excessive thermal stress. The system was designed to maintain the structural concrete temperature below 165 °F. The applicant further stated that reductions in excess of 10 percent in the compressive strength, tensile strength, and modulus of elasticity for concrete start to occur only when the temperature range of 180 °F to 200 °F is reached. The mix for the primary shield wall concrete was designed with 15 percent more compressive strength than the required design strength, so any reduction in concrete strength properties due to elevated temperatures is offset by the stronger mix. Therefore, the applicant's further evaluation concluded that change in material properties of concrete due to elevated temperature is not an AERM for Groups 1-5 Class 1 structures.

The staff's audit and review found the applicant's further evaluation acceptable because change in material properties due to elevated temperature is not an aging effect requiring management for Groups 1-5 Class 1 structures.

In LRA Section 3.5.2.2.1.1 the applicant provided a detailed description for the following containment aging effects:

In the LRA, the applicant stated:

- crack initiation and growth due to SCC and loss of material due to crevice corrosion of stainless steel liner for groups 7 and 8 structures

In the LRA, the applicant stated that the maximum temperature for fuel-related components in Groups 7-8 structures is 125 °F. A temperature threshold of 140 °F or a continuous temperature environment of 200 °F is necessary for SCC susceptibility. Therefore, the low temperature environment of stainless steel liners in Groups 7-8 structures warrants no AMP because the aging effects of crack initiation and growth and loss of material do not occur.

The staff's audit and review determined that no AMP is required for aging effects for stainless steel liners for Groups 7-8 structures.

Based on the programs identified above, the staff concludes that the applicant has met the criteria of SRP-LR Section 3.5.2.2.2.1. For those line items that apply to LRA Section 3.5.2.2.2.1, the staff determined that the applicant is consistent with the GALL Report and had demonstrated that the aging effects will be adequately managed so that the intended function(s) will be maintained consistent with the CLB during the period of extended operation, as required by 10 CFR 54.21(a)(3).

Aging Management of Inaccessible Areas. The staff reviewed LRA Section 3.5.2.2.2.2 against the criteria in SRP-LR Section 3.5.2.2.2.2.

In LRA Section 3.5.2.2.2.2, the applicant addressed aging management of inaccessible areas.

SRP-LR Section 3.5.2.2.2.2 states that cracking, spalling, and increases in porosity and permeability due to aggressive chemical attack and cracking, spalling, loss of bond, and loss of material due to corrosion of embedded steel could occur in below-grade inaccessible concrete areas. The GALL Report recommends further evaluation to manage these aging effects in inaccessible areas of Groups 1-3, 5, and 7-9 structures, if specific GALL Report criteria cannot be satisfied.

The applicant stated that for such inaccessible concrete components as exterior walls below grade and foundations for Groups 1-3, 5, and 7-9, the aging effects of cracking, spalling, and increases in porosity and permeability due to aggressive chemical attack and cracking, spalling, loss of bond, and loss of material due to corrosion of embedded steel are not applicable.

As documented in the Audit and Review Report, the staff noted that the applicant's Structural Monitoring Program did not discuss the need or lack of need for periodic ground water monitoring to ensure that the below-grade water chemistry does not become aggressive. The applicant was asked to justify not monitoring ground water periodically during the CLB and potential period of extended operation to check water chemistry for non-aggressiveness.

By letter dated September 2, 2005, the applicant stated that, as discussed in LRA Section 3.5.2.2.1.1, ground water chemistry records available for the current operating period, indicate that water in contact with below-grade concrete has been non-aggressive over at least the last 40 years. To ensure that ground water remains non-aggressive over the period of extended operation, sampling for pH, chlorides, and sulfates would be performed as part of the Structural Monitoring Program at least every 5 years. Accordingly, changes were made to the Structural Monitoring Program.

In addition, conforming changes were made where LRA Section 3.5.2.2 states that continued groundwater sampling was unnecessary. The new paragraph which replaced the former paragraph reads: "As part of the Structural Monitoring Program, Palisades will continue to monitor groundwater on a periodic basis to ensure it remains non-aggressive, such that the associated aging effects remain not applicable."

> On Page 3-297 replaced existing paragraph that began, 'It is also concluded that formal groundwater monitoring'
>
> On Page 3-298, replaced existing paragraph that began, 'It is also concluded that formal groundwater monitoring'

The staff's review found the applicant's response acceptable as the applicant had appropriately addressed the aging effect/mechanism, as recommended by the GALL Report.

The applicant stated that Groups 1-3, 5, and 7-9 Class 1 structures are designed in accordance with ACI 318-63 and meet ACI 201.2R-77 requirements for good quality, dense, well cured, and low permeability concrete with an acceptable degree of protection against aggressive environments. In addition, the applicant stated that the rate of groundwater flow is 650 feet per year, not an aggressive rate.

The staff determined through discussions with the applicant's technical personnel and review of the LRA that the recommendations of the GALL Report had been satisfied and that no plant-specific AMP for inaccessible concrete of Class 1 (Groups 1-3, 5, and 7-9) structures is required for these nonexistent aging effects and aging mechanisms.

On the basis that PNP does not have an aggressive environment aging mechanism for inaccessible concrete, the staff found that these aging effects (cracking, spalling, increases in porosity and permeability, loss of bond, loss of material) are not applicable to Groups 1-3, 5, 7-9 Class 1 structures. The staff concludes that the applicant has met the criteria of SRP-LR Section 3.5.2.2.2.2. For those line items that apply to LRA Section 3.5.2.2.2.2, the staff determines that the applicant is consistent with the GALL Report and has demonstrated that the effects of aging will be adequately managed so that the intended function(s) will be maintained consistent with the CLB during the period of extended operation, as required by 10 CFR 54.21(a)(3).

3.5.2.2.3 Component Supports

The staff reviewed LRA Section 3.5.2.2.3 against the criteria in SRP-LR Section 3.5.2.2.3, which addresses several areas discussed below.

Aging of Supports Not Covered by Structures Monitoring Program. The staff reviewed LRA Section 3.5.2.2.3.1 against the criteria in SRP-LR Section 3.5.2.2.3.1.

In LRA Section 3.5.2.2.3.1, the applicant addressed aging of support members (all groups) including anchor bolts, concrete surrounding anchor bolts, welds, grout pads, bolted connections, etc., not covered by the Structural Monitoring Program.

SRP-LR Section 3.5.2.2.3.1 states that the GALL Report recommends further evaluation of certain component support/aging effect combinations not managed by the Structures Monitoring Program, including (1) reduction in concrete anchor capacity due to degradation of the surrounding concrete for Groups B1-B5 supports, (2) loss of material due to environmental corrosion for Groups B2-B5 supports, and (3) reduction/loss of isolation function due to degradation of vibration isolation elements for Group B4 supports. Further evaluation is necessary only for structure/aging effect combinations not covered by the structures monitoring program.

In the LRA, the applicant stated that GALL Report Items III.B1.1.4-a, III.B1.2.3-a, III.B2.2-a, III.B3.2-a, III.B4.3-a, and III.B5.2-a discuss the aging effect/mechanism of reduction in anchor bolt capacity due to local concrete degradation/service-induced cracking. The GALL Report recommends further evaluation of this aging effect and mechanism if the Structures Monitoring Program does not have this aging effect and mechanism within the scope of license renewal. The Structural Monitoring Program includes local concrete degradation/service-induced cracking within the scope of license renewal. Therefore, the applicant stated that no further evaluation is required.

The applicant also stated that GALL Report Items III.B2.1-a, III.B3.1-a, III.B4.1-a, and III.B5.1-a discuss aging effect/mechanism, loss of material due to environmental corrosion (i.e., pitting corrosion, general corrosion, etc.) and the GALL Report recommends further evaluation of this aging effect and mechanism if the Structures Monitoring Program does not have this aging effect and mechanism within the scope of license renewal. The applicant's Structural Monitoring Program includes loss of material due to environmental corrosion (i.e., pitting corrosion, general corrosion, etc.) within the scope of license renewal. Therefore, the applicant stated that no further evaluation is required.

Additionally, in the LRA, the applicant stated that GALL Report Item III.B4.2-a discusses vibration isolation elements and the aging effect/mechanism of reduction or loss of isolation function due to radiation hardening, temperature, humidity, or sustained vibratory loading. The applicant's Structural Monitoring Program is credited with age managing vibration isolation elements for the emergency diesel generator. Therefore, the applicant stated that no further evaluation is required.

The staff finds that the applicant has included the aging effect combinations within the scope of its Structural Monitoring Program and that no further evaluation is required. The staff reviewed the applicant's Structural Monitoring Program as documented in SER Section 3.0.3.2.10. The staff finds the program acceptable for managing aging of component supports for all such GALL Report component support groups.

Based on the programs identified above, the staff concludes that the applicant has met the criteria of SRP-LR Section 3.5.2.2.3.1. For those line items that apply to LRA Section 3.5.2.2.3.1, the staff determined that the applicant is consistent with the GALL Report and had demonstrated that the aging effects will be adequately managed so that the intended function(s) will be maintained consistent with the CLB during the period of extended operation, as required by 10 CFR 54.21(a)(3).

Cumulative Fatigue Damage Due to Cyclic Loading. The staff reviewed LRA Section 3.5.2.2.3.2 against the criteria in SRP-LR Section 3.5.2.2.3.2.

SRP-LR Section 3.5.2.2.3.2 states that fatigue of component support members, anchor bolts, and welds for Groups B1.1, B1.2, and B1.3 component supports is a TLAA as defined in 10 CFR 54.3 only if a CLB fatigue analysis exists. TLAAs are required to be evaluated in accordance with 10 CFR 54.21(c).

In LRA Section 3.5.2.2.3.2, the applicant stated that a TLAA is required if, as part of the CLB, a fatigue analysis is performed for these supports. PNP does not perform fatigue analysis for ASME Class 1, 2, or 3 component supports (i.e., support members, welds, bolted connections, and support anchorage to building structure).

The staff finds this aging effect not applicable.

3.5.2.2.4 Quality Assurance for Aging Management of Nonsafety-Related Components

SER Section 3.0.4 provides the staff's evaluation of the applicant's QA program.

Conclusion. On the basis of its review, for component groups evaluated in the GALL Report for which the applicant claimed consistency with the GALL Report and for which further evaluation is recommended, the staff determined that the applicant has adequately addressed the issues that were further evaluated. The staff finds that the applicant has demonstrated that the aging effects will be adequately managed so that the intended function(s) will be maintained consistent with the CLB for the period of extended operation, as required by 10 CFR 54.21(a)(3).

3.5.2.3 AMR Results That Are Not Consistent with or Not Addressed in the GALL Report

Summary of Technical Information in the Application. In LRA Tables 3.5.2-1 through 3.5.2-10, the staff reviewed additional details of AMR results for material, environment, AERM, and AMP combinations not consistent with or not addressed in the GALL Report.

In LRA Tables 3.5.2-1 through 3.5.2-10, the applicant indicated, via Notes F through J, that the combination of component type, material, environment, and AERM does not correspond to a line item in the GALL Report. The applicant provided further information concerning how the aging effects will be managed. Specifically, Note F indicates that the material for the AMR line item component is not evaluated in the GALL Report. Note G indicates that the environment for the AMR line item component and material is not evaluated in the GALL Report. Note H indicates that the aging effect for the AMR line item component, material, and environment combination is not evaluated in the GALL Report. Note I indicates that the aging effect identified in the GALL Report for the line item component, material, and environment combination is not applicable. Note J indicates that neither the component nor the material and environment combination for the line item is evaluated in the GALL Report.

Staff Evaluation. For component type, material, and environment combinations not evaluated in the GALL Report, the staff reviewed the applicant's evaluation to determine whether the applicant had demonstrated that the aging effects will be adequately managed so that the intended function(s) will be maintained consistent with the CLB during the period of extended operation. The staff's evaluation is discussed in the following sections.

3.5.2.3.1 Structures and Component Supports – Auxiliary Building – Summary of Aging Management Evaluation – LRA Table 3.5.2-1

The staff reviewed LRA Table 3.5.2-1, which summarizes the results of AMR evaluations for the auxiliary building component groups. The staff determined that all AMR evaluation results in LRA Table 3.5.2-1 are consistent with the GALL Report, as discussed below.

As documented in the Audit and Review Report, the staff noted that LRA Table 3.5.2-1 for component type "Fuel Related Component - Carbon Steel, Protected," refers to GALL Report Item III.A5.2-b. The environment shown in the GALL Report is water. The AMR line item shows an environment of plant indoor air. The staff requested that the applicant explain how the applicant's Water Chemistry Program manages loss of material in a plant indoor air environment.

By letter dated August 27, 2005, the applicant stated that the aging effect/mechanism in question is loss of material/boric acid corrosion. Although the carbon steel anchor bolts are protected from boric acid by the liner plate, the applicant conservatively credited the AMP for the liner plate to prevent boric acid leakage past the liner. The following comment was included in the AMR evaluation: anchor bolts for the spent fuel pool gates, spent fuel pool liner, fuel tilt pool liner, and appurtenances for the fuel transfer tube are protected from exposure to corrosive environments by the stainless steel liners and transfer tube. The liners and transfer tube are age managed by the SFP Water Chemistry Program and TS surveillance of SFP water levels.

Ensuring minimal leakage from the liners ensures minimal potential boric acid wastage as well. Anchor bolt degradation could result in SFP liner damage which would be evident via SFP level monitoring. On further evaluation, it was evident that this component is redundant to component type "Building Framing - Concrete, Protected," that the full description of components for this concrete type in the scoping report included embedded steel reinforcements and shapes. Thus, this AMR line item component type was deleted from the LRA. There was a similar component in Table 3.5.2-4, "Fuel Related Component - Carbon Steel, Protected," also deleted from the LRA.

The staff's review of the applicant's response found that the deletion of the AMR line item eliminated the discrepancy. The staff finds this deletion acceptable.

As documented in the Audit and Review Report, the staff noted that LRA Table 3.5.2-1 for component type "Fuel Related Component - Stainless, Borated," refers to GALL Report Item III.A5.2-b. The environment shown in the GALL Report is water with an aging effect of crack initiation and growth. The GALL Report has no water temperature criteria for which no aging would occur. The staff requested that the applicant explain why no AMP was required to prevent cracking of the stainless steel liners as the GALL Report has no criteria for the temperature of the water to which the liners are exposed.

By letter dated August 27, 2005, the applicant stated that a temperature threshold of 140 °F is from industry guidance. The applicability criteria for cracking due to SCC is given as a temperature > 140 °F and chlorides, or fluorides, or sulfates > 150 ppb. Additionally, GALL Report, Revision 1, Table IX.D also specifies 140 °F as the threshold for SCC in treated water. It was noted that the loss of material due to the crevice corrosion portion of the same line item (III.A5.2-b) was considered an AERM with the Water Chemistry Program and monitoring of the fuel pool level per TS credited for aging management consistent with the GALL Report and that a more appropriate note for the AMR cracking line item would have been H rather than E in the LRA. Note E was changed to H in the response.

The staff's review found the applicant's response acceptable as the applicant appropriately addressed the aging effect/mechanism.

The staff's review of LRA Section 3.5.2.3.1 identified areas in which additional information was necessary to complete the review of the applicant's aging management results. The applicant responded to the staff's RAIs as discussed below.

In RAI 3.5.2-1-1, dated June 28, 2005, the staff stated:

> In Table 3.5.2-1 (Page 3-311), under the component type 'Flood Barrier-Carbon Steel, Protected,' PNP's Structural Monitoring Program is credited to manage the loss of leak tightness aging effect of flood doors. The Structural Monitoring Program is also credited to manage the loss of leak tightness effect in: (1) HELB doors (Table 3.5.2-1, Page 3-313); (2) Control room vestibule door (Table 3.5.2-2, Page 3-314); (3) Flood doors and hatch (Table 3.5.2-10, Page 3-389); and (4) Control room vestibule door (Table 3.5.2-10, Page 390). Summarize past PNP's operating/inspection experience in managing the leak tightness of the above listed PNP components, and discuss specific provision(s)

of the Structural Monitoring Program that are intended to maintain the leak tightness junction of the PNP components.

In its response, by letter dated July 28, 2005, the applicant stated:

> For the flood doors, hatch, and High Energy Line Break / Moderate Energy Line Break doors, the Structural Monitoring Program credits an existing watertight barrier inspection procedure, MSM-M-16, 'Inspection of Watertight Barriers.' The watertight barrier inspection procedure is currently performed on a yearly basis on all watertight barriers as well as twice every refueling outage on high use doors. Parameters inspected include seals (including performance of a chalk test), loose or missing parts, latch tightness, etc. A review of work order history identifies instances where the program has found barrier seals that failed chalk tests and latches that were discovered to be loose. Repairs were made and the barriers were retested satisfactorily.

> For the control room vestibule doors, the Structural Monitoring Program credits Palisades' monthly Technical Specification Test MO-33, 'Control Room Ventilation Emergency Operation.' One of the acceptance criteria is to ensure the control room pressure readings are equal to or greater than 0.125 inches of water as required by Technical Specification Surveillance Requirement SR 3.7.10.4. Review of action requests indicates that this test has found and repaired a degraded vestibule door closing mechanism that had impacted its leak tightness.

The staff reviewed the applicant's response and found that the applicant has adequately implemented its structural monitoring program to manage the loss of leak tightness aging effect of flood doors (HELB doors, Control Room vestibule doors, and flood doors and hatch). Therefore, the staff finds the applicant's aging management review regarding the flood doors is acceptable and RAI 3.5.2-1-1 is resolved.

In RAI 3.5.2-2-1(a), dated June 28, 2005, the staff stated:

> Table 3.5.2-2 (Page 3-318) of the LRA credits ASME Section XI IWB, IWC, IWD, IWF Inservice Inspection Programs to manage loss of material aging effect of Auxiliary Building cast iron components (ASME Class 2 & 3 Piping & Mechanical Component Support). Table 3.5.2-2 (Page 3-336) of the LRA credits Structural Monitoring Program to manage loss of material aging effect of Discharge Structure Cast Iron components (Non-ASME Piping & Mechanical Component Support). Note 582 referred to by the tables states that cast iron is considered consistent with carbon steel and is evaluated the same, but with the additional aging effect/mechanism of loss of material due to selective leaching also evaluated. Discuss PNP's past operating experience and inspection results related to selective leaching of PNP's in-scope cast iron components. Did any of these affected cast iron components experience cracking or loss of function as a result of leaching? If yes, summarize PNP's corrective action(s) taken to dispose the identified aging degradation.

In its response, by letter dated July 28, 2005, the applicant stated:

> The ASME and Non-ASME Cast Iron components in question are pump support skids located in the Auxiliary Building and in the Warm Water Recirculation Pump House above the Discharge Structure. Both are in an indoor air environment. As indicated in note 582, both sets of components were evaluated for loss of material due to selective leaching. The results of those evaluations are that loss of material due to selective leaching does not apply to these components in their plant indoor air environment. Thus, the ASME Section XI IWB, IWC, IWF Inservice Inspection Program and the Structure Monitoring Program are only being credited for age managing loss of material due to general corrosion, not for loss of material due to selective leaching.

> Where loss of material due to selective leaching does apply, namely in a wetted environment, the One Time Inspection Program is the appropriate program to manage the aging effect. Since the One Time Inspection Program is a new program (see LRA Section B2.1.13), there is no operating experience or inspection results related to selective leaching on in-scope cast iron components.

> It is noted, however, while reviewing the structural component types where loss of material due to selective leaching does apply in LRA Tables 3.5.2-5 (page 3-349) and 3.5.2-7 (page 3-355), the One Time Inspection Program was not listed. These line items should have credited the One Time Inspection Program for the aging management of loss of material due to selective leaching in addition to the Structural Monitoring Program for management of other loss of material mechanisms. The resulting line item revisions are provided below. The existing table line item is shown in gray. The added information is shown in bold text.

The staff reviewed the applicant's response and finds that the applicant has adequately evaluated the potential for loss of material due to selective leaching in an indoor air environment for the cast iron components mentioned in the RAI. The staff concurs with the applicant's determination that loss of material due to selective leaching does not apply to the above noted cast iron components located in an indoor air environment. Thus, the ASME Section XI IWB, IWC, IWF Inservice Inspection Program and the Structure Monitoring Program credited for aging management of loss of material due to general corrosion of the cast iron components in an indoor air environment are adequate and acceptable. The applicant indicated that where loss of material of cast iron components due to selective leaching does apply, namely in a wetted environment, the One Time Inspection Program is relied upon by the applicant to manage the aging effect. The applicant further noted that, while reviewing the structural component types where loss of material due to selective leaching does apply in LRA Tables 3.5.2-5 (page 3-349) and 3.5.2-7 (page 3-355), the One Time Inspection Program was not listed. These line items should have credited the One Time Inspection Program for the aging management of loss of material due to selective leaching in addition to the Structural Monitoring Program for management of other loss of material mechanisms. As part of its response to RAI 3.5.2-2-1(a), the applicant provided the revised LRA Tables 3.5.2-5 and 3.5.2-7 reflecting inclusion of the One Time Inspection Program to manage loss of material due to selective leaching of cast iron components. The staff reviewed the contents of the revised tables and found the proposed revisions acceptable, therefore, RAI 3.5.2-2-1(a) is resolved.

On the basis of its review, as discussed above, the staff concludes that the applicant has demonstrated that the aging effects associated with the components will be adequately managed so that the intended function(s) will be maintained consistent with the CLB for the period of extended operation, as required by 10 CFR 54.21(a)(3).

3.5.2.3.2 Structures and Component Supports – Component Supports – Summary of Aging Management Evaluation – LRA Table 3.5.2-2

The staff reviewed LRA Table 3.5.2-2, which summarizes the results of AMR evaluations for the component supports component groups.

The applicant proposed the Boric Acid Corrosion Program to manage loss of material from aluminum ASME Class 2 and 3 piping, mechanical, and non-ASME component supports exposed to atmosphere/weather, plant indoor air, or containment air (all with borated water leakage in air) environments.

The staff reviewed the applicant's Boric Acid Corrosion Program as documented in SER Section 3.0.3.2.2. The program monitors through periodic inspections component degradation due to boric acid leakage. The staff's audit and review of plant-specific and industry operating experience determined that the aging effect of loss of aluminum material exposed to an atmosphere/weather, plant indoor air, or containment air (all with borated water leakage in air) environment is adequately managed by the Boric Acid Corrosion Program and found management of loss of component supports material acceptable.

As documented in the Audit and Review Report, the staff noted that LRA Table 3.5.2-2 for component type "Non-ASME Component Support-Auxiliary Building, Aluminum, Protected," shows a GALL Report line item and a Table 1 item with Note F. The staff requested that the applicant explain why a GALL Report line item and a Table 1 item had been shown with Note F.

By letter dated August 27, 2005, the applicant stated that, as indicated in plant-specific Note 503, component type "Non-ASME Component Support-Auxiliary Bldg, Aluminum, Protected" represents the new fuel storage racks represented by GALL Report Item VII.A1.1-a. New fuel racks are aluminum whereas the GALL Report line item is for carbon steel racks. Hence, Note F (material not in the GALL Report for this component) was assigned. However, as neither the material or program match, alignment with the GALL Report line item was inappropriate. For this AMR line item, only the GALL Report and Table 1 line items were deleted, leaving Notes F and 503.

The staff's review found the applicant's response acceptable as the applicant appropriately addressed the aging effect/mechanism.

On the basis of its review, as discussed above, the staff concludes that the applicant has demonstrated that the aging effects associated with the components will be adequately managed so that the intended function(s) will be maintained consistent with the CLB for the period of extended operation, as required by 10 CFR 54.21(a)(3).

3.5.2.3.3 Structures and Component Supports – Containment – Summary of Aging Management Evaluation – LRA Table 3.5.2-3

The staff reviewed LRA Table 3.5.2-3, which summarizes the results of AMR evaluations for the containment component groups. The containment component groups include the containment shell and base, containment shell prestressing systems, and building frame, stall and concrete structures. The staff determined that all AMR evaluation results in LRA Table 3.5.2-3 are consistent with the GALL Report.

On the basis of its review, as discussed above, the staff concludes that the applicant has demonstrated that the aging effects associated with the components will be adequately managed so that the intended function(s) will be maintained consistent with the CLB for the period of extended operation, as required by 10 CFR 54.21(a)(3).

3.5.2.3.4 Structures and Component Supports – Containment Interior Structures – Summary of Aging Management Evaluation – LRA Table 3.5.2-4

The staff reviewed LRA Table 3.5.2-4, which summarizes the results of AMR evaluations for the containment interior structures component groups. The staff determined that, for the staff's scope of review, all AMR LRA Table 3.5.2-4 evaluation results are consistent with the GALL Report or, if not consistent, previously discussed in SER Section 3.5.2.1.

The staff's review of LRA Section 3.5.2.3.4 identified areas in which additional information was necessary to complete the review of the applicant's aging management results. The applicant responded to the staff's RAIs as discussed below.

In RAI 3.5.2-4-1, dated June 28, 2005, the staff stated:

> In Table 3.5.2-4, under the component type 'concrete protected,' a number of structural components (e.g., masonry walls, RC beams, columns, pedestals) are listed. It would be logical to have primary shield walls, secondary shield walls, and reactor pressure vessel (RPV) supports included under this component type. Section 3.5.2.2.2.1 describes the elevated temperature situation around the reactor vessel, and justifies the existence of the elevated temperatures in these areas, based on the estimated temperatures in the Palisades FSAR. The applicant is requested to provide the following information related to this component type.

In RAI 3.5.2-4-1(a), dated June 28, 2005, the staff requested the applicant to "provide the operating experience related to the effectiveness of the "shield cooling system." Are the shield wall temperatures, or any other parameter monitored, that would detect the malfunctioning of the cooling system?"

In its response, by letter dated July 28, 2005, the applicant stated:

> In the 1995 Refueling Outage, an array of temperature monitoring devices was installed in the annulus between the reactor vessel and the biological shield wall. Ten of these devices were installed on the shield wall itself. Measurements showed temperatures ranging between 164 °F and 202 °F at the shield wall steel

liner plate. These measured temperatures were used as input to the development of the revised biological shield wall temperature profiles shown in FSAR Figures 9-3, 9-4, 9-5 and 9-6. The results of this benchmarking analysis showed the structural concrete in the biological shield wall remained below 165 °F.

Shield wall temperatures are not continuously monitored. However, the shield cooling water temperatures are monitored and will alarm in the control room when temperature reaches 120 °F. Follow up actions to the alarm include commencing a plant shutdown if the alarm cannot be cleared and shield cooling return temperature exceeds 165 °F. Similarly, shield cooling pump breakers are monitored and should both shield cooling pumps trip, commencement of a plant shutdown is directed.

The staff reviewed the applicant's response to RAIs 3.5.2.4-1 and 3.5.2-4-1(a), and found that though the shield wall temperatures are not continuously monitored, there is an alarm system when the temperature is indicated to be above 120 °F. The applicant prepares to shutdown the plant when the temperature are indicated to be above 165 °F. Though this temperature is above the 150 °F threshold value, based on the licensee's planned action to shutdown the plant when needed, the staff finds it acceptable, because at 165 °F of occasional temperature, the effects on concrete properties are minimal.

In RAI 3.5.2-4-1(b), dated June 28, 2005, the staff stated:

Based on the discussion of the elevated temperature condition, in and around the primary shield wall in Section 3.5.2.2.2.1, the staff agrees with EPRI TR-103842, that the concrete properties will not be significantly affected, if the actual temperatures around the shield wall remain within the estimated limits. However, additional shrinkage and loss of moisture due to radiation could degrade the concrete on a long term basis. In this context, please provide a summary of the results of the last two inspections performed for: (1) the primary shield wall, (2) RPV supports, (3) grouted anchorages, and (4) masonry walls inside the containment.

In its response, by letter dated July 28, 2005, the applicant stated:

1 & 2) No inspection results are available. As discussed in section 2.4.4 of the LRA, the entire interior concrete surface of the Palisades Primary Shield Wall is lined with welded carbon steel plate. This includes the area around the reactor pressure vessel (RPV) supports. Accordingly, the shield wall concrete and the concrete around the RPV supports are not accessible for inspection.

3) The term 'grouted anchorage' in the description of 'Building Framing - Containment Cavity' in LRA Table 2.4.4-1 is used generically. The specific anchorage in the vicinity of the reactor shield wall is cast-in-place bolting or strap anchors, depending on elevation, for the liner plate. These anchors are not accessible for inspection.

4) There is one block wall inside containment on the 649' level, which is remote from the high temperature and radiation environment of the shield wall. Structural

Monitoring Program inspections were performed inside containment in 1996 and 1999. The top five courses of masonry wall blocks were found spalled at the northern most tip of the block wall. The existing condition was determined not to be damaging to the masonry wall integrity, which serves only as a partition wall. The condition was deemed acceptable as-is.

It should be noted that the concrete shielding blocks in the primary coolant pipe openings of the shield wall that are mentioned in LRA Section 3.5.2.2.2.1 are not masonry walls. These removable concrete blocks are held in place by external restraints without mortar, and are not inspected or evaluated as 'block walls.'

The staff reviewed the applicant's supplemental response to RAI 3.5.2-4-1(b), and found that the shield wall concrete and the concrete around the RPC supports are not accessible for inspections, as these areas are lined with carbon steel liner. The applicant also stated that as result of Structural Monitoring Inspections performed inside the containment in 1996 and 1999, the top five courses of masonry wall blocks were found spalled at the northern most tip of the block wall. The condition was determined not to be damaging to the masonry wall integrity, and the condition was deemed acceptable. After reviewing the applicant's RAI response, the staff requested that the applicant: (1) explain or identify under what program the carbon steel plate (liner) on primary shield wall is inspected, and (2) discuss the effects of degraded masonry wall courses on Class 1 components during the postulated accident or seismic event.

In the November 15, 2005 teleconference, the staff requested the following additional information from the applicant:

Based on the discussion of the elevated temperature condition in and around the primary shield wall in Section 3.5.2.2.2.1, the staff agrees with EPRI TR 103842, that the concrete properties will not be significantly affected, if the actual temperatures around the shield wall remain within the estimated limits. However, additional shrinkage and loss of moisture due to radiation could degrade the concrete on a long term basis. In this context, please provide a summary of the results of the last two inspections performed for: (1) the primary shield wall, (2) RPV supports, (3) grouted anchorages, and (4) masonry walls inside the containment.

In its supplemental response, dated December 16, 2005, the applicant stated that,

The NMC response to this RAI provided in a letter dated July 28, 2005, is hereby replaced in its entirety with the following:

1 & 2) No inspection results are available. As discussed in Section 2.4.4 of the LRA, the entire interior concrete surface of the Palisades Primary Shield Wall is lined with welded carbon steel plate. This includes the area around the RPV supports. Accordingly, the shield wall concrete and the concrete surrounding the RPV supports are not accessible for inspection. In Palisades FSAR Figure 6-6, which shows the reactor vessel, reactor cavity, bioshield, reactor supports, insulation, etc., it can be seen that these areas are inaccessible and are high radiation dose locations.

With regard to the potential for additional shrinkage and loss of moisture due to radiation that could degrade concrete on a long term basis, the shield cooling system maintains structural concrete at temperatures that mitigate concrete thermal heating due to radiation or conduction (see LRA page 2-147 for shield cooling system description). These embedded cooling system coils are installed more densely around the steel components that support the reactor vessel to maintain the steel and concrete around the supports at the design temperature.

Loss of concrete strength due to cumulative radiation exposure, is addressed in the Palisades LRA Page 3-7, Table 3.01, Service Environments, which provides the following discussion:

> Radiation - Plant radiation doses outside the reactor cavity are not of concern for aging management. Materials can be affected by cumulative radiation exposure. For concrete, neutron fluence above 10^{19} n/cm2 (>1 Mev) or gamma dose $>10^{10}$ rads is required to cause degradation

The Palisades' neutron fluence estimate through the end of the proposed extended operating period at the outside diameter (OD) of the reactor vessel, is 1.94×10^{18} n/cm^2. The fluence at the inside diameter (ID) of the biological shield can be assumed to be the same." This is less than the threshold value for degradation of 1.0×10^{19} n/cm^2. At a depth of approximately 11 inches into the biological shield (the approximate depth of the non-structural, non-reinforced, sacrificial concrete – See LRA page 2-228 for cavity description), the neutron fluence is equivalent to 10^{17} n/cm^2 (E > 1 Mev). This is less than the threshold value of 1.0×10^{19} n/cm^2 by two orders of magnitude. Palisades estimates that gamma dose at a depth of approximately 11' behind the biological shield liner plate is 4.74×10^8 rads (less than 1.0×10^{10} threshold). These values of neutron fluence and gamma dose are even lower above or below the core centerline. Therefore, the sacrificial concrete portion of the bioshield is not subject to degradation due to loss of concrete strength from cumulative radiation exposure.

Based on the above discussion, no aging management program is required for the interior reactor cavity steel liner plate, interior concrete primary shield wall or the RPV supports. However, as discussed, the accessible external portions of the reinforced concrete bioshield are included in the Palisades Structures Monitoring Program.

3) The term 'grouted anchorage' in the description of 'Building Framing - Containment Cavity' in LRA Table 2.4.4-1 is used generically. The specific anchorage in the vicinity of the reactor shield wall is cast-in-place bolting or strap anchors, depending on elevation, for the liner plate. These anchors are not accessible for inspection.

4) There is one block wall inside containment on the 649' level, which is remote from the high temperature and radiation environment of the shield wall. Structural Monitoring Program inspections were performed inside containment in 1996 and 1999. The top five courses of masonry wall blocks were found spalled at the northern most tip of the wall. The existing condition was determined not

damaging to the masonry wall integrity, which serves only as a partition wall. The condition was deemed acceptable as-is.

This masonry wall (with vertical and horizontal steel reinforcement) partially surrounds the shield cooling system surge tank and associated piping components which are not safety related. Palisades FSAR Figure 1-6 (E2) shows the location at Elevation 649' of the shield cooling surge tank, that is surrounded on three sides by the masonry block wall. FSAR Table 5.2-3 shows the component classification of the surge tank, piping and valves, as Class 3. On the West side, these two walls are "qualified", indicating that they were analyzed to ensure that failure would not occur due to a design basis earthquake load in accordance with NRC IEB 80-11 (See Palisades FSAR Section 5.10.3.2 Masonry Walls). The North and East side walls are identified as 'unqualified,' since their failure would not impact any safety related equipment or components, since that area is used for lay-down/storage purposes, as shown on FSAR Figure 1-6.

The staff found that the significant portion of the masonry walls is seismically qualified. The impact of the five top courses of the masonry wall in a localized area is insignificant, and the applicant asserts that their failure will not affect any Class I components during the postulated accident or seismic event. Therefore, the staff finds the applicant's response acceptable and the concerns described in RAI 3.5.2-4-1(b) are resolved.

On the basis of its review, as discussed above, the staff concludes that the applicant has demonstrated that the aging effects associated with the components will be adequately managed so that the intended function(s) will be maintained consistent with the CLB for the period of extended operation, as required by 10 CFR 54.21(a)(3).

3.5.2.3.5 Structures and Component Supports – Discharge Structure – Summary of Aging Management Evaluation – LRA Table 3.5.2-5

The staff reviewed LRA Table 3.5.2-5, which summarizes the results of AMR evaluations for the discharge structure component groups. The staff determined that, for the staff's scope of review, all AMR evaluation results in LRA Table 3.5.2-5 are consistent with the GALL Report.

On the basis of its review, as discussed above, the staff concludes that the applicant has demonstrated that the aging effects associated with the components will be adequately managed so that the intended function(s) will be maintained consistent with the CLB for the period of extended operation, as required by 10 CFR 54.21(a)(3).

3.5.2.3.6 Structures and Component Supports – Feedwater Purity Building – Summary of Aging Management Evaluation – LRA Table 3.5.2-6

The staff reviewed LRA Table 3.5.2-6, which summarizes the results of AMR evaluations for the feedwater purity building component groups. The staff determined that all AMR evaluation results in LRA Table 3.5.2-6 are consistent with the GALL Report.

On the basis of its review, as discussed above, the staff concludes that the applicant has demonstrated that the aging effects associated with the components will be adequately managed so that the intended function(s) will be maintained consistent with the CLB for the period of extended operation, as required by 10 CFR 54.21(a)(3).

3.5.2.3.7 Structures and Component Supports – Intake Structure – Summary of Aging Management Evaluation – LRA Table 3.5.2-7

The staff reviewed LRA Table 3.5.2-7, which summarizes the results of AMR evaluations for the intake structure component groups. The staff determined that for the staff's scope of review, all AMR evaluation results in LRA Table 3.5.2-7 are consistent with the GALL Report.

On the basis of its review, as discussed above, the staff concludes that the applicant has demonstrated that the aging effects associated with the components will be adequately managed so that the intended function(s) will be maintained consistent with the CLB for the period of extended operation, as required by 10 CFR 54.21(a)(3).

3.5.2.3.8 Structures and Component Supports – Miscellaneous Structural and Bulk Commodities – Summary of Aging Management Evaluation – LRA Table 3.5.2-8

The staff reviewed LRA Table 3.5.2-8, which summarizes the results of AMR evaluations for the miscellaneous structural and bulk commodities component groups.

The applicant proposed the Structural Monitoring Program to manage change in material properties, loss of form, and loss of built-up roofing materials for roofing component types exposed to an atmosphere/weather environment.

The staff reviewed the applicant's Structural Monitoring Program as documented in SER Section 3.0.3.2.10. The program is designed to ensure that age-related (as well as other) deterioration of plant structures (including masonry walls) and components within its scope is adequately managed to ensure that each structure or component retains the ability to perform its intended functions. The staff accepted the position that change in material properties, loss of form, and loss of material exhibited by built-up roofing in an atmosphere/weather environment is adequately managed by the Structural Monitoring Program, which through visual examination inspects the built-up roofing for any sign of aging degradation. The staff's audit and review of plant-specific and industry operating experience determined that the aging effects of change in material properties, loss of form, and loss of built-up roofing material for roofing component types exposed to an atmosphere/weather environment are adequately managed by the Structural Monitoring Program and that management of change in material properties, loss of form, and loss of built-up roofing material in miscellaneous structural and bulk commodities is acceptable.

In LRA Table 3.5.2-8, the applicant proposed the Structural Monitoring Program in conjunction with the Fire Protection Program to manage loss of concrete material for fire barrier component types exposed to a plant indoor air environment.

The staff reviewed the applicant's Structural Monitoring and Fire Protection Programs as documented in SER Sections 3.0.3.2.10 and 3.0.3.2.5, respectively. The Fire Protection Program includes (a) fire barrier inspections, (b) electric and diesel-driven fire pump tests, and (c) periodic maintenance, testing, and inspections of water-based fire protection systems. The staff's audit and review of plant-specific and industry operating experience determined that the aging effect of loss of concrete materials for fire barrier component types exposed to a plant indoor air environment is adequately managed by the Structural Monitoring and Fire Protection Programs and that management of loss of concrete materials in miscellaneous structural and bulk commodities is acceptable.

In LRA Table 3.5.2-8, the applicant proposed the Fire Protection Program to manage loss of fire stop (sealant/maranite) materials for fire barrier component types exposed to a plant indoor air environment.

The staff reviewed the applicant's Fire Protection Program as documented in SER Section 3.0.3.2.5. The staff's audit and review of plant-specific and industry operating experience determined that the aging effects of loss of material of fire stop (sealant/maranite) material for fire barrier component types exposed to a plant indoor air environment are adequately managed by the Fire Protection Program and that management of loss of material of fire stop (sealant/maranite) material in miscellaneous structural and bulk commodities is acceptable.

In LRA Table 3.5.2-8, the applicant proposed the Fire Protection Program to manage cracking and loss of fire wrap material for fire barrier component types exposed to an plant indoor air environment.

The staff reviewed the applicant's Fire Protection Program as documented in SER Section 3.0.3.2.5. The staff accepted the position that cracking and loss of material exhibited by fire wraps in a plant indoor environment is adequately managed by the Fire Protection Program, which through visual examination inspects the fire wraps for any sign of aging degradation. The staff's audit and review of plant-specific and industry operating experience determined that the aging effects of cracking and loss of material of fire wraps for fire barrier component types exposed to a plant indoor air environment are adequately managed by the Fire Protection Program and that management of cracking and loss of fire wrap material in miscellaneous structural and bulk commodities is acceptable.

In LRA Table 3.5.2-8, the applicant proposed the Structural Monitoring Program to manage change in material properties and cracking of elastomer materials for "Flood Barrier," "HELB/MELB & EQ Civil/Structural Components," and "Seal, Gasket or Filler" component types exposed to a plant indoor air, containment air, or atmosphere/weather environment.

The staff reviewed the applicant's Structural Monitoring Program as documented in SER Section 3.0.3.2.10. The staff accepted the position that change in material properties and cracking exhibited by elastomers in a plant indoor air, containment air, or atmosphere/weather environment is adequately managed by the Structural Monitoring Program, which through visual examination inspects the elastomers for any sign of aging degradation. The staff's audit and review of plant-specific and industry operating experience determined that the aging effects of change in material properties and cracking for "Flood Barrier," "HELB/MELB & EQ Civil/Structural Components," and "Seal, Gasket or Filler" exposed to a plant indoor air, containment air or atmosphere/weather environment are adequately managed by the Structural Monitoring Program and determined that management of change in material properties and cracking of elastomer materials in miscellaneous structural and bulk commodities is acceptable.

As documented in the Audit and Review Report, the staff noted that LRA Table 3.5.2-8 shows for component type "Seal, Gasket or Filler - Auxiliary Building - Elastomer, Exposed," cracking as the only aging effect. The staff requested that the applicant explain why change in material properties is not another aging effect for this component as for identical component types in other buildings.

By letter dated August 27, 2005, the applicant stated that the AMR basis document indicates that change of material properties had been evaluated as an AERM with the Structural Monitoring Program credited with managing it and a standard Note J applied, but that the LRA should have included the change in material properties AERM with Note J. For this AMR line item, the subject AERM was added to LRA Table 3.5.2-8 with Note J.

The staff's review found the applicant's response acceptable as the applicant had appropriately addressed the aging effect/mechanism.

As documented in the Audit and Review Report, the staff noted that LRA Table 3.5.2-8 shows for component type "Seal Gasket or Filler - Auxiliary Building - Elastomer, Protected," the two aging effects change in material properties and cracking with GALL Report Item III.B4.2-a and Table 3.5.1, Item 3.5.1-29. GALL Report Item III.B4.2-a lists an aging effect of reduction or loss of isolation function with the component vibration isolation elements. As documented in the Audit and Review Report, the staff requested that the applicant explain why this AMR line item had a Note A (consistent with the GALL Report) when the component type and aging effects shown differed from the GALL Report line item. The intended function of vibration isolation also was not shown in the AMR line item.

By letter dated August 27, 2005, the applicant stated that the answer to this question was in plant-specific Note 593 included in the LRA Table 3.5.1 line item. Restated here: "Aging effect terminology used in the GALL Report for the Emergency Diesel Generators vibration isolation elements is slightly different, but overall deterioration is the same (e.g., cracking and change in material properties due to thermal exposure, etc.). Other elements included in this component (thermal expansion/seismic separation joint filler, gap or crack seal, etc.) are not addressed in the GALL." As to intended function, expansion/separation was considered applicable to vibration isolation. Additional clarification was required to explain the consistency between the GALL Report loss of vibration isolation aging effect and the evaluated aging effects that support the expansion/separation intended function summarized in the LRA. Additionally, better clarification was required to describe the other components included in the component type that did not align with GALL Report Item III.B.4a and required a Note J consistent with other elastomers in the table. A revision to Note 593 made the clarifications. The response included the addition of revised Note J to the line item in the table, in addition to Note A, for the other components in the group.

The staff's review found the applicant's response acceptable as the applicant appropriately addressed the aging effect/mechanism.

As documented in the Audit and Review Report, the staff noted that LRA Table 3.5.2-8 shows for component type "Fire Barrier-Auxiliary Building-Concrete, Protected," a GALL Report line item and a Table 1 item with Note H for the aging effect loss of material. The staff requested that the applicant explain why a GALL Report line item and a Table 1 item were shown with Note H.

By letter dated September 2, 2005, the applicant stated that in LRA Table 3.5.2-8 for the loss of material AERM of component type "Fire Barrier - Auxiliary Building - Concrete, Protected," the GALL Report, Volume 2, and Table 1 entries for the Fire Protection and Structural Monitoring Programs had been deleted.

The staff's review found the applicant's response acceptable as the applicant appropriately addressed the aging effect/mechanism.

As documented in the Audit and Review Report, the staff noted that LRA Table 3.5.2-8 shows for component type "Fire Barrier-Auxiliary Building - Fire Stop, Protected," a GALL Report line item and a Table 1 item with Note H for the aging effect loss of material. The staff requested that the applicant explain why a GALL Report line item and a Table 1 item were shown with Note H. This question also applied to LRA Table 3.5.2-8 for component type "Fire Barrier - Intake Structure Building - Fire Stop, Protected" for aging effect loss of material; to LRA Table 3.5.2-8 for component type "Fire Barrier - Turbine Building - Fire Stop, Protected" for aging effect loss of material; and to LRA Table 3.5.2-8 for component type "Fire Barrier - Water Treatment Building - Fire Stop, Protected" for aging effect loss of material.

By letter dated September 2, 2005, the applicant stated that in LRA Table 3.5.2-8 for the loss of material AERM of component type "Fire Barrier - Auxiliary Building - Fire Stop, Protected," the GALL Report Volume 2 and Table 1 entries for the Fire Protection Program had been deleted.

The same deletions of the GALL Report Volume 2 and Table 1 information were also made to the loss of material AERMs of the following additional line items of LRA Table 3.5.2-8, which had standard Note H:

- Page 3-366, "Fire Barrier - Intake Structure Bldg - Fire Stop, Protected"
- Page 3-368, "Fire Barrier - Turbine Bldg - Fire Stop, Protected"
- Page 3-370, "Fire Barrier - Water Treatment Bldg - Fire Stop, Protected"

The staff's review found the applicant's response acceptable as the applicant appropriately addressed the aging effect/mechanism.

The staff's review of LRA Table 3.5.2-8 identified an area in which additional information was necessary to complete the review of the applicant's results. The applicant responded to the staff's RAI as discussed below.

In RAI 3.3.2.7-1, dated August 31, 2005, the staff stated:

> LRA Table 3.5.2-8 referenced the Fire Protection Program, Section B2.1.10, as the aging management program addressing these barriers. However, Section B2.1.10, did not specifically address radiant energy shields, and refers to the Structural Monitoring Program, Section B2.1.19, for fire barriers, such as walls, floors, and ceilings. The staff requested that the applicant identify where the radiant energy shields referenced in the FHA are included in the AMR and which program manages their aging effects. The applicant was also requested to verify that the Table 3.5.2-8 should include a reference to the Structural Monitoring Program, Section B2.1.19, for these barriers or identify where in the Fire Protection Program, Section B2.1.10, they are addressed.

In its response, by letter dated September 16, 2005, the applicant stated that the radiant energy shields in question are included in component types "Fire Barrier - Auxiliary Building - Carbon Steel, Protected," "Fire Barrier - Intake Structure - Carbon Steel, Protected," and "Fire Barrier - Containment Building - Carbon Steel, Protected" as indicated in LRA Table 3.5.2-8, pages 3-362 and 3-365. They include the radiant energy shield between the diesel fire pumps in the intake structure, and radiant energy shields in cable trays in the auxiliary building and in containment. They are age managed by the Fire Protection Program as indicated in Table 3.5.2-8. In LRA Section B2.1.10, on page B-71, among others, fire barrier inspections are described as being part of the Fire Protection Program. The radiant energy shields are included in the group of fire barriers subject to inspection by this program.

Based on the above discussion, the staff finds the applicant's response to RAI 3.3.2.7-1 acceptable. The reference for the radiant energy shield in the FHA was clarified. Also, fire barrier inspections were clarified as being part of the Fire Protection Program. Therefore, the staff's concern described in RAI 3.3.2.7-1 is resolved.

On the basis of its review, as discussed above, the staff concludes that the applicant has demonstrated that the aging effects associated with the components will be adequately managed so that the intended function(s) will be maintained consistent with the CLB for the period of extended operation, as required by 10 CFR 54.21(a)(3).

3.5.2.3.9 Structures and Component Supports – Switchyard and Yard Structures – Summary of Aging Management Evaluation – LRA Table 3.5.2-9

The staff reviewed LRA Table 3.5.2-9, which summarizes the results of AMR evaluations for the switchyard and yard structures component groups. The staff determined that all AMR evaluation results in LRA Table 3.5.2-9 are consistent with the GALL Report.

On the basis of its review, as discussed above, the staff concludes that the applicant has demonstrated that the aging effects associated with the components will be adequately managed so that the intended function(s) will be maintained consistent with the CLB for the period of extended operation, as required by 10 CFR 54.21(a)(3).

3.5.2.3.10 Structures and Component Supports – Turbine Building – Summary of Aging Management Evaluation – LRA Table 3.5.2-10

The staff reviewed LRA Table 3.5.2-10, which summarizes the results of AMR evaluations for the turbine building component groups. The staff determined that all AMR evaluation results in LRA Table 3.5.2-10 are consistent with the GALL Report or, if not consistent, previously discussed in SER Section 3.5.2.1.

The staff reviewed the AMR line items for items not consistent with the GALL Report and identified loss of leak tightness for control room vestibule door, carbon steel components exposed to plant indoor air. The applicant stated that aging effects are manage with the Structural Monitoring Program. The staff review concluded that the Structural Monitoring Program is an acceptable program to monitor the aging effects of the control room vestibule door exposed to plant indoor air.

On the basis of its review, as discussed above, the staff concludes that the applicant has demonstrated that the aging effects associated with the components will be adequately managed so that the intended function(s) will be maintained consistent with the CLB for the period of extended operation, as required by 10 CFR 54.21(a)(3).

Conclusion. On the basis of its review, the staff finds that the applicant has appropriately evaluated AMR results involving material, environment, AERMs, and AMP combinations not evaluated in the GALL Report. The staff finds that the applicant has demonstrated that the aging effects will be adequately managed so that the intended functions will be maintained consistent with the CLB for the period of extended operation, as required by 10 CFR 54.21(a)(3).

3.5.3 Conclusion

The staff concludes that the applicant has demonstrated that the aging effects associated with the containments, structures, and component supports components will be adequately managed so that the intended function(s) will be maintained consistent with the CLB for the period of extended operation, as required by 10 CFR 54.21(a)(3).

The staff also reviewed the applicable FSAR supplement program summaries and concludes that they adequately describe the AMPs credited for managing aging of the containments, structures, and component supports, as required by 10 CFR 54.21(d).

3.6 Aging Management of Electrical and Instrumentation and Controls

This section of the SER documents the staff's review of the applicant's AMR results for the electrical and I&C systems components and component groups of the electrical commodity groups.

3.6.1 Summary of Technical Information in the Application

In LRA Section 3.6, the applicant provided AMR results for the electrical and I&C systems components and component groups. In LRA Table 3.6.1, "Summary of Aging Management Evaluations in Chapter VI of NUREG-1801 for Electrical Components," the applicant provided a summary comparison of its AMRs to those evaluated in the GALL Report for electrical and I&C systems components and component groups.

The applicant's AMRs incorporated applicable operating experience in the determination of AERMs. These reviews included evaluation of plant-specific and industry operating experience. The plant-specific evaluation included reviews of condition reports and discussions with appropriate site personnel to identify AERMs. The applicant's review of industry operating experience included a review of the GALL Report and issues identified since the issuance of the GALL Report.

3.6.2 Staff Evaluation

The staff reviewed LRA Section 3.6 to determine whether the applicant had provided sufficient information to demonstrate that the aging effects for the electrical and I&C systems components within the scope of license renewal and subject to an AMR will be adequately managed so that the intended function(s) will be maintained consistent with the CLB for the period of extended operation, as required by 10 CFR 54.21(a)(3).

The staff conducted an onsite audit of AMRs to confirm the applicant's claim that certain identified AMRs were consistent with the GALL Report. The staff did not repeat its review of the matters described in the GALL Report; however, the staff did verify that the material presented in the LRA was applicable and that the applicant had identified the appropriate GALL Report AMRs. The staff's evaluations of the AMPs are documented in SER Section 3.0.3. Details of the staff's audit evaluation are documented in the Audit and Review Report and summarized in SER Section 3.6.2.1.

In the onsite audit, the staff also selected AMRs consistent with the GALL Report for which further evaluation is recommended. The staff confirmed that the applicant's further evaluations were consistent with the acceptance criteria in SRP-LR Section 3.6.2.2. The staff's audit evaluations are documented in the Audit and Review Report and summarized in SER Section 3.6.2.2.

In the onsite audit, the staff also conducted a technical review of the those remaining AMRs not consistent with or not addressed in the GALL Report. The audit and technical review evaluated whether all plausible aging effects were identified and whether the aging effects listed were appropriate for the combination of materials and environments specified. The staff's audit evaluations documented in the Audit and Review Report are summarized in SER Section 3.6.2.3. The staff's evaluation of its technical review is also documented in SER Section 3.6.2.3.

Finally, the staff reviewed the AMP summary descriptions in the FSAR supplement to ensure that they adequately describe the programs credited with managing or monitoring electrical and I&C systems component aging.

SER Table 3.6-1 below summarizes the staff's evaluation of components, aging effects/mechanisms, and AMPs listed in LRA Section 3.6 and addressed in the GALL Report.

Table 3.6-1 Staff Evaluation for Electrical and Instrumentation and Controls in the GALL Report

Component Group	Aging Effect/ Mechanism	AMP in GALL Report	AMP in LRA	Staff Evaluation
Electrical equipment subject to 10 CFR 50.49 environmental qualification (EQ) requirements (Item Number 3.6.1-01)	Degradation due to various aging mechanisms	Environmental qualification of electric components	TLAA	This TLAA is evaluated in SER Section 4.4, "Environmental Qualification of Electrical Equipment"
Electrical cables and connections not subject to 10 CFR 50.49 EQ requirements (Item Number 3.6.1-02)	Embrittlement, cracking, melting, discoloration, swelling, or loss of dielectric strength leading to reduced insulation resistance (IR); electrical failure caused by thermal/ thermoxidative degradation of organics; radiolysis and photolysis [ultraviolet (UV) sensitive materials only] of organics; radiation-induced oxidation; moisture intrusion	Aging management program for electrical cables and connections not subject to 10 CFR 50.49 EQ requirements	Non-EQ Electrical Commodities Condition Monitoring Program (B2.1.12)	Consistent with GALL, which recommends no further evaluation (See SER Section 3.6.2.1)
Electrical cables used in instrumentation circuits not subject to 10 CFR 50.49 EQ requirements that are sensitive to reduction in conductor insulation resistance (IR) (Item Number 3.6.1-03)	Embrittlement, cracking, melting, discoloration, swelling, or loss of dielectric strength leading to reduced IR; electrical failure caused by thermal/ thermoxidative degradation of organics; radiation-induced oxidation; moisture intrusion	Aging management program for electrical cables used in instrumentation circuits not subject to 10 CFR 50.49 EQ requirements	Non-EQ Electrical Commodities Condition Monitoring Program (B2.1.12)	Consistent with GALL, which recommends no further evaluation (See SER Section 3.6.2.1)

Component Group	Aging Effect/ Mechanism	AMP in GALL Report	AMP in LRA	Staff Evaluation
Inaccessible medium-voltage (2K VAC to 15K VAC) cables (e.g., installed in conduit or direct buried) not subject to 10 CFR 50.49 EQ requirements (Item Number 3.6.1-04)	Formation of water trees; localized damage leading to electrical failure (breakdown of insulation); water tress caused by moisture intrusion	Aging management program for inaccessible medium-voltage cables not subject to 10 CFR 50.49 EQ requirements	Non-EQ Electrical Commodities Condition Monitoring Program (B2.1.12)	Consistent with GALL, which recommends no further evaluation (See SER Section 3.6.2.1)
Electrical connectors not subject to 10 CFR 50.49 EQ requirements that are exposed to borated water leakage (Item Number 3.6.1-05)	Corrosion of connector contact surfaces caused by intrusion of borated water	Boric acid corrosion	Boric Acid Corrosion Program (B2.1.4)	Consistent with GALL, which recommends no further evaluation (See SER Section 3.6.2.1)

The staff's review of the electrical and I&C systems component groups followed several approaches. One approach, documented in SER Section 3.6.2.1, discusses the staff's review of AMR results for components the applicant indicated are consistent with the GALL Report and require no further evaluation. Another approach, documented in SER Section 3.6.2.2, discusses the staff's review of AMR results for components the applicant indicated are consistent with the GALL Report and for which further evaluation is recommended. A third approach, documented in SER Section 3.6.2.3, discusses the staff's review of AMR results for components the applicant indicated are not consistent with or not addressed in the GALL Report. The staff's review of AMPs credited to manage or monitor aging effects of the electrical and I&C systems components is documented in SER Section 3.0.3.

3.6.2.1 AMR Results That Are Consistent with the GALL Report

Summary of Technical Information in the Application. In LRA Section 3.6.2.1, the applicant identified the materials, environments, and AERMs. The applicant identified the following programs that manage the aging effects of the electrical and I&C systems components:

- Boric Acid Corrosion Program
- Non-EQ Electrical Commodities Condition Monitoring Program

Staff Evaluation. In LRA Table 3.6.2-1, the applicant summarized AMRs for the electrical and I&C systems components and identified which AMRs it considered consistent with the GALL Report.

For component groups evaluated in the GALL Report for which the applicant had claimed consistency and for which the GALL Report does not recommend further evaluation, the staff performed an audit and review to determine whether the plant-specific components in these GALL Report component groups were bounded by the GALL Report evaluation.

The applicant provided a note for each AMR line item. The notes describe how the information in the tables aligns with the information in the GALL Report. The staff audited those AMRs with Notes A through E, which indicate how the AMR was consistent with the GALL Report.

Note A indicates that the AMR line item is consistent with the GALL Report for component, material, environment, and aging effect. In addition, the AMP is consistent with the GALL Report AMP. The staff audited these line items to verify consistency with the GALL Report and the validity of the AMR for the site-specific conditions.

Note B indicates that the AMR line item is consistent with the GALL Report for component, material, environment, and aging effect. In addition, the AMP takes some exceptions to the AMP identified in the GALL Report. The staff audited these line items to verify consistency with the GALL Report. The staff verified that the identified exceptions to the GALL Report AMPs had been reviewed and accepted by the staff. The staff also determined whether the AMP identified by the applicant was consistent with the AMP identified in the GALL Report and whether the AMR was valid for the site-specific conditions.

Note C indicates that the component for the AMR line item, although different from, is consistent with the GALL Report for material, environment, and aging effect. In addition, the AMP is consistent with the AMP identified by the GALL Report. This note indicates that the applicant was unable to find a listing of some system components in the GALL Report; however, the applicant identified a different component in the GALL Report that had the same material, environment, aging effect, and AMP as the component under review. The staff audited these line items to verify consistency with the GALL Report. The staff also determined whether the AMR line item of the different component applied to the component under review and whether the AMR was valid for the site-specific conditions.

Note D indicates that the component for the AMR line item, although different from, is consistent with the GALL Report for material, environment, and aging effect. In addition, the AMP takes some exceptions to the AMP identified in the GALL Report. The staff audited these line items to verify consistency with the GALL Report. The staff verified whether the AMR line item of the different component was applicable to the component under review. The staff verified whether the exceptions to the GALL Report AMPs had been reviewed and accepted by the staff. The staff also determined whether the AMP identified by the applicant was consistent with the AMP identified in the GALL Report and whether the AMR was valid for the site-specific conditions.

Note E indicates that the AMR line item is consistent with the GALL Report for material, environment, and aging effect, but a different AMP is credited. The staff audited these line items to verify consistency with the GALL Report. The staff also determined whether the identified AMP would manage the aging effect consistent with the AMP identified in the GALL Report and whether the AMR was valid for the site-specific conditions.

The staff's review of LRA Section 3.6 identified an area in which additional information was necessary to complete the review of the applicant's AMR results. The applicant responded to the staff's RAI as discussed below.

In RAI-3.6.2-1, dated June 3, 2005, the staff stated that various components described in the LRA Notes for Table 3.6.2-1 (and in notes for other tables) are not specified in the table. For example, it is not apparent whether neutron monitoring cables and uninsulated ground connectors are within the scope of license renewal and specified in the associated LRA tables. Therefore, the staff requested the applicant to clarify this information.

In its response, by letter dated July 1, 2005, the applicant stated:

> The referenced plant specific notes are associated with a particular line item in the Table 3.6.2-1. The plant specific note applicable to neutron monitoring cables is note 602 (on page 3-418 of the LRA), which is referenced in the second row of Table 3.6.2-1 on page 3-415. Neutron monitoring cables are in-scope and are included in the listed commodity, 'Electrical cables and connections used in instrumentation circuits not subject to 10 CFR 50.49 EQ requirements that are sensitive to reduction in conductor IR. (ISG-15) (Nuclear Instrumentation and Radiation monitoring systems).'

> Uninsulated ground conductors are not referenced in Table 3.6.2-1, as they are not in scope for license renewal. The Palisades plant uninsulated grounding cables are installed to provide personnel safety and economic equipment protection, and are not associated with supporting any system or component License Renewal intended function. This is described in the fifth paragraph of FSAR section 8.3.1.2 which states 'The 4,160 volt switchgear is provided with relay protection, grounding and the mechanical safeguards necessary to assure adequate personnel protection and to prevent or limit equipment damage during system fault conditions.'

Based on its review, the staff finds the applicant's response to RAI-3.6.2-1 acceptable because the applicant provided the requested clarification; therefore, the staff's concern described in RAI-3.6.2-1 is resolved.

The staff did not repeat its review of the matters described in the GALL Report; however, the staff did verify that the material presented in the LRA was applicable and that the applicant had identified the appropriate GALL Report AMRs. The staff's evaluation is discussed in the following section.

3.6.2.1.1 Radiolysis and Photolysis (Ultraviolet Sensitive Materials Only) of Organics

GALL Report, Volume 2, Item VI.A.1-a, identifies radiolysis and photolysis (ultraviolet (UV) sensitive materials only) of organics as aging effects/mechanisms of cable and connection insulations. As documented in the Audit and Review Report, the staff asked the applicant why those aging mechanisms were not applicable. By letter dated August 27, 2005, the applicant stated that they were applicable and would be added to LRA Table 3.6.2-1 for the component type "Electrical Cables and Connections not Subject to 10 CFR 50.49 EQ Requirements." The staff finds the applicant's response acceptable because it identified the aging effect consistent with the GALL Report.

GALL Report, Volume 2, Item VI.A.1-b, identifies radiolysis and photolysis (UV sensitive materials only) of organics as aging effects/mechanisms of instrumentation circuit cable and connection insulations sensitive to reduction in conductor insulation resistance. As documented in the Audit and Review Report, the staff asked why those aging mechanisms were not applicable. By letter dated August 27, 2005, the applicant stated that they were applicable and would be added to LRA Table 3.6.2-1 for the component type "Electrical Cables and Connections not Subject to 10 CFR 50.49 EQ Requirements Used in Instrumentation Circuits." The staff finds the applicant's response acceptable because the aging effects identified were consistent with the GALL Report.

Conclusion. The staff evaluated the applicant's claim of consistency with the GALL Report. The staff also reviewed information pertaining to the applicant's consideration of recent operating experience and proposals for managing the associated aging effects. On the basis of its review, the staff concludes that the AMR results, which the applicant claimed to be consistent with the GALL Report, are consistent with the GALL Report AMRs. Therefore, the staff concludes that the applicant has demonstrated that the aging effects for these components will be adequately managed so that their intended function(s) will be maintained consistent with the CLB for the period of extended operation, as required by 10 CFR 54.21(a)(3).

3.6.2.2 AMR Results That Are Consistent with the GALL Report, for Which Further Evaluation is Recommended

Summary of Technical Information in the Application. In LRA Section 3.6.2.2, the applicant provided further evaluation of aging management as recommended by the GALL Report for electrical and I&C systems components. The applicant provided information concerning how it will manage aging effects for electrical equipment subject to EQ.

Staff Evaluation. For component groups evaluated in the GALL Report for which the applicant had claimed consistency with the GALL Report and for which the GALL Report recommends further evaluation, the staff audited and reviewed the applicant's evaluations to determine whether they adequately address those issues. In addition, the staff reviewed the applicant's further evaluations against the criteria in SRP-LR Section 3.6.2.2. Details of the staff's audit are documented in the Audit and Review Report. The staff's evaluation of the aging effects is discussed in the following sections.

3.6.2.2.1 Electrical Equipment Subject to Environmental Qualification

In LRA Section 3.6.2.2.1, the applicant stated that EQ is a TLAA, as defined in 10 CFR 54.3. Applicants must evaluate TLAAs in accordance with 10 CFR 54.21(c)(1). SER Section 4.4 documents the staff's review of the applicant's evaluation of this TLAA.

3.6.2.2.2 Quality Assurance for Aging Management of Nonsafety-Related Components

SER Section 3.0.4 provides the staff's evaluation of the applicant's QA program.

Conclusion. On the basis of its review, for component groups evaluated in the GALL Report for which the applicant claimed consistency with the GALL Report and for which further evaluation is recommended, the staff determined that the applicant has adequately addressed the issues that were further evaluated. The staff finds that the applicant has demonstrated that the aging effects will be adequately managed so that the intended function(s) will be maintained consistent with the CLB for the period of extended operation, as required by 10 CFR 54.21(a)(3).

3.6.2.3 AMR Results That Are Not Consistent with or Not Addressed in the GALL Report

Summary of Technical Information in the Application. In LRA Table 3.6.2-1, the staff reviewed additional details of AMR results for material, environment, AERM, and AMP combinations not consistent with or not addressed in the GALL Report.

In LRA Table 3.6.2-1, the applicant indicated, via Notes F through J, that the combination of component type, material, environment, and AERM does not correspond to a line item in the GALL Report. The applicant provided further information concerning how the aging effects will be managed. Specifically, Note F indicates that the material for the AMR line item component is not evaluated in the GALL Report. Note G indicates that the environment for the AMR line item component and material is not evaluated in the GALL Report. Note H indicates that the aging effect for the AMR line item component, material, and environment combination is not evaluated in the GALL Report. Note I indicates that the aging effect identified in the GALL Report for the line item component, material, and environment combination is not applicable. Note J indicates that neither the component nor the material and environment combination for the line item is evaluated in the GALL Report.

Staff Evaluation. For component type, material, and environment combinations not evaluated in the GALL Report, the staff reviewed the applicant's evaluation to determine whether the applicant had demonstrated that the aging effects will be adequately managed so that the intended function(s) will be maintained consistent with the CLB during the period of extended operation. The staff's evaluation is discussed in the following section.

3.6.2.3.1 Electrical Components – Electrical Commodity Groups – Summary of Aging Management Evaluation – LRA Table 3.6.2-1

The staff reviewed LRA Table 3.6.2-1, which summarizes the results of AMR evaluations for the electrical commodity groups component types. The staff's evaluation of each of the component types follows.

Fuse Holder. LRA Table 3.6.2-1 identifies AMR results for line items where no AMPs were proposed even though aging effects were identified. Specifically, instances in which the applicant stated that no aging effects had been identified occurred for components fabricated from phenolic, copper, and aluminum materials exposed to adverse localized environments of heat, radiation, or moisture in the presence of oxygen.

ISG-5, "Interim Staff Guidance on the Identification and Treatment of Electrical Fuse Holders for License Renewal," identifies aging effects of metallic portions of fuse holders including fatigue/ohmic heating, thermal cycling, electrical transients, frequent manipulation, vibration, chemical contamination, corrosion, and oxidation. Fuse holders have an indoor air adverse localized environment. During the audit, the staff requested that the applicant identify environments and aging effects for fuse holders and propose an AMP with the ten program elements or justify why no AMP is required for fuse holders. By letter dated August 27, 2005, the applicant stated that the aging effects of fatigue/ohmic heating, thermal cycling, electrical transients, frequent manipulation, vibration, chemical contamination, corrosion, and oxidation and the indoor air adverse localized environment were applicable and would be added to LRA Table 3.6.2-1 for fuse holders.

By letter dated August 27, 2005, the applicant stated that of the fuse holders not inside active equipment, 36 fuses were installed in junction boxes and that from this number, 12 bolted fuse holders (installed in 1981) are cycled once per refueling outage and not susceptible to the relaxation or fatigue experienced by fuse clips, the other 24 fuses with clips are not cycled with any frequency or routinely removed for maintenance and/or surveillance. Therefore, the applicant did not consider fatigue due to mechanical stress an AERM.

In the August 27, 2005, letter, the applicant further stated that all of the fuse holders are installed in metal junction boxes seismically mounted on their support structure, separate from sources of vibration. Therefore, the applicant did not consider vibration an applicable aging mechanism.

The applicant also stated, in the August 27, 2005, letter, that the junction boxes are located inside rooms with a controlled environment that protects the panels from the weather and that no sources of potential mechanical system leakage are located in proximity to the junction boxes. Therefore, the applicant did not consider corrosion an applicable aging mechanism.

In addition, the applicant stated in the August 27, 2005, letter, that as to internal moisture, plant-specific operating experience revealed no aging from the formation of condensation in the panels. All the junction boxes were inspected and the surface condition of the fuse clips showed no signs of corrosion or moisture. Therefore, the applicant did not consider corrosion an applicable aging mechanism.

In conclusion, the applicant, in its August 27, 2005, letter, indicated that it had found no fatigue/ohmic heating, thermal cycling, frequent manipulation, vibration, chemical contamination, corrosion, and oxidation AERMs for fuse holders within the scope of license renewal.

For aging effects of fatigue/ohmic heating, thermal cycling, electrical transients, frequent manipulation, vibration, chemical contamination, corrosion, and oxidation, the staff reviewed the applicant's finding that they do not occur at PNP and concludes that no AMP was needed and that the applicant had addressed the aging effects of fuse holders, as identified in ISG-5.

On the basis of its review, the staff found that air indoor/adverse localized environment on copper and aluminum would not cause aging of concern during the period of extended operation. The applicant adequately addressed why aging effects of metallic portions of fuse holders, as identified in ISG-5, are not applicable. Therefore, the staff concludes that there are no applicable AERMs for metallic portions of fuse holder components exposed to indoor air or adverse localized environments.

Non-Segregated Phase Bus and Connections. LRA Table 3.6.2-1 identifies AMR results for line items where no AMPs were proposed even though aging effects were identified. The applicant identified various metals and porcelain as construction materials of the non-segregated phase bus and connections. The adverse localized environment is caused by heat, radiation, or moisture in the presence of oxygen. The aging effects are oxidation, loosening of bolted connections due to thermal cycling, and corrosion due to moisture to be managed by the applicant's Non-EQ Electrical Commodities Condition Monitoring Program.

As documented in the Audit and Review Report, the staff observed that bus/connections, insulation/insulators, and enclosure assemblies are the SCs of the metal enclosed bus. The bus material is various metals, porcelain, xenon, and thermo-plastic organic polymers. The metal-enclosed bus environment is indoor and outdoor air. AERMs include embrittlement, cracking, melting, discoloration, swelling, or loss of dielectric strength leading to reduced insulation resistance, loss of material/general corrosion, loosening of bolted connections, thermal cycling and ohmic heating, hardening, and loss of strength/elastomer degradation. Therefore, the staff asked the applicant for appropriate material, environment, aging effects, and AMPs for each SC of the metal-enclosed bus.

By letter dated August 27, 2005, the applicant stated the aging effects (embrittlement, cracking, melting, discoloration, swelling, or loss of dielectric strength leading to reduced insulation resistance, loss of material/general corrosion, loosening of bolted connections, thermal cycling and ohmic heating, hardening, and loss of strength/elastomer degradation) and materials (various metals, porcelain, and glass). The applicant stated that PNP has no xenon or thermo-plastic organic polymers and that environment (indoor and outdoor air) would be added to LRA Table 3.6.2-1 for the component type "Non-Segregated Phase Bus and Connections." In the component type column below "Non-Segregated Phase Bus and Connections," "Bus/Connections, Insulation/Insulators, and Enclosure Assembly" would be added. The staff's review found the applicant's response acceptable because it identified all materials and aging effects of metal-enclosed bus.

In LRA Table 3.6.2-1, the applicant proposed its Non-EQ Electrical Commodities Condition Monitoring Program to manage embrittlement, cracking, melting, discoloration, swelling, or loss of dielectric strength leading to reduced insulation resistance, loss of material/general corrosion, loosening of bolted connections due to thermal cycling and ohmic heating of various metals, porcelain, and glass for non-segregated phase bus and connections exposed to indoor and outdoor air environments.

The staff reviewed the applicant's Non-EQ Electrical Commodities Condition Monitoring Program as documented in SER Section 3.0.3.1.7. This new program manages aging in selected non-EQ commodity groups within the scope of license renewal. Program activities are responsive to the staff guidance of the GALL Report and to industry standards. The staff's review of plant-specific and industry operating experience found the aging effects of embrittlement, cracking, melting, discoloration, swelling, or loss of dielectric strength leading to reduced insulation resistance, loss of material/general corrosion, loosening of bolted connections due to thermal cycling and ohmic heating of various metals, porcelain, and glass materials exposed to indoor and outdoor air environments adequately managed by the Non-EQ Electrical Commodities Condition Monitoring Program. On this basis, the staff finds management of the aging effects acceptable.

High-Voltage Transmission Conductors. LRA Table 3.6.2-1 identifies AMR results for line items where no AMPs were proposed even though aging effects were identified. Specifically, the applicant stated that no aging effects were identified for components fabricated from aluminum and steel materials exposed to atmosphere and weather environments. The applicant identified loss of conductor strength and vibration as aging effects of high-voltage transmission, but stated that no AMP was required.

As documented in the Audit and Review Report, the staff observed that the most prevalent mechanism contributing to loss of high-voltage transmission conductor strength is corrosion, including corrosion of steel core and aluminum strand pitting, and that other aging effects include loss of material/wind-induced abrasion and fatigue, increased resistance of

3-305

connection/oxidation, or loss of preload. The staff requested that the applicant explain why transmission conductors require no AMP. In the LRA, the applicant stated that routine switchyard inspection detects loose connections and it appeared to have a plant procedure to manage loose connections, but not credited with managing aging effects for high-voltage connections. The staff requested that the applicant explain why this procedure is not credited with managing the aging effects for high-voltage connections.

By letter dated August 27, 2005, the applicant stated that the component type would be changed to "High-Voltage Transmission Conductors and Connections" and the aging effects (loss of material/wind-induced abrasion and fatigue, loss of conductor strength/corrosion, increased resistance of connection/oxidation or loss of preload) would be added to LRA Table 3.6.2-1 for the high-voltage transmission conductors. The plant experience in the LRA was a site-specific experience in a corrective action document. The program basis document indicates no other site-specific experience with any other problems of high-voltage connections. The corrective action document, with a discovery time of August 7, 2001, noted that there was a loose connection in the switchyard disconnect for 29R8 on the east side of the Z phase. A review of the work order history for disconnect 29R8 determined that the contact on the east side of the Z phase for 29R8 was worked on March 31, 2001. Therefore, the problem noted was due to poor workmanship and not to any aging mechanisms.

In the August 27, 2005, letter, the applicant also stated:

> The Palisades transmission conductor component type includes both the transmission conductors and the hardware used to secure the conductors to the insulators. The materials for aluminum cable [conductor]-steel reinforced (ACSR) transmission conductors are aluminum and steel, and the environment is outdoor weather. Based on industry guidance, potential aging effects and aging mechanisms are loss of conductor strength due to general corrosion (atmospheric oxidation of metals) and loss of material due to wear from wind loading.

> Corrosion in ACSR conductors is a very slow acting mechanism. Corrosion rates are dependent on air quality. Palisades is located in a mostly agricultural area with no significant nearby industries that could contribute to corrosive air quality. Corrosion testing of transmission conductors at Ontario Hydroelectric showed a 30 percent loss of composite conductor strength of an 80-year-old ACSR conductor. The Institute of Electrical and Electronic Engineers National Electrical Safety Code (NESC) requires that tension on installed conductors be a maximum of 60% of the ultimate conductor strength. Therefore, assuming a 30% loss of strength, there would still be significant margin between what is required by the NESC and the actual conductor strength. In determining actual conductor tension, the NESC considers various loads imposed by ice, wind, and temperature as well as length of conductor span. The transmission conductors in scope for license renewal are short spans located within the high voltage switchyard. The Palisades line near the plant is designed for heavy loading; therefore, the Ontario Hydroelectric heavy loading zone study is aligned with respect to loads imposed by weather conditions.

The Ontario Hydroelectric test envelops the conductors at Palisades, demonstrating that the material loss on the Palisades ACSR transmission conductors is acceptable for the period of extended operation. This illustrates with reasonable assurance that transmission conductors at Palisades will have ample strength to perform their intended function throughout the renewal term; therefore, loss of conductor strength due to corrosion of the transmission conductors is not an aging effect requiring management.

Loss of material due to mechanical wear can be an aging effect for strain and suspension insulators that are subject to movement. Experience has shown that transmission conductors do not normally swing and that when they do swing because of substantial wind, they do not continue to swing for very long once the wind has subsided. Wear has not been identified during routine inspection. Therefore, loss of material due to wear is not an aging effect requiring management for transmission conductors.

NMC reviewed industry operating experience and NRC generic communications related to the aging of transmission conductors in order to ensure that no additional aging effects exist beyond those identified above. NMC also reviewed plant-specific operating experience, including nonconformance reports, licensee event reports, and condition reports, and documented interviews with transmission engineering personnel. This review did not identify unique aging effects for transmission conductors beyond those identified above.

In conclusion, no aging management program is required for the Palisades transmission conductors and connections aging effects of loss of conductor strength and loss of material (mechanical wear).

For the aging effects of loss of conductor strength and vibration, the staff reviewed the applicant's basis for concluding that these aging effects do not occur at PNP and concludes that no AMP was needed.

The staff's review of current industry research and operating experience found that atmosphere and weather environments on aluminum and steel would not cause significant aging of concern during the period of extended operation. The applicant adequately addressed why aging effects of loss of material/wind-induced abrasion and fatigue, increased resistance of connection/oxidation, or loss of preload are not significant. Therefore, the staff concludes that there are no AERMs for high-voltage transmission conductor and connection components exposed to atmosphere and weather environments.

High Voltage Switchyard Bus and Connections. LRA Table 3.6.2-1 identifies AMR results for line items where no AMPs were proposed even though aging effects were identified. Specifically, the applicant stated that no aging effects were identified for components fabricated from aluminum, stainless steel (bolting), and copper materials exposed to atmosphere and weather environments. The applicant identified connection surface oxidation and vibration as aging effects of high-voltage switchyard bus and connections, but stated that no AMP was required.

As documented in the Audit and Review Report, the staff noted that the LRA states that routine inspections detect loose connections in the switchyard. The applicant appeared to have a plant procedure to manage switchyard bus loose connections. Therefore, the staff requested that the applicant explain why this procedure was not credited with managing the aging effects for high-voltage connections.

By letter dated August 27, 2005, the applicant stated:

> The plant experience documented in the Palisades LRA on page B-95 was based on the fact that there was a corrective action document that documented a site-specific experience. Note: there was no other site-specific experience to document other problems with high-voltage connections at Palisades. This document, with a time of discovery of August 7, 2001, noted that there was a loose connection in the switchyard disconnect for 29R8 on the east side of the Z phase. A review of the work order history, for disconnect 29R8, determined that the contact on the east side of the Z phase for 29R8 was worked on March 31, 2001. Therefore, the problem noted was due to poor workmanship from the work performed by work order and not due to any aging mechanisms.
>
> As stated in Table 3.6.2-1 of the LRA the switchyard bus and connections subject to an AMR (1) are constructed of aluminum, copper, and stainless steel (bolting), (2) are exposed to an atmosphere/ weather (same as Air- Outdoor) environment consisting of temperatures up to 40 deg. C (105 deg. F), precipitation, and negligible radiation, (3) provide electrical connections to specific sections of an electrical circuit to deliver voltage, current or signals, and (4) require no AMP. There are no aging effects from the outdoor environment (consisting of temperatures up to 40 deg. C (105 deg. F) and precipitation) that would cause the loss of the capability to provide electrical connections to specified sections of an electrical circuit to deliver voltage, current, or signals.
>
> In conclusion Palisades determined that an environment consisting of temperatures up to 40 deg. C (105 deg F) and precipitation has no significant aging effect on aluminum, copper, and stainless steel from (the component parts from which the switchyard bus and connections are constructed. Therefore, no AMP is required for High-Voltage Switchyard Bus and Connections.

The staff's review of the applicant's response in the August 25, 2005, letter concluded that the site-specific experience was the result of poor workmanship and not of aging mechanisms. Also, the staff reviewed the applicant's basis for concluding that the aging effects of connection surface oxidation and vibration do not exist at PNP and concludes that no AMP is needed.

The staff's review of current industry research and operating experience found that atmosphere and weather environments on aluminum, stainless steel, and copper would not cause significant aging of concern during the period of extended operation. Therefore, the staff concludes that there are no AERMs for high-voltage switchyard bus and connections exposed to atmosphere and weather environments.

High-Voltage Insulators. LRA Table 3.6.2-1 identifies AMR results for line items where no AMPs were proposed even though aging effects were identified. Specifically, the applicant stated that no aging effects were identified for components fabricated from porcelain, cement, and metal materials exposed to atmosphere and weather environments. The applicant identified surface contamination and cracking as aging effects of high-voltage insulators, but stated that no AMP was required.

As documented in the Audit and Review Report, the staff noted that various airborne materials like dust, salt, and industrial effluents can contaminate insulator surfaces. A large buildup of contamination enables the conductor voltage to track along the surface more easily and can lead to insulator flashover. Surface contamination can be a problem in areas with greater concentrations of airborne particles as near facilities that discharge soot or near the seacoast where salt spray is prevalent. Cracks have been known to occur with insulators when the cement that binds the parts together expands enough to crack the porcelain. Mechanical wear is another aging effect for strain and suspension insulators subject to movement from wind causing the supported transmission conductor to swing from side to side. If frequent enough, such swinging could wear the metal contact points of the insulator string and between the insulator and supporting hardware. Therefore, staff requested that the applicant identify all aging effects of high-voltage insulators and explain why no AMP is required.

By letter dated August 27, 2005, the applicant stated:

> These aging effects (loss of material/mechanical wear due to wind blowing on transmission conductors) are hereby added to the Palisades Table 3.6.2-1 for the High-Voltage Insulators, on LRA page 3-417. A conforming change is also made to Section 3.6.2.1.9 on page 3-408 under Aging Effects Requiring Management.

> The high-voltage insulators (including high voltage strain and suspension insulators), that perform the function of insulating and supporting electrical transmission conductors and are subject to an AMR, (1) are constructed of porcelain, galvanized metal, and cement, (2) are exposed to an outdoor weather environment consisting of temperatures up to 40 deg. C (105 deg. F), precipitation, and negligible radiation, (3) insulate and support an electrical conductor, and (4) require no AMP. NMC did not identify any aging effects from the outside environment (consisting of temperatures up to 40 deg. C (105 deg. F) and precipitation) that would cause the loss of the capability to insulate or support its associated electrical conductor.

> Regarding the potential for contamination of insulators, the buildup of surface contamination is gradual and in most areas such contamination is washed away by rain. Surface contamination can be a problem in areas where there are high concentrations of airborne particles, such as near facilities that discharge soot, or near the seacoast where salt spray is prevalent. Palisades is located in an area with moderate rainfall where airborne particle concentrations are comparatively low; consequently, the rate of contamination buildup on the insulators is not significant. At Palisades, as in most areas of the Michigan transmission system, contamination build-up on insulators is not a problem due to rainfall periodically 'washing' the insulators. The glazed insulator surface aids this contamination removal. Additionally, there is no nearby heavy industry or other producers of industrial effluents, which could cause excessive contamination. There is no salt spray at Palisades as the plant is far from any ocean. Therefore, surface contamination is not an applicable aging effect for the insulators in the service conditions they are exposed to at Palisades.

Regarding high voltage porcelain insulator cracking, porcelain is essentially a hardened, opaque glass. As with any glass, if subjected to enough force, it will crack or break. The most common cause for cracking or breaking of an insulator is being struck by an object (e.g., a rock or bullet). Cracking and breaking caused by physical damage is not an aging effect and is not subject to an AMR. Cracks have been known to occur with insulators when the cement that binds the parts together expands enough to crack the porcelain. This phenomenon, known as cement growth, occurs mainly because of improper manufacturing processes or materials, which make the cement more susceptible to moisture penetration, and the specific design and application of the insulator. The string insulators which have experienced porcelain cracking caused by cement growth are isolated to bad batches (specific, known brands and manufacture dates) of string insulators used in strain application. The post insulators most susceptible to this aging effect are multicone (post) insulators used in cantilever applications. Research of Palisades corrective action documents revealed no instance of insulator cracking or failure related to cement growth in the Palisades switchyard. Accordingly, cracking due to cement growth is not an applicable aging effect for the high voltage insulators in the service conditions they are exposed to at Palisades.

Regarding mechanical wear, this is an aging effect for strain and suspension insulators in that they are subject to movement. Movement of the insulators can be caused by wind blowing the supported transmission conductor, causing it to swing from side to side. If this swinging is frequent enough, it could cause wear in the metal contact points of the insulator string and between an insulator and the supporting hardware. Although this mechanism is possible, experience has shown that the transmission conductors do not normally swing and that when they do, due to a substantial wind, do not continue to swing for very long once the wind has subsided. Wind loading that can cause a transmission line and insulators to vibrate or sway is considered in the design and installation. The concern of wear due to loss of material will not cause a loss of intended function of the insulators at Palisades; therefore, loss of material due to wear is not an applicable aging effect for insulators.

Palisades operating experience was reviewed to validate aging effects for switchyard insulators. This review included corrective action documents for any documented instances of switchyard insulator aging, in addition to interviews with Palisades engineering and maintenance personnel. No instance of aging related problems with in-scope switchyard insulators due to contaminants, cracking, cement growth, or mechanical wear was uncovered.

For the aging effects of surface contamination, cracking, and mechanical wear, the staff reviewed the applicant's basis for concluding that these aging effects do not exist at PNP and concludes that no AMP is needed.

The staff's review of current industry research and operating experience found that atmosphere and weather environments would not cause significant aging of porcelain, cement, and metal of concern during the period of extended operation. The applicant adequately addressed why aging effects of loss of material/mechanical wear, surface contamination, and cracking are not applicable aging effects. Therefore, the staff concludes that there are no AERMs for high-voltage insulator components exposed to atmosphere and weather environments.

Cable Connections (Metallic Portions). LRA Table 3.6.2-1 does not address the aging effects of metallic cable connections. The aging effects of metallic cable connections include loosening of bolted connections due to thermal cycling, ohmic heating, electrical transient, vibrations, chemical contamination, corrosion, and oxidation. The environments are indoor and outdoor air. As documented in the Audit and Review Report, the staff asked the applicant for the environment and aging effects of cable connections and to propose an AMP or justify why no AMP is required for metallic cable connections.

By letter dated August 27, 2005, the applicant stated that the material (various metals used for electrical contacts), environment (air-indoor and outdoor) and aging effects (including loosening of bolted connections due to thermal cycling, ohmic heating, electrical transient, vibrations, chemical contamination, corrosion, and oxidation) would be added to LRA Table 3.6.2-1 (Page 3-414) for the component type "Electrical Cables and Connections not Subject to 10 CFR 50.49 EQ Requirement."

In addition, by letter dated August 25, 2005, the applicant added the following information to the Non-EQ Electrical Commodities Condition Monitoring Program:

> Parameters Monitored/Inspected: This program will focus on the metallic parts of the connection. The monitoring includes loosening of bolted connections due to thermal cycling, ohmic heating, electrical transients, vibration, chemical contamination, corrosion, and oxidation. A representative sample of electrical cable connections is tested. The following factors are considered for sampling: application (high, medium and low voltage), circuit loading, and location (high temperature, high humidity, vibration, etc.). The technical basis for the sample selected is to be documented.

> Detection of Aging Effects: Electrical connections within the scope of license renewal are tested at least once every 10 years. Testing may include thermography, contact resistance testing, or other appropriate testing methods justified in the application. This test frequency is adequate to preclude failures of the electrical connections since experience has shown that aging degradation of electrical connections is a slow process. A 10-year testing frequency will provide two data points during a 20-year period, which can be used to characterize the degradation rate. The first tests for license renewal are to be completed before the period of extended operation.

> Acceptance Criteria: The acceptance criteria for each connection test are defined by the specific type of test performed and the specific type of cable connections tested.

> Corrective Actions: Pursuant to 10 CFR Part 50, Appendix B, an engineering evaluation is performed when the test acceptance criteria are not met in order to ensure that the intended functions of the cable connections can be maintained consistent with the current licensing basis. Such an evaluation is to consider the significance of the test results, the operability of the component, the reportability

of the event, the extent of the concern, the potential root causes for not meeting the test acceptance criteria, the corrective action warranted, and the likelihood of recurrence. When an unacceptable condition or situation is identified, a determination is made on whether the same condition or situation is applicable to other within the scope of license renewal cable connections not tested.

The staff reviewed the applicant's Non-EQ Electrical Commodities Condition Monitoring Program as documented in SER Section 3.0.3.1.7. This new program manages aging in selected non-EQ commodity groups within the scope of license renewal. Program activities are responsive to the staff guidance of the GALL Report and to industry standards. The staff's review of plant-specific and industry operating experience found the aging effect of loosening of bolted connection due to thermal cycling, ohmic heating, electrical transient, vibrations, chemical contamination, corrosion, and oxidation of various metals in cable connections exposed to indoor and outdoor air environments adequately managed by the Non-EQ Electrical Commodities Condition Monitoring Program. As discussed in the appendix to GALL Report, Volume 2, the staff finds the requirements of 10 CFR Part 50, Appendix B, acceptable to address corrective actions. On this basis, the staff finds management of the aging effects acceptable.

Conclusion. On the basis of its review, the staff finds that the applicant has appropriately evaluated AMR results involving material, environment, AERMs, and AMP combinations not evaluated in the GALL Report. The staff finds that the applicant has demonstrated that the aging effects will be adequately managed so that the intended functions will be maintained consistent with the CLB for the period of extended operation, as required by 10 CFR 54.21(a)(3).

3.6.3 Conclusion

The staff concludes that the applicant has demonstrated that the aging effects associated with the electrical and I&C systems components will be adequately managed so that the intended function(s) will be maintained consistent with the CLB for the period of extended operation, as required by 10 CFR 54.21(a)(3).

The staff also reviewed the applicable FSAR supplement program summaries and concludes that they adequately describe the AMPs credited for managing aging of the electrical and I&C systems, as required by 10 CFR 54.21(d).

3.7 Conclusion for Aging Management Review Results

The staff reviewed the information in LRA Section 3, "Aging Management Review Results," and Appendix B, "Aging Management Programs." On the basis of its review of the AMR results and AMPs, the staff concludes that the applicant has demonstrated that the aging effects will be adequately managed so that the intended functions will be maintained consistent with the CLB for the period of extended operation, as required by 10 CFR 54.21(a)(3). The staff also reviewed the applicable FSAR supplement program summaries and concludes that the FSAR supplement adequately describes the AMPs credited for managing aging as required by 10 CFR 54.21(d).

With regard to these matters, the staff concludes that there is reasonable assurance that the activities authorized by the renewed license will continue to be conducted in accordance with the

CLB, and any changes made to the CLB, in order to comply with 10 CFR 54.21(a)(3), are in accordance with the NRC's regulations.

SECTION 4

TIME-LIMITED AGING ANALYSES

4.1 Identification of Time-Limited Aging Analyses

This section discusses the identification of time-limited aging analyses (TLAAs). Nuclear Management Company, LLC (NMC or the applicant) discussed the TLAAs in Sections 4.2 through 4.7 of its license renewal application (LRA). Sections 4.2 through 4.8 of this safety evaluation report (SER) document the review of the TLAAs, conducted by the staff of the U.S. Nuclear Regulatory Commission (NRC or the staff).

TLAAs are certain plant-specific safety analyses that involve time-limited assumptions defined by the current operating term. Pursuant to Title 10, Section 54.21(c)(1), of the *Code of Federal Regulations* (10 CFR 54.21(c)(1)), the applicant for license renewal must provide a list of TLAAs, as defined in 10 CFR 54.3.

In addition, pursuant to 10 CFR 54.21(c)(2), an applicant must provide a list of plant-specific exemptions granted under 10 CFR 50.12 that are based on TLAAs. For any such exemptions, the applicant must provide an evaluation that justifies the continuation of the exemptions for the period of extended operation.

4.1.1 Summary of Technical Information in the Application

To identify the TLAAs, the applicant evaluated calculations for the Palisades Nuclear Plant (PNP) against the six criteria specified in 10 CFR 54.3. The applicant indicated that it had identified the calculations that met the six criteria by searching the current licensing basis (CLB). The CLB includes the final safety analysis report (FSAR), engineering calculations, technical reports, engineering work requests, licensing correspondence, and applicable vendor reports. In LRA Table 4.1-1, "Time Limited Aging Analyses," the applicant listed the applicable TLAAs in the following categories:

- reactor vessel neutron embrittlement
- metal fatigue
- environmental qualification of electrical equipment
- concrete containment tendon prestress analysis
- containment liner plate and penetrations load cycles
- other plant-specific time-limited aging analyses

Pursuant to 10 CFR 54.21(c)(2), the applicant stated that it did not identify any exemptions granted under 10 CFR 50.12 that were based on a TLAA, as defined in 10 CFR 54.3.

4.1.2 Staff Evaluation

In LRA Section 4.1, the applicant identified the TLAAs applicable to PNP. The staff reviewed the information to determine if the applicant had provided adequate information to meet the requirements of 10 CFR 54.21(c)(1) and 10 CFR 54.21(c)(2).

As defined in 10 CFR 54.3, TLAAs are analyses that meet the following six criteria:

(1) involve systems, structures, and components that are within the scope of license renewal, as delineated in 10 CFR 54.4(a)

(2) consider the effects of aging

(3) involve time-limited assumptions defined by the current operating term, for example, 40 years

(4) are determined to be relevant by the applicant in making a safety determination

(5) involve conclusions, or provide the basis for conclusions, related to the capability of the system, structure, and component to perform its intended functions, as delineated in 10 CFR 54.4(b)

(6) are contained or incorporated by reference in the CLB

The applicant provided a list of common TLAAs from NUREG-1800, "Standard Review Plan for Review of License Renewal Applications for Nuclear Power Plants," (SRP-LR) dated July 2001. The applicant listed those TLAAs applicable to PNP in LRA Table 4.1-1, "Time Limited Aging Analyses."

As required by 10 CFR 54.21(c)(2), an applicant must list all exemptions granted under 10 CFR 50.12 based on a TLAA and evaluated and justified for continuation through the period of extended operation. In its LRA, the applicant stated that each active exemption was reviewed to determine if it was based on a TLAA. The applicant identified no TLAA-based exemptions. From the information provided by the applicant about the process used to identify TLAA-based exemptions, as well as the applicant's search results, the staff concludes that the applicant identified no TLAA-based exemptions justified for continuation through the period of extended operation in accordance with 10 CFR 54.21(c)(2).

4.1.3 Conclusion

On the basis of its review, the staff concludes that the applicant has provided an acceptable list of TLAAs, as required by 10 CFR 54.21(c)(1). The staff also confirmed that no exemptions under 10 CFR 50.12 have been granted on the basis of a TLAA, as required by 10 CFR 54.21(c)(2).

4.2 Reactor Vessel Neutron Embrittlement

The reactor pressure vessel (RPV) is subject to neutron irradiation from the core that results in embrittlement of reactor vessel materials. The NRC approved methodology used for neutron fluence calculations is consistent with Regulatory Guide (RG) 1.190 and described in Westinghouse Commercial Atomic Power (WCAP)-15353, "Palisades Reactor Pressure Vessel Fluence Evaluation." Analyses have addressed upper shelf energy (USE), pressurized thermal shock (PTS), primary coolant system (PCS) pressure-temperature (P-T) operating limits, and low temperature overpressure protection (LTOP) pressure-operated relief valve (PORV) setpoints. PNP began use of low leakage core designs when RG 1.99, Revision 2, "Radiation Embrittlement of Reactor Vessel Materials," was issued May 1988. Gradual improvements to the low leakage core design and in the analytical methods used to estimate the amount of fast fluence accumulated at the RPV beltline resulted in the current ultra low leakage core design approved by the NRC on November 14, 2000. This core design consists of third and fourth cycle assemblies loaded in peripheral locations with specially designed shield assemblies placed to reduce flux at the limiting RPV beltline axial welds. The limiting welds are estimated to reach the PTS screening criterion in 2014.

Neutron embrittlement is a significant aging mechanism for all ferritic materials that have a neutron fluence greater than 10^{17} n/cm^2 (E > 1.0 MeV). The relevant calculations, using predictions of the cumulative damage to the RPV from neutron embrittlement, were originally based on the 40-year plant license. The RPV containing the core fuel assemblies is made of thick steel plates or forgings welded together. Neutrons from the reactor fuel irradiate the vessel as the reactor is operated and change the material properties of the steel. The most pronounced and significant changes occur in the material property known as fracture toughness.

Fracture toughness is a measure of the resistance to crack extension in response to stresses. A reduction in this material property due to irradiation is referred to as embrittlement. The largest amount of embrittlement usually occurs at the section of the vessel's wall closest to the reactor fuel, otherwise referred to as the vessel's beltline. The rate at which the vessel's steel embrittles also depends on its chemical composition. The amounts of two elements in the steel, specifically copper and nickel, are the most important chemical elements in determining how sensitive the steel is to neutron irradiation.

The applicant identified three analyses affected by irradiation embrittlement. These TLAAs are discussed in LRA Section 4.2 and include:

- RPV upper shelf energy
- pressurized thermal shock
- pressure-temperature (P-T) limits

The staff review findings on the RPV irradiation embrittlement TLAAs in LRA Sections 4.2.1 through 4.2.4 are presented in SER Sections 4.2.1 through 4.2.4, respectively.

4.2.1 Upper Shelf Energy

Appendix G of 10 CFR Part 50 requires RPV beltline materials to have initial Charpy USE values in the transverse direction for the base metal and for the weld material along the weld of no less than 75 ft-lb (102 J). Throughout the life of the RPV the beltline materials must maintain Charpy USE values of no less than 50 ft-lb (68 J). However, in accordance with Appendix G paragraph IV.A.1.a, Charpy USE values below these criteria may be acceptable if it is demonstrated, in a manner approved by the Director of the Office of Nuclear Reactor Regulation, that lower Charpy USE values will provide margins of safety against fracture equivalent to those required by American Society of Mechanical Engineers (ASME) Code Section XI, Appendix G.

RG 1.99, Revision 2, provides an expanded discussion of the calculations of Charpy USE values and describes two methods for determining them for RPV beltline materials depending on if a given RPV beltline material is represented in the plant's Reactor Vessel Material Surveillance Program (i.e., 10 CFR Part 50, Appendix H, program). If surveillance data is not available, the Charpy USE is determined in accordance with position 1.2 in RG 1.99, Revision 2. If two or more surveillance data are available, the Charpy USE should be determined in accordance with position 2.2 in RG 1.99, Revision 2. These methods refer to Figure 2 in RG 1.99, Revision 2, which indicates that the percentage drop in Charpy USE depends upon the amount of copper and the neutron fluence. Since the analyses performed in accordance with 10 CFR Part 50, Appendix G, are based on a flaw with a depth equal to one-quarter of the RPV wall thickness (1/4t), the neutron fluence used in the Charpy USE analysis is the neutron fluence at the 1/4t depth location.

4.2.1.1 Summary of Technical Information in the Application

In LRA Section 4.2.1, the applicant summarized the evaluation of USE for the period of extended operation. The applicant indicated that calculations have shown that the RPV beltline Charpy USE for the limiting plate will be less than 50 ft-lb, based on RG 1.99, Revision 2. LRA Table 4.2.1-1 lists the estimated USE values for the RPV beltline plates and welds based on their estimated fluence on March 24, 2031, the end of 60 years of operation. These fluence values and the material properties needed for USE calculations are also listed in LRA Table 4.2.1-1. The USE for the limiting beltline material, plate D-3804-1 (heat No. C-1308A), is 48.97 ft-lb (66 J), which is below the USE of 50 ft-lb (68 J), as required by 10 CFR Part 50, Appendix G.

Since the end of extended license Charpy USE for the limiting beltline material is below 50 ft-lb (68 J), LRA Section 4.2.1 states that the applicant will submit an equivalent margins analysis (EMA) for the staff's review and approval at least three years before the date when the USE of the limiting material decreases to less than 50 ft-lb (68 J). Therefore, the applicant concluded that, pursuant to 10 CFR 54.21(c)(1)(iii), the effects of aging with respect to Charpy USE on the intended function of RPV beltline materials will be adequately managed for the period of extended operation through the plant's Reactor Vessel Integrity Surveillance Program.

4.2.1.2 Staff Evaluation

The staff reviewed LRA Section 4.2.1, pursuant to 10 CFR 54.21(c)(1)(iii), to verify that the effects of aging on the intended function(s) will be adequately managed for the period of extended operation.

The applicant summarized the Charpy USE analyses for the RPV beltline materials in LRA Table 4.2.1-1. The applicant originally documented its analyses on USE for 32 effective full-power years (EFPYs) in its responses to Generic Letter (GL) 92-01, Revision 1, "Reactor Vessel Structural Integrity." The number of EFPYs at the end of the period of initial operation (40 years) is assumed to be 32, using an assumed 80 percent capacity factor. The information related to this GL, as modified by the staff evaluation, is compiled in the NRC's Reactor Vessel Integrity Database (RVID). The staff compared the copper content and unirradiated Charpy USE values for the RPV beltline materials listed in LRA Table 4.2.1-1 with the values reported in the RVID and found that, except for the unirradiated Charpy USE values for two beltline materials, they were identical. For the two exceptions, the applicant's unirradiated USE values are lower (more conservative) than those in the reactor vessel integrity database and have no impact on the conclusion in the LRA that the USE value for the limiting beltline material is 48.97 ft-lb (66 J).

The applicant calculated the estimated USE values for the RPV beltline materials using the RPV fluence predicted for March 24, 2031. This end of extended operation fluence was based on the fluence calculation methodology documented in Westinghouse Report WCAP-15353, "Palisades

Reactor Pressure Vessel Neutron Fluence Evaluation," forwarded to the NRC by letter dated February 21, 2000. This WCAP, along with a separate amendment request, dated April 27, 2000, to change the operating license expiration date from March 14, 2007, to March 24, 2011, was approved by the NRC in a safety evaluation dated November 14, 2000. In that safety evaluation, the staff concludes that the calculation of the RPV fluence and dosimetry in WCAP-15353 was consistent with the methods of draft RG DG-1053, "Calculational and Dosimetry Methods for Determining Pressure Vessel Neutron Fluence," which eventually became RG 1.190 issued in 2001 with the same title. The staff also concludes that the calculation and update of the RPV fluence were acceptable.

The staff's review of LRA Section 4.2.1 identified areas in which additional information was necessary to complete the review of the applicant's USE evaluation. The applicant responded to the staff's requests for additional information (RAIs) as discussed below.

In RAI 4.2-1, dated November 30, 2005, the staff stated that LRA Section 4.2.1 does not provide the EFPYs for the proposed 60 calendar years of plant operation. Therefore, the staff requested that the applicant provide the EFPYs for 40 calendar years and for 60 calendar years of operation.

In its response, by letter dated January 13, 2006, the applicant stated that the projected EFPY are 24.17 and 42.37 for 40 and 60 calendar years of operation, respectively; therefore, the capacity factor is assumed to be 0.91 for the period of extended operation.

The staff's review of the applicant's calculations verified the values; therefore, the staff's concern described in RAI 4.2-1 is resolved.

In RAI 4.2-2, dated November 30, 2005, the staff stated that the LRA's end of extended license Charpy USE values for all beltline materials are based on position 1.2 of RG 1.99, Revision 2. The staff's independent calculation indicated that position 1.2 is not conservative for the lower shell axial weld 3-112A/C (fabricated with weld wire heat No. 34B009), for which a surveillance weld of the same heat is available from Millstone Power Station Unit 1. Therefore, the staff requested that the applicant provide the end of extended license USE values based on surveillance data for the beltline materials for which surveillance data are available from PNP, or from a sister plant, and demonstrate that this revision of USE values based on surveillance data will not change the conclusion in LRA Section 4.2.1.

In its response, by letter dated January 13, 2006, the applicant provided the end of extended license USE values using position 2.2 of RG 1.99, Revision 2 for the beltline materials having valid surveillance data from Millstone Power Station, Unit 1 as reported in NUREG/CR-6551, "Improved Embrittlement Correlations for Reactor Pressure Vessel Steels." The applicant demonstrated that the USE value reported in LRA Table 4.2-1 is conservative when compared to the USE value that would be established for weld wire heat No. 34B009 from the Millstone Power Station, Unit 1 surveillance data.

The staff's review found the applicant's response to RAI 4.2-2 acceptable. Since the information in LRA Table 4.2-1 is based on position 1.2 of RG 1.99 and considered conservative, the applicant's response to RAI 4.2-2 does not change the original conclusions in the LRA. The limiting beltline material is still plate D-3804-1. Therefore, the staffs concern described in RAI 4.2-2 is resolved.

Limiting Plate and Weld for USE	Acceptance Criterion	Calculated USE Value for 42.37 EFPY	Conclusion
Lower shell plate (D-3804-1)	Projected USE > 50 ft-lbs	48.97 ft-lbs (Confirmed by Staff)	Acceptance criterion is exceeded in 2021
Intermediate to lower shell circ. Weld (9-112)	Projected USE > 50 ft-lbs	50.83 ft-lbs (Confirmed by Staff)	Acceptable TLAA satisfies §54.21(c)(1)(ii)]

To resolve the issue of the RPV not meeting the 10 CFR Part 50, Appendix G requirement to maintain all RPV beltline material USE values above 50 ft-lb (68 J) at the end of the period of extended operation, the applicant stated in LRA Section 4.2.1 that it will submit an EMA for review and approval at least three years before the date the USE of the limiting material decreases to less than 50 ft-lb (68 J). The staff noted that the proposed date for the EMA submittal is consistent with the 10 CFR Part 50, Appendix G requirements for operating plants. Further, guidelines and criteria for the EMA for low USE RPV materials were established in RG 1.161, "Evaluation of Reactor Pressure Vessels with Charpy Upper-Shelf Energy Less than 50 ft-lb," dated 1995. The staff has reviewed many owners group and plant-specific EMA applications with predicted USE values much lower than the applicant's projected value. Based on this review experience, the staff determined that the applicant need not submit the EMA results until three years before the date the USE of the limiting material decreases to less than 50 ft-lb (68 J). The applicant's proposal is consistent with 10 CFR Part 50, Appendix G.

The staff determined that the applicant had demonstrated that the effects of aging on the intended function of the RPV beltline materials will be adequately managed under the Reactor Vessel Integrity Surveillance Program for the period of extended operation.

4.2.1.3 FSAR Supplement

The applicant provided an FSAR supplement summary description of its TLAA evaluation of USE in LRA Section A4.1.1. On the basis of its review of the FSAR supplement, the staff concludes that the summary description of the applicant's actions to address the USE analysis is adequate.

4.2.1.4 Conclusion

On the basis of its review, as discussed above, the staff concludes that the applicant has demonstrated, pursuant to 10 CFR 54.21(c)(1)(iii), that, for the USE TLAA, the effects of aging on the intended function(s) will be adequately managed for the period of extended operation. The staff also concludes that the FSAR supplement contains an appropriate summary description of the TLAA evaluation, as required by 10 CFR 54.21(d).

4.2.2 Pressurized Thermal Shock

The PTS Rule, 10 CFR 50.61, provides the fracture toughness requirements for protecting the RPVs of pressurized water reactors (PWRs) against PTS consequences. Applicants are required to assess the RPV materials projected values of the PTS reference temperature (RT_{PTS}) through the end of the operating licenses. The PTS rule requires that each applicant calculate the end-of-license RT_{PTS} value for each material within the RPV beltline. The RT_{PTS} value for each beltline material is the sum of the unirradiated nil ductility reference temperature (RT_{NDT}) value, a shift in the RT_{NDT} value caused by exposure to high-energy neutron irradiation of the material (i.e., ΔRT_{NDT} value), and an additional margin value to account for uncertainties (i.e., M value). Section 50.61 of 10 CFR also provides screening criteria against which the calculated end-of-license RT_{PTS} values are evaluated. RPV beltline base-metal materials (forging or plate materials) and longitudinal (axial) weld materials are considered as adequate protection against PTS events if the calculated RT_{PTS} values are less than or equal to 270 °F. RPV beltline circumferential weld materials are considered as adequate protection against PTS events if the calculated RT_{PTS} values are less than or equal to 300 °F.

RG 1.99, Revision 2, provides an expanded discussion of the calculations of the shift in the RT_{NDT} value caused by exposure to high-energy neutron irradiation and the margin value to account for uncertainties. In this RG, the shift in the RT_{NDT} value caused by exposure to high-energy neutron irradiation is the product of a chemistry factor (CF) and a fluence factor. The fluence factor depends upon the neutron fluence and the CF may be determined from surveillance material or from tables in the RG. If the RPV beltline material is not represented by surveillance material, its CF and the shift in the RT_{NDT} value caused by exposure to high-energy neutron irradiation may be determined by the methodology documented in position 1.1 and tables in the RG. The CF determined from tables in the RG depends upon the amounts of copper and nickel in the beltline. If the RPV beltline material is represented by surveillance material, its CF may be determined from the surveillance data with the methodology documented in position 2.1 of the RG. Section 50.61 of 10 CFR provides methods for determining RT_{NDT} values equivalent to RG 1.99, Revision 2.

4.2.2.1 Summary of Technical Information in the Application

In LRA Section 4.2.2, the applicant discussed its PTS analysis using 10 CFR 50.61 criteria. RT_{PTS} values were calculated at the inside surface of the beltline region materials of the RPV using Charpy-based fracture toughness evaluations in accordance with 10 CFR 50.61 methods for a 42.37 EFPY operating period. The RT_{PTS} values for the beltline region materials at the end of the period of extended operation were calculated to be lower than the applicable 10 CFR 50.61 screening criteria values with the exception of the intermediate and lower shell axial welds. The axial welds fabricated with weld wires of heat No. W5214 are the limiting RPV welds.

Limiting Material for PTS	Screening Criteria	Calculated 42.37 EFPY RT_{PTS} Value	Conclusion
Intermediate shell and lower shell axial welds (W5214)	270 °F	Applicant: 287 °F (Confirmed by Staff)	Screening Criterion is exceeded in 2014

In its LRA, the applicant provided limited information about its plan for addressing PTS concerns prior to exceeding the screening criteria. Section 54.21(c)(1)(iii) of 10 CFR allows the applicant to demonstrate that the effects of aging on the intended function of the system, structure, or component (SSC) will be adequately managed for the period of extended operation. This option permits an applicant to elect not to extend the existing TLAA.

4.2.2.2 Staff Evaluation

The staff reviewed LRA Section 4.2.2, pursuant to 10 CFR 54.21(c)(1)(iii), to verify that the effects of aging on the intended function(s) will be adequately managed for the period of extended operation.

In the LRA, the applicant provided the RT_{PTS} analyses for the materials in the RPV beltline. LRA Table 4.2.2-1 provides the chemistry, fluence values, and predicted RT_{PTS} value through 42.37 EFPY for each plate and weld in the RPV beltline. The staff verified that the applicant's RPV beltline material chemistry and fluence were based on the information in WCAP-15353 and the staff's associated November 14, 2000, safety evaluation, as discussed in SER Section 4.2.1.2. Further, the staff's independent calculations confirmed the applicant's end of extended license RT_{PTS} values, which exceed the 10 CFR 50.61 screening criteria. The staff's calculations indicate that the RT_{PTS} of the limiting weld would reach 132 °C (270 °F) in 2014, the same year predicted by the applicant in the LRA, at a neutron fluence of 1.59×10^{19} n/cm^2 (E > 1.0 MeV). For the two beltline plates having surveillance data, the staff applied position 2.1 in RG 1.99, Revision 2 in estimating their end of extended license RT_{PTS} values.

If an RPV is projected to exceed the PTS screening criteria, 10 CFR 50.61(b)(3) requires that the applicant implement a flux reduction program reasonably practicable to avoid exceeding the PTS screening criteria. If the flux reduction program does not prevent the RPV from exceeding the PTS screening criteria before the end of the operating license, the applicant, to meet PTS requirements, can choose in 10 CFR 50.61 between two options:

(1) The applicant can submit a plant-specific safety analysis, pursuant to 10 CFR 50.61(b)(4), to determine what, if any, modifications to equipment, systems, and plant operation are necessary to prevent failure of the RPV from a postulated PTS event.

(2) The other option is to perform a thermal annealing treatment of the RPV, pursuant 10 CFR 50.61(b)(7), to recover fracture toughness. Section 50.61 of 10 CFR requires submission to the NRC of details of the approach selected for approval at least three years before the RPV is projected to exceed the PTS screening criteria.

An applicant choosing to use the 10 CFR 54.21(c)(1)(iii) option for managing the RPV PTS TLAA must provide (a) an assessment of the CLB TLAA for PTS, (b) a discussion of the flux reduction program implemented in accordance with 10 CFR 50.61(b)(3), and (c) an identification of viable options for managing the aging effect in the future ("Pressurized Thermal Shock Analyses for Renewal of Certain Nuclear Power Plant Operating Licenses," Executive Director Memo to Commissioners, dated May 27, 2004, ML041190564).

Recognizing that the RPV welds are expected to exceed 10 CFR 50.61 PTS screening criteria during the period of extended operation, the applicant chose to use the 10 CFR 54.21(c)(1)(iii) option for managing the PTS TLAA. Accordingly, the applicant discussed the flux reduction program implemented in accordance with 10 CFR 50.61(b)(3) and also identified other viable options for managing the aging effect.

The staff's review of LRA Section 4.2.2 identified an area in which additional information was necessary to complete the review of the applicant's PTS evaluation. The applicant responded to the staff's RAI as discussed below.

In RAI 4.2-3, dated November 30, 2005, the staff stated that:

(1) The applicant provided a TLAA which determines that the limiting material for the RPV will exceed the PTS screening criterion in the year 2014. Therefore, the staff requested that the applicant describe any current or planned flux reduction program required to support the determination that the RPV limiting material will comply with 10 CFR 50.61 requirements until the year 2014.

(2) LRA Section 4.2.2 states that the applicant will select the optimum alternative to manage PTS in accordance with 10 CFR 50.61 and will provide applicable submittals for NRC review and approval prior to exceeding the PTS screening criteria during the period of extended operation. Therefore, the staff requested that the applicant provide additional information regarding specific plant equipment modifications, operational modifications, revised PTS analysis, or thermal annealing which could be implemented to allow the RPV to comply with 10 CFR 50.61 requirements through the period of extended operation.

In its response, by letter dated January 13, 2006, the applicant addressed to each area of the RAI and is summarized as follows:

(1) The determination that the RPV material will remain below the 10 CFR 50.61 PTS screening criteria until year 2014 is based on its implementation of an ultra low leakage core design. The current ultra low leakage core design was developed from a series of changes in core reload design resulting in flux reductions. The reduction in RPV neutron fluence due to the core design changes was summarized most recently in a Consumers Energy Company letter dated August 31, 2000.

(2) The applicant identified the options it is contemplating for resolving the PTS issue. The applicant's options include flux reduction, preheating of the safety injection water, provision of a compressive prestress in the beltline region, and thermal annealing of the RPV.

Therefore, the staff's review of the applicant's future submittals applying these options will be manageable. The staff determined, consistent with the requirements of 10 CFR 50.61, that the applicant need not submit its resolution of the PTS issue until three years before the limiting RPV material is projected to exceed the specified screening criteria. The staff determined, in accordance with 10 CFR 54.21(c)(1)(iii) and the guidelines and criteria for thermal annealing and plant-specific PTS analysis were established in RG 1.162, "Format and Content of Report for Thermal Annealing of Reactor Pressure Vessels," and RG 1.154, "Format and Content of Plant-Specific Pressurized Thermal Shock Safety Analysis Reports for Pressurized Water Reactors," respectively, that the applicant had demonstrated that the effects of aging on the intended function of the RPV beltline materials will be adequately managed for the extended period of operation under the Reactor Vessel Integrity Surveillance Program and through future analyses, plant modifications, or thermal annealing of the RPV.

Based on its review, the staff finds the applicant's response to RAI 4.2-3 acceptable. The applicant's plant-specific analyses, plant modifications, or thermal annealing plan for the RPV must be submitted in accordance with 10 CFR 50.61 to ensure that the aging effects will be managed during the period of extended operation. Therefore, the staff's concerns described in RAI 4.2-3 have been adequately addressed.

The staff reviewed the applicant's TLAA on PTS as summarized in LRA Section 4.2.2, and determines that the applicant's TLAA will continue to comply with the staff's requirements for PTS, in accordance with 10 CFR 50.61 throughout the period of extended operation. Although RPV welds in the beltline region exceed PTS screening criteria, the applicant proposed to manage the PTS issue through future analyses, plant modifications, or thermal annealing of the RPV, permitted by and in accordance with 10 CFR 50.61, to resolve this PTS issue so that it can maintain and monitor the entire RPV. The staff finds this position acceptable. The staff determines that the applicant has demonstrated that PTS would be adequately managed for the period of extended operation as required by 10 CFR 54.21(c)(1)(iii).

4.2.2.3 FSAR Supplement

The applicant provided an FSAR supplement summary description of its TLAA evaluation of PTS in LRA Section A4.1.2. On the basis of its review of the FSAR supplement, the staff concludes that the summary description of the applicant's actions to address PTS is adequate.

4.2.2.4 Conclusion

On the basis of its review, as discussed above, the staff concludes that the applicant has demonstrated, pursuant to 10 CFR 54.21(c)(1)(iii), that, for the PTS TLAA, the effects of aging on the intended function(s) will be adequately managed for the period of extended operation. The staff also concludes that the FSAR supplement contains an appropriate summary description of the TLAA evaluation, as required by 10 CFR 54.21(d).

4.2.3 Pressure-Temperature (P-T) limits

Section 50.60 of 10 CFR, "Acceptance Criteria for Fracture Prevention Measures for Lightwater Nuclear Power Reactors for Normal Operation," and 10 CFR Part 50, Appendix G, "Fracture Toughness Requirements," set forth the staff's requirements and criteria for generating the P-T limits required for U.S. commercial light-water reactors.

4.2.3.1 Summary of Technical Information in the Application

In LRA Section 4.2.3, the applicant summarized the evaluation of the pressure-temperature (P-T) limits for the period of extended operation. Appendix G of 10 CFR Part 50 requires that the RPV be maintained within established P-T limits including during heatup and cooldown. These limits specify the maximum allowable pressure as a function of reactor coolant temperature. As the RPV becomes embrittled and its fracture toughness is reduced, the allowable pressure (given the required minimum temperature) is reduced. Heatup and cooldown limit curves are calculated using the adjusted RT_{NDT} corresponding to the limiting beltline region material of the RPV. The adjusted RT_{NDT} of the limiting material in the core region of the RPV is determined by using the unirradiated RPV material fracture toughness properties, estimating the radiation-induced RT_{NDT}, and adding a margin. LTOP limits and setpoints are determined as part of the calculation of P-T operating limit curves.

The applicant concluded that the P-T limits were TLAAs that needed to be assessed against the acceptance criteria of 10 CFR 54.21(c)(1)(iii). The P-T limit curves that apply for the current operating conditions are included in the plant technical specifications (TSs). The amendment request regarding the current P-T limit curves was approved by the staff and these requirements are incorporated into the PNP technical specifications. LRA Section 4.2.3 states, "The current pressure/temperature analyses are valid beyond the current operating license period, but not to the end of the period of extended operation. These analyses are estimated to expire in 2014." LRA Section 4.2.3 further states that 10 CFR Part 50, Appendices G and H, require that the P-T limits contained in the TSs be updated, as needed, to remain valid.

4.2.3.2 Staff Evaluation

The staff reviewed LRA Section 4.2.3, pursuant to 10 CFR 54.21(c)(1)(iii), to verify that the effects of aging on the intended function(s) will be adequately managed for the period of extended operation.

Section 50.60 of 10 CFR provides the acceptance criteria for fracture prevention measures for light water nuclear power reactors for normal operation and invokes the application of 10 CFR Part 50, Appendices G or H, as applicable. The staff found from reviewing the TSs that the current P-T limits, which the NRC approved on March 2, 1995, via License Amendment No. 163, are valid for an RPV wall fluence up to 2.192×10^{19} n/cm^2 (E > 1.0 MeV). Based on the latest fluence estimates reported in WCAP-15353, the current P-T limit analyses would be valid until 2014; however, under the 40-year license, the P-T limits would expire on March 24, 2011, at the end of the current license. To support operation beyond the current license, LRA Section A2.16 states, in the first of four improvements made to Reactor Vessel Integrity Surveillance Program, that the applicant will update P-T limits and LTOP curves to reflect the additional neutron fluence accumulated during the extended period of operation and submit the updated curves for staff approval prior to the period of extended operation (Commitment 29).

The TSs will continue to be updated, as required by either Appendices G or H of 10 CFR Part 50, or as operational needs dictate. The staff agreed that this updating will assure that the operational limits remain valid for current and projected cumulative neutron fluence levels and that additional P-T limit analysis is not required at this time.

4.2.3.3 FSAR Supplement

The applicant provided an FSAR supplement summary description of its TLAA evaluation of P-T limits in LRA Section A4.1.3. On the basis of its review of the FSAR supplement, the staff concludes that the summary description of the applicant's actions to address the P-T limits is adequate.

4.2.3.4 Conclusion

On the basis of its review, as discussed above, the staff concludes that the applicant has demonstrated, pursuant to 10 CFR 54.21(c)(1)(iii), that, for the P-T limits TLAA, the effects of aging on the intended function(s) will be adequately managed for the period of extended operation. The staff also concludes that the FSAR supplement contains an appropriate summary description of the TLAA evaluation, as required by 10 CFR 54.21(d).

4.2.4 Low Temperature Overpressure Protection (LTOP) PORV Setpoints

4.2.4.1 Summary of Technical Information in the Application

In LRA Section 4.2.4, the applicant summarized the evaluation of the LTOP PORV setpoints TLAA for the period of extended operation. LTOP limits and setpoints are determined as part of the calculation of P-T operating limit curves in LRA Section 4.2.3.

4.2.4.2 Staff Evaluation

The staff reviewed LRA Section 4.2.4, pursuant to 10 CFR 54.21(c)(1)(iii), to verify that the effects of aging on the intended function(s) will be adequately managed for the period of extended operation. The staff's review of the LTOP PORV setpoints TLAA is included within SER Section 4.2.3.1.

4.2.4.3 FSAR Supplement

The applicant provided an FSAR supplement summary description of its TLAA evaluation of the LTOP PORV setpoints in LRA Section A4.1.4. On the basis of its review of the FSAR supplement, the staff concludes that the summary description of the applicant's actions to address the LTOP PORV setpoints is adequate.

4.2.4.4 Conclusion

On the basis of its review, as discussed above, the staff concludes that the applicant has demonstrated, pursuant to 10 CFR 54.21(c)(1)(iii), that, for the LTOP PORV setpoints TLAA, the effects of aging on the intended function(s) will be adequately managed for the period of extended operation. The staff also concludes that the FSAR supplement contains an appropriate summary description of the TLAA evaluation, as required by 10 CFR 54.21(d).

4.3 Metal Fatigue

A metal component subjected to cyclic loading at loads less than the static design load may fail due to fatigue. Metal fatigue of components may have been evaluated based on an assumed number of transients or cycles for the current operating term. The validity of such metal fatigue analyses is reviewed for the period of extended operation. The GALL Report identifies fatigue aging related effects that require evaluation as possible TLAAs pursuant to 10 CFR 54.21(c). Each of these TLAAs is summarized in the SRP-LR and presented in LRA Section 4.

4.3.1 Evaluation of Fatigue in Vessels, Piping and Components

In LRA Section 4.3.1, the applicant summarized the evaluation of fatigue in vessels, piping and components for the period of extended operation. Design analyses of vessels and piping predict fatigue effects by constructing a set of normal operating and abnormal cyclic load events, each of which is assumed to occur no more than an assumed number of times during the design life or licensed operating period. The applicant maintains a record of the occurrence, and cumulative number of occurrences, of the subset of normal and abnormal cyclic events which contribute significantly to the predicted cumulative usage factor (CUF) of vessels, piping, and components. The design-basis number of cycles, the estimated number of cycles experienced to date, and the estimated number of cycles expected to be experienced by the end of the extended operating period are shown in LRA Table 4.3.1-1.

At PNP, the cumulative effect of alternating stresses produced by these events (the CUF) was originally calculated under ASME Section III Class A rules, 1965 Edition, and some Addenda. ASME Section III Class 1 rules were used for later replacements (e.g., the steam generators) and for certain piping modifications. Re-analyses, to incorporate events and conditions not considered in the original design, typically used the original design code for the component. The code of record for each component is listed in LRA Table 4.3.1-2. At almost all PNP locations, for which a fatigue analysis was performed, the CUF predicted by the original analysis of record, based on a 40-year life, is not expected to approach the allowable limit of 1.0 in a 60-year life.

Fatigue issues will be managed through cycle counting. The program will record plant transients for a bounding subset of RPV, pressurizer, steam generator, other reactor coolant system (RCS) components, and Consumers Design Class I piping to ensure that fatigue effects are adequately managed for Consumers Design Class I systems and components. Cycle counting will assure that either (a) the analysis and licensing basis design cycle count assumptions are not exceeded during the period of extended operation or (b) corrective action is taken to ensure that the basis for the safety determination remains valid upon reaching the assumed cycle count for a given event.

4.3.2 Reactor Vessel Fatigue Analyses

4.3.2.1 Summary of Technical Information in the Application

In LRA Section 4.3.2, the applicant summarized the evaluation of the reactor vessel fatigue analyses for the period of extended operation. The reactor vessel was designed, constructed, and analyzed to the ASME Boiler and Pressure Vessel (B&PV) Code, Section III, Subsection 4, for Class A vessels, 1965, with Addenda through Winter, 1965. ASME III-1965 Class A vessels require a Subsection N-415.2 fatigue analysis for parts subject to stress ranges which exceed the criteria of Subsection N-415.1. The original analyses have been amended to address issues that have arisen since fabrication. The current highest calculated fatigue usage factors, based on the number of design-basis load cycles assumed by the vessel analyses, have been determined. The number of design-basis load cycles for each event was selected to be adequate for the originally licensed 40-year design life. Of the high usage factor vessel locations, the vessel studs have the highest usage factor, but are replaceable, and the control rod drive mechanism (CRDM) housings and appurtenances have been replaced. For these reasons, fatigue in these components is addressed separately - the vessel studs in LRA Section 4.3.3 and the replaced CRDM housings and appurtenances in LRA Section 4.3.4.

4.3.2.2 Staff Evaluation

The staff reviewed LRA Section 4.3.2, pursuant to 10 CFR 54.21(c)(1)(i), to verify that the analyses remain valid for the period of extended operation and, pursuant to 10 CFR 54.21(c)(1)(iii), to verify that the effects of aging on the intended function(s) will be adequately managed for the period of extended operation.

The applicant stated that the fatigue analyses at various reactor vessel limiting locations were evaluated for extended operation in accordance with the ASME Code Section III Class A with

Addenda through Winter, 1965, Subsection N-415.2, for components with stress ranges exceeding Subsection N-415.1 criteria. The fatigue analyses were based on the nuclear steam supply system design transient conditions and corresponding design cycles, as shown in LRA Table 4.3.1-1. The limiting locations which were evaluated consisted of the shell and bottom head, inlet and outlet nozzles, internal welded attachments, instrument nozzle shroud tubes, vessel head CRDM nozzles, and instrument flange bolts. The highest CUF calculated at these locations was 0.4516 on the outlet nozzle. The applicant indicated that the CUF for the period of extended operation was not expected to exceed the ASME Code Section III limit value of 1.0 because the number of design basis transient events is not expected to approach the number assumed in the design-basis analysis. However, in the unlikely event that a design basis cycle count limit is projected to be reached at any time during the period of extended operation, the applicant stated that it intends to use the Fatigue Monitoring Program, described in LRA Appendix B, to take corrective action or re-analyze to ensure that the CUF does not exceed the ASME Section III limiting value before the end of the period of extended operation.

The staff review of LRA Section 4.3.2 identified areas in which additional information was necessary to complete the review of the reactor vessel structural integrity evaluation. The applicant responded to the staff's RAIs as follows.

In RAI 4.3-1, dated July 21, 2005, the staff requested that the applicant justify the difference in plant heat-ups (134) and plant cool-downs (119) listed in LRA Table 4.3.1-1.

In its response, by letter dated August 19, 2005, the applicant attributed the difference to misinterpretation of logging criteria between 1978 and 1982. The numbers of heat-ups and cool-downs have coincided since 1982. The applicant also stated that the number of cool-downs should have been listed as 134.

Based on its review, the staff finds the applicant's response to RAI 4.3-1 acceptable because the applicant clarified that the plant has experienced 134 cool-down consistent with the 134 plant heat-ups described in the LRA. The staff's concern described in RAI 4.3-1 is resolved.

In RAI 4.3-2, dated July 21, 2005, the staff requested that the applicant provide the basis for not counting the safety valve operation cycles and for assuming the design value of 200 cycles in LRA Table 4.3.1-1 will never be approached.

In its response, by letter dated August 19, 2005, the applicant stated that the PCS safety valves have not operated during the plant life to date and are not expected to operate in the future. The 2005 and 60-year values for safety valve operation in the table should be zero instead of considered as not counted. The applicant also stated that the PCS pressure increases during startup and other periods of low-temperature operation will cause the PORVs to open prior to the safety valves, precluding the need for safety valve operation. Pressure increases during plant operation at power have been managed by the pressurizer and the chemical and volume control system. These controls have been adequate to control primary system volume and to maintain the pressure below the PCS code safety valve set points.

Based on its review, the staff finds the applicant's response to RAI 4.3-2 acceptable because it conforms with current industry practice; therefore, the staff's concern described in RAI 4.3-2 is resolved.

In RAI 4.3-3, dated July 21, 2005, the staff requested that the applicant show that the limiting components of the reactor vessels evaluated for extended operation correspond to the components listed in the GALL Report, Volume 2, Table IV.A2, for PWR reactor vessels.

In its response, by letter dated August 19, 2005, the applicant provided a table showing the components evaluated in the fatigue TLAAs. Based on its review the staff finds that locations found in the RAI response correspond to those in the GALL Report, Volume 2. The highest CUF for 60-year operation was appropriately identified in the LRA. Therefore, the staff's concern in RAI 4.3-3 is resolved.

In RAI 4.3-4, dated July 21, 2005, the staff requested that the applicant provide the basis for stating that the number of design basis transients is not expected to approach the number assumed by the analysis for the period of extended operation.

In its response, by letter dated August 19, 2005, the applicant stated that the number of design transients projected for 60-year operation was determined based on actual plant data transients. This information is listed in LRA Table 4.3.1-1 The applicant also stated that the reactor vessel head closure stud design basis assumes 50 cycles for 60 years. Projecting this rate into the future results in an estimated 32 cycles to be experienced in 60 years.

Based on its review, the staff finds that the numbers of transients experienced to date and the maximum total numbers of transients estimated for 60 years are below those allowed by design; therefore, the staff's concern described in RAI 4.3-4 is resolved.

In RAI 4.3-5, dated July 21, 2005, the staff requested that the applicant provide the highest 60-year CUFs and disposition for all structures or components in the reactor vessel internals, such as those listed in Table IV B3, "Reactor Vessel Internals (PWR) - Combustion Engineering," Items B3.2f, B3.4d and B3.5g, of the GALL Report, Volume 2, which require further evaluation as TLAAs for the period of extended operation.

In its response, by letter dated August 19, 2005, the applicant stated that fatigue analyses for the core support barrel or the upper guide structure were not performed because they are not parts of the pressure boundary. These removable assemblies contain the components equivalent to the GALL items (control rod shroud, core shroud core support plate, and fuel alignment pins). The reactor vessel fatigue evaluation of LRA Section 4.3.2 did include the core stabilizing lugs, the core stop lugs, and the flow baffle because they are integral parts of the reactor vessel. These were shown to have projected 60-year CUFs much lower than the ASME Section III fatigue limit.

Based on its review, the staff finds the applicant's response to RAI 4.3-5 acceptable because the projected 60-year CUFs are lower than the ASME Section III fatigue limit. Therefore, the staff's concern described in RAI 4.3-5 is resolved.

On the basis of its review of LRA Section 4.3.3 and the RAI responses, the staff concludes that the reactor vessel fatigue analyses TLAA will remain valid for the period of extended operation in accordance with 10 CFR 54.21(c)(1)(i) and will be adequately managed for the period of extended operation in accordance with 10 CFR 54.21(c)(1)(iii).

4.3.2.3 FSAR Supplement

The applicant provided an FSAR supplement summary description of its TLAA evaluation of the reactor vessel fatigue analyses in LRA Section A4.2.1. On the basis of its review of the FSAR supplement, the staff concludes that the summary description of the applicant's actions to address the reactor vessel fatigue analyses is adequate.

4.3.2.4 Conclusion

On the basis of its review, as discussed above, the staff concludes that the applicant has demonstrated, pursuant to 10 CFR 54.21(c)(1)(i), that, for the reactor vessel fatigue analyses TLAA, the analyses remain valid for the period of extended operation. The applicant had also demonstrated, pursuant to 10 CFR 54.21(c)(1)(iii), that, for the TLAA, the effects of aging on the intended function(s) will be adequately managed for the period of extended operation. The staff also concludes that the FSAR supplement contains an appropriate summary description of the TLAA evaluation, as required by 10 CFR 54.21(d).

4.3.3 Reactor Vessel Head Closure Stud Fatigue Analysis

4.3.3.1 Summary of Technical Information in the Application

In LRA Section 4.3.3, the applicant summarized the evaluation of the reactor vessel head closure stud fatigue analysis for the period of extended operation. The highest fatigue usage factor calculated by the reactor vessel fatigue analysis is in the vessel head studs.

4.3.3.2 Staff Evaluation

The staff reviewed LRA Section 4.3.3, pursuant to 10 CFR 54.21(c)(1)(i), to verify that the analyses remain valid for the period of extended operation and, pursuant to 10 CFR 54.21(c)(1)(iii), to verify that the effects of aging on the intended function(s) will be adequately managed for the period of extended operation.

The applicant stated that, based on the number of design-basis load events, the 40-year CUF in the head closure studs was 0.835. It also stated that the design-basis events (DBEs) that affect the RPV and the head closure studs is not expected to approach the number of design-basis load events during the period of extended operation, and the CUF in the head closure studs will remain below the ASME Section III limiting value of 1.0. However, the applicant also committed to monitor the number of design basis cycles via the Fatigue Monitoring Program and perform a reanalysis or other corrective actions to ensure that the ASME Section III fatigue limit is not exceeded during the period of extended operation (Commitment 5).

Based on its review, the staff concludes that the reactor vessel closure studs fatigue TLAA will remain valid to the end of the period of extended operation because the CUF for head closure studs will remain below the ASME Section III limiting value of 1.0.

4.3.3.3 FSAR Supplement

The applicant provided an FSAR supplement summary description of its TLAA evaluation of the reactor vessel head closure stud fatigue analysis in LRA Section A4.2.2. On the basis of its review of the FSAR supplement, the staff concludes that the summary description of the applicant's actions to address the reactor vessel head closure stud fatigue analysis is adequate.

4.3.3.4 Conclusion

On the basis of its review, as discussed above, the staff concludes that the applicant has demonstrated, pursuant to 10 CFR 54.21(c)(1)(i), that, for the reactor vessel head closure stud fatigue analysis TLAA, the analyses remain valid for the period of extended operation. The applicant had also demonstrated, pursuant to 10 CFR 54.21(c)(1)(iii), that, for the TLAA, the effects of aging on the intended function(s) will be adequately managed for the period of extended operation. The staff also concludes that the FSAR supplement contains an appropriate summary description of the TLAA evaluation, as required by 10 CFR 54.21(d).

4.3.4 Control Rod Drive Mechanism (CRDM) Housing Fatigue Analyses

4.3.4.1 Summary of Technical Information in the Application

In LRA Section 4.3.4, the applicant summarized the evaluation of the CRDM housing fatigue analyses for the period of extended operation. The reactor CRDMs are enclosed in pressure housings bolted and seal-welded to the RPV CRDM nozzle flanges. The CRDM housings, their seal housings, their instrument and vent tube nozzles, the flange bolts, and the Omega seal welds between the CRDM housing flanges and the reactor vessel CRDM nozzle flanges were all replaced in 2001. Extension of the operating license to March 24, 2031, therefore requires a 31-year design life. The replacements are ASME III (1989) - Class 1, NPT stamped, with a reconciliation to the 1965 Code.

4.3.4.2 Staff Evaluation

The staff reviewed LRA Section 4.3.4, pursuant to 10 CFR 54.21(c)(1)(i), to verify that the analyses remain valid for the period of extended operation.

The applicant stated that an evaluation of the replaced CRDM housings for fatigue did not require a fatigue analysis because the evaluation of these components determined that the fatigue criteria of ASME Section III, 1989 Edition, Paragraph NB-3222.4(d), "Components not Requiring Analysis for Cyclic Operation," were met. For other replaced components like CRDM housing bolts, nozzle flange bolts, and the Omega seal welds the applicant stated that their expected installed design life, which includes the period of extended operation, will be 30 years, less than the 40-year period on which the DBE cycles were based. The two highest calculated 40-year CUFs for these components, 0.624 for the flange bolts and 0.5621 for the flange Omega seal welds, meet the ASME Section III Class 1 fatigue limit. The applicant also stated that the actual fatigue CUFs are not expected to reach their calculated values because the number of DBEs that affect the RPV, the CRDM housings, and appurtenances is not expected to approach the design basis limits for these events during the period of extended operation.

Based on its review, the staff concludes that the reactor vessel CRDM housing fatigue TLAAs will remain valid to the end of the period of extended operation because the two highest calculated 40-year CUFs meet the ASME Section III Class 1 fatigue limit and actual fatigue CUFs are not expected to approach the design basis limits since components were replaced in 2001.

4.3.4.3 FSAR Supplement

The applicant provided an FSAR supplement summary description of its TLAA evaluation of CRDM housing fatigue analyses in LRA Section A4.2.3. On the basis of its review of the FSAR supplement, the staff concludes that the summary description of the applicant's actions to address the CRDM housing fatigue analyses is adequate.

4.3.4.4 Conclusion

On the basis of its review, as discussed above, the staff concludes that the applicant has demonstrated, pursuant to 10 CFR 54.21(c)(1)(i), that, for the CRDM housing fatigue analyses TLAA, the analyses remain valid for the period of extended operation. The staff also concludes that the FSAR supplement contains an appropriate summary description of the TLAA evaluation, as required by 10 CFR 54.21(d).

4.3.5 Steam Generator Fatigue Analyses

4.3.5.1 Summary of Technical Information in the Application

In LRA Section 4.3.5, the applicant summarized the evaluation of the steam generator fatigue analyses for the period of extended operation. The steam generators were replaced in 1990 through 1991. Extension of the operating license to March 24, 2031, therefore, requires a 40-year design life. The replacement steam generators were designed to the ASME B&PV Code, Section III, 1977. The primary coolant pressure boundary (tube side) of the steam generators is designed to Section III Class 1 rules. Critical components of the Class 2 design secondary side (e.g., the feedwater nozzles) were also analyzed using Class 1 methods.

4.3.5.2 Staff Evaluation

The staff reviewed LRA Section 4.3.5, pursuant to 10 CFR 54.21(c)(1)(i), to verify that the analyses remain valid for the period of extended operation.

The staff review of LRA Section 4.3.5 identified areas in which additional information was necessary to complete the review of the steam generator fatigue analyses TLAA. The applicant responded to the staff's RAIs as follows.

The applicant stated that ASME Section III Class 1 fatigue analyses were performed for the vessels and components, except for the manway studs, of the replacement steam generators. The analyses were based on the same number of DBE cycles assumed in the original plant design for a 40-year licensing operating period. The maximum CUF was determined to be 0.9158 on the main feedwater nozzle, less than the ASME Section III Class 1 fatigue limit of 1.0. Therefore, in RAI 4.3-6, dated July 21, 2005, the staff requested that the applicant verify that the CUF value of 0.9158 represents the highest 60-year value for structures or components in the steam generators like those listed in GALL Report, Volume 2, Table IV D1, "Steam Generators (Recirculating)," Items D1.1 and D1.2.

In its response, by letter dated August 19, 2005, the applicant provided a table for the steam generator components with corresponding CUFs showing that the CUF value of 0.9158 represents the highest 60-year CUF for all steam generator components.

Based on its review, the staff finds the applicant's response to RAI 4.3-6 acceptable because the highest CUF value is representative of all SG components and is below the ASME Section III Class 1 fatigue limit of 1.0. Therefore, the staff's concern described in RAI 4.3-6 is resolved.

In the LRA, the applicant stated that prior fatigue analysis of the manway studs indicated that the CUF will exceed the ASME Section III Class 1 fatigue limit in about 200 heat-up and cool-down cycles, less than the expected 60-year design basis number of 240 heat-up and cool-down cycles in LRA Table 4.3.1-1. The fatigue analysis of the manway studs is, therefore, not a TLAA. The applicant stated that plant procedures require periodic monitoring of the PCS heat-up and cool-down cycles to ensure that the manway studs are replaced before they experience 200 heat-up and cool-down cycles.

In RAI 4.3-7, dated July 21 2005, the staff requested that the applicant provide the dates when 200 heat-up and cool-down cycles are projected to occur from the installation date of the replacement steam generators.

In its response, by letter dated August 19, 2005, the applicant stated that 200 cycles are expected to occur in the year 2046, which is past March 24, 2031, the expected expiration date of the renewed operating license.

Based on its review, the staff finds the applicant's response to RAI 4.3-7 acceptable because the 200 cycle are expected to occur after the expected expiration date of the renewed operating license; therefore, the staff's concern described in RAI 4.3-7 is resolved.

4.3.5.3 FSAR Supplement

The applicant provided an FSAR supplement summary description of its TLAA evaluation of the steam generator fatigue analyses in LRA Section A4.2.4. On the basis of its review of the FSAR supplement, the staff concludes that the summary description of the applicant's actions to address the steam generator fatigue analyses is adequate.

4.3.5.4 Conclusion

On the basis of its review, the staff concludes that the applicant has demonstrated, pursuant to 10 CFR 54.21(c)(1)(i), that, for the steam generator fatigue analyses TLAA, the analyses remain valid for the period of extended operation. The staff also concludes that the FSAR supplement contains an appropriate summary description of the TLAA evaluation, as required by 10 CFR 54.21(d).

4.3.6 Pressurizer Fatigue Analyses

4.3.6.1 Summary of Technical Information in the Application

In LRA Section 4.3.6, the applicant summarized the evaluation of the pressurizer fatigue analyses for the period of extended operation. The pressurizer is a vertical, hemispherical-head vessel, supported from its lower head by a flanged skirt, connected to the PCS by a 12-inch surge nozzle in the center of the lower head. The lower head also has 120 electric heater nozzles and four level instrument nozzles. There is a liquid-temperature element nozzle in the lower shell. The upper head has a manway, a spray nozzle for pressure control, four level instrument nozzles, a PORV and three code relief valve nozzles, and a vapor-space-temperature-element nozzle. The surge and spray nozzles have thermal sleeves. The vessel shell, heads, skirt, surge nozzle, and manway are carbon steel. The shell and upper head are clad with 304 stainless steel. The lower head and surge nozzles are clad with Ni-Cr-Fe alloy (Alloy 600). The heater sleeves and the remaining nozzles are Alloy 600 or have Alloy 600 safe ends (or flanges, for the code safety valves), except the PORV nozzle safe end, which was modified. The pressurizer was designed to the ASME B&PV Code, Section III, Class A, 1965, Winter 1966, Addenda. The code design calculation includes a fatigue analysis for those nozzles or other parts which do not meet the fatigue analysis exemption criteria of Section III, Paragraph N-415.1. These include all nozzles attached by J-welds and other nozzles and parts subject to more-severe thermal transients.

Thermal stratification phenomena in the surge line have required re-analysis of the surge nozzle, and concerns for high differential temperatures with auxiliary spray have required re-analysis of the spray nozzle. Primary water stress-corrosion cracking (PWSCC) of the Alloy 600 temperature nozzles required repair, a revised fatigue analysis, and analyses of the PWSCC effects. These failures, failure of the PORV nozzle safe end, and industry-wide cracking of Alloy 600 components have required evaluation of PWSCC effects in all Alloy 600 components.

4.3.6.2 Staff Evaluation

The staff reviewed LRA Section 4.3.6, pursuant to 10 CFR 54.21(c)(1)(i), to verify that the analyses remain valid for the period of extended operation and, pursuant to 10 CFR 54.21(c)(1)(iii), to verify that the effects of aging on the intended function(s) will be adequately managed for the period of extended operation.

The staff's review of LRA Section 4.3.6 identified an area in which additional information was necessary to complete the review of the pressurizer fatigue analyses TLAA. The applicant responded to the staff's RAI as follows.

In the LRA, the applicant stated that a safe-end-to-pipe Alloy 600 weld at the PORV nozzle was found cracked and leaking in 1993. This weld was repaired by two welds and a short section of stainless steel pipe, and the modified nozzle was re-analyzed for fatigue over a 40-year operation. The highest calculated CUF was determined as 0.7572, well within the ASME Section III Class 1 fatigue limit of 1.0. In 1994, the safe end material was replaced with 316 stainless steel material and Alloy 690 welds. The fatigue analysis for the currently installed safe end determined a worst-case CUF of 0.084 at the inside wall of the safe end transition.

In RAI 4.3-8, dated July 21, 2005, the staff requested that the applicant verify that the 0.7572 CUF represents the highest 60-year value for structures or components in the pressurizer like those listed in GALL Report, Volume 2, Table IV C2.5, "Pressurizers."

In its response, by letter dated August 19, 2005, the applicant provided a table of the pressurizer components with corresponding CUFs showing that the 0.7572 value represents the highest 60-year CUF for all pressurizer components except for pressurizer surge nozzle, the pressurizer spray nozzle, and a repaired pressurizer upper head temperature instrument nozzle. The applicant also stated that the number of DBEs is not expected to approach the design-basis cycle limits during the period of extended operation, and the actual CUFs are not expected to approach their calculated values in this period. In LRA Section 4.3.6, the applicant also stated that the Fatigue Monitoring Program will ensure that the bases for the modified PORV nozzle reanalysis, and for those portions of the original pressurizer analysis not superseded, remain valid. The fatigue analyses and the actual CUFs are, therefore, expected to remain valid for the period of extended operation.

Based on its review the staff finds that the calculated CUF value is not expected to exceed of the ASME Section III Class 1 fatigue limit of 1.0. The staff, therefore, found the applicant's response acceptable because the implementation of the Fatigue Monitoring Program will ensure that the CUF analysis remains valid for the period of extended operation. The staff finds the applicant's response acceptable; therefore, the staff's concern described in RAI 4.3-8 is resolved.

4.3.6.3 FSAR Supplement

The applicant provided an FSAR supplement summary description of its TLAA evaluation of the pressurizer fatigue analyses in LRA Section A4.2.5. On the basis of its review of the FSAR supplement, the staff concludes that the summary description of the applicant's actions to address the pressurizer fatigue analyses is adequate.

4.3.6.4 Conclusion

On the basis of its review, the staff concludes that the applicant has demonstrated, pursuant to 10 CFR 54.21(c)(1)(i), that, for the pressurizer fatigue analyses TLAA, the analyses remain valid for the period of extended operation. The applicant had also demonstrated, pursuant to 10 CFR 54.21(c)(1)(iii), that, for the TLAA, the effects of aging on the intended function(s) will be adequately managed for the period of extended operation. The staff also concludes that the FSAR supplement contains an appropriate summary description of the TLAA evaluation, as required by 10 CFR 54.21(d).

4.3.7 Regenerative Heat Exchanger Fatigue Analyses

4.3.7.1 Summary of Technical Information in the Application

In LRA Section 4.3.7, the applicant summarized the evaluation of the regenerative heat exchanger fatigue analyses for the period of extended operation. The regenerative heat exchanger recovers energy from the letdown line to heat chemical and volume control system charging water for reactor system makeup (charging) and auxiliary pressurizer spray. The heat exchanger letdown or tube side is separated from the PCS by a normally-open manual valve and a control valve and is therefore a Consumers Design Class 1 component. The charging or shell side is separated from the PCS by a check valve and control valve on each of the charging and auxiliary spray lines. The charging system has three positive-displacement pumps. The pumps cannot be throttled without lifting discharge safety valves, but one of them is variable speed. The charging system was designed to operate continuously, with flow from the fluid-drive variable-speed pump controlled by a primary coolant volume control signal. If the variable-speed pump is out of service, the system controls PCS makeup by cycling a constant-speed pump on and off. This cycling of cold makeup water against the (approximately constant) hot letdown flow produces significant thermal transients in the regenerative heat exchanger. Isolation of letdown flow introduces a similar differential thermal load and has a similar effect.

4.3.7.2 Staff Evaluation

The staff reviewed LRA Section 4.3.7, pursuant to 10 CFR 54.21(c)(1)(ii), to verify that the analyses have been projected to the end of the period of extended operation.

The applicant stated that the original design of the regenerative heat exchanger included a fatigue evaluation based on ASME Section III, 1965 Edition, Paragraph N-415. Revisions were made in 1993 and 1995 for changed operating conditions and procedures. The 1995 revision showed that the highest design-basis CUF was determined as 0.88 at the inner ligament of the shell side of the tube sheet. After an evaluation of plant events in the first 20 years of operation, the number of design transient events was projected for 60-year operation. The revised fatigue analysis shows that the highest 60-year CUF at the two most critical locations in the regenerative heat exchanger (tube sheet and tube sheet-to-shell junction) was calculated as 0.439, well below the ASME Section III Class 1 fatigue limit.

Based on its review, the staff finds that highest 60-year CUF at the two most critical locations in the regenerative heat exchanger will remain below the ASME Section III Class 1 fatigue limit. The staff concludes that the regenerative heat exchanger fatigue TLAAs have been projected to the end of the period of extended operation.

4.3.7.3 FSAR Supplement

The applicant provided an FSAR supplement summary description of its TLAA evaluation of the regenerative heat exchanger fatigue analyses in LRA Section A4.2.6. On the basis of its review of the FSAR supplement, the staff concludes that the summary description of the applicant's actions to address the regenerative heat exchanger fatigue analyses is adequate.

4.3.7.4 Conclusion

On the basis of its review, as discussed above, the staff concludes that the applicant has demonstrated, pursuant to 10 CFR 54.21(c)(1)(ii), that, for the regenerative heat exchanger fatigue analyses TLAA, the analyses have been projected to the end of the period of extended operation. The staff also concludes that the FSAR supplement contains an appropriate summary description of the TLAA evaluation, as required by 10 CFR 54.21(d).

4.3.8 ASME III Class A Primary Coolant Piping Fatigue Analyses

4.3.8.1 Summary of Technical Information in the Application

In LRA Section 4.3.8, the applicant summarized the evaluation of the ASME III Class A primary coolant piping fatigue analyses for the period of extended operation. A piping fatigue analysis was originally applied only to the main loops of the PCS, the two 42-inch hot legs and the four 30-inch cold legs, and to the connecting nozzles for smaller piping. The hot and cold legs are seam-welded from ASTM A516 Grade 70 plate, clad by roll-bonding with 1/4-inch (nominal) 304L stainless steel. The hot and cold legs are supported only by the reactor and steam generator nozzles and by the primary coolant pumps. The original analyses calculated fatigue usage factors for: hot legs, cold legs, safety injection-shutdown cooling nozzles, hot leg to surge line nozzle, charging inlet nozzles, hot leg temperature nozzle, shutdown cooling outlet nozzle, and cold leg temperature nozzles. The hot-leg-to-surge-line nozzle was re-analyzed to address transients not contemplated in the original analysis. The cold-leg-to-pressurizer spray nozzles and others, not listed above, were exempt from a fatigue calculation in the original analysis because all then-known cyclic stress ranges were below the endurance limit; however, the cold-leg-to-pressurizer-spray nozzles have since been evaluated for additional transients not contemplated in the original analysis.

4.3.8.2 Staff Evaluation

The staff reviewed LRA Section 4.3.8, pursuant to 10 CFR 54.21(c)(1)(i), to verify that the analyses remain valid for the period of extended operation and, pursuant to 10 CFR 54.21(c)(1)(iii), to verify that the effects of aging on the intended function(s) will be adequately managed for the period of extended operation.

The staff's review of LRA Section 4.3.8 identified areas in which additional information was necessary to complete the review of the primary coolant piping fatigue analyses TLAA. The applicant responded to the staff's RAIs as follows.

The applicant stated that the fatigue analysis of record for the hot and cold legs is based on the worst-possible-case stress ranges calculated at all locations on a typical hot leg and a typical cold leg. The 40-year calculated CUFs were 0.07551 and 0.7531, respectively. These CUFs are considered higher than those that would be calculated using location-specific fatigue analyses. The cold-leg CUF value of 0.7531 was also found to be higher than the nozzle CUFs in the original calculations. The number of DBEs that affect the hot and cold legs is not expected to approach the limits of the DBEs during the period of extended operation and, therefore, the actual fatigue CUFs are also not expected to approach their calculated values during this period.

The applicant intends to use the Fatigue Monitoring Program to monitor the cycles of the operating transients in these systems. The applicant stated that, if the cycle count reaches or exceeds the design-basis count limit during the period of extended operation, a re-analysis or other corrective action will ensure that the hot and cold leg CUFs meet the ASME Section III fatigue limit.

In RAI 4.3-9, dated July 21, 2005, the staff requested that the applicant provide the highest 60-year CUFs and disposition for all structures or components in the RCS and connected lines like those listed in GALL Report, Volume 2, Table IV C2 "Reactor Coolant System and Connected Lines (PWR)," Items C2.2-a, C2.2-b and C2.2-c, "Connected Systems Piping and Fittings."

In its response, by letter dated August 19, 2005, the applicant stated that, with the exception of the regenerative heat exchanger and the charging inlet nozzle, no ASME Section III fatigue analyses were performed because these systems were designed to the United States of America Standard (USAS) B31.1 Power Piping Code. The fatigue analysis results for the connected systems piping and fittings are summarized in LRA Section 4.3.13. The USAS B31.1 Power Piping Code addresses piping fatigue indirectly, through a series of reduction factors which depend on the number of specified design cycles.

The applicant stated that the Fatigue Monitoring Program will be applied to monitor the cycles of the operating transients in these systems. The applicant stated that, if the cycle count reaches or exceeds the design-basis count limit during the period of extended operation, a re-analysis or other corrective action will ensure that the stresses meet the USAS B31.1 Power Piping Code design criteria for the period of extended operation.

Based on its review the staff finds that these systems were designed according to USAS B31.1 Power Piping Code. The staff finds the applicant's response acceptable because the implementation of the corrective action program and the Fatigue Monitoring Program will ensure that the stresses meet the design criteria; therefore, the staff's concern described in RAI 4.3-9 is resolved

In RAI 4.3-10, dated July 21, 2005, the staff requested that the applicant provide the basis for stating that the number of each of the DBEs that affect the hot and cold legs is not expected to approach its design-basis limit during the period of extended operation.

In its response, by letter dated August 19, 2005, the applicant stated that the projected 60-year transients that affect the hot legs and the cold legs are based on the estimated actual transients to January 9, 2005 (34 years of operation) listed in LRA Table 4.3.1-1. Using the actual transients, the projected 60-year transients are shown to be lower than the initial design-basis transients.

Based on its review, the staff finds the applicant's response to RAI 4.3-10 acceptable because the projected 60-year transients are lower than the initial design-basis transients; therefore, the staff's concern described in RAI 4.3-10 is resolved.

4.3.8.3 FSAR Supplement

The applicant provided an FSAR supplement summary description of its TLAA evaluation of the ASME III Class A primary coolant piping fatigue analyses in LRA Section A4.2.7. On the basis of its review of the FSAR supplement, the staff concludes that the summary description of the applicant's actions to address the ASME III Class A primary coolant piping fatigue analyses is adequate.

4.3.8.4 Conclusion

On the basis of its review, as discussed above, the staff concludes that the applicant has demonstrated, pursuant to 10 CFR 54.21(c)(1)(i), that, for the ASME III Class A primary coolant piping fatigue analyses TLAA, that the analyses remain valid for the period of extended operation. The applicant had also demonstrated, pursuant to 10 CFR 54.21(c)(1)(iii), that, for the TLAA, the effects of aging on the intended function(s) will be adequately managed for the period of extended operation. The staff also concludes that the FSAR supplement contains an appropriate summary description of the TLAA evaluation, as required by 10 CFR 54.21(d).

4.3.9 Revised NRC Bulletin 88-11 Fatigue Analysis of the Hot Leg to Pressurizer Surge Line Nozzle, Surge Line, and Pressurizer Surge Nozzle

4.3.9.1 Summary of Technical Information in the Application

In LRA Section 4.3.9, the applicant summarized the evaluation of the revised NRC Bulletin 88-11 fatigue analysis of the hot leg to pressurizer surge line nozzle, surge line, and pressurizer surge nozzle for the period of extended operation. NRC Bulletin 88-11, dated

December 1988, was issued to address pressurizer surge line temperature stratification concerns. The effects of thermal stratification were evaluated by the Combustion Engineering Owners Group (CEOG). The CEOG report concluded the structural integrity of the pressurizer surge line is acceptable for the 40-year life of the plant. The NRC issued an safety evaluation on September 13, 1993, concluding that the CEOG analysis adequately demonstrates that the bounding surge line and nozzles meet ASME Section III stress and fatigue requirements for the 40-year design. Consumers Energy provided additional information detailing completion of the required actions of Bulletin 88-11, including the requirement to update the pressurizer surge line stress and fatigue analyses, as stated in FSAR Section 4.3.7.

4.3.9.2 Staff Evaluation

The staff reviewed LRA Section 4.3.9, pursuant to 10 CFR 54.21(c)(1)(i), to verify that the analyses remain valid for the period of extended operation; pursuant to 10 CFR 54.21(c)(1)(ii), to verify that the analyses have been projected to the end of the period of extended operation; and, pursuant to 10 CFR 54.21(c)(1)(iii), to verify that the effects of aging on the intended function(s) will be adequately managed for the period of extended operation.

The staff's review of LRA Section 4.3.9 identified areas in which additional information was necessary to complete the review of the fatigue analysis of the hot leg to pressurizer surge line nozzle, surge line, and pressurizer surge nozzle TLAA. The applicant responded to the staff's RAIs as follows.

The applicant stated that the effects of thermal stratification in the pressurizer surge line were evaluated for fatigue by the CEOG in response to NRC Bulletin 88-11, dated December 1988. The locations with the highest fatigue CUFs were determined to be the surge elbow, the hot-leg to pressurizer surge line nozzle, and the pressurizer surge nozzle. The effects of thermal stratification were considered in the fatigue analyses in addition to the original design-basis load events.

The fatigue evaluation of the surge line elbow determined a 40-year CUF of 0.937. This analysis was based on thermal stratification conditions under initial intermittent pressurizer spray operation. The fatigue evaluation was revised to reflect thermal stratification conditions under continuous pressurizer spray. This calculation determined a lower CUF of 0.0447 for the expected number of cycles at the end of the period of extended operation.

The fatigue analysis of the hot leg-to-surge line and the pressurizer surge line nozzle with intermittent pressurizer spray and the revised set of design basis load events, which included the thermal stratification transients, determined CUFs of 0.3818 and 0.9611, respectively. The fatigue evaluation was revised to reflect thermal stratification conditions under continuous pressurizer spray, which determined much lower CUFs. These CUFs were projected to remain below the ASME Section III fatigue limit for the period of extended operation.

In RAI 4.3-11, dated July 21, 2005, the staff requested that the applicant clarify whether PNP has operated with continuous pressurizer spray operation since the start of plant operation. If not, the staff requested that the applicant indicate when continuous spray operation was implemented.

In its response, by letter dated August 19, 2005, the applicant stated that intermittent pressurizer spray operation occurred in hot functional testing performed prior to plant operation. The intermittent operation resulted in PCS pressure oscillations. The operation was revised to permit continuous spray operation, which eliminated these pressure oscillations. This change was implemented prior to the beginning of commercial operation in 1971. However, the fatigue evaluation of a typical combustion engineering plant was based at that time on intermittent spray operation, which is what the applicant reported in the LRA.

Based on its review, the staff finds the applicant's response to RAI 4.3-11 acceptable because it clarified the basis of the pessurizer surge line fatigue analyses; therefore, the staff's concern described in RAI 4.3-11 is resolved.

In RAI 4.3-12, dated July 21, 2005, the staff requested that the applicant provide the basis for its statement that, "The number of transient events which might be expected to initiate these thermal stratification events will not exceed the design basis limits for the extended licensed operating period."

In its response, by letter dated August 19, 2005, the applicant stated that the "transient events" refer to the limiting cycles for the revised design-basis heat-up and cool-down transient. The expected cycles are listed as 240. The thermal stratification transient will also, therefore, not exceed 240 for 60-year operation.

Based on its review, the staff finds the applicant's response acceptable because it agreed with previous RAI responses; therefore, the staff's concern described in RAI 4.3-12 is resolved.

Based on its review the staff finds that the implementation of the Fatigue Monitoring Program will monitor the cycles of the operating transients in the pressurizer surge line elbow, the hot leg and surge line nozzles to ensure that appropriate action is taken in the event that a design-basis cycle count limit is reached during the period of extended operation and that the CUF analysis remain valid.

4.3.9.3 FSAR Supplement

The applicant provided an FSAR supplement summary description of its TLAA evaluation of revised NRC Bulletin 88-11 fatigue analysis of the hot leg to pressurizer surge line nozzle, surge line, and pressurizer surge nozzle in LRA Section A4.2.8. On the basis of its review of the FSAR supplement, the staff concludes that the summary description of the applicant's actions to address the revised NRC Bulletin 88-11 fatigue analysis of the hot leg to pressurizer surge line nozzle, surge line, and pressurizer surge nozzle is adequate.

4.3.9.4 Conclusion

On the basis of its review, as discussed above, the staff concludes that the applicant has demonstrated, pursuant to 10 CFR 54.21(c)(1)(ii), that, for the surge line elbow, the analyses have been projected to the end of the period of extended operation. The applicant has also demonstrated, pursuant to 10 CFR 54.21(c)(1)(iii), that the effects of aging on the intended function(s) will be adequately managed for the period of extended operation. The staff concludes that the applicant has demonstrated, pursuant to 10 CFR 54.21(c)(1)(i), that, for the hot leg and pressurizer surge nozzles, the analyses remain valid for the period of extended operation. The applicant had also demonstrated, pursuant to 10 CFR 54.21(c)(1)(iii), that the effects of aging on the intended function(s) will be adequately managed for the period of extended operation. The staff also concludes that the FSAR supplement contains an appropriate summary description of the TLAA evaluation, as required by 10 CFR 54.21(d).

4.3.10 Revised Fatigue Analysis of Nozzles from PCS Cold Legs 1B and 2A to Pressurizer Spray and of the Pressurizer Spray Nozzle

4.3.10.1 Summary of Technical Information in the Application

In LRA Section 4.3.10, the applicant summarized the evaluation of the revised fatigue analysis of nozzles from PCS cold legs 1B and 2A to pressurizer spray and of the pressurizer spray nozzle for the period of extended operation. Pressurizer spray is normally supplied by the reactor coolant pump head through three-inch nozzles on two of the four 30-inch PCS cold legs. Normal spray flow in each of these three-inch lines is continuous, through a normally throttled 3/4-inch main spray bypass valve, and through a three-inch main spray control valve. The charging line downstream of the regenerative heat exchanger supplies auxiliary spray from the chemical and volume control system through a two-inch control valve. All three of these sources supply a single pressurizer spray nozzle. The original design of the pressurizer included a fatigue analysis of the pressurizer spray nozzle. However, the normal spray piping and the auxiliary spray piping were designed to the American National Standards Institute (ANSI) B31.1 Code. Revised operating conditions and pressurizer cooldown rate prompted addition of a fatigue analysis for the two cold leg nozzles and for the auxiliary spray piping.

4.3.10.2 Staff Evaluation

The staff reviewed LRA Section 4.3.10, pursuant to 10 CFR 54.21(c)(1)(i), to verify that the analyses remain valid for the period of extended operation and, pursuant to 10 CFR 54.21(c)(1)(iii), to verify that the effects of aging on the intended function(s) will be adequately managed for the period of extended operation.

The staff's review of LRA Section 4.3.10 identified areas in which additional information was necessary to complete the review of the revised fatigue analysis of nozzles from PCS cold legs 1B and 2A to pressurizer spray and of the pressurizer spray nozzle TLAA. The applicant responded to the staff's RAIs as follows.

The applicant stated that the highest 40-year CUFs in the cold leg to pressurizer spray nozzles and for the pressurizer spray nozzle were 0.66 and 0.8214, respectively. These analyses were based on the design-basis transients and cycles. On this basis, the 60-year CUF for the cold leg-to-pressurizer spray nozzles was projected to the end of the period of extended operation as 0.99.

The revised analysis also estimated the maximum CUF in the spray nozzle to October 1991 as 0.353. Based on the trends at that time, the maximum 40-year CUF was estimated as 0.435. On that basis, the 60-year CUF to the end of the period of extended operation was projected as 0.517. However, the applicant also stated that the projected number of cycles of the DBEs is not expected to exceed the design-basis cycle limits during the period of extended operation. On this basis, the applicant concluded that the actual CUF is not expected to approach the calculated CUF by the end of the period of extended operation.

In RAI 4.3-13, dated July 21, 2005, the staff requested that the applicant provide the basis for the 60-year projected CUF of 0.517.

In its response, by letter dated August 19, 2005, the applicant stated that the estimated frequency of spray during the 20-year period subsequent to 1991 would increase the CUF by 0.082 to 0.435 for 40-year operation. On the assumption that operation will continue to be the same during the period of extended operation, the applicant demonstrated that the 60-year CUF would be 0.517.

Based on its review, the staff finds the applicant's response acceptable because it clarified the basis for the fatigue analysis; therefore, the staff's concern described in RAI 4.3-13 is resolved.

In RAI 4.3-14, dated July 21, 2005, the staff requested that the applicant provide the basis for the statement that, "The projected number of cycles of the design basis events does not exceed the design basis limit during the extended licensed operating period."

In its response, by letter dated August 19, 2005, the applicant provided the design-basis cycles for high-differential-temperature spray events. The response also included the number of cycles that had actually been counted to January 9, 2005. The counted cycles were substantially lower than the design-basis number of cycles. On this basis, the applicant extrapolated the expected number of cycles for 60 years and demonstrated that it would be lower than the projected number of design-basis cycles.

Based on its review, the staff finds the applicant's response to RAI 4.3-14 acceptable because it clarified the basis of the fatigue analysis; therefore, the staff's concern described in RAI 4.3-14 is resolved.

Based on its review the staff finds that the implementation of the Fatigue Monitoring Program will monitor the cycles of the operating transients in the cold leg-to-pressurizer spray nozzles and the pressurizer spray nozzle to ensure that appropriate action is taken in the event that a design-basis cycle count limit is reached during the period of extended operation and that the CUF analysis remain valid.

4.3.10.3 FSAR Supplement

The applicant provided an FSAR supplement summary description of its TLAA evaluation of the revised fatigue analysis of nozzles from PCS cold legs 1B and 2A to pressurizer spray and of the pressurizer spray nozzle in LRA Section A4.2.9. On the basis of its review of the FSAR supplement, the staff concludes that the summary description of the applicant's actions to address the revised fatigue analysis of nozzles from PCS cold legs 1B and 2A to pressurizer spray and of the pressurizer spray nozzle is adequate.

4.3.10.4 Conclusion

On the basis of its review, as discussed above, the staff concludes that the applicant has demonstrated, pursuant to 10 CFR 54.21(c)(1)(i), that, for the cold leg to pressurizer spray nozzles and the pressurizer spray nozzles, the analyses remain valid for the period of extended operation. The applicant has also demonstrated, pursuant to 10 CFR 54.21(c)(1)(iii), that, for the pressurizer spray nozzles, the effects of aging on the intended function(s) will be adequately managed for the period of extended operation. The staff also concludes that the FSAR supplement contains an appropriate summary description of the TLAA evaluation, as required by 10 CFR 54.21(d).

4.3.11 Pressurizer Auxiliary Spray Line Tee Fatigue Analysis in Response to NRC Bulletin 88-08

4.3.11.1 Summary of Technical Information in the Application

In LRA Section 4.3.11, the applicant summarized the evaluation of the pressurizer auxiliary spray line tee fatigue analysis in response to NRC Bulletin 88-08 for the period of extended operation. NRC Bulletin 88-08 and supplements describe observed effects of thermal cycling and thermal stratification in RCS pressure boundary components due to thermally-driven cyclic in-leakage at isolation valves and similar phenomena. In 1989, a conservative, bounding analysis of the section of the auxiliary spray line from check valve CK2118 to the pressurizer spray line tee demonstrated that fatigue due to these effects would be acceptable for the then-remaining 30-year licensed operating life.

The piping material is A 376 Type 316. The auxiliary spray line connects to the normal spray line vertically and below so that the cooler auxiliary spray water does not cause a thermal stratification effect. The analysis assumed 500 lifetime operating basis earthquake cycles and 500 lifetime full-range thermal expansion (heat-up and cool-down) cycles. It modeled the NRC Bulletin 88-08 phenomena (due to in-leakage through the check valve) as a thermal cycle every two minutes, between 536 and 400 °F, or 1.84×10^5 per year at a 70 percent availability factor.

The 500 lifetime operating bases earthquake cycles are assumed to occur only once in a design lifetime, independent of its length. The analysis also conservatively attributed the entire 500 full-range thermal expansion to the remaining 30-year life. These operating basis earthquake (OBE) and full-range thermal cycles contributed a negligible 0.0063 usage factor. The assumed NRC Bulletin 88-08 cycles contributed 0.4245 in 30 years for an end-of-life CUF of 0.43.

4.3.11.2 Staff Evaluation

The staff reviewed LRA Section 4.3.11, pursuant to 10 CFR 54.21(c)(1)(i), to verify that the analyses remain valid for the period of extended operation.

The applicant stated that the 1989 bounding fatigue analysis of the section between check valve CK2118 to the pressurizer spray line tee of the auxiliary spray line was based on the design-basis heat-up and cool-down transient cycles and the design-basis OBE cycles taken over the remaining 30 years of operation. In addition, the NRC Bulletin 88-08 check valve in-leakage phenomena were modeled as a thermal transient ranging from 400 °F to 536 °F at a rate of 1.84×10^5 cycles per year with a 70 percent availability factor. On this basis, the fatigue analysis of the tee determined a CUF of 0.4245 at the end of the 40-year licensing period.

The applicant stated that recent experience indicated an availability factor of about 90 percent. Increasing the number of design-basis thermal transients, OBE cycles, and NRC Bulletin 88-08 thermal cycles resulted in a CUF of 0.92 to the end of the period of extended operation, below the ASME Section III Class 1 fatigue limit of 1.0. The staff finds this acceptable for those components where the CUF did not exceed 1.0, and concurred with the applicant that the effects of aging of the pressurizer auxiliary spray line will remain valid for the period of extended operation.

Based on its review, the staff concludes that the applicant has demonstrated that the pressurizer auxiliary spray line tee TLAA will remain valid to the end of the period of extended operation.

4.3.11.3 FSAR Supplement

The applicant provided an FSAR supplement summary description of its TLAA evaluation of the pressurizer auxiliary spray line tee fatigue analysis in response to NRC Bulletin 88-08 in LRA Section A4.2.10. On the basis of its review of the FSAR supplement, the staff concludes that the summary description of the applicant's actions to address the pressurizer auxiliary spray line tee fatigue analysis in response to NRC Bulletin 88-08 is adequate.

4.3.11.4 Conclusion

On the basis of its review, as discussed above, the staff concludes that the applicant has demonstrated, pursuant to 10 CFR 54.21(c)(1)(i), that, for the pressurizer auxiliary spray line tee TLAA, the analyses remain valid for the period of extended operation. The staff also concludes that the FSAR supplement contains an appropriate summary description of the TLAA evaluation, as required by 10 CFR 54.21(d).

4.3.12 Absence of a TLAA for ASME III Class 1 HELB Locations and Leak-Before-Break Analyses Based on Fatigue Usage Factor

4.3.12.1 Summary of Technical Information in the Application

In LRA Section 4.3.12, the applicant summarized the evaluation of TLAAs for ASME III Class 1 high-energy line break (HELB) locations and leak-before-break analyses based on fatigue usage factor criteria for the period of extended operation. Review of the licensing basis and the associated HELB reports revealed that selected break locations, either inside or outside containment, were not dependent on aging factors. Therefore, HELB analyses are not TLAAs.

4.3.12.2 Staff Evaluation

The staff reviewed LRA Section 4.3.12, pursuant to 10 CFR 54.3(a), to verify that the absence of TLAAs for ASME III Class 1 HELB locations and leak-before-break analyses.

The staff reviewed FSAR Section 5.6.3.1, which shows the requirement that pipe breaks be postulated at all intermediate locations throughout a piping system where the CUF exceeds 0.10. This criterion is also stated in GL 87-11, which is referenced in the FSAR. The applicant stated that HELB locations in ASME Section III Class 1 piping at are not dependent on aging factors. The staff interpreted this statement as indicating that this criterion has not been considered in the postulation of intermediate HELB in ASME Section III Class 1 piping. The applicant did not provided any basis for this statement.

By letter dated March 30, 2006, the applicant, provided clarification regarding the postulation of pipe breaks in ASME III Class 1 high energy lines. The applicant stated that the HELB locations inside containment were postulated based primarily on an effects oriented approach. This procedure assumed that pipe breaks could occur at any location within high energy piping that is in the vicinity of safety related equipment or components. Pipe breaks were postulated at terminal ends and intermediate locations, chosen as follows: a longitudinal break was postulated at a point which produces the greatest impingement loading on each component of each essential system; a circumferential break was postulated at the point which produces the greatest pipe whip loading on each component of each essential system.

The applicant stated that a few high-energy lines (PCS hot and cold legs) had ASME III Class 1 fatigue analyses. However, the CUFs for these lines were calculated based on enveloped stress ranges for the entire line, a procedure that does not permit the calculation of CUFs at individual locations. Thus, postulation of breaks at particular locations based on the fatigue criterion of GL 87-11, CUF ≤ 0.10, was not possible, since location-based CUFs did not exist. Breaks in these lines were, therefore, stress-based (i.e., intermediate) breaks postulated at locations where the pipe stress exceeded the GL 87-11 stress requirement of 2.4 S_m.

The staff reviewed the applicants response, and concurred that, since the postulation of HELB did not involve time-limited assumptions defined by the current operating term (i.e., fatigue CUFs were not used as a basis for postulating HELB in Class 1 lines), the postulation of pipe breaks based on fatigue CUFs is not a TLAA, in accordance with 10 CFR 54.3.

4.3.12.3 FSAR Supplement

The staff concludes that HELB is not a TLAA; therefore, an FSAR supplement is not required.

4.3.12.4 Conclusion

The staff concludes that the applicant had provided an acceptable demonstration, pursuant to 10 CFR 54.3(a), regarding the absence of TLAAs for ASME III Class 1 HELB locations and leak-before-break analyses. The staff also concludes that an FSAR supplement is not necessary to satisfy the requirements of 10 CFR 54.21(d).

4.3.13 Assumed Thermal Cycle Count for Allowable Secondary Stress Range Reduction Factor in Piping and Components

4.3.13.1 Summary of Technical Information in the Application

In LRA Section 4.3.13, the applicant summarized the evaluation of the assumed thermal cycle count for allowable secondary stress range reduction factor in piping and components for the period of extended operation. This section addresses the issue of assumed thermal cycle counts which determine an allowable secondary stress range reduction factor for some Consumers Power (CP Co) Design Class 1 piping and components and for non-CP Co Design Class 1 piping and components.

Only the PCS piping and components have an ASME Class 1 fatigue analysis. Other CP Co Design Class 1 piping and components were designed to the ANSI B31.1 Power Piping Code, ASME Section VIII, or ASME III Classes 2 and 3, which require a stress range reduction factor to the allowable stress range for secondary (expansion and displacement) stresses to account for thermal cycling. For ANSI B31.1, the allowable secondary stress range is 1.0 S_A for

7,000 equivalent full-range thermal cycles or less. The allowable secondary stress range is reduced to 0.5 S_A for thermal cycles greater than 100,000. Components designed to other codes, such as ASME VIII, have identical or very similar provisions. An increase in design life could increase the number of full-range thermal cycles; therefore, design analyses under these codes are TLAAs.

With regard to the stress range reduction factors and corresponding thermal cycle count assumptions, the Consumers Power Class 2 and 3 piping systems and components are designed to the same requirements as CP Co Design Class 1 piping. The review of possible TLAAs found no piping and components design analyses to the ANSI B31.1 rules, which invoke lower stress range reduction factors for an increase in the equivalent full-range thermal and displacement cycles.

4.3.13.2 Staff Evaluation

The staff reviewed LRA Section 4.3.13, pursuant to 10 CFR 54.21(c)(1)(i), to verify that the analyses remain valid for the period of extended operation; pursuant to 10 CFR 54.21(c)(1)(ii), to verify that the analyses have been projected to the end of the period of extended operation; and, pursuant to 10 CFR 54.21(c)(1)(iii), to verify that the effects of aging on the intended function(s) will be adequately managed for the period of extended operation.

In the LRA, the applicant stated that the number of lifetime (full-range or equivalent) thermal and displacement cycles applicable to most of the ANSI B31.1 piping and components are not expected to exceed the number of plant DBE cycles. As long as these are not exceeded, the ANSI B31.1 stress range reduction factors assumed in the analyses will remain valid for the period of extended operation.

The applicant stated that the TLAA fatigue review of piping designed to the ANSI B31.1 Power Piping Code revealed two piping systems where the projected number of full-range and equivalent thermal cycles exceed the minimum 7,000 cycle limit of the code. Thermal cycles greater than the limit of 7,000 cycles require reduction factors in calculating the allowable secondary stress-range limit of the code. One system consists of the charging lines inboard of the regenerative heat exchanger. For this system, the original design basis 6,000 events increased to 18,000 for a 40-year life and 27,000 for a 60-year life as a result of an increase in partial-range thermal cycling due to cycling of the charging pumps. The evaluation of the effects of these additional cycles on the regenerative heat exchanger and the charging inlet nozzle were reported in LRA Sections 4.3.7 and 4.3.8. The applicant stated that the calculation of the charging lines will be revised to reflect the effects of the additional cycles on these lines. The other system is the PCS hot leg sampling piping, which may exceed 7,000 cycles during the period of extended operation.

The applicant stated that the analysis of the charging lines inboard of the regenerative heat exchanger will be revised to evaluate the effect that the increase in variable speed charging pump events may have on these lines and will take actions to ensure that these lines meet the licensing basis design criteria for the period of extended operation. The applicant will complete this evaluation and advise the staff of the results, and of any necessary corrective actions, before the end of the current licensed operating period. The other system is the PCS hot leg sampling piping, which may experience cycling exceeding the ANSI B31.1 limit of 7,000 cycles.

The applicant will perform a calculation to justify PCS sampling at any reasonable frequency for 60 years of operation without exceeding the 7,000 cycle limit.

The applicant also stated that the Fatigue Monitoring Program will monitor the cycles of the operating transients in the cold leg to pressurizer spray nozzles and the pressurizer spray nozzle. The program will ensure that a re-analysis or other corrective action will be taken in the event that a design-basis cycle count limit is reached during the period of extended operation.

The applicant committed to evaluate the effect that the increase in variable speed charging pump events may have on these lines and will take actions to ensure that these lines meet the licensing basis design criteria. The applicant also committed to submit the results of the evaluation to the staff prior to entering the period of extended operation. Therefore, the staff concludes that with this commitment and the implementation of the Fatigue Monitoring Program, the TLAAs associated with the charging lines inboard of the regenerative heat exchanger will remain valid through the end of the period of extended operation (Commitment 37).

4.3.13.3 FSAR Supplement

The applicant provided an FSAR supplement summary description of its TLAA evaluation of the assumed thermal cycle count for allowable secondary stress range reduction factor in piping and components in LRA Section A4.2.11. On the basis of its review of the FSAR supplement, the staff concludes that the summary description of the applicant's actions to address the assumed thermal cycle count for allowable secondary stress range reduction factor in piping and components is adequate.

4.3.13.4 Conclusion

On the basis of its review, as discussed above, the staff concludes that the applicant has demonstrated, pursuant 10 CFR 54.21(c)(1)(ii), that, for the charging lines inboard of the regenerative heat exchanger, the analyses have been projected to the end of the period of extended operation. The staff concludes that the applicant has demonstrated, pursuant to 10 CFR 54.21(c)(1)(i), that, for piping and components, the analyses remain valid for the period of extended operation. In addition, the staff concludes that the applicant has demonstrated, pursuant to 10 CFR 54.21(c)(1)(iii), that, for the piping and components, the effects of aging on the intended function(s) will be adequately managed for the period of extended operation. The staff also concludes that the FSAR supplement contains an appropriate summary description of the TLAA evaluation, as required by 10 CFR 54.21(d).

4.3.14 Effects of Reactor Coolant System Environment on Fatigue Life of Piping and Components (Generic Safety Issue 190)

4.3.14.1 Summary of Technical Information in the Application

In LRA Section 4.3.14, the applicant summarized the evaluation of the effects of the RCS environment on the fatigue life of piping and components for the period of extended operation. The effects of the reactor coolant environment may need to be included in the calculated fatigue life of components; Generic Safety Issue (GSI)-190, "Fatigue Evaluation of Metal Components for 60-year Plant Life," addressed this issue. Although the parent GSI-190 safety issue has been resolved, SRP-LR Section 4.3.1.2, states that, "The applicant's consideration of the effects of coolant environment on component fatigue life for license renewal is an area of review." The GSI-190 review requirements are therefore imposed by the standard review plan and do not depend on the individual plant licensing basis.

4.3.14.2 Staff Evaluation

The staff reviewed LRA Section 4.3.14, pursuant to 10 CFR 54.21(c)(1)(ii), to verify that the analyses have been projected to the end of the period of extended operation and, pursuant to 10 CFR 54.21(c)(1)(iii), to verify that the effects of aging on the intended function(s) will be adequately managed for the period of extended operation.

GSI-166, "Adequacy of the Fatigue Life of Metal Components," raised concerns regarding the conservatism of the fatigue curves used in the design of the RCS components. Although GSI-166 was resolved for the current 40-year design life of operating components, the staff identified GSI-190, to address license renewal. The NRC closed GSI-190 in December 1999, concluding that:

> The results of the probabilistic analyses, along with the sensitivity studies performed, the iterations with industry (NEI and EPRI), and the different approaches available to the licensees to manage the effects of aging, lead to the conclusion that no generic regulatory action is required, and that GSI-190 is closed. This conclusion is based primarily on the negligible calculated increases in core damage frequency in going from 40- to 60-year lives. However, the calculations supporting resolution of this issue, which included consideration of environmental effects, and the nature of age-related degradation indicate the potential for an increase in the frequency of pipe breaks as plants continue to operate. Thus, the staff concluded that, consistent with existing requirements in 10 CFR 54.21, licensees should address the effects of the coolant environment on component fatigue life as aging management programs are formulated in support of license renewal.

The applicant stated that plant-specific calculations were performed for the six sample locations applicable to PNP adapted from the seven locations identified in NUREG/CR-6260, "Application of NUREG/CR-5999 Interim Fatigue Curves to Selected Nuclear Power Plant Components," for older-vintage Combustion Engineering plants. The identified locations consisted of the reactor vessel lower head to shell transition, the primary coolant inlet nozzle, the primary coolant outlet nozzle, the surge line elbow, the charging nozzles, the safety injection nozzles, and the shut-down cooling line inlet transition. Of these, PNP has no shut-down cooling line inlet transition. The safety injection and shut-down cooling functions share a common nozzle.

The detailed environmental fatigue calculations used the appropriate environmental fatigue factor (F_{en}) relationships from NUREG/CR-6583, "Effects of LWR Coolant Environments on Fatigue Design Curves of Carbon and Low-Alloy Steels," for carbon and alloy steels, and from NUREG/CR 5704, "Effects of LWR Coolant Environments on Fatigue Design Curves of Austenitic Stainless Steels," for stainless steel as appropriate for the material. The analyses showed that the CUFs at all locations, based on the inclusion of environmental effects in the fatigue analyses, will remain less than the ASME Section III Class 1 fatigue limit with the exception of the charging nozzle. The CUF for this location was calculated as 4.428.

The staff's review of LRA Section 4.3.14 identified an area in which additional information was necessary to complete the review of the analyses of the effects of the RCS environment on fatigue life of piping and components TLAA. The applicant responded to the staff's RAI as follows.

In RAI 4.3-15, dated July 21, 2005, the staff requested that the applicant provide the environmentally-based highest CUF at the nozzle shared by the safety injection and shutdown cooling functions.

In its response, by letter dated August 19, 2005, the applicant stated that preliminary analysis results indicate that the limiting location in this area is at the end of the cladding near the safe end on the safety injection nozzle. This nozzle supports both safety injection and shutdown cooling. The non-environmental CUF has been determined to be 0.0308. The environmentally corrected CUF has been determined to be 0.472, based on applying the environmental factor of 15.35 for stainless steel.

In its supplemental response, by letter dated October 28, 2005, the applicant indicated that the limiting location of concern occurs at the acute angle of the intersection at the inside surfaces of the safety injection nozzle and the primary coolant system pipe. The 60-year fatigue usage factor is 0.036. After application of the environmental factor of 1.79 for carbon steel, the environmentally corrected usage factor is 0.065.

Based on its review, the staff finds the applicant's response to RAI 4.3-15 acceptable because it conformed to the staff position on environmental effects on fatigue; therefore, the staff's concern described in RAI 4.3-15 is resolved.

In the LRA, the applicant stated that the charging nozzle has an Alloy 600 safe end and is one of the components in the Alloy 600 Inspection Program, which specifies inspection methods and inspection frequency. The applicant also stated that the Fatigue Monitoring Program will ensure that the original design-basis number of load cycles for each loading event is not exceeded during the period of extended operation and that appropriate action is taken if exceeded.

The staff found that the environmental fatigue TLAAs of locations listed in NUREG/CR 6260 for older-vintage Combustion Engineering plants will remain valid to the end of the period of extended operation. The staff also found acceptable the applicant's use of the Fatigue Monitoring Program to ensure that the original design-basis number of load cycles for each loading event is not exceeded during the period of extended operation, and that appropriate action is taken if exceeded.

4.3.14.3 FSAR Supplement

The applicant provided an FSAR supplement summary description of its TLAA evaluation of the analyses of the effects of the RCS environment on fatigue life of piping and components in LRA Section A4.2.12. On the basis of its review of the FSAR supplement, the staff concludes that the summary description of the applicant's actions to address the analyses of the effects of the RCS environment on fatigue life of piping and components is adequate.

4.3.14.4 Conclusion

On the basis of its review, as discussed above, the staff concludes that the applicant has demonstrated, pursuant to 10 CFR 54.21(c)(1)(ii), that, for the effects of the RCS environment on fatigue life of piping and components TLAA, the analyses have been projected to the end of the period of extended operation. The applicant had also demonstrated, pursuant to 10 CFR 54.21(c)(1)(iii), that, for the TLAA, the effects of aging on the intended function(s) will be adequately managed for the period of extended operation. The staff also concludes that the FSAR supplement contains an appropriate summary description of the TLAA evaluation, as required by 10 CFR 54.21(d).

4.4 Environmental Qualification of Electrical Equipment

The staff established nuclear station environmental qualification (EQ) requirements in 10 CFR Part 50, Appendix A, Criterion 4, and in 10 CFR 50.49, which specifically requires that an EQ program be established to demonstrate that certain electrical components located in "harsh" plant environments are qualified to perform their safety function in those environments after the effects of in-service aging. "Harsh" environments are defined as those areas of the plant that could be subject to the environmental effects of a loss-of-coolant accident (LOCA), HELB, or post-LOCA radiation. The effects of significant aging mechanisms are required by 10 CFR 50.49 to be addressed as part of an EQ program. For the purpose of license renewal, only those components with a qualified life of 40 years or greater would be TLAAs.

As required by 10 CFR 54.21(c)(1), the applicant must provide a list of EQ TLAAs in the LRA. The applicant shall demonstrate that one of the following is true for each type of EQ equipment: (1) the analyses remain valid for the period of extended operation, (2) the analyses have been projected to the end of the period of extended operation, or (3) the effects of aging on the intended function(s) will be adequately managed for the period of extended operation.

4.4.1 Summary of Technical Information in the Application

In LRA Section 4.4, the applicant summarized the evaluation of EQ of electrical equipment for the period of extended operation. Section 50.49(e)(5) of 10 CFR contains provisions for aging that require, in part, consideration of all significant types of aging degradation that can affect component functional capability. Section 50.49(e)(5) also requires component replacement or maintenance prior to the end of designated life unless additional life is established through ongoing qualification. Sections 50.49(k) and (l) permit different qualification criteria to apply based on plant vintage. Supplemental EQ regulatory guidance for compliance with these different qualification criteria is provided in RG 1.89, Revision 1, "Environmental Qualification of Certain Electric Equipment Important to Safety for Nuclear Power Plants," Division of Operating Reactors (DOR) Guidelines, and NUREG-0588, "Interim Staff Position on Environmental Qualification of Safety-Related Equipment," dated July 1981.

The Electrical Equipment Qualification (EEQ) Program was established to demonstrate that certain electrical components, located in harsh plant environment (that is, those areas of the plant that could be subject to the harsh environmental effects of a LOCA, HELB or post-LOCA radiation), are qualified to perform their safety function operation in those harsh environments after the effects of in-service aging. The EEQ Program manages applicable component thermal, radiation, and cyclic aging effects for the current operating license period using the qualification methods established by 10 CFR 50.49(f). Maintaining qualification through the extended license renewal period requires that existing EQ evaluations be re-analyzed.

Under 10 CFR 54.21(c)(1)(iii), a plant EQ program, which implements the requirements of 10 CFR 50.49 (as further defined and clarified by the DOR Guidelines, NUREG-0588, and RG 1.89, Revision 1), is viewed as an aging management program (AMP) for license renewal. Re-analysis of an aging evaluation to extend the qualification of components under 10 CFR 50.49(f) is performed on a routine basis as part of the EEQ program. Important attributes of the re-analysis of an aging evaluation include: (a) analytical methods, (b) data collection and reduction methods, (c) underlying assumptions, (d) acceptance criteria, and (e) corrective actions (if acceptance criteria are not met). Each of these attributes of the re-analysis is described in detail below.

Analytical Methods. The analytical models used in the re-analysis of an aging evaluation are the same as those applied during the first evaluation. The applicant stated that the Arrhenius methodology is an acceptable thermal model for performing an aging evaluation. The analytical method used for a radiation aging evaluation involves a demonstration of qualification for the total integrated dose (i.e., normal radiation dose for the projected installed life plus accident-radiation dose). For license renewal, one acceptable method of establishing the 60-year normal radiation dose is to multiply the 40-year normal radiation dose by 1.5 (i.e., 60 years/40 years). The result is added to the accident-radiation dose to obtain the total integrated dose for the component. For cyclical aging, a similar approach may be used. Other models may be justified on a case-by-case basis.

Data Collection and Reduction Methods. Reducing excess conservatism in the component service conditions (e.g., temperature, radiation, cycles) used in the prior aging evaluation is the chief method used for a re-analysis. Temperature data used in an aging evaluation are to be conservative and based on plant design temperatures or actual plant temperature data. Plant temperature data can be obtained in several ways, including monitors used for TS compliance, other installed monitors, measurements made by plant operators during rounds, and temperature sensors on large motors (while the motor is not running). A representative number of temperature measurements is conservatively evaluated to establish the temperatures used in an aging evaluation. Plant temperature data may be used in an aging evaluation in different ways, such as directly applying the plant temperature data or by using the plant temperature data to demonstrate conservatism when using plant design temperatures. Any changes to material activation energy values, as part of a re-analysis, are justified on a case-specific basis. Similar methods of reducing excess conservatism in the component service conditions from prior aging evaluations may be used for radiation and cyclical aging.

Underlying Assumptions. EQ component aging evaluations contain sufficient conservatism to account for most environmental changes occurring due to plant modifications and events. When unexpected adverse conditions are identified during operational or maintenance activities that affect the normal operating environment of a qualified component, the affected EQ component is evaluated and appropriate corrective actions are taken that may include changes to the qualification bases and conclusions.

Acceptance Criteria and Corrective Actions. The reanalysis of an aging evaluation could extend the qualification of a component. If the qualification cannot be extended by re-analysis, the component is maintained, replaced, or re-qualified prior to exceeding the period for which the current qualification remains valid. A reanalysis is performed in a timely manner (i.e., sufficient time must be available to maintain, replace, or requalify the component if the re-analysis is unsuccessful).

4.4.2 Staff Evaluation

The staff reviewed LRA Section 4.4, pursuant to 10 CFR 54.21(c)(1)(iii), to verify that the effects of aging on the intended function(s) will be adequately managed for the period of extended operation.

The staff reviewed LRA Section 4.4 where the applicant described the technical bases and justification for its EEQ program to adequately manage the effects of aging on the intended function(s) of electrical components for the period of extended operation. The staff reviewed this section to determine whether the applicant demonstrated that the effects of aging on the electrical equipment intended function(s) will be adequately managed through the EEQ Program, together with other programs and processes, during the period of extended operation, as required by 10 CFR 54.21(c)(1)(iii).

The staff reviewed LRA Section 4.4 to determine whether the applicant had submitted adequate information to meet the 10 CFR 54.21(c)(1) requirement. For the electrical equipment identified in LRA Table 4.4, the applicant used 10 CFR 54.21(c)(1)(iii) in its TLAA evaluation to demonstrate that the aging effects of EQ equipment will be adequately managed during the period of extended operation. The staff reviewed the EEQ Program to determine whether it would assure that the electrical and I&C components covered would continue to perform their intended functions consistently with the CLB for the period of extended operation. The staff's evaluation of the components qualification focused on how the EEQ Program manages the aging effects to meet 10 CFR 50.49 requirements.

The staff conducted an audit of the information provided in LRA Section B3.1 and of the program bases documents, which are available at the applicant's engineering office. The staff audit and review found that the EEQ Program, which the applicant claimed to be consistent with GALL AMP X.E1, "Environment Qualification of Electrical Components," is indeed consistent with GALL AMP X.E1. Therefore, the staff found that the EEQ Program is capable of programmatically managing the qualified life of components within the scope of the program for license renewal. The continued implementation of the EEQ Program assures that the aging effects will be managed and that components within the scope of the EEQ Program will continue to perform their intended functions for the period of extended operation.

4.4.3 FSAR Supplement

The applicant provided an FSAR supplement summary description of its TLAA evaluation of environmental qualification of electrical equipment in LRA Section A4.3. On the basis of its review of the FSAR supplement, the staff concludes that the summary description of the applicant's actions to address environmental qualification of electrical equipment is adequate.

4.4.4 Conclusion

On the basis of its review, as discussed above, the staff concludes that the applicant has demonstrated, pursuant to 10 CFR 54.21(c)(1)(iii), that, for the environmental qualification of electrical equipment TLAA, the effects of aging on the intended function(s) will be adequately managed for the period of extended operation. The staff also concludes that the FSAR supplement contains an appropriate summary description of the TLAA evaluation, as required by 10 CFR 54.21(d).

4.5 Concrete Containment Tendon Prestress Analysis

4.5.1 Summary of Technical Information in the Application

In LRA Section 4.5, the applicant summarized the evaluation of the concrete containment tendon prestress analysis for the period of extended operation. The containment is a Consumers Design Class 1 structure. It consists of a post-tensioned, reinforced concrete cylinder and dome connected to and supported by a massive, reinforced concrete foundation slab. It is designed to American Concrete Institute (ACI) 318-63, with some adaptations of the design rules and equations, as described in FSAR Section 5.8. The post-tensioning system consists of three groups of tendons: (1) three sets of 55 dome tendons (165 total), spaced 120 degrees apart and anchored at the vertical face of the dome ring girder; (2) 178 vertical tendons anchored at the top surface of the ring girder and at the bottom of the base slab; and (3) six sets of hoop tendons (502 total), spaced 60 degrees apart, each spanning 120 degrees of arc, anchored at six vertical buttresses (the midpoint of each hoop tendon therefore passes under the anchor buttress for two other sets of hoop tendons). PNP tendons are ungrouted. The tendon sheaths (glands, conduits) are filled with a corrosion preventive medium. Some sheaths were blocked. Tendons were not installed in two of the 180 vertical sheaths, nor in 20 of the 522 hoop sheaths. Evaluation of the original containment design determined that these tendons were not required.

For the TLAA, the applicant noted that the original design included a calculation of expected loss of prestress for the plant design life in accordance with ACI 318-63. The calculation evaluated loss of prestress due to friction and initial seating loss, tendon relaxation, concrete elasticity, concrete shrinkage, and concrete creep. Furthermore, the applicant noted that the original analysis was conservative, as demonstrated by a regression analysis of tendon surveillance data from the 20th and 25th-year tendon surveillances. This regression analysis indicated that the effective dome, hoop, and vertical tendon forces would remain significantly higher than values predicted by the original relaxation estimates beyond the 40-year licensed operating period.

The Containment Inservice Inspection Program predicts time-dependent lower limits of the lift-off force (predicted lower limits (PLLs)) for each tendon subgroup by regression analysis of individual tendon surveillance data and maintains trend lines of the data for each tendon surveyed. The program inspects a sample of tendons from each group (dome, vertical, and hoop) in each inspection interval to confirm that the trend lines remain within the tolerances of the PLLs and therefore that tendon prestresses will remain above their respective minimum required values (MRVs) for the succeeding inspection interval. The program provides for appropriate actions if surveillance data indicate that a trend line may cross its MRV.

4.5.2 Staff Evaluation

The staff reviewed LRA Section 4.5, pursuant to 10 CFR 54.21(c)(1)(iii), to verify that the effects of aging on the intended function(s) will be adequately managed for the period of extended operation.

In its review, the staff noted that the applicant did not provide quantitative comparison of the measured prestressing force trend line with the PLL or with MRV. In the description of the TLAA, the applicant provides references to the Containment In-service Inspection Program and various sections of FSAR Chapter 5. The staff reviewed these references and identified areas in which additional information was necessary to complete the review of the applicant's concrete containment tendon prestress analysis TLAA. The applicant responded to the staff's RAIs as follows.

In RAI 4.5-1, dated June 22, 2005, the staff stated that FSAR Section 5.8.5.3.1 provides the final prestress values as 140.6 KIP per square inch (ksi), 138 ksi, and 143 ksi for the dome, hoop, and vertical tendons, respectively. It appears that the applicant had used these values as the MRVs. As these values are not at the anchorages, they should not be used as MRVs without modifying them for appropriate friction and elastic shortening losses. Therefore, the staff requested that the applicant provide MRVs in terms of force per tendon.

In its response, by letter dated July 25, 2005, the applicant stated:

> FSAR Section 5.8.5.3.1 provides the initial estimate of the final wire prestress for the containment tendons. The calculated forces in this section of the FSAR do not represent the minimum required values (MRV) for the containment tendons. FSAR Section 5.8.8.3.4 lists the acceptance criteria of the lift-off force per tendon as no less than 584 KIPs per tendon for dome tendons and no less than 615 KIPs per tendon for hoop and vertical tendons.

Based on its review, the staff finds the applicant's response to RAI 4.5-1 acceptable because the applicant provided a clarification; therefore, the staff's concern described in RAI 4.5-1 is resolved.

In RAI 4.5-2, dated June 22, 2005, the staff noted that, in describing the scope of the program related to GALL AMP X.S1, in LRA Section B2.1.7, the applicant stated that, "The prestressing force in the common tendon is measured during each surveillance and is used to establish the trend of prestressing loss of the group." This practice is not consistent with GALL AMP X.S1 and Information Notice (IN) 99-10, "Degradation of Prestressing Tendon Systems in Prestressed Concrete Containments." The IN recommends the use of all available measured prestressing forces for developing trend lines for each group of tendons. The applicant started using random sampling of tendons since the 15-year tendon surveillance; therefore, the applicant has at least three sets of measured prestressing forces for performing regression analyses and developing trend lines. The staff requested that the applicant provide trend lines for each group of tendons for comparison with the PLLs and MRVs. The staff also requested that the applicant tabulate the measured prestressing forces (not individual wire forces) used in the corrected regression analyses.

In its response, by letter dated July 25, 2005, the applicant stated that regression analysis results are contained in the Precision Surveillance Corporation report entitled "30th Year Physical Tendon Surveillance of Palisades Nuclear Plant, 2002." In Enclosure 4 of the response letter, the applicant provided a discussion in Section VIII, "Comparison with Original Installation Data."

The staff reviewed the information in Enclosure 4 of the applicant's response letter and determined that additional information was necessary to complete its review.

In RAI 4.5.2(a), dated September 28, 2005, the staff noted that Section VIII in Enclosure 4 of the applicant's letter dated July 25, 2005, states that "as a result of the generator change and the re-tensioning of a large number of vertical tendons these must now be excluded from this analysis." As a result, the staff requested that the applicant provide the following information:

(a)   when the generator change and re-tensioning of a large number of vertical tendons were performed

(b)   an explanation of how the TLAA for excluded tendons is performed as a large number of vertical tendons was excluded from the analysis

In its response, by letter dated October 28, 2005, the applicant stated that the steam generator replacement was completed in 1991 between the 15-year and 20-year surveillance. In addressing the number of vertical tendons excluded from the analysis, the applicant stated that the measurement of tendon liftoff force had been performed on previously de-tensioned and re-tensioned tendons. Tendons V72, V128, V126, V116, 48AE and 52AE were all retensioned during the steam generator replacement project. Since re-tensioning, these tendons have been selected and tested during a subsequent surveillance (1992, 1997, or 2002). The applicant stated that surveillance test results for each of these tendons were acceptable (above the MRV). The tendons were excluded from the regression analysis because they had been previously de-tensioned and re-tensioned. The tendons are not excluded from future testing under the Containment In-service Inspection Program, which assures the continued acceptability of these tendons in accordance with 10 CFR 54.21(c)(1)(iii) requirements.

Based on its review, the staff finds the applicant's response to RAI 4.5.2(a) acceptable because the applicant provided the dates for the SG replacement and re-tensioning. The re-tensioned tendons have been tested in three subsequent surveillances and are not excluded from future testing. Therefore, the staff's concerns described in RAI 4.5.2(a) are resolved.

In RAI 4.5.2(b), dated September 28, 2005, the staff requested, based on the information provided in response to RAI 2.5-2, that the applicant explain how the 95 percent confidence curves in Enclosure 4 of letter dated July 25, 2005, were established.

In its response, by letter dated October 28, 2005, the applicant provided a description of the method used for determining the confidence interval in the tendon regression analyses. The procedure consisted of inserting the error band in the regression analysis equation with a standard "t-statistic" with n-2 degrees of freedom. The applicant also explained that the regression analysis for each group of tendons using the data from the last surveillance report has been revised to project out to 60 years. The results for the vertical and horizontal regressions previously provided were changed because the test results from tendon V334 and 84DF were incorrectly excluded from the data.

Based on its review, the staff finds the applicant's response acceptable because the applicant explained how the regression analyses curves are determined; therefore, the staff's concern described in RAI 4.5.2(b) is resolved.

The staff review of LRA Section 4.5, and the relevant references cited in the TLAA, found that the applicant has adequate procedures in place for monitoring and trending the containment prestressing forces, and the applicant's choice to manage this TLAA pursuant to

10 CFR 54.21(c)(1)(iii) is acceptable. The staff, therefore, concluded that in conjunction with the Boric Acid Corrosion Program, the prestressing tendon forces in containment will be adequately managed.

4.5.3 FSAR Supplement

The applicant provided an FSAR supplement summary description of its TLAA evaluation of the concrete containment tendon prestress analysis in LRA Section A4.4.1. In this section, the applicant summarized the TLAA by describing the method used to estimate the predicted prestressing forces at 40 years and emphasized that, because of the conservatism in the prestressing force loss calculations, the trend lines are considerably above the minimum estimated values. In conclusion, the applicant stated that, "This issue is dispositioned under 10 CFR 54.21(c)(1)(iii), the effects of aging on the intended function(s) will be adequately managed for the extended operating period, in that the existing Palisades tendon surveillance program activities will be continued in accordance with 10 CFR 50.55a."

The staff's review of LRA Section A4.4.1 identified areas in which additional information was necessary to complete the review of the applicant's FSAR supplement. The applicant responded to the staff's RAIs as follows.

In RAI 4.5-3, dated June 22, 2005, the staff requested that the applicant provide a tabulated comparison of the MRVs for each group of tendons and the projected prestressing forces at 40 and 60 years, based on the regression analyses, utilizing the procedure recommended in GALL AMP X.S1 and IN 99-10.

In its response, by letter dated July 25, 2005, the applicant stated:

> Based on the discussion contained in FSAR Chapter 5.8, and specifically 5.8.5.3, "Prestressing System", the Palisades containment tendons are expected to remain above minimum required force levels beyond the 40-year plant life. FSAR 5.8.5.3 states the following conclusion; "At the conclusion of the Twentieth and Twenty-Fifth year tendon surveillances, regression analyses were performed utilizing the surveillance data (References 36, 37, 44, and 45). The results consistently indicated that the effective group tendon forces, dome/hoop/vertical, were significantly higher than predicted values beyond the 40-year time period." This conclusion is also supported by the report extract discussed in the response to RAI 4.5-2, and provided as Enclosure 4 to this letter. By inspection of the graphs provided in this report, when the current trend curves are extrapolated, it is apparent that tendon forces will satisfy their minimum required values, with margin, beyond 60 years.

The staff reviewed the applicant's response to RAI 4.5-3 and determined that additional information was necessary to complete the review.

By letter dated October 28, 2005, the applicant provided the results of the analysis performed to project the expected containment tendon prestressing forces to 60 years, utilizing, to the extent practical with available surveillance data, the procedures recommended in GALL Program X.S1 and IN 99-10. The following table summarizes the post-tension surveillance test results at 40 and 60 years, as provided by the applicant on November 29, 2005, which were extracted from the regression analyses tables and graphs provided in letter dated October 28, 2005.

Tendons	Minimum Required Value (kips)	40-Year Projection (kips)	60-Year Projection (kips)
Dome	584	643	640
Vertical	615	670	669
Horizontal	615	646	642

In RAI 4.5.3(a), dated September 28, 2005, the staff stated that, in its response to RAI 4.5-3, the applicant provided a summary of the regression analysis data through tables and graphics in Enclosure 4 of the applicant's letter dated July 25, 2005. The staff noted that a number of tendons appears to have been excluded from the analysis in these tables. The specific tendons are numbered as D1-38, V334, and 65BF. Therefore, the staff requested that the applicant provide the basis excluding these tendons from the analysis.

In its response, by letter dated October 28, 2005, the applicant stated:

> Tendons D1-38 and 65BF were excluded because they were de-tensioned and re-tensioned during the first surveillance; therefore, any data after that time is excluded as unrepresentative of the general population.

> Tendon V334 was incorrectly excluded from the analysis. Apparently, V334 was mistakenly identified as V324, which was de-tensioned and re-tensioned during the first surveillance. The regression analysis prepared in response to RAI 4.5.2(b) includes tendon V334.

> It was also noted that tendon 84DF was de-tensioned and re-tensioned during the first surveillance, and should have been excluded, along with D1-38 and 65BF. The regression analysis prepared in response to RAI 4.5.2(b) excludes 84DF.

The staff finds the applicant's response to RAI 4.5.3(a) acceptable based on the applicant's clarifications. The staff finds the content of the FSAR supplement in LRA Section A4.4.1 acceptable and the concern described in RAI 4.5.3(a) is resolved.

The staff review of the FSAR supplement concluded that the summary description of the applicant's actions to address the concrete containment tendon prestress analysis is adequate.

4.5.4 Conclusion

On the basis of its review, as discussed above, the staff concludes that the applicant has demonstrated, pursuant to 10 CFR 54.21(c)(1)(iii), that, for the concrete containment tendon prestress analysis, the effects of aging on the intended function(s) will be adequately managed for the period of extended operation in conjunction with AMP B2.1.4. The staff also concludes that the FSAR supplement contains an appropriate summary description of the TLAA evaluation, as required by 10 CFR 54.21(d).

4.6 Containment Liner Plate and Penetrations Load Cycles

The containment structure is a Consumers Design Class 1 structure. The entire interior surface of the containment structure is lined with 1/4-inch thick steel plate to ensure a high degree of leak tightness. Numerous mechanical and electrical systems penetrate the containment wall through steel penetrations welded to the containment liner plate. The containment liner plate and penetrations were conservatively designed, in part, to the rules of the ASME B&PV Code, Section III, 1965. This code edition classifies containment as a Class B vessel. A fatigue analysis is required under this code edition only for Class A vessels (reactor coolant pressure boundary (RCPB), etc.); however, the PNP containment liner and penetration designs use some of the methods and data from Section III, Article 4, for design of Class A vessels for fatigue loads.

4.6.1 Containment Liner Plate Load Cycles

4.6.1.1 Summary of Technical Information in the Application

In LRA Section 4.6.1, the applicant summarized the evaluation of the containment liner plate load cycles for the period of extended operation. The containment liner plate is stitch welded to a gridwork of structural steel angles embedded in the concrete. The anchoring system is designed to prevent significant distortion of the liner plate during accident conditions and to ensure that the liner maintains its leak-tight integrity. The liner plate has been coated on the inside for corrosion protection. There is no paint on the side in contact with the concrete.

The containment design relies on the liner only to maintain a leak-tight containment. There are no design conditions under which the liner plate is relied upon to assist the concrete in maintaining the integrity of the structure. At times, the liner plate provides assistance in order to maintain deformation compatibility. Stress concentrations around penetration openings in the liner plate are calculated using the theory of elasticity. These stress concentrations were then reduced by thickening the liner plate around each penetration in accordance with the ASME B&PV Code, Section III, 1965.

FSAR Section 5.8.3.2.1 states that the following fatigue loads were considered in the design of the liner plate: thermal cycling due to annual outdoor temperature variations, thermal cycling due to containment interior temperature varying during the start-up and shutdown of the reactor system, and thermal cycling due to the design-basis accident (DBA) was assumed to be one cycle. The liner plate was analyzed for these fatigue loadings using Figure N-415(a) of the ASME B&PV Code, Section III, Article 4, 1965.

4.6.1.2 Staff Evaluation

The staff reviewed LRA Section 4.6.1, pursuant to 10 CFR 54.21(c)(1)(i), to verify that the analyses remain valid for the period of extended operation.

The staff review of LRA Section 4.6.1 identified an area in which additional information was necessary to complete the review of the containment liner plate load cycles TLAA. The applicant responded to the staff's RAI as follows.

The applicant stated that the containment liner plate was evaluated for metal fatigue on the basis of ASME Section III, Article 4, 1965. The fatigue loadings used in the analysis were stated in FSAR Section 5.3.8.2.1 and consist of thermal cycling due to 40 cycles of annual outdoor temperature variation, thermal cycling due to containment interior temperature variation during start-up and shutdown, assumed as 500 cycles, and thermal cycling due to DBA, consisting of one cycle. Based on these design load cycles, the applicant evaluated the requirements for fatigue analysis in accordance with ASME Section III, Article 4, 1965, Paragraph N-415.1, "Vessels not Requiring Analysis for Cyclic Operation," and Figure N-415(a), and determined that the conditions stated in this paragraph for not requiring an ASME Section III fatigue analysis were satisfied. For the period of extended operation, the applicant stated that the effect of increasing the annual cycles is negligible. The applicant also stated that the design-basis 500 cycles of interior containment heat-up and cool-down is conservative and can be considered to remain applicable for 60-year operation. On this basis, the conditions for satisfying the requirements of Paragraph N-415.1 of the code remain applicable for the period of extended operation.

In RAI 4.6.1, dated July 20, 2005, the staff requested that the applicant provide an editorial correction in the summary description in LRA Section 4.6-1.

In its response, by letter dated August 19, 2005, the applicant stated:

> The missing text in the summary description on page 4-46 is shown underlined below:
>
> The Palisades containment design relies on the liner only to maintain a leak-tight containment. There are no design conditions under which the liner plate is relied upon to assist the concrete in maintaining the integrity of the structure even though the liner will, at times, provide assistance in order to maintain deformation compatibility.

Based on its review, the staff finds the applicant's response to RAI 4.6-1 acceptable because the applicant provided a correction to the LRA that helped the staff understand the design of the containment liner; therefore, the staff's concern described in RAI 4.6-1 is resolved.

4.6.1.3 FSAR Supplement

The applicant provided an FSAR supplement summary description of its TLAA evaluation of the containment liner plate load cycles in LRA Section A4.4.2. On the basis of its review of the FSAR supplement, the staff concludes that the summary description of the applicant's actions to address the containment liner plate load cycles is adequate.

4.6.1.4 Conclusion

On the basis of its review, as discussed above, the staff concludes that the applicant has demonstrated, pursuant to 10 CFR 54.21(c)(1)(i), that, for the containment liner plate load cycles TLAA, the analyses remain valid for the period of extended operation. The staff also concludes that the FSAR supplement contains an appropriate summary description of the TLAA evaluation, as required by 10 CFR 54.21(d).

4.6.2 Containment Penetration Load Cycles

4.6.2.1 *Summary of Technical Information in the Application*

In LRA Section 4.6.2, the applicant summarized the evaluation of the containment penetration load cycles for the period of extended operation. FSAR Sections 5.8.3 and 5.8.6.4 describe design of both the "large penetrations" (personnel air lock and equipment hatch) and the "small penetrations" (escape air lock, piping, ventilation, and electrical penetrations) to limiting strain criteria for a set of design loads, including thermal loads and DBA loads. FSAR Section 5.8.6.4.1 describes the design of the pipe penetrations to acceptance criteria for combined primary and secondary stresses under the ASME B&PV Code, Section III, 1965, Article 4. Figure N-415(a) of that article is a fatigue S-N diagram (stress versus number of allowed cycles or vice-versa) for alternating stress intensity Sa versus N for carbon steel. Numerous mechanical and electrical subsystems penetrate the containment wall through steel penetrations welded to the containment liner plate. Containment penetrations are designed to maintain the leak tightness of the containment structure under normal and accident conditions. They are designated as CPCo Design Class 1 components of the containment structure. Piping, heating, ventilating, and air conditioning (HVAC), and electrical penetrations are designed, fabricated, inspected, and installed in accordance with the ASME B&PV Code, Section III, Subsection B. The in-service functionality of each individual penetration is monitored by means of local leak-rate testing (also referred to as Type B Test in 10 CFR Part 50, Appendix J), in accordance with plant TSs, at a frequency of at least every refueling, but not exceeding a two-year interval.

4.6.2.2 *Staff Evaluation*

The staff reviewed LRA Section 4.6.2, pursuant to 10 CFR 54.21(c)(1)(i), to verify that the analyses remain valid for the period of extended operation.

The applicant stated that, of the design-basis fatigue loads on the containment penetrations, only the environmental and operational load cycles would increase due to the period of extended operation. Of these two events, the effect of the assumed summer-winter annual cycles is and will remain negligible for the period of extended operation. The number of design basis interior heat-up and cool-down cycles is 500. Based on the counted cycles shown in LRA Table 4.3.1-1, the applicant determined that 240 cycles may be expected for 60 years of operation. Therefore, the applicant determined that the design basis containment penetration fatigue evaluation will remain valid for the period of extended operation.

In RAI 4.6-2, dated July 20, 2005, the staff stated:

> The Analysis subsection in Section 4.6.2 reproduces the corresponding paragraphs of the Analysis subsection in Section 4.6.1. Provide confirmation that this conforms with the corresponding section in FSAR 5.8.6.4.1.

In its, response, by letter dated August 19, 2005, the applicant stated:

> The Analysis subsection in Section 4.6-2 has been revised to focus on to containment penetrations, deleting the inadvertent reference to the containment liner. The revised text conforms with FSAR 5.8.6.4.1.

Revise the first two paragraphs to read as follows:

The allowable stress was conservatively based on the ASME B&PV Code, Section III, Article 4, 1965. Specifically, the following sections were adopted as guides in establishing allowable stress limits:

1. Paragraph N-412(m) - Thermal Stress, Subparagraph 2

2. Paragraph N-412(n) - Operational Cycle

3. Paragraph N-414.5, Table N-413, Figures N-414 and N-415(a) - Peak Stress Intensity

4. Paragraph N-415.1 - Vessels Not Requiring Analysis for Cyclic Operation

The number of design load cycles is relatively insignificant when compared to the allowable number of cycles on the fatigue curve in code Figure N-415 (a) for the peak design stress in the penetrations, and when compared to the allowable number of cycles for 3 times the S_m value of code Table N-421 at the operating temperature. The results of the analysis confirm that the design of the containment penetrations complies with the provisions of code paragraph N-415.1, and a fatigue analysis for design load cycles is not required.

Based on its review, the staff finds the applicant's response to RAI 4.6-2 acceptable because the applicant revised the analysis subsection in LRA Section 4.6-2 to conform with FSAR Section 5.8.6.4.1; therefore, the staff's concern described in RAI 4.6-2 is resolved.

The staff review of LRA Section 4.6.2 and the relevant references cited in the TLAA, found that the design load cycles will not be reached by the end of the period of extended operation, therefore, pursuant to 10 CFR 54.21(c)(1)(i), the containment penetration load cycles TLAA analyses remain valid for the period of extended operation.

4.6.2.3 FSAR Supplement

The applicant provided an FSAR supplement summary description of its TLAA evaluation of the containment penetration load cycles in LRA Section A4.4.2. On the basis of its review of the FSAR supplement, the staff concludes that the summary description of the applicant's actions to address the containment penetration load cycles is adequate.

4.6.2.4 Conclusion

On the basis of its review, as discussed above, the staff concludes that the applicant has demonstrated, pursuant to 10 CFR 54.21(c)(1)(i), that, for the containment penetration load cycles TLAA, the analyses remain valid for the period of extended operation. The staff also concludes that the FSAR supplement contains an appropriate summary description of the TLAA evaluation, as required by 10 CFR 54.21(d).

4.7 Other Plant-Specific Time-Limited Aging Analyses

In LRA Section 4.7, the applicant provided its evaluation of plant-specific TLAAs. The TLAAs evaluated include the following:

- crane load cycles

- Alloy 600 nozzle safe end life assessment analyses

- ASME Code Case N-481 relaxation of the primary coolant pump weld category B-L-1 inspection interval from 10 years to 40 years (not used)

- risk-informed in-service inspection program calculations (not used)

- absence of a TLAA for reactor coolant pump flywheel fatigue or crack growth analysis

4.7.1 Crane Load Cycles

4.7.1.1 Summary of Technical Information in the Application

In LRA Section 4.7.1, the applicant summarized the evaluation of the crane load cycles for the period of extended operation. A crane evaluation to the Crane Manufacturers Association of America (CMAA) Standard CMAA-70 assumes a number of rated lifts in the design lifetime in order to establish the design service level, and hence the allowable stresses. At PNP, two cranes have been re-analyzed to CMAA-70 design criteria. The NUREG-0612 heavy loads evaluation of the reactor building polar crane was performed to CMAA-70. A re-design of the spent fuel pool crane for dry fuel storage also included a NUREG-0612 evaluation to CMAA-70 design criteria. The limiting components of the containment polar crane (135 tons) and the redesigned spent fuel pool crane (110 tons) are now rated for CMAA-70, "Service Level A - Standby or Infrequent Service." Therefore, the analyses of the polar crane and the spent fuel pool crane are TLAAs for license renewal purposes.

4.7.1.2 Staff Evaluation

The staff reviewed LRA Section 4.7.1, pursuant to 10 CFR 54.21(c)(1)(i), to verify that the analyses remain valid for the period of extended operation.

The staff's review identified areas in which additional information was necessary to complete the review of the applicant's analysis of crane load cycles TLAA. The applicant responded to the staff's RAIs as follows.

LRA Section B2.1.15 indicates that there are no fatigue concerns for the containment polar crane and the spent fuel pool crane as these cranes cannot realistically approach the 20,000 to 100,000 rated lifts assumed for their design evaluation during the period of extended operation. LRA Section B2.1.15 further indicates that the applicant does not track the number and magnitude of all the lifts made by cranes.

In LRA Section 4.7.1, the applicant stated that the polar crane cannot realistically approach the 20,000 to 100,000 rated lifts assumed for components evaluated to CMAA 70 (1975) Service Level A during a 60-year licensed operating period. For the spent fuel pool crane, the applicant identified the expected lifts between rerating and the end of the 60-year extended license and concluded that this machine cannot realistically approach the 20,000 to 100,000 rated lifts assumed for its evaluation during a 60-year licensed operating period.

In RAI 4.7.1-1(a), dated June 22, 2005, the staff stated that the LRA does not characterize the estimated number of lifts for the polar crane. LRA Section A4.5.1 indicates that polar crane planned engineering lifts (over-rated capacity) have been evaluated separately and approved up to 140 tons, less than 4 percent over the 135 ton rating. According to NUREG-1744, "A Survey of Crane Operating Experience at U.S. Nuclear Power Plants from 1968 through 2002," the maximum load weight for the reactor building crane was exceeded twice (greater than 125 percent) when lifting the reactor vessel head with additional lead shielding intact. Therefore, the staff requested that the applicant identify the number of lifts considered in the fatigue evaluation and clarify how the over-rated capacity lifts have been considered in combination with the other lifts for the fatigue evaluation of the polar crane.

In its response, by letter dated July 25, 2005, the applicant stated that the NUREG-1744 characterization of the planned engineering lift is incorrect. The review of operating experience indicated no loads, including test loads, have exceeded 125 percent of the rated capacity of the crane. The applicant further stated that certain lifts have been evaluated and approved up to 140 tons, which is less than 4 percent over the 135 ton rating. Regarding the number of lifts, the applicant stated that there have been fewer than 10 lifts exceeding 50 percent of the crane's rated capacity in each 12-month period, including a limited number of planned engineering lifts. At that rate, only 600 lifts of that magnitude would be expected for the 60 years of operation. The applicant concluded that 600 lifts is substantially less than the 20,000 full load cycles that would require fatigue analysis in accordance with CMAA 70.

Based on its review, the staff finds the applicant's response to RAI 4.7.1-1(a) acceptable because the applicant clarified that over capacity lifts have been considered and the number of near rated capacity expected lifts have been identified and evaluated according to CMAA 70. Therefore, the staff's concern described in RAI 4.7.1-1 is resolved.

In RAI 4.7.1-1(b), dated June 22, 2005, the staff stated that industry experience in documents like NUREG-1744 indicates that crane vibrations have occurred during crane operation. Therefore, the staff requested a review of the applicant's plant-specific operating experience for whether crane vibrations have occurred during crane operation and, if so, clarification of how vibrations are considered in the fatigue evaluation. The staff requested technical justification if crane vibrations are not considered for license renewal.

In its response, by letter dated July 25, 2005, the applicant stated that:

> Review of plant operating experience did not identify any instances of significant vibration noted during the operation of the overhead cranes. Crane vibration would be indicative of abnormal crane operation that would be identified and evaluated separately on a plant-specific basis under the corrective action process in order to determine its significance with respect to the crane.

The staff review of the applicant's analysis in the LRA and additional clarifications in its response found the applicant's response to RAI 4.7.1-1(b) acceptable. The staff concurred that the polar crane and spent fuel pool crane have been evaluated and are qualified for the period of extended operation. The cranes are qualified for a 100,000 cycle design life, which exceeds the estimated load cycles for the life of the cranes including life extension. A fatigue analysis is not required for the polar crane or spent fuel pool crane because estimated load cycles are well below the limits established by CMAA 70. Fatigue life is not significant in the operation of this equipment and the analysis is valid for the period of extended operation. Therefore, the staff's concern described in RAI 4.7.1-1(b) is resolved.

The staff review of LRA Section 4.7.1 and the relevant references cited in the TLAA, found that the design crane load cycles limits will not be reached by the end of the period of extended operation, therefore, pursuant to 10 CFR 54.21(c)(1)(i), the polar and spent fuel cranes TLAA analyses remain valid for the period of extended operation.

4.7.1.3 FSAR Supplement

The applicant provided an FSAR supplement summary description of its TLAA evaluation of the crane load cycles in LRA Section A4.5.1. On the basis of its review of the FSAR supplement, the staff concludes that the summary description of the applicant's actions to address the crane load cycles is adequate.

4.7.1.4 Conclusion

On the basis of its review, as discussed above, the staff concludes that the applicant has demonstrated, pursuant to 10 CFR 54.21(c)(1)(i), that, for the crane load cycles TLAA, the analyses remain valid for the period of extended operation. The staff also concludes that the FSAR supplement contains an appropriate summary description of the TLAA evaluation, as required by 10 CFR 54.21(d).

4.7.2 Alloy 600 Nozzle Safe End Life Assessment Analyses

4.7.2.1 Summary of Technical Information in the Application

In LRA Section 4.7.2, the applicant summarized the evaluation of the Alloy 600 nozzle safe end life assessment analyses for the period of extended operation. Alloy 600 (Ni-Cr-Fe alloy) was used to clad the pressurizer lower head and surge nozzle for the pressurizer heater sleeves and for smaller nozzles and safe ends and flanges on larger nozzles of the reactor vessel head, PCS loop piping, and pressurizer. There were originally 251 Alloy 600 heater sleeves, nozzles, safe ends, and flanges in the PCS.

By the 1990's, Alloy 600 had become recognized as susceptible to a cracking phenomenon identified as primary water stress-corrosion cracking (PWSCC). A significant contributing factor to the initiation and propagation of PWSCC was determined to be temperature. Thus, since the pressurizer has the highest operating temperature of any location in the PCS, a reasonable assumption is that the Alloy 600 nozzles would develop PWSCC first.

Evidence of PWSCC was first noted at PNP in September 1993 when a through-wall crack was found in the pressurizer PORV nozzle safe end to pipe weld heat affected zone. The crack was found after the PCS leak rate increased while the plant was in hot shutdown. The plant was returned to cold shutdown and the leak was repaired. The cracked section of field weld was removed, the safe end modified for a new weld prep, and a short new section of stainless steel pipe installed to replace the material removed. The failed section was saved for thorough analysis of the failure mechanism. The failure mechanism was attributed to PWSCC.

LRA Section 4.7.2 also reviews past component integrity analyses approved by the NRC. The applicant's review concluded that: (a) a fracture mechanics assessment of the most susceptible Alloy 600 locations in pressurizer and PCS loop penetrations and nozzle safe ends is a TLAA; (b) although the fatigue analysis for the repaired pressurizer temperature element nozzles is a TLAA, the fracture mechanics analysis for them is not; and (c) although the service time assessment for a pressurizer surge line nozzle is a TLAA, the corresponding assessment for a pressurizer spray line nozzle is not.

4.7.2.2 Staff Evaluation

The staff reviewed LRA Section 4.7.2, pursuant to 10 CFR 54.21(c)(1)(i), to verify that the analyses remain valid for the period of extended operation and, pursuant to 10 CFR 54.21(c)(1)(iii), to verify that the effects of aging on the intended function(s) will be adequately managed for the period of extended operation.

The information provided by the applicant is that the Alloy 600 nozzle and safe end assessment analyses for certain components in the PCS meet the 10 CFR 54.3(a) definition of TLAAs. These components are within the scope of 10 CFR 54.4 SSCs, which ensure the integrity of the RCPB. The staff determined that, with the PCS Alloy 600 nozzles and safe ends currently in service, the TLAA does not support continued operation pursuant to 10 CFR 54.21(c)(1)(i) and (ii) due to Alloy 600 weld penetration cracking.

The applicant presented a set of identified TLAA results in LRA Table 4.7.2-1, which lists pressurizer and PCS Alloy 600 locations (components) susceptible to PWSCC along with their estimated service time based on an initial flaw depth of 0.010 inch. LRA Table 4.7.2-1 is based on plant-specific Report 32-1238965-00, "Fracture Mechanics Assessment of Palisades Alloy 600 Components," which was approved by the NRC in a safety evaluation dated June 27, 1995. In LRA Table 4.7.2-1, locations with a calculated service time of 40 years or more are identified as TLAA items and those of less than 40 years are not so identified.

The staff's review of LRA Section 4.7.2 identified areas in which additional information about the applicant's identification of these TLAA items was necessary. The applicant responded to the staff's RAIs as follows.

In RAI 4.7.2-1, dated November 30, 2005, the staff stated that LRA Table 4.7.2-1 summarizes the fracture mechanics assessment of the most susceptible pressurizer and PCS Alloy 600 locations. Therefore, the staff requested that the applicant:

(1)   Confirm that all the components in LRA Table 4.7.2-1 are inspection items under the Alloy 600 AMP.

(2)   Clarify how this fracture mechanics assessment supports the inspection methods and intervals of the pressurizer and PCS Alloy 600 locations and how the fracture mechanics assessment supports the inspection methods and intervals of the Alloy 600 AMP.

(3) Justify 0.01 inch as the postulated initial flaw depth in the fracture mechanics analyses and provide the flaw depth considered to be detectable by approved methods and the basis of the determination.

In RAI 4.7.2-1, the staff further stated that the fracture mechanics assessment summary appeared to be based on Report 32-1238965-00. In recent years, the industry started a systematic approach to manage reactor vessel head and pressurizer Alloy 600 penetration nozzles (e.g., the work related to crack growth rate in MRP-55, "Crack Growth Rates for Evaluating Primary Water Stress Corrosion Cracking (PWSCC) of Thick-Wall Alloy 600 Material," dated July 18, 2002, and the fatigue growth rate of ASME B&PV Code, Appendix O, "Evaluation of Flaws in PWR Reactor Vessel Upper Head Penetration Nozzles"). Therefore, the staff also requested that the applicant assess the impact of this new information on the fracture mechanics assessment of the Alloy 600 components, as summarized in LRA Table 4.7.2-1.

In its response, by letter dated January 13, 2006, the applicant stated:

(1) All the components in Table 4.7.2-1 are inspection items under Palisades' Alloy 600 aging management program (AMP). These items were described in the Alloy 600 project plan approved by NRC on June 27, 1995.

(2) The subject fracture mechanics assessment is discussed in the NRC's June 27, 1995 safety evaluation. Significant items affected by the analysis are that the pressurizer spray nozzle safe end is inspected every other refueling outage and pressurizer temperature element nozzles are inspected each refueling outage.

(3) A value of 0.01 inch was selected as the postulated initial flaw depth in the fracture mechanics analyses. This is smaller than detectable by approved NDE methods. The NRC's June 27, 1995 SER, on page 8, estimates that the ultrasonic examination detection limit would be 2 mm (0.079 in.). Because of this, the inspection frequencies have not been extended based on some predicted propagation rate. The pressurizer spray nozzle safe end is inspected every other refueling outage and pressurizer temperature element nozzles are inspected each refueling outage.

In 2004, the Combustion Engineering Owners Group performed a generic fatigue crack growth evaluation associated with small diameter nozzles. This evaluation looked at limiting pressurizer and hot leg nozzle locations. Only two transients were determined to contribute to fatigue crack growth. The transients and the numbers of cycles assumed in the analysis for each are:

Start-up/Shut-down (Normal)   500 Events
OBE (Hot Leg only) (Upset)   200 Events

The initial crack depths evaluated were 0.938" for the hot leg, 0.719" for the pressurizer lower side shell and 1.125" for the pressurizer lower head. None of these cracks propagated substantially for the entire operating period as defined by the above transients. The disposition at the bottom of page 4-60 reads, 'The Palisades plant-specific bounding fracture mechanics analysis demonstrates the validity of the cycle-dependent aspects of the generic bounding fracture mechanics analysis (WCAP-15973-P) by demonstrating that the plant-specific load and thermal events are within those assumed by the generic bounding analysis. The basis for the safety determination of the fracture mechanics evaluation calculation will therefore remain valid so long as the numbers of these events do not exceed the design basis values.'

The staff concludes that the applicant's response to subparts 2 and 3 of RAI 4.7.2-1 inadequately addresses the staff's concerns.

By letter dated March 30, 2006, the applicant provided its supplemental response. The applicant proposed to replace the discussion provided in the paragraph, "In 2004, the Combustion Engineering Owners Group....design basis values." with a paragraph that indicated that the conclusions of the 1995 report, "Fracture Mechanics Assessment of Palisades Alloy 600 Components" remained valid. The revised paragraph identified two reports: WCAP-15937-P, Rev. 1, "Low-Alloy Steel Component Corrosion Analysis Supporting Small-Diameter Alloy 600/690 Nozzle Repair/Replacement Programs (May 2004), and Westinghouse Calculation Note CN-CI-02-71, Rev. 1, "Summary of Fatigue Crack Growth Evaluation Associated with Small Diameter Nozzles in CEOG Plants." The applicant stated that these later analyses confirmed that the Palisades cycle-dependent information developed in the earlier analyses and reported in LRA Table 4.7.2-1 remain valid. The applicant stated that these later analysis demonstrate that the plant-specific loading and thermal events are enveloped by those assumed by the generic bounding analyses.

In original RAI 4.7.2-1, dated November 30, 2005, regarding LRA Table 4.7.2-1, the staff requested information regarding whether and how the applicant applied more recent knowledge regarding PWSCC crack growth rates to the TLAA for Alloy 600 components. The applicant's response of March 30 makes reference to two documents that included fatigue crack growth analyses of alloy steel components, where the fatigue crack started at the tip of an assumed flaw at an interface between the alloy steel and the Alloy 600 (or Alloy 82/182 weld metal). Since the references discuss fatigue crack growth of alloy steel and the subject of the TLAA is PWSCC of Alloy 600, the staff found that the applicant's supplemental response still did not adequately address the concerns.

The applicant subsequently proposed to revise LRA Table 4.7.2-1 to indicate the effects of aging on the intended function of the Alloy 600 components would be adequately managed for the period of extended operation in accordance with CFR 54.21(c)(1)(iii). The applicant added a note to LRA Table 4.7.2-1 to identify the table as being for information only and to indicate that the table describes the combined effects of fatigue and PWSCC, with both mechanisms being managed by AMPs.

The applicant also committed to revise the Alloy 600 Program to update the PWSCC crack growth rate assessments and inspection program consistent with the latest NRC requirements and industry commitments (e.g., EPRI Report 1010087 "Materials Reliability Program: Primary System Piping System Butt Weld Inspection and Evaluation Guidelines [MRP-139]," (August 2005)). The updated program will be submitted for NRC review and approval at least three years prior to entering the period of extended operation (Commitment 6).

With these clarifications and revisions, the staff finds the applicant's evaluation of PWSCC of all Alloy 600 heater sleeves, nozzles, safe ends, and flanges to be acceptable in accordance with 10 CFR 54.21(c)(1)(iii). Therefore, the staff's concerns described in RAI 4.7.2-1 are resolved.

For the repaired pressurizer temperature element nozzles, the applicant determined that the fatigue analysis of the weld repairs is a TLAA; however, the fracture mechanics and corrosion evaluations of the weld repairs are not. The fatigue analysis of the weld repairs is based on half of the 40-year pressurizer DBE cycles. The applicant proposed to monitor the cumulative number of pressurizer temperature element nozzle fatigue cycles within the Fatigue Monitoring Program and maintain a special action level to ensure that appropriate actions are taken if, at any time, the cycle count for any DBE since 1993 reaches the number assumed by the analysis. Therefore, in accordance with 10 CFR 54.21(c)(1)(iii), the fatigue analysis is acceptable for the period of extended operation. The applicant did not consider the fracture mechanics analysis of the weld pad repair a TLAA because the 7.68-year crack growth assessment of Report 32-1238965-00 was not used as the basis for the safety determination for continued operation. Further, the staff questioned the applicant's conclusion that the corrosion evaluation of the weld pad repairs is not a TLAA based on its evaluation time of 60 years. The evaluation time of 60 years, is the basis to conclude that the corrosion evaluation is acceptable for the extended period of operation in accordance with 10 CFR 54.21(c)(1)(i). The staff determined that additional information about inspections of the weld pad repairs during the period of extended operation was necessary.

In RAI 4.7.2-2, dated November 30, 2005, the staff noted that LRA Section 4.7.2 states, as to the pressurizer temperature element nozzles repaired in 1993, that the fatigue analysis of the weld pad repairs and the ASME XI, Appendix A, crack growth evaluation are TLAAs. Therefore, the staff requested that the applicant:

(1) Provide the actual cycle count recorded in the fatigue monitoring program for the pressurizer events from 1993 to date as a fraction (e.g., 7/10) of the 20-year pressurizer DBE cycles assumed in the weld pad fatigue analysis to determine whether action levels for cycles will be reached by major events early in the extended period of operation and, if reached, describe the actions to be taken.

(2) Provide information about the pad material and pad weld material and confirm whether the repair was in accordance with ASME Code Section XI and the welded pad was analyzed in accordance with ASME Code, Section III.

(3) Provide information about the inspection results for the repair since 1993 and discuss the consistency of the inspections of the repair with those of pressurizer temperature element nozzles in the Alloy 600 Program.

(4) Identify the submittal and the NRC safety evaluation of corrosion for the extended 60-year operating period and provide a consequence assessment of the pressurizer temperature element nozzle having a predicted bore diameter increase of 0.28 inch due to corrosion.

In its response, by letter dated January 13, 2006, the applicant provided additional information related to each area in the staff's RAI:

(1) The license renewal application did not discuss the fatigue crack growth analysis of the weld pad repairs that had been prepared in 2003. The analysis determined that only pressurizer cool-down and system leak tests contribute to fatigue crack growth at these locations. The analysis assumed 500 cool-downs and 320 leak tests. In LRA Table 4.3.1-1, the maximum number of cool-downs for 60 years is determined to be 240 (240/500 = 0.48) and the maximum number of leak tests for 60 years is 10 (10/320 = 0.03125). Fatigue failure during the extended operating period of the weld pad repairs to the pressurizer temperature element nozzles is judged not to be a concern.

(2) The weld pad is deposited with SFA 5.11 ENiCrFe-3 welding electrodes as specified in ASME Section II, Part C. This filler material is compatible with both the alloy 600 (P43) nozzle and the vessel base material, SA-533 Grade B Class 1 for the top head nozzle and the A-10 composition weld pad of the side shell nozzle. The welding filler material, SFA 5.11 ENiCrFe-3, conforms to the requirements of Sections II and III 1986 Edition and is compatible with the materials being joined. This weld material (ENiCrFe-3) provides corrosion resistance in the primary water environment and meets the strength requirements for design under ASME Section III. Repair of the temperature element nozzles was performed in accordance with the requirements of ASME Section XI 1983 with Summer 1983 addenda. The welded pad was analyzed in accordance with ASME Section III, 1965 with Winter 1966 addenda.

(3) The most recent inspection of the upper temperature element nozzle was the bare metal visual performed during the 2004 refueling outage. This inspection was acceptable and is performed each refueling outage. There have been no corrective actions related to the condition of this nozzle since the 1993 repair. The inspection will be continued through the license renewal period.

(4) This corrosion analysis has been superseded. The revised analysis was submitted in a letter dated on August 9, 2004 (ML042240249) entitled "Request for Relief from ASME Section XI Code Requirements for Repair of Pressurizer Nozzle Penetrations." The supporting calculations determined the allowable maximum hole size in the vessel wall, under ASME code rules for the particular design of the lower TE-0101 nozzle, and the allowable corrosion as the difference between that value and the vessel wall bore for the nozzle. This, with an estimated corrosion rate (from WCAP-15973-P), resulted in an estimated repair lifetime of 52.3 years; that is, an estimated corrosion life of 52.3 years following initiation of leakage.

Based on its review, the staff finds the applicant's response to RAI 4.7.2-2 acceptable; therefore, staff's concerns described in RAI 4.7.2-2 are resolved.

For the pressurizer spray and surge nozzle service time assessment, the applicant concluded that the spray nozzle safe end is not a TLAA, but the surge nozzle safe end is a TLAA, if their calculated service time is less or greater than 40 years. The staff did not agree with the applicant's criterion for determining the TLAA. In addition, the service time assessments for both the pressurizer spray and surge nozzle should be considered TLAAs. The staff noted that neither the spray nozzle TLAA nor the surge nozzle TLAA is included in the applicant's disposition for the 40-year fracture mechanics analysis.

In RAI 4.7.2-3, dated November 30, 2005, the staff stated that in the pressurizer spray and surge nozzle service time assessment the applicant concluded that the predicted service time for the surge nozzle is 40 years. Therefore, the staff requested that the applicant provide the disposition for the surge line nozzle TLAA because it was not clearly included within the applicant's "Disposition for all Alloy 600 Heater Sleeves, Nozzles, Safe Ends, and Flanges: 10 CFR 54.21(c)(1)(iii)." For the spray nozzle, which has a predicted service time of 5.36 years at 640 °F, the staff requested that the applicant provide the inspection results for the spray nozzle since 1995 and discuss the consistency of the current inspections with those in the Alloy 600 Program.

In its response, by letter dated January 13, 2006, the applicant stated:

> Disposition for the pressurizer surge line nozzle is in accordance with 10 CFR 54.21(c)(I)(iii). The Palisades Alloy 600 Program includes the surge line nozzle in the inspection program.
>
> For the spray nozzle, the Alloy 600 Program requires a bare-metal VT-2, volumetric, or penetrant inspection every other refueling outage. Volumetric examination was performed during the most recent inspection during the 2004 refueling outage. No corrective action documents have been identified related to leakage or structural indication issues regarding this nozzle.

Based on its review, the staff finds the applicant's response to RAI 4.7.2-3 acceptable; therefore, the staff's concerns described in RAI 4.7.2-3 are resolved.

In RAI 4.7.2-4, dated November 30, 2005, the staff noted that, on bounding fracture mechanics analysis of the hot leg, piping resistence temperature detector, and sampling nozzles, pressurizer instrument nozzles, and pressurizer heater sleeves, LRA Section 4.7.2 states "bounding fracture mechanics portion of the analysis employs elastic-plastic methods with IWB 3600 and Regulatory Guide 1.161 acceptance criteria." The staff believed that this bounding fracture mechanics analysis refers to that of WCAP-15973, "Low-Alloy Steel Component Corrosion Analysis Supporting Small-Diameter Alloy 600/690 Nozzle Repair/Replacement Programs." Therefore, the staff requested that the applicant modify the referenced sentence in LRA Section 4.7.2 by adding the following phrase to the end: "as modified by the NRC SE dated January 12, 2005." RG 1.161 acceptance criteria, especially the structural factors, are not acceptable for analyzing detected flaws, including flaws with indication of leakage.

In its response, by letter dated January 13, 2006, the applicant stated:

> The first sentence of the last paragraph on page 4-59 is in error. The sentence is hereby revised to read, 'The bounding fracture mechanics portion of the analysis employs elastic-plastic methods with IWB 3600 and Regulatory Guide 1.161

acceptance criteria, as modified by the NRC Safety Evaluation dated January 12, 2005.'

Based on its review, the staff finds the applicant's response to RAI 4.7.2-4 acceptable; therefore, the staff's concern described in RAI 4.7.2-4 is resolved.

In RAI 4.7.2-5, dated November 30, 2005, the staff noted that LRA Section 4.7.2 states for disposition for all Alloy 600 heater sleeves, nozzles, safe ends, and flanges that, "Locations which are more susceptible to PWSCC, or whose failure could result in a more-significant safety hazard, are also subject to initial or periodic bare-metal VT-2, volumetric, or penetrant inspections." Therefore, the staff requested the applicant to list the "more susceptible" locations and state the criteria for determining these locations developed from the fracture mechanics assessment in Report 32-1238965-00.

In its response, by letter dated January 13, 2006, the applicant stated that the locations most susceptible to PWSCC are the surge nozzles safe ends, pressurizer temperature element nozzles, pressurizer relief valve safe ends, and heater sleeves. The applicant further stated that PWSCC susceptibility is the leading criterion in determining susceptible locations along with failure consequences, leakage detection margin, and radiation dose rate.

Based on its review, the staff finds the applicant's response to RAI 4.7.2-5 acceptable; therefore, the staff's concerns described in RAI 4.7.2-5 are resolved.

In RAI 4.7.2-6, dated November 30, 2005, the staff noted that LRA Section 4.7.2 states, "NMC will re-evaluate effects of primary water stress corrosion cracking for all Alloy 600 components for which the current analyses found acceptable crack sizes at 40 years to identify those for which the analysis would predict unacceptable crack sizes at 60 years, and to identify appropriate additional inspections for them. NMC will complete these re-evaluations before the period of extended operation." From newly available information on PWSCC and fatigue crack growth rates, the staff determined that this reevaluation should be performed three years prior to instead of just before entering the period of extended operation. Therefore, the staff requested that the applicant revise its commitment letter to reflect the change in the submittal date for this re-evaluation.

In its response, by letter dated January 13, 2006, the applicant revised its commitment to state that re-evaluations will be completed three years prior to entering the period of extended operation.

Based on its review, the staff finds the applicant's response to RAI 4.7.2-6 acceptable; therefore, the staff's concern in RAI 4.7.2-6 is resolved.

The staff reviewed the applicant's TLAA on Alloy 600 nozzle and safe end life assessment analyses, as summarized in LRA Section 4.7.2, and determined that the relevant fatigue, fracture mechanics, and corrosion analyses, as modified in the responses to the staff's RAI, would continue to comply with the staff's requirements throughout the period of extended operation.

4.7.2.3 FSAR Supplement

The applicant provided an FSAR supplement summary description of its TLAA evaluation of the Alloy 600 nozzle safe end life assessment analyses in LRA Section A4.5.2. On the basis of its review of the FSAR supplement, the staff concludes that the summary description of the applicant's actions to address the Alloy 600 nozzle safe end life assessment analyses is adequate.

4.7.2.4 Conclusion

On the basis of its review, as discussed above, the staff concludes that the applicant has demonstrated, pursuant to 10 CFR 54.21(c)(1)(i), that, for the Alloy 600 nozzle safe end life assessment analyses TLAA, the analyses, as modified in the responses to the staff's RAI, remain valid for the period of extended operation. The applicant had also demonstrated, pursuant to 10 CFR 54.21(c)(1)(iii), that, for the TLAA, the effects of aging on the intended function(s) will be adequately managed for the period of extended operation. The staff also concludes that the FSAR supplement contains an appropriate summary description of the TLAA evaluation, as required by 10 CFR 54.21(d).

4.7.3 ASME Code Case N-481 Relaxation of the Primary Coolant Pump Weld Category B-L-1 Inspection Interval from 10 Years to 40 Years (Not Used)

By letter dated May 5, 2005, the applicant provided supplementary information to support its LRA. In this letter, the applicant stated that it had revisited the TLAA discussion in LRA Section 4.7.3 of the need for the ASME Code Case N-481 relaxation of the primary coolant pump weld category B-L-1 inspection interval from 10 years to 40 years. The applicant stated that this discussion very conservatively interpreted 10 CFR 54.3(a) and, upon further review, it had concluded that the issue did not meet 10 CFR 54.3(a) TLAA criteria. The applicant stated that, "The Section 4.7.3 discussion of the Code Case N-481 is unnecessary since this code case will no longer be in use when the plant enters the period of extended operation." In the letter, the applicant withdrew LRA Section 4.7.3 and, to conform with this change, the applicant also withdrew the corresponding FSAR supplement summary description of the TLAA in LRA Section A4.5.3.

By letter dated December 16, 2005, the applicant provided additional information about the deletion of this TLAA discussion, stating that, "NMC has concluded that the ASME Section XI IWB, IWC, IWD, IWF Inservice Inspection Program for aging management should be revised to reflect the 2001 edition through 2003 addenda as the Section XI code of record." In the letter the applicant revised the ASME Section XI IWB, IWC, IWD, IWF Inservice Inspection Program description in LRA Section B2.1.2. Under the heading "Program Description," the applicant replaced the second paragraph, in its entirety, with the following:

> The Palisades ASME Section XI IWB, IWC, IWD, IWF Inservice Inspection aging management program is based on the ASME B&PV Code, Section XI, 2001 edition, including 2002 and 2003 addenda. The IWB-2500 Category B-Q requirements to perform volumetric examinations of steam generator tubes is addressed by the Steam Generator Tube Integrity Program.

During the period of extended operation, the program will be maintained consistent with the requirements of 10 CFR 50.55a. 10 CFR 50.55a currently requires that inservice inspection of Class 1, 2, and 3 pressure retaining components, their integral attachments, and supports, be conducted in accordance with the latest edition of ASME Section XI approved by the NRC twelve months prior to the start of a ten year interval. 10 CFR 50.55a also provides for the use of NRC-approved relief requests. Therefore, any relief requests or other alternatives to the ISI code of record would be submitted for NRC review and approval at least twelve months prior to the start of each inspection interval in accordance with existing regulations.

Based on its review, the staff finds deletion of LRA Section 4.7.3 acceptable because the 2001 Edition of the ASME Code does not require pump casing weld, volumetric, or routine internal visual examinations during the period of extended operation; however, it does require external surface examinations of the casing welds and internal visual examinations when the reactor coolant pump (RCP) is disassembled for other reasons. In addition, the staff finds that the RCP casings are not considered susceptible to thermal aging; therefore, a structural integrity analysis is not necessary. As such, the ASME Code Case N-481 relaxation of the primary coolant pump weld category B-L-1 inspection interval from 10 years to 40 years discussion is not considered a TLAA in accordance with 10 CFR 54.3 criteria. As the RCP casing analysis is no longer considered a TLAA, the staff also concludes that a corresponding FSAR supplement summary description is not required by 10 CFR 54.21(d). Therefore, deletion of LRA Sections 4.7.3 and A4.5.3 is acceptable.

4.7.4 Risk-Informed Inservice Inspection Program Calculations (Not Used)

By letter dated May 5, 2005, the applicant provided supplementary information to support its LRA. In this letter, the applicant stated that it had revisited the need for the Risk-Informed In-service Inspection Program Calculations TLAA discussion in LRA Section 4.7.4. The applicant stated that this discussion very conservatively interpreted 10 CFR 54.3(a) and, upon further review, it had concluded that the issue did not meet the 10 CFR 54.3(a) TLAA criteria. The applicant stated that, "The Risk Informed Inservice Inspection Program calculations discussed in Section 4.7.4 were used for determining inspection intervals. The risk at any point in time is unaffected by this evaluation, and is not time-limited." In the letter, the applicant withdrew LRA Section 4.7.4 and, to conform with this change, the applicant also withdrew the corresponding FSAR supplement summary description of the TLAA in LRA Section A4.5.4.

The staff review concluded that risk is not time-limited and, as such, Risk-Informed In-service Inspection Program calculations are not considered a TLAA in accordance with 10 CFR 54.3 criteria. As the calculations are no longer considered a TLAA, the staff also concludes that a corresponding FSAR supplement summary description is not required by 10 CFR 54.21(d). Therefore, deletion of LRA Sections 4.7.4 and A4.5.4 is acceptable.

4.7.5 Absence of a TLAA for Reactor Coolant Pump Flywheel Fatigue or Crack Growth Analysis

The function of the RCP in the reactor coolant system (RCS) of a PWR is to maintain an adequate cooling flow rate by circulating a large volume of primary coolant water at high temperature and pressure through the RCS. A concern about overspeed of the RCP and its potential for failure led to the issuance of RG 1.14, "Reactor Coolant Pump Flywheel Integrity," dated 1971. The regulatory position of RG 1.14 on ISI calls for an in-place ultrasonic volumetric examination of the areas of higher stress concentration at the bore and keyway at approximately three-year intervals and an examination of all exposed surfaces and complete ultrasonic volumetric examination at approximately 10-year intervals. The flywheel inspection schedule is to coincide with the individual plant's ISI schedule, as required by ASME Code Section XI.

CEOG report (SIR)-94-080, "Relaxation of Reactor Coolant Pump Flywheel Inspection Requirements," was submitted on April 4, 1995, to propose a change to the RG 1.14 inspection requirements. This report, which provides an engineering analysis based on fracture mechanics, is intended to eliminate RCP flywheel ISI requirements for certain CEOG plants, including PNP; however, the NRC safety evaluation of SIR-94-080, dated May 22, 1997, states that "the staff believes that even for flywheels meeting all the design criteria of RG 1.14, as modified in this SER, inspections should not be completely eliminated." In the NRC safety evaluation, the staff concluded:

(1)　Licensees for ANO-2, Palisades, Millstone 2, Waterford 3, and St. Lucie 1 & 2 who plan to submit a plant-specific application of this topical report need to verify the reference temperature RT_{NDT} for their flywheels. Also, if these licensees have flywheels made of materials other than SA 533 B and SA 508, they need to justify the use of the K_{IC} v.s. $(T-RT_{NDT})$ curve in Appendix A of Section XI of the ASME Code to derive their respective K_{IC} values. In both cases, they should report any existing plant-specific test results which are directly or indirectly related to fracture toughness of the RCP flywheel material.

(2)　Since ANO-1 already has a unique flywheel inspection program of 10-year intervals, this SER does not affect its status regarding flywheel inspections.

(3)　Licensees meeting (1) above should either conduct a qualified in-place UT examination over the volume from the inner bore of the flywheel to the circle of one-half the outer radius or conduct a surface examination (MT and/or PT) of exposed surfaces defined by the volume of the disassembled flywheels once every 10 years. The staff considers this 10-year inspection requirement not burdensome when the flywheel inspection is conducted during scheduled ISI inspection or RCP motor maintenance. This would provide an appropriate level of defense in depth.

4.7.5.1　Summary of Technical Information in the Application

In LRA Section 4.7.5, the applicant summarized the evaluation of the need for a TLAA for RCP flywheel fatigue or crack growth analysis for the period of extended operation. The four original RCP motors were built by Allis-Chalmers. Westinghouse built an additional motor. One of the five is maintained as a spare, and any combination of four may be installed. All have flywheels at the top of the motor to provide additional rotational inertia for gradual coastdown and continued circulation, in case of a power supply loss or inadvertent trip.

An RCP flywheel could theoretically burst because of centrifugal stresses, which could produce missiles inside containment and could also damage pump seals or other pressure boundary components. This concern is the subject of RG 1.14. The flywheels may therefore be subject to crack growth or fatigue analyses, ISI, or both to assess and reduce the probability of a failure.

Early TSs required periodic, relatively frequent, inspections of RCP motor flywheels. To justify a longer inspection interval, the CEOG prepared report SIR-94-080. This report used a crack growth analysis of PNP RCP flywheels to establish acceptable limits for the flywheel inspection intervals. The analysis does not depend on the licensed operating period (10 CFR 54.3 Criterion 3) and is not a TLAA. The evaluation determined that the primary coolant pump would be subject to approximately 500 startup/shutdown cycles, and the crack growth fatigue analysis assumed 4000 cycles. It was concluded that a ten-year inspection interval was acceptable since an assumed pre-service flaw would not grow to a critical flaw size during the period between inspections.

4.7.5.2 Staff Evaluation

The staff reviewed LRA Section 4.7.5, pursuant to 10 CFR 54.21(c)(1)(i), to verify that the analyses remain valid for the period of extended operation.

RG 1.14, Revision 1, provides the staff's recommended acceptance criteria for material and minimum fracture toughness properties of SA 508, Classes 2 and 3, materials and SA 533 Grade B, Class 2, materials used in the fabrication of U.S. RCP flywheels. RG 1.14, Revision 1, also provides guidelines for performing structural integrity assessments of the RCP flywheels in U.S. light-water reactors, including assessments for ensuring the integrity of the flywheels against unacceptable fatigue-induced crack growth failures. The fatigue crack growth assessments are based on the number of start/stop cycles assumed in the design-basis for the pumps. Therefore, to meet the 10 CFR 54.21(c)(1)(i) acceptance criterion, the applicant is required to demonstrate that the total number of RCP start/stop cycles, projected through the end of the extended periods of operation, will be bounded by the number of RCP start/stop cycles assumed in the 60-year fatigue-induced crack growth analysis for the RCP flywheels.

A bounding fatigue crack growth evaluation of the RCP flywheels over the period of extended operation was performed in SIR-94-080, dated March 1995, for certain operating CEOG plants. It demonstrated that the flywheel design had an adequate structural reliability with high flaw tolerance and negligible flaw crack extension over a 60-year service life. The staff reviewed and approved the evaluation with certain conditions and limitations for application in reducing the inspection frequency for the period of the current license. In 1998, the applicant submitted a license amendment request to reduce its RCP flywheel inspection frequency and scope based on SIR-94-080. The staff approved the PNP license amendment request in License Amendment 182.

In the LRA, the applicant stated that the SIR-94-080 analysis does not depend on the licensed operating period and is not a TLAA. Topical Report SIR-94-080 includes a stress and fracture evaluation that addresses fatigue crack growth for 4,000 startup/shutdown cycles. In the safety evaluation of SIR-94-080, the staff accepted the fatigue crack growth evaluation only as a basis to increase the flywheel inspection interval from approximately three to ten years for the period of the current license. This acceptance implied that the fatigue crack growth analysis in the SIR-94-080 safety evaluation was approved by the staff for the CLB and that further justification is needed to extend its license for an additional 20 years. Consequently, this analysis is a TLAA.

The staff review of LRA Section 4.7.5 identified an area in which additional information was necessary to complete the review of the RCP flywheel analysis evaluation. The applicant responded to the staff's RAI as follows.

In RAI 4.7.5-1, dated June 3, 2005, the staff noted that the LRA states that the RCP flywheel evaluation is not considered as a TLAA; however, from past review experience the staff determined that the evaluation is a TLAA. Therefore, the staff requested that the applicant provide the 10 CFR 54.21(c)(1) TLAA evaluations or further technical justification for not providing them.

In its response, by letter dated July 1, 2005, the applicant revised its LRA to consider the fatigue crack growth analysis for RCP flywheels a TLAA. The revised LRA Section 4.7.5 states that, "The primary coolant pump is assumed to experience approximately 500 startup/shutdown cycles, and the crack growth fatigue analysis assumed 4000 cycles. The expected number of cycles for the 60-year extended licensed operating period, as listed in Table 4.3.1-1, is substantially less than the 500 assumed cycles."

Based on its review, the staff finds the applicant's response to RAI 4.7.5-1 acceptable because the staff determined that the fatigue crack growth analysis of SIR-94-080 for certain CEOG plant RCP flywheels is valid for 60 years of operation and, under 10 CFR 54.21(c)(1)(i), acceptable for the period of extended operation. Therefore, the staff's concern described in RAI 4.7.5-1 is resolved.

The staff reviewed the applicant's TLAA on RCP flywheel analysis, as summarized in LRA Section 4.7.5, and determined that the fracture mechanics analyses would continue to comply with the staff's requirements throughout the period of extended operation.

4.7.5.3 FSAR Supplement

Under 10 CFR 54.21(d), applicants for license renewal must include a supplement summary description of the programs and activities for managing the effects of aging and the evaluation of TLAAs for the period of extended operation. In its response to RAI 4.7.5-1, by letter dated July 1, 2005, the applicant provided an FSAR supplement summary description for the fatigue-induced crack growth of RCP flywheels TLAA by adding LRA Section A4.5.5. In the letter, the applicant stated that an evaluation of the probability of failure over the period of extended operation in SIR-94-080 for certain operating CEOG plants demonstrated that flywheel design has an adequate structural reliability with a high flaw tolerance and negligible flaw crack extension over a 60-year service life. The applicant demonstrated that the RCP flywheels met the topical report conditions in a plant-specific application which invoked this evaluation as the basis for reducing the frequency of RCP flywheel inspections. The applicant further stated that the analysis associated with the structural integrity of the RCP flywheel had been determined to remain valid for the period of extended operation, in accordance with 10 CFR 54.21(c)(1)(i).

The applicant provided an FSAR supplement summary description for the RCP flywheels analysis in LRA Section A4.5.5. The staff review of the FSAR supplement concluded that the applicant's actions to address the RCP flywheel fatigue or crack growth analysis is adequate.

4.7.5.4 Conclusion

On the basis of its review, as discussed above, the staff concludes that the applicant has demonstrated, pursuant to 10 CFR 54.21(c)(1)(i), that, for the RCP flywheel fatigue or crack growth analysis TLAA, the analyses remain valid for the period of extended operation. The staff also concludes that the FSAR supplement contains an appropriate summary description of the TLAA evaluation, as required by 10 CFR 54.21(d).

4.7.6 Reactor Vessel Underclad Cracking

By letter dated April 26, 2006, the applicant indicated it had determined that the preferred way to disposition the issue for license renewal was to classify it as a TLAA, prepare a technical report that dispositions the TLAA for the full 60-year extended operating period, and update the pertinent LRA sections to reflect this new information. To provide a more detailed discussion of the issue, the applicant also provided the LRA changes that reflect the expected technical conclusions for the issue. These LRA changes were prepared based on preliminary information from the nuclear steam supply system vendor. Upon completion of the final technical report on the effects of potential underclad cracking at PNP, the applicant will notify the staff that the technical report for the final disposition of the issue has been completed and the associated LRA changes submitted in the April 26 letter have been confirmed. If the final report identifies a need for any additional LRA revisions, the revised information would be provided at that time for staff review and approval. The applicant will submit this information no later than September 1, 2006.

The staff reviewed the proposed LRA changes related to potential underclad cracking. The applicant proposed to add Section 4.7.6 to the LRA that described the mechanisms responsible for underclad cracking and indicated the effects of aging could be managed by performing a fracture mechanics analysis. The staff agreed that the effects of aging on underclad cracks can be evaluated and managed by a fracture mechanics analysis. The applicant indicated that the staff had reviewed and approved a bounding fracture mechanics analysis that assessed underclad cracking for Westinghouse RPVs. The applicant stated that although PNP is not a Westinghouse plant, it had been determined that if underclad cracking were postulated to exist in the RPV, the bounding fracture mechanics analysis in "A Review of Cracking Associated with Weld Deposited Cladding in Operating PWR Plants," Westinghouse WCAP 15338-A, October 2002, would be applicable to the issue and would be used as a reference to support a 60 year life. The applicant proposed to make similar changes to LRA Sections 3.1.2.2.5, A4.5.6, and 4.7 references to add the references to WCAP-15338-A and the staff approval letter. The staff determined the changes were acceptable pending verification of the applicability of the WCAP-15388-A.

The applicant made a commitment (Commitment 55) to validate the applicability of WCAP-15338-A as follows:

> Upon completion of the final technical report on the effects of potential underclad cracking at Palisades, NMC will notify NRC that the technical report for the final disposition of the issue has been completed, and the associated LRA changes submitted on April 26, 2006, have been confirmed. If the final report identifies a need for any additional LRA revisions, the revised information will be provided at that time for NRC review and approval. NMC will submit this information no later than September 1, 2006.

To close Commitment 55, by letter dated July 5, 2006, the applicant provided an updated response to the staff's question about the underclad cracking TLAA based on plant-specific report WCAP-16605-NP, "A Review of Cracking Associated with Weld Deposited Cladding at Palisades Nuclear Plant," dated June 2006.

4.7.6.1 Summary of Technical Information in the Application

In LRA Section 4.7.6, the applicant summarized the evaluation of reactor vessel underclad cracking for the period of extended operation. Underclad cracking has occurred in the low alloy steel base metal heat-affected zone (HAZ) beneath the austenitic stainless steel weld overlay that is deposited to protect the ferritic material from corrosion. Two types of underclad cracking have been identified. Reheat cracking has occurred as a result of postweld heat treatment of austenitic stainless steel cladding applied using high-heat-input welding processes on ASME SA-508, Class 2 forgings. Cold cracking has occurred in ASME SA-508, Class 3 forgings after deposition of the second and third layers of cladding, when no pre-heating or post-heating was applied during the cladding procedure. The cold cracking was determined to be attributable to residual stresses near the yield strength in the weld metal/base metal interface after cladding deposition, combined with a crack-sensitive microstructure in the HAZ, and high levels of diffusible hydrogen in the austenitic stainless steel or Inconel weld metals. The hydrogen diffused into the HAZ and caused cold (hydrogen-induced) cracking as the HAZ cooled.

A generic fracture mechanics evaluation of Westinghouse plants initially demonstrated that the growth of underclad cracks during a 40-year plant life was insignificant. The evaluation was extended to 60 years, using fracture mechanics analysis based on a representative set of design transients, with the occurrences extrapolated to cover 60 years of service life. The 60-year evaluation, WCAP-15338-A, showed insignificant growth of the underclad cracks, and concluded that the cracks were of no concern relative to structural integrity of the reactor vessel. The staff reviewed and approved the evaluation for application to all Westinghouse RPVs, and identified two plant-specific Applicant Action Items to be completed by each applicant as a condition for referencing WCAP-15338-A. These action items include verifying that the design transients and operating conditions assumed in the report are applicable to the applicant's plant, and providing an description of the issue as a TLAA to be incorporated into the FSAR.

Although PNP is not a Westinghouse plant, the reactor vessel was fabricated using similar processes and materials as those used in some reactor vessels fabricated by Combustion Engineering for Westinghouse. The applicant has determined that if underclad cracking were postulated to exist in the PNP reactor vessel, WCAP-15338-A is applicable to the issue, and provides a valid technical reference to support a 60-year service life.

By letter dated July 5, 2006, the applicant provided additional information on the underclad cracking TLAA based on plant-specific report WCAP-16605-NP, which documents fatigue crack growth analysis of underclad flaws in the RPV in accordance with Section XI of the ASME Code and concludes that "underclad cracks in the Palisades reactor vessel are of no concern relative to the structural integrity of the vessel for continued plant operation, even through 60 years of operation."

4.7.6.2 Staff Evaluation

The staff reviewed the LRA changes related to potential underclad cracking. By letter, dated April 26, 2006, the applicant added LRA Section 4.7.6 that described the mechanisms responsible for underclad cracking and indicated the effects of aging could be managed by performing a fracture mechanics analysis. The staff agrees that the effects of aging on underclad cracks can be evaluated and managed by a fracture mechanics analysis. The applicant indicated that the staff had reviewed and approved a bounding fracture mechanics analysis that assessed underclad cracking for Westinghouse RPVs. The applicant stated that although PNP is not a Westinghouse plant, it had been determined that if underclad cracking were postulated to exist in the PNP RPV, the bounding fracture mechanics analysis in "A Review of Cracking Associated with Weld Deposited Cladding in Operating PWR Plants," Westinghouse WCAP 15338-A, October 2002, would be applicable to the issue and would be used as a reference to support a 60 year life. The applicant made similar changes to LRA Sections 3.1.2.2.5 and A4.5.6 and to the LRA Section 4.7 references to add the references to WCAP-15338-A and the staff approval letter. The staff determined the changes were acceptable pending verification of the applicability of the WCAP-15388-A. This was Confirmatory Item CI 4.7.6-1.

The staff reviewed the July 5, 2006, information for closing this confirmatory item and determined that WCAP-16605-NP is a plant-specific version of WCAP-15338-A, but with PNP plant-specific data (including design transients) for the fatigue crack growth analysis. The staff found that for underclad flaws with aspect ratio (flaw length to flaw depth) of 2 to 6, the amounts of flaw growth cited in WCAP-15338-A are about the same as, or bound, those of WCAP-16605-NP for all assumed flaw locations. As fatigue crack growth analysis uses design transients and applied stress intensity factors as input, the bounding nature of this analysis in WCAP-15338-A indicates that the ASME Code Section XI flaw evaluation provided by Westinghouse in 2001 to supplement WCAP-15338-A also applies to PNP. This flaw evaluation demonstrates that after 60 years of fatigue crack growth, significant margins (2 for normal, upset, and test conditions and 3 for emergency and faulted conditions) remain for underclad flaws of realistic shapes at different locations in addition to the ASME Code-specified structural factors. Hence, this TLAA is acceptable and Confirmatory Item CI 4.7.6-1 is closed.

4.7.6.3 FSAR Supplement

The applicant provided an FSAR supplement summary description of its TLAA evaluation of reactor vessel underclad cracking in LRA Section A4.5.6. On the basis of its review of the FSAR supplement, the staff concluded that the summary description of the applicant's actions to address the underclad cracking issue is adequate.

4.7.6.4 Conclusion

On the basis of its review, as discussed above, the staff concludes that the applicant has demonstrated, pursuant to 10 CFR 54.21(c)(1)(ii), that, for the reactor vessel underclad cracking TLAA, the effects of aging on the intended function(s) will be adequately managed for the period of extended operation. The staff also concludes that the FSAR supplement contains an appropriate summary description of the TLAA evaluation, as required by 10 CFR 54.21(d).

4.8 Conclusion for Time-Limited Aging Analyses

The staff reviewed the information in LRA Section 4, "Time-Limited Aging Analyses." On the basis of its review, the staff concludes that the applicant has provided an adequate list of TLAAs, as defined in 10 CFR 54.3. Further, the staff concludes that the applicant demonstrated that: (1) the TLAAs will remain valid for the period of extended operation, as required by 10 CFR 54.21(c)(1)(i); (2) the TLAAs have been projected to the end of the period of extended operation, as required by 10 CFR 54.21(c)(1)(ii); or (3) that the aging effects will be adequately managed for the period of extended operation, as required by 10 CFR 54.21(c)(1)(iii). The staff also reviewed the FSAR supplement for the TLAAs and found that the FSAR supplement contains descriptions of the TLAAs sufficient to satisfy the requirements of 10 CFR 54.21(d). In addition, the staff concludes that no plant-specific exemptions are in effect that are based on TLAAs, as required by 10 CFR 54.21(c)(2).

With regard to these matters, the staff concludes that there is reasonable assurance that the activities authorized by the renewed license will continue to be conducted in accordance with the CLB, and that any changes made to the CLB, in order to comply with 10 CFR 54.21(c), are in accordance with the Atomic Energy Act of 1954 and the NRC's regulations.

SECTION 5

REVIEW BY THE ADVISORY COMMITTEE ON REACTOR SAFEGUARDS

The NRC staff issued its safety evaluation report (SER) with confirmatory item related to the renewal of operating licenses for Palisades Nuclear Plant (PNP) on June 1, 2006. On July 11, 2006, the applicant presented its license renewal application, and the staff presented its review findings to the ACRS Plant License Renewal Subcommittee. The staff reviewed the applicant's comments on the SER and completed its review of the license renewal application. The staff's evaluation is documented in an SER that was issued by letter dated September 28, 2006.

During the 537th meeting of the ACRS, November 1, 2006, the ACRS completed its review of the PNP license renewal application and the NRC staff's SER. The ACRS documented its findings in a letter to the Commission dated November 17, 2006. A copy of this letter is provided on the following pages of this SER Section.

November 17, 2006

The Honorable Dale E. Klein
Chairman
U. S. Nuclear Regulatory Commission
Washington, DC 20555-0001

Subject: REPORT ON THE SAFETY ASPECTS OF THE LICENSE RENEWAL
            APPLICATION FOR THE PALISADES NUCLEAR POWER PLANT

Dear Chairman Klein:

During the 537th meeting of the Advisory Committee on Reactor Safeguards, November 1-3,
2006, we completed our review of the license renewal application for the Palisades Nuclear
Plant (PNP) and the final Safety Evaluation Report (SER) prepared by the NRC staff. Our Plant
License Renewal Subcommittee also reviewed this matter during a meeting on July 11, 2006.
During our review, we had the benefit of discussions with representatives of the NRC staff and
the applicant, Nuclear Management Company, LLC (NMC). In addition, we had the benefit of
input from the public. We also had the benefit of the documents referenced. This report fulfills
the requirements of 10 CFR 54.25 that the ACRS review and report on all license renewal
applications.

CONCLUSION AND RECOMMENDATION

The programs established and committed to by the applicant to manage age-related degradation
provide reasonable assurance that PNP can be operated in accordance with its current licensing
basis for the period of extended operation without undue risk to the health and safety of the
public.

The NMC application for renewal of the operating license for PNP should be approved.
Continued operation during the entire period of extended operation is contingent on the
resolution of the issues associated with three Time-Limited Aging Analyses (TLAAs) related to
reactor pressure vessel (RPV) integrity.

BACKGROUND AND DISCUSSION

PNP is a Combustion Engineering 2-loop pressurized water nuclear plant with a large, dry,
ambient-pressure containment. PNP is located five miles south of South Haven, Michigan, on
the eastern shore of Lake Michigan. The current power rating of the PNP is 2566 MWt, for a
gross electrical output of 767 MWe. PNP was originally licensed to operate on February 21,
1971. NMC requested renewal of the PNP operating license for 20 years beyond the current
license term, which expires on February 20, 2011.

In the final SER, the staff documented its review of the license renewal application and other
information submitted by NMC and obtained during the audit and inspection conducted at the
plant site. The staff reviewed the completeness of the applicant's identification of structures,

systems, and components (SSCs) that are within the scope of license renewal; the integrated plant assessment process; the applicant's identification of the plausible aging mechanisms associated with passive long-lived components; the adequacy of the applicant's Aging Management Programs (AMPs); and the identification and assessment of TLAAs requiring review.

The NMC application is largely consistent with NUREG-1801, "Generic Aging Lessons Learned (GALL) Report," issued in July 2001. All deviations from the GALL Report are documented in the application. The applicant identified the SSCs that fall within the scope of license renewal and performed a comprehensive aging management review for these SSCs. Based on the results of this review, the applicant will implement 24 AMPs for license renewal including existing, enhanced, and new programs. In the final SER, the staff concluded that the applicant has appropriately identified the SSCs within the scope of license renewal and that the AMPs described by the applicant are appropriate and sufficient to manage aging of long-lived passive components that are within the scope of license renewal. We concur with this conclusion.

The staff conducted an inspection and an audit. The inspection verified that the scoping and screening methodologies are consistent with the regulations and are adequately reflected in the application. The audit verified the appropriateness of the AMPs and the aging management reviews. Based on the inspection and audit, the staff concluded that these programs are consistent with the descriptions contained in the NMC license renewal application. The staff also concluded that the existing programs, to be credited as AMPs for license renewal, are generally functioning well and that an implementation plan has been established in the applicant's commitment tracking system to ensure timely completion of the license renewal commitments.

During our meetings with the staff and the applicant, we discussed the adequacy of programs proposed by NMC to manage aging of certain components that are projected to exceed acceptance limits during the period of extended operation.

The applicant identified the systems and components requiring TLAAs and reevaluated them for 20 additional years of operation. As required by 10 CFR Part 54, the applicant must identify any exemptions granted under 10 CFR 50.12 which rely on a TLAA and determine if that exemption should be continued for an additional 20 years of operation. No such exemption currently exists in the PNP licensing basis. The applicant reexamined 23 TLAAs. All of these TLAAs are valid, without restriction, for 20 more years of operation, except for three TLAAs associated with reactor vessel neutron embrittlement, namely: reactor vessel upper shelf energy, reactor vessel pressurized thermal shock, and reactor vessel pressure-temperature curves. In each of these cases, PNP will exceed the acceptance limits prior to the end of the extended period of operation.

To analyze the reactor vessel neutron fluence for purposes of RPV integrity evaluations, the applicant uses the methodology described in WCAP-15353, which is consistent with Regulatory Guide 1.190.

The applicant began using low neutron leakage cores in 1988 to reduce the neutron embrittlement of the reactor vessel to extend the time before exceeding the acceptance limits. However, the applicant predicts that the following acceptance limits will be exceeded:

- Upper Shelf Energy limit – exceed in 2021.
- Reactor Vessel Pressurized Thermal Shock (PTS) screening criterion – exceed in 2014.
- Pressure-Temperature limit curves – expire in 2014.

The staff's confirmatory calculations show reasonable agreement with the applicant's findings.

Upper Shelf Energy Limit. The applicant predicts this criterion will be exceeded in 2021. Appendix G of 10 CFR 50 requires RPV beltline materials to have Charpy upper shelf energy values no less than 50 ft-lb in the transverse direction in the base metal and along a weld for weld material. However, in accordance with Appendix G, Charpy upper shelf energy values below 50 ft-lb may be acceptable if it is demonstrated that lower Charpy upper shelf energy values will provide margins of safety against fracture (ductile tearing) equivalent to those required by ASME Code, Section XI, Appendix G. Regulatory Guide 1.99 describes two acceptable methods for determining the upper shelf energy values for RPV beltline materials.

Because the reactor vessel upper shelf energy limit will be exceeded prior to the end of the extended period of operation, the applicant must provide an analysis in accordance with 10 CFR Part 50, Appendix G at least three years prior to exceeding the upper shelf energy limit.

PTS Screening Criterion. The applicant predicts the criterion for axial welds and plates will be exceeded in 2014. 10 CFR 50.61 provides the fracture toughness requirements for protecting reactor vessels from the effects of PTS events. The end of life reference temperature (RTPTS) value is the sum of a reference value for an unirradiated material, a shift in the reference value caused by exposure to high-energy neutron irradiation, and an additional margin to account for uncertainties.

If an applicant determines that the RPV will not meet the PTS screening criterion through the end of the facility's current license term, several actions must be taken. 10 CFR 50.61(b)(3), requires that an applicant implement a reasonably practicable flux reduction program in an effort to avoid exceeding the PTS screening criterion. If no reasonably practicable flux reduction program will meet this objective (as is true in the case of PNP) the applicant has several options. The applicant may submit a safety analysis in accordance with 10 CFR 50.61(b)(4) to demonstrate that the RPV can be operated beyond the 10 CFR 50.61 screening criterion. This safety analysis may include plant modifications. Such an analysis must be submitted three years prior to the time the RPV is projected to exceed the PTS screening criterion. In accordance with 10 CFR 50.61(b)(7), the applicant could propose to anneal the RPV in order to improve its material properties and permit continued operation. In accordance with 10 CFR 50.66, the applicant's thermal annealing plan would have to be submitted three years prior to when the facility's RPV is projected to exceed the PTS screening criterion.

Pressure-Temperature Limit Curves. Pressure-temperature limit curves are contained in the PNP technical specifications and are assessed against the limits in 10 CFR 50.60, Appendix G to 10 CFR 50, and Appendix G to Section XI of the ASME Code. The current pressure-temperature limits approved by the staff are valid beyond the current license term, but not through the extended period of operation. Based on the neutron fluence expected to be accumulated, the pressure-temperature limit curves will expire in 2014. Prior to entering the period of extended operation, the applicant must submit an amendment requesting a technical specification change and approval of new limits covering the period of extended operation beyond 2014.

The staff has concluded that the applicant has provided an adequate list of TLAAs. Further, the staff has concluded that the applicant has met the license renewal rule by demonstrating that the TLAAs have been projected to the end of the period of extended operation. In those cases where the current TLAAs do not cover the entire period of extended operation, the applicant must provide additional information in a timely manner and submit a license amendment for a technical specification change to extend these three TLAAs to cover the entire period of extended operation. We concur with the staff that the applicant has properly identified the applicable TLAAs, reviewed the associated analyses and licensing bases, and identified those instances where additional measures are needed to modify the TLAAs to cover the entire period of extended operation. We concur with the staff's conclusions and the resulting license conditions and commitments.

During our Plant License Renewal Subcommittee meeting on July 11, 2006, members of the Public provided comments and raised several questions. These comments and questions were recorded and are contained in the transcript of that meeting. The reference to the transcript that contains these comments and questions was provided to the Executive Director for Operations. Subsequently, the staff has responded to these questions and comments.

We agree with the staff that there are no issues related to the matters described in 10 CFR 54.29(a)(1) and (a)(2) that preclude renewal of the operating license for PNP. The programs established and committed to by NMC provide reasonable assurance that PNP can be operated in accordance with its current licensing basis for the period of extended operation without undue risk to the health and safety of the public. Continued operation during the entire period of extended operation is contingent on the resolution of the issues associated with three TLAAs related to RPV integrity. The NMC application for renewal of the operating license for PNP should be approved.

Sincerely,

/RA/

Graham B. Wallis
Chairman

References:
1. Safety Evaluation Report Related to the License Renewal of the Palisades Nuclear Power Plant, September 2006.
2. Palisades Nuclear Power Plant - Application for Renewed Operating Licenses, March 22, 2005
3. Safety Evaluation Report with Open Items Related to the License Renewal of the Palisades Nuclear Power Plant, June 2006
4. Audit and Review Report for Plant Aging Management Reviews and Programs (AMPs) (AMRs) - Palisades Nuclear Power Plant, October 20, 2005
5. Palisades Nuclear Power Plant, Inspection Report 05000255/2005009, December 28, 2005
6. Memorandum dated September 13, 2006, from John T. Larkins, Executive Director, ACRS, to Luis A. Reyes, Executive Director for Operations, Subject: Questions Raised

by Members of the Public During the ACRS Subcommittee Meeting on Palisades
Nuclear Plant License Renewal Application

7. Regulatory Guide 1.99 Revision 2, Radiation Embrittlement of Reactor Vessel Materials, May 1988

8. Regulatory Guide 1.190, Calculational and Dosimetry Methods for Determining Pressure Vessel Neutron Fluence, March 2001

9. Palisades Reactor Pressure Vessel Fluence Evaluation, WCAP-15353, January 2000

SECTION 6

CONCLUSION

The staff of the U.S. Nuclear Regulatory Commission (staff) reviewed the license renewal application (LRA) for the Palisades Nuclear Plant (PNP), in accordance with the NRC regulations and NUREG-1800, "Standard Review Plan for Review of License Renewal Applications for Nuclear Power Plants," dated July 2001. Title 10, Section 54.29, of the *Code of Federal Regulations* (10 CFR 54.29) provides the standards for issuance of a renewed license.

On the basis of its evaluation of the license renewal applications, the NRC staff concludes that the requirements of 10 CFR 54.29(a) has been met and that the confirmatory item has been resolved.

The staff notes that any requirements of Subpart A of 10 CFR Part 51 are documented in Supplement 27 to NUREG-1437, "Generic Environmental Impact Statement for License Renewal of Nuclear Plants: Regarding Palisades Nuclear Plant Final Report," dated October 12, 2006.

APPENDIX A

COMMITMENTS FOR LICENSE RENEWAL OF PNP

During the review of the Palisades Nuclear Plant (PNP), license renewal application (LRA) by the staff of the U.S. Nuclear Regulatory Commission (staff), the applicant made commitments related to aging management programs (AMPs) to manage aging effects of structures and components (SCs) prior to the period of extended operation. The following table lists these commitments, along with the implementation schedules and the sources of the commitment.

APPENDIX A: LONG TERM COMMITMENTS FOR LICENSE RENEWAL OF PNP

No.	Commitment	Implementation Schedule	Source
1	Each year, following the submittal of the Palisades License Renewal Application and at least three months before the scheduled completion of the NRC review, NMC will submit an amendment to the application pursuant to 10 CFR 54.21 (b). This amendment will identify any changes to the Current Licensing Basis of the facility that materially affect the contents of the License Renewal Application, including the FSAR supplement, that have not already been submitted.	Annually	March 22, 2005 Letter
2	NMC will submit an equivalent margins analysis, completed in accordance with 10 CFR 50 Appendix G Section IV.A.1, for NRC approval, at least three years before any reactor vessel beltline material upper shelf energy decreases to less than 50 ft-lb.	Three years before limit is reached	March 22, 2005 Letter
3	At the appropriate time, prior to exceeding the PTS screening criteria, Palisades will select the optimum alternative to manage PTS in accordance with NRC regulations and make relevant submittals to obtain NRC review and approval.	As required	March 22, 2005 Letter
4	NMC will evaluate the effect the increase in variable speed charging pump out-of-service events may have on these lines (Charging Lines Inboard of the Regenerative Heat Exchanger), and will take actions necessary to ensure these lines meet licensing basis design criteria for the extended operating period. NMC will complete this evaluation and will advise the NRC of the results, and of any necessary corrective actions, before the end of the current licensed operating period.	Prior to Period of Extended Operation	March 22, 2005 Letter

A-3

	APPENDIX A: LONG TERM COMMITMENTS FOR LICENSE RENEWAL OF PNP		
No.	**Commitment**	**Implementation Schedule**	**Source**
5	NMC will monitor the cumulative number of pressurizer temperature element nozzle fatigue cycles within the Fatigue Monitoring Program, and maintain a special action level to ensure that appropriate actions are taken if at any time the cycle count for any design basis event since 1993 reaches the number assumed by these analyses.	Prior to Period of Extended Operation	March 22, 2005 Letter
6	NMC will revise the Alloy 600 Program to update the PWSCC corrosion rate assessments and inspection program consistent with the latest NRC requirements and industry commitments (e.g., EPRI Report 1010087 "Materials Reliability Program: Primary System Piping System Butt Weld Inspection and Evaluation Guidelines [MRP-139]," (August 2005)). The updated program will be submitted for NRC review and approval by March 24, 2008.	March 24, 2008	April 26, 2006 Letter
7	The supporting calculations for the Palisades RI-ISI program will be reviewed, and updated as needed, to reflect a 60-year operating period; and the program inspection scope will be updated accordingly, before the period of extended operation.	Prior to Period of Extended Operation	March 22, 2005 Letter
8	In a periodic FSAR update following NRC issuance of the renewed operating license, in accordance with 10 CFR 50.71 (e), the summary descriptions of Aging Management Programs and Time Limited Aging Analyses, provided in Appendix A, will be incorporated into appropriate sections of the FSAR. Analyses, provided in Appendix A, will be incorporated into appropriate sections of the FSAR.	First periodic FSAR update following issuance of renewed license	March 22, 2005 Letter

APPENDIX A: LONG TERM COMMITMENTS FOR LICENSE RENEWAL OF PNP

No.	Commitment	Implementation Schedule	Source
9	The Quality Program implementation procedures will be expanded to apply the elements of corrective action, confirmation process, and administrative controls to both safety related and non-safety related systems, structures, and components that are subject to aging management review for license renewal.	Prior to Period of Extended Operation	March 22, 2005 Letter
10	Review and revise ASME ISI Master Plan, procedures that implement credited License Renewal Programs, and plant maintenance procedures to reflect and reference the applicable guidance provided in NUREG-1339 and EPRI TR-104213 for safety and non-safety related bolting. These revisions should also include instructions for selection of bolting material and use of lubricants and sealants, in accordance with the guidelines of EPRI NP-5769 and the additional recommendations of NUREG-1339 to prevent or mitigate degradation and failure of safety-related bolting.	Prior to Period of Extended Operation	March 22, 2005 Letter
11	Evaluate the high strength bolting used for component supports for susceptibility to cracking as described in NUREG-1 801, Section XI.M.18, "Parameters Monitored/Inspected," and implement appropriate inspection requirements to provide adequate age-management for these bolts. This is to be completed prior to the end of the current operating license.	Prior to Period of Extended Operation	March 22, 2005 Letter
12	Revise applicable plant procedures to include criteria for observing susceptible SSC, within the scope of license renewal, for boric acid leakage and degradation, during system walkdown inspections.	Prior to Period of Extended Operation	March 22, 2005 Letter

A-5

APPENDIX A: LONG TERM COMMITMENTS FOR LICENSE RENEWAL OF PNP

No.	Commitment	Implementation Schedule	Source
13	Revise applicable plant procedure(s) to include explicit acceptance criteria for boric acid inspections.	Prior to Period of Extended Operation	March 22, 2005 Letter
14	Revise applicable plant procedures to include inspection of structural steel and non-ASME component supports for evidence of boric acid residue and boric acid wastage/corrosion on a periodic frequency.	Prior to Period of Extended Operation	March 22, 2005 Letter
15	A Buried Services Corrosion Monitoring Program will be developed and implemented. Features of the program will include development and implementation of procedures for inspection of selected buried SSCs for corrosion, pitting and MIC. The periodicity of these inspections will be based on opportunities for inspection such as scheduled excavation and maintenance work.	Prior to Period of Extended Operation	March 22, 2005 Letter
16	Develop and implement procedures for periodic draining and cleaning of diesel fuel oil storage tanks, Emergency Diesel Generator day tanks, and Diesel Fire Pump Day Tanks. These procedures shall include steps to perform a visual inspection of interior tank surfaces for signs of degradation or corrosion, with acceptance criteria, corrective actions, and documentation of inspection results.	Prior to Period of Extended Operation	March 22, 2005 Letter
17	Develop and implement procedures for periodic draining of water accumulated in the bottom of the fuel oil storage tanks and fuel oil day tanks for the Diesel Generators and Diesel Fire Pumps.	Prior to Period of Extended Operation	March 22, 2005 Letter
18	Develop and implement procedures for periodic ultrasonic measurement of thickness of the bottom of Fuel Oil Storage Tanks, Emergency Diesel Generator Day Tanks, and Diesel Fire Pump Day Tanks.	Prior to Period of Extended Operation	March 22, 2005 Letter

A-6

APPENDIX A: LONG TERM COMMITMENTS FOR LICENSE RENEWAL OF PNP

No.	Commitment	Implementation Schedule	Source
19	The Structures Monitoring Program shall be revised to include specific inspection criteria and documentation requirements for verifying that walls, ceilings and floors that serve as Fire Protection Program fire barriers are verified to be free from aging related degradation that would impact the fire barrier's intended function.	Prior to Period of Extended Operation	March 22, 2005 Letter
20	Plant procedures shall be revised to more specifically address aging related degradation and expectations for documentation of fire door condition.	Prior to Period of Extended Operation	March 22, 2005 Letter
21	Develop and implement procedures to perform visual inspections for fire door clearances.	Prior to Period of Extended Operation	March 22, 2005 Letter
22	Revise diesel-driven fire pump performance test procedures to more specifically address requirement to inspect and monitor fuel oil supply line for aging related degradation, and to document inspection results.	Prior to Period of Extended Operation	March 22, 2005 Letter
23	Develop and implement procedures for inspection of below grade fire protection system piping. Inspections shall occur when below grade piping is excavated for maintenance, and shall include pipe wall thickness (NDE or direct measurement)and documentation of aging related degradation of pipes. Procedures shall include acceptance criteria, and criteria for further corrective actions if acceptance criteria are not met.	Prior to Period of Extended Operation	March 22, 2005 Letter

A-7

No.	APPENDIX A: LONG TERM COMMITMENTS FOR LICENSE RENEWAL OF PNP		
	Commitment	Implementation Schedule	Source
24	Plant procedures shall be revised to more specifically address identification of aging related degradation and expectations for documentation of fire hydrant condition. Also, these revisions shall include provisions to perform flow testing for fire hydrants within the scope of License Renewal that are credited for fire suppression in the Palisades current licensing basis.	Prior to Period of Extended Operation	March 22, 2005 Letter
25	Develop and implement procedures to replace all sprinkler heads prior to the end of the 50 year service life, or for testing of a representative sample of sprinkler heads prior to the end of the 50 year service life and at 10 year intervals thereafter, per requirements of NFPA 25, Section 5.3.	Prior to Period of Extended Operation	March 22, 2005 Letter
26	A Non-EQ Electrical Commodities Condition Monitoring Program will be developed and implemented. Features of the program will include development and implementation of procedures to conduct periodic inspection of insulated cables and connectors, test sensitive instrumentation circuits, test medium voltage cables, and inspect manhole water levels.	Prior to Period of Extended Operation	March 22, 2005 Letter
27	A One Time Inspection Program will be developed and implemented. Features of the program are as described in the Enhancement section of LRA Section B2.1.13.	Prior to Period of Extended Operation	March 22, 2005 Letter
28	Revise crane and fuel handling machine inspection procedures to specifically inspect for general corrosion on passive components making up the bridge, trolley, girders, etc., and to inspect rails of Bridge Cranes for wear. Revision should also include documentation of results of these inspections, acceptance criteria, and qualification requirements for inspectors and crane supervisors.	Prior to Period of Extended Operation	March 22, 2005 Letter

APPENDIX A: LONG TERM COMMITMENTS FOR LICENSE RENEWAL OF PNP

No.	Commitment	Implementation Schedule	Source
29	The Reactor Vessel Integrity Surveillance Program will ensure that additional neutron fluence due to the power uprate will be accounted for when calculating fluence and developing pressure-temperature and LTOP curves for the extended operating period.	Prior to Period of Extended Operation	March 22, 2005 Letter
30	Document and establish requirement to save and store all pulled and tested reactor vessel surveillance capsules for future reconstitution use.	Prior to Period of Extended Operation	March 22, 2005 Letter
31	Evaluate and revise as necessary, the surveillance capsule withdrawal and testing schedule of FSAR Table 4-20 such that at least one capsule remains in the reactor vessel and is tested during the period of extended operation to monitor the effects of long-term exposure to neutron irradiation.	Prior to Period of Extended Operation	March 22, 2005 Letter
32	Develop a program level procedure to implement and control Technical Specification and FSAR activities associated with the Reactor Vessel Integrity Surveillance Program, including activities associated with surveillance capsules, pressure-temperature limit curves, LTOP setpoints, neutron embrittlement calculation methodology, neutron fluence calculations and control, and documentation requirements. Title of procedure should be, "Reactor Vessel Integrity Surveillance Program."	Prior to Period of Extended Operation	March 22, 2005 Letter

A-9

No.	Commitment	Implementation Schedule	Source
33	NMC will participate in industry initiatives that will generate additional data on aging mechanisms relevant to reactor vessel internals (RVI), including void swelling, and develop appropriate inspection techniques to permit detection and characterization of features of interest. Recommendations for augmented inspections and techniques resulting from this effort will be incorporated into the Reactor Vessel Internals Program as applicable. The revised Reactor Vessel Internals Program will be submitted for NRC review and approval by March 24, 2009.	March 24, 2009	August 25, 2005 Letter
34	Withdrawn		August 25, 2005 Letter
35	Incorporate into the Structural Monitoring Program all structural members listed in Tables 3.5.2-1 through 3.5.2-10 that will use the Structural Monitoring Program as an AMR	Prior to Period of Extended Operation	March 22, 2005 Letter
36	Enhance system walkdown procedures to more specifically address the types of components to be inspected, and to specifically describe the relevant degradation mechanisms and effects of interest, and for use of the Corrective Action Program to document related aging degradation, identified during the inspections, that may affect the ability of the SSC to perform its intended function.	Prior to Period of Extended Operation	March 22, 2005 Letter
37	A Fatigue Monitoring Program will be developed and implemented. Features of the program will include monitoring and tracking selected cyclic loading transients (cycle counting) and their effects on critical reactor pressure boundary components and other selected components.	Prior to Period of Extended Operation	March 22, 2005 Letter

APPENDIX A: LONG TERM COMMITMENTS FOR LICENSE RENEWAL OF PNP

No.	Commitment	Implementation Schedule	Source
38	The final text and schedule of licensee commitments that are confirmed by NRC in the final SER for the Palisades Renewed Operating License will be incorporated into appropriate locations of the FSAR in the first regular FSAR update under 10 CFR 50.71 (e) following NRC issuance of the renewed operating license.	Prior to Period of Extended Operation	July 1, 2005 Letter
39	Visual inspections of a sample of buried carbon, low-alloy, and stainless steel components will be performed within ten years prior to entering, and within ten years after entering, the period of extended operation. Prior to the tenth year of each period, NMC will perform an evaluation of available data to determine if sufficient opportunistic inspections have been performed within that period to assess the condition of the components. If insufficient data exists, focused inspection(s) will be performed as needed.	Prior to Period of Extended Operation and Within 10 Years of Entering the Period of Extended Operation	September 16, 2005
40	NMC will enhance the preventive maintenance program to periodically inspect, and replace as necessary, the expansion joints/flexible connections in the portions of the Heating, Ventilation and Air Conditioning System, and the Service Water System, that are in-scope for license renewal.	Prior to Period of Extended Operation	July 25, 2005 Letter
41	NMC will identify specific methods of inspection for individual components as part of the System Monitoring Program implementation procedure development. Industry documents such as EPRI 1009743, EPRI GS-7086, and API 575 will be used as source documents to define tank testing and inspection requirements.	Prior to Period of Extended Operation	July 25, 2005 Letter

APPENDIX A: LONG TERM COMMITMENTS FOR LICENSE RENEWAL OF PNP

No.	Commitment	Implementation Schedule	Source
42	NMC will revise the governing procedure for the Flow Accelerated Corrosion Program to include the value of 87.5% of nominal wall thickness for non safety related piping as a trigger point to initiate engineering analysis to confirm that remaining wall thickness is acceptable to support the intended function or to determine corrective action, as applicable. This requirement will be implemented by March 24, 2009.	March 24, 2009	August 25, 2005 Letter
43	NMC will perform a neutron absorption ("blackness") test of selected cells in the NUS spent fuel racks prior to March 24, 2011, to validate that there is no significant degradation of the neutron absorption capability. An additional test will be performed within the first 10 years following the start of the period of extended operation. If degradation is identified in either test, an evaluation of the condition will be performed under the NMC Corrective Action Program. If applicable, this evaluation will consider the potential need for additional or more frequent testing.	March 24, 2011	November 18, 2005
44	Palisades procedures will be enhanced to inspect and document the internal condition of applicable components, in-scope for license renewal, when maintenance provides an opportunity. Applicable components are those that have an internal environment of water, are constructed of materials that are potentially susceptible to internal aging degradation in a wetted environment, but are not subject to another Aging Management Program (e.g., Water Chemistry, Open Cycle Cooling) that would manage the internal environment such that aging degradation of the internal surfaces would not be expected.	Prior to Period of Extended Operation	March 30, 2006

A-12

APPENDIX A: LONG TERM COMMITMENTS FOR LICENSE RENEWAL OF PNP			
No.	Commitment	Implementation Schedule	Source
45	To verify that Corrosion Under Insulation (CUI) is not causing excessive corrosion of insulated piping and components, inspections of opportunity will be performed to assess the external surface condition when insulation is removed for maintenance or surveillance. The piping and components of interest are those within the scope of the System Monitoring Program, constructed of carbon or low alloy steel, with low normal operating temperatures in an indoor or outdoor environment such that the piping could be wetted under its insulation (e.g., from condensation or rain water) for extended periods without being detected. The System Monitoring Program will be enhanced to require a periodic review of documented under-insulation inspection results to verify that there were a sufficient number of inspection opportunities to provide a representative indication of system condition, and to assess the need for further action. If there were insufficient opportunities for inspection, insulation will be removed from additional sample locations to assess system condition under insulation. This program requirement will be implemented prior to March 24, 2011.	March 24, 2011	April 26, 2006

A-13

APPENDIX A: SHORT TERM COMMITMENTS FOR LICENSE RENEWAL OF PNP

No.	Commitment	Implementation Schedule	Source
46	NMC will submit, for NRC review and approval, a comparison of EPRI TR-105714 revision 5 with revision 3 to identify the material changes that impact aging management and justify their acceptability by October 31, 2005. If necessary, the submittal will include a Water Chemistry Program description, revised to identify and justify use of TR-105714, Revision 5, as an exception to the NUREG 1801 program description.	Completed October 31, 2005	August 25, 2005 Letter
47	NMC will submit, for NRC review and approval, a comparison of TR-102134 revision 6 with revision 3 to identify the material changes that impact aging management and justify their acceptability by October 31, 2005. If necessary, the submittal will include a Water Chemistry Program description, revised to identify and justify use of TR-1 02134, Revision 6, as an exception to the NUREG1801 program description.	Completed October 31, 2005	August 25, 2005 Letter
48	NMC will submit, for NRC review and approval, a comparison of TR-107396 revision I with revision 0 to identify the material changes that impact aging management and justify their acceptability by October 31, 2005. If necessary, the submittal will include a Closed Cycle Cooling Water Program description, revised to identify and justify use of TR-1 07396, Revision 1, as an exception to the NUREG 1801 program description.	Completed October 31, 2005	August 25, 2005 Letter

APPENDIX A: SHORT TERM COMMITMENTS FOR LICENSE RENEWAL OF PNP

No.	Commitment	Implementation Schedule	Source
49	NMC will revise the ASME Section XI IWB, IWC, IWD, IWF Aging Management Program descriptions in LRA Appendices A and B to reflect the 2001 edition including the 2002 and 2003 addenda of ASME Section XI. The revised program descriptions will identify exceptions to this code taken by the program, if any, that impact aging management effectiveness. Appropriate justification will also be provided to show that the exceptions, if any, still provide an acceptable level of aging management. The revised program descriptions will be submitted for NRC review and approval by October 31, 2005.	Completed October 31, 2005	August 25, 2005 Letter
50	NMC will revise the Containment Inservice Inspection Program description in the LRA to identify use of the 1998 edition as an exception to GALL. Exceptions taken to the 1998 edition, if any, will be identified and justified as part of the program description. A comparison of the 1998 edition with the 1995 edition/1996 addendum referenced in NUREG 1801, revision 0, or the 2001 edition, including the 2002 and 2003 addenda, referenced in NUREG 1801, draft revision 1 (publicly released on August 12, 2005), will also be developed to support the adequacy of the 1998 edition of IWE and IWL for aging management. The revised program description and comparison will be submitted for NRC review and approval by October 31, 2005.	Completed October 31, 2005	August 25, 2005 Letter

A-15

APPENDIX A: SHORT-TERM COMMITMENTS FOR LICENSE RENEWAL OF PNP

No.	Commitment	Implementation Schedule	Source
51	NMC will develop a new Compressed Air Program for Palisades. This program will manage aging in carbon steel components within the compressed, saturated or moist air environments of the Compressed air systems. Compressed Air System descriptions for LRA Appendices A and B will be submitted for NRC review and approval by October 31, 2005. In addition, LRA Appendix A and B descriptions of the One-Time Inspection Program, revised to delete reference to management of compressed air components, will be provided.	Completed October 31, 2005	August 27, 2005 Letter
52	NMC will update the appropriate sections of the License Renewal Application to reflect inclusion of the Auxiliary Feedwater (AFW) Pump steam supply line insulation within the AFW Pump Room in scope for license renewal, and provide the results of the aging management review. This information will be submitted for NRC review and approval by October 31, 2005.	Completed October 31, 2005	August 27, 2005 Letter
53	NMC will have an analysis performed to project the expected containment tendon pre-stressing forces out to 60 years, utilizing, to the extent practical with available surveillance data, the procedures recommended in NUREG 1801, Generic Aging Lessons Learned (GALL) Report, Program X.S1, and Regulatory Guide 1.99, Radiation Embrittlement of Reactor Vessel Materials. The results of the analysis will be submitted to NRC when available.	Completed October 28, 2005	July 25, 2005 Letter

APPENDIX A: SHORT TERM COMMITMENTS FOR LICENSE RENEWAL OF PNP

No.	Commitment	Implementation Schedule	Source
54	NMC will submit for NRC review either a technical discussion that provides a more detailed basis for concluding that underclad cracking is not a TLAA at Palisades, or a description to be incorporated into LRA Section 4 and Appendix A which describes underclad cracking as a TLAA and identifies the appropriate disposition under 10 CFR 54.21(c)(1). This information will be submitted by September 1, 2006.	Completed April 26, 2006	March 30, 2006 Letter
55	Upon completion of the final technical report on the effects of potential underclad cracking at Palisades, NMC will notify NRC that the technical report for the final disposition of the issue has been completed, and the associated LRA changes submitted on April 26, 2006, have been confirmed. If the final report identifies a need for any additional LRA revisions, the revised information will be provided at that time for NRC review and approval. NMC will submit this information no later than September 1, 2006.	Completed July 5, 2006	April 26, 2006 Letter

A-17

APPENDIX B

CHRONOLOGY

This appendix contains a chronological listing of the routine correspondence between the staff of the U.S. Nuclear Regulatory Commission (NRC or the staff) and the Nuclear Management Company, LLC (NMC), and other correspondence regarding the staff's reviews of the Palisades Nuclear Plant (PNP), Docket Number 50-255, license renewal application (LRA).

March 22, 2005	Letter Transmitting Palisades Nuclear Plant Application for Renewed Operating License. (Accession No. ML050940434)
March 22, 2005	Palisades Nuclear Plant, Application for Renewed Operating License. Cover Page through Appendix D. (Accession No. ML050940446)
March 22, 2005	Palisades Nuclear Plant, Applicant's Environmental Report Operating License Renewal Stage (Accession No. ML050940449)
March 23, 2005	Transmittal of Information set of License Renewal Scoping Boundary Drawings for Palisades Nuclear Plant (Accession No. ML050940480)
March 23, 2005	Information set of License Renewal Scoping Boundary Drawings for Palisades Nuclear Plant, Drawings LR-M-206-1A, Rev. 24 through LR-M-210-1C, Rev. 14 (Accession No. ML050940481)
March 23, 2005	Information set of License Renewal Scoping Boundary Drawings for Palisades Nuclear Plant, Drawings LR-M-210-2, Rev. 34 through LR-M-218-4, Rev. 22 (Accession No. ML050940483)
March 23, 2005	Information set of License Renewal Scoping Boundary Drawings for Palisades Nuclear Plant, Drawings LR-M-218-5, Rev. 27 through LR-M-225-2, Rev. 44 (Accession No. ML050940484)
March 23, 2005	Information set of License Renewal Scoping Boundary Drawings for Palisades Nuclear Plant, Drawings LR-M-226-1, Rev. 56 through LR-M-200-1, Rev. 32 (Accession No. ML050940485)
March 31, 2005	Mech 3.1, Original Section 3 Tables for the Palisades Nuclear Plant, Units 1 and 2, License Renewal Application (Accession No. ML051450274)
March 31, 2005	Palisades Mech 3.2, Original Section 3 Tables for the Palisades Nuclear Plant, Units 1 and 2, License Renewal Application (Accession No. ML051450298)

March 31, 2005	Palisades Mech 3.3, Original Section 3 Tables for the Palisades Nuclear Plant, Units 1 and 2, License Renewal Application (Accession No. ML051450318)
March 31, 2005	Palisades Mech 3.4, Original Section 3 Tables for the Palisades Nuclear Plant, Units 1 and 2, License Renewal Application (Accession No. ML051450323)
March 31, 2005	Palisades Civil 3.5, Original Section 3 Tables for the Palisades Nuclear Plant, Units 1 and 2, License Renewal Application (Accession No. ML051450396)
March 31, 2005	Palisades Elec. 3.6, Original Section 3 Tables for the Palisades Nuclear Plant, Units 1 and 2, License Renewal Application (Accession No. ML051450399)
March 31, 2005	Palisades Precedent Table, Civil Section 3.5. (Accession No. ML051380092)
March 31, 2005	Palisades Precedent Table, Elec. Section 3.6. (Accession No. ML051380086)
March 31, 2005	Palisades Precedent Table, Mech Section 3.1 (Accession No. ML051380094)
March 31, 2005	Palisades Precedent Table, Mech Section 3.2. (Accession No. ML051380095)
March 31, 2005	Palisades Precedent Table, Mech Section 3.3 (Accession No. ML051380098)
March 31, 2005	Palisades Precedent Table, Mech Section 3.4 (Accession No. ML051380099)
April 1, 2005	Precedents for Section 3.5 Exceptions to GALL. (Accession No. ML052000440)
April 1, 2005	Precedents for Section 3.6 Exceptions to GALL. (Accession No. ML052000454)
April 1, 2005	Palisades, License Amendment Request: One - Time Extension to Technical Specification Action Completion Time For Restoration of A Service Water Train to Operable Status (Accession No. ML051020349)
April 5, 2005	Press Release-05-060: NRC Announces Availability of License Renewal Application for Palisades (Accession No. ML050950137)

April 6, 2005	Receipt and Availability of the License Renewal Application for the Palisades Nuclear Plant and Federal Register Notice (Accession No. ML050960344)
April 8, 2005	4/8/05 - Maintenance of Reference Material at the South Haven Memorial Library at the Palisades Nuclear Plant, License Renewal Application (Accession No. ML051100210)
April 12, 2005	Palisades - Section 3.2 ESF Precedent Tables (E-Mail) (Accession No. ML061720311)
April 12, 2006	Palisades - Section 3.2 ESF Tables (Accession No. ML061720318)
April 12, 2006	Palisades - ESF Drawings (Accession No. ML061720283)
April 12, 2006	Safety Injection System Drawing (VISIO_SOD _SI_02_r01.pdf) (Accession No. ML061720292)
April 12, 2006	Safety Injection and Containment Spray System Drawing (Accession No. ML061720298)
April 13, 2005	04/28/2005, Forthcoming Public Information Session for the U.S. Nuclear Regulatory Commission (NRC) Staff to Describe the License Renewal Process (Accession No. ML051030194)
April 15, 2005	Palisades - Supplemental ESF Drawings. (Accession No. ML061800319)
April 15, 2005	SAFETY INJECTION AND CONTAINMENT SPRAY SYSTEM DRAWING WITH SI HIGHLIGHTED (s1_01_r02) (Accession No. ML061800338)
April 15, 2005	SAFETY INJECTION AND CONTAINMENT SPRAY SYSTEM DRAWING WITH SI HIGHLIGHTED (s1_02_r01). (Accession No. ML061800343)
April 15, 2005	SAFETY INJECTION AND CONTAINMENT SPRAY SYSTEM DRAWING WITH CONTAINMENT SPRAY HIGHLIGHTED. (Accession No. ML061800349)
April 15, 2005	SAFETY INJECTION AND CONTAINMENT SPRAY SYSTEM DRAWING WITH CONTAINMENT SPRAY HIGHLIGHTED. (Accession No. ML061800353)
April 16, 2005	Precedents for Section 3.3 Exceptions to GALL. (Accession No. ML052000241)
April 25, 2005	Press Release-III-05-020: NRC Staff Schedules Public Meeting for April 28 to Discuss License Renewal Process for Palisades Nuclear Plant (Accession No. ML051150147)

April 26, 2005	G20050315/LTR-05-0245 - Rep. Fred Upton Ltr re: License Renewal Application of the Palisades Nuclear Plant (Accession No. ML051220248)
April 26, 2005	Letter from Congressman Fred Upton to Chairman Nils Diaz Pertaining to Palisades Nuclear Power Plant. (Accession No. ML053530137)
April 27, 2005	TRAINING PRESENTATION ON PALISADES SAFETY INJECTION & CONTAINMENT SPRAY (Accession No. ML061800363)
April 27, 2005	Training Presentation on Palisades Safety Injection & Containment Spray (Accession No. ML061800370)
April 27, 2005	Palisades - Additions to ESF Tables 2.3.2-1 & 3.2.2-1. (Accession No. ML061800379)
April 27, 2005	Palisades - Additions to ESF Tables (Accession No. ML061800389)
May 5, 2005	Supplementary Information to Support the Application for Renewed Operating License for the Palisades Nuclear Plant. (Accession No. ML051300128)
May 5, 2005	Supplementary Information to Support the Application for Renewed Operating License for the Palisades Nuclear Plant. (Accession No. ML051320235)
May 25, 2005	Summary of Public Information Session for the U.S. Nuclear Regulatory Commission (NRC) Staff to Describe its License Renewal Process (Accession No. ML051450441)
June 2, 2005	Federal Register Notice, Palisades, Notice of Acceptance for Docketing of Application, Notice of Opportunity for Hearing (Accession No. ML051530122)
June 3, 2005	Requests for Additional Information for the review of the Palisades Nuclear Plant, License Renewal Application (Accession No. ML051540237)
June 3, 2005	07/01/05 Notice of Forthcoming Exit Meeting with Nuclear Management Company, LLC on License Renewal Scoping and Screening Methodology Audit for Palisades Nuclear Plant (Accession No. ML051540449)
June 3, 2005	Summary of Telephone Conference held on April 12, 2005, between the U.S. Nuclear Regulatory Commission and Nuclear Management Company, LLC, concerning information pertaining to the Palisades Nuclear Plant License Renewal Application (Accession No. ML051560003)
June 6, 2005	Precedents for Section 3.1 Exceptions to GALL. (Accession No. ML052000234)

June 7, 2005	Precedents for Section 3.2 Exceptions to GALL. (Accession No. ML052000237)
June 8, 2005	· G20050315/LTR-05-0245 - Fred Upton Ltr. Re License Renewal Application of the Palisades Nuclear Plant (Accession No. ML051260065)
June 9, 2005	Press Release-05-091: NRC Announces Opportunity For Hearing On Application To Renew Operating License For Palisades Nuclear Power Plant (Accession No. ML051600436)
June 22, 2005	06/16/2005 Summary of Telephone Conference with Nuclear Management Company, LLC, Concerning Draft Requests for Additional Information Pertaining to the Palisades Nuclear Plant License Renewal Application (Accession No. ML051730200)
June 22, 2005	Requests for Additional Information for the review of the Palisades Nuclear Plant, License Renewal Application (Accession No. ML051730291)
June 22, 2005	06/20/2005 Summary of Telephone Conference Held on Between U.S. Nuclear Regulatory Commission (NRC) &Nuclear Management Company, LLC, Concerning Draft Requests for Additional Information Pertaining to Palisades Nuclear Plant License Renewal Application (Accession No. ML051730343)
June 22, 2005	Requests for Additional Information for the review of the Palisades Nuclear Plant, License Renewal Application (Accession No. ML051730577)
June 22, 2005	Request for Additional Information (RAIs) for the Review of the Palisades Nuclear Plant, License Renewal Application (TAC No. MC6433) (Accession No. ML051730597)
June 22, 2005	Request for Additional Information (RAIs) for the Review of the Palisades Nuclear Plant, License Renewal Application (TAC No.MC6433) (Accession No. ML051730597)
June 22, 2005	Summary of Telephone Conference Held on June 20, 2005, Between the U.S. Nuclear Regulatory Commission (NRC) and Nuclear Management Company, LLC (NMC) Concerning Draft Requests for Additional Information Pertaining to the Palisades NP LR Application. (Accession No. ML051730518)
June 28, 2005	Requests for Additional Information (RAIs) for the review of the Palisades Nuclear Plant, License Renewal Application (Accession No. ML051790133)

June 28, 2005	Requests for Additional Information for the review of the Palisades Nuclear Plant, License Renewal Application (Accession No. ML051790142)
June 28, 2005	Requests for Additional Information for the review of the Palisades Nuclear Plant, License Renewal Application (Accession No. ML051790157)
June 30, 2005	Acknowledgment of Receipt of Your Letter on the Applications for Renewal of the Operating Licenses for Palisades and Donald C. Cook, Units 1 and 2 Nuclear Plants (Accession No. ML051820578)
June 30, 2005	Palisades Nuclear Plant License Renewal Review. (Accession No. ML051860359)
July 1, 2005	Palisades, Response to NRC Request for Additional Information Relating to License Renewal dated June 3, 2005 (Accession No. ML051960390)
July 7, 2005	06/22/2005, Summary of Telephone Conference Between the NRC and NMC, LLC, Concerning Draft Requests for Additional Information Pertaining to the Palisades Nuclear Plant License Renewal Application (TAC No. MC6433) (Accession No. ML051880341)
July 7, 2005	Summary of Telephone Conference Held on June 22, 2005, Between the U.S. NRC and NMC, LLC, Concerning Draft RAIs pertaining to the Palisades Nuclear Plant License Renewal Application (Accession No. ML051880435)
July 7, 2005	Summary of Telecon Held on June 22, 2005, Between the NRC and NMC, LLC, Concerning Draft RAIs Pertaining to the Palisades Nuclear Plant License Renewal Application (TAC No. MC6433) (Accession No. ML051880462)
July 11, 2005	Summary of Telephone Conference Held on June 28, 2005, Between the NRC and NMC, Concerning Draft Requests for Additional Information Pertaining to the Palisades Nuclear Plant License Renewal Application (Accession No. ML051940323)
July 12, 2005	Palisades, Request for Additional Information for the Review of the Palisades Nuclear Plant, License Renewal Application (Accession No. ML051940189)
July 14, 2005	Summary of Meeting Held on July 1, 2005, Between the NRC and NMC Representatives to Discuss the Results of the Scoping and Screening Methodology Audit for the Palisades Nuclear Plant License Renewal Application (Accession No. ML051950164)

July 19, 2005	Press Release-III-05-034: NRC Seeks Public Input on Palisades Nuclear Plant Environmental Issues for License Renewal (Accession No. ML052000242)
July 20, 2005	Request for Additional Information for the review of the Palisades Nuclear Plant, License Renewal Application (Accession No. ML052010471)
July 21, 2005	Palisades Nuclear Plant, Request for Additional Information (RAI) for the Review of the License Renewal Application (TAC NO. MC6433) (Accession No. ML052030025)
July 21, 2005	Palisades Nuclear Plant, Request for Additional Information (RAI) for the Review of the License Renewal Application (TAC NO. MC6433) (Accession No. ML052030025)
July 25, 2005	NMC Response to NRC Requests for Additional Information dated June 22, 2005 Relating to License Renewal for the Palisades Nuclear Plant (Accession No. ML052090229)
July 26, 2005	Request for Additional Information for the Review of the Palisades Nuclear Plant, License Renewal Application (Accession No. ML052070727)
July 26, 2005	Request for Additional Information for the Review of the Palisades Nuclear Plant, License Renewal Application (Accession No. ML052070788)
July 27, 2005	Request for Additional Information for the Review of the Palisades Nuclear Plant, License Renewal Application (Accession No. ML052080205)
July 27, 2005	09/02/05, Forthcoming Exit Meeting with Nuclear Management Company to Present Results of the Palisades Nuclear Plant Aging Management Program and Aging Management Review Audits. (Accession No. ML052080548)
July 28, 2005	Palisades, Response to NRC Requests for Additional Information Relating to License Renewal dated June 28, 2005 (Accession No. ML052130137)
August 2, 2005	Summary of a Telephone Conference Held on July 20, 2005, Between the NRC and Nuclear Management Company, LLC Concerning Responses to Open Questions from the Site Aging Management Program Audit. (Accession No. ML052140634)
August 8, 2005	2005/08/08-Palisades- Request for Hearing and Petition to Intervene (Accession No. ML052940221)

August 12, 2005	NMC Response to NRC Requests for Additional Information Dated July 12, 2005 Relating to License Renewal for the Palisades Nuclear Plant (Accession No. ML052290273)
August 19, 2005	Palisades, Response to NRC Requests for Additional Information Relating to License Renewal dtd July 20 and 21, 2005 (Accession No. ML052360305)
August 23, 2005	Request for Additional Information (RAIs) for the Review of the Palisades Nuclear Plant, License Renewal Application (TAC No. MC6433) (Accession No. ML052360026)
August 23, 2005	Request for Additional Information (RAI) for the Review of the Palisades Nuclear Plant, License Renewal Application (Pkg. 20) (Accession No. ML052360061)
August 25, 2005	Palisades Nuclear Plant Response to Request for Additional Information Related to License Amendment Request for One-Time Extension to the Technical Specification Action Completion Time for Restoration of a Service Water Train to Operable Status (Accession No. ML052450148)
August 25, 2005	Supplementary Information for the Palisades Application for Renewed Operating License Resulting from Aging Management Programs Audit. (Accession No. ML052410206)
August 26, 2005	Request for Additional Information (RAI) for the Review of the Palisades Nuclear Plant, License Renewal Application (TAC No. MC6433) (Accession No. ML052380035)
August 26, 2005	Summary of Telephone conference held on August 11, 2005, Between the NRC and Nuclear Management Company Co. LLC, Concerning Draft RAI Pertaining to the Palisades Nuclear Plant, License Renewal Application (Accession No. ML052430773)
August 27, 2005	Palisades, Response to NRC Requests for Additional Information Relating to License Renewal dated July 26 and 27, 2005 (Accession No. ML052440284)
August 27, 2005	Palisades - Supplementary Information for the Plant Application for Renewed Operating License Resulting from Aging Management Review Audit. (Accession No. ML052440392)
August 31, 2005	Request for Additional Information (RAIs) for the Review of the Palisades Nuclear Plant, License Renewal Application (TAC No. MC6433) (Accession No. ML052430767)
September 2, 2005	NMC Responses to Follow Up Questions From License Renewal Aging Management Review and Aging Management Program Audits (Accession No. ML052500207)

September 2, 2005   2005/09/02-Palisades - NRC Staff Response Opposing Petition to Intervene and Request for Hearing and Notices of Appearance's for Darani M. Reddick and Susan L. Uttal on behalf of the NRC Staff (Accession No. ML052500344)

September 2, 2005   2005/09/02-Nuclear Management Company's Answer to the August 8, 2005 Request for Hearing and Petition to Intervene (Accession No. ML052510507)

September 8, 2005   Summary of Telecon held on August 24, 2005 between the NRC and Nuclear Management Company, LLC Regarding Draft Request for Additional Information Pertaining to the Palisades Nuclear Plant, License Renewal Application (Accession No. ML052510606)

September 8, 2005   Request for Additional Information for the Review of the Palisades Nuclear Plant, License Renewal Application (Tac No. MC6433) (Accession No. ML052520310)

September 14, 2005   Request for Additional Information (RAI) for the Review of the Palisades Nuclear Plant, License Renewal Application (Accession No. ML052580267)

September 15, 2005   2005/09/15-Letter from David Lewis to Administrative Judges providing a compact disc of the Palisades license renewal application (Accession No. ML052650222)

September 16, 2005   Palisades Responses to NRC Requests for Additional Information Relating to License Renewal dated August 23, August 26, and August 31, 2005 (Accession No. ML052650369)

September 21, 2005   Summary of Public Scoping Meetings to support review of Palisades Nuclear Plant License Renewal Application (TAC MC6434) (Accession No. ML052630426)

September 21, 2005   Palisades, Request for Additional Information (RAI) for the Review of the Palisades Nuclear Plant, License Renewal Application (TAC No. MC6433) (Accession No. ML052650077)

September 23, 2005   NMC Response to NRC Requests for Additional Information Dated August 23, 2005 Relating to License Renewal for the Palisades Nuclear Plant (Accession No. ML052710377)

September 26, 2005   2005/09/26-Palisades - NRC Staff Motion to Strike Petitioners' Combined Reply to NRC Staff and NMC Answers to Petition to Intervene and Request for Hearing (Accession No. ML052690301)

September 27, 2005   AMP and AMR Audit Q & A Report (Accession No. ML052720222)

September 27, 2005   Summary Report of License Renewal Review Questions for AMP Audit (Accession No. ML052720250)

September 27, 2005   Summary Report of License Renewal Review Questions for AMR Audit (Accession No. ML052720254)

September 28, 2005   Request for Additional Information for the Review of the Palisades Nuclear Plant, License Renewal Application (Accession No. ML052730468)

September 28, 2005   Summary of a Telephone Conference Call Held on August 24, 2005, Between the NRC and NMC, Concerning Responses to Follow-Up Questions from the Site Aging Management Program and Aging Management Review Audit. (Accession No. ML052730460)

September 29, 2005   Summary of meeting held on September 22, 2005, between NRC and NMC representatives to discuss the results of the Palisades Nuclear Plant Aging Management Program and Aging Management Review Audits. (Accession No. ML052730440)

October 6, 2005   NMC Response to NRC Requests for Additional Information Dated September 8, 2005 Relating to License Renewal for the Palisades Nuclear Plant (Accession No. ML052850022)

October 11, 2005   Summary of telephone conference held on 9/14/05 between the NRC and the Nuclear Management Company, LLC concerning the applicant's response to the Request for Additional Information pertaining to the Palisades Nuclear Plant, License Renewal Application (Accession No. ML052850109)

October 14, 2005   NMC Response to NRC Requests for Additional Information Dated September 14, 2005 Relating to License Renewal for the Palisades Nuclear Plant (Accession No. ML052920252)

October 20, 2005   Audit and Review Report for Plant Aging Management Reviews and Programs for the Palisades Nuclear Plant License Renewal Application (Accession No. ML052980168)

October 21, 2005   Response to NRC Requests for Additional Information Dated August 24, 2005, Relating to License Renewal for the Palisades Nuclear Plant (Accession No. ML052990316)

October 24, 2005   Request for Additional Information for the review of the Palisades Nuclear Plant, License Renewal Application (Accession No. ML052980033)

October 24, 2005   Request for Additional Information for the review of the Palisades Nuclear Plant, License Renewal Application (Accession No. ML052980038)

October 28, 2005   Press Release-III-05-041: NRC Licensing Board to Hear Oral Argument and Receive Public Comments Nov. 3 in Palisades License Renewal Proceeding (Accession No. ML053010166)

October 28, 2005   Summary of a Telephone Conference Call Held on October 17, 2005, Between the NRC and Nuclear Management Company, LLC, Concerning Draft Request for Additional Information Pertaining to the Palisades License Renewal Application (Accession No. ML053040379)

October 28, 2005   Palisades, NMC Response to NRC Requests for Additional Information Dated September 28, 2005 Relating to License Renewal (Accession No. ML053070017)

October 28, 2005   Summary of Telephone Conference Call held on September 7, 2005, between the U.S. Nuclear Regulatory Commission (NRC) and Nuclear Management Company, LLC, concerning Draft Request for Additional Information Pertaining to the Palisades Nuclear Plant. (Accession No. ML053040425)

October 31, 2005   Supplementary Information for the Application for Renewal of the Palisades Plant Operating License (Accession No. ML053070561)

November 8, 2005   Letter to Paul Harden, Re: Audit and Review for Plant Aging Management Reviews and Programs for the Palisades Nuclear Plant License Renewal Application (Tac MC6433) (Accession No. ML053140415)

November 10, 2005   12/01/2005 Notice of Meeting with Nuclear Management Co, LLC to Discuss the License Renewal Project for Palisades Nuclear Plant (Accession No. ML053180497)

November 10, 2005   Ltr. 11/10/05 Palisades re: Public Meeting to discuss the Palisades Nuclear Plant License Renewal Project (Accession No. ML053180535)

November 18, 2005   Palisades, Response to NRC Request for Additional Information Relating to the License Renewal Dated October 24, 2005 (Accession No. ML053270257)

November 18, 2005   Supplement to "Response to NRC Request for Additional Information Dated August 24, 2005, dated October 21, 2005, and telecon on November 10, 2005 for Palisades." (Accession No. ML053470426)

November 29, 2005   Summary of a Telephone Conference Call held on September 8, 2005, between the NRC and Nuclear Management Company, LLC (NMC) Concerning Draft Request for Additional Information Pertaining to Palisades License Renewal (Accession No. ML053340021)

November 30, 2005   Request for Additional Information for the review of the Palisades Nuclear Plant, License Renewal Application (TAC No. MC6433) (Accession No. ML053340149)

December 1, 2005	Slides/Handouts - Palisades Nuclear Plant - Scoping, Screening and Aging Management License Renewal Inspection Exit Meeting (Accession No. ML053610105)
December 12, 2005	Summary of a telephone conference call held on November 15, 2005, between the NRC and Nuclear Management Company, LLC concerning responses to Request for Additional Information. (Accession No. ML053470339)
December 14, 2005	12/14/05 - Issuance of Environmental Scoping Summary Report Associated with the Staff's Review of the Application by Nuclear Management Company, LLC, for Renewal of the Operating License for Palisades Nuclear Plant (Accession No. ML053490390)
December 16, 2005	Palisades, Supplemental Responses to NRC Questions Relating to License Renewal (Accession No. ML053530082)
December 20, 2005	Ltr 12/01/05, Palisades Summary with NRC to Discuss the Results of the License Renewal Inspection (Accession No. ML053610091)
December 22, 2005	12/22/05 - Summary of Teleconference Conducted on November 10, 2005, with Nuclear Management Company, LLC (NMC) to Discuss the Severe Accident Mitigation Alternatives (SAMA) Request for Additional Information (RAI) for Palisades Nuclear Plant (Accession No. ML053570348)
December 28, 2005	IR 05000255-05-009(DRS); 10/24/2005 - 12/01/2005; Palisades Nuclear Plant; License Renewal Inspection (Accession No. ML053630216)
January 4, 2006	Note to File: Docketing of Additional Information pertaining to License Renewal Application of the Palisades Nuclear Plant (TAC MC6433) (Accession No. ML060060075)
January 4, 2006	Note to File: Docketing of Additional Information pertaining to License Renewal Application of the Palisades Nuclear Plant (TAC MC6433) (Accession No. ML060100035)
January 5, 2006	Withdrawal of License Amendment Request for One-Time Extension to the Technical Specification Action Completion Time for Restoration of a Service Water Train to Operable Status (Accession No. ML060050372)
January 13, 2006	NMC Response to NRC Requests for Additional Information Dated November 30, 2005 Relating to License Renewal for the Palisades Nuclear Plant (Accession No. ML060170211)
January 26, 2006	Palisades, Supplementary Copies of License Renewal Drawings Related to Application for Renewed Operating License (Accession No. ML060260438)

January 26, 2006   Palisades Nuclear Plant, Supplementary Copies of License Renewal Drawings Related to Application for Renewed Operating License (Accession No. ML060310646)

January 26, 2006   Supplementary Precedent Information Related to the Palisades Application for Renewed Operating License. (Accession No. ML060260287)

January 27, 2006   02/14/2006 Notice of Meeting Between the U.S. Nuclear Regulatory Commission (NRC) Staff and Nuclear Management Company, LLC (NMC) Regarding the License Renewal Application for the Palisades Nuclear Plant (Accession No. ML060270074)

March 8, 2006   Request for Additional Information for the Review of the Palisades Nuclear Plant, License Renewal Application (Accession No. ML060670288)

March 21, 2006   Annual Update of the Application for Renewed Operating License for the Palisades Nuclear Plant. (Accession No. ML060870166)

March 30, 2006   NMC Responses to NRC Requests for Additional information Relating to License Renewal Dated March 8, 2006 (Accession No. ML060900288)

April 26, 2006   Palisades NMC Responses to NRC Follow Up Questions Relating to License Renewal (Accession No. ML061170214)

July 05, 2006   Palisades, Updated NMC Response to NRC Question Relating to License Renewal. (Accession No. ML061860565)

September 13, 2006   Questions Raised by Members of the Public During the ACRS Subcommittee Meeting on Palisades Nuclear Plant License Renewal Application(Accession No. ML062640268)

September 19, 2006   Revision of Schedule for the Conduct of the Review of the Palisades Nuclear Plant, License Renewal Application (TAC NO. MC6433) (Accession No. ML062620482)

September 28, 2006   Safety Evaluation Report (SER) Related to the License Renewal of Palidades Nuclear Plant (Accession No. ML062710074)

October 16, 2006   Response to Questions Raised by Members of the Public During the Advisory Committee on Reactor Safeguards Subcommittee Meeting on Palisades Nuclear Plant License Renewal Application (Accession No. ML062640588)

November 17, 2006   Report on the Safety Aspects of the License Renewal Application for the Palisades Nuclear Power Plant (Accession No. ML063210400) b

APPENDIX C

PRINCIPAL CONTRIBUTORS

NAME	RESPONSIBILITY
F. Akstulewicz	Management Oversight
H. Asher	Structural Engineering
D. Ashley	Project Support
J. Ayala	Lead Project Manager
W. Bateman	Management Oversight
B. Boger	Management Oversight
T. Chan	Management Oversight
K. Chang	Management Oversight
M. Concepcion	Quality Assurance
K. Cozens	GALL Audit and Review
G. Cranston	Management Oversight
Z. Cruz Perez	Mechanical Engineering
R. De La Garza	SER Support
Y. Diaz-Castillo	Material Engineering
R. Denning	Management Oversight
J. Downs	Fire Protection
B. Elliot	Material Engineering
T. Ford	Reactor Systems
G. Galleti	Quality Assurance
F. Gillespie	Management Oversight
R. Goel	Mechanical Engineering
J. Grobe	Management Oversight
J. Hannon	Management Oversight
R. Hardies	Material Engineering

NAME	RESPONSIBILITY
M. Hartzman	Mechanical Engineering
M. Heath	GALL Audit and Review
A. Hiser	Management Oversight
K. Hsu	GALL Audit and Review
G. Imbro	Management Oversight
N. Iqbal	Fire Protection
D. Jeng	Mechanical Engineering
R. Jenkins	Management Oversight
S. Jones	Management Oversight
R. Karas	Management Oversight
P. Kuo	Management Oversight
C. Lauron	Chemical Engineering
S. Lee	Management Oversight
C. Li	Plant Systems
M. Li	Project Support
Y. Li	Mechanical Engineering
P. Lougheed	Region III Inspection
L. Lund	Management Oversight
J. Lyons	Management Oversight
K. Manoly	Management Oversight
T. Martin	Management Oversight
D. Mathews	Management Oversight
M. Mayfield	Management Oversight
R. McIntyre	Quality Assurance
R. McNally	Mechanical Engineering
M. Morgan	Project Manager
M. Mitchell	Management Oversight

NAME	RESPONSIBILITY
D. Nguyen	Electrical Engineering
A. Pal	Electrical Engineering
K. Parczewski	Chemical Engineering
P. Prescott	Quality Assurance
V. Rodriguez	SER Support
J. Rowley	Material Engineering
C. Sheng	Material Engineering
D. Solorio	Management Oversight
F. Talbot	Quality Assurance
D. Thatcher	Management Oversight
L. Tran	GALL Audit and Review
J. Tsao	Mechanical Engineering
M. Tschiltz	Management Oversight
H. Walker	Mechanical Engineering
S. Weerakkody	Management Oversight
P. Wen	GALL Audit and Review
G. Wilson	Management Oversight
O. Yee	Project Support
R. Young	Plant Systems
J. Zimmerman	Management Oversight

CONTRACTORS

CONTRACTOR	TECHNICAL AREA
Legin Group, Inc.	SER Support
Information Systems Laboratories	Plant Systems/GALL Audit

APPENDIX D

REFERENCES

This appendix contains a listing of the references used in the preparation of the Safety Evaluation Report prepared during the review of the license renewal application (LRA) for Palisades Nuclear Plant (PNP), Docket Number 50-255.

(1)   Title 10, Code of Federal Regulations, Part 54, "Requirements for Renewal of Operating Licenses for Nuclear Power Plants"

(2)   Title 10, Code of Federal Regulations, Part 51, "Environmental Protection Regulations for Domestic Licensing and Related Regulatory Functions"

(3)   NEI 95-10, "Industry Guideline for Implementing the Requirements of 10 CFR Part 54 -The License Renewal Rule," Rev. 4, Nuclear Energy Institute, October 2004.

(4)   NUREG-1800, "Standard Review Plan for Review of License Renewal Applications for Nuclear Power Plants, U.S. Nuclear Regulatory Commission," July 2001.

(5)   NUREG-1801, "Generic Aging Lessons Learned (GALL) Report, U.S. Nuclear Regulatory Commission," July 2001.

(6)   Title 10, Code of Federal Regulations, Part 140, Appendix B, "Form of Indemnity Agreement With Licensees Furnishing Insurance Policies As Proof of Financial Protection"

(7)   Title 10, Code of Federal Regulations, Part 50, "Domestic Licensing of Production and Utilization Facilities"

(8)   Palisades Nuclear Plant Final Safety Analysis Report (FSAR)

(9)   NUREG 1437, Supplement 27, "Generic Environmental Impact Statement of Nuclear Plants: Regarding Palisades Nuclear Plant"

(10)   RG 1.188, "Standard Format and Content for Applications To Renew Nuclear Power Plant Operating Licenses"

(11)   ASTM E 185-82, "Standard Practice for Conducting Surveillance Tests for Light-Water Cooled Nuclear Power Reactor Vessels"

(12)   Information Notice (IN)-2001-09, "Main Feedwater System Degradation in Safety Related ASME Code Class 2 Piping Inside the Containment of a Pressurized Water Reactor"

(13)   License Renewal Project Guideline (LRPG) 1, Revision 2, "License Renewal Project Guidance"

(14)   LRPG 3, Revision 3, "IPA Scoping and Screening"

(15)   LR-TR-012, Revision 2, "Mechanical and Electrical Scoping and Screening Methodology and Summary Report"

(16)   LR-TR-002-NSAS, Revision 3, "Components Identification and Data Processing for Non-Safety Related Affecting Related SSCs within the Scope of License Renewal."

(17)   LR-TR-015-FP-APPR Motors, Revision 0, "Fire Protection - APPR Spare Motors"

(18)   LR-TR-003-FP, Revision 3, "Component Identification and Data Processing for SSCs Within Scope of 10 CFR 54.4(a)(3) for Fire Protection"

(19)   LR-TR-004-EQ, Revision 2, "Component Identification and Data Processing for SSCs Within Scope of 10 CFR 54.4(a)(3) for Environmental Qualification"

(20)   LR-TR-005-PTS, Revision 2, "Component Identification and Data Processing for SSCs Within Scope of 10 CFR 54.4(a)(3) for Pressurized Thermal Shock"

(21)   LR-TR-006-ATW, Revision 2, "Component Identification and Data Processing for SSCs Within Scope of 10 CFR 54.4(a)(3) for Anticipated Transients Without SCRAM"

(22)   LR-TR-007-SBO, "Component Identification and Data Processing for SSC within Scope of 10 CFR 54.4(a)(3) for Station Blackout"

(23)   LR-TR-022-CS, Revision 2, "Civil Structural (C/S) Integrated Plant Assessment-Scoping/Screening and Aging Management Review Methodology and Results"

(24)   LR-SS-ESS, Revision 2, "Palisades Nuclear Plant System/Structure Scoping and Screening Results for Engineering Safeguards"

(25)   LRPG 2, Revision 2, "Staff Training Requirements and Qualifications"

(26)   NUREG/CR-6260, "Application of NUREG/CR-5999 Interim Fatigue Curves to Selected Nuclear Power Plant Components.," March 1995

(27)   NUREG/CR-6412, "Aging and loss-of-coolant accident (LOCA) testing of electrical connections," January 1998.

(28)   NUREG-0696, "Functional Criteria for Emergency Response Facilities," February 1981

(29)   NUREG-0737, "Clarification of TMI Action Plan Requirements," November 1980

(30)   ASME Boiler & Pressure Vessel Code, Section III

NRC FORM 335 (9-2004) NRCMD 3.7	U.S. NUCLEAR REGULATORY COMMISSION	1. REPORT NUMBER (Assigned by NRC, Add Vol., Supp., Rev., and Addendum Numbers, if any.)
BIBLIOGRAPHIC DATA SHEET *(See instructions on the reverse)*		NUREG-1871

2. TITLE AND SUBTITLE	3. DATE REPORT PUBLISHED	
Safety Evaluation Report Related to the License Renewal of Palisades Nuclear Plant	MONTH	YEAR
	January	2007
	4. FIN OR GRANT NUMBER	

5. AUTHOR(S)	6. TYPE OF REPORT
Juan Ayala	
	7. PERIOD COVERED *(Inclusive Dates)*
	3/22/2005 - 11/1/2006

8. PERFORMING ORGANIZATION - NAME AND ADDRESS *(If NRC, provide Division, Office or Region, U.S. Nuclear Regulatory Commission, and mailing address; if contractor, provide name and mailing address.)*

Division of License Renewal
Office of Nuclear Reactor Regulation
Washington, DC 20555-0001

9. SPONSORING ORGANIZATION - NAME AND ADDRESS *(If NRC, type "Same as above"; if contractor, provide NRC Division, Office or Region, U.S. Nuclear Regulatory Commission, and mailing address.)*

Division of License Renewal
Office of Nuclear Reactor Regulation
Washington, DC 20555-0001

10. SUPPLEMENTARY NOTES

11. ABSTRACT *(200 words or less)*

This safety evaluation report (SER), documents the technical review of the Palisades Nuclear Plant (PNP) license renewal application (LRA) by the U.S. Nuclear Regulatory Commission (NRC) staff (the staff). By letter dated March 22, 2005, Nuclear Management Company, LLC (NMC or the applicant) submitted the LRA in accordance with Title 10, Part 54, of the Code of Federal Regulations (10 CFR Part 54). NMC requests renewal of the operating license for Palisades Nuclear Plant (Facility Operating License Number DPR-20) for a period of 20 years beyond the current expiration date at midnight March 24, 2011.

This SER presents the status of the staff's review of information submitted to the NRC through July 5, 2006, the cutoff date for consideration in the SER. The staff identified no open items and one confirmatory item that had to be resolved before the staff could make a final determination on the application. Sections 1.5 and 1.6 of this report summarize these items. Section 6 provides the staff's final conclusion on the review of the PNP LRA.

The NRC Palisades license renewal project manager is Juan Ayala. Mr. Ayala may be reached at 301-415-4063. Written correspondence should be addressed to the Division of License Renewal, US Nuclear Regulatory Commission, Washington, DC 20555-0001

12. KEY WORDS/DESCRIPTORS *(List words or phrases that will assist researchers in locating the report.)*	13. AVAILABILITY STATEMENT
10 CFR Part 54, license renewal, Palisades, scoping and screening, aging management, time-limited aging analysis, safety evaluation report	unlimited
	14. SECURITY CLASSIFICATION
	(This Page) unclassified
	(This Report) unclassified
	15. NUMBER OF PAGES
	16. PRICE

(98) NRC Information Notice 99-10, "Degradation of Prestressing Tendon Systems in Prestressed Concrete Containments"

(99) NUREG-1744, "A Survey of Crane Operating Experience at U.S. Nuclear Power Plants from 1968 through 2002"

(100) MRP-55, "Crack Growth Rates for Evaluating Primary Water Stress Corrosion Cracking (PWSCC) of Thick-Wall Alloy 600 Material," July 18, 2002

(101) ASME B&PV Code, Appendix O, "Evaluation of Flaws in PWR Reactor Vessel Upper Head Penetration Nozzles"

(102) NRC Regulatory Guide 1.14, "Reactor Coolant Pump Flywheel Integrity," August 1975

(103) SIR-94-080, "Relaxation of Reactor Coolant Pump Flywheel Inspection Requirements," April 4, 1995

(104) WCAP-15973, "Low-Alloy Steel Component Corrosion Analysis Supporting Small-May 24, 2006 Diameter Alloy 600/690 Nozzle Repair/Replacement Programs"

(105) NMC, LLC, License Renewal Application for Palisades Nuclear Plant, dated March 22, 2005

(82)   NUREG/CR-6551, "Improved Embrittlement Correlations for Reactor Pressure Vessel Steels"

(83)   NRC Regulatory Guide 1.161, "Evaluation of Reactor Pressure Vessels with Charpy Upper-Shelf Energy Less Than 50 Ft-Lb," June 1995

(84)   Executive Memo to Commissioners, "Pressurized Thermal Shock Analyses for Renewal of Certain Nuclear Power Plant Operating Licenses," May 27, 2004

(85)   American Society of Mechanical Engineers (ASME) Boiler and Pressure Vessel Code (Code), Section XI, Appendix G

(86)   NRC Regulatory Guide 1.162, "Format and Content of Report for Thermal Annealing of Reactor Pressure Vessels," February 1996

(87)   NRC Regulatory Guide 1.162, "Format and Content of Plant-Specific Pressurized Thermal Shock Safety Analysis Reports for Pressurized Water Reactors," January 1987

(88)   ASTM A 516, "Standard Specification for Pressure Vessel Plates, Carbon Steel, for Moderate- and Lower-Temperature Service," ASTM International

(89)   NRC Bulletin 1988-11, "Pressurizer Surge Line Thermal Stratification," December 20, 1988

(90)   NRC Bulletin 1988-08, "Thermal Stresses in Piping Connected to Reactor Coolant Systems," August 8, 1988

(91)   NRC Generic Letter 1987-11, "Relaxation in Arbitrary Intermediate Pipe Rupture Requirements," June 19, 1987

(92)   NRC Generic Safety Issue (GSI)-190, "Fatigue Evaluation of Metal Components for 60-year Plant Life"

(93)   NRC Generic Safety Issue (GSI)-166, "Adequacy of the Fatigue Life of Metal Components"

(94)   NUREG/CR-6583, "Effects of LWR Coolant Environments on Fatigue Design Curves of Carbon and Low-Alloy Steels"

(95)   NUREG/CR-5704, "Effects of LWR Coolant Environments on Fatigue Design Curves of Austenitic Stainless Steels"

(96)   NRC Regulatory Guide 1.89, "Environmental Qualification of Certain Electric Equipment Important to Safety for Nuclear Power Plants," June 1984

(97)   NUREG-0588, "Interim Staff Position on Environmental Qualification of Safety-Related Equipment," July 1981

(65)   EPRI TR-107396, Revision 0, "Closed Cycle Cooling Water Chemistry Guideline,"
         November 1997

(66)   EPRI Technical Report 1009743 "Aging Identification and Assessment Checklist,"
         August 27,2004.

(67)   EPRI-GS-7086, "Testing, monitoring, and maintenance of aboveground storage
         tanks," December 1990

(68)   American Petroleum Institute (API) 575, "Inspection of Atmospheric and
         Low-Pressure Storage Tanks"

(69)   Topical Report CPC-2A, "Quality Program Description for Nuclear Power Plants
         (Part 2) - Palisades Nuclear Plant"

(70)   Title 10, Code of Federal Regulations, Part 50, Appendix R, "Fire Protection
         Program for Nuclear Power Facilities Operating Prior to January 1, 1979"

(71)   Branch Technical Position (BTP) 9.5-1, "Guidelines for Fire Protection for Nuclear
         Power Plants," August 1976

(72)   SAND96-0344, UC-523, "Aging Management Guideline for Commercial Nuclear
         Power Plants - Electrical Cables and Terminations," September 1996

(73)   NUMARC 93-01, Rev. 2, "Industry Guideline for Monitoring the Effectiveness of
         Maintenance at Nuclear Power Plants," April 1996.

(74)   ASTM D 4057, "Standard Practice for Manual Sampling of Petroleum and
         Petroleum Products," ASTM International

(75)   ASTM D 1796, "Standard Test Method for Water and Sediment in Fuel Oils by the
         Centrifuge Method (Laboratory Procedure)," ASTM International

(76)   ASTM D 2709, "Standard Test Method for Water and Sediment in Middle
         Distillate Fuels by Centrifuge," ASTM International

(77)   ASTM D 2276, "Standard Test Method for Particulate Contaminant in Aviation
         Fuel by Line Sampling," ASTM International

(78)   EPRI NP-7077, Revision 2, Project 2493, "PWR Primary Water Chemistry
         Guideline," November 1990

(79)   Westinghouse Commercial Atomic Power (WCAP)-15353, "Palisades Reactor
         Pressure Vessel Fluence Evaluation"

(80)   NRC Regulatory Guide 1.190, "Calculational and Dosimetry Methods for
         Determining Pressure Vessel Neutron Fluence," March 2001

(81)   NRC Generic Letter 92-01, Revision 1, "Reactor Vessel Structural Integrity"

(48) NRC Regulatory Guide 1.99, Revision 2. "Radiation Embrittlement of Reactor Vessel Materials," May 1988

(49) NRC Regulatory Guide 1.160, "Monitoring the Effectiveness of Maintenance at Nuclear Power Plants," March 1997

(50) NEI 94-01, "Industry Guideline for Implementing Performance-Based Option of 10 CFR 50 Appendix J," July 26, 1995

(51) NEI 97-06, "Steam Generator Program Guidelines," January 2001

(52) NEI 96-03, "Guideline for Monitoring the Condition of Structures at Nuclear Power Plants"

(53) NSAC-202L-R2, "Recommendations for an Effective Flow Accelerated Corrosion Program," April 8, 1999.

(54) EPRI TR-103834-P1-2, "Effects of Moisture on the Life of Power Plant Cables, Electric Power," August 1994.

(55) EPRI TR-105714, "PWR Primary Water Chemistry Guidelines, Revision 5," 2003

(56) EPRI TR-102134, "PWR Secondary Water Chemistry Guidelines, Revision 5," December 1999

(57) EPRI TR-105714, "PWR Primary Water Chemistry Guidelines, Revision 3" May 1993

(58) EPRI TR-102134, "PWR Secondary Water Chemistry Guidelines, Revision 3 " November 1995

(59) EPRI 102884, "Pressurized Water Reactor Primary Water Chemistry Guidelines, Revision 5" dated September 2003 (revised October 2003)

(60) EPRI NP-7079, "Instrument Air Systems: A Guide for Power Plant Maintenance Personnel," 1990

(61) EPRI TR-108147, "Compressor and Instrument Air System Maintenance Guide," March 1998

(62) EPRI NP-5769, "Degradation and failure of bolting in nuclear power plants: Volume 2: Final Report," April 1988

(63) EPRI TR-104213, "Bolted Joint Maintenance & Application Guide," December 1995

(64) EPRI TR-1007820, "Closed Cooling Water Chemistry Guideline"

(31)   Branch Technical Position (BTP) Auxiliary and Power Conversion Systems
         Branch (APCSB) 9.5-1, Appendix A, August 1976.

(32)   NUREG-1493, "Primary Reactor Containment Leakage Testing for Water-Cooled"

(33)   NUREG-0612, "Control of Heavy Loads at Nuclear Power Plants," 1980

(34)   NUREG-1825, "Safety Evaluation Report Related to the License Renewal of the
         Joseph M. Farley Nuclear Plant, Units 1 and 2"

(35)   NUREG-0313, "BWR Coolant Pressure Boundary," January 1988.

(36)   NUREG-1339, "Resolution of Generic Safety Issue 29: Bolting Degradation or Failure in
         Nuclear Power Plants," June 1990.

(37)   Title 10, Code of Federal Regulations, Part 50, Appendix H, "Reactor Vessel Material
         Surveillance Program Requirements"

(38)   Title 10, Code of Federal Regulations, Part 50, Appendix B, "Quality Assurance Criteria
         for Nuclear Power Plants and Fuel Reprocessing Plants"

(39)   Title 10, Code of Federal Regulations, Part 50, Appendix G, "Fracture Toughness
         Requirements"

(40)   NRC Order EA-03-009, "Issuance of Order Establishing Interim Inspection Requirements
         for Reactor Pressure Vessel Heads at Pressurized Water Reactors," February 11, 2003.

(41)   NRC Bulletin 2001-01, "Circumferential Cracking of Reactor Pressure Vessel Head
         Penetration Nozzles," August 3, 2001

(42)   NRC Bulletin 2002-01, "Reactor Pressure Vessel Head and Vessel Head Penetration
         Nozzle Inspection Programs," March 18, 2002

(43)   NRC Bulletin 2002-02, "Reactor Pressure Vessel Head Degradation and Reactor
         Coolant Pressure Boundary Integrity," August 9, 2002

(44)   NRC Bulletin 2004-01, "Inspection of Alloy 82/182/600 Materials Used in the Fabrication
         of Pressurizer Penetrations and Steam Space Piping Connection at Pressurized-Water
         Reactors, "May 28, 2004

(45)   ASME Boiler & Pressure Vessel Code, Section XI, Subsection IWB, IWC, IWD, and IWF

(46)   NRC Regulatory Guide 1.163, "Performance-Based Containment Leak Test Program,"
         September 1995

(47)   NEI 94-01, "Industry Guideline for Implementing Performance-Based Option of
         10 CFR 50 Appendix J"